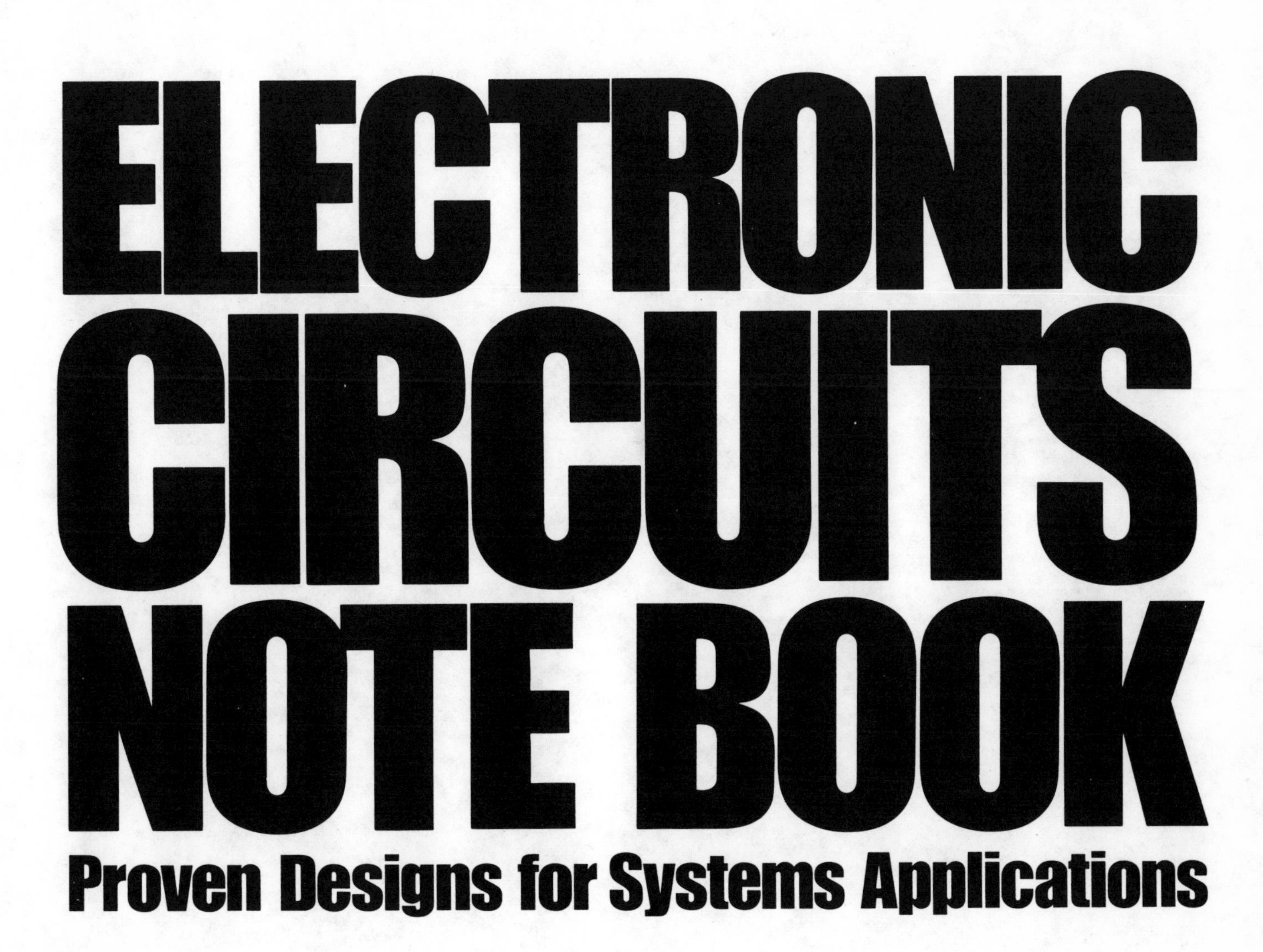

ELECTRONIC CIRCUITS NOTE BOOK

Proven Designs for Systems Applications

ELECTRONIC CIRCUITS NOTE BOOK

Proven Designs for Systems Applications

Edited by
SAMUEL WEBER
Editor-in-chief
Electronics

McGraw-Hill Publications Co.
1221 Avenue of the Americas
New York, New York 10020

ELECTRONICS BOOK SERIES

Also published by *Electronics*

- **Microprocessors**
- **Basics of data communications**
- **Large scale integration**
- **Applying microprocessors**
- **Circuits for electronics engineers**
- **Design techniques for electronics engineers**
- **Memory design: microcomputers to mainframes**
- **Personal computing: hardware and software basics**
- **Microelectronics interconnection and packaging**
- **Practical applications of data communications**
- **Microprocessors and microcomputers**

Library of Congress Cataloging in Publication Data

Main entry under title:
Electronic Circuits Notebook
(Electronics magazine books)
The circuits in this book originally appeared in Electronics from late 1977 to mid-1980.
1. Electronic circuits.
I. Weber, Samuel II. Electronics
TK7867.E42 621.3815'3 80-29479
McGraw-Hill Book Company ISBN 0-07-019244-8
McGraw-Hill Publications Company ISBN 0-07-606720-3

Contents

Preface

Although the circuit designer's job has changed radically with the introduction of integrated circuits of ever-higher complexity, his need for fresh ideas and new approaches to realizing electronic functions is as strong as ever. As electronic technology expands into ver more applications, the designer must continually break new ground in performance, cost reduction and reliability.

This probably accounts for the popularity of Designer's Casebook, the regular section of *Electronics* magazine on which this volume is based. Because of its frequency and the high standards maintained by its editors, Designer's Casebook always reflects the state of the art in circuit design, utilizing the latest components and techniques.

This volume is a companion to the highly successful "Circuits for Electronics Engineers," an earlier compilation of Designer's Casebook articles which was published in 1977. The circuits in this book originally appeared in *Electronics* from late 1977 to mid-1980.

The book contains 268 circuits arranged alphabetically by function in 39 chapters for convenient access. Like its predecessor, this collection is intended to provide solutions to recurrent engineering problems that can be used immediately and to stimulate ideas by presenting approaches that can be adapted or built upon, thus saving valuable engineering time.

—Samuel Weber

Chapter 1
AMPLIFIER CIRCUITS

Bi-FET devices improve absolute-value amplifier

by Dan L. Vogler
Lintech Electronics, Albuquerque, N.M.

An absolute-value amplifier, also known as a precision full-wave rectifier, which features wide bandwidth and dynamic range, can be built with high-impedance operational amplifiers to produce a circuit that is more reliable than those implementing the usual phase-cancellation technique. The low input current and wide frequency range of the CA3140 bipolar/field-effect-transistor op amps eliminate the gain and phase-shift errors encountered in other designs.

As shown in the figure, op amp A_1 serves as a unity-gain buffer, op amp A_2 has a gain of +1 during the positive half-cycle of the input wave and a gain of −1 during the negative portions, and A_3 in association with the diode forms a precision clamp.

During the positive portion of the input signal equal voltage is present at both inputs of A_2. The op amp behaves as a unity-gain follower, as determined by the feedback elements.

During the negative portions, however, the clamping action of A_3 with the diode prevents the voltage at the noninverting input of A_2 from going negative, effectively tying the pin to ground. Op amp A_2 therefore either operates in the inverting mode or else multiplies the signal by a factor of −1.

Precision resistors for the gain-controlling elements of op amp A_2 assure no greater than 2% deviation from the desired gain. The clamping circuit of A_3 can accurately process signals down to −0.3 volt below the negative supply rail of the amplifier, which in this case is ground.

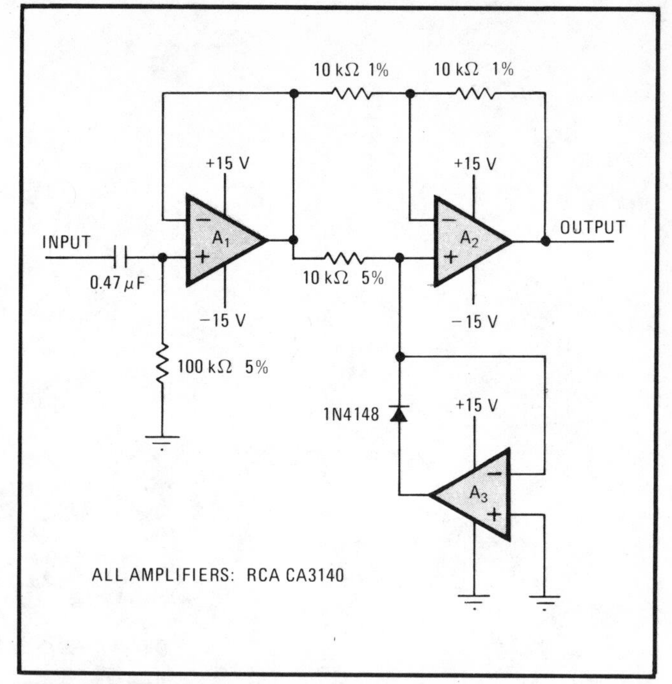

Precision full-wave rectifier. Op amp A_3, which ensures A_2 follows positive voltages and inverts negative ones, has single-ended power supply to minimize slew time and maximize stability. Power-supply pins are decoupled with 0.47-μF capacitors.

The result is an absolute-value amplifier which has a dynamic range exceeding 90 decibels and a bandwidth exceeding 1 megahertz. When this circuit is used in conjunction with a peak detector or integrator network, it becomes an invaluable building block in ac-to-dc conversion applications. □

Unity-gain buffer amplifier is ultrafast

by James B. Knitter and Eugene L. Zuch
Datel Systems Inc., Canton, Mass.

Applications where transmission-line drivers, active voltage probes, or buffers for ultrahigh-speed analog-to-digital converters are needed can use a stable buffer amplifier capable of driving a relatively low-resistance, moderate-capacitance load over a wide range of frequencies. The circuit shown in (a) fulfills these requirements. With a bandwidth of 300 megahertz, it exhibits no peaking of its response curve, having a gain of virtually 1 (0.995) under no-load conditions and 0.9 under a maximum load of 90 ohms.

The circuit is a variation on a basic emitter-follower network, which is inherently capable of wideband performance. However, no feedback loops are needed anywhere within the circuit to boost the gain at the high frequencies, and dispensing with them contributes to the stability of the circuit. Also, using two matched npn-pnp transistor pairs ensures close tracking between input and output voltages (a task normally addressed by suitable feedback circuitry) as well as low offset-voltage drift (20 microvolts/°C).

The complementary-transistor pairs are 2N4854s wired for active current sourcing and sinking so that bipolar input signals can be processed. Each transistor has a typical β of 100. With the npn and pnp input-bias currents tending to cancel each other, the resultant input-bias current of the amplifier is ±5 microamperes.

Layout is critical to the stability of the circuit. The buffer should be constructed as shown in (b). The two transistor pairs are mounted close together, in holes drilled in a copper-clad circuit board as shown. The flanges on the TO-99 cases encapsulating the 2N485s should be soldered to the copper, which serves as a ground plane. The collector of each transistor must be bypassed by a 0.1-microfarad ceramic-chip capacitor mounted close to the transistor. This is done by standing the capacitors on end, with the bottom contact lead soldered to the ground plane and the top contact lead soldered to the collector.

All leads must be less than ½ inch in length and be as

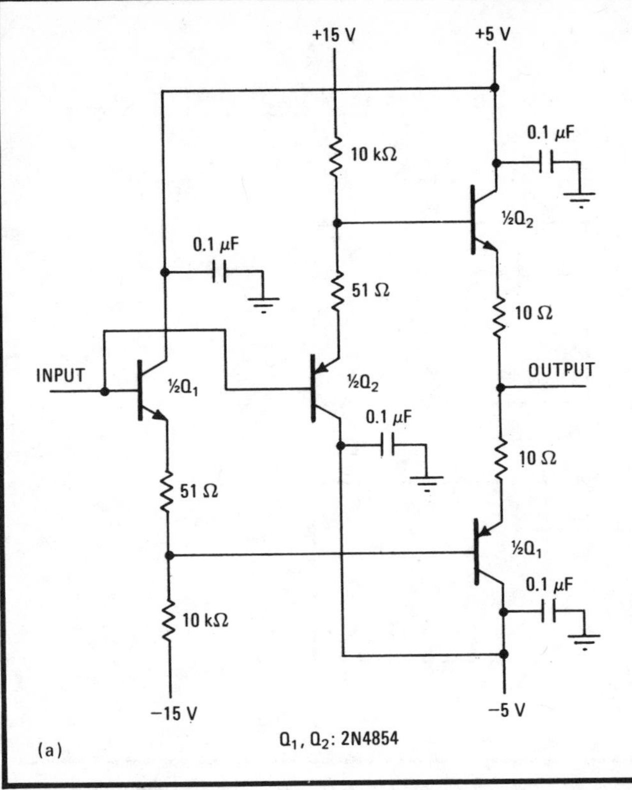

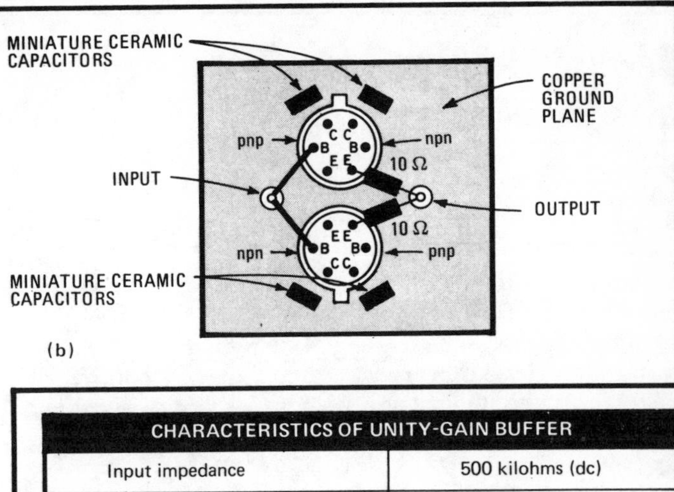

Wideband buffer. Emitter-follower configuration yields unity gain from dc to 300 megahertz. Absence of feedback in circuit contributes to buffer stability. Use of matched npn-pnp transistor pairs ensures almost perfect input/output signal tracking (a). Component layout is critical for circuit stability (b).

CHARACTERISTICS OF UNITY-GAIN BUFFER	
Input impedance	500 kilohms (dc)
Input bias current	±5 μA
Input capacitance	16 pF max
Input/output voltage range	±3 V
Output offset-voltage drift	±20 μV/°C
Output impedance	10 ohms
Load resistance	90 ohms max
Gain, no load	+0.995
Bandwidth, −3 dB	300 MHz
Power supply, quiescent	±15 V dc at 1.5 mA ± 5 V dc at 4.5 mA
Power consumption	90 mW

directly wired as possible. One-eighth-watt resistors are used throughout and are soldered to the transistor leads as close as possible to the case. For clarity, not all components are shown. For coupling to or from the amplifier, subminiature radio-frequency connectors can be mounted at the input and output ports of the buffer.

Typical characteristics of the unity-gain buffer circuit are listed in the table. □

Demand-switched supply boosts amplifier efficiency

by Jerome Leiner
Loral Electronic Systems, Yonkers, N. Y.

The efficiency of high-power audio amplifiers operating in class B will be improved by up to 80% at low power levels if the supply voltage can be switched from a low to a high value as the power demands on the amplifier increase. Using the automatic-switching circuit described here ensures a lower heat dissipation than would be possible with an amplifier that delivers a low-power output from a single high-voltage supply, which is the most common situation. The increased efficiency of this power amplifier can produce considerable savings in its weight and size and also reduces the amplifier's heat sink requirements.

The amplifier is designed to switch from the low- to the high-voltage supply as the audio signal level passes up through the low-voltage supply level. The switchover is made with virtually no perturbation in the output current up to as high as 25 kilohertz. At higher speeds, any waveshape distortion can be reduced by implementing negative feedback around the amplifier. Both supply voltages can be derived from one source, with the low voltage taken from a selected tap on the power transformer.

Several circuit configurations were tried, among them a cascaded emitter follower, a series transistor configuration, and the parallel transistor arrangement shown in the figure. The cascaded emitter follower and the series configuration performed adequately below 10 kHz. At higher frequencies, however, the effects of carrier storage produced by the first two arrangements caused large perturbations in the output current. The parallel arrangement finally adopted has virtually no storage problem and appears to be the most useful at higher frequencies.

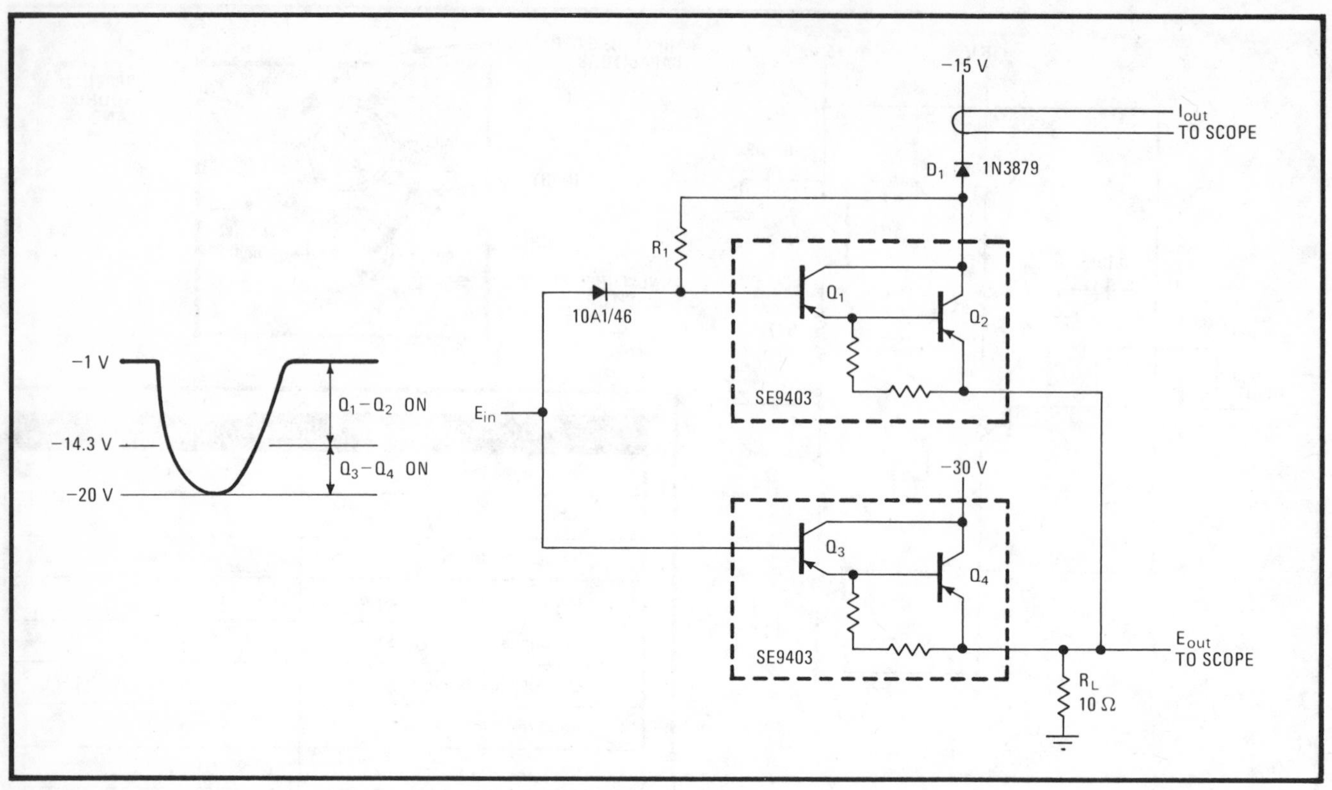

High efficiency. Audio power amplifier switches from Darlington pair Q_1–Q_2 and low-voltage supply to Q_3–Q_4 and high-voltage supply only when power output demands increase. Circuit thereby eliminates the need for amplifier to dissipate excessive heat, a condition that occurs when an amp with a single high-voltage supply is used to process a low-amplitude input signal.

For simplicity, the operation of one half of a complementary-output stage (see figure) is described. A half-sine wave that swings from −1 to −20 volts is the input-signal source in this case.

When the input level is at −1 v, current flows through D_2, R_1, and D_1. Thus the input base of Darlington pair Q_1–Q_2 is one diode drop lower than the base of Q_3–Q_4. As a result, Q_1–Q_2, which uses the −15-v supply, is on and Q_3–Q_4 is off.

When the input signal reaches within one diode drop of the low-voltage power supply, D_2 begins to turn off and Q_3–Q_4, which uses the −30-v supply, starts to turn on. As Q_3–Q_4 moves into the linear region, Q_1–Q_2 begins to turn off, so there is a smooth transition of current in the load. Q_3–Q_4 stays on until the signal polarity reverses; when the signal passes through the low-voltage supply level, Q_1–Q_2 goes on and Q_3–Q_4 goes off. □

Voltage-controlled amplifier phase-adjusts wave generator

by G. B. Clayton
Liverpool Polytechnic, Liverpool, England

When added to a generator that produces two triangular and two square waves in quadrature, a voltage-controlled, gain-switching amplifier makes it easy to adjust the phase difference of each pair of signals. The entire circuit—that is, the generator and the controller—requires only two chips and one field-effect transistor for providing phase differences from 0° to 180°.

In the arrangement shown, a quad operational amplifier (A_1–A_4) serves as the quadrature oscillator, and a dual op amp (A_5–A_6) is the control section. Amps A_1 and A_2 form an integrator and comparator, needed for generating the triangular and square waves. A_3 is a zero-crossing detector, used to produce a square wave from the triangular input of A_1. A_4 produces a second triangular wave from A_3's output. Note that the feedback resistor R_3 in the A_1–A_4 loop will prevent A_4 from drifting into saturation, even if offset voltages from the op amps are high.

In the phase-control section, A_5 acts as the switched-gain element. A_2 and Q_1 control the gain of A_5. When Q_1 turns on, A_5 has a gain of −1; otherwise, its gain is +1. This element, appropriately biased at its input with a control voltage, V_c, thus turns on and off sooner or later than usual, depending on the magnitude of the control voltage. This acts to advance or retard A_3's on-off transitions on both the rising and the falling edge of A_2's square-wave signal. As a result, the signals from A_3 and A_4 lead their respective counterparts at A_2 and A_1 by a value almost linearly proportional to V_c.

A_5's output is inverted by A_6, which is in turn connected to potentiometer R_4 and A_2. R_4, included to overcome

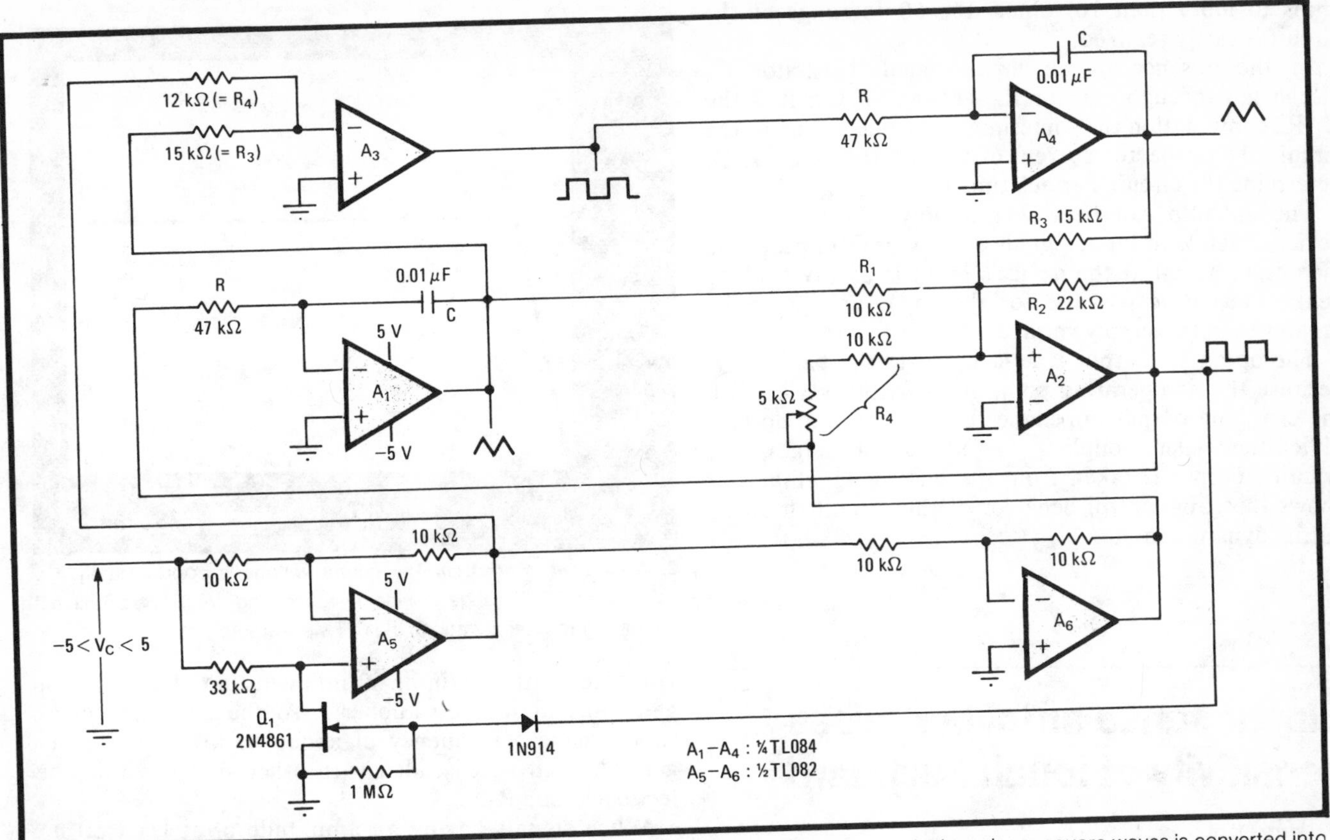

Quadrature variance. Waveform generator that normally produces in-quadrature (90° departure) triangular or square waves is converted into variable phase-delay circuit when gain-switching amps A_5–A_6 are added. A_5–A_6 act to advance A_3's turn-on transition, so that signals at A_3 and A_4 lead those at A_1 and A_2. Phase shift between both sets of waves is controlled by V_c.

the effects of component mismatch, is placed strategically, so that it will not interfere with the generation of waves produced by A_1 and A_2.

If the components and amplifiers are matched, the frequency of oscillation for all waveforms will be $f = R_2/4R_1CR$. The relation between control voltage, resistor R_4's value, and phase shift is given by:

$$\theta = 90°[(V_cR_2R_3/V_{o\,sat}R_1R_4) - 1]$$

where $V_{o\,sat}$ is 0.7 v below the supply voltage. □

Automatic gain control has 60-decibel range

by Neil Heckt
The Boeing Co., Seattle, Wash.

An automatic-gain-control circuit with an input range of 60 decibels (20 millivolts to 20 volts) can be built using a junction FET as a voltage-controlled resistor in a peak-detecting control loop. The circuit exhibits a quick response of 1 to 2 milliseconds and a delay time of 0.4 second.

As shown in the schematic of Fig. 1, the 2N4861 n-channel field-effect transistor Q_2, connecting the noninverting input of the operational amplifier to ground, determines the closed-loop gain of the system. Negative base-voltage peaks from the output of the op amp, beyond V_{BE} of Q_1, turns Q_1 on, and its collector current then charges capacitor C_1.

The voltage across C_1 determines the channel resistance of Q_2. Since the range of this resistance is 120

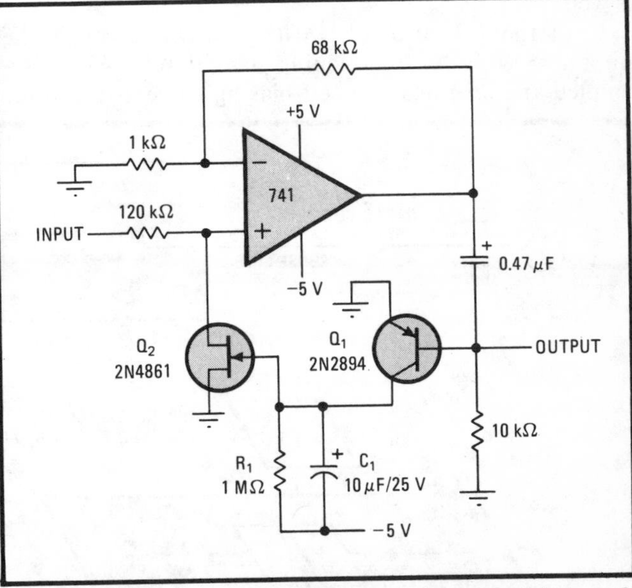

1. Audio AGC. Input voltage for distortionless output is from 20 millivolts to 20 volts in this quick-response AGC circuit. The input signal can have greater magnitude than the supply voltage because the maximum signal across the FET is 25 millivolts.

ohms to more than 10^8 ohms, the 60-dB range of the circuit is easily realized.

In the absence of an input signal, capacitor C_1 discharges through resistor R_4, cutting off Q_2. It is the C_1-R_4 combination that determines the delay time of the circuit. The collector current of Q_1 and the value of C_1 determine the circuit's attack time.

The op amp can be a 741 or any general-purpose device. Even with input signals of 20 v peak to peak, the maximum signal at the device's input is 25 mv peak to peak. Thus it is possible for the input voltage to be greater than the supply voltage.

The op amp's output is ac-coupled to the base of Q_1 because the dc operating point of its output varies with the changing output impedance of Q_2. To avoid dc bias difficulties when coupling to subsequent stages, the circuit's output is taken from the base of Q_1. Figure 2 shows the gain-control behavior of this circuit throughout its dynamic range. □

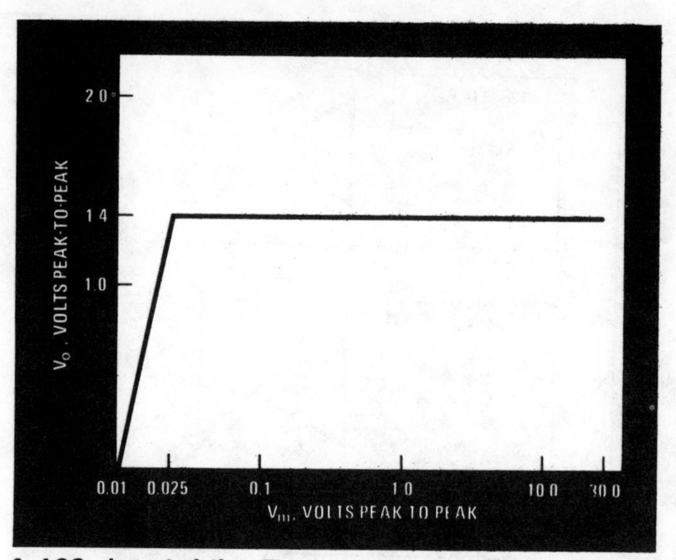

2. AGC characteristics. The output voltage is approximately 1.4 volts over a 60-dB range. A wider dynamic range would be possible if the off/on-resistance ratio of the FET were greater.

Linear sense amplifier raises sensitivity of touch keyboard

by Jerry Dahl

IBM Corp., Research Triangle Park, Raleigh, N. C.

Keyboards relying on hand capacitance to simulate contact closure require a sense amplifier to detect the capacitive changes and thus determine when a key is depressed. Using one half of a complementary-MOS gate array, where one gate is operated in a linear mode to detect currents as low as 50 microamperes, this sense amplifier is not only simple and inexpensive but sensitive as well.

A_1 of the 74C00 quad NAND gate (a) serves as the amplifier, with A_2 functioning as a latch. A_1 is ac-coupled and operates as a self-biasing current-to-voltage converter with a gain of 50 millivolts/μA. Its open-loop gain falls above 100 kilohertz, so the drive-line clock should have a frequency of about 10 kHz. For higher gain, A_1 can be cascaded with other stages within the feedback loop R_1.

When signals having an amplitude of at least 50 μA are coupled to the sense line via the coupling capacitor connected to the keybutton, A_1 goes high and triggers A_2 for about 70 μs. A_2 operates as a one-shot and thus it does not need to be reset.

The construction of the key module capacitor is shown in (b). The coupling capacitor is connected to the keybutton directly. The circuit-card pads are coated with a thin insulating epoxy covering that serves as the dielectric. When the keybutton is depressed, the clock pulse on the drive line will therefore be coupled to the sense line through the electric field of the capacitor. □

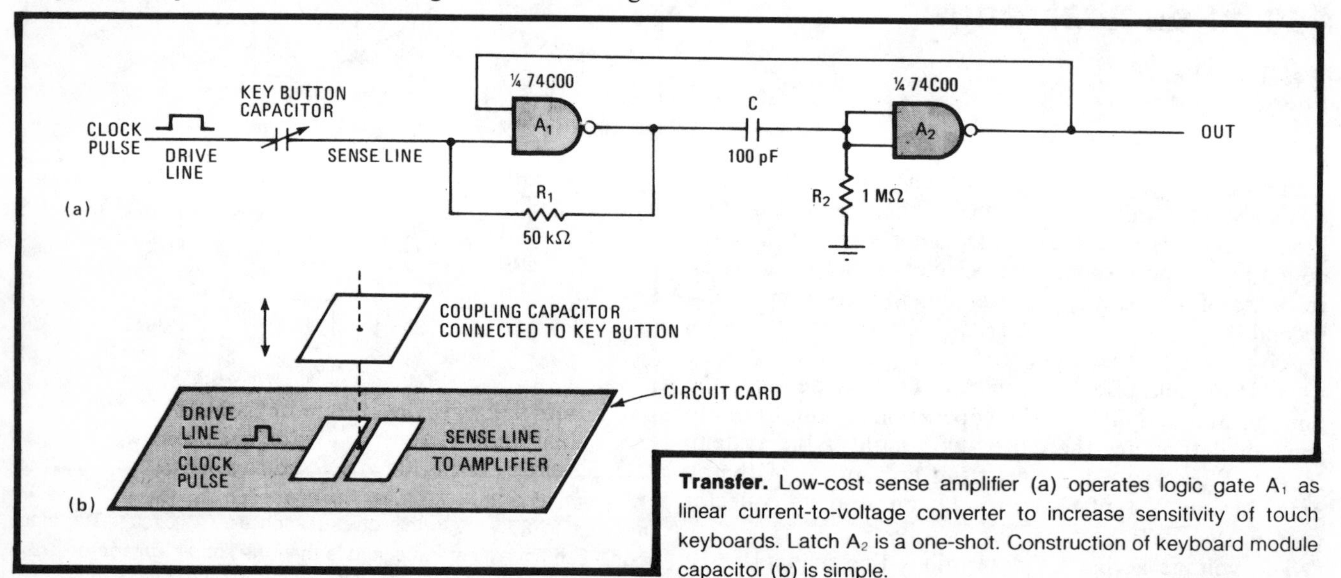

Transfer. Low-cost sense amplifier (a) operates logic gate A_1 as linear current-to-voltage converter to increase sensitivity of touch keyboards. Latch A_2 is a one-shot. Construction of keyboard module capacitor (b) is simple.

Level shifter builds high-voltage op-amp block

by Leon C. Webb
Ball Corp., Aerospace Systems Division, Boulder, Colo.

Placing a level-shifting network inside the major loop of an operational amplifier adapts it for high-voltage applications. The output swing of the circuit, which can be in the hundreds of volts, is limited only by the breakdown voltage of the active devices (in this case, transistors) used. At the same time, the op amp is isolated from high potentials, even in the absence of its ±15-volt supply voltages, by the attenuator formed by the amp's gain-controlling resistor and the input resistance of the circuit and by the clamping action of the circuit's common-base stages.

The method can be applied to op amps that will generate either bipolar or unipolar voltage swings. As shown in (a) for the general voltage amplifier, which generates a bipolar swing, the op amp's output is trans-

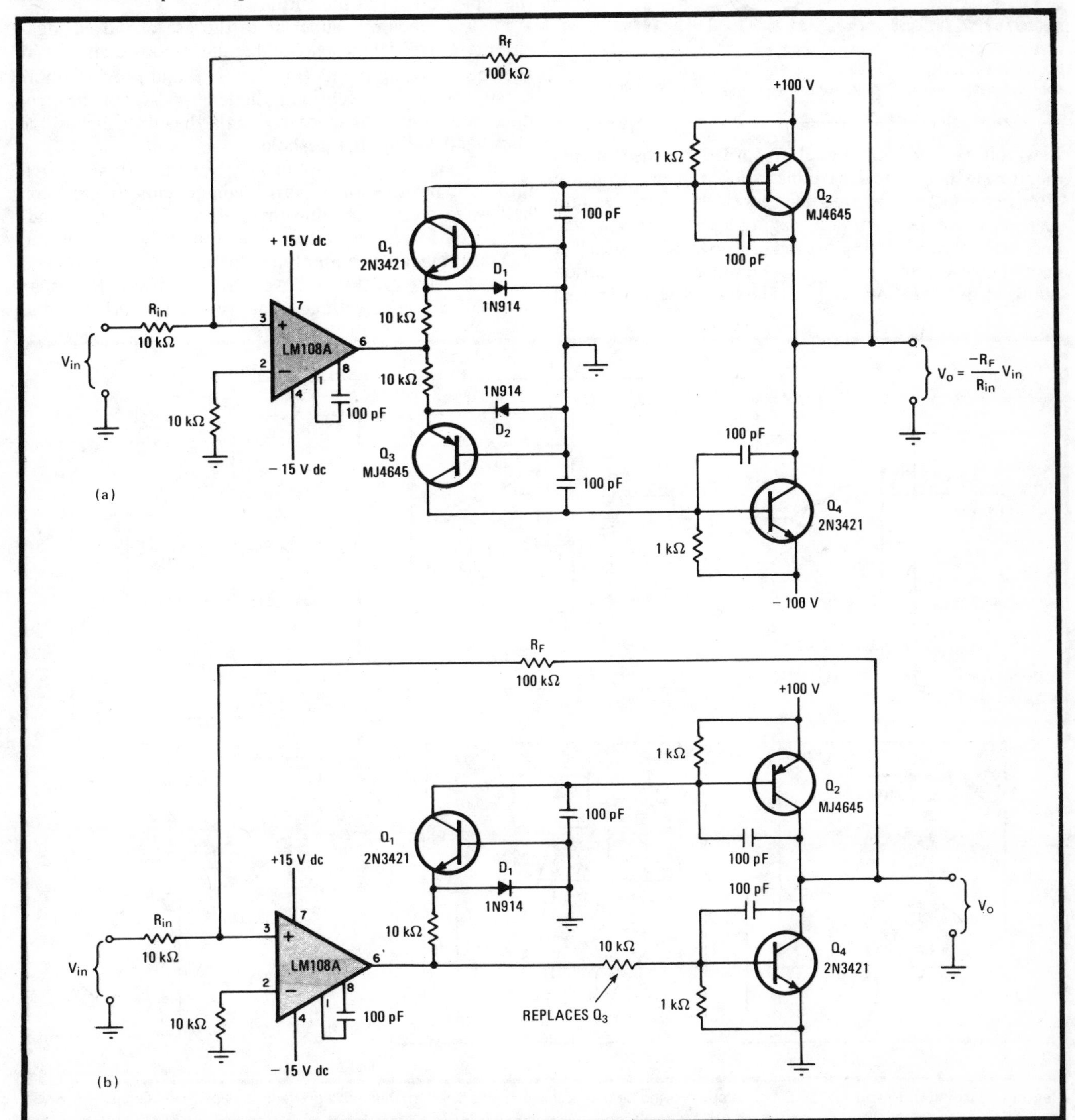

Translation. Level-shifter adapts operational voltage amplifier (a) for high-voltage duty. Output swing of circuit is limited only by the shifter's supply voltage and the breakdown voltage of transistors used. For unidirectional output swings (b), simplified circuit will suffice.

formed into an emitter current that flows through either transistor Q_1 or Q_3, depending upon the polarity of the output voltage from the LM108A. As a result, a corresponding base current is applied to either transistor Q_2 or transistor Q_4, respectively, thereby turning it on to a greater or lesser degree. Thus, V_o assumes a value equal to a $-10V_{in}$, where the voltage multiplication holds true for $-10\ v < V_{in} < 10\ v$.

If only a unidirectional output is desired, the configuration shown in (b) will suffice. This circuit, which delivers only positive output voltages, has the same transfer function (that is, $V_{out} = -10\ V_{in}$) for $-10\ v < V_{in} < 0\ v$. If a negative-only voltage output is required, stage Q_1 is replaced by a 10-kΩ resistor. The level-shifter's supply voltage must also be negative. □

Antilog amplifiers improve biomedical signal's S/N ratio

by T. G. Barnett and D. L. Wingate
London Hospital Medical College, Department of Physiology, England

Low-voltage, biphasic signals recorded by instruments monitoring biomedical variables such as heart rate are often accompanied by high noise levels due to inadequate sensing, movement artefact, paging systems and power-line interference. Using paired antilogarithmic amplifiers, however, to provide the nonlinear amplification required, the level of the biphasic signals can be raised well above the amplitude of the interfering signals. The signal-to-noise ratio can thus be improved from 2:1 at the input to 10:1 at the output.

Such a scheme is superior to the use of paired logarithmic amps, which cannot handle biphasic signals at the zero-crossing points ($\log 0 = \infty$), and provides more sensitivity than conventional diode clippers, which introduce noise and cannot pass signals that drop below the circuit's 0.7 clipping threshold.

Input signals are amplified by A_1 and are separated into their positive and negative components by precision half-wave rectifier A_2 and inverter A_3. The corresponding outputs are then introduced into the AD759N and AD759P log/antilog amplifiers, which are wired to yield $e_o = E_{ref}10^{-e_{in1}}/K$ for $-2 \le e_{in1}/K \le 2$, where E_{ref} is an internal reference voltage of approximately 0.1 volt and

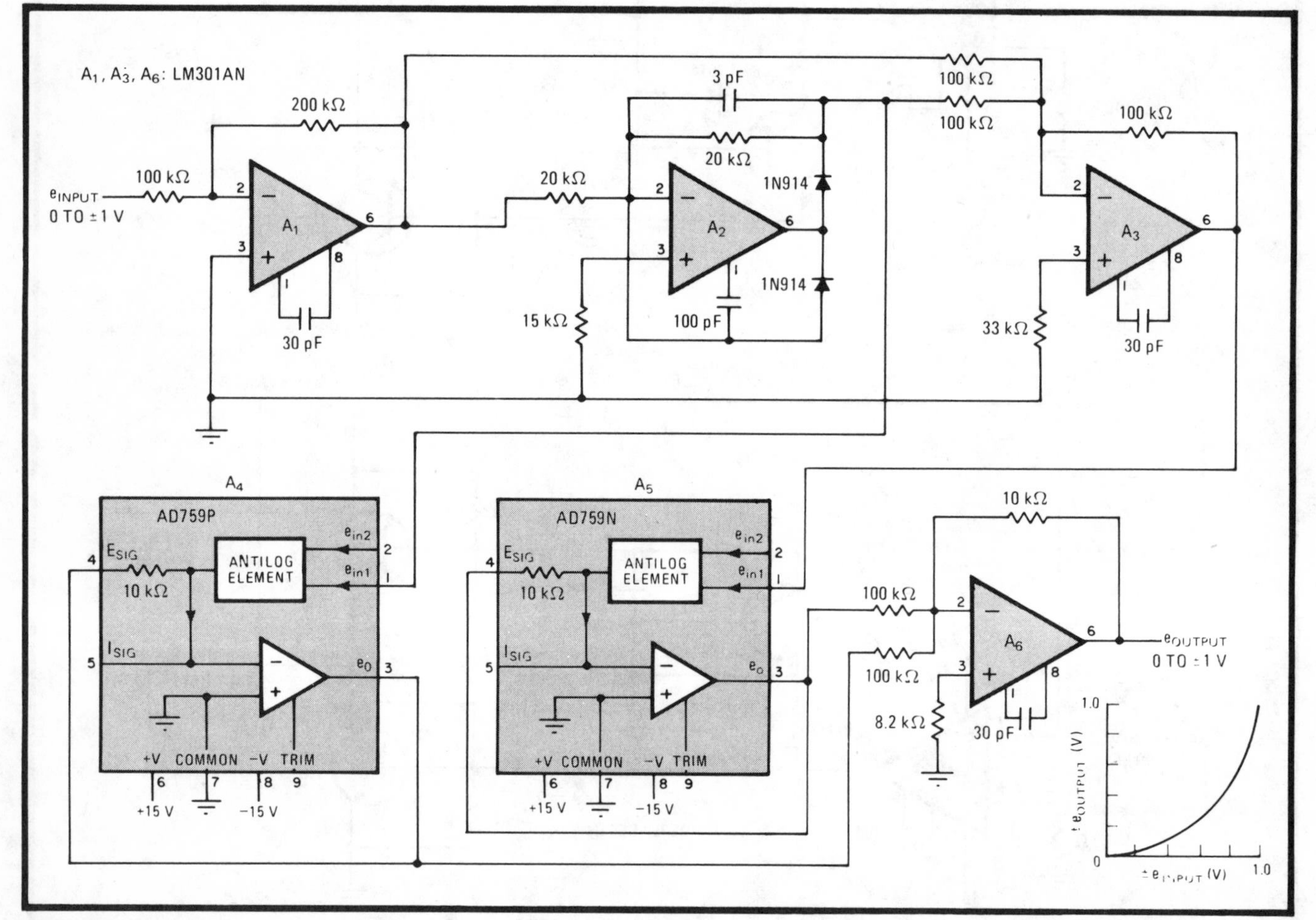

Biomedical booster. Paired AD759 antilog amps provide bipolar, nonlinear amplification, thus raising level of biphasic signals such as EKGs with respect to noise and so increasing S/N ratio. Circuit is superior to those using log amps, which cannot provide accurate output at zero crossings of signals, and is more sensitive than diode clippers, provides greater noise rejection than filters, and introduces no phase shift.

K is a multiplying constant that has been set at 1 as a consequence of utilizing input e_{in1}. The output voltages generated by A_4 and A_5 are of negative and positive polarity, respectively.

These components are then summed by A_6, whose output yields a bipolar, antilogged signal that can be introduced to appropriate trigger circuits. If desired, the original signal can be reconstructed by passing it through paired logarithmic amplifiers. □

Analog current switch makes gain-programmable amplifier

by John Maxwell
National Semiconductor Corp., Santa Clara, Calif.

Moderate-cost binary or binary-coded-decimal gain-programmed amplifiers (GPAs) can be built with monolithic current-mode analog switches such as the National AH5010 or the AM97C10. GPAs are useful in audio and other systems that require logic control for signal preconditioning, leveling, and dynamic-range expansion.

The logic-controlled GPA is actually a multiplying digital-to-analog converter. The analog input is the reference node, which is multiplied by the digital input word. Although multiplying d-a converters have been available for some time in module, hybrid, and monolithic form, most are either prohibitively expensive or have poor signal-handling capabilities.

A 4-bit binary GPA with an input voltage swing of ±25 V can be built with a quad current-mode switch (of the multiplexing type, which has all FET drains tied together), four binary-weighted resistors, and an operational amplifier, as shown in Fig. 1. The output voltage—which is a function of the feedback resistor, input resistors, and the logic states of the field-effect-transistor gates—behaves according to the equation shown in the figure.

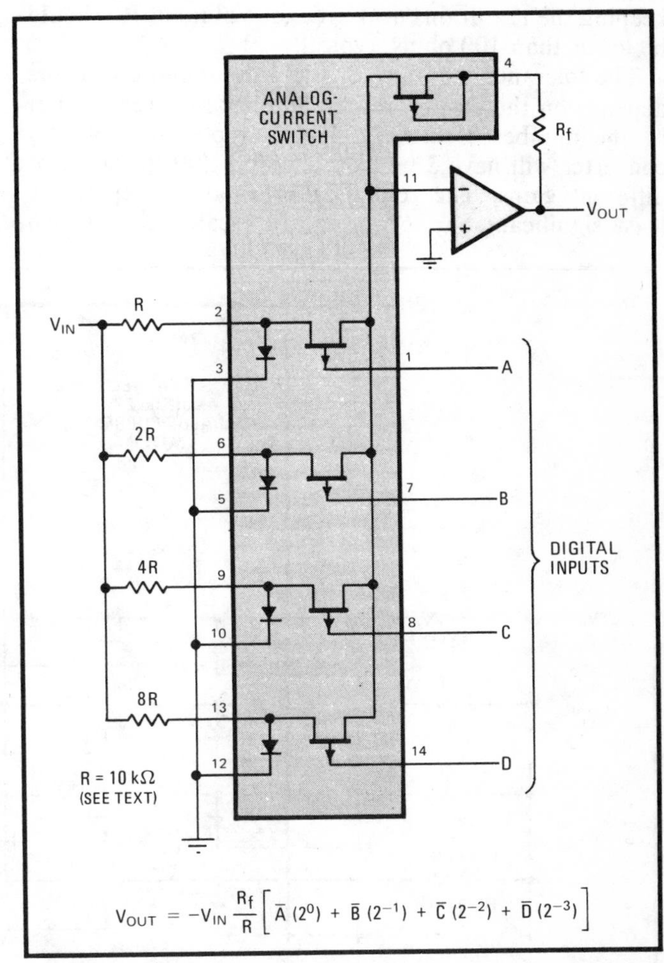

1. Gain-programmable amplifier. A monolithic current-mode analog switch connected to an operational amplifier becomes a moderate-cost GPA, or multiplying d-a converter. The multiplex switch depends on the controlling logic: for TTL, an AH5010 serves, while for C-MOS logic, the AM97C10 should be used. The op amp may be any general-purpose unit such as the LM118.

Current-mode analog switches, unlike conventional analog switches, control only the signal current at the virtual ground of an op amp. And since the voltage across each of the FET switches is clamped by a diode to a few hundred millivolts, the switches can be driven by standard logic levels, without the need for power supplies, logic interfaces, and level-translator circuits.

A logic 0 turns the switch on because the FET passes current until the channel is pinched off by a gate voltage, and the voltage of a logic 1 is sufficient to pinch the FET off. The built-in clamping diodes hold the source-to-drain voltage to about 0.7 V. A fifth FET is incorporated in the multiplexing-type chip to facilitate op-amp compensation for the on resistance of the FET switches, and, as such, it is placed in series with the feedback resistor.

The number of bits can be expanded by cascading another quad-current switch and resistor array to the first, as shown in Fig. 2. Instead of continuing the binary progression of the input resistors (16R, 32R, etc.), current-splitting resistors (R_{CS}) and shunts (R_S) are used so that the same four-resistor array may be used for additional bits, minimizing the number of different values required for higher-order converters. Binary weighting requires a 1/16 current split for the second quad switch, while BCD weighting requires a 1/10 split.

Certain practical limitations must be considered in selecting values of the gain-programming resistors. If high values are used, switch resistance becomes negligible, but leakage at elevated temperatures can cause trouble, and the signal bandwidth is decreased as well.

Using programming resistors that are too small increases switch-resistance errors. In addition, the signal current could saturate the FET as I_{DSS} is approached. High signal currents may even forward-bias the diode and the FET source-gate junction.

An input resistor value of R = 10 kilohms limits the switch current to less than 2 milliamperes, minimizing both leakage and switch-resistance problems. The accuracy at unity gain (including the compensation FET in the feedback loop) is within less than 0.05%, with $R = R_f = 10\ k\Omega$.

When cascading switches, the current-shunt resistor (R_S) should be small to minimize the voltage drop, keeping the FET drains near ground. Values of R_S should be lower than 100 ohms, typically 20 Ω.

The tolerance required for the programming resistors depends on the desired resolution of the converter; that is, the number of bits, N. For example, an 8-bit d-a converter will have $2^N - 1$ or 255 steps (99 for BCD), or different gains. The resolution or smallest step is the least significant bit, $\frac{1}{2}^N$ of the full-scale value, which works out to .0039. Error for d-a converters is usually specified as 1 LSB, or ±½ LSB, which would be ±0.2% for the 8-bit binary unit. The feedback resistor, which is the most critical, should be a 0.1%-tolerance unit. The first resistor, R, contributes one half the full-scale error, and similarly, the second contributes one fourth, and so on. Therefore, using a 0.2%-tolerance resistor for the two most significant resistors, R and 2R, 0.5% resistors for the third resistor, and 1% resistors for the fourth and fifth, allows 5% resistors to be used in the sixth, seventh, and eighth positions, still producing an overall accuracy within 0.2% Thus, high accuracy joins the GPA's other assets of high speed and large signal-handling capability. □

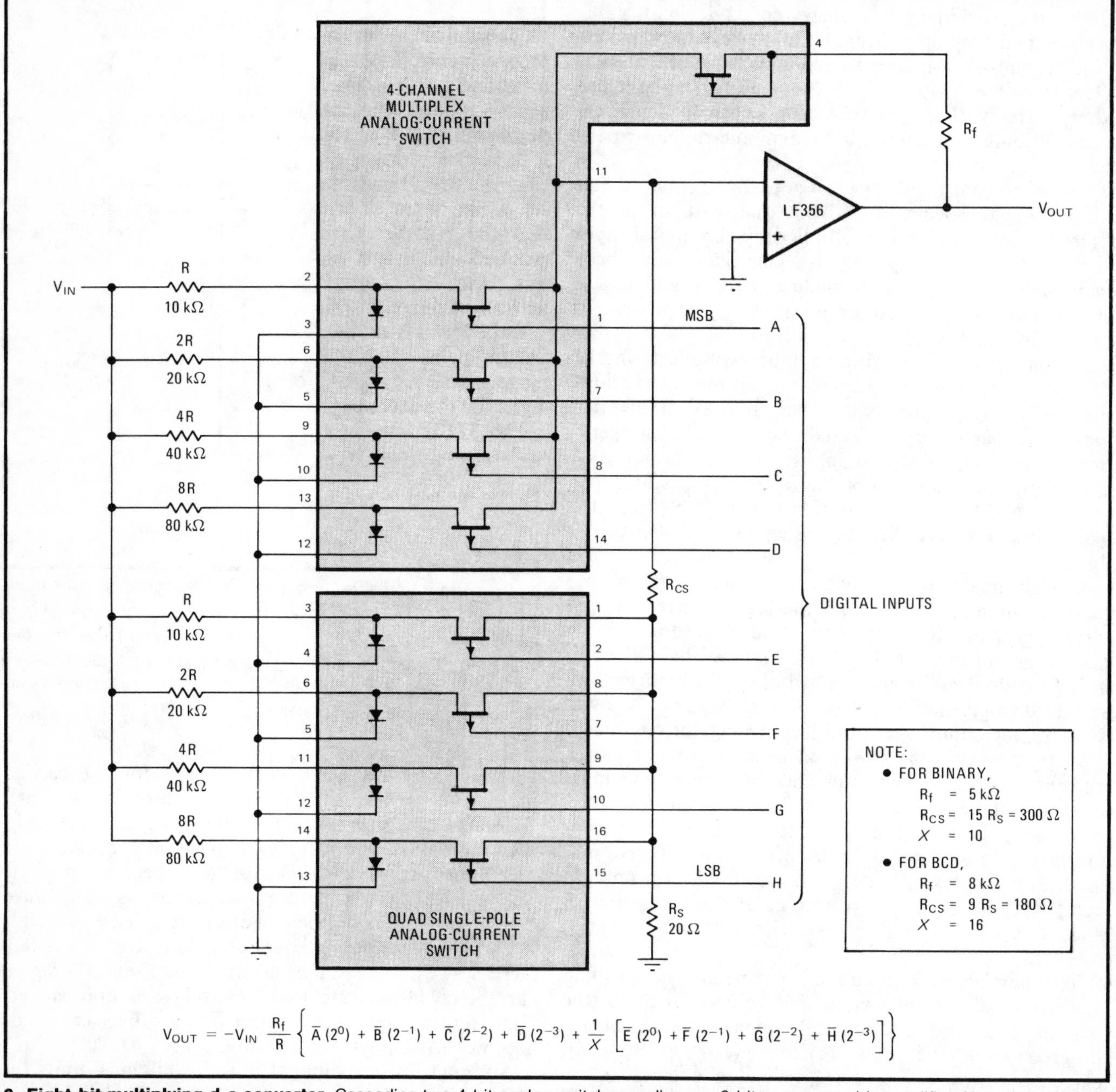

2. Eight-bit multiplying d-a converter. Cascading two 4-bit analog switches realizes an 8-bit programmable amplifier. Note that while the first monolithic switch is the multiplexing type, as in Fig. 1, the second switch is a quad spst type, with drains uncommitted and no compensating FET built in. Suitable types are AH5012 for TTL and AM97C12 for C-MOS digital control. The LF356 op amp settles fast.

Chapter 2
ANALOG SIGNAL PROCESSING

Improved analog divider finds large-signal quotients

by Umesh Kumar
New Delhi, India

The linearity and signal-handling capability of the divider circuit proposed by Kraus for finding the quotient of two analog voltages [*Electronics*, Aug. 5, 1976] can be easily improved. What's more, the cost of the updated unit is virtually equal to that of the original design.

As shown in the voltage-to-time converter portion of the system, input signal V_y drives constant-current source A_1–Q_1. Note that V_y is capable of assuming a peak value of at least 10 volts, in contrast to the original circuit where A_1's gain of 15 limits V_y's maximum input value to 1 V for a 15-V supply voltage. In addition, the constant current source affords a virtually linear charging of C, unlike the scheme employed initially, where a field-effect transistor operates as a voltage-controlled resistor (R_{ds}) through which C is charged.

Upon application of a trigger pulse at pin 2 of the 555, the output (pin 3) of the one-shot goes to logic 1 and C charges toward $2/3\ V_{cc}$. When the voltage across C indeed reaches $2/3\ V_{cc}$, the capacitor discharges and the output returns to its low state. The process thereupon repeats, with the oscillating output of the 555 serving as a modulating signal at Q_2–A_2.

Thus, during the charging of C, $2/3\ V_{cc} = (I/C)\ t_c = (V_y/RC)\ t_c$, or $t_c = (2V_{cc}RC)/3V_y$, where I is the current emanating from the current source and t_c is the charge time of the capacitor. At this time, Q_2 is on and so the output voltage is $V_{out} = -V_x$.

If τ denotes the period of the trigger pulse, the timer will be low for the fraction of the period when $(\tau - t_c)/\tau)$, during which time Q_2 is off and the output voltage $V_{out} = 0$. Therefore, the average voltage over a cycle is:

$$\overline{V}_{out} = -V_x t_c/\tau = -2/3(V_{cc}RC/\tau)\ (V_x/V_y)$$

where $\overline{V}_{out}$ can be recovered by an RC network placed at the output of the circuit or read with an average-responding meter. If RC/τ is made equal to 10 and $V_{cc} = 15$, then $V_{out} = -V_x/V_y$. □

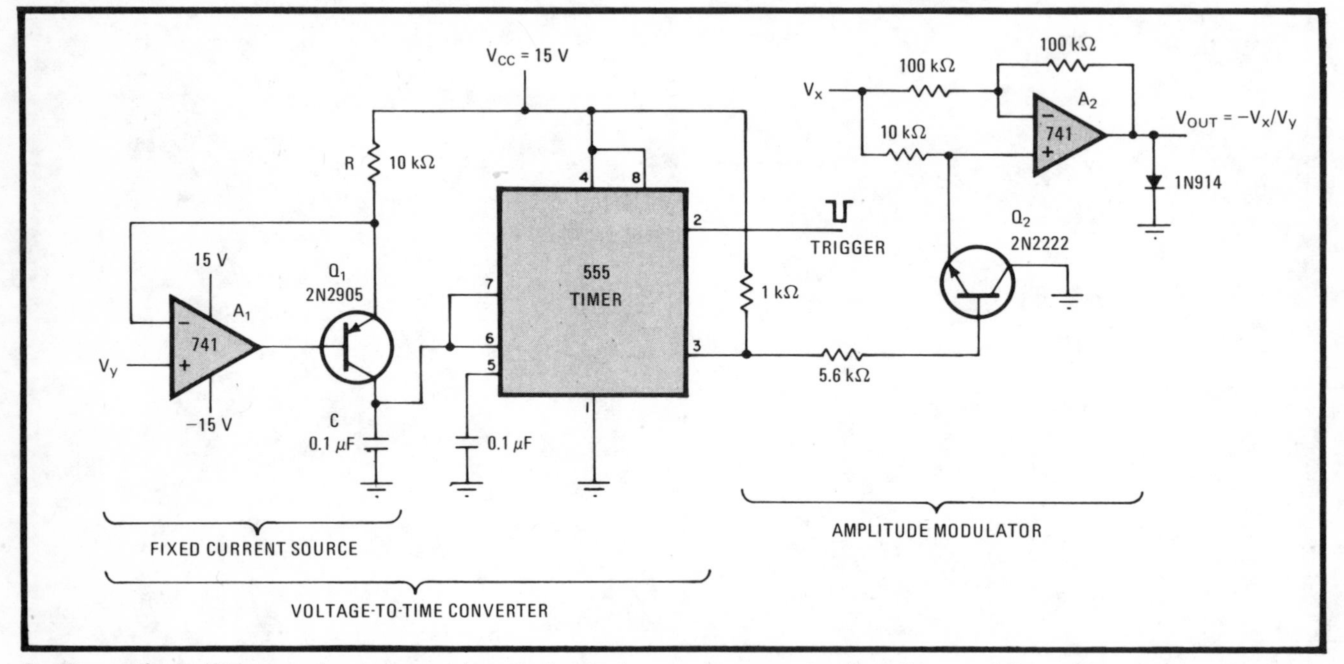

Quality quotients. Wide-range input stage and constant-current source increases signal-handling capability and linearity of analog divider Output of 555, which is a function of V_y, R, C, and τ, modulates V_x so that $\overline{V}_{out} = -V_x/V_y$ for $1 < V_y < 10$ and $0 < V_x < 15$.

Switched V-f converter linearizes analog multiplier

by Kamil Kraus
Rokycany, Czechoslovakia

This analog voltage multiplier provides a degree of linearity not attainable with circuits that use rudimentary voltage-to-frequency converters. And when modified, it is a more versatile analog divider than the circuit proposed by Kumar[1], which finds the quotient for only one fixed reference voltage because of the way its current source is configured. The much improved performance is achieved with a one-chip unit designed specifically for V-f conversion duties, and a simple but accurate switched-capacitor arrangement to do the actual multiplication.

The product of two voltages is found by utilizing the TL604 single-pole, double-throw switches, thereby sampling input voltage V_2 periodically and placing a corresponding charge on capacitor C_u. The average current that flows to charge C_u during these intervals will thus be proportional to V_2, C_u, and the sampling frequency, which is a function of input V_1. The equivalent resistance corresponding to the average current that flows will therefore be $R_{eq} = k/C_u f$, where k is a constant and f is the sampling frequency. Assuming a high sampling frequency, a pure resistance equal to R_{eq} may be considered to be in series between the output of the switch and the inverting input of the TL071 operational amplifier, as shown.

Thus the output voltage V_{out} may be expressed as some function of V_2 multiplied by the switching function, or, more clearly, as $V_{out} = -V_2(R_2/R_{eq}) = -V_2R_2C_uf/k$, where the sampling frequency, on the order of 50 to 500 kHz for $0 < V_1 < 25$ V, is generated by the V-f converter. But the sampling frequency is given by $f = V_1/2R_1Ck_1$, and so $V_{out} = V_1V_2R_2C_u/R_1Ckk_1$, which is proportional to $-V_1V_2$ (constant k_1 is introduced by the VCO).

If the output circuit is modified slightly (see inset), the circuit will function as a divider. Then, $V_{out} = V_2(R_{eq}/R_2) = R_2V_2k/C_uR_1C_1k_1 \sim -V_2/V_1$. □

References
1. Umesh Kumar, "Improved analog divider finds large-signal quotients," p. 12

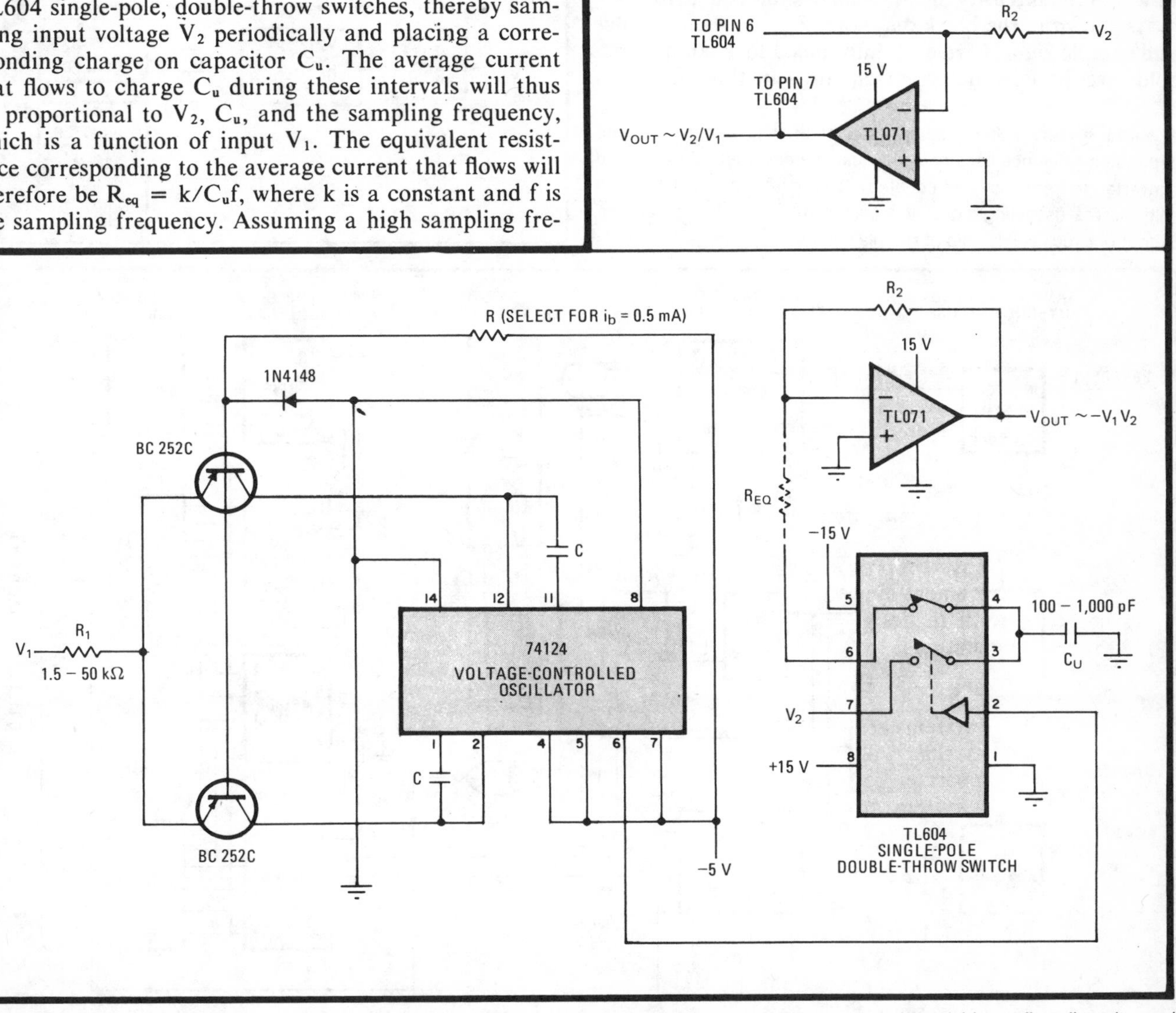

Mathematics. One-chip V-f converter and switched capacitor arrangement provided by solid-state switches yield excellent linearity and wide-range voltage-handling performance in an analog multiplier. Interchanging R_{eq} and R_2 by modifying output circuit slightly (see inset), where R_{eq} represents the average current flowing in capacitor C_u during sampling period, converts unit into two-input analog divider.

Digital normalizer derives ratio of two analog signals

by James H. McQuaid
University of California, Lawrence Livermore Laboratory, Livermore, Calif.

The absolute value of a voltage or current at a circuit point is frequently less important than the ratio of that quantity to a reference. This circuit compares two analog signals by using a high-accuracy digital technique for normalizing the reference voltage, thus simplifying circuitry and avoiding the use of analog dividers or microprocessors. It is invaluable in many light-chopping applications, such as lasers, where the measurement of light intensity at a specific frequency must take into account total source-intensity variations. It is also useful in atmospheric physics when measuring the concentration of a specific gas in a mixture by infrared techniques in which the intensity of the beam is subject to drift.

As shown in the block diagram (Fig. 1), the reference and sample signals are each introduced to a sample-and-hold circuit. Peak-detection circuitry in this device, in conjunction with track-and-hold logic, which controls the sample rate, produces a voltage output that is presented to their respective voltage-to-frequency converters. Each converter's output is a pulse train with a frequency directly proportional to the input voltage.

The track-and-hold logic and associated gating circuits simultaneously initialize both scalar circuits and allow the output pulse train of both converters to be counted. When the count in the reference scaler exceeds its capacity, an overflow pulse is generated that closes the gating circuit; at this time, the contents of the sample scaler are clocked into the holding register and then sent to the display. The contents of the reference scaler are regarded as a unit voltage, and the reference scaler

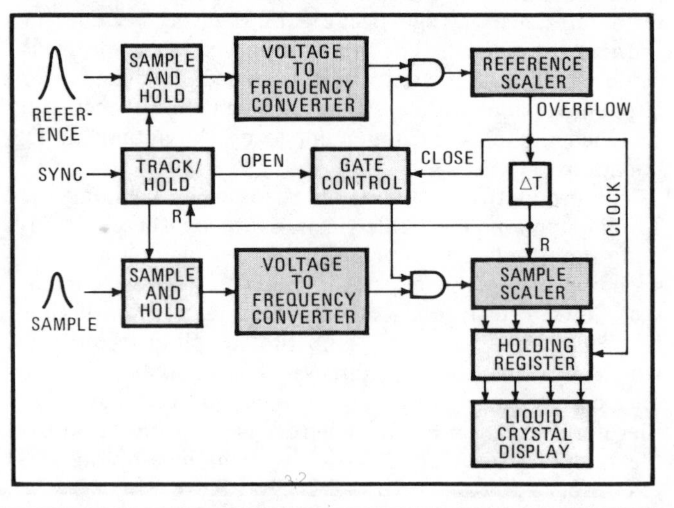

1. Ratio meter. Voltage comparison of analog sample to time-dependent reference uses analog-to-digital converters. Voltages are converted to frequency and counted. Reference count normalized to unity serves as gating signal to count number in sample counter. Output of circuit yields ratio of sample to reference voltage.

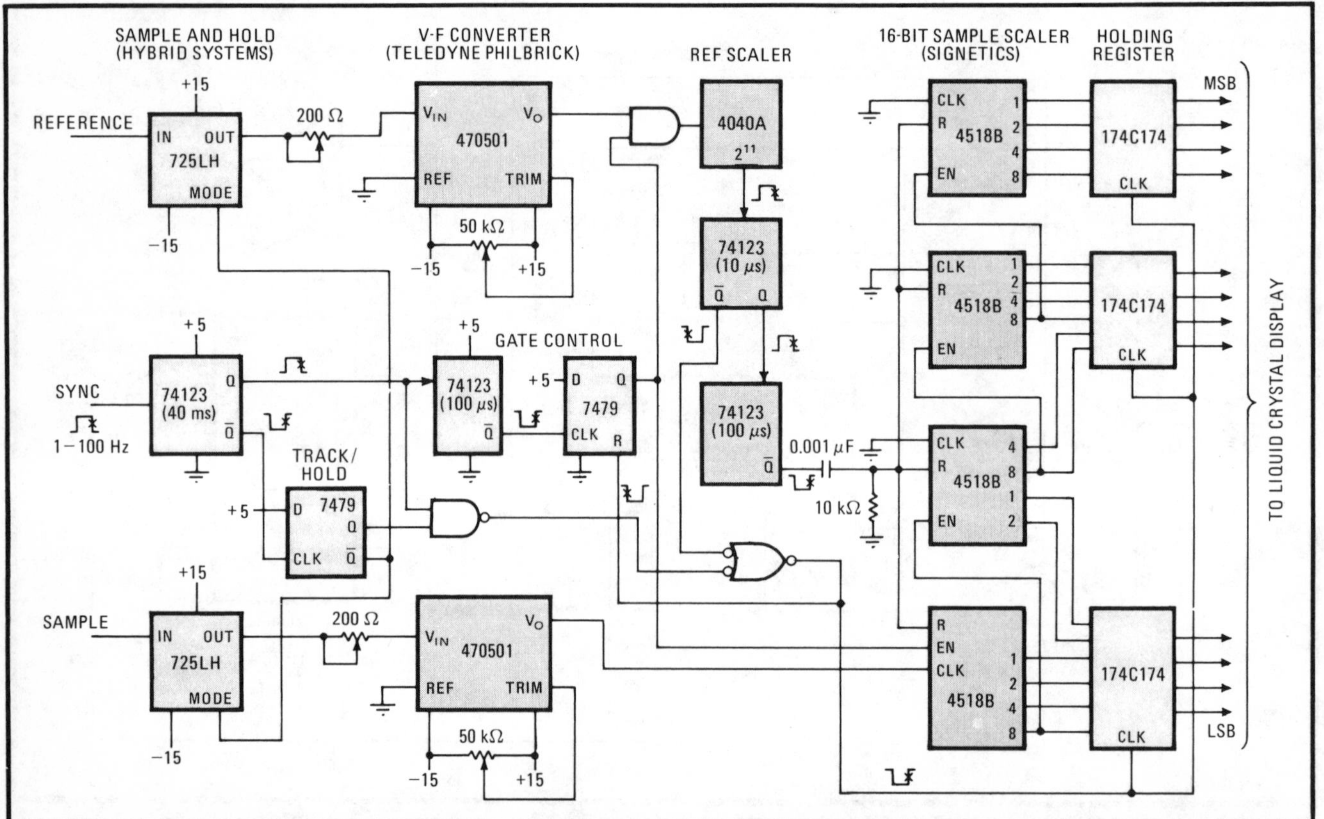

2. Digital normalization. Circuit detects relationship of analog sample amplitude to reference voltage by digital means. Several one-shots are used to obtain proper timing and gating. Circuit uses C-MOS devices where possible to reduce power consumption.

controls the gating time to the sample scaler. Thus the contents of the holding register will normally be some fraction of the unity-set voltage. The sample scaler and sampling circuits are then reset, ready to process the next sample.

As shown in Fig. 2, the sample-and-hold devices are Hybrid Systems 725LH devices, which are accurate to within 0.01% and have a droop rate of 15 millivolts per second using their internal holding capacitor. The sync input used to control the sampling period is a 5-volt, 1-to-100-hertz pulsed voltage. The converters are Teledyne Philbrick 470501 devices with an upper limiting frequency of about 1 megahertz. This frequency is produced at an input of 10 v, and the device's voltage-frequency characteristic is linear to 0.005%. The converter may be easily calibrated with its 50-kilohm trimmer potentiometer and the 200-ohm rheostat at the output of the 725LH device.

The 4040 reference scaler is a 12-bit binary counter. After reaching its counting capacity (1,024) during a given sampling period, it clocks the contents of the Signetics 4518B into the 174C174 C-MOS holding register while resetting the gating circuits. The sample scaler capacity (16 bits) and the large reference scaler capacity (12 bits) ensure a high counting accuracy, typically to within 0.1%. In addition, the larger capacity of the sample scaler allows the signal amplitude to exceed the reference amplitude while the correct ratio is still displayed.

The ΔT function in the block diagram is a small but important part of the circuit. It is implemented as shown in Fig. 2 with a number of one shots to achieve correct timing and edge triggering for data transfer. The digitizing time for the circuit shown is 4 milliseconds. Greater speed (with less accuracy) can be easily achieved by reducing the number of bits in the reference scaler. For instance, a digitizing time of 250 microseconds may be achieved with an 8-bit reference scaler. □

Balanced modulator chip multiplies three signals

by Henrique Sarmento Malvar
Department of Electrical Engineering, University of Brazilia, Brazil

Three signals can be multiplied by a one-chip double-balanced modulator, a device normally capable of mixing only two. In this case, the third input is introduced at the bias port of Fairchild's μA796, enabling the unit's transconductance to be varied at an audio rate, which effects modulation. Although the technique reduces the bandwidth and dynamic range over which the device can operate, the three-input mixer will still be useful in many applications, notably for generating discrete sidebands and synthesizing music.

The output voltage from the mixer, V_o, is the product of a carrier switching function, V_c, and a modulated signal, V_s. Because the bias signal, V_b, controls the conductivity of transistors that are effectively in series with the usual modulating signal, it is simply regarded as a second modulating signal such that $V_o = kV_bV_cV_s$, where $k = 0.00064R_L/R_I$ over a small dynamic range and the voltages are in millivolts.

Note that V_c and V_s may be positive or negative but that V_b must always be positive to prevent the reverse-biasing of the internal transistors connected to pin 5. When V_b is negative, A_1 blocks the application of negative voltage by turning off the BC 178 transistor.

Note also that k will vary nonlinearly with the amplitude of V_c and V_s, and in order to keep k within 1% of its given value, both voltages must not exceed 8 millivolts root mean square. V_b will not affect k if $I_{bias} << 1$ milliampere. Unfortunately, k also is sensitive to temperature changes ($-0.67\%/°C$), and this factor can limit the circuit's effectiveness in high-accuracy applications.

The output frequencies generated by the mixing

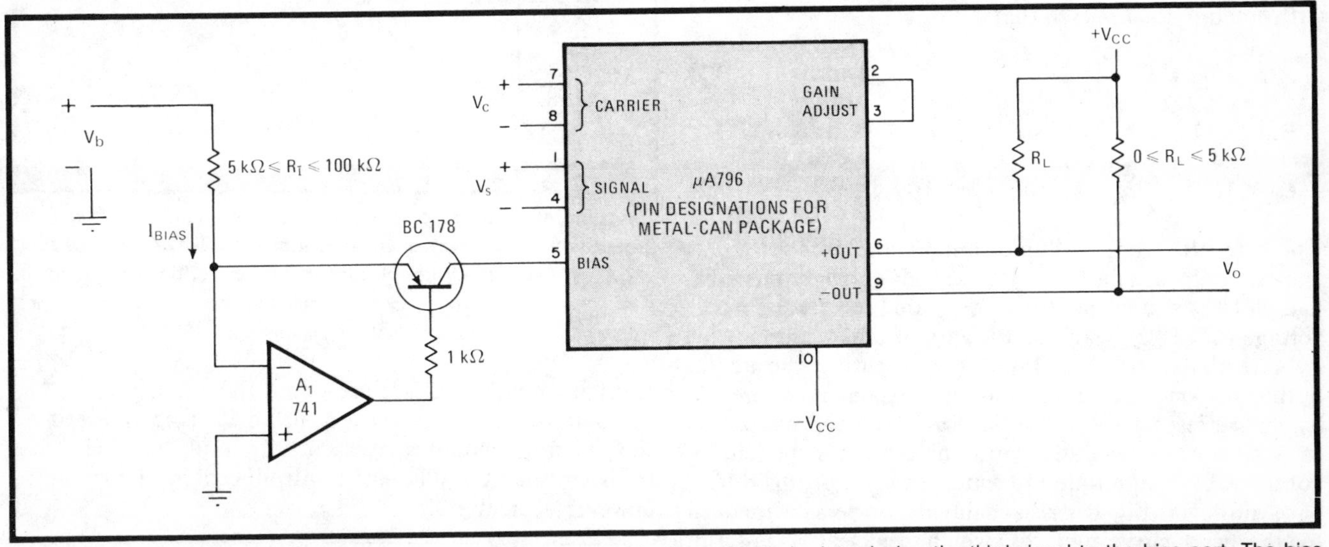

Biased mixer. A balanced modulator can be configured to multiply three signals by introducing the third signal to the bias port. The bias current, which normally sets the dc operating point of device, is varied at an audio rate to effect linear modulation over a small dynamic range.

process will be $|f_c \pm f_s|$, and $|f_c \pm f_s \pm f_b|$, where the f subscripts correspond to their voltage counterparts. There will be no output if any of the driving signals are disconnected.

Signal inputs V_c and V_s may have an f_c or f_s of one megahertz at the maximum. The third input, V_b, will be band-limited to 20 kilohertz, however, because of the relatively poor frequency response of the 741. □

Multiplier, op amp generate sine for producing vectors

by Jerald Graeme
Burr-Brown, Tucson, Ariz.

To compute trigonometric functions of the kind that produce a vector from its X and Y components, a system needs to be supplied with signals that are proportional to the sines of the applied X and Y dc-voltage inputs. A multiplier, operating as a signal squarer, and an operational amplifier, serving as a subtracter, will process the X or Y component more simply than other circuits now generally used. With these components, the desired waveform response for either input—the positive half of a sine function—can be approximated to within 5% of a perfect waveform, a value well within the limits that yield good computational accuracy.

Using the multiplier and the op amp to generate the sinusoidal transfer function is much easier than using diodes in complex circuits to synthesize the nonlinear response by producing a piecewise-linear approximation. Also, this circuit produces the sine response over the required −90° to +90° quadrant using one less multiplier than existing analog circuits,[1] which generate the approximation by means of a long mathematical series or an equivalent method.

The circuit shown in (a) processes the dc input voltage that corresponds to either the X or Y component of the vector. The BB4213AM has a transfer function of $G = X'Y'/10$, where X' and Y' are the inputs; it is made to square the input voltage, e_i, when the X' and Y' ports are connected. The output of the squarer is then introduced into the inverting port of the BB3500 op amp, with e_i fed to its input, so that:

$$e_o = 3.86\left(e_i - \frac{e_i^2}{10}\right) \qquad (1)$$

But Eq. 1 can be expressed by:

$$e_o \doteq 10 \sin\left(\frac{e_i}{10}\pi\right), \quad 0 \leq e_i \leq 10 \qquad (2)$$

which results from a simple series approximation. This can be confirmed in the actual output-versus-input voltage plot (b). Equation 2 may be scaled for other input-voltage ranges by changing the gain of the op amp.

Note that term 2 in Eq. 1 (representing the actual output response) has a second-order exponent, whereas a Taylor-series, which would yield a closer approximation to a sine wave, would have a third-order exponent of considerable magnitude in term 2. In spite of this difference, however, Eq. 1 is reasonably accurate as evidenced by the error-curve plot (c), which peaks at 4% of full scale. Dominant distortion is related to the second and third harmonics that are present at the output.

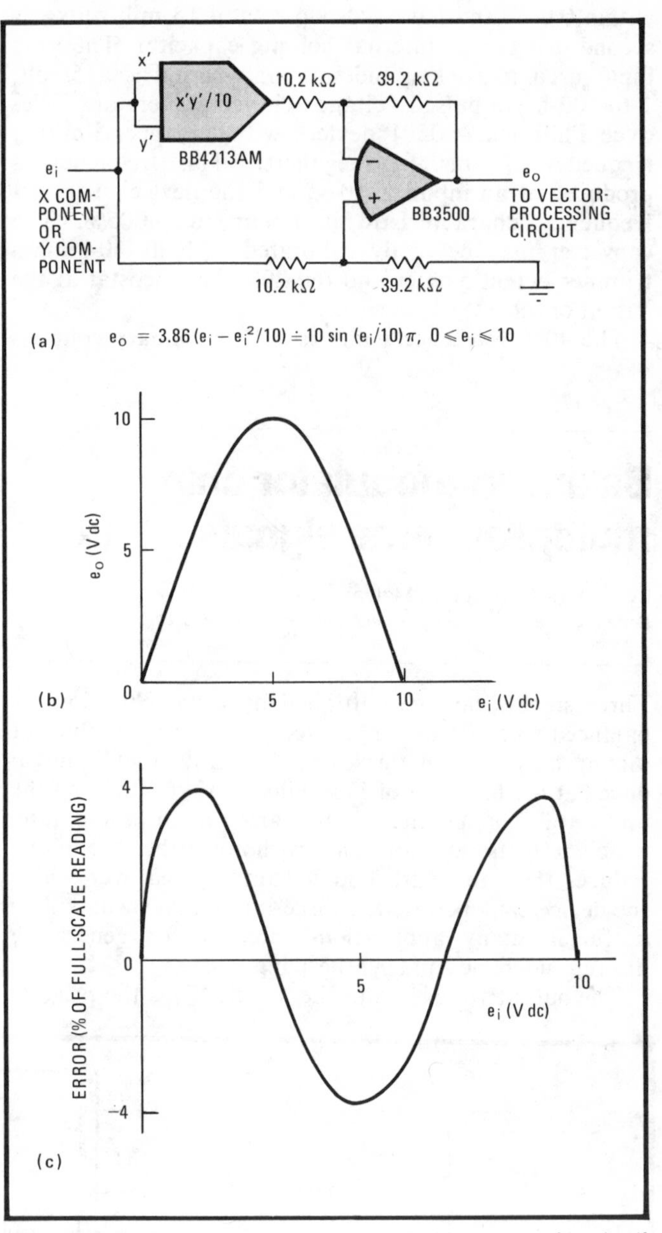

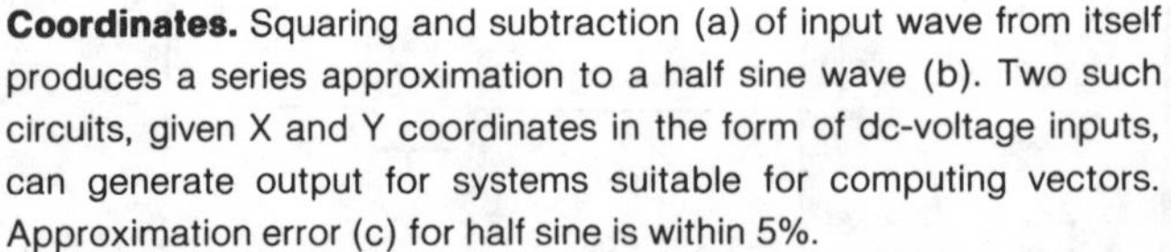

Coordinates. Squaring and subtraction (a) of input wave from itself produces a series approximation to a half sine wave (b). Two such circuits, given X and Y coordinates in the form of dc-voltage inputs, can generate output for systems suitable for computing vectors. Approximation error (c) for half sine is within 5%.

An additional error term of up to 1% is introduced by the inherent nonlinearities in the multiplier circuit's transfer function. The error introduced by the op amp, however, is negligible. □

References

1. Burr-Brown, Model 4213 Product Data Sheet, PDS366, 1976.

Chapter 3
AUDIO CIRCUITS

Adjustable e^x generator colors synthesizer's sounds

by Randall K. Kirschman
Mountain View, Calif.

Providing the control signals for voltage-controlled amplifiers, oscillators and filters in order to modulate sound parameters such as loudness, pitch and timbre, this adjustable e^x generator is the indispensable ingredient required to attain superior performance in a music synthesizer. Only four integrated circuits and a few passive components are needed in the inexpensive unit, which costs under $6.

When gated or triggered, the generator produces a waveform that passes through four states:

- An exponential attack.
- An initial decay, or fallback.
- A sustain, or steady dc level.
- A final decay, or release.

Each of these four parameters is continuously variable, so that waveforms having a large variety of shapes can be generated.

The waveforms are generated by the sequential charging and discharging of capacitor C_1 (see figure). In general operation, C_1 is connected to a current source or sink as required, through the 4016 complementary-MOS analog switches. These switches are controlled by simple logic set into action by the gate-input pulse. Triggered operation is made possible by adding a monostable multivibrator to the circuit.

In the dormant state (gate input low), analog switch C is on, switches A and B are off and the RS flip-flop formed by two 4001 NOR gates is reset. The onset of a gate pulse turns on switch A and turns C off. Consequently, C_1 charges through R_3 and R_4, producing the attack segment of the waveform. Note that the LM356 buffer protects C_1 from excessive loading.

When the voltage across C_1 reaches V_{max} (determined by voltage divider R_1–R_2), the LM311 comparator sets the RS flip-flop. This action in turn switches B on and A off. Thus the initial decay segment is generated as C_1 discharges through R_5 and R_6 to reach the sustain voltage, the level of which is determined by the setting of potentiometer R_2.

Concurrently, the comparator's output has gone low, but the RS flip-flop remains set until the gate pulse moves to logic 0, at which time switch C turns on. Thus C_1 discharges through R_7 and R_8 to produce the final-decay portion of the wave, after which the circuit reverts to its dormant state. □

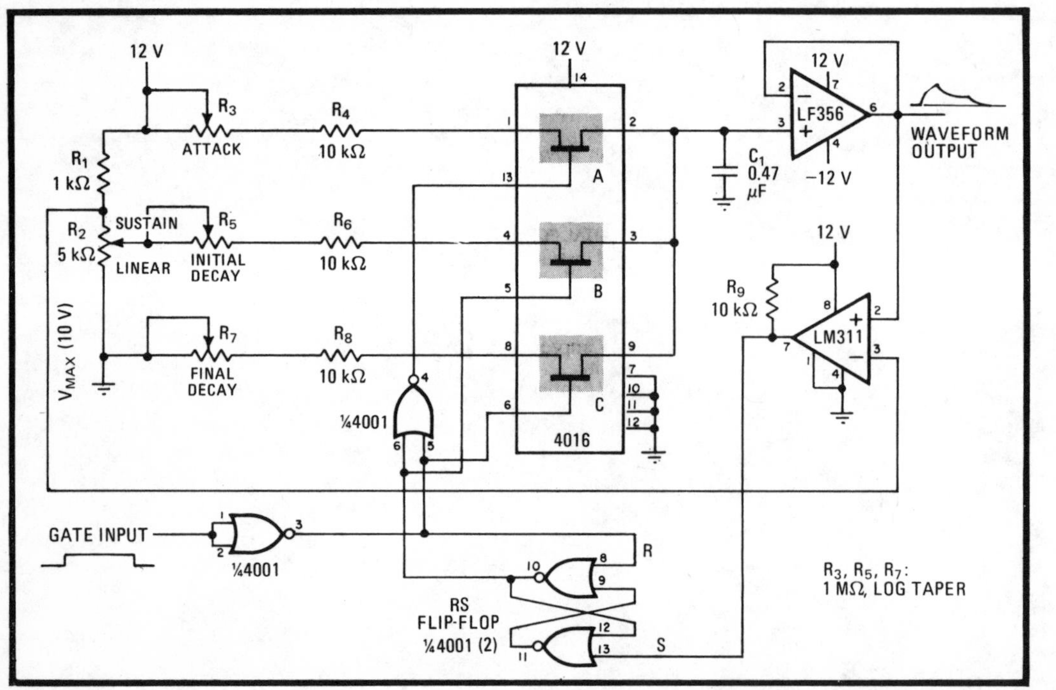

Musical tint. Four-state generator provides myriad control waveforms for modulating voltage-controlled amplifiers, oscillators, and filters in a music synthesizer, and thus is useful for coloring loudness, pitch, and timbre. Attack and decay times are variable from 5 to 500 milliseconds; sustain level is adjustable from 0 to 10 volts.

Two-chip generator shapes synthesizer's sounds

by Jonathan Jacky
Seattle, Wash.

Generating the same adjustable modulating waveforms for a music synthesizer as the circuit proposed by Kirschman[1], but using only two integrated circuits, this generator also works from a single supply. It has, in addition, separate gate and trigger inputs for providing a more realistic keyboard response.

When gated or triggered, the generator, which is built around Intersil Inc.'s C-MOS 7555 timer, produces a waveform that passes through four states:

- An exponential attack.
- An initial decay, or fallback.
- A sustain, or steady dc level.
- A final decay, or release.

Each of these four parameters is continuously variable, so that waveforms having a wide variety of shapes can be generated.

The waveforms are generated by the sequential charging and discharging of capacitor C_1. Here, the 7555 controls the sequencing while diodes switch the currents, unlike Kirschman's circuit where comparators and flip-flops control the stepping and analog switches steer the currents. Furthermore, the 7555 is well suited for handling the two logic signals provided by most synthesizer keyboards—the gate, which is high as long as any key is depressed, and the trigger, which provides a negative pulse as each key is struck. The gate and trigger features eliminate the need to release each key before striking the next to initiate an attack phase.

In the dormant state (the gate input at pin 4 of the 7555 is low), capacitor C_2 is discharged. When the gate goes high and a trigger pulse appears at pin 2, the 7555 output (pin 3) goes high and charges C_1 through R_3, R_4, and D_1, producing the attack segment of the waveform. Note that diode D_2 is reverse-biased because pin 7 of the 7555 is high and that diode D_3 is back-biased by logic 1 signal applied to the gate input.

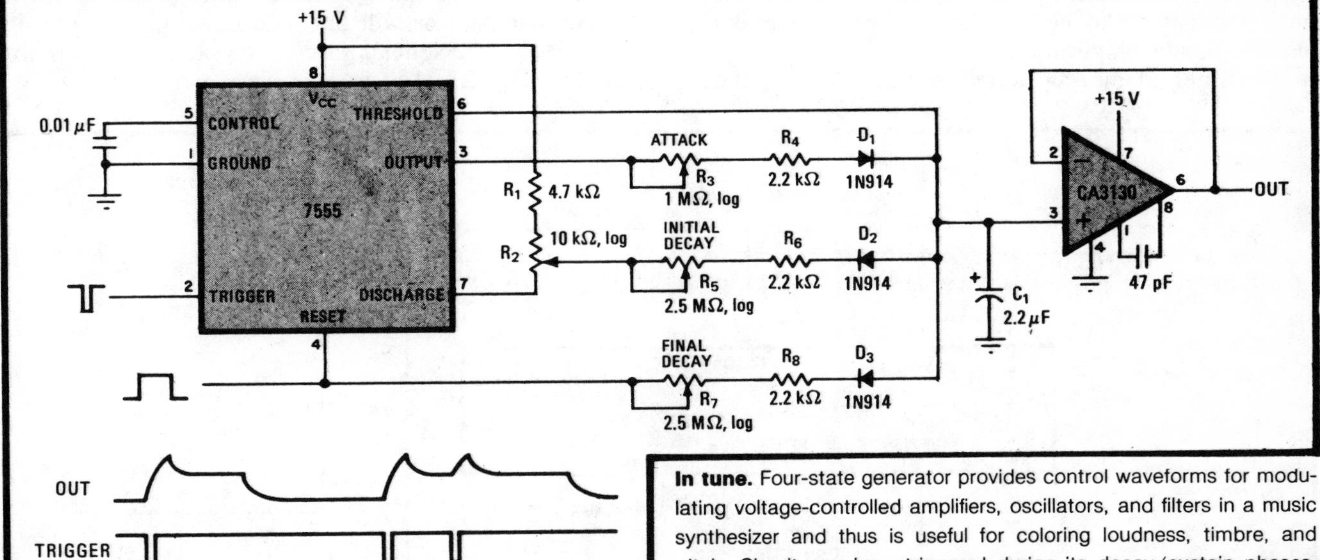

In tune. Four-state generator provides control waveforms for modulating voltage-controlled amplifiers, oscillators, and filters in a music synthesizer and thus is useful for coloring loudness, timbre, and pitch. Circuit can be retriggered during its decay/sustain phases. Attack time is variable from 5 milliseconds to 2 seconds. Initial and final decay times can range from 5 ms to 5 s. The sustain level has a dynamic range of 0 to 10 V.

When the voltage across C_1 reaches 10 volts, pin 3 of the 7555 goes low and pin 7 is grounded, terminating the attack phase. D_1 and D_3 are now reverse-biased and C_1 discharges through D_2, R_5, and R_6 to produce the initial decay. The sustain level reached is determined by the voltage divider formed by resistor R_1 and potentiometer R_2. During this phase, a second attack can be obtained by striking another key (see timing diagram). When the last key is released, the gate goes low and C_1 will discharge through D_3, R_7, and R_8 to produce the final decay. The CA3130 operational amplifier serves as a buffer to protect C_1 from excessive loading. □

References
1. Randall K. Kirschman, "Adjustable e^x generator colors synthesizer's sounds," p. 18

Tuning-meter muting improves receiver's squelch response

by Albert Helfrick
Aircraft Radio and Control Division, Cessna Aircraft Co., Boonton, N. J.

Although the CA3089 FM/IF system offers the advantages of one-chip simplicity, good limiting capability, high gain, and excellent linearity when used in wideband fm-broadcast receivers,[1] its audio-channel muting performance, and thus its squelch response, suffer in narrow-band applications, especially at low signal levels. Utilizing the output voltage from the device's tuning-meter port to drive the audio-muting control circuit through an operational amplifier greatly improves squelching, particularly for signals whose amplitudes are barely detectable.

The circuit configuration of a typical limiter-discriminator designed for a modulation deviation of ±5 kilohertz is shown in the figure. As in many discriminators, a crystal serves for the high-Q tuned circuit and so makes possible the high audio recovery required in a narrow-band configuration.

The internal muting action of the CA3089, though sufficient for wideband service, lacks the speed and precision necessary for narrowband operation, because the system's effectiveness is a function of the characteristics of the detector's frequency-determining elements connected to pins 8, 9, and 10, as well as the gain distribution of the entire receiver. Narrowband receivers usually make full use of the system's available sensitivity by having as much gain as possible before the detector so that limiting occurs on noise, and the small-bandwidth characteristics of the CA3089 circuit are similar. At low signal levels, this limiting causes the squelch circuit to be almost useless. In some of the recommended circuits for frequency discriminators, the squelch circuit will not operate at all.

Driving the mute-control amplifier from the tuning-meter port (pin 13) instead ensures that the tuned circuit and the chip's gain distribution have no effect on squelch operation. The meter-output voltage, taken from the unit's three intermediate-frequency amplifiers and their level detectors, has a constant characteristic—that is, it is independent of the tuned circuit used. In fact, the response is virtually linear for input signals ranging from 5 to 10,000 microvolts.

The high input impedance of the low-noise CA3140 op-amp comparator will not load down pin 13, and its gain enables the squelch circuit to operate in a surefire manner. The CA3140 is used as a comparator with a

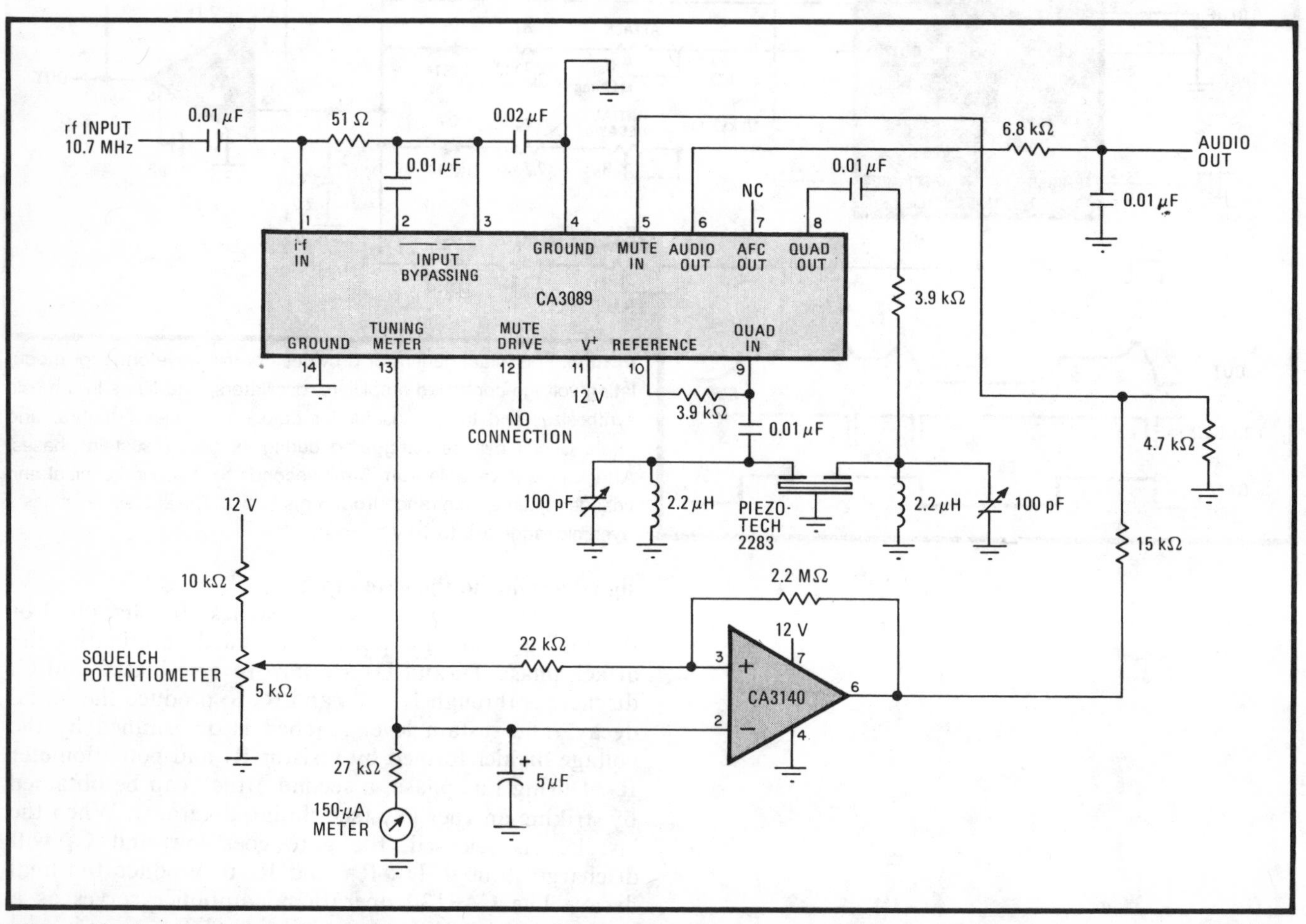

Silence. Circuit derives voltage for squelch-control amplifier from CA3089's tuning-meter port, whose output is linear over 5 to 10,000 μV. Op amp provides gain for surefire operation. Configuration provides positive squelch response, even at low signal levels, by bypassing the combined nonlinear response of tuned circuit and mute-drive circuit internal to the CA3089.

variable threshold set by the squelch potentiometer, as shown. A voltage divider at its output ensures that no more than about 5 volts can be applied to port 5—anything higher would cause latchup in the CA3089, which might cause excessive power dissipation.

The CA3140 can operate with a common-mode voltage equal to that of the negative supply, and it may therefore be operated from the same power source as the CA3089. □

References
REFERENCES
1. J. Brian Dance, "One chip fm demodulator has improved response." p. 230

Resistor matrix orchestrates electronic piano/tone generator

by Hsi Jue Tsi
National Taiwan University, Taipei, Taiwan

Combining an eight-by-eight-resistor matrix that is programmed to generate 64 different frequencies with control circuits that send each tone in sequence for half-second intervals, this unit makes a tuneful electronic music box. It can also serve as a programmable tone generator for testing purposes, in which case the number of tones may be extended to 512.

In this circuit, the resistor timing element in an astable multivibrator is periodically switched to a new value every half second with the aid of a multiplexer circuit. Serving as the multivibrator (or tone generator) is the versatile 555 timer, which can generate frequencies over a range of 100 hertz to 5 kilohertz.

Timing resistors are switched into the R_iC network

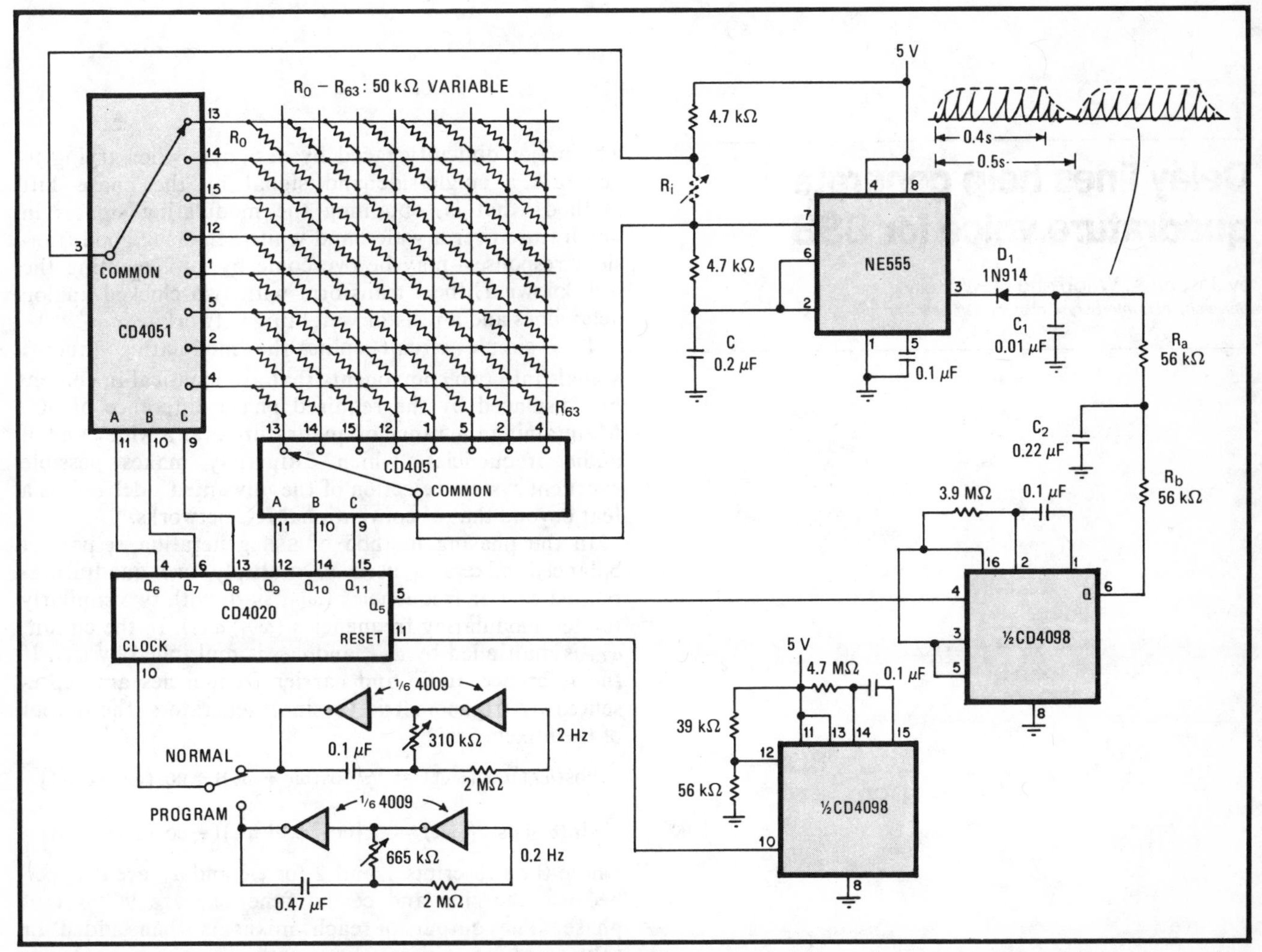

Tuning up. Music box/tone generator uses programmable resistor matrix to generate 64 tones in sequence, each for a 0.5-second period. Oscillator clocks 4051s, which in turn place each of 64 resistors in timing network of 555. Output of timer is shaped by a 4098 one-shot and suitable integrator network for click-free output during switching. Matrix can be programmed in 10 minutes if counter is available.

with the aid of two CD4051 multiplexers, as shown. Each device has three address ports that are updated by a 4020 counter, which is in turn stepped by an oscillator normally running at 2 Hz. If, for example, all outputs of the 4020 are low, the common ports of the 4051s will be connected to row 1, column 1, of the matrix and R_0 will be in the 555's timing loop. As the counter steps through all locations, R_1–R_{63} in turn set the frequency of the 555.

The output of the 555 is a train of square waves. To avoid the key clicks that occur each time the frequency of a tone is changed, the square wave is rounded off, being converted to a sawtooth wave with D_1, R_aC_1, and R_bC_2. Note that the 4098 monostable multivibrator connected to R_b aids greatly in controlling the attack-delay characteristic of each waveform at the output; it applies a gradual turn-on bias to D_1 initially and then a turn-off bias after about 0.4 second. This 4098 is driven by the 4020. R_b and C_2 integrate the 4098's output pulse, then C_2 discharges, enabling the device to bias the diode as described. A second 4098 one-shot is used to reset the 4020 so that resistor R_0 will be immediately accessed on power-up.

Programming of the resistor values can be tedious, but with practice it can be done in 10 minutes. A frequency counter connected to the output of the 555 is helpful. Despite the harmonics in the 555's output, the counter will read the fundamental frequency of a given tone.

First, it is necessary to switch to the program clock. This clock will advance the 4020 counter once every 5 seconds and gives the user time to adjust each R_i for the particular output frequency desired during that span. Two passes over the 64 tones should be adequate.

If a counter is not available, a piano or tuning fork will be needed and tuning will have to be done by ear, requiring an extremely long programming time. Means will also have to be found to single-step the 4020.

For more demanding applications, the number of tones can be expanded to 512 by adding another CD4051 and the appropriate number of resistors. □

Delay lines help generate quadrature voice for SSB

by Joseph A. Webb and M. W. Kelly
University of Canterbury, Christchurch, New Zealand

The major difficulty faced by designers when trying to generate a single-sideband signal by the phase-shift method—that is, obtaining the modulating signals in quadrature over a wide band while achieving good transient response—may be overcome by implementing the well-known Hilbert transform with two clocked analog delay lines and a resistor weighting network.

This simple circuit splits the modulating (audio) signals into two components that are identical in content but displaced by the required phase difference of 90°. Maintaining the range of quadrature over a wide band of audio frequencies, which ultimately makes possible excellent system rejection of the unwanted sideband, is a feat beyond that of conventional RC networks.

In the phasing method of SSB generation, a pair of balanced mixers is used to multiply two quadrature-related carrier frequencies (ω_{C1}, ω_{C2}), with two similarly related modulating frequencies (ω_{V1}, ω_{V2}). In the circuit, ω_{C1} is multiplied by ω_{V2}, and ω_{C2} is multiplied by ω_{V1}. If the reference audio and carrier frequencies are represented by trigonometric (cosine) generators, the output of the mixers are:

$$\cos(\omega_C t)\cos(\omega_V t) = \tfrac{1}{2}[\cos(\omega_C+\omega_V)t+\cos(\omega_C-\omega_V)t]$$

$$\sin(\omega_C t)\sin(\omega_V t) = \tfrac{1}{2}[\cos(\omega_C+\omega_V)t-\cos(\omega_C-\omega_V)t]$$

where the subscripts 1 and 2 for ω_V and ω_C are dropped because the sine and cosine functions are 90° out of phase. The output of each mixer is then added or subtracted to obtain the upper ($\omega_C+\omega_V$) or lower ($\omega_C-\omega_V$) sideband, as desired. Remember, however, that quadrature between the audio and carrier frequencies

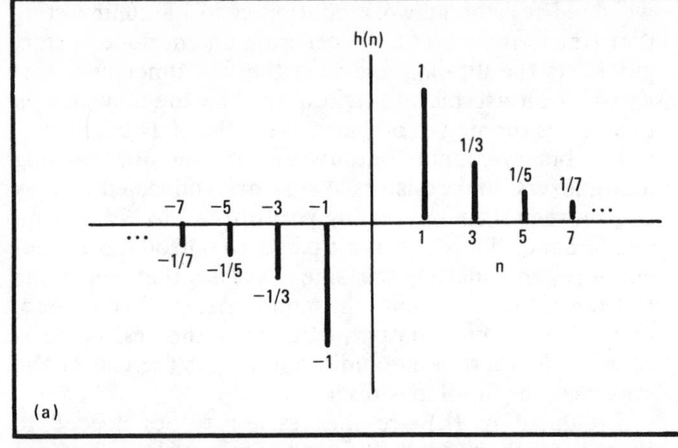

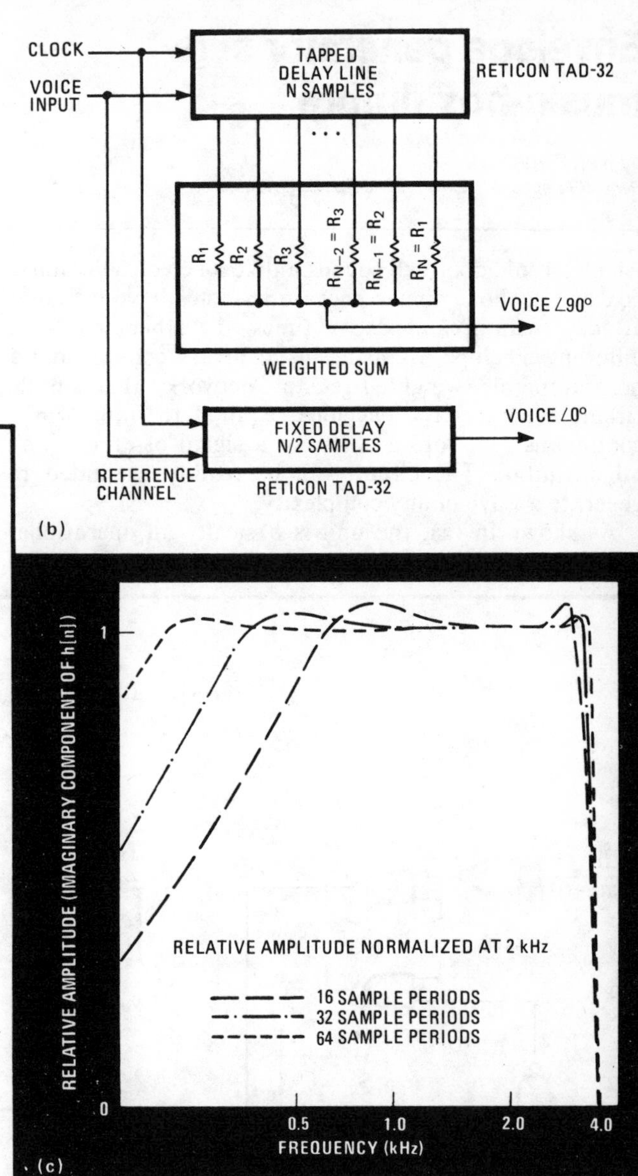

Constant phase. Hilbert transform function shown in (a) is implemented by delay-line circuit shown in (b) in order to keep modulating signals in phase-modulated single-sideband system in true quadrature. Plot of imaginary component of circuit's generated Hilbert transform, h(n), indicates good transient response (c). Audio signals remain in quadrature over entire frequency range shown.

must be maintained for optimum response.

The discrete Hilbert transform of any signal, that is:

$$h(n) = \frac{1 - e^{j\pi n}}{\pi n} = \frac{1 - \cos\pi n}{\pi n}$$

corresponds to a 90° phase shift of all its frequency components, and thus by implementing this function the quadrature relationship for the audio channels is maintained. Attaining quadrature for carrier signals is simple, since the ω_C signal has virtually zero bandwidth.

The discrete Hilbert transform is defined from plus to minus infinity, although truncation is needed for physical realization of the function. The truncated impulse response of this function is illustrated in (a).

The required response may be generated with the delay-line circuit shown in (b). A Reticon TAD-32 charge-coupled device is used for the delay line. The weighting resistors are selected so that the circuit will generate the product of the truncated function, h(n), and a smoothing or weighted function, W(n), where $W(n) = \cos^2 n\pi/N$. Each resistor is selected so that $R(n) = h(n)W(n)$. Note that the $\cos^2$ function is defined from +90° to −90°, not from plus to minus infinity.

The reference voice channel is delayed by N/2 samples for the audio channels to remain in true quadrature. At a clock frequency of 8 kilohertz, the delay amounts to 4 milliseconds for 64 samples.

The plot of the imaginary component of h(n) in (c) of the figure illustrates the excellent transient response of the circuit. As can be seen, relatively few samples are needed for good performance. In these tests, the clock frequency was 8 kHz. For telephone-quality voice signals, N = 32 is sufficient, and N = 64 represents excellent performance. Since the Hilbert transform is symmetrical, that is, $f(t) = -f(t)$, quadrature is perfect over the entire frequency range shown. □

Envelope generator sets music-box timbre

by Ken Dugan
General Telephone & Electronics Corp., Clearwater, Fla.

An electronic door bell sounds unlike an electronic music box or telephone ringer because its notes have different attack, sustain, and decay times—in other words, a different envelope. By using just a binary counter and a programmable weighted-resistor network, this simple circuit generates the envelope required to transform a continuous tone into a chime or a signal of almost any other timbre. The circuit can be readily expanded to generate a wave of any complexity.

As shown in (a), the unit is basically an operational amplifier that operates as a subtracter, with the weighted-resistor network connected to its noninverting port (the switched leg). To generate an envelope, a start pulse sets the flip-flop and fires the 555 timer, which is wired as an astable multivibrator. The timer, which in this case is running at 60 hertz, steps the 4024 counter.

The binary-counter outputs address the 4051 analog multiplexers, and resistors A – N are connected one by one between the noninverting port of the LM324 op amp and ground. Thus the multiplexers control the output envelope, modulating the sine wave so that when the resistance switched into the noninverting port is zero, there is maximum output, but when the resistance is equal to R, there is no audio output. At the end of the sequence, the flip-flop is reset.

Tabulated in (b) are the resistor values needed to generate a chime, or bell sound. The envelope required for a perfect chime is logarithmic (fast attack, no sustain, long delay), but the envelope is approximated by a simple sloping line as shown; otherwise many resistors and multiplexers would be needed. □

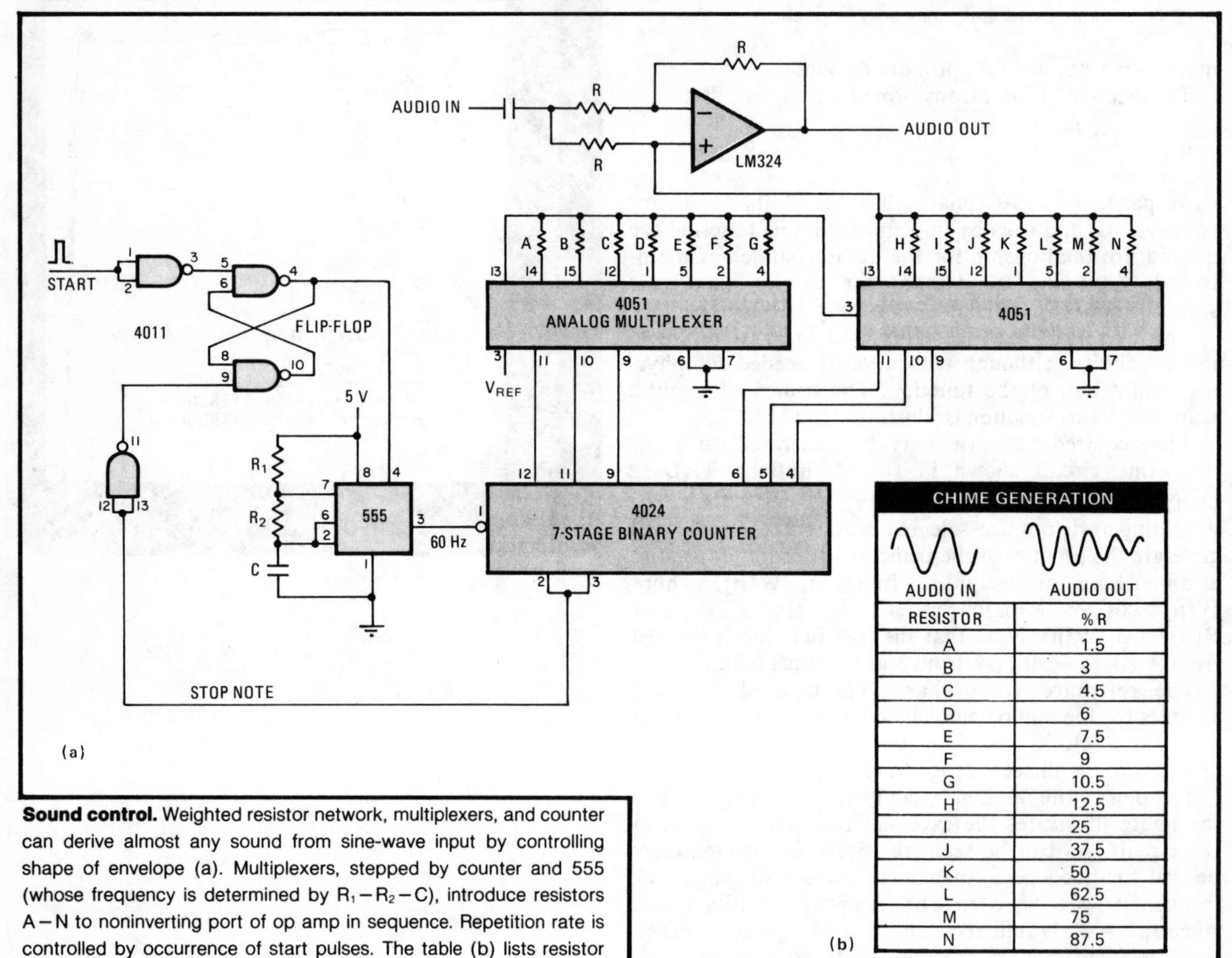

RESISTOR	% R
A	1.5
B	3
C	4.5
D	6
E	7.5
F	9
G	10.5
H	12.5
I	25
J	37.5
K	50
L	62.5
M	75
N	87.5

Sound control. Weighted resistor network, multiplexers, and counter can derive almost any sound from sine-wave input by controlling shape of envelope (a). Multiplexers, stepped by counter and 555 (whose frequency is determined by $R_1 - R_2 - C$), introduce resistors A – N to noninverting port of op amp in sequence. Repetition rate is controlled by occurrence of start pulses. The table (b) lists resistor values required for generation of chimes, or bell sounds.

Controller halts playback when taped voice pauses

by N. Bhaskara Rao
U.V.C.E., Department of Electrical Engineering, Bangalore, India

Few typists can transcribe a dictated message or speech without stopping the tape recorder from time to time. The illustrated circuit stops the recorder automatically at the end of a sentence or other pause. Its programmable halt time is proportional to the length of the preceding playback segment.

The circuit is preset on power up by R_1 and C_1. Audio signals from the output of the tape recorder may then be fed into a buffer amplifier having a low output impedance, so that a dc voltage proportional to the audio input is produced by the full-wave rectifier, D_1 and D_2.

The 7413 Schmitt triggers and an RC network define the time delay, T. The voltage at x(t), which is initially set high by the audio signal, goes low when V_o is low for a period greater than T, so that any pause in the audio signal triggers the recorder-halting circuit. The delay time selected may be varied for the particular application by adjusting one or both elements of the RC network.

As one-shot A_1 is triggered by x(t), A_2 is cleared and starts to count up. If the time during which x(t) is high is T_{on}, then the output of A_2 at the end of that period is given by $N = f_1T_{on}$, where f_1 is derived from divider A_3 and the master clock frequency, f.

As x(t) goes low at the end of a phrase, one-shot A_4 is triggered, flip-flop A_5 clears, and the recorder's motor is braked to a halt. At the same time, A_2 starts to count down at a rate, f_2, which is determined by divider A_6 and the master clock. As A_2 goes through zero after a time equal to $T_{off} = T_{on}(L/M)$, where L and M are the divider ratios, it generates a borrow pulse that sets A_5 and restarts the recorder's motor. The audio output from the recorder then sets x(t) high, and the cycle repeats.

Because the actual interface between A_5 and the tape machine varies widely with the recorder used, the wiring details of this portion of the circuit are not shown. Other parts of the circuit may be easily modified to suit the application. For instance, the 74193 ratio counters each provide divisor ratios to 15, but they may be replaced by dividers that provide any value of L and M. □

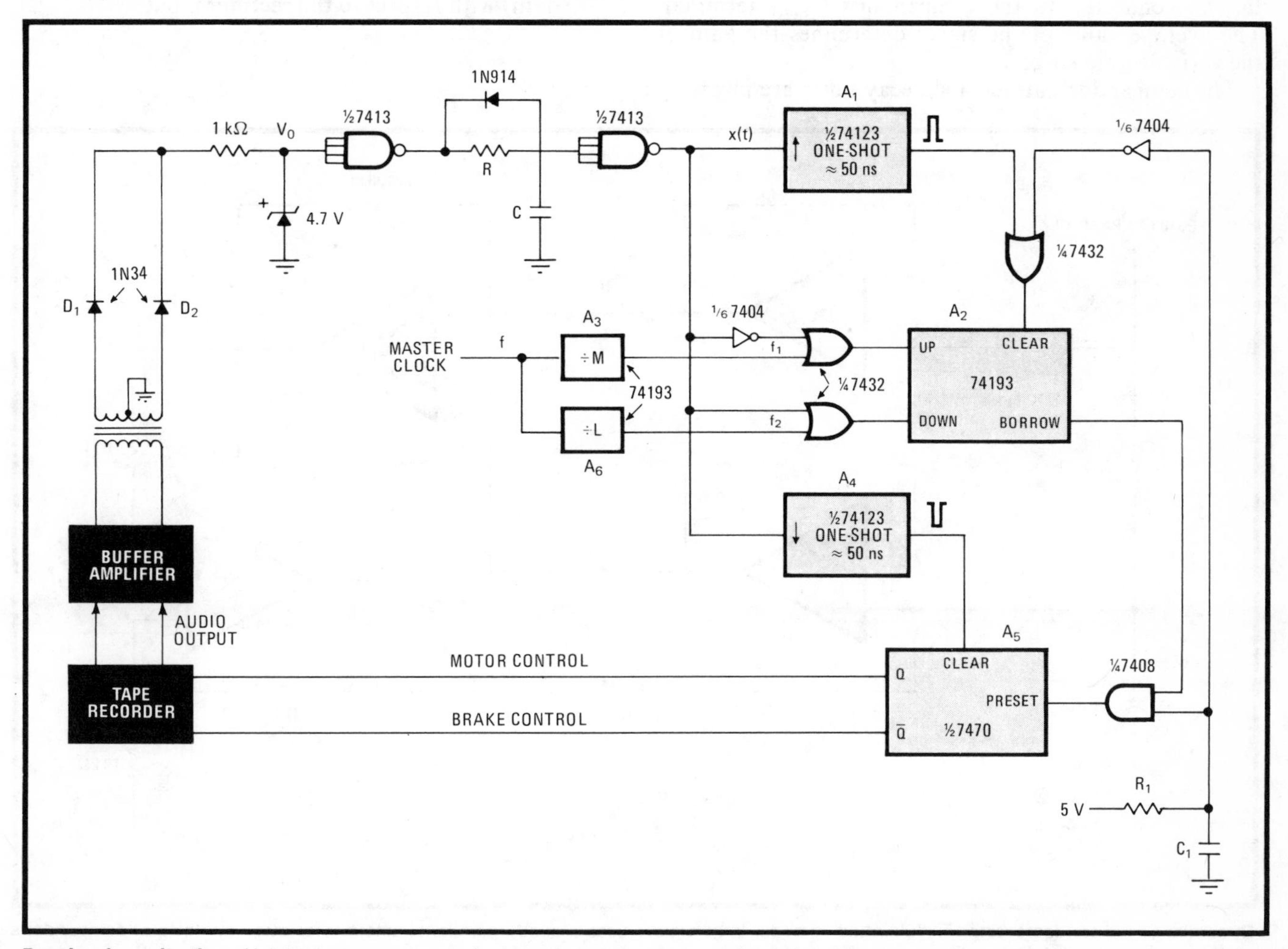

Pausing for write time. Unit brakes tape recorder's drive after each sentence of taped message to provide transcription secretary with time to write information. Circuit uses up-down counter to derive a halt time proportional to the length of the preceding playback period.

Charge pump cuts compandor's attack time

by Devlin M. Gualtieri
Allied Chemical Corp., Morristown, N. J.

Integrated-circuit compandors such as the Signetics NE570 could be more effective in high-fidelity noise-reduction schemes if it were not for one built-in shortcoming: their slow attack time permits large input signals to overdrive the device's compressor and thereby create distortion. But by adding a quad operational amplifier, a diode, and a few resistors, the compandor's attack-to-decay ratio (which is internally set at 1:5) can be dynamically controlled. Specifically, the attack time can be decreased for a given decay period. This concept can be extended to any charge-storage circuit, such as a sample-and-hold module, to speed voltage-level acquisition for a given set of circuit parameters.

The NE570 contains a full-wave rectifier, a variable-gain stage, and other peripheral circuits. The rectifier converts an audio-input signal into a pulsating direct current, which is averaged by an external filter capacitor, C, connected to the compandor's C_{RECT} terminal. The average value of the signal determines the gain of the variable-gain stage.

The compandor's attack and decay times are inversely proportional to the value of the filter capacitor. Thus attack time could be reduced simply by substituting a smaller capacitor for C. But the circuit's purpose is to reduce attack time without using a smaller C, because the third-order harmonic distortion generated when the compandor processes low- and medium-amplitude signals increases as the value of C decreases. The effective capacitance of C during charging is reduced by using this circuit as a pump to charge C more quickly for large input signals, thereby shortening attack time without appreciably reducing the average capacitance of C.

The averaging capacitor is connected to the compandor's rectifier through a 150-ohm resistor, R. At low signal levels, the LM324 quad op amp is not active, so that R contributes only 0.2% additional distortion to what would normally be expected with C alone.

When the rectifier processes a large signal, the relatively large voltage drop across R activates the circuit. Differential amplifier A_1–A_3 generates a large voltage at the input to A_4. This causes A_4 to charge C through D_1 at its short-circuit value of 40 milliamperes. C is effectively charged at 40 volts per millisecond, which corresponds to an attack time of less than 0.1 ms.

The threshold of the enhanced attack rate, which is set by the quiescent 1.3-V drop across C and the 0.7-V drop across D_1, is approximately 0.1 V root mean square (−20 dBm) with respect to the rectifier input. □

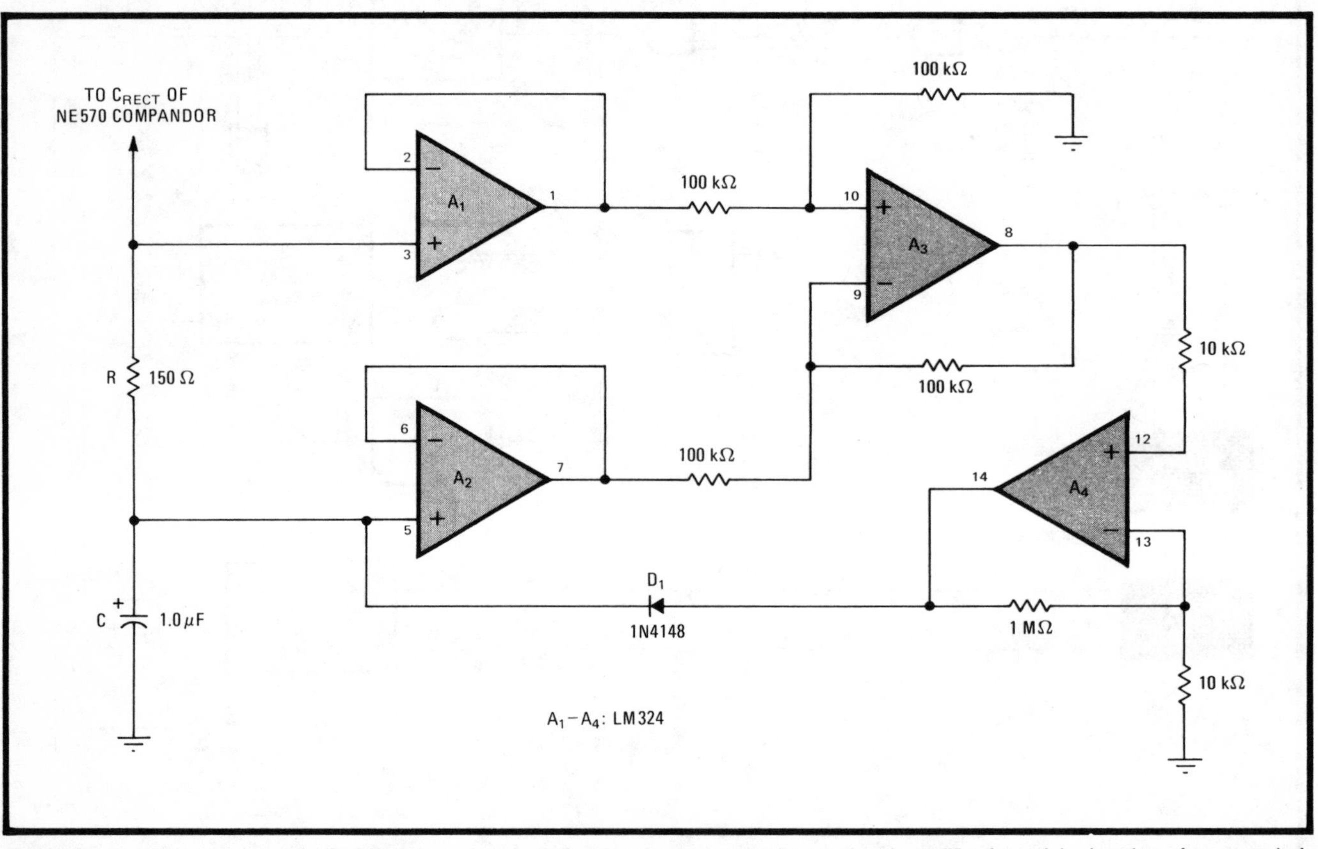

Dynamic. Circuit for decreasing NE570's attack time for large signals uses a quad operational amplifier for quick charging of compandor's averaging capacitor, C, in this way reducing its effective value. The fundamental waveform distortion from the compandor output is substantially reduced as a result, but third-harmonic distortion for low- and medium-amplitude signals is not substantially increased.

Bilateral speaker networks form switchless intercom

by Frank Kasparec
St. Poelten, Austria

Only one transducer—a dynamic loudspeaker—is required at each station of this intercom to permit transfer of audio information in both directions simultaneously. Having no need for push-to-talk switches, the circuit is less costly and less bulky than conventional transceiver units. Undesired acoustic distortion normally encountered in this type of transmission system is eliminated by using a simple phase-compensation network.

As described for the receiver portion of the unit, audio signals at the input are converted into their acoustical equivalent, S_1, by the speaker after passing through the RC network made up of C_1 and resistors R_3 to R_7, which is required to offset the frequency-dependent phase shift created by the speaker's inductance. When properly set, potentiometer R_6 also cancels the feedback (talkback) voltage appearing at the noninverting input of the 741 differential amplifier, which is normally used to amplify the electrical equivalent of the acoustic vibrations, S_2, hitting the cone of the speaker in the transmit mode.

As expected, the compensation network responds similarly to voltages emanating from the speaker's coil, acting to minimize phase distortion at V_{out}. In this case, however, input signals to the differential amplifier are in the millivolt range. The 741 op amp provides a gain of R_9/R_8 and thus the needed amplification at V_{out}. Note that in most cases a power stage will be required following the 741 op amp to energize a loudspeaker.

The design of the phase-shift network, although easy, must be done with care if acoustic feedback is to be reduced to a minimum. The phase differences as seen by the differential amplifier must be canceled, and thus $(R_1+R_2)/Z_1 = Z_{C1}/(R_3+R_4/2)$, where $Z_1 = jX_{L1}$, Z_{C1} is the reactance of capacitor C_1, and L_1 is the speaker's inductance. It is also assumed that R_2 is approximately equal to R_1, and R_3 and R_4 are much smaller than R_5–R_7. For the general purpose speaker, R_1 will be about 4 or 8 ohms. The above equation reduces to $(R_1+R_2)(R_3+R_4/2) = L_1/C_1$. At the same time, for direct current balance at the output of the op amp, $R_2/R_1 = (R_7+R_6/2)/(R_5+R_6/2)$.

To determine the element values in the phase-compensation network, and to estimate the frequency response of the unit, it is necessary to measure both the loudspeaker's dc resistance and its inductance. The best way to find the dc resistance is to use a digital ohmmeter. To determine coil inductance, it is necessary to apply an audio signal (preferably at f = 1 kHz) to the loudspeaker, as shown in the inset, and to measure A, the ratio of the root-mean-square output voltage to the rms input voltage. The coil inductance is then given by:

$$L_1 = [R_1^2 - A^2(R_1+R_2)^2/4\pi^2f^2(A^2-1)]^{1/2}$$

Rearranging the equation:

$$A = [R_1^2 + 4\pi^2f^2L^2/(R_1+R_2)^2 + 4\pi^2f^2L^2]^{1/2}$$

and it can be seen that the speaker response tends toward a high-pass characteristic. Because L_1 is generally negligible, the frequency response is largely flat over the audio range.

Adjustment of the circuit is simple. R_4 is initially set to null the output of the differential amp under no-signal conditions. A square wave is then applied at V_{in} and R_6 set to minimize V_{out}. □

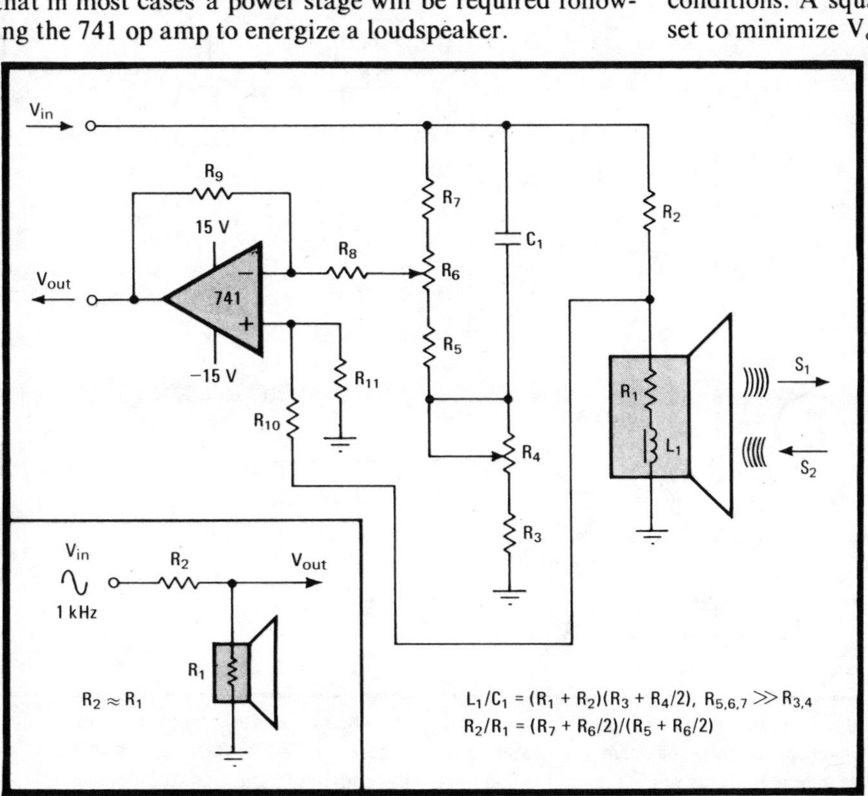

Double duty. One-transducer intercom station using dynamic loudspeaker minimizes cost and bulk of conventional units. Compensation network $(R_3\text{–}R_7)C_1$ reduces phase distortion created by speaker. Speaker inductance may be easily measured (see inset) in order to set component values in compensation network.

Audio blanker suppresses radar-pulse and ignition noise

by Carl Andren
E-Systems Inc., St. Petersburg, Fla.

This simple audio-stage noise blanker will reject most repetitious, pulse-type interference, like radar and automobile ignition spikes, that often plagues a-m receivers. The circuit is both less costly and far less complex than the radio-frequency stage blankers employed in some of the more sophisticated receivers, and though not as effective in eliminating interference, it outperforms the more commonly used noise-limiting audio-clipping circuits.

The blanker shown in the figure detects whether the amplitude of an offending pulse train at the output of the receiver's envelope detector exceeds a set threshold and then disables the output stage if necessary. Waveform diagrams are shown at several circuit points to help clarify operation of the blanker.

A typical amplitude-modulated signal might appear at the input of an a-m receiver as shown in the upper left of the figure, where a 20-megahertz radio-frequency wave, modulated 30%, is overridden by radar pulses 20 decibels greater in amplitude. A time-magnified portion of the a-m detector output, after passing through an inverting operational-amplifier stage, would appear as shown, where the maximum amplitude of the pulse would be limited by the saturating level of the intermediate-frequency amplifier. Only two offending pulses are shown for clarity, but this detected signal contains a pulse train of sufficient amplitude and repetition rate to generate a substantial pulse noise and so impair the readability of the signal.

The interfering spikes increase the effective modulation percentage to well over 100%. The blanker is triggered into operation when the modulation peak exceeds 140%, whereupon Q_1 and Q_2 switch on and disable signal-gate Q_3 for the duration of each spike. The 140% threshold has been experimentally determined as the point at which the interference caused by the blanking operation itself is still less than the interference generated by the offending pulse train. Note that to ensure that the blanking action occurs at the set modulation peak independently of signal-level changes, the receiver's automatic-gain-control signal is introduced at the threshold bias point at the emitter of Q_1.

Q_3 operates with no applied dc voltage so that no switching transients will be generated by the blanking action to impair circuit performance. Q_2, R_1, and C_1 have a fast-attack, slow-decay characteristic. Q_3 is thus gently turned on after a spike has passed so that the popping and clicking sounds that often accompany the operation of a blanking circuit that processes a randomly occurring train of spikes will be further suppressed.

The results of the blanking action are shown at the output of Q_3, where it is seen that only brief transients appear. The signal is slightly distorted, but the distortion is barely audible. There is a great improvement in noise reduction, however. □

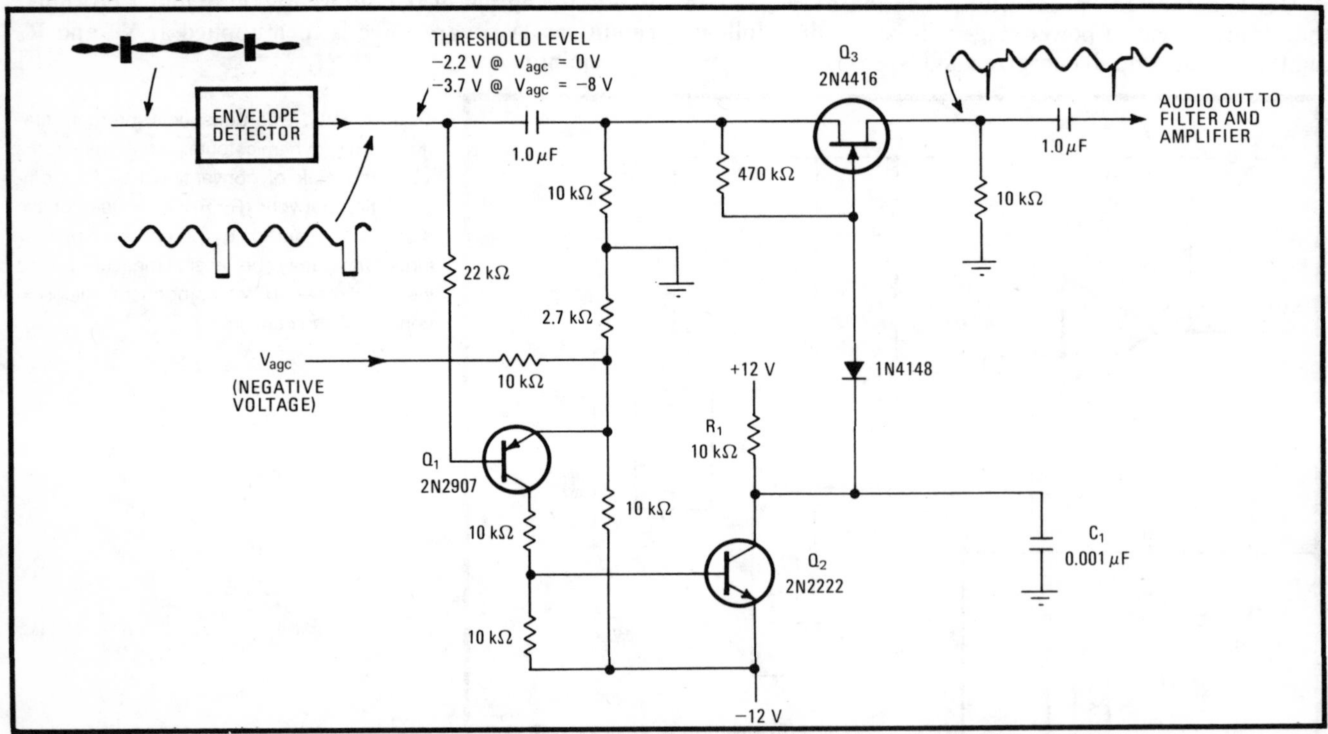

Spike eater. Audio-stage noise blanker, although not as effective at eliminating pulse-type interference as rf-stage blankers, outperforms noise-limiter/clipper circuits. Blanking occurs when spikes raise effective modulation percentage over 140%. Receiver's agc signal is introduced to emitter of Q_1 to ensure that the blanking action occurs at the set modulation point independent of input signal amplitude.

Chapter 4
AUTOMOTIVE CIRCUITS

Darlington-switched relays link car and trailer signal lights

by M. E. Gilmore, and C. W. Snipes
Florence, Ala.

New cars with separate turn and brake signals—a safety feature—require a special circuit to properly drive the combination turn-and-brake lights on a trailer; otherwise, if the trailer lights are connected to the brake command, the turn signal will not work, and connecting the lights to the turn command will not yield a brake signal. But two relays and low-cost transistors will combine the signals onto a common bus again, ensuring that the trailer's lights respond to both commands.

As shown in the figure, the brake-command line is normally connected to the trailer lights through relays K_1 and K_2 during normal operation. However, a left- or right-turn command will turn on the respective Darlington amplifier, Q_1Q_2 or Q_3Q_4, thus activating K_1 or K_2. The turn signal is then routed to the lights.

Capacitors C_1 and C_2 charge to the peak amplitude of the turn signal, which flashes at one to two times per second. C_1 and C_2 should therefore be selected to hold the relay closed between these flash intervals (0.5 to 1.0 second), but no longer. If the capacitance is too large, the brake signal cannot immediately activate the trailer lights after the turn signal is canceled. Diodes D_1 and D_2 prevent capacitor discharge through the left or right turn-signal lines, respectively. □

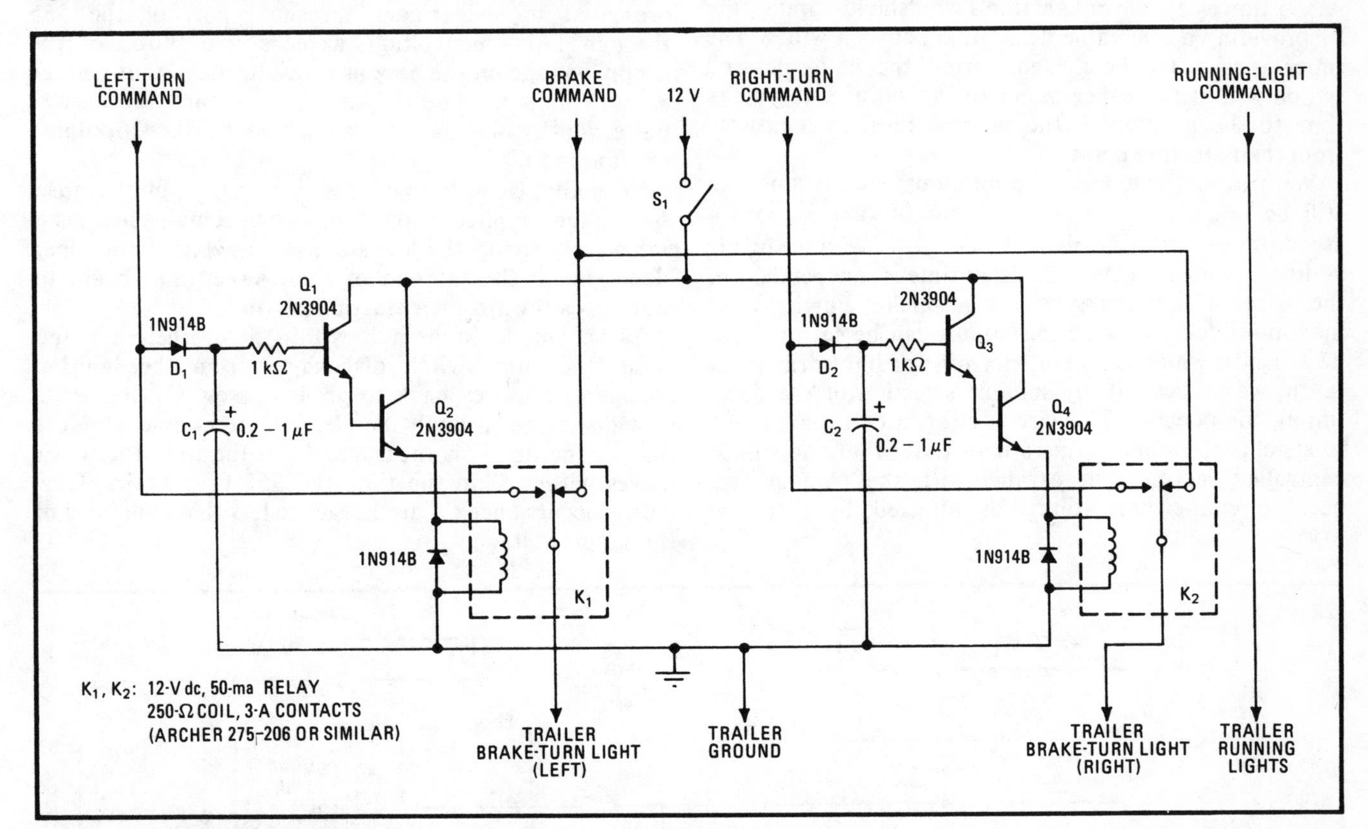

Auto-to-trailer interface. Relays multiplex brake and turn commands onto common bus line, permit control of brake-turn lights on trailer. Darlington amplifiers provide high command-line isolation and sufficient drive for the relays.

Synchronous timing loop controls windshield wiper delay

by John Okolowicz
Honeywell Inc., Fort Washington, Pa.

A 555 timer can control an auto's windshield-wiping rate by providing a selectable delay time between wipes. The timer uses a feedback signal from the cam-operated switch within the wiper motor to synchronize the delay time to the position of the wiper blades, as measured from their starting point.

With synchronization, the minimum delay time can still be reliably kept to nearly zero (normal delay for standard systems), which is best for heavy rain. In addition, it ensures that the delay time is independent of the wiper speed across the windshield. Further, the maximum delay time in this circuit can be set to about 22 seconds, which is suitable for mist or light drizzle, or to any value desired, by suitable selection of the 555's timing components. This circuit offers a better approach to synchronous-delay wipers than those that use silicon controlled rectifiers in parallel with the cam switch, because cam-switch voltage is affected by dirt and grease.

The circuit shown in Fig. 1 is for a Volkswagen Rabbit, but it can be used in any car that has one end of the wiper motor always grounded (which requires a positive energizing voltage). As shown, the 555 assumes the high state on power-up (S_1 initialized), driving the MJ1000 Darlington amplifier, which in turn energizes the wiper motor. As the blades traverse through an angle of approximately 5°, the cam switch is engaged to the 12-volt ignition-line voltage. Thus a feedback voltage is presented to the trigger threshold port of the 555 through D_1. This voltage exceeds two thirds of the supply voltage on the 555; as a result, the output voltage at pin 3 falls at time t_1, which is less than the normal pulse width time usually controlled by the 20-kilohm resistor and C_1.

Normally, C_1 would now begin to charge, but because the voltage supplied to the cam switch remains high, the 555 simply sits in the low state. Meanwhile, both wiper blades reach the far end of their sweep and begin to move back toward their starting position.

As the blades approach within 5° of their starting point, the cam switch disengages from the ignition voltage and moves back to ground. Now C_1 begins its slow discharge through the 300-kΩ potentiometer. Note that t_2, the delay, is measured from the time the wiper moves within 5° to the time the 555 fires again. Triggering occurs once C_1 discharges to less than one third of the supply voltage.

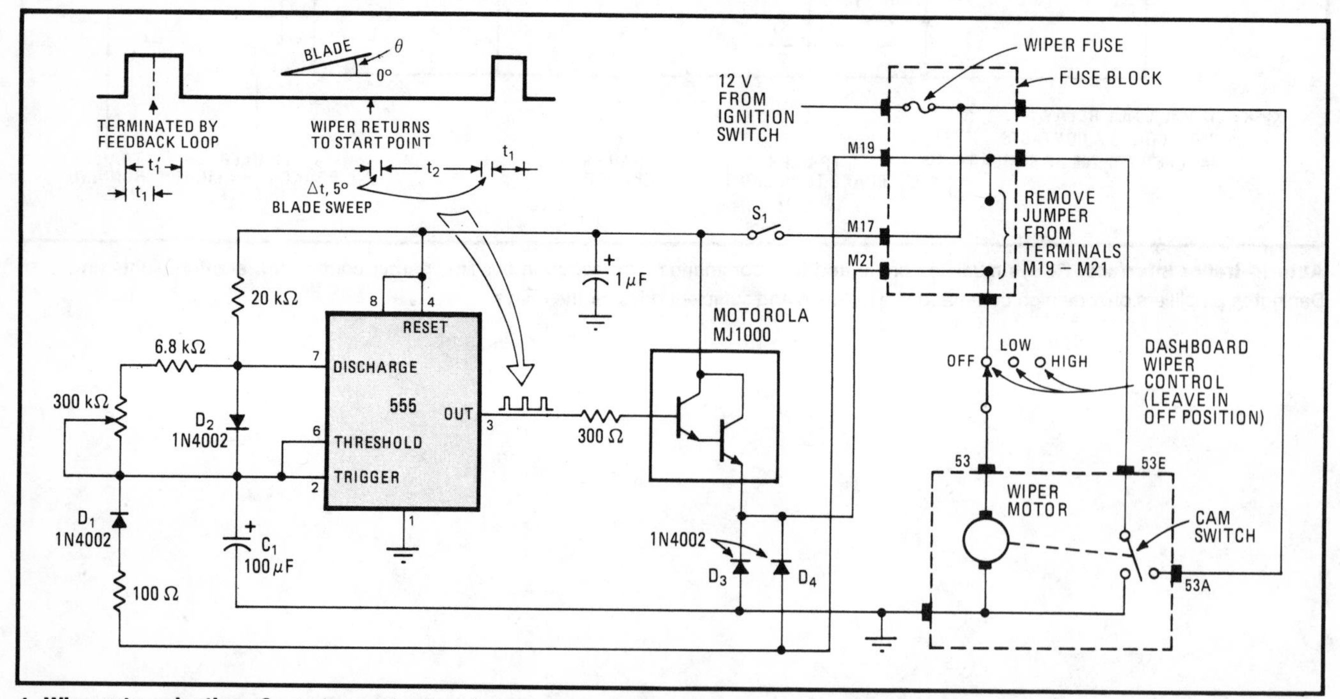

1. Wipe-rate selection. Control over the time between sweeps in negative-ground windshield wiper systems is done by the 555 and a synchronous feedback loop. Delay time can be varied from 0 to 22 seconds. Synchronizing the feedback to the wiper blades' position ensures that sweep rate will be independent of variations in wiper speed. Terminal block notations are for Volkswagen Rabbit.

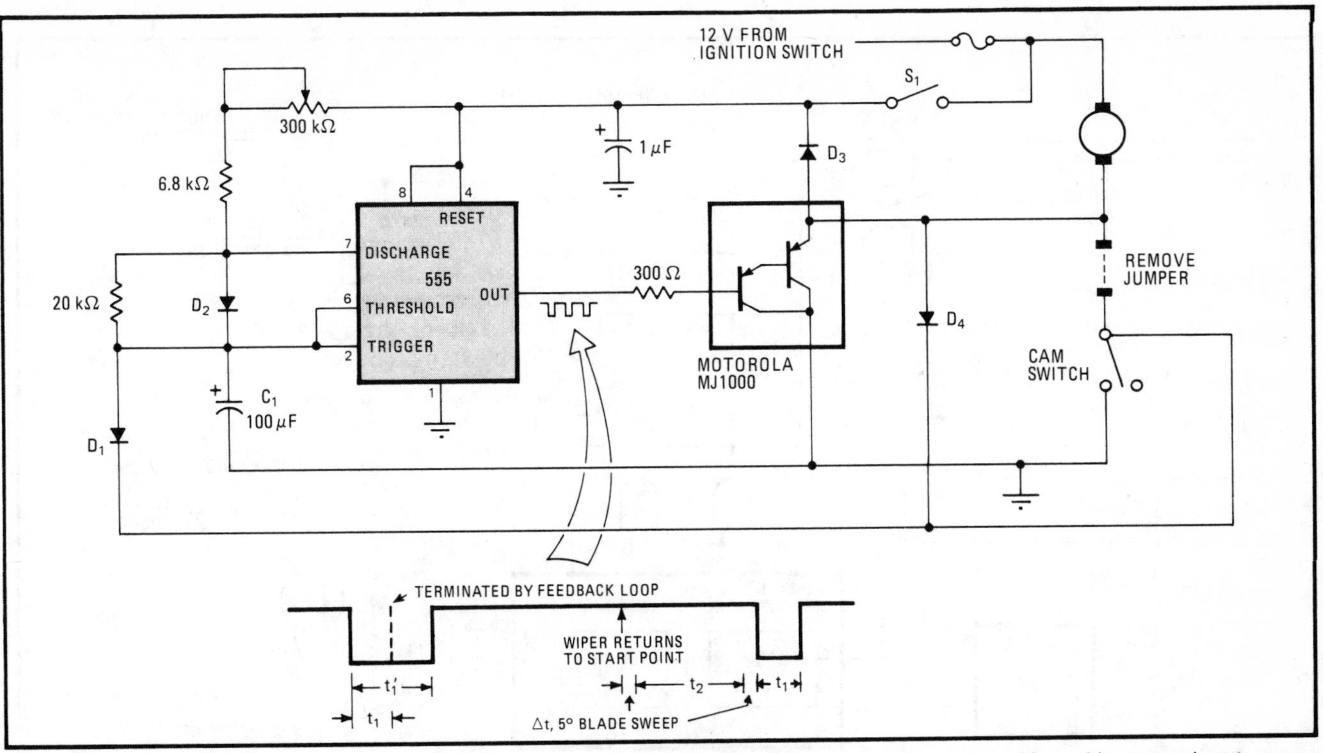

2. Positive-ground system. Suggested modification for many General Motors vehicles and other autos with positive-ground system uses same number of components as standard system. Wiper motor is energized by completing ground connection. Note that diodes D_1, D_3, and D_4 are reversed with respect to the polarities shown in Fig. 1 and that pulse-train polarity from 555 output is inverted.

D_1 prevents C_1 from discharging through the cam switch, and D_2 allows independent selection of times t_1 and t_2. D_3 provides suppression of the back voltage produced when the MJ1000 turns off; without the diode, the 555 will be retriggered falsely. D_4 allows normal operation when the circuit is not activated and also prevents the MJ1000 from shorting to ground through the cam switch.

Some cars have a reversed wiper-motor configuration. Thus one end of the wiper motor is always connected to the positive ignition-line voltage and so requires a connection to ground to energize it. Figure 2 shows a suggested modification of the wiper-delay circuit to handle such cases. □

Engine staller thwarts car thieves

by Gary L. Grundy
Apple Valley, Minn.

Car thieves are becoming surprisingly astute at finding and disabling theft-prevention devices in automobiles. But what if the vehicle starts and then stalls repeatedly? Chances are the would-be thief won't suspect an anti-theft device but will instead hunt for a new victim, if the car is equipped with a unit that simulates engine malfunction.

When in operation, the device permits the car to start normally and after 12 seconds opens the circuit to the ignition coil, stalling the engine. Four seconds later, the circuit again closes, enabling the thief to restart the engine. The cycle repeats and then after an additional 12 seconds, the engine stalls, and will not start again. By this time, the thief will probably abandon the car a short distance at most from where it was parked.

The circuit of the anti-theft device, shown in Fig. 1, makes use of a 555 timer and complementary-metal-

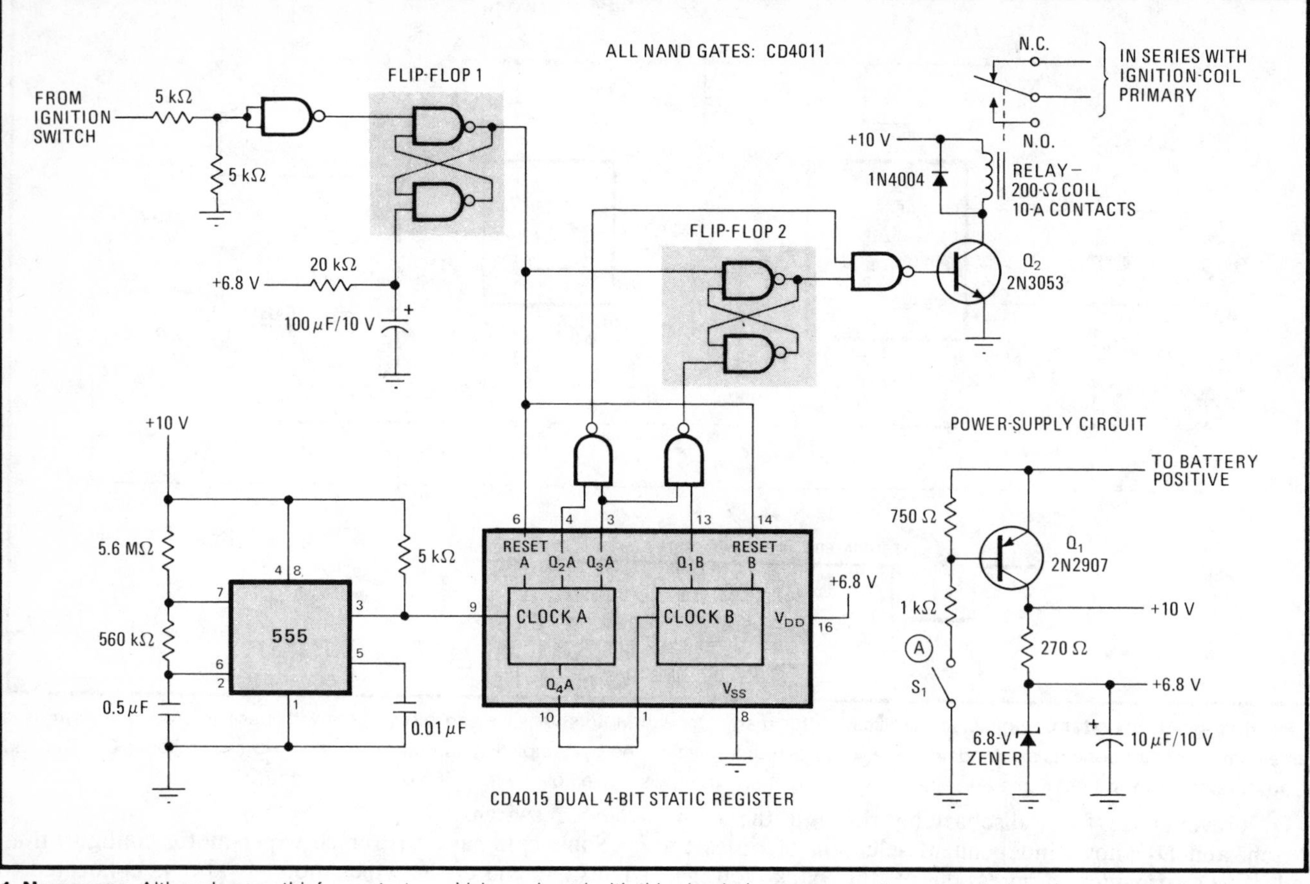

1. No escape. Although a car thief can start a vehicle equipped with this circuit, he soon gets discouraged when the engine stalls repeatedly. Timer circuit periodically removes power from the ignition coil, simulating engine malfunction.

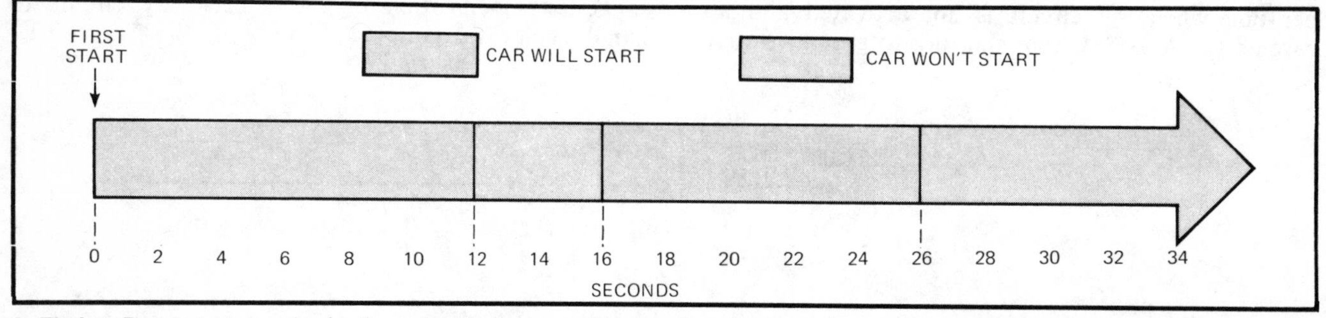

2. Timing. The start-stop cycles for the anti-theft device are 12 seconds and 4 seconds, respectively. After the engine stalls the third time, it can not be restarted until the driver opens the enable switch S_1, which is concealed within the car's interior.

oxide-semiconductor logic for low battery drain. Enable switch, S_1, which should be concealed, is set by the driver to activate the circuit whenever he leaves his car. A good way for a nonsmoker to conceal S_1 is to use the cigarette lighter—its wire to the battery is simply disconnected and routed to the circuit at point A. A key switch may be placed in series with S_1 to safeguard against tampering.

Once the circuit is armed by closing S_1, turning on the car's ignition toggles flip-flop 1, made up of two NAND gates, permitting 0.5-hertz pulses from the 555 timer to enter the CD4015 shift register. The CD4015 contains two 4-bit shift registers, which, in this case, are cascaded. After 12 seconds (six pulses) the NAND gating turns on transistor Q_2, opening the normally closed contacts of the relay and stalling the engine. After two more timer pulses enter the shift register, the NAND gating turns off Q_2, enabling engine to be restarted. The cycle is repeated, except that the third application of ignition voltage toggles flip-flop 2, and the car can not be started again until the driver returns and opens S_1. The timing sequence is detailed in Fig. 2.

To make device unrecognizable as an anti-theft unit, it can be built on a small circuit board and housed in a small box such as those used for pollution controls.

An optional alarm that actuates after 60 seconds can be added by connecting pins 5 and 12 of the shift register to an additional two-input NAND gate. The gate output can be made to drive a relay activating the horn, lights, or a siren to draw attention to the abandoned vehicle, if desired. □

Chapter 5
CLOCK CIRCUITS

Decoders drive flip-flops for clean multiphase clock

by Craig Bolon
Massachusetts Institute of Technology, Cambridge, Mass.

A multiphase clock suitable for driving circuits with strict timing requirements, such as charge-coupled-device memories, can be built with a counter, decoder, and set/reset flip-flops. This clock can generate any number of outputs at any duty cycle, yet never suffers from the drift and "glitches" encountered in most multiphase designs. Although each signal is phase-locked to a master clock, its timing and duty cycle can be set independently of the others, permitting great flexibility.

The figure shows a typical application—an eight-phase clock designed for driving parallel banks of Intel 2416 CCD memories. A master clock drives the 74163 synchronous 4-bit binary counter. The binary output is then presented to the combinational logic of two 74138 1-of-8 decoders.

The decoders count one pulse on the rising edge of each clock. The first decoder counts eight pulses (0 – 7) before its outputs are held high by the Q_D output of the binary counter. The second decoder is then enabled and counts an additional eight pulses, after which the 16-count sequence is repeated.

Each decoder output controls an R/S flip-flop by setting or resetting it at the desired moment. Thus, the

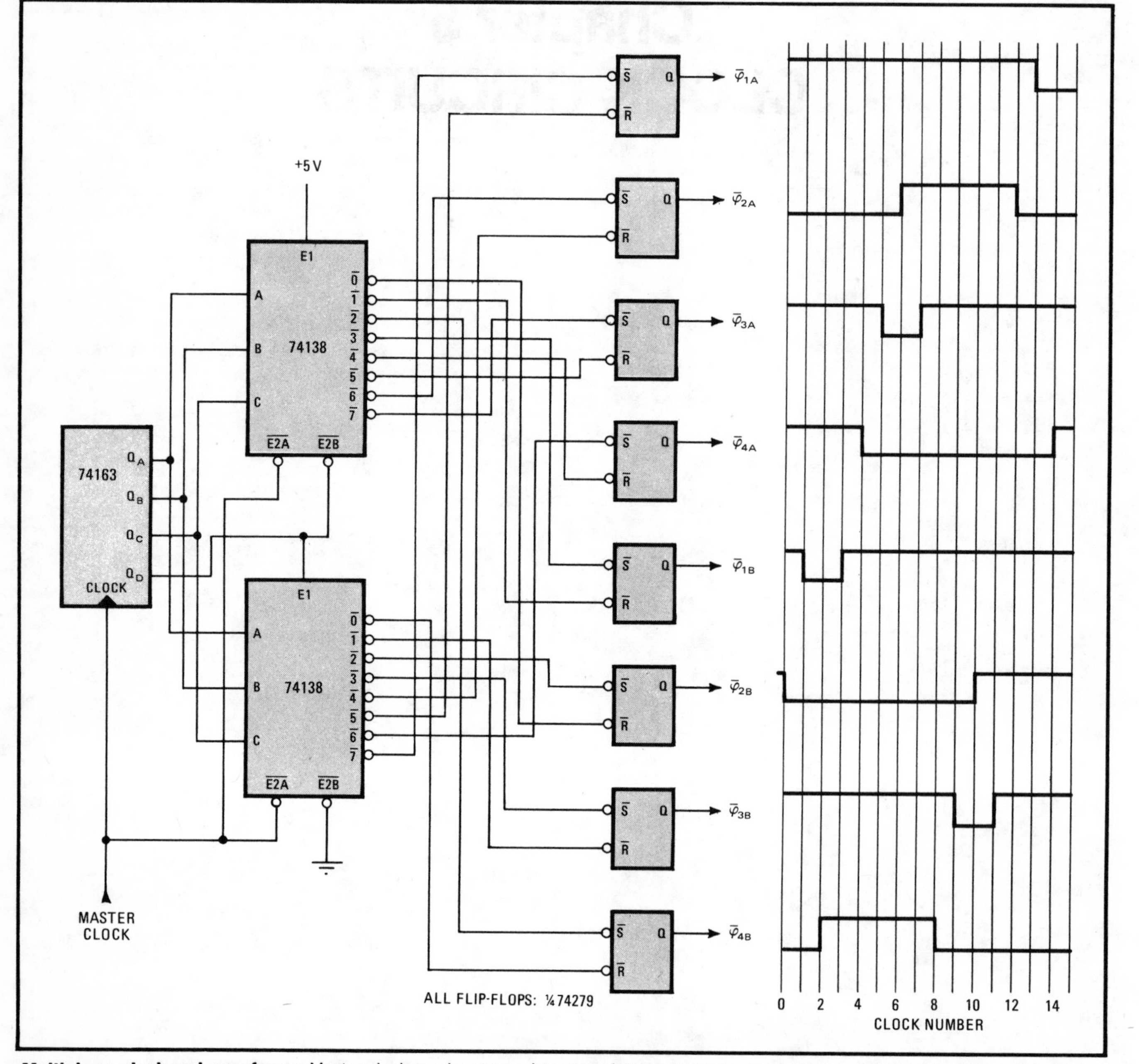

Multiphase clock and waveforms. Master clock can be stopped or started at any point in cycle without affecting the phase relationship at all. Skew in output waveforms can be limited to propagation delay of one gate, provided that edge-triggered flip-flops are used.

master clock assures synchronous operation, and the strictly sequential nature of the decoder output keeps it glitch-free for all time.

The clock waveforms are shown to the right of the circuit. Phase 2A of the clock, for example, is generated by setting a flip-flop on count 6 of the 16-count cycle ($\overline{6}$ of the first decoder) and resetting it on count 12 of the cycle ($\overline{4}$ of the second decoder). The number of phases can easily be increased by expanding the binary counter and adding decoders and flip-flops. □

Two-phase clock features nonoverlapping outputs

by Neil Heckt
The Boeing Co., Seattle, Wash.

A reliable, two-phase clock signal with nonoverlapping outputs—the kind that is an absolute must for the Motorola 6800 and other microprocessing units—can easily be derived from a pulse train input. This design avoids overlap by exploiting the propagation delays inherent in transistor-transistor logic, and it uses only one integrated circuit.

As shown in the schematic of Fig. 1, a pair of two-input NOR gates in the 7402 chip are wired as an R-S flip-flop to provide the split-phase outputs. The propagation delay of the gates depends on their capacitive loading; it is typically 10 nanoseconds with a 15-picofarad load, increasing to 20 ns with a 150-pF load. As specified in the Motorola 6800 applications manual, the clock inputs of the central processing unit are capacitive, with maximum values of 160 pF but typically 110 pF.

With a 1-megahertz, 50%-duty-cycle input pulse train, this circuit produces a ϕ_1 output 470 ns in duration, a ϕ_2 output 490 ns in duration, and 20 ns of nonoverlap, as shown in Fig. 2. The duration of the nonoverlap is independent of the input duty cycle. □

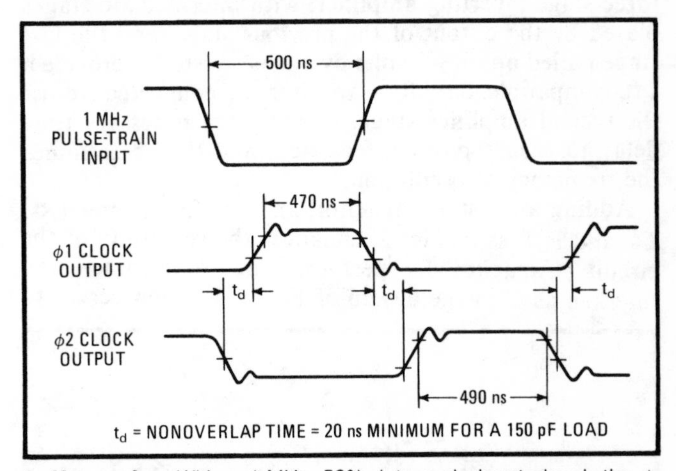

2. No overlap. With a 1-MHz, 50%-duty-cycle input signal, the ϕ_1 and ϕ_2 outputs have durations of 470 ns and 490 ns, respectively. The nonoverlap, which is dependent upon the propagation delay of the gates, is a function of load capacitance and varies from about 10 ns with a 15-pF load to 20 ns with a 150-pF load.

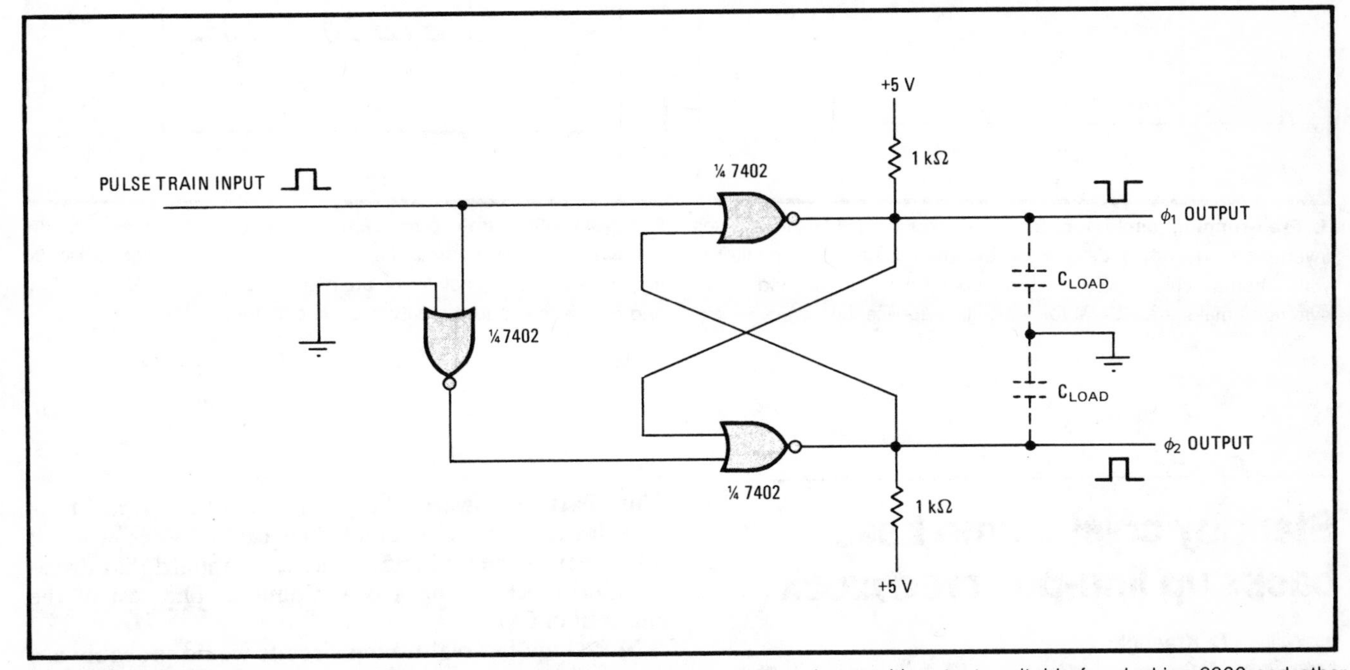

1. Phase splitter. Simple circuit derives two out-of-phase signals from oscillator input, with outputs suitable for clocking 6800 and other microprocessors that have strict timing requirements. Nonoverlapping of outputs is constant, regardless of input duty cycle.

Adjustable TTL clock maintains 50% duty cycle

by Wilton Helm
Seventh-day Adventist Radio-TV-Film Center, Thousand Oaks, Calif.

The utility of the basic free-running transistor-transistor-logic clock can be greatly increased by adding a few components to provide a variable-frequency output over a 5:1 range or better, while still maintaining a 50% duty cycle of the square-wave output.

A common design of the free-running oscillator is shown in Fig. 1. Three inverters are connected as a three-stage inverting amplifier, with intermediate stages biased by the output of the previous stage, and the last stage pulled up to +5 volts by a bias resistor to provide a TTL-compatible output. Capacitor C, connected across the second amplifier stage, provides the necessary time delay to ensure positive feedback, and thus determines the frequency of oscillation.

Adding a resistor, capacitor, and potentiomenter (R_2, C_2, and R_3), as in Fig. 2, enhances the versatility of the circuit. Capacitor C_1 performs the same time-delay function as C in the design of Fig. 1, but now serves to determine the upper limit of oscillation frequency, since it is paralleled by C_2.

The combination of C_2 and R_2 adds a time delay to the transitions of the second amplifier stage, additional to the delay of the upper-limit capacitor C_1. But the resistance in series with C_2 could tend to upset the symmetry of the square wave, as is usually the case when attempts are made to vary the frequency of the free-running-oscillator design of Fig. 1.

Symmetry of the square-wave output is maintained by connecting the right side of R_2 through resistor R_3 to the output of the third amplifier stage. This changes the charging current to the capacitors in proportion to the setting of frequency-adjusting potentiometer R_2. Thus, a duty cycle of 50% is constant over the entire range of oscillation.

The lower frequency limit is set by capacitor C_2. With the components shown, the frequency of oscillation can be varied by R_2 from about 4 to 20 hertz. Other frequency ranges can be obtained by changing the values of C_1 and R_3, which control the upper limit of oscillation, or C_2, which limits the low-frequency end.

Note that the inverters used in the adjustable oscillator are open-collector types, such as the 7405. The inherent low impedance of other types of inverters would swamp the effect of charging-current resistor R_3 and should not be used in this application. □

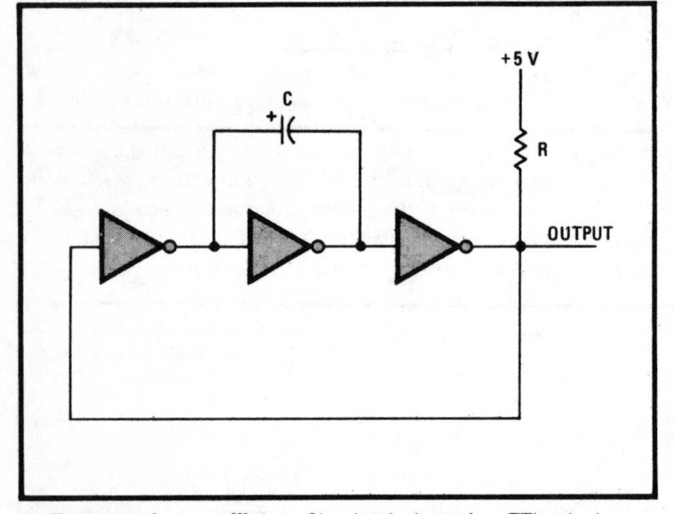

1. Free-running oscillator. Simple design of a TTL clock uses inverters. Frequency is determined by time constant of capacitor C with internal chip resistors. Low component count and good symmetry make it a natural for noncritical applications.

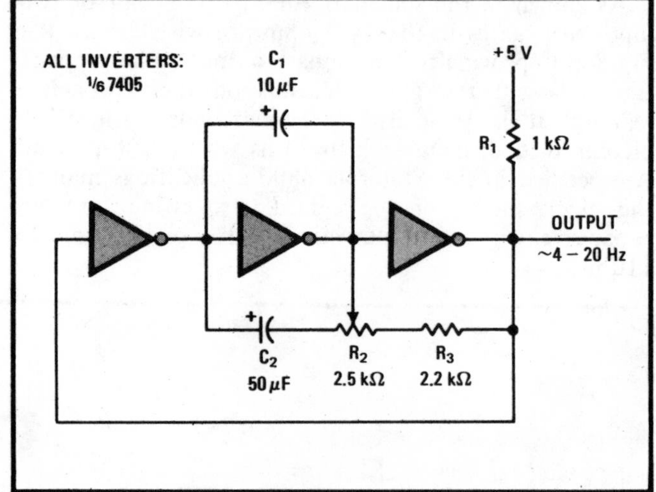

2. Adjustable counterpart. With the addition of three parts, the frequency of oscillation can be adjusted over a 5:1 range, while the attributes of the original design are maintained: the component count and cost is low, and the duty cycle is constant at 50%.

Standby crystal time base backs up line-powered clock

by William D. Kraengel, Jr.
Valley Stream, N.Y.

This battery-powered, crystal-controlled time base provides accurate and glitch-free performance when it takes over as the 60-hertz frequency standard that drives a digital clock during a power outage. The cost of the unit is about $7.

More long-interval timing circuits would probably use the ac power line as a time base because of its long-term average-frequency accuracy (1 part in 10^7), were it not

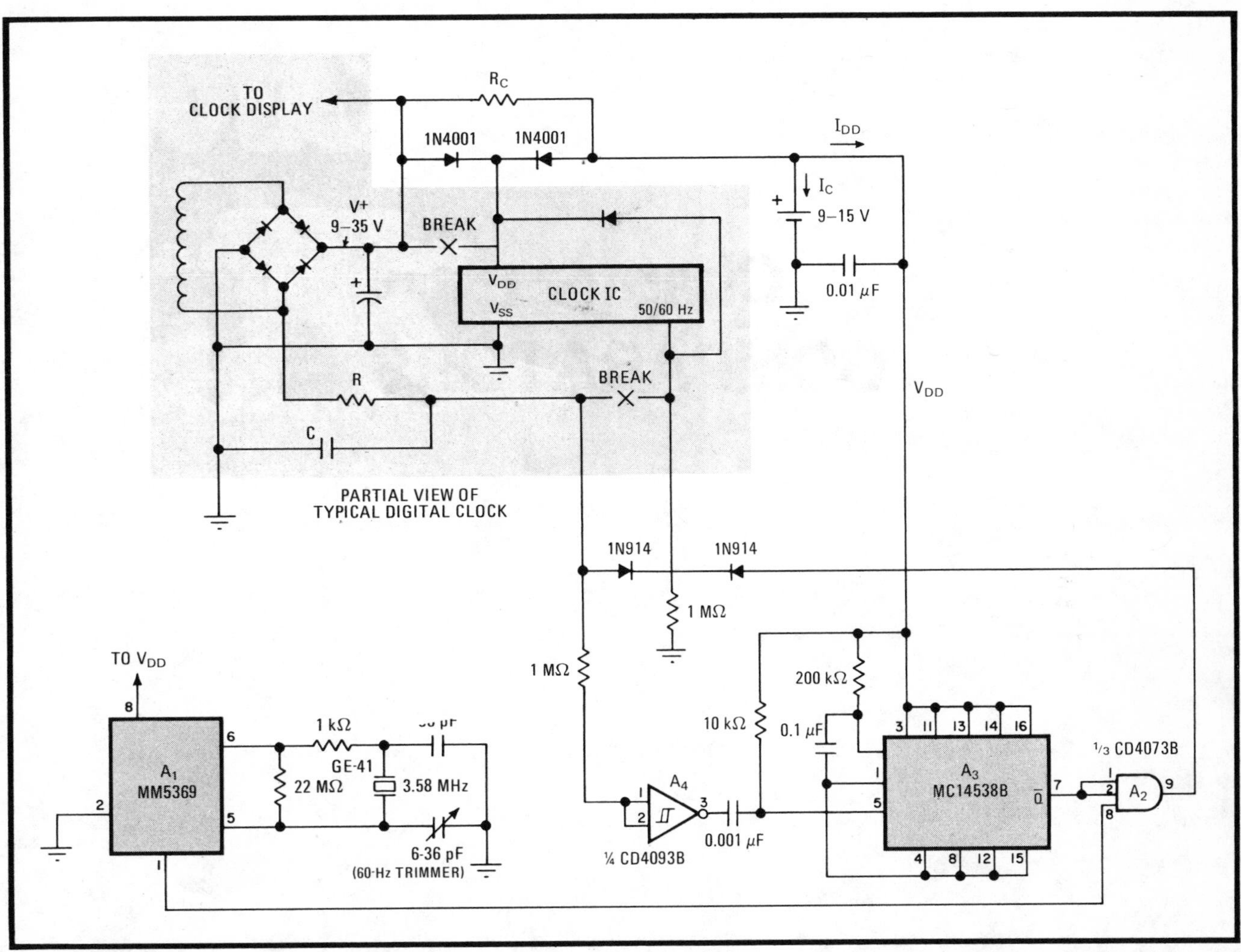

Standby standard. Battery-powered, crystal-controlled time base, having sufficient accuracy for most short-term applications, takes over clock-driving duties of digital chronometer in event of ac power loss. Unit uses 3.58-MHz oscillator, which is divided down to 60 hertz.

for the transients and blackouts that occur frequently. This back-up time base takes over smoothly in such instances and has sufficient accuracy over a period of several hours to satisfy all but the most demanding applications.

The standby time base uses a low-cost crystal oscillating at 3.58 megahertz, which is generally the frequency required for the color-burst circuits in standard television receivers. The frequency produced by the crystal's programmable oscillator-divider chip, A_1, is 60 hertz. This signal is fed to one input of an AND gate, A_2, which is activated if line power is lost.

During normal operation, the battery is trickle-charged (I_c) by the clock's supply through R_c, at a rate of 0.01 C, where C is the capacity of the battery in ampere-hours. R_c is equal to $(V^+ - V_{bat})/(I_c + I_{DD})$, where I_{DD} = 2.5 milliamperes. The digital clock must be modified slightly, as shown, in order to lengthen the charge life of the battery. Thus the digital clock's display will be blanked while the battery is the power source.

Meanwhile, one-shot A_3, configured as a missing-pulse detector, is triggered by Schmitt trigger A_4 at the beginning of each cycle of the ac input.

The one-shot's pulse width is 20 milliseconds, slightly longer than the period of the 60-Hz line input. Thus, A_3 is continually retriggered, and so A_2 is disabled.

With a loss of line power, the battery takes over the supply chores. A_3 times out, and then A_2 is enabled, so that the 60-Hz signal derived by the crystal circuit drives the digital clock's timing chip. The maximum length of time between the power outage and the first clock pulse from the standby unit is 8.3 milliseconds.

Almost the reverse action occurs when the ac line power is restored. When the filter capacitor in the clock's power supply recharges enough for the line pulses to rise above the set threshold of the Schmitt trigger, the one-shot is triggered, and the AND gate is disabled. As the voltage across the filter capacitor rises further, the power source duties revert back to the digital clock's power supply. □

Chapter 6
COMPARATORS

Digital comparator saves demultiplexing hardware

by V. L. Patil and Rahul Varma
Central Electronics Engineering Research Institute, Pilani, India

Comparing two *m*-digit numbers, where each digit comprises *n* bits, by conventional means requires the services of a demultiplexer for separating the data into two corresponding sets, *2mn* storage elements that convert the data into bit-parallel, digit-parallel form, and *m* magnitude comparators for performing the actual comparison. The demultiplexing can be simplified, however, and the number of storage elements reduced to *3m* with this technique, which utilizes strobed memory elements in the form of D flip-flops and combinational logic to ascertain the relationship of the two numbers.

The method is illustrated for an example where two 4-bit, 3-digit numbers are compared. As seen, the corresponding digits of both numbers are simultaneously introduced to the 7485 4-bit comparator, A_0, with the least significant bits being introduced first. The result of the comparison is then strobed into the 7475 quad latch, A_1, by digit strobe D_1.

Similarly, the second-most significant bits (SSB) and the most significant bits are then strobed into A_2 and A_3, respectively, by strobes D_2 and D_3. The combinational logic that follows then evaluates the three-digit (MSD, SSD, and LSD) comparison from:

$$OL = L_3 + E_3L_2 + E_3E_2L_1$$
$$OG = G_3 + E_3G_2 + E_3E_2G_1$$
$$OE = E_3E_2E_1$$

where OL = 1, OE = 1, and OG = 1 signify that $A < B$, $A = B$ and $A > B$, respectively, and L_i, E_i, and G_i are the individual corresponding outputs of flip-flops A_i.

The truth table outlines circuit operation. □

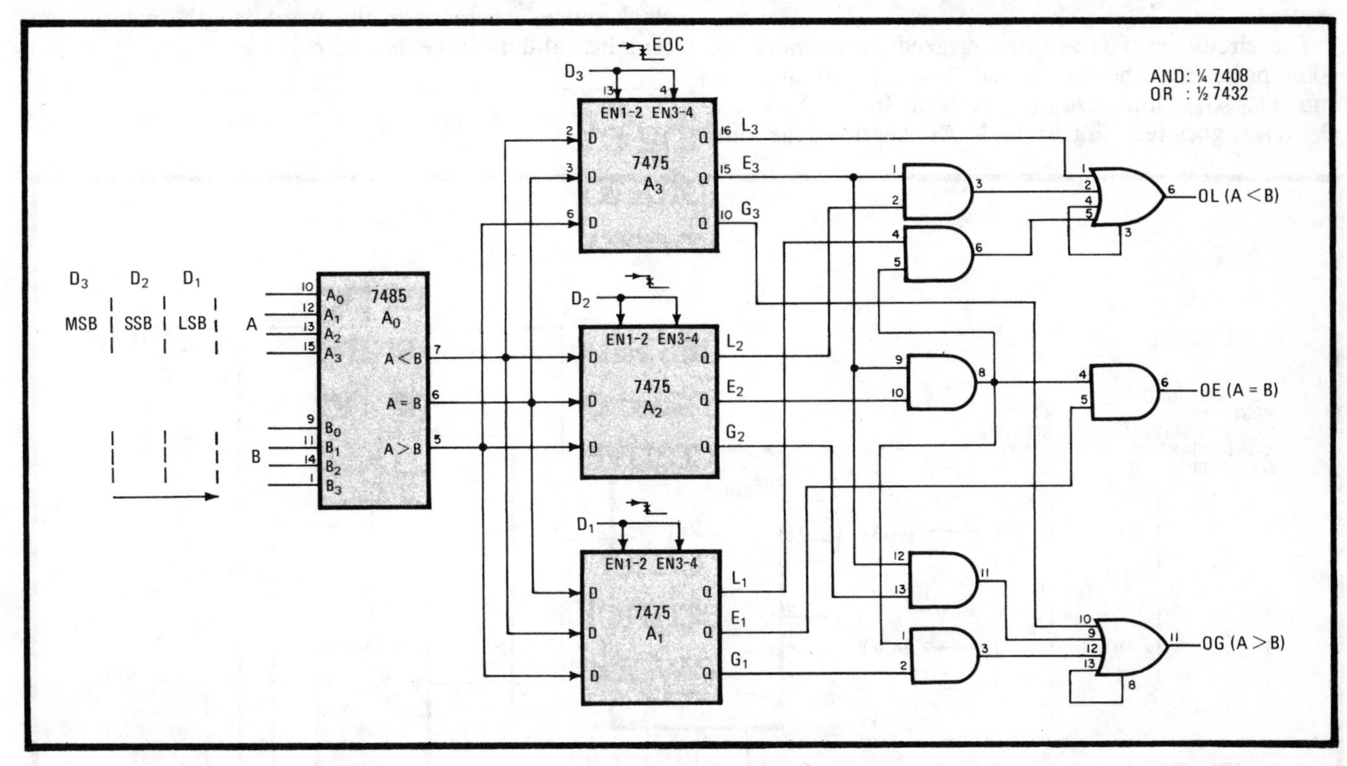

MSD			SSD			LSD			OUTPUT		
G_3	E_3	L_3	G_2	E_2	L_2	G_1	E_1	L_1	OG	OE	OL
>	=	<	>	=	<	>	=	<	>	=	<
0	0	1	X	X	X	X	X	X	0	0	1
0	1	0	0	0	1	X	X	X	0	0	1
0	1	0	0	1	0	0	0	1	0	0	1
1	0	0	X	X	X	X	X	X	1	0	0
0	1	0	1	0	0	X	X	X	1	0	0
0	1	0	0	1	0	1	0	0	1	0	0
0	1	0	0	1	0	0	1	0	0	1	0

Reduction. Circuit performs n-bit, m-digit comparison of two numbers without a conventional IC demultiplexer, reduces number of memory elements normally required. Simplified decoding technique utilizes combinational logic. Truth table outlines circuit operation in comparing most, second-most, and least significant digits.

Digital comparator minimizes serial decoding circuitry

by Harland Harrison
Memorex Inc., Communications Division, Cupertino, Calif.

Using significantly fewer chips than the comparator proposed by Patil and Varma[1], this two-word, 4-bit comparator offers other advantages as well—it accommodates any word length and is more easily modified to handle any bit width. The control signals needed to facilitate the comparison can also be more conveniently applied.

The circuit outputs are first cleared with a negative-going pulse from the start signal. This sets outputs O_{LESS} and $O_{GREATER}$ low. O_{EQUAL}, derived from O_{LESS} and $O_{GREATER}$, goes temporarily high. At this time, the numbers can be presented to data buses A and B for comparison, with the least significant bit pair introduced first. As a consequence of the configuration, any number of bit pairs per word can be compared without modifying the circuit at all. The data buses each accept up to 4 bits, but this number may be expanded simply by cascading 7485 comparators.

The result of each bit-pair comparison is then latched into the 7474 flip-flops by the D_i clock pulse, with the results of each bit-test being fed to the cascade inputs of the 7485. As a result, the comparator keeps track of the previous bit-pair check while continuing to update its results as each succeeding bit pair is introduced. Thus, the need for additional memory and logic elements is eliminated. The final result becomes valid after the D_m clock pulse, where m is the word length in bits, and remains valid until the next start pulse. □

References
1. V. L. Patil and R. Varma, "Digital comparator saves demultiplexing hardware," p. 41

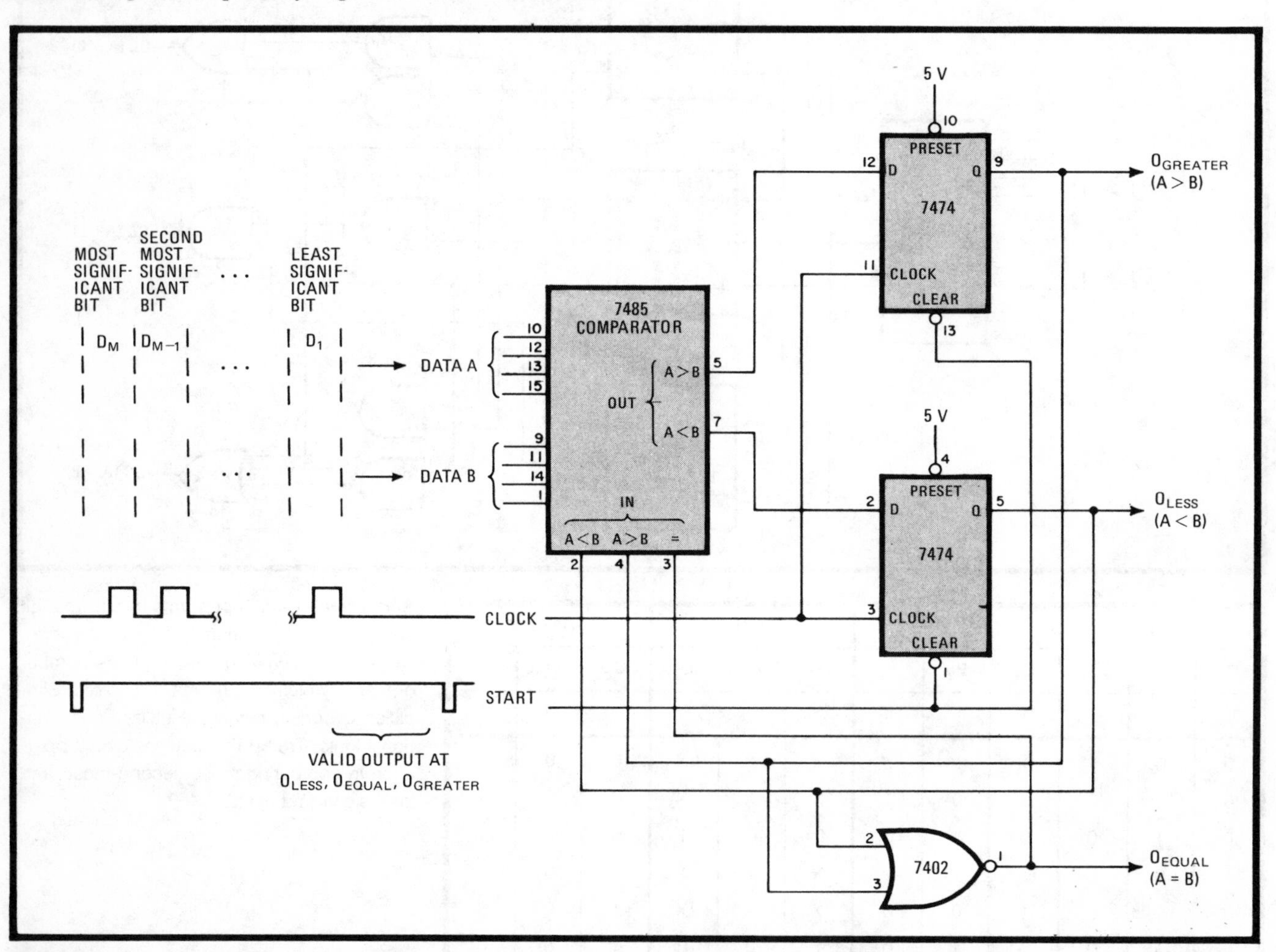

Less memory. Circuit performs 4-bit comparison of two numbers with minimal circuitry. 7485 comparator replaces large numbers of flip-flop–type memories and logic elements by keeping track of previous bit-pair checks in real time as each pair is introduced. Circuit accommodates any word length; bit width is expandable simply by cascading 7485 comparators.

Comparators replace mechanical set-point meter

by Louis A. Perretta
HNU Systems Inc., Newton, Mass.

A standard ammeter and a simple comparator circuit can replace the expensive and bulky mechanical dual-set-point meters used for most process control applications. This electronic circuit provides higher efficiency at low cost and can be built on a printed-circuit card. With a slight modification to the basic circuit, a double high set point and a double low set point may be established—a feature that is not available from the mechanical meter.

The mechanical meter used for industrial purposes contains internal relays and a 115-volt power supply to drive them, as well as a meter that triggers when its pointer or indicating device comes into contact with the high or low set points on the meter dial.

In the electronic circuit shown in the figure, the mechanical set control is replaced by two potentiometers, transistors Q_1 through Q_5, and the LM319 operational amplifier. A 1-milliampere meter indicates conditions at the output of the monitored circuit. The meter is no longer part of the relay triggering circuit, and the relays no longer need a power supply. If the circuit is to control heavy loads, relays must be employed, but they draw only little power from output transistors Q_4 and Q_5 and can be placed far from the metering circuit.

In normal operation, an input signal is applied to each of the two op amps in the LM319, which compares it to a previously set (high- and low-point) reference. In the circuit, the reference voltages are adjustable from 0 to -5.6 v. This is made possible by the biasing arrangement at Q_1, in which diodes Z_1 and D_1 are configured to supply a constant -5.6 v at its emitter. The voltage across potentiometers R_3 and R_6 thus may vary between two references, one at -5.6 v and the other at -0.6 v, so that the input to the comparator can vary between -5 v and ground.

When the input voltage, as set by the meter, exceeds the high or low reference, the appropriate output of the LM319 fires and switches either transistor Q_4 or Q_5, which closes the proper relay.

The high and low reference points may be set by using the meter. Switching switch S_1 to P_1 removes the metering circuit from the input signal, and the meter then displays the high-point reference voltage, which can be adjusted by R_3. With S_1 moved to position P_2, the desired low-level voltage may be adjusted with potentiometer R_6.

The voltage drop across zener diode Z_1 and the

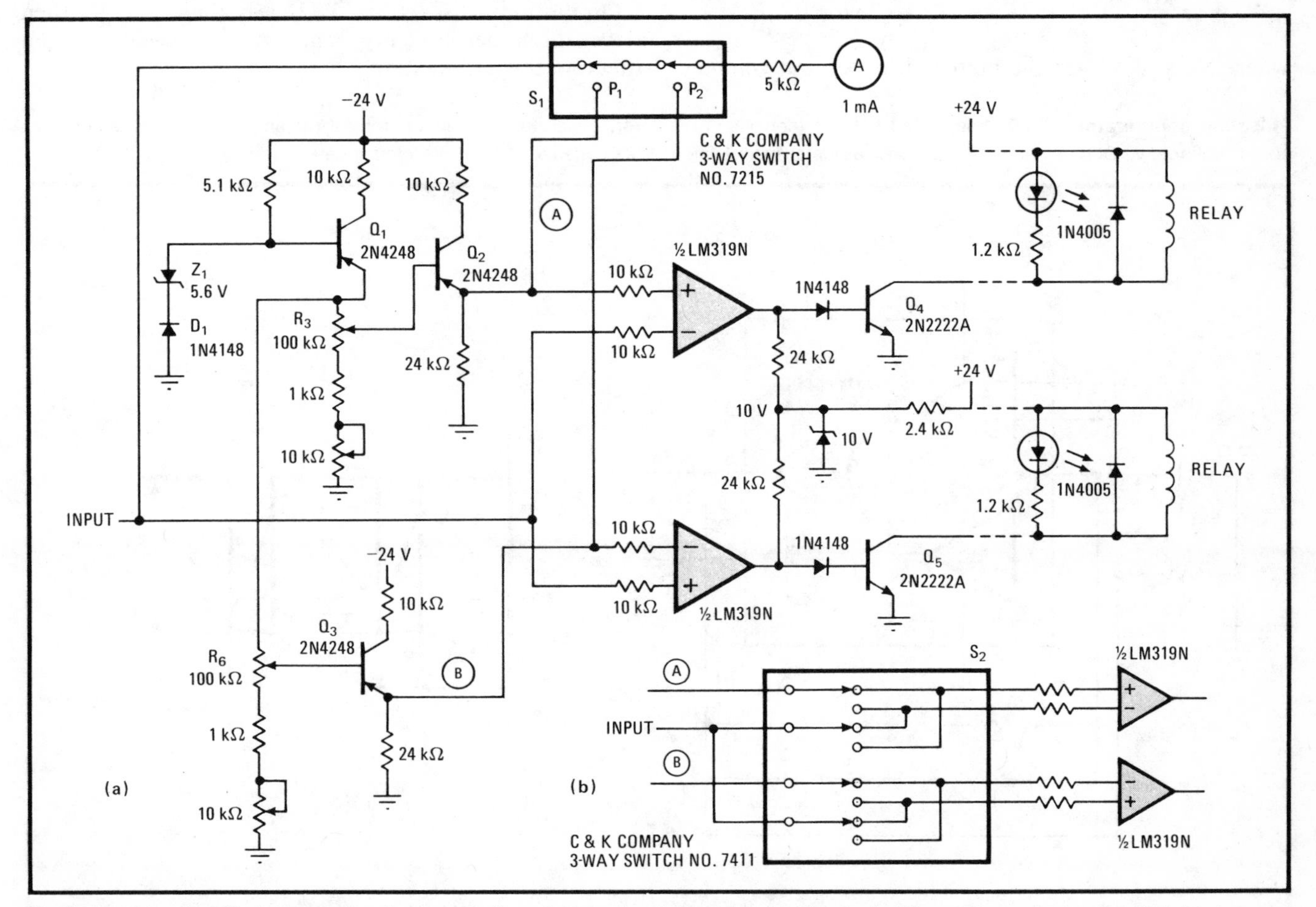

Mechanical to electrical conversion. Implementation of dual-set-point meter using standard meter and comparator circuits (a) is low-cost and reliable. When a three-way switch is added to circuit (b), double high and low set points can be set.

number of standard diodes (D_1) determines the range of reference voltages possible. Reversing diode polarities and changing Q_1, Q_2, and Q_3 to npn devices produces positive reference voltages.

To establish double high set and low set points, a four-pole, three-way toggle switch is placed in the path of transistors Q_2 and Q_3 and the input of the op amp, as the lower part of the figure shows. □

One-chip comparator circuit generates pulsed output

by Virgil Tiponut and Daniel Stoiciu
Timisoara, Romania

Generating a steady stream of pulses rather than the usual logic 0 or logic 1 output when triggered, this special comparator circuit is useful in many control applications. The one-chip circuit is perhaps the simplest way to build what is essentially a low-cost switched oscillator.

A 711 dual comparator serves as a Schmitt trigger and rectangular-wave generator, as shown in the figure. The Schmitt trigger (A_1, R_1, and R_2) has switched levels V_h and V_l, given by:

$$V_h = \frac{R_1}{R_1+R_2} V_H$$

$$V_l = \frac{R_1}{R_1+R_2} V_L$$

where V_H and V_L are the high and low output voltages, respectively (at pin 10), $V_H = V_s - 0.75$, and V_s is the comparator's strobe voltage.

The outputs of each comparator are connected in the wired-OR configuration. When the control voltage, V_i, is greater than the user-set reference, V_{ref}, the output of A_2 is low. Thus the output voltage, V_o, is determined by the signals at the input to comparator A_1.

A_1 will then begin to oscillate. The output V_o will move high to V_H because of the small differential voltage that exists across the input of A_1, charging C through D_1 and R_3. When the voltage across C, V_f, approaches V_H, A_1 switches, bringing V_o low. C then discharges through D_2 and R_4 until the voltage across C drops below that at the noninverting input. V_o then moves high again, charging C, and the process repeats.

Circuit operation may be visualized with the aid of the waveforms shown. A_1's switching times, T_1 and T_2, are given by:

$$T_1 = C_1R_3 \ln [(V_l - V_H)/(V_h - V_H)]$$
$$T_2 = C_1R_4 \ln [(V_h - V_L)/(V_l - V_L)]$$

neglecting the output resistance of the comparator.

Oscillation ceases when V_i is less than V_{ref}. V_o then moves high permanently, since A_1 is prevented from affecting the output state. □

Pulsating comparator. One 711, wired as Schmitt trigger and oscillator, generates continuous train of rectangular waves when fired by control voltage, V_i. Output voltage V_o otherwise assumes logic 1 (high) state (when $V_i < V_{ref}$). Waveforms detail circuit operation.

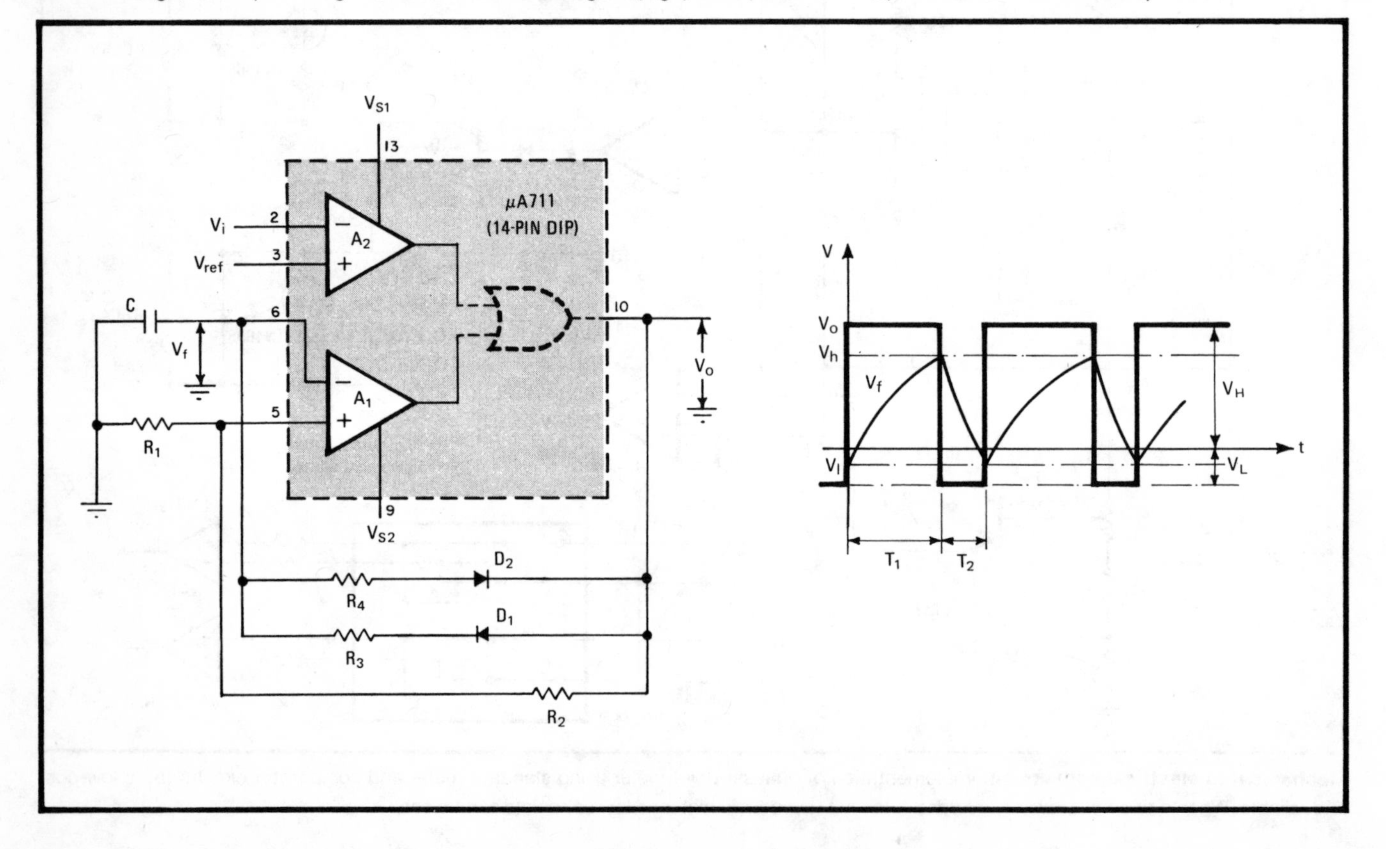

Dual-feedback amplifier zeros comparator hysteresis

by Svein Olsen
Royal Institute of Technology, Stockholm, Sweden

Amplifiers with positive feedback may be combined to create voltage comparators and zero-crossing detectors devoid of hysteresis. Alternatively, the amount of hysteresis, either positive or negative, may be selected. In both cases, feedback ensures that true bistable (switching) operation is achieved without undue sacrifice of noise immunity—a necessary condition for optimum comparator and zero-detector performance.

The ideal voltage comparator cannot be realized with a single amplifier because bistable operation does not occur until hysteresis starts. Witness today's typical comparator—a fast differential amp and a transistor-switch output stage that actually operates as a linear amplifier within a small region about the transition level. Achieving a step transition for slowly varying input signals is difficult with these high-gain, wideband devices, too, because of the radio-frequency oscillations and multiple transitions that occur in association with very small noise signals.

Introducing positive feedback to increase loop gain and thus ensure bistable operation, as some have tried, will yield clean switching independent of input slope. But hysteresis also is introduced, and, worst of all, Δt, a varying input/output delay—which depends on the slope of the input signal and the instantaneous value of hysteresis—comes into play.

The block diagram (a) shows how to achieve bistable operation while eliminating all of these problems. Amplifiers 1 and 2, each having positive feedback (α, β, respectively), are applied to their individual summing junctions, where they are combined with the input signal. Amplifier 1 also drives the second summing junction with a dc-level shift signal ($\pm k$) that is a function of the amp's hysteresis. Note that this feedback signal can be derived by either a switching or a linear stage.

Depending upon its polarity, the signal may add to or subtract from the amount of hysteresis inherent in amplifier 2. In the special case, total circuit hysteresis may be eliminated with little loss of noise immunity. At the same time, the circuit will retain high gain for true bistable operation. (The lengthy mathematical analysis of the circuit may be found elsewhere.[1])

A practical circuit having TTL-compatible outputs is shown in (b). Feedback in both amplifiers is determined by resistors r and R. In this application, r is 150 ohms and R is 15 kilohms, so that V_{H1} = 40 millivolts ($V_{in\ min}$ = 15 mV root mean square) and V_x = 20 mV, where V_{H1} is the hysteresis for amplifier 1 and V_x is the noise immunity.

Amplifier Q_1–Q_2 provides an inverted feedback signal to the second summing junction, with the magnitude of

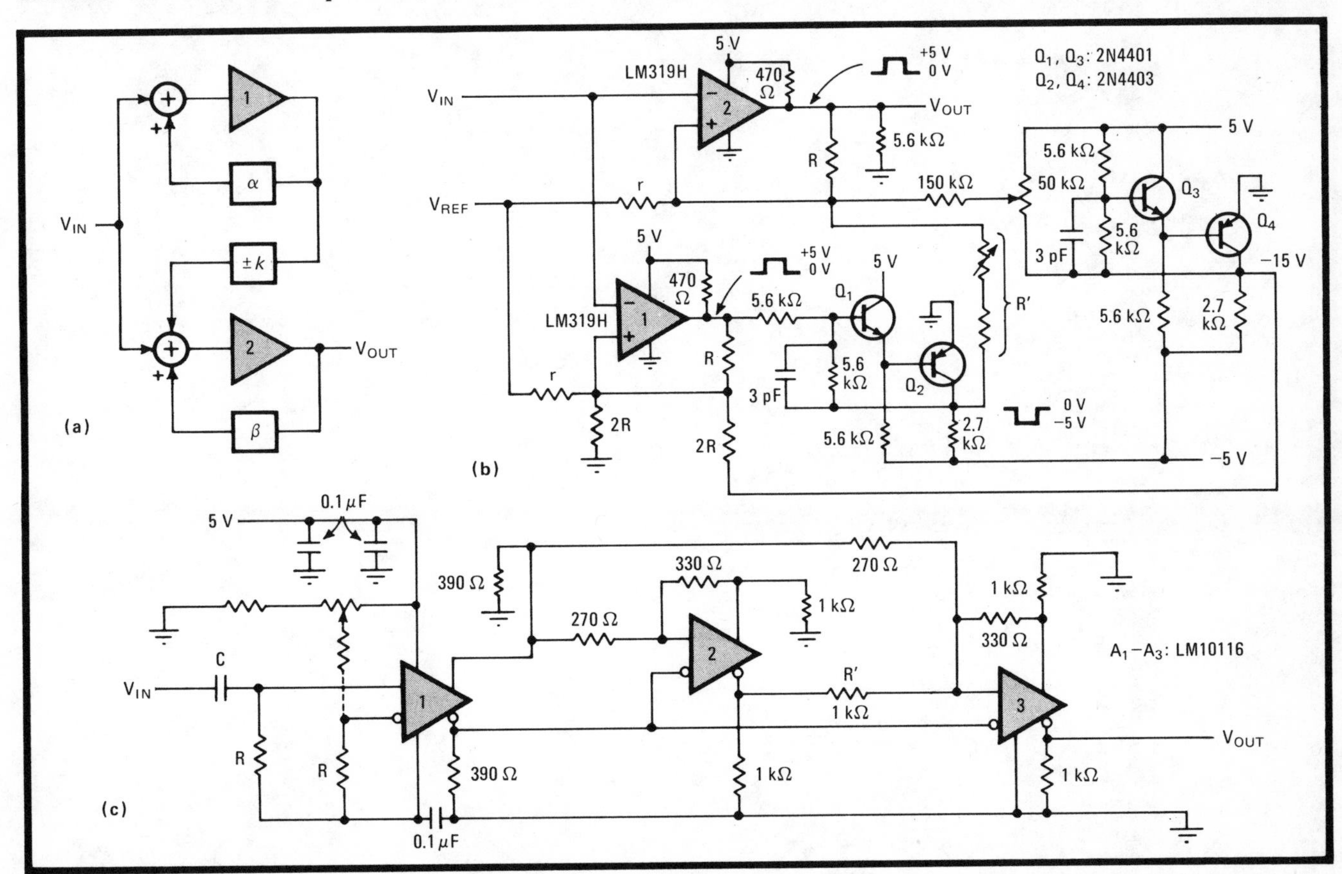

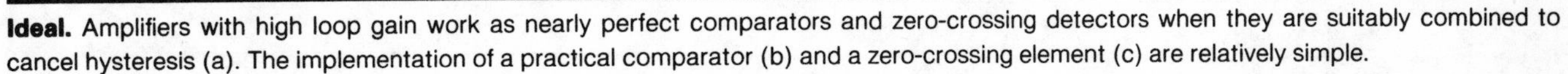

Ideal. Amplifiers with high loop gain work as nearly perfect comparators and zero-crossing detectors when they are suitably combined to cancel hysteresis (a). The implementation of a practical comparator (b) and a zero-crossing element (c) are relatively simple.

the signal set by potentiometer R′. The negative voltage at the junction of amplifier 1 required to establish a level-shift voltage at amplifier 2 is provided by Q_3–Q_4. The output hysteresis is adjustable to zero.

A fast (3-nanosecond) zero-crossing detector with zero hysteresis is shown in (c). This application requires an LM10116 emitter-coupled-logic receiver to be used, and although its low amplification factor makes it a little more difficult to achieve high loop gain, three sections are used to make up for the shortcoming.

Amplifier 1 is the input stage biased for Class A amplification. The input RC values are selected according to the impedance-matching requirements and to provide the required low-frequency response. Amplifiers 2 and 3 serve the functions previously mentioned. □

References

1. Svein Olsen, "The Zero Hysteresis Comparator," RVK-78 Conference Notes, Stockholm, March 29, 1978.

Chapter 7
CONTROL CIRCUITS

Train speed controller ignores track resistance

by Stephen H. Burns
U. S. Naval Academy, Electrical Engineering Department, Annapolis, Md.

Because it keeps the voltage applied to any universal motor constant and independent of line resistance for a given line voltage, this circuit will be especially attractive for use with model trains, for which it will maintain speed independently of the resistive joints in the tracks. No modification of the speed-controlling transformer used with the trains is required.

This circuit rides within or near the locomotive and is inserted in the electrical path between the motor and the track pickup and return. Thus it is necessary to connect the ac input of this circuit to the track pickup and frame of the locomotive and to connect the dc output of the circuit across the motor windings. This way the circuit develops a constant average voltage across the motor for a given transformer setting.

This is accomplished with a four-section circuit, with block A containing the supply for powering the unit. Blocks B and C derive a gate pulse inversely proportional to the magnitude of the input voltage and a control voltage proportional to the input signal so that motor speed can be set. Block D handles the power-switching function.

Diode bridge B_1, diodes D_1 and D_2, capacitor C_1, and Norton amplifier A_1 constitute block A, which is actually a three-stage supply. B_1 provides full-wave rectified dc for power stage Q_3 and D_1 and C_1 extract and filter power for A_1–A_4. Zener diode D_2 and A_1 provide a regulated 5 volts for current source Q_1 and to establish a switching reference voltage for amplifier A_2.

In operation, B_1 and Q_3 electrically disconnect the motor from the power source near each zero crossing of the ac input. At this time, a pulse is developed at the output of A_2, its width equal to $P = 2V_r/\omega V_i$, where V_r is the switching reference voltage, ω is 120π radians/sec, and V_i is the input voltage (see timing diagram). Note that by sampling near the zero crossings, the width of the pulse is made independent of the load on the ac line. One sense diode, D_3 or D_4, reference diodes D_5 and D_6, and one bridge diode set the switching threshold, V_r. This reference voltage is thus equal to four diode drops, or approximately 2 V.

A_3, Q_1, and Q_2 develop from the pulse a control voltage, V_c, whose average value can be expressed as $V_c = \pi I_o R_o V_i/(2V_r)$. A_3 and Q_1 constitute a 0.3-milliam-

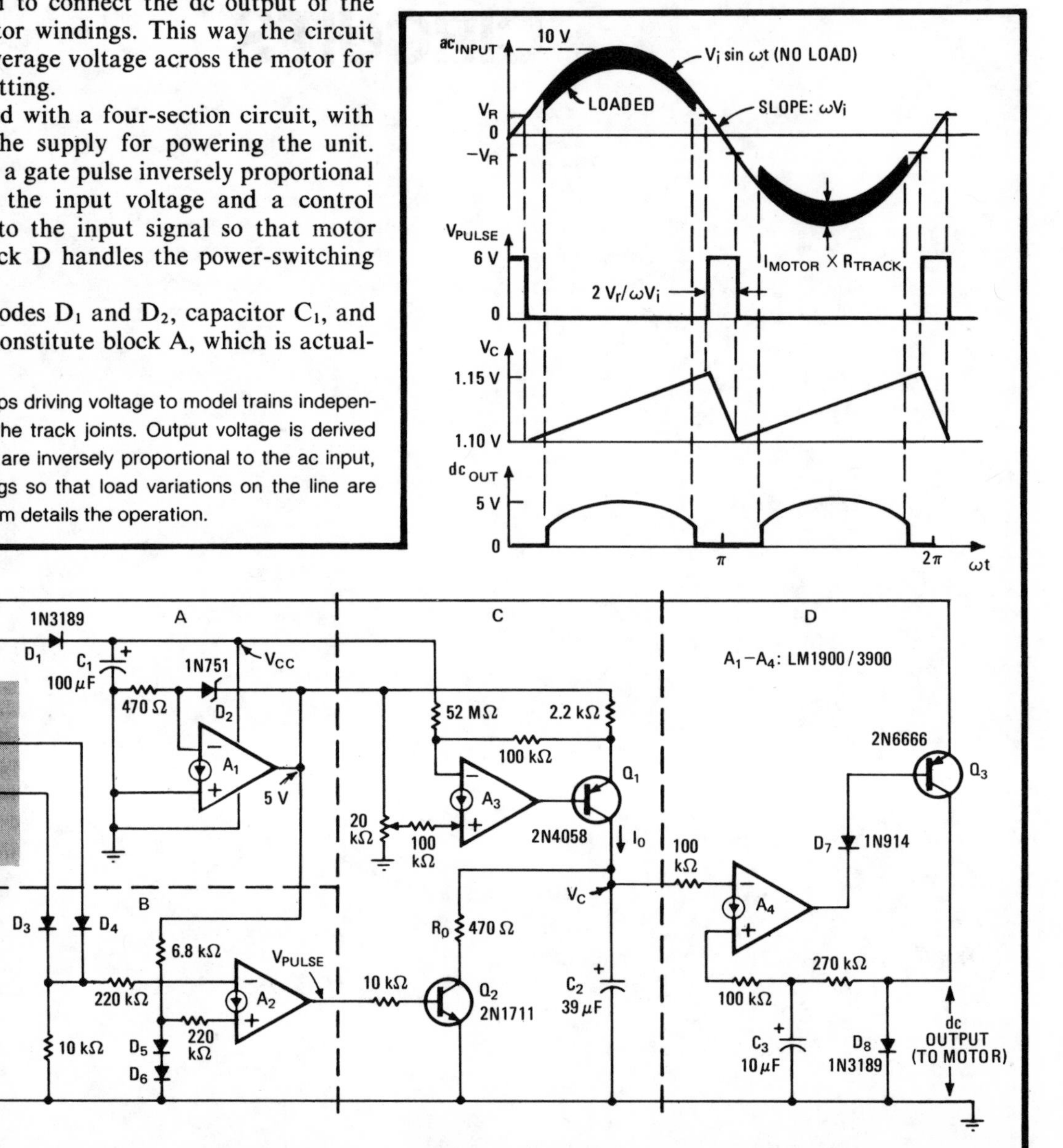

Rail tamer. Controller keeps driving voltage to model trains independent of the resistance in the track joints. Output voltage is derived from pulses whose widths are inversely proportional to the ac input, generated at zero crossings so that load variations on the line are disregarded. Timing diagram details the operation.

pere (I_o) constant-current source. A total resistance of 52 megohms placed in series between V_{cc} and the inverting input of A_3 was found best to compensate for changes in I_o versus supply variations.

The power switching stage includes feedback capacitor C_3, amplifier A_4, isolation diode D_7, damper diode D_8, and Q_3. C_2 and C_3 were selected so that Q_3 switches efficiently under all load conditions. Smaller values of C_2 result in more efficient switching but increase output resistance. Larger values of C_3 will result in greater overshooting in response to a transient.

With a source voltage of from 7 to 25 V rms and a load of 10 ohms, the circuit will generate an output of from 3 to 12 V. Increasing the line resistance from 0 to 4 Ω will cause a typical change in output voltage of only 0.1 V for a line drop of several volts. □

Switching-mode controller boosts dc motor efficiency

by Jay C. Sinnett
U. S. Environmental Research Laboratory, Narragansett, R. I.

On-site monitoring equipment that makes use of a variable-speed dc motor places a special premium on the efficient use of the instrument's battery supply, because current drain is often high. This motor-speed controller circuit, which works on the principle of the highly efficient switching-mode power supply, saves energy and thus reduces circuit losses associated with the motor.

In this circuit, large, low-duty cycle pulses of supply current set up continuous currents in a small (0.01-horsepower) motor that are almost equal in magnitude to the peak current drawn by the supply, thereby contributing to circuit efficiency. As a typical example, almost 200 milliamperes of continuous motor current can flow when the average battery current drain is 100 mA, for an output voltage of 3.5 volts.

A_1, a voltage comparator, serves as both an oscillator and a duty cycle element in the controller, as shown in the figure. C_1 and R_1 provide positive feedback to A_1, enabling it to oscillate at about 20 kilohertz. The duty cycle, which can be from 10% to 70% of one 20-kHz period, is controlled by the negative feedback loop formed by Q_1, R_1, C_3, and R_3.

When the system's control signal, $\overline{\text{MOTOR RUN}}$, is asserted low, Q_2 turns on and applies power to the entire circuit. Pulses emanating from A_1 are amplified and inverted through Q_1 and pass through the motor, M. R_4 and D_1 set the average voltage supplied to the motor and thus largely determine the motor speed.

Note the absence of a capacitor at the output, which would normally be required to filter the pulsed signals and enable the motor to run smoothly. If a capacitor were used, it would have to be large in value and therefore large in size and costly as well. Instead, diode D_2 is placed in the circuit for filtering, enabling the pulsed energy to be stored by the motor's inductance in the field surrounding the windings. Between pulses, when Q_1 is off, little battery current is drawn, but the motor current is relatively large, since the amplitude of the current decays slowly through D_2.

Note also that although D_1 provides a stable, accurate reference, the average voltage fed back from the motor's terminals is affected by the forward-voltage drop of D_2. The drop varies with temperature and the current drawn through it and so reduces the absolute accuracy with which the output voltage can be set. However, resettability and stability are both very good with respect to battery voltage variations, and in applications where the temperature variations are minimal, the drawback will be unimportant. For example, the current variations due to even a 4-V supply-voltage change will be less than 2%. □

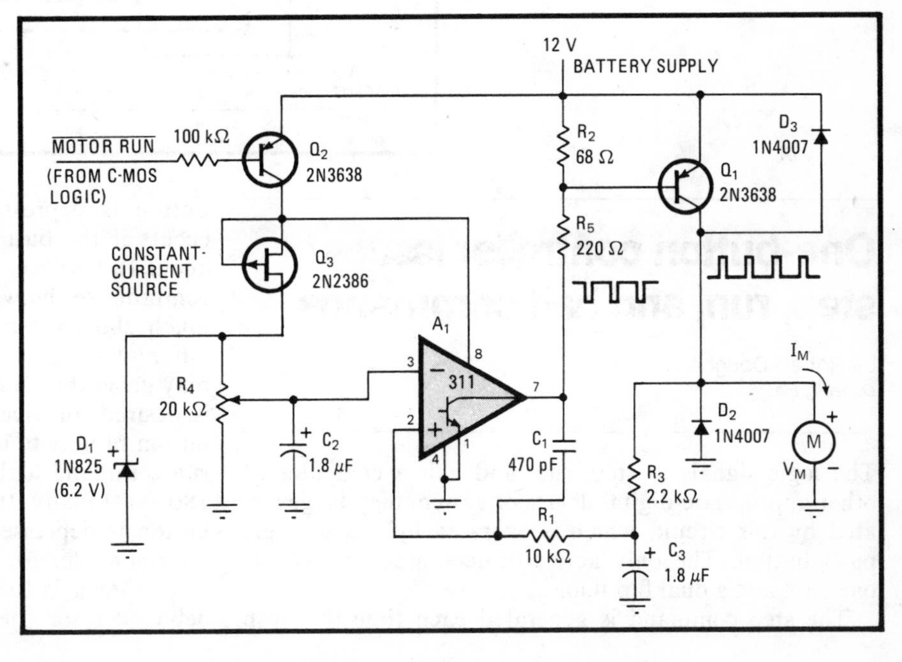

Less drain. Dc motor-speed control, which works on principle of switching-mode power supply, ensures minimum circuit losses. Duty cycle of pulsed output, 10% to 70% of one cycle at 20 kHz, drives motor, keeping battery current to minimum. Motor's inductance stores and filters pulses, essentially replacing filter capacitor normally used.

Diodes and integrator brake small motors dynamically

by Stephen Wardlaw
Yale-New Haven Hospital, Dept. of Laboratory Medicine, New Haven, Conn.

Alternating-current motors used in position-sensing circuits must be quickly braked and stopped if the system is to retain its positional accuracy. In the case of a small shaded-pole motor, a dc source connected directly to its field winding brakes it dynamically by rapidly dissipating its kinetic energy. But if not turned off in time, the source will overheat the motor.

A safer way is to derive the dc voltage through a silicon controlled rectifier, a diode, and a resistance-capacitance network. Moreover, such a circuit costs less than an electromechanical switch and is simpler than a thermal-delay or momentary-contact switch.

As shown in the figure, the braking unit (within the dotted lines) must be placed in parallel with a manual electronic switch, S_1, that is used to trigger the braking of motor M. With S_1 in the normally closed position, no voltage appears across the braking unit, and R_1 bleeds off any charge being stored in capacitor C_1.

When braking is desired, S_1 is activated and thus opened, so that the positive half-cycle of the line voltage will appear across D_1, C_1, and R_1–R_3 and the SCR will be triggered. This action, in addition to enabling a strong pulse of direct current to flow through the motor windings, partly charges C_1.

When the line current drops through zero and into its negative half-cycle, the SCR turns off and remains in that state until the ac input reaches its positive half-cycle again. The process is repeated until C_1 is charged to near the peak value of the line voltage, at which time direct current will cease to flow. The SCR will not turn on again, because D_1 will be permanently back-biased.

The 150-volt varistor helps to suppress line spikes. The fuse, F, is included as a safety precaution and will open if for some reason the braking unit continues to enable the power line to feed a relatively high direct current through the motor winding. Using the component values shown, the braking unit will enable the line to supply a pulsating dc to the motor for approximately 1 second—more than enough time to completely brake any small motor with a rating of up to ¼ horsepower or so. □

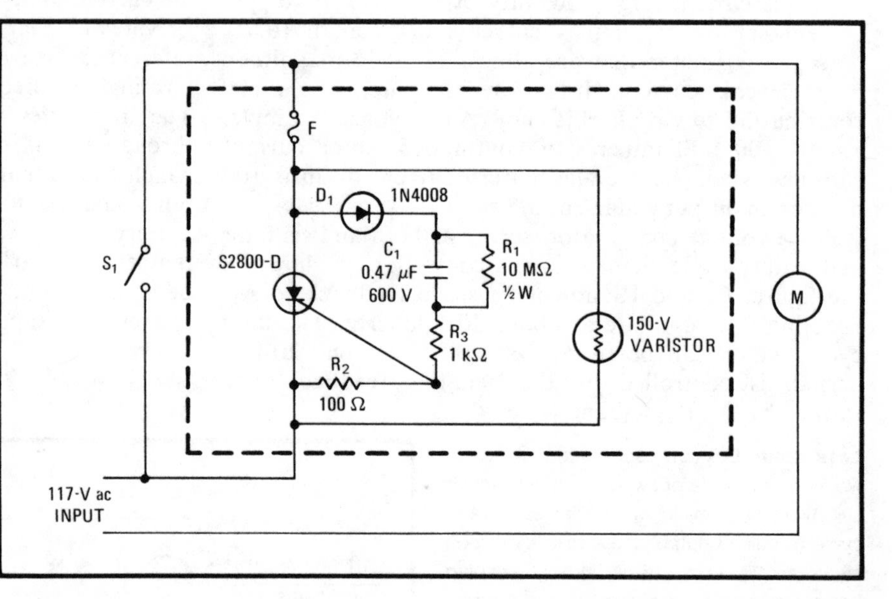

Fast reaction. S_1 initiates motor braking. Positive half-cycle of input voltage appears across D_1, C_1, R_1–R_3, firing SCR and enabling direct current to flow through small shaded-pole motor. C_1 charges to nearly peak value of input voltage during succeeding positive half-cycles, terminating process.

One-button controller issues step, run, and halt commands

by Robert Dougherty
Dunedin, Fla.

The logic signals to step, run, and halt a computer or other appropriate digital device or system may be generated by this circuit, which is operated by just a single push button. The only active devices used are a dual one-shot and a dual flip-flop.

The step command is generated each time the push button is depressed momentarily. The run command occurs if the button is held down for a time exceeding about 180 milliseconds. This time represents an excellent compromise between circuit speed and accuracy. A much shorter duration means the circuit may fail to differentiate between the step and run commands and may generate the run command when the step command is desired, or vice versa. Also, repeatedly pressing the button rapidly to initiate step functions will generate the run command if the duration is set for much more than 180 ms. Finally, the computer will be halted if the push button is depressed momentarily when the circuit is in the run mode.

As shown in the figure, A_1 acts as an effective switch debouncer for the push button. For a step command, poking the button quickly will cause the Q output of A_1 to go high and fire A_2, the run-and-stop one-shot. A_1 is also fed to A_3, the D input of the run-and-idle latch. At the same time, the $\overline{Q}$ output of A_1, which moves low, will fire the step one-shot A_4, yielding the step function.

The sequence of events just discussed also describes the initial portion of the run command, whereby the step pulse is used to manually advance the computer's program counter by one. The run pulse then commands the computer to rapidly execute succeeding steps automatically. The $\overline{Q}$ output of A_2 moves high 180 ms after the push button is depressed. The positive-going, or trailing, edge of this pulse then clocks the state of the push button (as detected by A_1) into A_3.

If the button has been released before time-out, a zero appears at the Q output of A_3. But if the button is activated, A_3 moves high and the run command is executed by the computer.

A press of the button will cause the circuit to halt the machine if it is in the run mode, by clocking in a logic 0 to the run-and-idle latch. Note that the step pulse generated at the start of the halt sequence, as shown in the timing diagram, is of no consequence, since when the step is received, the machine is already in the run mode and will override that command. □

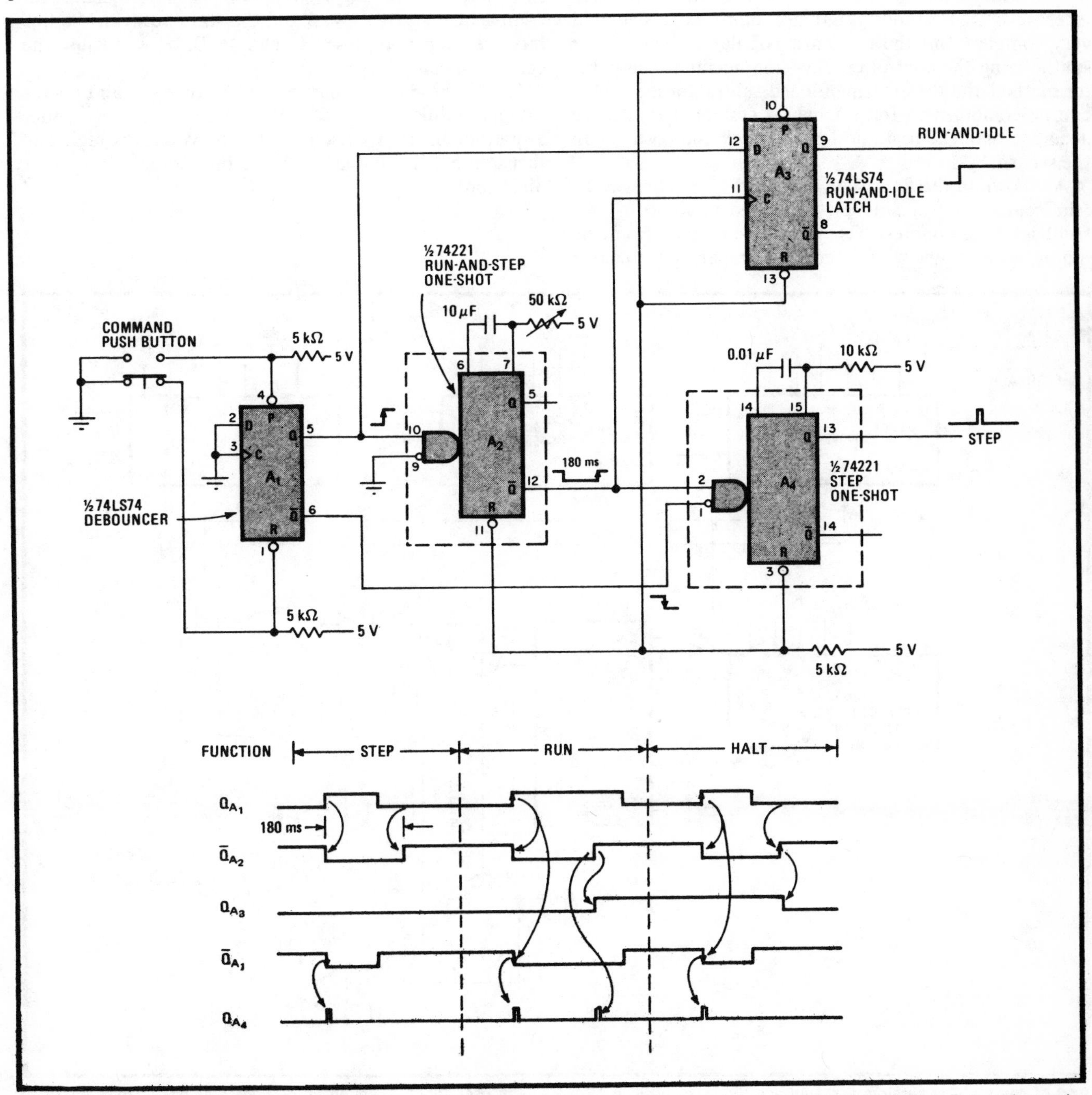

Touch control. One push button and two ICs single-step a computer's program counter or control run and halt operations. Button-depression time and present mode of controller determine the command generated. Timing diagram details circuit operation.

Hybrid servo system minimizes hunting

by C. V. Rajaraman
ISRO, Trivandrum, India

Although the shunt comparator circuit of Vojnovic[1] is simple and tends to reduce the overshoot and hunting problems inherent in a high-speed digital servo system, difficulties may arise when the servomechanism is continually called upon to follow small changes in position. Digital subtraction circuits, on the other hand, will increase system stability but are more expensive and very complex. But the two most popular techniques for synthesizing the control circuit—the no-shunt comparator method and the aforementioned subtractor method—can be combined to form a hybrid system that is more accurate than the first and less complex and costly than the second.

As shown in the figure, A_1 and A_2 compare lines 5–12 (the coarse bits) of a 12-bit command input with their feedback-data counterparts, which are derived from the motor position by a shaft encoder. Comparing the coarse bits in this manner enables the system to converge quickly on the desired position.

At the same time, the low-order (fine) bits of the command word are compared by A_3, and A_4 is programmed to find a~b, the difference between the two 4-bit binary words. The a~b result is then transformed into an equivalent analog signal by the digital-to-analog converter, A_5. Using the d-a converter allows a precise voltage to be applied to the motor, instead of the constant-magnitude (logic 1) signal that a comparator-type circuit would generate whenever there was an a~b offset of any value. Thus the motor has little tendency to overshoot its intended mark.

As for transferring the voltage from the converter to the motor, V moves high either when the coarse-bit comparison yields $A = B$ and the state between the fine bits are such that $A = B$ or $A > B$, or when just the coarse-bit comparison yields $A > B$.

Under these conditions, S_1 and S_4 turn on, and operational amplifier B_1 moves high to drive the motor. Under any other bit-comparison condition, W moves high and turns on S_2, S_3, and B_2, to drive the motor in the other direction. □

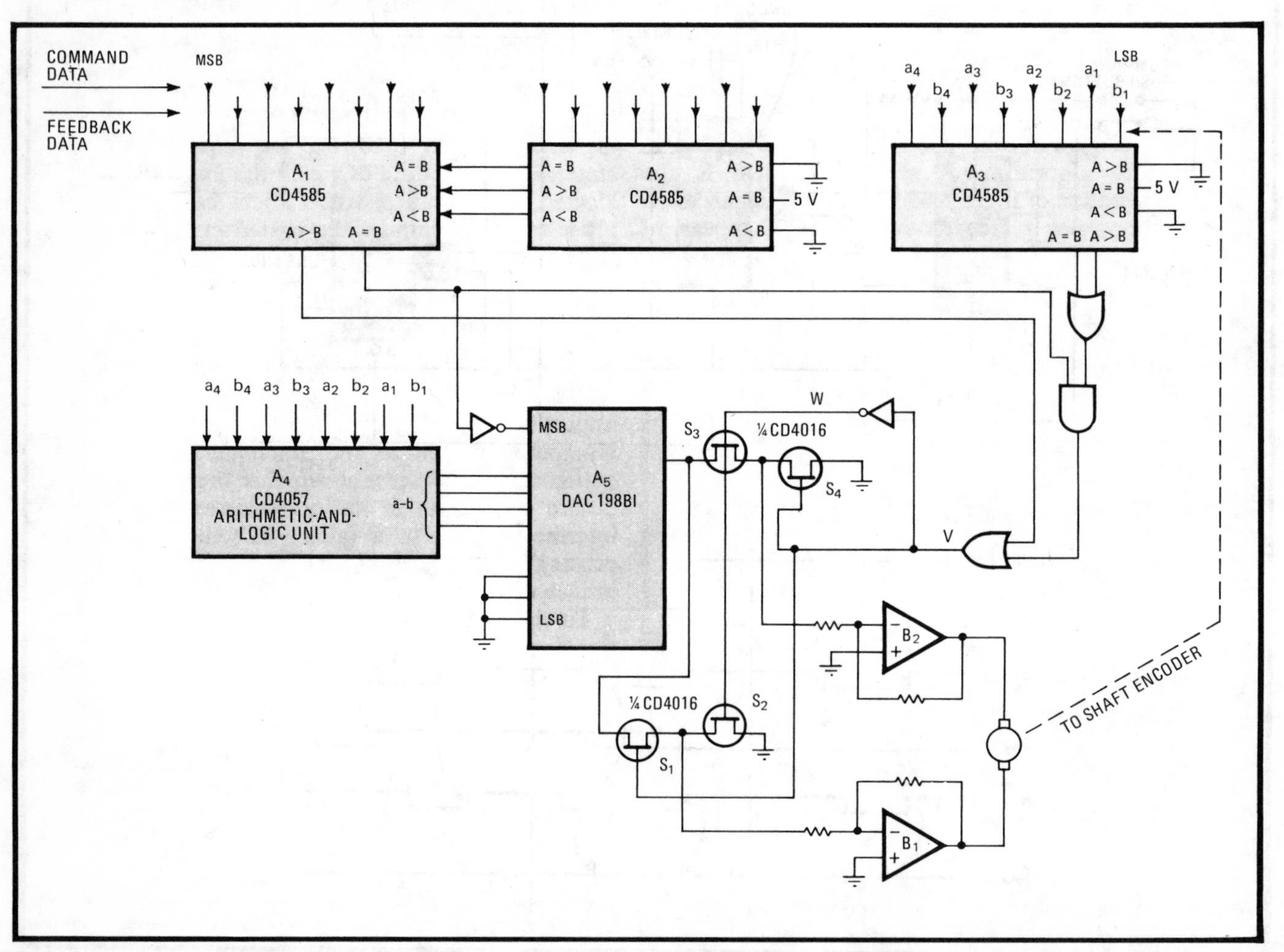

End of search. Comparators and subtractor in servo position motor without oscillations. Comparison of high-order bits enables motor to converge on desired location. Fine-bit comparison with a d-a converter resolves precise feedback voltages to minimize overshooting.

Annunciator control uses only passive components

by John A. Haase
Fort Collins, Colo.

Combining relays with other passive components, this circuit provides time delays of anywhere from milliseconds to minutes without the need for the polarizing potentials normally required by monostable multivibrators. Such an electromechanical arrangement is ideally suited to the implementation of a practical call controller, or annunciator.

Delays are produced by first generating a staircase voltage. Pushing button PB_1 momentarily latches relay K_1 and energizes its bell circuit. The bridge circuit used with each relay permits the use of an inexpensive dc relay.

Call-indicator lamp L_1 then lights, and supply voltage is applied to C_1, C_2, D_1, and D_2, which make up the staircase generator. The step interval of the rising output voltage, V, at the junction of D_2 and C_2 becomes $\Delta V_o = C_1 E(10^3)/C_2$ millivolts per cycle, where $E = 12.6(1.414) = 18$ volts. Note that the diodes are considered passive elements in this application, as their characteristic curve, per se, is not utilized in the generation of the staircase waveform.

Circuit constants are selected so that D_5, which serves as a comparator, breaks down after 155 cycles (2 seconds), when output voltage V reaches 28 v. Relay K_2 is then energized, enabling K_3 to close and lamp L_2 to light, because the alternating voltage is applied to a second delay circuit through one of the normally open contacts of K_2. At the same time, voltage to relay K_1 is removed, since the normally closed relay contact of K_3 (in K_1's energizing path) opens. Simultaneously, C_2 is discharged by K_3's closed relay contact (in series with the 1-kilohm resistor).

Relay K_3 remains closed for 80 seconds to prevent repeated (and most times annoying) calls. At that time, the output voltage across C_4, generated by the second staircase waveform, steps to 28, whereupon D_6 breaks down and relay K_4 is energized. Relay K_3 then opens and the path of relay K_1 is reactivated to accommodate call requests, while C_4 discharges.

Call requests may be extended to 4 seconds by placing switch S_1 in the hold position. In this mode, relay K_4 does not come into play, so the user must wait an infinite inhibiting time before initiating a second call request. Under this condition, push button PB_2 must be depressed to clear the circuit so that it can respond to calls. □

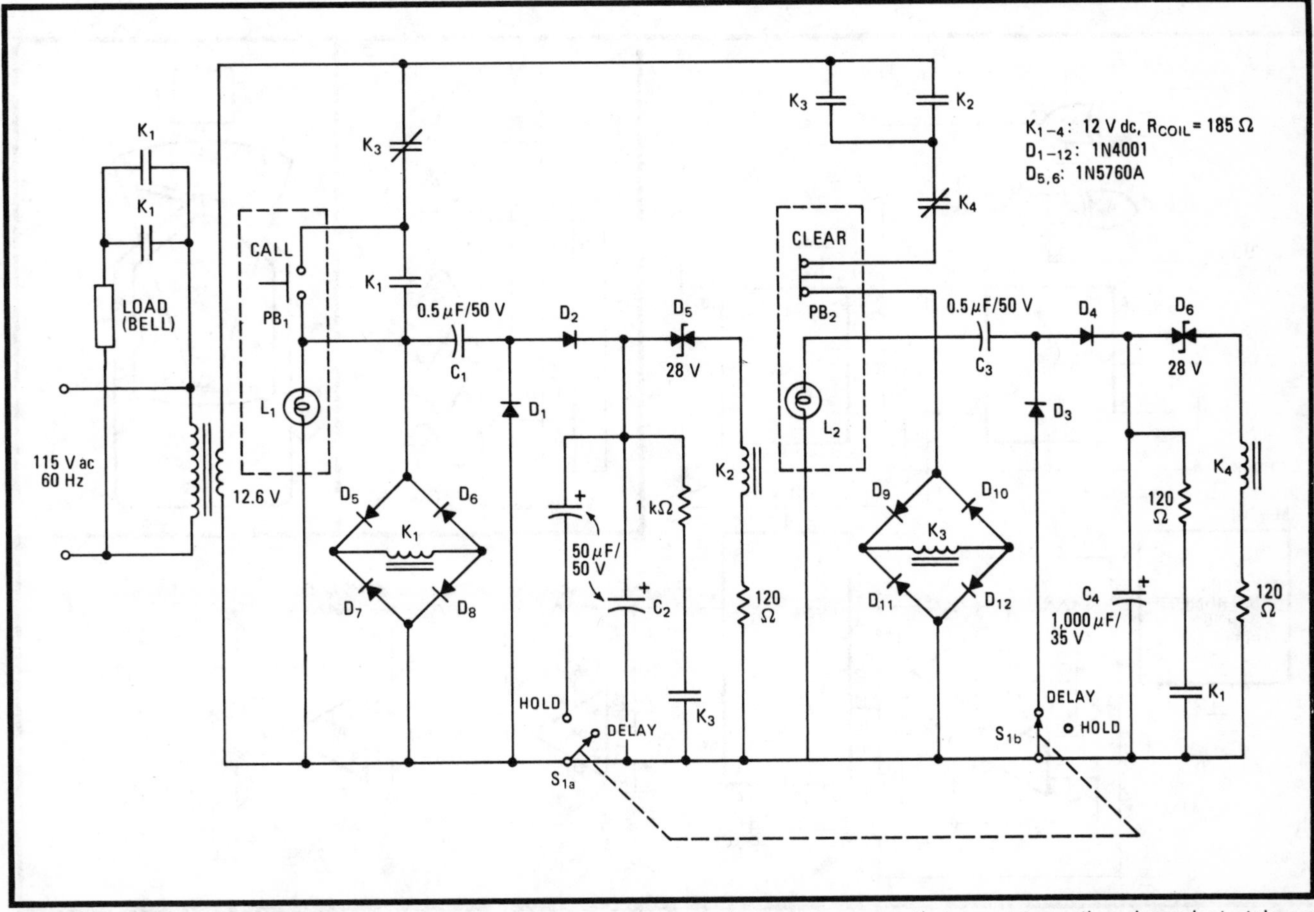

Charged delay. Electromechanical relays and other passive components form charge pump that generates a time-dependent staircase voltage, V, for call controller. By suitably selecting diodes D_5 and D_6 so that their breakdown voltages conform to some preset value of V, the circuit provides delay. Depressing PB_1 rings bell for 2 seconds, inhibits circuit for 80 seconds to prevent quick second requests.

Gray-code counter steps torque motor

by Thomas L. Clarke
Miami, Fla.

The positional accuracy of a simple stepping-motor system is limited by the response of its mechanical drive. This drawback can be eliminated electronically by using a position sensor and a counter working in Gray code to control the motor. The mechanical-drive circuit can be simplified with digital logic to reduce system errors and nonlinearities. A four-state Gray-code counter enables the system to move smoothly from its starting point to the desired position.

In this circuit, the summed quadrature outputs of a photoelectric sensor and the counter (see inset) set the position of the system. With suitable clock signals, the counter is advanced one location, causing the motor's position to change and the output from the sensor to vary accordingly. Thus the system is rotated 90° for each clock signal.

This circuit is intended for visual setting of a desired position through manual control of the clock and up/down inputs. For automatic tracking, the sensor's output must be compared to the desired position with additional circuitry in order to generate those signals.

Flip-flops Q_1 and Q_2 and exclusive-OR gates A_1–A_3 comprise the up-down Gray-code counter. The direction of the counting is determined by the logic state at the up/down input.

The output of the counter changes on the positive transition of each clock pulse. Depending upon the state of the counter, either the normal or inverted sine-wave outputs of the sensor are summed at the output of the 4052 four-input multiplexer. As a consequence, the output from A_4 forces the system to a new position, which is reflected at the sensor as its output steps a quarter cycle. The motor is driven through Q_3 and Q_4 by a positional signal that progressively advances or recedes (depending upon the state of the up/down counter) by a quarter cycle.

A minimum settling time of a few milliseconds is set for the system by the lead-compensation components between stages A_4 and A_5. Lead compensation is required in this situation because the system response is that of a double integration network that acts to saturate Q_3 and Q_4. The open loop would tend to be sluggish without the lead compensation, which reduces the effective system gain at low frequencies. □

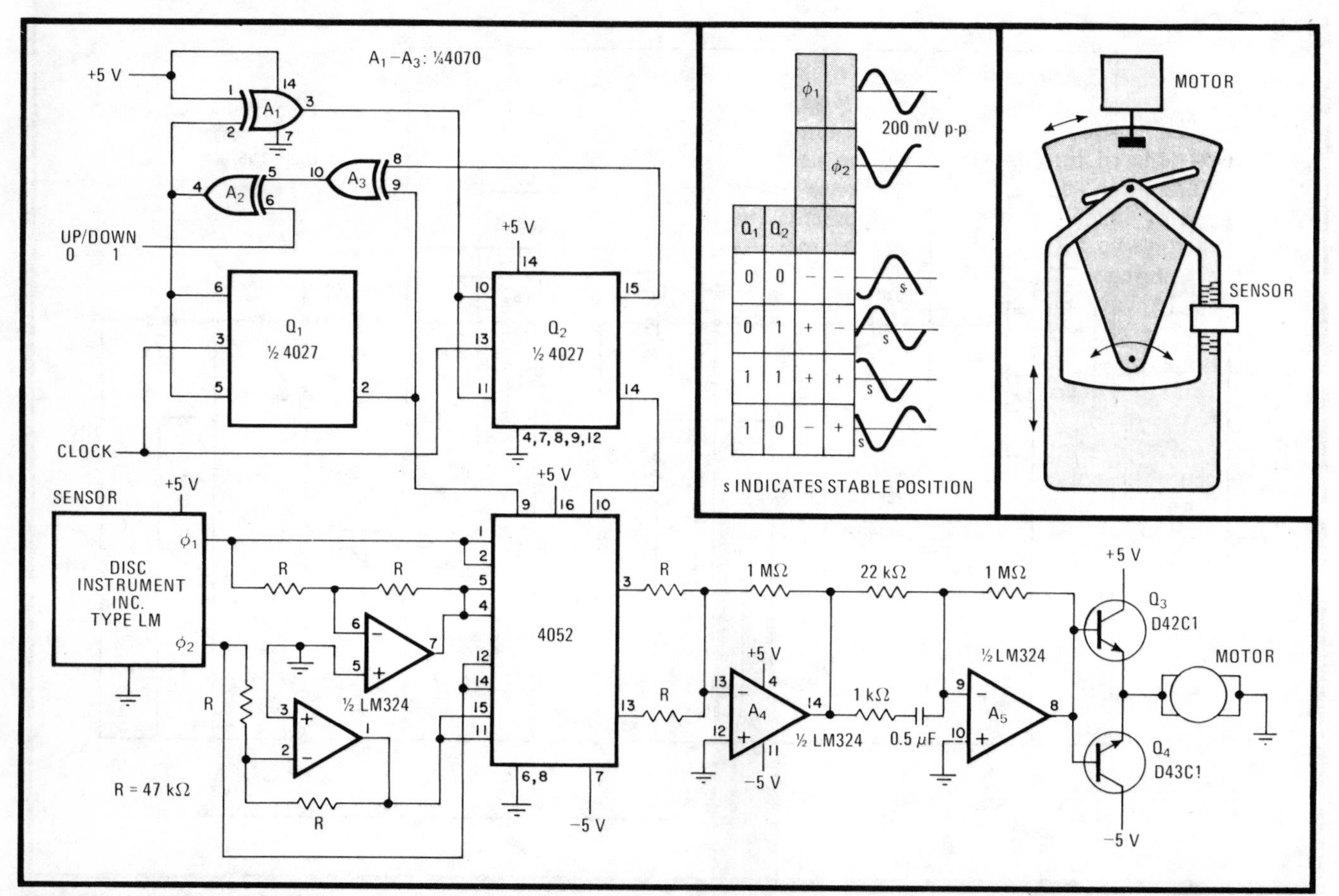

Smooth. Four-state Gray-code counter Q_1–Q_2 provides signal that, when summed with output of optical sensor by 4052 multiplexer and op amp A_4, generates quadrature output for smooth stepping of motor in quarter-cycle increments. Compensation network between A_4 and A_5 prevents saturation of Q_3 and Q_4, eliminating sluggish system response by reducing effective gain at low frequencies.

One-chip power amplifier controls dc motor's speed

by Kuang-Lu Lee and Dennis Monticelli
National Semiconductor Corp., Santa Clara, Calif.

Circuits for regulating the speed of small dc motors need not be expensive or complicated now that one-chip power operational amplifiers are available. In fact, using the power device (such as the LM13080) in a simple negative-feedback configuration provides better regulation than many speed controllers now on the market. In addition, common-mode rejection of power-supply transients is large.

As shown in (a), the circuit's reference voltage is established by D_2 and R_3 and filtered by R_5 and C_1. D_1 simply serves as a common-mode level shifter for the inputs of the op amp. Negative feedback around the op amp provides the contolled-voltage drive to the motor. Thus:

$$V_{motor} = (V_{D2} + I_m R_3)(R_2/R_1) + V_{D2}$$

where V_{D2} is the forward voltage drop of diode D_2 and I_m is the current through the motor.

As the motor load increases, I_m increases, and this results in a corresponding increase in V_{motor}. To accommodate large changes in load, V_{motor} varies considerably. The amp therefore needs a 10-volt source voltage to provide sufficient swing, current, and power dissipation for most small motors. Powered by such a source, the LM13080 will handle up to 2 watts in free air and can deliver 0.5 ampere.

The optimum settings for potentiometers P_1 and R_3 are those that provide stable regulation. They are found empirically with the actual motor to be used. P_1 is first adjusted experimentally so that the motor will provide slightly fewer than the desired number of revolutions per minute. R_3 is then increased until a minimal loss in speed is observed for a substantial increase in motor load. Note that excessive positive feedback via R_3 will cause instability. Because the adjustments of P_1 and R_3 interact, it will be necessary to readjust both until the best settings are obtained.

The circuit's performance for a small motor is shown in (b). Note its superior performance with respect to a popular configuration that drives the motor from a constant-voltage source. □

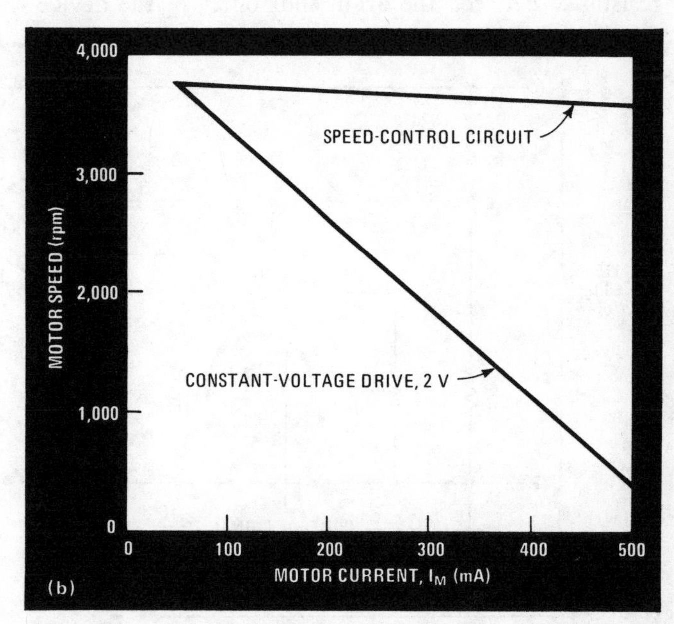

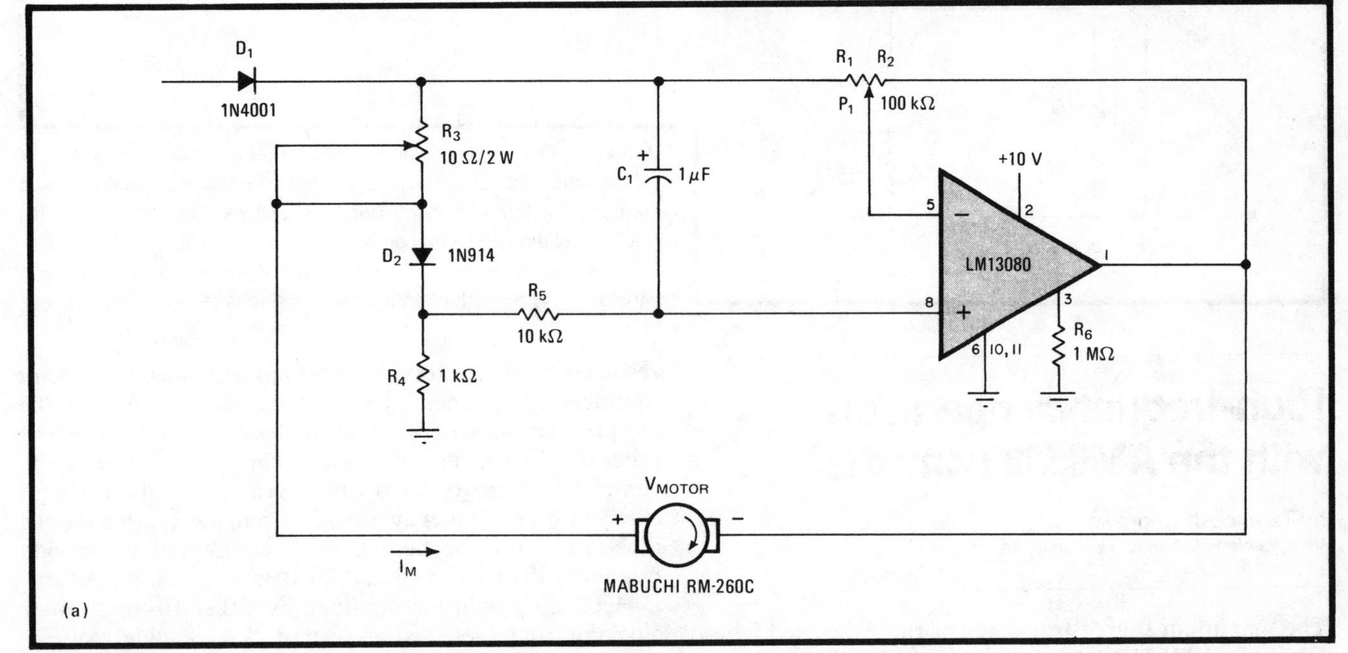

Speedy solution. One-chip power op-amp circuit (a) makes simple, low-cost speed control for small dc motors. Circuit affords excellent common-mode rejection. Controller's rpm-vs-load performance (b) is superior to that of circuits utilizing a constant-voltage drive.

Remote controller sets universal motor's speed

by Hari Herscovici
Cordis Corp., Miami, Fla.

The speed of an ac-dc motor is easily set with this circuit. Millivolt-level input voltages drive its variable-speed control amplifier through an optocoupler that is isolated from the rest of the circuit to permit its use in remotely controlled applications.

Control signals in the range of 0 to 3 volts are applied to the optoisolator (GE H11F2) as shown (a). The resistance between the drain and source of the device's field-effect transistor varies with input voltage V_{in}, and so the gain of the 741 operational amplifier, which amplifies the rectified 60-hertz power-line input, is controlled accordingly.

When the instantaneous output of the op amp is greater than the motor's counter electromotive-force voltage, diode D_2 conducts and thus the silicon controlled rectifier is switched on. Power is thereby applied to the motor. The greater the difference between the op amp's output and the counter emf voltage at any instant, which indicates motor speed is lower than programmed, the earlier in the cycle the trigger pulse to the SCR occurs.

Diode D_1 and resistors R_1 and R_2 have been selected so that the circuit will withstand a reverse voltage of 200 V. If a transformer-based input circuit (b) is substituted, however, it is only necessary for D_1 to have a reverse-breakdown value of 20 V. □

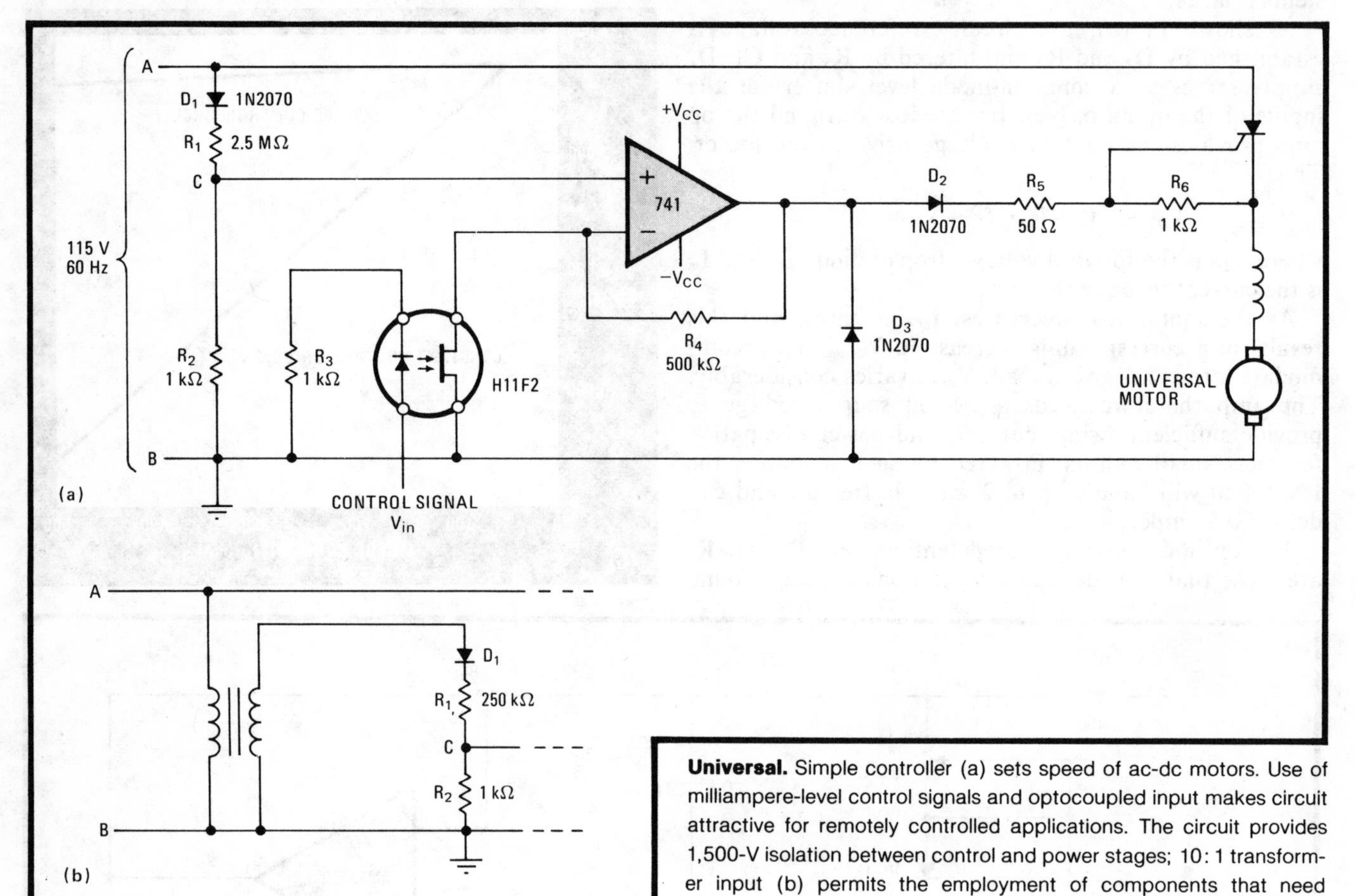

Universal. Simple controller (a) sets speed of ac-dc motors. Use of milliampere-level control signals and optocoupled input makes circuit attractive for remotely controlled applications. The circuit provides 1,500-V isolation between control and power stages; 10:1 transformer input (b) permits the employment of components that need withstand only a fraction of the power-line voltage.

High-frequency operation with the AM9513 controller

by Terence J. Andrews
Vega Precision Laboratories, Vienna, Va.

The maximum source frequency of the Advanced Micro Devices' AM9513 system timing controller may be extended from 7 MHz to 20 MHz with this circuit. By resolving timing intervals to 50 nanoseconds, the circuit significantly enhances the 9513's usefulness as a multiple programmable frequency divider. The improvement relies on the concept of swallow counting,[1] whereby the controller is made to operate synchronously with an external high-frequency clock. In particular, this circuit illustrates how the controller is configured to provide frequency division at 10 and 20 MHz as an example.

Here, the resolution obtained by either 10- or 20-MHz clocking can be selected as desired. A_2–A_4 achieve operation at 10 MHz. Register 1 of the 9513 is first loaded via its 16-bit data bus with $01F5_{16}$, which instructs the

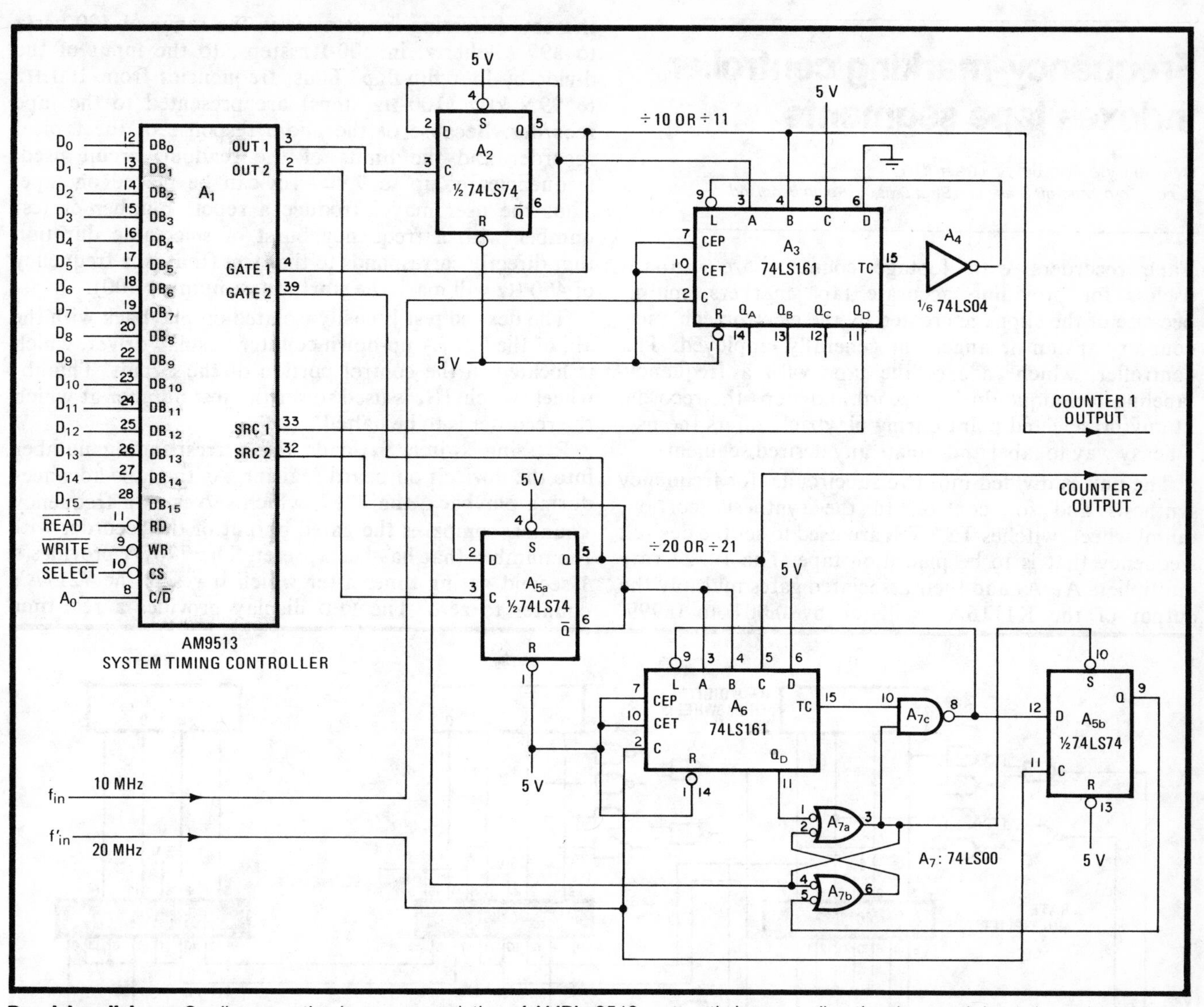

Resolving divisors. Swallow-counting improves resolution of AMD's 9513 system timing controller, thereby permitting pulse-train division at high frequency. Contents of the controller's load and hold registers set the divisor. Operation at 10 or 20 MHz can be chosen.

device to utilize incoming signals at pin 33 (SRC_1) as its system clock and pin 4 (gate 1) to select either the device's hold or its load register for reloading each of the unit's five 16-bit counters upon receipt of a terminal count.[2] Counter 1 counts down in binary-coded-decimal fashion, with the contents of the load and hold registers determining the divisor.

In this application, only counters 1 and 2 of the device are used. Counter 1 begins to count down from a preset number on the rising edge of SRC_1. SRC_1 in turn is driven by the 10-MHz clock through frequency divider A_3. A_3 is programmed as a 10's complement counter and divides f_{in} by either 10 or 11.

However, counter 1 is loaded with the contents of the load register if Q of toggled flip-flop A_2 (and thus gate 1) is low; otherwise, the contents of the hold register are loaded into counter 1. Thus when counter 1, which holds the contents of the load register, steps down to zero, A_2 orders A_3 to divide by 10; when the counter steps down from a value preset by the hold register, division is by 11. Because in this feedback arrangement there is a change of state at OUT 1 each time the counter reaches 0, it is seen that the contents of both the load and hold registers determine when A_2 is toggled and thereby determine the divisor. Although the numbers loaded into the hold and load registers of the timing controller are in the BCD form, the AM9513 can be programmed to count binary numbers as well.

To determine the divisor, N, multiply the contents of the hold register by 11 and the load register by 10 and add the results. Because the loading of 0001 into either register will cause improper operation, 0011 must be used instead. For example, if division by 301 is desired, the hold register should be loaded with 0011 (not 0001) and the load register with 0018 (not 0029). If 0000 is loaded into either register, a divisor of 10,000 will be the result. □

References

1. The TTL Applications Handbook, "Swallow Counters," Fairchild Camera and Instrument, August 1973.
2. AM9513 Applications Sheet, Advanced Micro Devices, 1979.

Frequency-marking controller indexes tape segments

by Joe Lyle and Jerry Titsworth
Bendix Corp., Aircraft Brake and Strut Division, South Bend, Ind.

Tape recorders with footage counters are virtually useless for providing accurate tape markers, mainly because of the slippage created by the electromechanical counter-capstan arrangement generally employed. This controller, which indexes the tape with a frequency-labeling technique during record and stops the recorder at any preselected point during playback, offers the user an easy way to label and isolate any desired segment.

The unit is divided into two subcircuits, for frequency synthesis and for control. In the synthesis portion, thumbwheel switches TS_1–TS_3 are used to set the desired frequency that is to be placed on tape. The 14527 rate multipliers A_1–A_3 and their associated gates multiply the output of the K1116A oscillator by 0.001 to 0.999, thereby delivering frequencies in the range of 400 hertz to 399 kilohertz, in 400-Hz steps, to the input of the divide-by-four flip-flop. Thus, frequencies from 100 Hz to 99.9 kHz (100-Hz steps) are presented to the tape recorder. Because of the audio response of the typical recorder and the limits of the readout circuit used, frequencies of up to 9,900 Hz can be placed on tape. Thus the user may introduce a report number or test number with a frequency burst of selectable duration that directly corresponds to that test (that is, a frequency of 400 Hz will mark the start of test number 400).

The desired test is easily isolated on playback with the aid of the 7217A up-down counter/display driver, which is located in the control portion of the circuit. Thumbwheel switch TS_4 is used to set the test number at which the recorder is to be halted.

Pressing switch S_4 loads the corresponding number into the device's on-board register. As the tape advances during playback, the 7217, which serves as a frequency counter, compares the gated output of the recorder with the number that has been preset. The 7207A provides a 1-second gating time, after which it resets the 7217A's counter to zero. The LED display provides a real-time

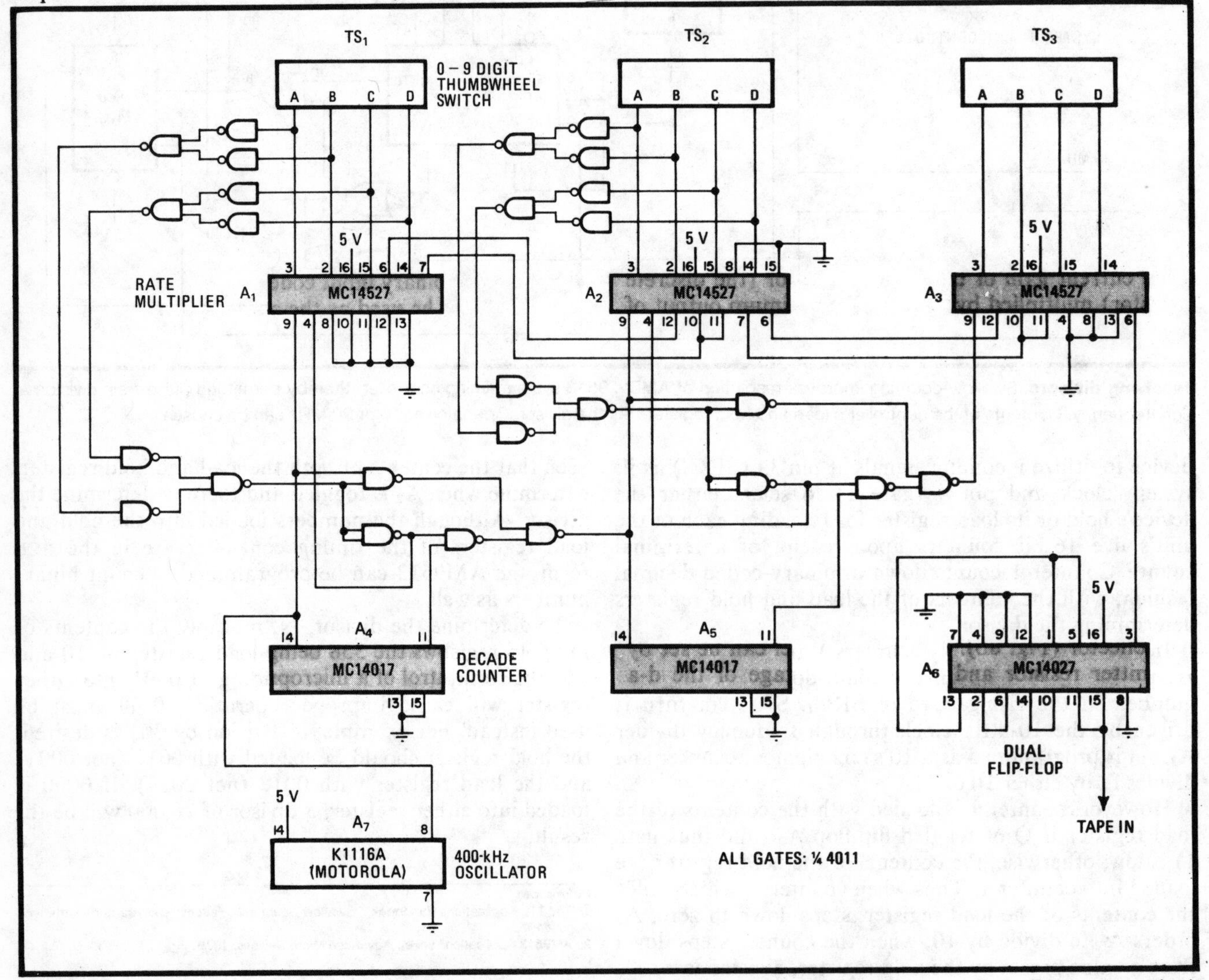

1. Search. Controller performs frequency labeling and searching functions to mark, isolate tape segments, respectively. Frequency synthesis portion injects tape markers over range of 100 Hz to 99.9 kHz, in 100-Hz steps.

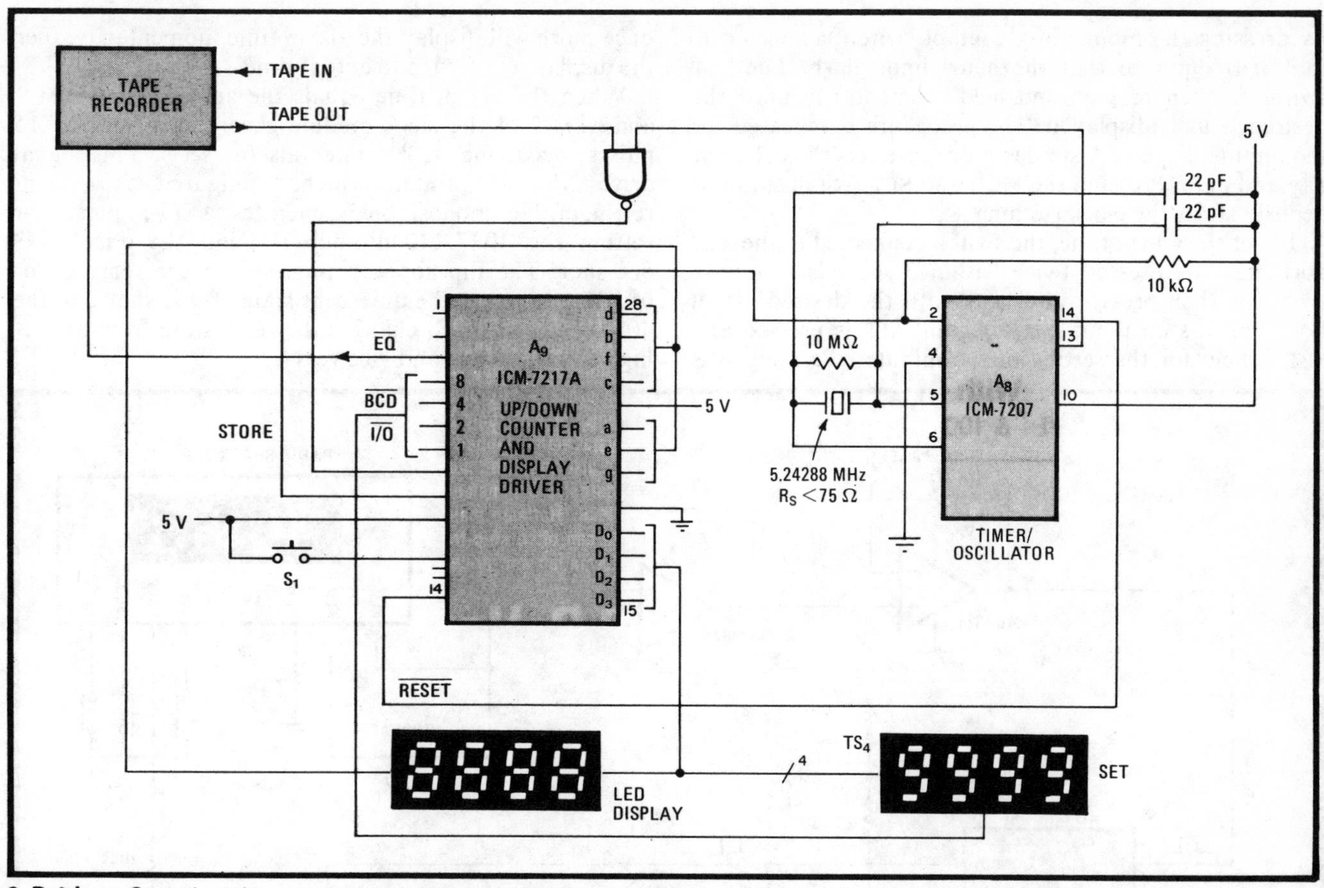

2. Retrieve. Control portion compares number corresponding to frequency selected to that generated by tape on playback, halts recorder when both are equal. TS_4 sets the test number to be retrieved. LED display provides real-time readout of frequency measured.

readout of the frequency measured.

Thus the equal (pin 3) output of the counter moves low if the tone matches the preset number. This signal is used (with appropriate logic) to halt the recorder. □

Module activates appliances at preset clock time

by Leslie D. Paul
Madison, Wis.

An inexpensive alarm clock module and an appropriate interface let this circuit activate or shut off any system at a time preselected by the user. The hours and minutes readout of its clocked liquid-crystal display allows direct and precise setting of the desired time, a major advantage over many commercial electromechanical units.

The module used is the Archer 277-1005, available from Radio Shack for about $20. The time of day is set by pressing the momentary-contact switch connected to the MOD input so that the hour digits flash. The SET switch is then pressed and held momentarily until the desired hour is displayed. The procedure is repeated for the minute display. A similar procedure sets the calendar day and date. Pressing the MOD and SET switch simultaneously starts the clock running.

To set the alarm time, the switch connected to the ALS port must be pressed twice within 3 seconds. The SET switch is then pressed and held until the desired alarm hour appears on the display. Again, ALS is pressed and SET is held for the setting of the minutes. Pressing ALS once more will display the alarm time momentarily, then the display will return to actual time.

When the alarm time equals the actual time, ALM 1 and ALM 2 of the clock module generates a burst of 15 pulses, occurring at 1-s intervals for 15 s. This signal drives the 555 timer, which, configured as a non-retriggerable monostable, generates a 17-s pulse for setting the 4027 JK flip-flop through the dual 4098 one-shot. The flip-flop can then switch the relay on or off, depending on the quiescent state of one-shot 2 of the 4098. Depressing S_1 changes the relay state from active-high to active-low, and vice versa. □

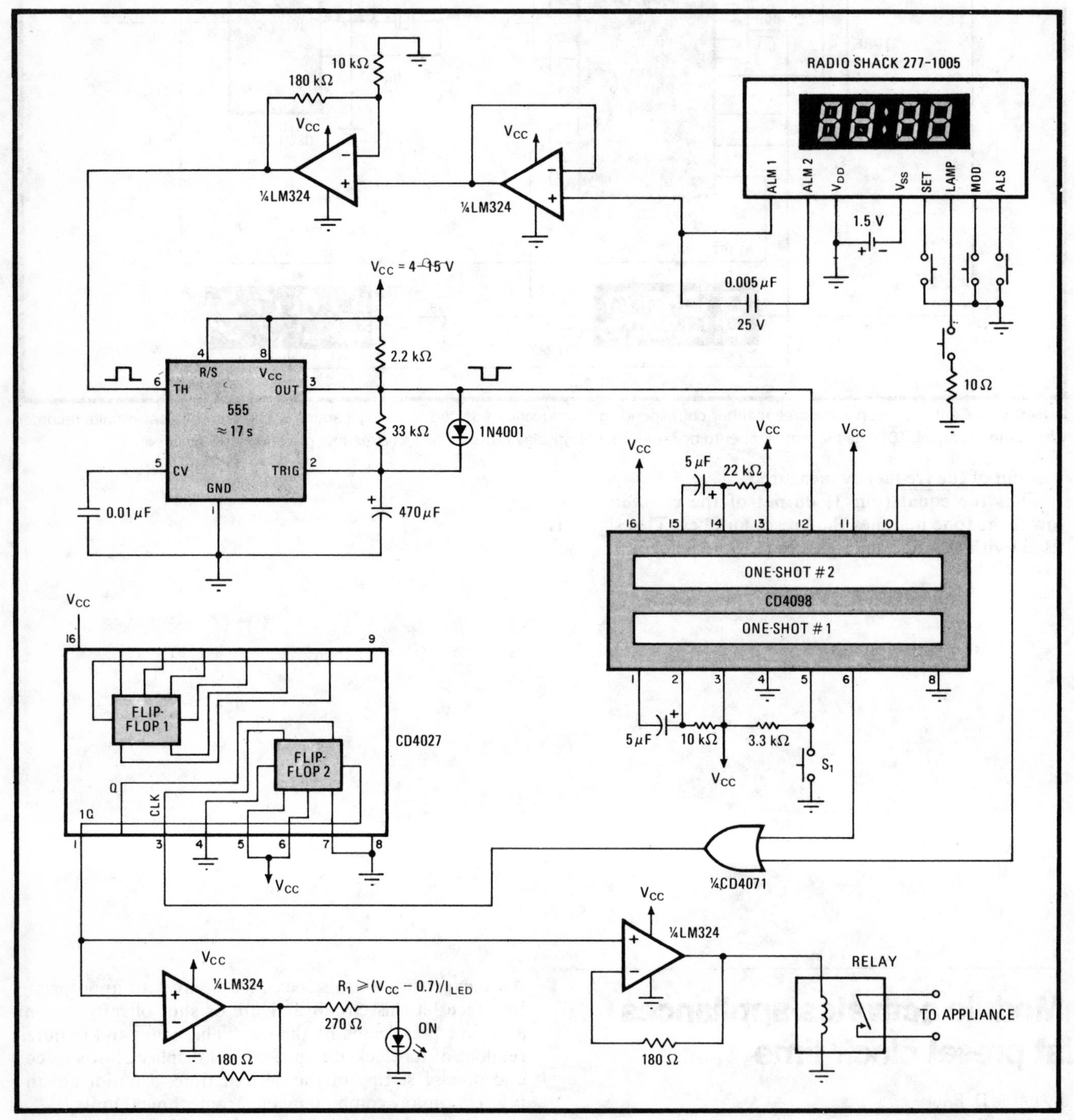

On time. Archer LCD alarm-clock module allows direct and precise setting of time to activate or shut off appliances. Pulsed alarm-signal output, not directly suitable for turning external devices on or off, passes through C-MOS interface so that relay is switched.

Magnetic levitator suspends small objects

by Bob Leser
Desert Technology, Las Cruces, N. M.

This circuit is a modern solution to the problem of securing frictionless bearings for small rotors and levitating small magnetic objects a few millimeters in space. Operational amplifiers replace the tubes used in earlier approaches, and an optical arrangement replaces the radio-frequency induction circuit originally used to position the object.

Potentiometer R_1 (a) sets the current through the PR9 lamp and thus its brightness and the gain of the position-sensing circuit. R_1 thus provides a fine adjustment of the position of the magnetic object that is suspended beneath the levitation coil L_1. The optical position-sensing circuitry (b), which should be mounted horizontally under L_1 if possible, includes two lenses to focus the beam via the levitated load to solar cell (photodetector) D_1. The light shield, with an aperture of approximately 3 millimeters, effectively eliminates background light. The suggested focal lengths and lens diameters are shown; as a check on the optics system, the beam should be aligned to yield a short-circuit current of 4 to 25 microamperes in D_1.

As for the basic circuitry, D_1's output is amplified by about 5 by operational amplifier A_1 and is then introduced to A_2, which is the all-important servo-loop stage. C_1, R_8, and R_9 provide positive feedback of the high-frequency components of the positioning signal. The stage thus generates the voltage derivative of the amp's output, preventing oscillations in the closed loop that would otherwise occur because of the lack of damping in

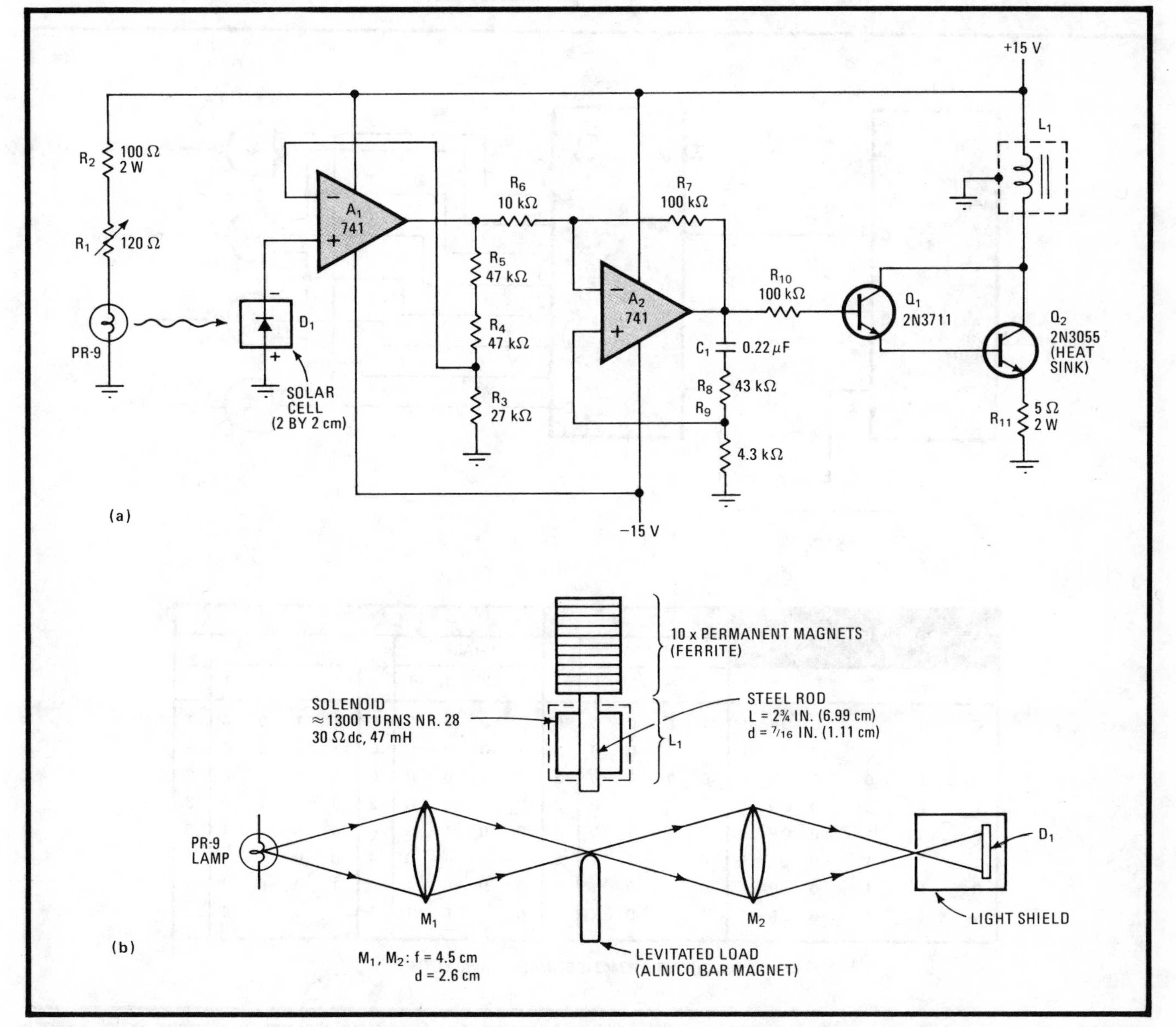

Rising rotors. Levitator circuit (a) suspends 1-in. steel spheres up to 2½ millimeters off reference surface. Optical arrangement (b) sets object distance. Details of levitation coil construction are outlined. Permanent magnets set ultimate levitation range.

the position servo portion of the circuit. Any closed-loop oscillation will be manifest as vibration of the levitated object.

Output stage Q_1 to Q_2 is a discrete darlington pair that drives L_1. The coil itself has 1,300 turns around a steel rod 2.75 inches long and 7/16 in. in diameter. A stack of 10 small permanent magnets atop the coil provides a bias field extending the range of levitation beyond that which would be normally attained. The coil is surrounded by a grounded shield to reduce the amount of stray coupling to the op amp inputs.

The most stable closed-loop condition is set by adjusting Q_2's collector voltage to about 7.5 volts by altering the levitation distance between the sensing optics and L_1. Levitation distances in this circuit range from about 20 millimeters for a small Alnico bar magnet to 2½ mm for a steel ball with a diameter of 1 in. □

Controller selects mode for multiphase stepping motor

by Oldrich Podzimek
Electrical Engineering Research Institute, Prague, Czechoslovakia

Offering a selection of the most common stepping modes, these circuits are an inexpensive solution to the problem of torque control in four- and five-phase motors. The mode can be changed simply by the flip of a switch.

The basic circuit is the same for either stepping motor. It consists of a 4-bit binary-coded-decimal counter, a BCD-to-decimal converter, and several gates that serve as phase detectors.

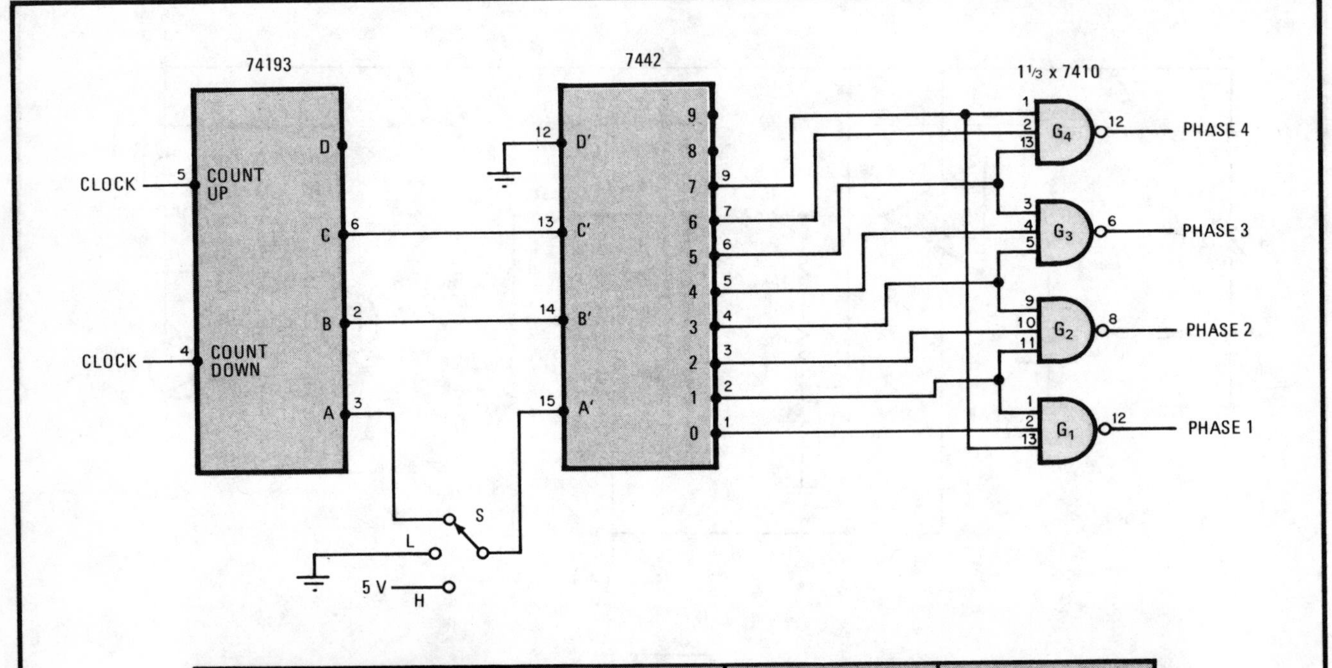

COUNTER OUTPUT				A					L					H				
				PHASE				Z	PHASE				Z	PHASE				Z
A	B	C	D	1	2	3	4		1	2	3	4		1	2	3	4	
0	0	0	0 OR 1	1	0	0	0	0	1	0	0	0	0	1	1	0	0	1
1	0	0	0 OR 1	1	1	0	0	1	1	0	0	0	0	1	1	0	0	1
0	1	0	0 OR 1	0	1	0	0	2	0	1	0	0	2	0	1	1	0	3
1	1	0	0 OR 1	0	1	1	0	3	0	1	0	0	2	0	1	1	0	3
0	0	1	0 OR 1	0	0	1	0	4	0	0	1	0	4	0	0	1	1	5
1	0	1	0 OR 1	0	0	1	1	5	0	0	1	0	4	0	0	1	1	5
0	1	1	0 OR 1	0	0	0	1	6	0	0	0	1	6	1	0	0	1	7
1	1	1	0 OR 1	1	0	0	1	7	0	0	0	1	6	1	0	0	1	7

Z = SWITCHED OUTPUT OF 7442 (DECIMAL EQUIVALENT)

1. Multimode. Step controller uses up-down counter, decoder, and logic to control excitation of four-phase motor windings. Switch S selects one of three possible operating modes, ranging from smooth-stepping to high-torque.

either one or two windings of the motor will be active at any given time.

The motor will step most smoothly when S is connected to A. When S is connected to L, the step angle will be doubled, and only one of four motor windings will be excited at any given instant, thereby improving the efficiency of the step operation for a given torque. Note that when S is in the H position, the step angle will be the same as in the preceding case, but two motor windings will be active at any time, power input will be doubled, and 41% greater torque will be obtained.

A similar circuit suitable for stepping five-phase motors is shown below. In this case, either two or three phases of the motor are excited simultaneously. This circuit can be extended to solve a general m-phase motor problem. □

The four-phase controller is shown on p. 62. The 74193 counter advances with each input-clock pulse at a frequency determined by individual requirements. Note that the 74193 can count up or down and so may be used to step the motor in the opposite direction if desired.

As the counter increments or decrements, the output of the 7442 4-to-10-line decoder switches in a manner dependent on switch S. If S connects port A′ of the 7442 to the A port of the 74193, the decoder's output will move from 0 to 7 in sequence. Otherwise, the output will switch to even values every other count (S connected to logic 0) or switch to odd values (S connected to logic 1).

A combinational logic circuit using gates G_1–G_4 converts the 7442's output to phase information in order to drive the motor. As can be seen in the truth table,

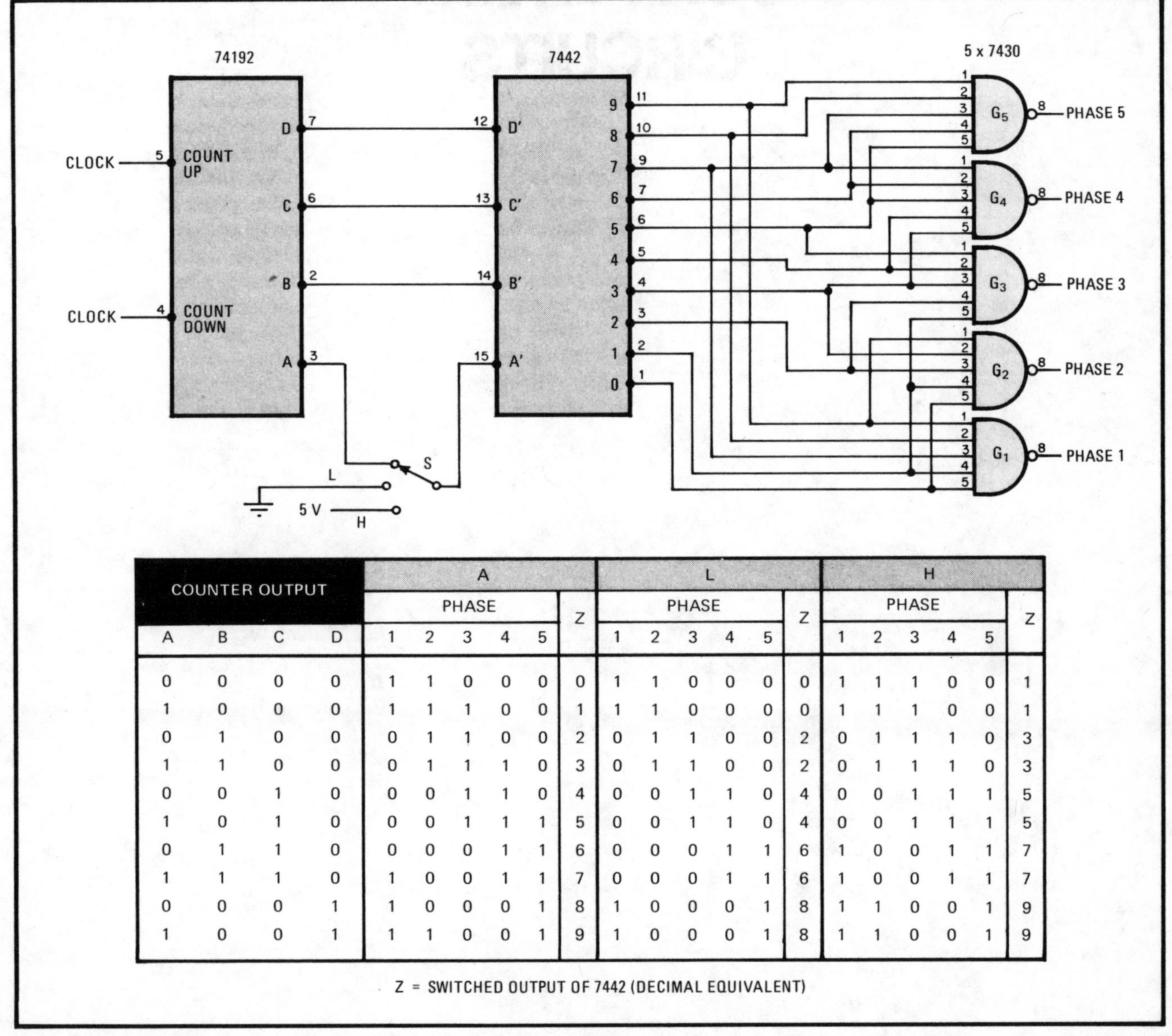

COUNTER OUTPUT				A						L						H					
				PHASE					Z	PHASE					Z	PHASE					Z
A	B	C	D	1	2	3	4	5		1	2	3	4	5		1	2	3	4	5	
0	0	0	0	1	1	0	0	0	0	1	1	0	0	0	0	1	1	1	0	0	1
1	0	0	0	1	1	1	0	0	1	1	1	0	0	0	0	1	1	1	0	0	1
0	1	0	0	0	1	1	0	0	2	0	1	1	0	0	2	0	1	1	1	0	3
1	1	0	0	0	1	1	1	0	3	0	1	1	0	0	2	0	1	1	1	0	3
0	0	1	0	0	0	1	1	0	4	0	0	1	1	0	4	0	0	1	1	1	5
1	0	1	0	0	0	1	1	1	5	0	0	1	1	0	4	0	0	1	1	1	5
0	1	1	0	0	0	0	1	1	6	0	0	0	1	1	6	1	0	0	1	1	7
1	1	1	0	1	0	0	1	1	7	0	0	0	1	1	6	1	0	0	1	1	7
0	0	0	1	1	0	0	0	1	8	1	0	0	0	1	8	1	1	0	0	1	9
1	0	0	1	1	1	0	0	1	9	1	0	0	0	1	8	1	1	0	0	1	9

Z = SWITCHED OUTPUT OF 7442 (DECIMAL EQUIVALENT)

2. Multiphase. Controller for stepping five-phase motors is similar to that for four-phase case. Five 5-input NAND gates and some additional wiring are the only new changes required for enabling up to three phases of a motor to be excited simultaneously.

Chapter 8
CONVERTER CIRCUITS

Hyperbolic clock inverts time

by Keith Baxter
New Haven, Conn.

Instruments designed to measure speed must contain circuits for converting a time function, t, to units of 1/t in order to calculate rate. This circuit aids in plotting the hyperbolic curve (1/t) using relatively few parts and, notably, having no need for logarithmic dividers.

As shown in the figure, a start pulse triggers transmission gate S_1, initiating the measurement cycle. At that time, A_1 and A_2 are reset. Current I_1 thereupon charges C_1 linearly, and thus a ramp voltage is applied at the noninverting input of A_3.

When the ramp voltage reaches 1/16 volt, the potential at the lowermost tap on the resistor voltage divider A_3 moves high and advances counter A_1. At this instant, the total elapsed time is 1/16 (C_1/I_1) V.

Switch S_2 is then activated, so that the second resistor in the ladder is shorted. Thus the voltage at the inverting port of A_3 increases to 1/15 V. When the ramp voltage reaches this value, A_3 again moves high and fires A_2. This action occurs at an elapsed time of 1/15 (C_1/I_1) V. The process continues as A_1 counts to 15 in a binary sequence and either all combinations of the resistor ladder are shorted or the measurement cycle ends (start pulse held at logic 1). It is assumed the data will be stored in a latch prior to the start pulse, because A_2 will be reset. At that time, A_2 will have counted down with each clock from A_3 to provide an output corresponding to an elapsed time of 1/t (C_1/I_1) V, where t may assume integer values from 1 to 15. Note that comparable circuit action cannot be realized easily with an astable multivibrator operating as a fixed-frequency source for stepping A_2; that is, the circuit is not performing a time-to-frequency conversion.

The contents of A_2 at the termination of the measurement cycle provide a direct indication of an object's speed. The factor (C_1/I_1)V may be adjusted to 1 by suitable choice of component values or set to a multiplicative constant as required. Also, to reduce the voltage divider increment, the resistor ladder may be expanded by cascading counters and their appropriate circuitry. □

Time twist. Circuit inverts time function, t, to 1/t in order to measure speed. Note that unit does not perform a standard time-to-frequency conversion. Ramp voltage derived from current generator is compared with resistive-ladder voltage at A_3, and A_1–A_2 are clocked each time the ramp exceeds changing ladder potential. Output of down counter, A_2, yields rate.

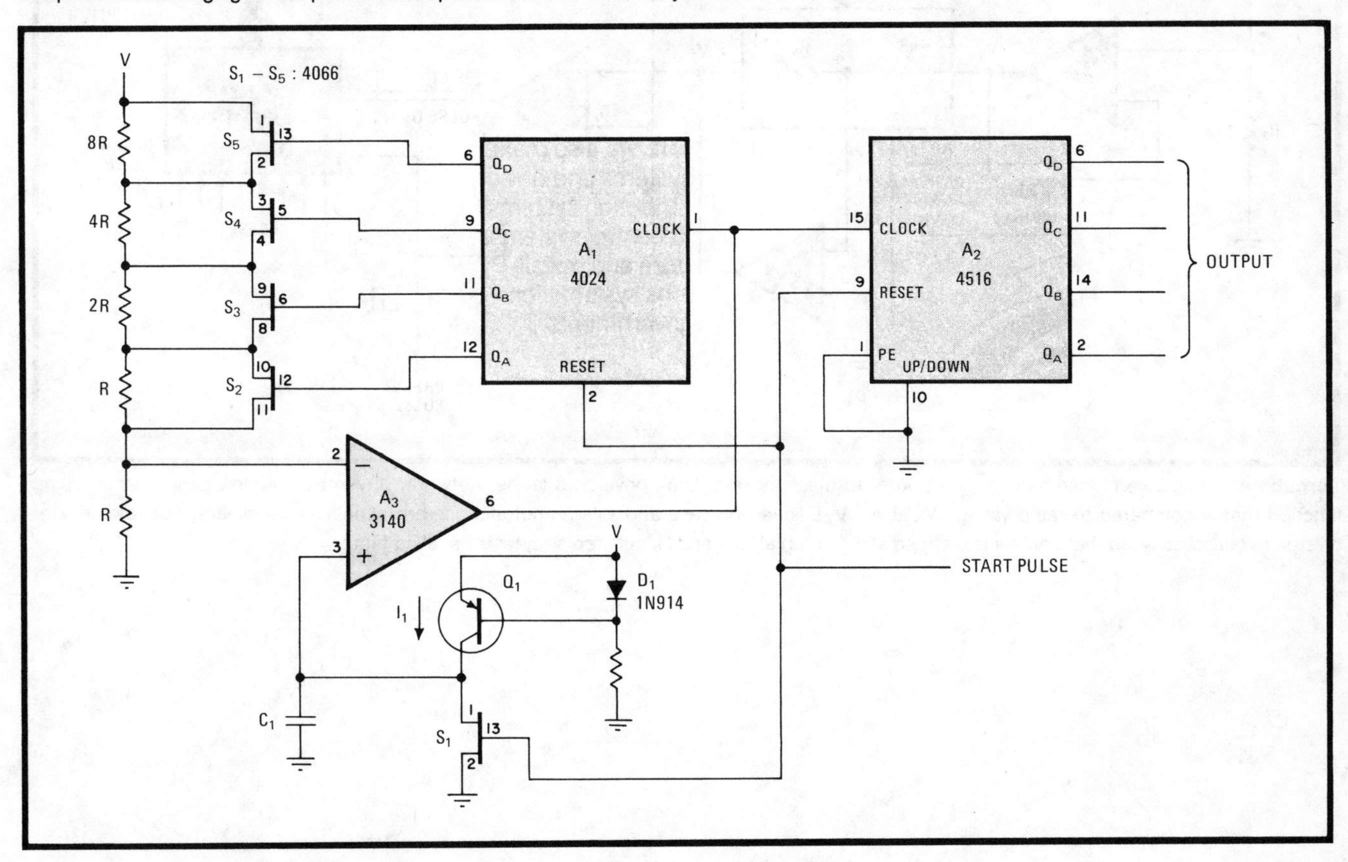

D-a converter simplifies hyperbolic clock

by R. H. Riordan
Cybec Electronics, Bentleigh, Australia

The hyperbolic clock circuit proposed by Baxter [in the preceding article], which transforms a time function, t, into units of 1/t in order to measure rate, can be made more compact by employing a one-chip digital-to-analog converter. Using the converter's monolithic ladder of 255 equivalent resistors in place of Baxter's discrete network for scaling provides greater resolution for a circuit of a given size and is easily modified for applications requiring a decimal output.

Initially, voltage V_1 is set to zero and the 8-bit binary counter formed by cascading two C-MOS 4516 chips is preset to a count of 255. The converter is equivalent to a fixed resistor R_f and a variable resistor R_v, where $R_v/R_f = 256/N$ and N is the output count of the 4516. The converter is connected in the reverse of the normal configuration, so that $V_2/V_3 = -R_v/R_f = -V_3(256/N)$. But $V_3 = -V_{ref}/256$, so $V_2 = V_{ref}/N$.

At first, $V_1 = 0$ and $V_2 = V_{ref}/255$. At $t = 0^+$, current generator I_1 starts to charge capacitor C_1, and V_1 begins to rise linearly with time. Whenever V_1 climbs above V_2, comparator A_1 generates a clock pulse, decrementing the counter and causing V_2 to rise above V_1 again. For any given count N, a clock pulse will be generated at a time t when $V_1 = I_1t/C = V_{ref}/N$. Thus, until the counter reaches zero, the count at any instant will be proportional to the reciprocal of the elapsed time.

This circuit, as well as Baxter's, has a potential weakness in that if a single-step cycle does not result in V_2 rising above V_1, a lock-up condition will occur. This danger is eliminated if a separate clock signal is provided for the counter, with A_1 used only to enable the 4516s.

If decimal timing signals are required, it is a relatively simple matter to replace the d-a unit and the counter with their decimal-output equivalents. If a different step range is required, it can be selected accordingly by changing only R_1 and R_2. R_1 and R_2 are also used to set the desired count range. □

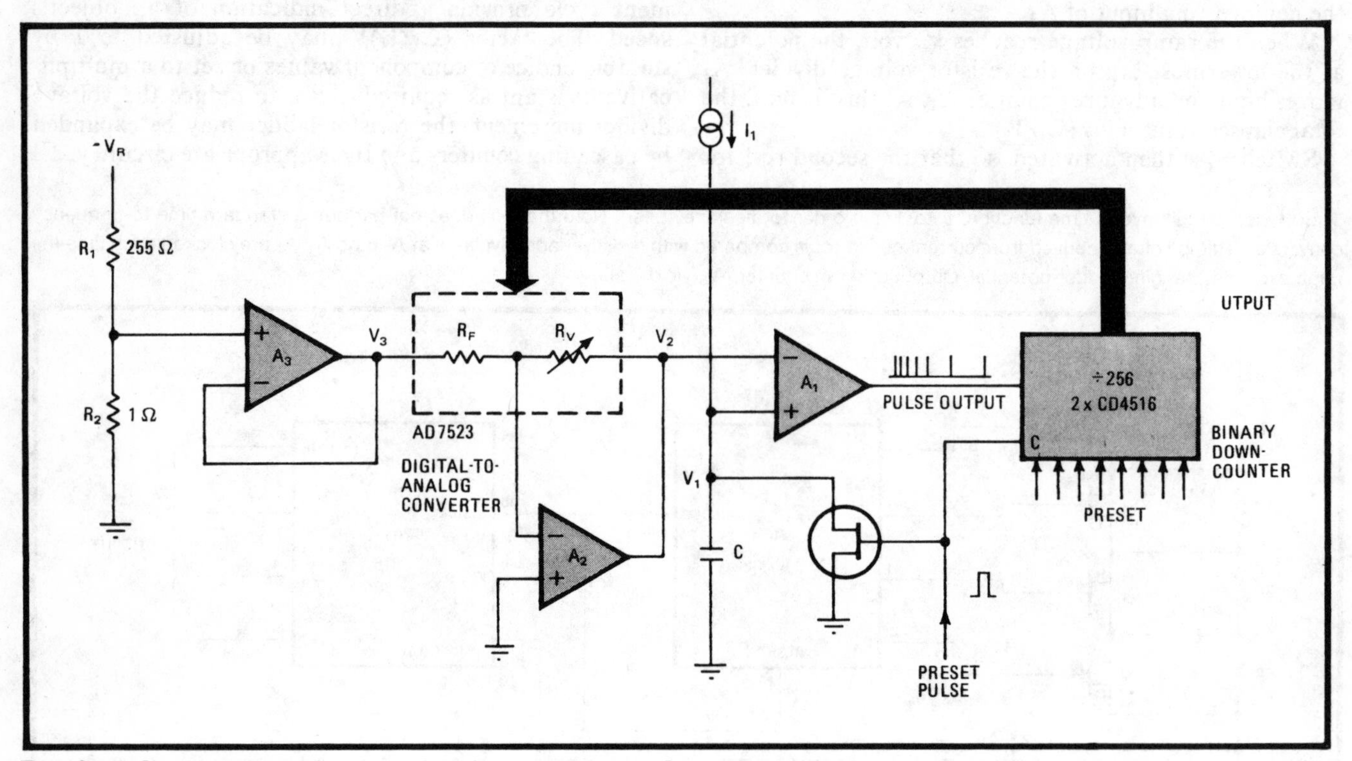

Turnabout. Clock inverts time function, t, in order to measure rate. D-a converter's ladder network, driven by counter, generates stepping function that is compared to ramp voltage V_1 at A_1. V_1's linear increase and V_2's monotonic rise are almost equal initially, but rates of rise diverge hyperbolically, so that counter is stepped at 1/t intervals. R_1 and R_2 set step size; I_1 sets scaling factor.

PROM converts push-button command to binary number

by Marco A. Brandestrini
University of Washington, Center for Bioengineering, Seattle, Wash.

A programmable read-only memory can be used to convert a decimal-input command from a one-of-five momentary-contact-switch array to its equivalent binary number, thus forming a circuit with countless uses in logic- and microprocessor-control applications. The circuit is superior to systems using a thumbwheel switch to digitally set the binary number and is more reliable than single-switch arrays using an all-mechanical arrangement. Variations and extensions of this idea are limited only by the size of PROMs available.

The basic idea is to connect the PROM's output lines, B_0–B_4, to their respective address inputs, A_0–A_4 so that any input signal may be latched and the resulting signal at the output will remain active after a given key is released (see figure).

The PROM, here the Signetics 82S23, should be programmed as shown in the table if the output code is to be the binary equivalent number of the decimal input signal. In order to force the output to a given state after the circuit has been turned on (power up), inputs A_0–A_4, usually all high, can be programmed to actuate any of the five output combinations. The truth table shows how the PROM is programmed for state 0 after power up. Of course, any input-output relation may be programmed into the PROM as desired.

Circuit operation is simple. If, for example, the 0 key is depressed, a logic 0 appears at address A_0 (all input and output ports are active low). The resulting output at B_0 moves low, and this signal is fed back to its corresponding input, keeping A_0 low after key 0 has been released. A light-emitting diode monitoring the A_0 port glows, indicating that the line has been activated.

If a second key is depressed, all of the PROM's open-collector output lines move high temporarily, clearing the output (see table), and the new key position is latched. The temporary logic 1 state is sensed by the NAND gate shown and produces an interrupt request—which is useful in microprocessor applications. □

Key-stroke commands. Circuit converts decimal-input command to its equivalent binary number. Momentarily depressed key is sensed by PROM and displayed by light-emitting diode. Outputs of PROM are wired to inputs and actuate latching. Interrupt generation is provided for microprocessor applications.

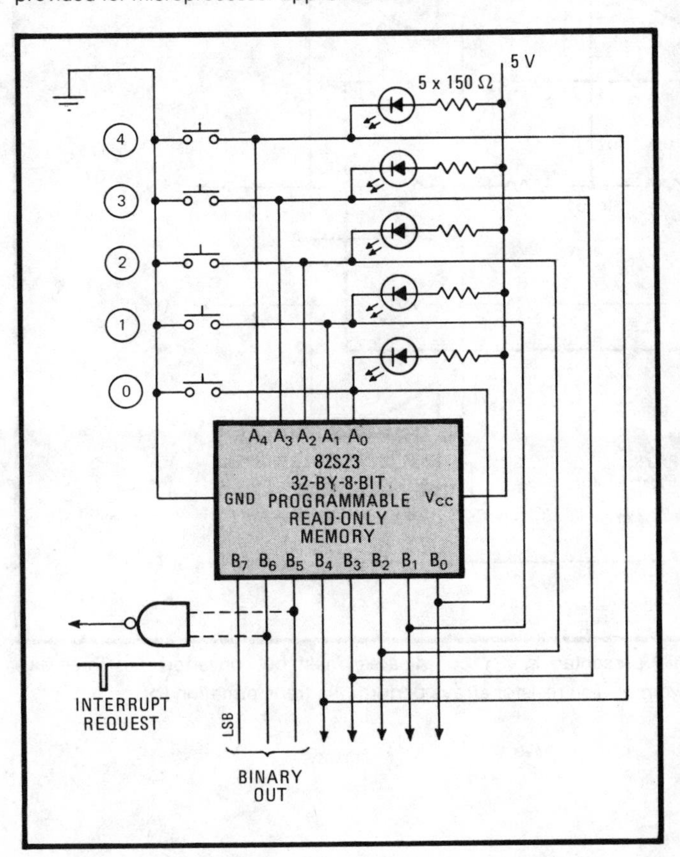

PROGRAMMING OF THE PROM

DECIMAL INPUT NUMBER	INPUTS A_4 A_3 A_2 A_1 A_0	OUTPUTS B_7 B_6 B_5 B_4 B_3 B_2 B_1 B_0
	0 0 0 0 0	1 1 1 1 1 1 1 1
	0 0 0 0 1	
	0 0 0 1 0	
	0 0 0 1 1	
	0 0 1 0 0	
	0 0 1 0 1	
	0 0 1 1 0	
	0 0 1 1 1	
	0 1 0 0 0	
	0 1 0 0 1	
	0 1 0 1 0	
	0 1 0 1 1	
	0 1 1 0 0	
	0 1 1 0 1	
	0 1 1 1 0	
4	0 1 1 1 1	0 0 1 0 1 1 1 1
	1 0 0 0 0	1 1 1 1 1 1 1 1
	1 0 0 0 1	
	1 0 0 1 0	
	1 0 0 1 1	
	1 0 1 0 0	
	1 0 1 0 1	
	1 0 1 1 0	
3	1 0 1 1 1	1 1 0 1 0 1 1 1
	1 1 0 0 0	1 1 1 1 1 1 1 1
	1 1 0 0 1	
	1 1 0 1 0	
2	1 1 0 1 1	0 1 0 1 1 0 1 1
	1 1 1 0 0	1 1 1 1 1 1 1 1
1	1 1 1 0 1	1 0 0 1 1 1 0 1
0	1 1 1 1 0	0 0 0 1 1 1 1 0
POWER UP	1 1 1 1 1	0 0 0 1 1 1 1 0

OUTPUT PORTS

Coordinate converter aligns piezoelectric positioner

by Lawrence E. Schmutz
Adaptive Optics Associates, Cambridge, Mass.

Piezoelectric tilt elements of the kind used to position laser beams and optical scanners can be aligned with the help of this circuit, which converts the transducer's high input driving voltages, normally resolved in X-Y coordinates, into corresponding coordinates (a, b, c) in a nonorthogonal three-axis system. Only one quad operational amplifier and two resistor-array packages are used for the transformation.

The geometry of many popular piezoelectric positioning elements (a), as for example the Burleigh Instruments' PZ-80, is such that:

$$x = c - a \qquad y = b - \tfrac{1}{2}(a + c) \qquad 0 = a + b + c$$

Solving these equations simultaneously for a, b, and c yields:

$$a = \tfrac{1}{2}x - \tfrac{1}{2}y \qquad b = \tfrac{2}{3}y \qquad c = -\tfrac{1}{2}x - \tfrac{1}{3}y$$

Simplifying further:

$$a = c + x \qquad b = -(a + c) \qquad c = -(x/2 + y/3)$$

The last set of equations is easily implemented by using precision resistors to set the gain of several op amps (b).

Resistor arrays in dual in-line packages will perform the transformation accurately, for their elements have a tolerance of ±0.5%. The overall circuit uncertainty becomes ±1% when two arrays are configured as shown. To ensure that the circuit occupies no more space than that taken by three dual in-line packages, the resistors are grouped in their respective arrays as shown by the dotted lines.

The TL084 op amp is more than adequate for the circuit accuracy desired, since in most applications the piezoelectric devices operate in the lower audio-frequency range. □

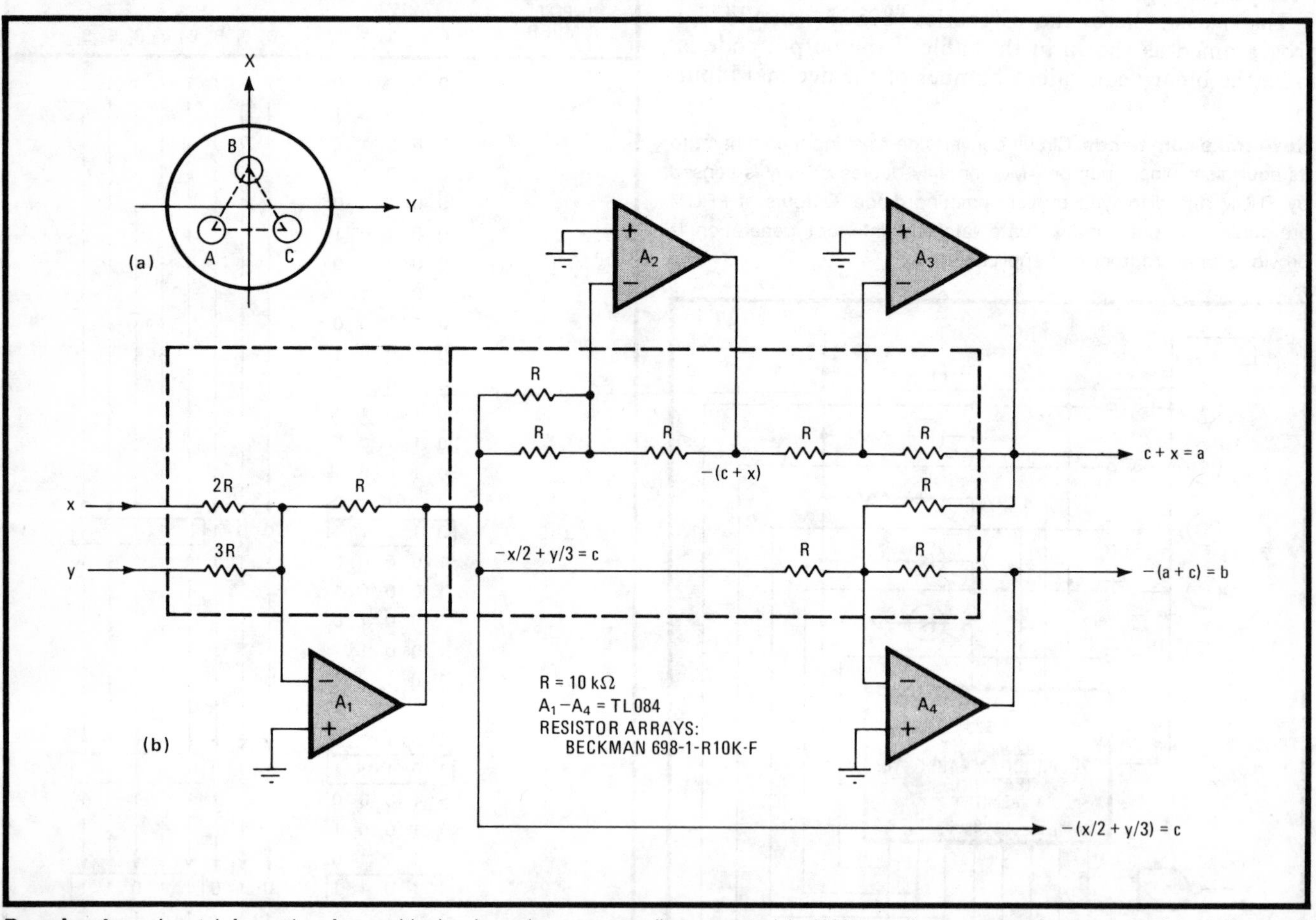

Beaming true. Input information for positioning laser beam, normally presented in X-Y coordinates, must be converted into three-axis coordinates for many piezoelectric transducers (a). One op amp and two precision resistor arrays perform the transformation (b).

Serial-to-parallel converter decodes width-modulated BCD

by William D. Kraengel Jr.
Valley Stream, N. Y.

Converting a binary-coded-decimal pulse train into its parallel equivalent is normally straightforward, unless of course, the signal assumes the form of a pulse-width–, pulse-code–, or pulse-position–modulated data stream. In data-processing cases where the stream is encoded by means of pulse-width modulation, however, this simple 4-bit decoder will serve well in performing the serial-to-parallel transformation.

The MC14538 one-shot, A_1, and the CD4015 4-bit shift register, A_4, form the central part of the decoder, serving as the timing and storage elements. As can be seen from the schematic and the timing diagram, the BCD input data is grouped into 5-bit cycles, with bit 5 being the framing pulse. A_1 must generate a pulse width of $t_{pw} = RC = \frac{1}{2}(t_1 + t_2)$ when triggered, and one-shot A_2 has a very small triggering time of $t_s = R_1C_1 = 10$ microseconds, where both times are defined as shown in the timing diagram.

The firing time for A_1 must be selected to correspond to the input frequency of the serial BCD data. Under this condition, the one-shot is triggered by the negative edge of each serial input data bit, and if the serial input data line is low at the time the one-shot times out, a logic 0 is introduced into the shift register. On the other hand, a logic 1 will be read into the shift register if the input data goes high by the time A_1 times out.

Shift register A_3 serves as a counter, acting to disable A_4 and enable quad latch A_5 and one-shot A_2 during the framing pulse. The valid data from A_4 (which is inverted) is latched into A_5 as A_1 times out during the framing pulse. Simultaneously, A_2 is triggered so as to generate a data available/reset strobe signal for any peripheral control circuitry. The parallel equivalent BCD data and its inverse are available at the outputs of quad latch A_5. □

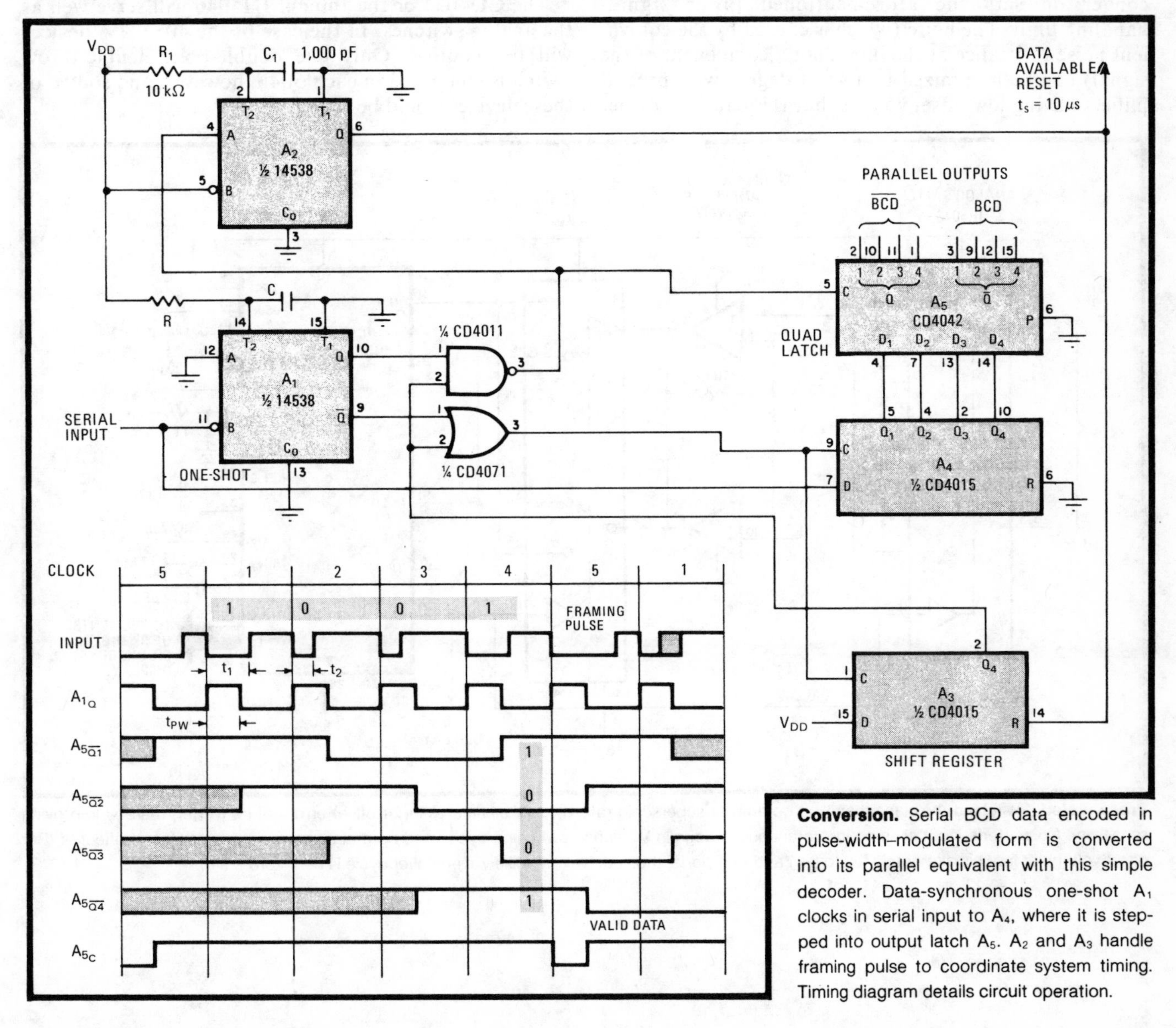

Conversion. Serial BCD data encoded in pulse-width–modulated form is converted into its parallel equivalent with this simple decoder. Data-synchronous one-shot A_1 clocks in serial input to A_4, where it is stepped into output latch A_5. A_2 and A_3 handle framing pulse to coordinate system timing. Timing diagram details circuit operation.

Switching preamp improves a-d converter sensitivity

by Peter Bradshaw
Intersil Inc., Cupertino, Calif.

The low-signal resolution of even the best analog-to-digital converters, including those equipped to eliminate input-offset disturbances, is limited by the noise generated at the inputs. But the resolution, and thus the true sensitivity, of a converter can be inexpensively improved by an order of magnitude if a switched, differential amplifier is employed at the input to precancel offset errors without affecting the normal conversion process.

The technique is illustrated for Intersil's ICL7106 3½-digit autozero converter/display driver, which normally handles signals over the range of 100 millivolts to 2 volts. In such a converter, the small input noise voltage trapped on the autozero capacitor during a conversion sets the aforementioned lower signal-handling limit. The noise (which is caused by the equivalent noise resistance at the input, not a component of the signal) can be minimized to a great degree with preamplifiers having low offset voltage, but this can be a rather expensive solution to the problem.

As shown, an alternative approach is to use the liquid-crystal-display backplane (BP) drive output of the ICL7106 to synchronously switch one half of the low-cost LM348 quad operational amplifier via analog switches so that, over a switching cycle, the input of the converter sees no instantaneous change in the magnitude or polarity of sample voltage V_{in}. Offset voltages, including that of the op amp, on the other hand, are virtually canceled because an equal but opposite noise component (average value is near zero) is applied to the IN HI (and IN LO) ports of the converter over a given interval. In this case, the switching (BP) signal is set at 45 kilohertz, but this can be varied by suitable selection of the RC components at pins 38 and 39 of the ICL7106. In this configuration, excellent performance is obtained for input signals ranging from 1 to 20 mV full scale.

Most dual (matched) op amps will be suitable for the switching task, but it is important that both the positive and negative slew rates of the device be reasonably close. Op amps having significant crossover distortion (such as the LM124/324) should not be used.

The CD4053 or the Intersil IH5046 will serve well as the analog switches. In the case of the 4053, 1⅓ devices will be required. Only one double-pole, double-throw switch is contained in each 5046, however, and so two of these devices would be required. □

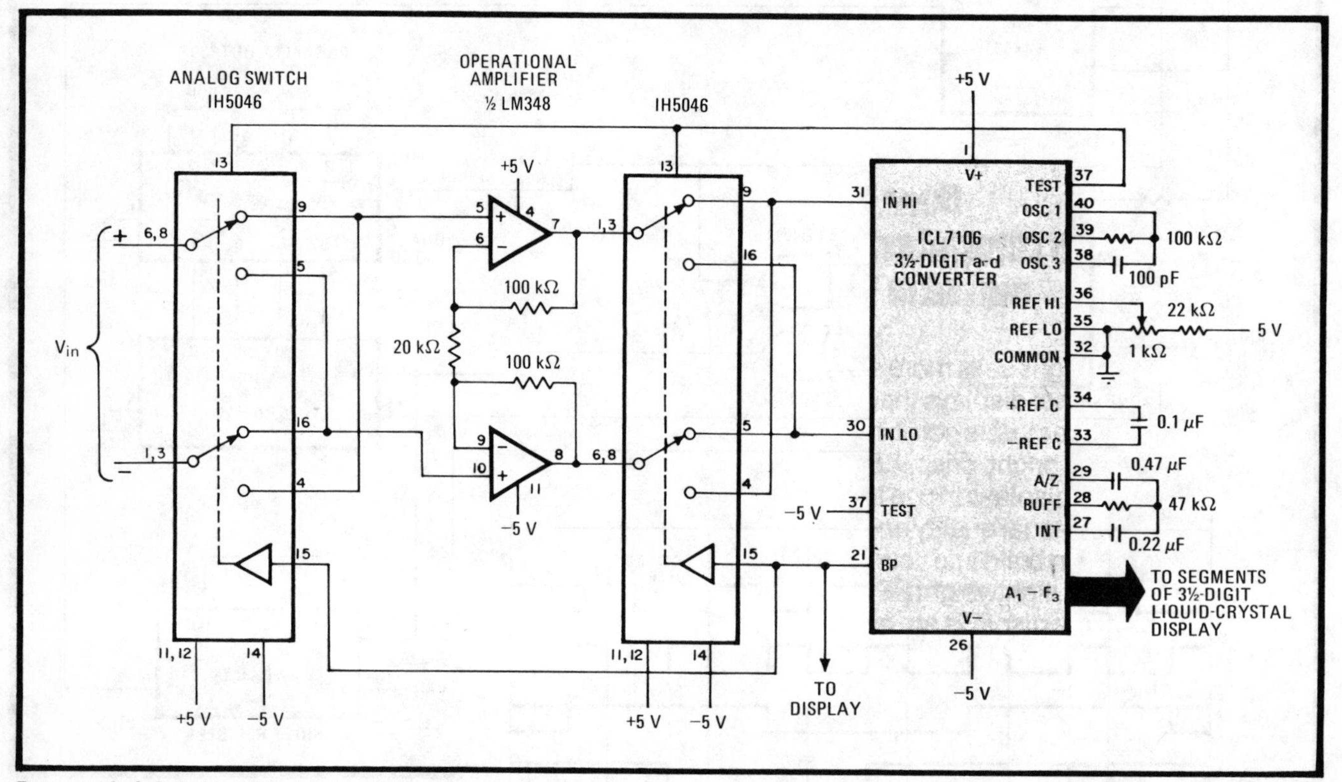

Precanceled. Switching the signal-handling op amp at supersonic rate virtually eliminates input-offset errors of converter, thereby improving sensitivity. Converter's input sees no instantaneous change in V_{in} during switching cycle, and normal conversion process is not affected. But offset voltage at input is alternately fed to (+) and (−) ports; thus equivalent noise voltage over cycle is near zero.

Single a-d converter cuts cost of droopless sample-and-hold

by Carl Andren
Harris Corp., Electronics Systems Division, Melbourne, Fla.

Because leakage currents cause droop, sample-and-hold circuits with capacitors as storage elements cannot retain a sampled voltage indefinitely. This is the major reason designers, to improve sample-and-hold performance, have resorted to converters combining analog-to-digital and digital-to-analog converter functions. But a single a-d device can be made to perform both functions alternately, thus cutting the cost and complexity of the two-converter scheme. Only one operational amplifier and a solid-state switch are needed in addition.

The more popular forms of a-d converter use a successive-approximation register that—with the aid of a self-contained comparator and a d-a converter—generates a digital estimate of the sampled analog voltage. When the comparison has been approximated to the least significant bit, the measurement is ended and an end-of-conversion signal is generated. If the unit is then configured as a latched d-a converter, the sampled analog voltage may be recovered and held indefinitely (assuming that one input of the comparator is accessible).

As shown, in the normal a-d conversion mode of a representative device like the AD582, a start-convert pulse initiates the measurement. An analog voltage, applied across resistor R_1 in the summing junction of comparator A_1, can then be sampled.

The successive-approximation register generates a 12-bit equivalent of the analog voltage and also drives the d-a converter that is connected to A_1's summing junction. The d-a converter then attempts to null A_1's output, whereupon the end-of-conversion signal (EOC) is generated. The sampling period takes a nominal time of 2.5 microseconds.

In this circuit, the EOC signal energizes a solid-state relay (CAG13) so that the converter can be switched to the holding mode. R_1 is then placed at the output of an op amp, A_2.

A_2's output maintains A_1's summing junction at a voltage null so that the output voltage becomes the potential across R_1—that is, the sampled voltage. Note that the switch resistance and A_2's input-bias current are taken into consideration for both modes and therefore they are not, for all practical purposes, sources of error in the measurement. □

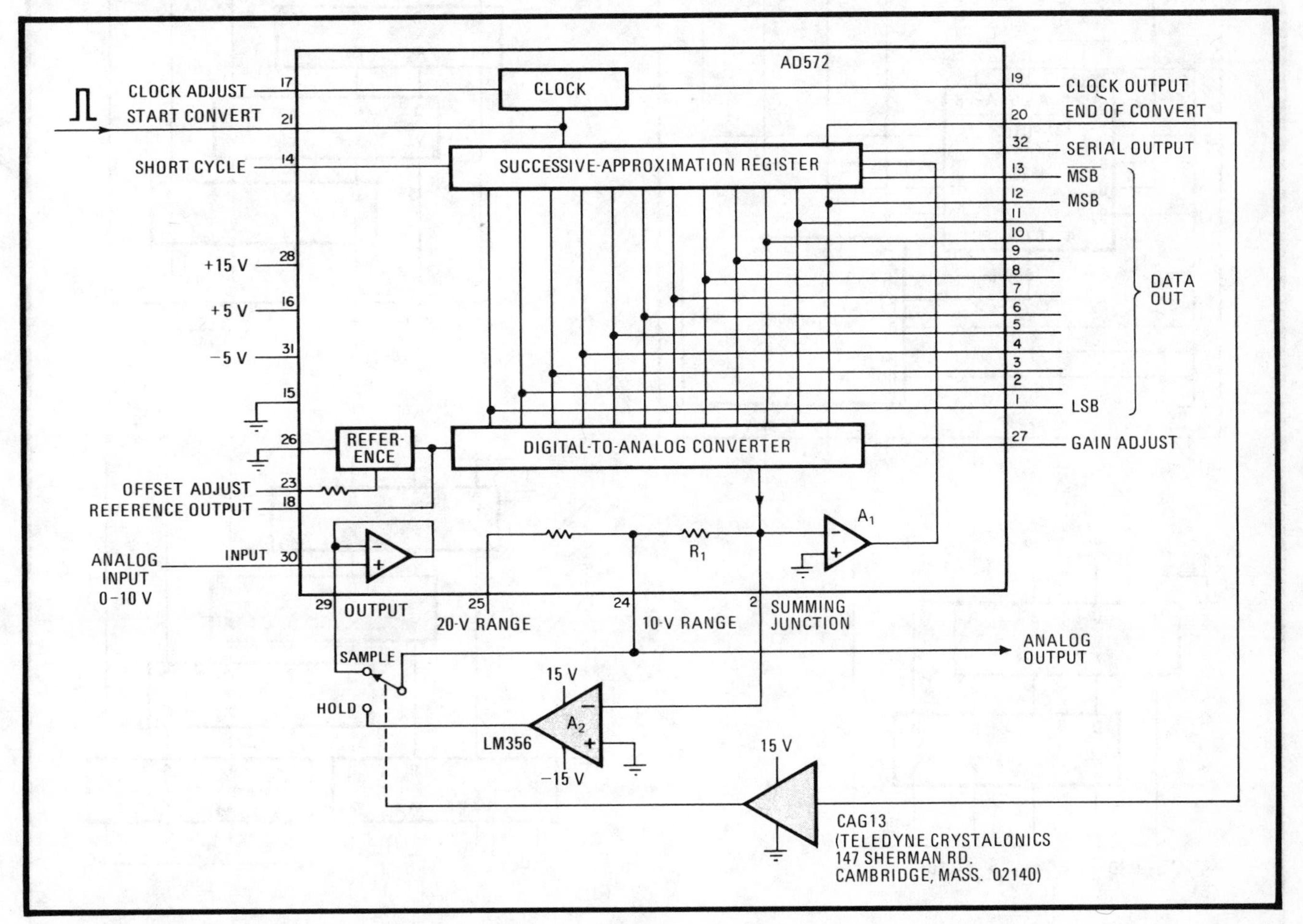

Inverting the converting. This analog-to-digital converter, when it is combined with an op amp and switch, can provide d-a function on the hold portion of sample-and-hold cycle, thereby reducing cost and complexity of the usual two-converter (a-d–d-a) scheme. No sampling-peak capacitor is required in the converter sampling technique, so that the sample-voltage droop is eliminated.

Cascaded C-MOS blocks form binary-to-BCD converters

by Haim Bitner
Seforad-Applied Radiation Ltd., Emek Hayarden, Israel

Low-power complementary-metal-oxide-semiconductor adders and comparators are easily combined to form this 4-bit binary-to-BCD converter. When the basic adder-comparator blocks are cascaded, the converter can be expanded to turn n binary input bits into a binary-coded-decimal output. The circuit is simpler than one using counters, and read-only memories are eliminated.

Comprising the basic 4-bit converter block (a) are the 4008 full adder and the 4585 comparator. The binary inputs are introduced at A–C, with A being the next to least significant bit, and D grounded. The BCD output appears at X_1–X_3 and Y of the 4008. The LSB input bypasses the unit and becomes the LSB output.

The 4585 compares the input bits to a binary number (0100) which is hard-wired to pins B_0–B_3. Thus the output of the 4585 is low if the number at A–D is less or equal to 4. The X_1–X_3 outputs of the 4008 are then identical to the input bits.

If the input number becomes greater than 4 (that is, greater than 0100), the 4585's output moves high, and so binary number 0011 is placed on the B_1 and B_2 inputs of the 4008 adder. Thus, 3 is added to the input number and the Y output of the 4008 goes high, indicating the most significant digit is active.

By cascading units, 5-bit (b), 6-bit (c) and 7-bit (d) converters can be built. The method can be used to extend indefinitely the number of bits processed. Note that the value of the least significant input bit (a_0) is numerically equal to its BCD-equivalent and so passes straight from input to output in all cases. □

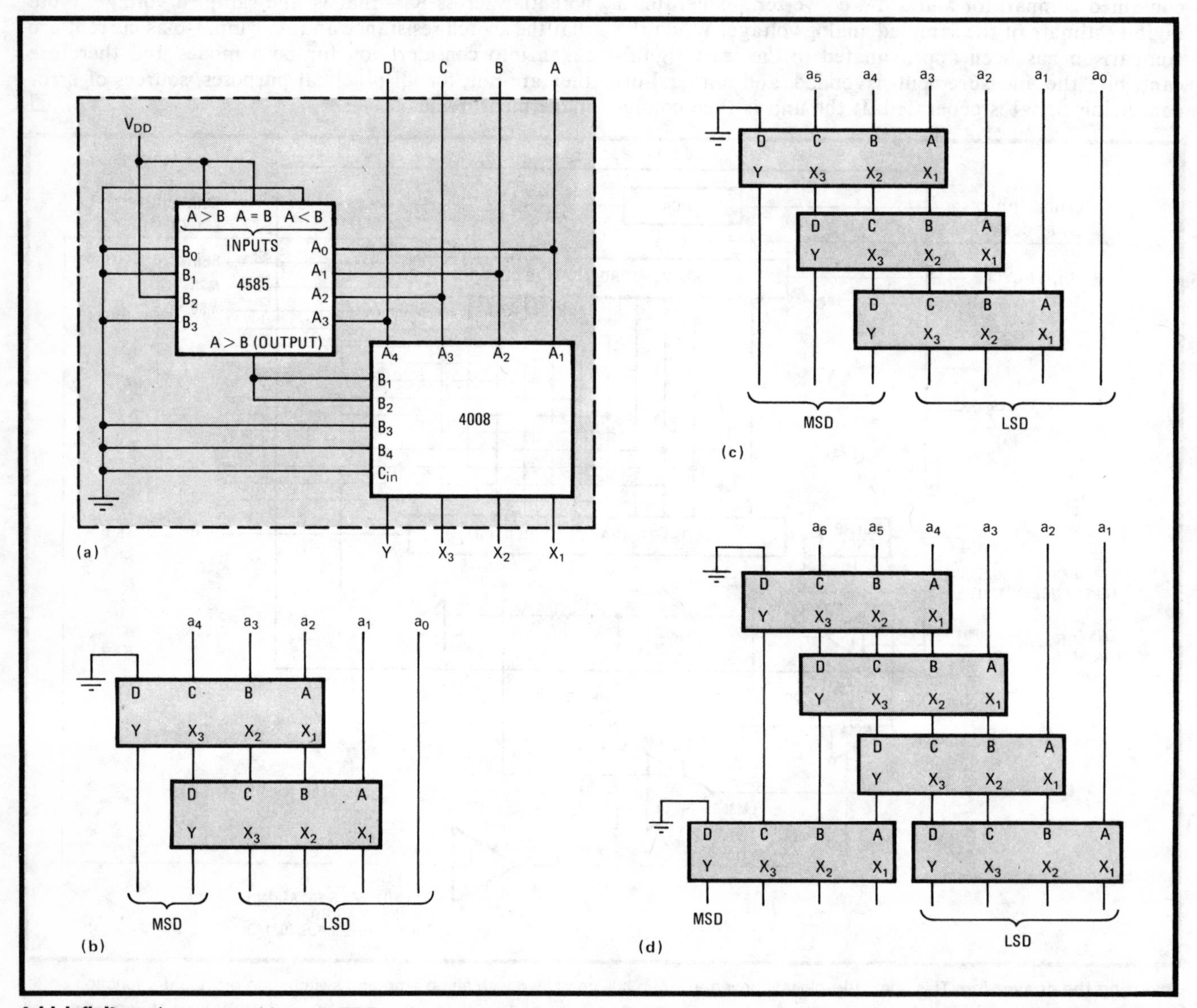

Add infinitum. Low-power binary-to-BCD converter (a) requires only C-MOS adder and comparator for processing 4 bits. Unit is so configured that basic building blocks can be easily combined to form 5-bit (b), 6-bit (c), and 7-bit (d) converters.

Voltage-to-current converter handles bilateral signals

by Kelvin Shih
General Motors Proving Ground, Milford, Mich.

Sending an analog voltage from one point to another via a slip ring and brush assembly often causes disaster in instrumentation applications, because of the attendant voltage drops across the variable resistances encountered and also because of induced noise. By converting the voltage to a current before transmission (and back to a voltage at the receiving end), however, these circuits eliminate those problems. And unlike converters that generate a problem of their own—their inability to handle bilateral input signals—this one will transform a ±10-volt signal into a ±10 milliampere current at the transmitter and recover the ±10-V signal at the receiver.

At the transmitter, input voltage V_1 is applied to one branch of a summing-amplifier circuit consisting of A_1, R_1–R_3, and R_7. The other branch is driven by V_2, which is the transmitter's output voltage. As a consequence of this feedback arrangement, the voltage from inverting amplifier A_2 is $V_1 + V_2$.

Current booster Q_1–Q_2, which is part of A_2's feedback resistor network, provides a low-impedance source for generating a current, I_o, with Q_1 becoming active for positive input voltages and Q_2 active for negative input voltages. Thus, $I_o = [(V_1 + V_2) - V_2]/R_6 = V_1/R_6$, and therefore the output current is a function of input voltage V_1 only.

At the receiver, A_4, Q_3, and Q_4 detect input current I_o and convert it to a voltage, $-V_1$. Note that the two-transistor arrangement similar to that employed in the transmitter is again required to process the bilateral input currents encountered. A_5 acts as an inverting amplifier so that the output signal, V_1, is recovered. □

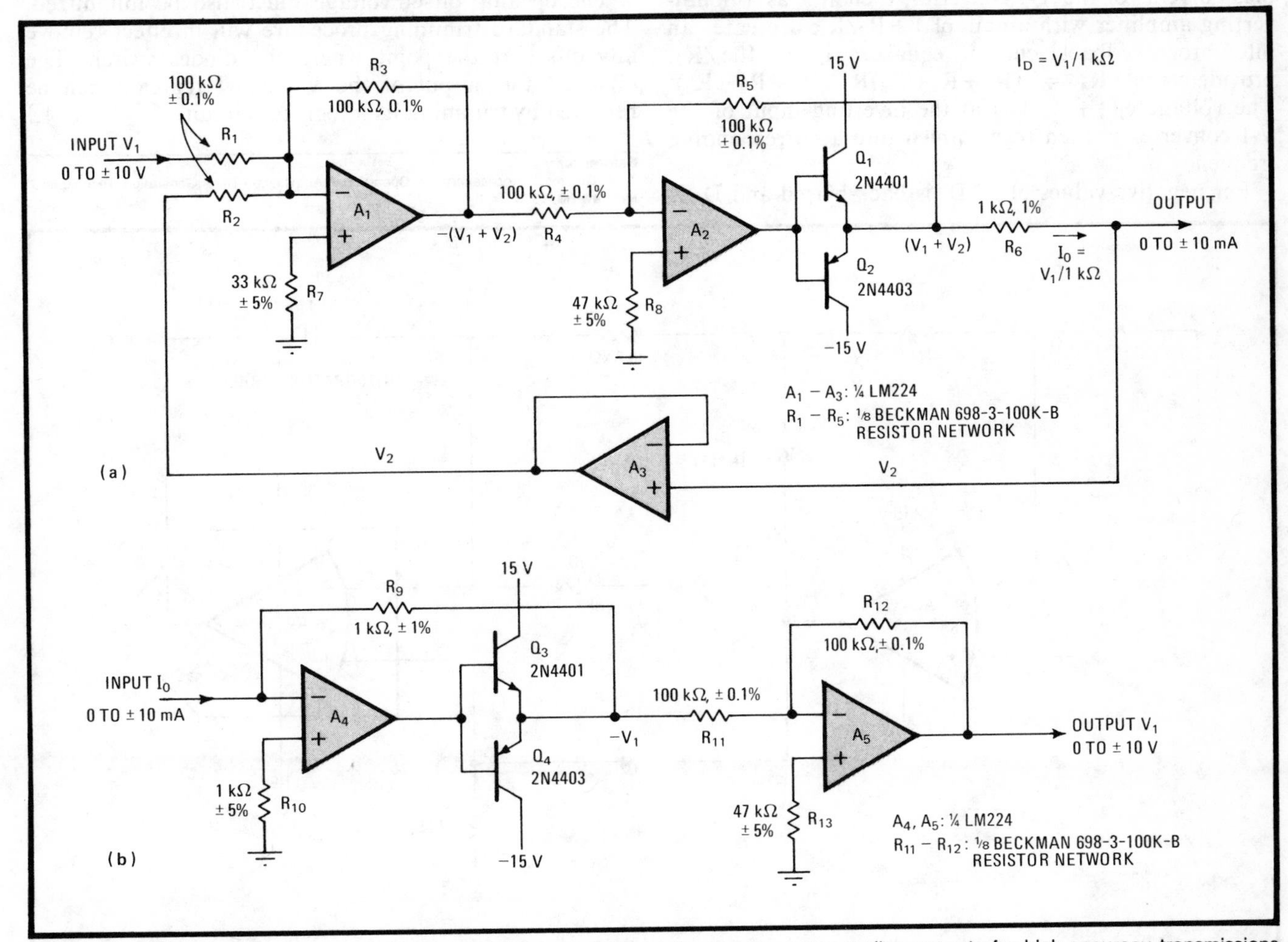

Communicating current. Converter (a) transforms bilateral analog voltages into corresponding currents for high-accuracy transmissions over high-resistance networks. Receiving converter (b) performs inverse operation to recover input voltage. Low-cost op-amps and transistors are used throughout. Beckman package in (a) and (b) provides low-cost source of matched resistors required for very precise conversions.

Diodes adapt V-f converter for processing bipolar signals

by Jerald Graeme
Burr-Brown Research Corp., Tucson, Ariz

Two diodes and one operational amplifier will enable a voltage-to-frequency converter to process bipolar input signals, thus adapting it for operation in absolute-value circuits. Using the integrator of the converter eliminates at least one of the op amps normally required for such absolute-value converters. Moreover, the approach is simpler overall than ones that bias the converter's inputs at a value midway between supply voltage and ground.

When input signal e_i is positive (see figure), diode D_1 becomes forward-biased and D_2 is reverse-biased. Op amp A_1, which isolates the signal from the offset and bias currents of the V-f converter, then acts as a noninverting amplifier with a gain of $1+R_2/R_1$; it creates an integrator feedback current equal to $i_f = 10e_i/R_3$, provided that $R_4 = (R_1+R_2+R_3)R_3/(R_1+R_2-R_3)$. The voltage $e_i(1+R_2/R_1)$ at the inverting input of the V-f converter is then transformed into a corresponding frequency.

For negative values of e_i, D_1 is back-biased and D_2 is forward-biased, enabling the op amp's output signal to be applied to the noninverting input of the V-f converter. In this configuration, the gain is negative, so that the integrator current generated has the same polarity as before. Thus the V-f converter cannot distinguish positive and negative voltages having the same magnitude, and so generates the same frequency for both signals.

If D_1 and D_2 are replaced with the emitter-base junctions of any general-purpose transistors and the transistors' collectors are used to drive lamps or other indicators, the polarity of the input signal can be displayed. Care should be taken to avoid reverse emitter-base breakdown caused by large input-signal levels by placing diodes in series with the base of each transistor.

The accuracy of the converter is determined by the same factors as affect conventional absolute-value circuits: resistor-ratio matching and the op amp's input offset voltage.[1] It is most important that the R_1–R_4 resistor values be correct for a given gain, as they have a part in equalizing circuit gain for both signal polarities.

The op amp offset voltage must also be minimized. The standard trimming procedure will in effect remove any offset at the point where the diodes switch. The offset at the output of the V-f converter can then be removed by trimming its integrator circuit. □

References

1. J. Graeme, "Applications of Operational Amplifiers—Third Generation Techniques," McGraw-Hill, 1973.

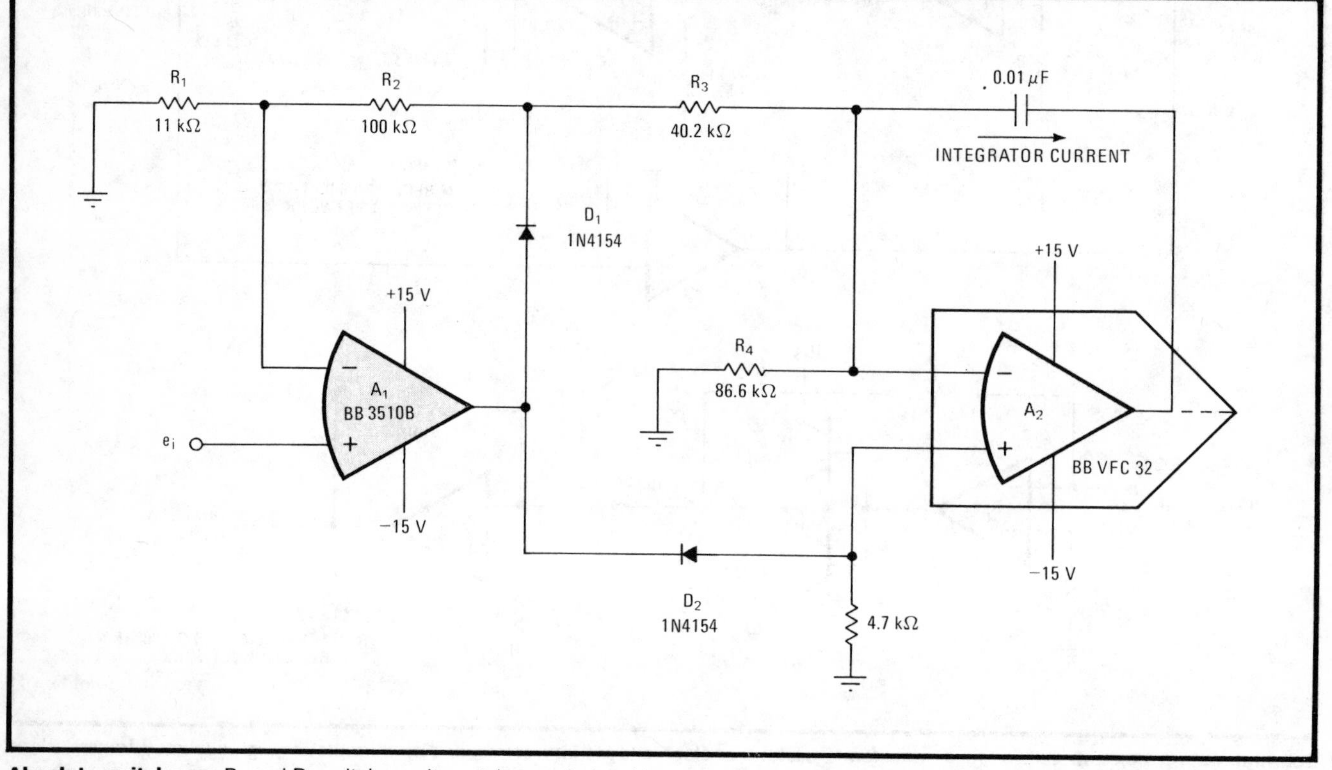

Absolute switchover. D_1 and D_2 switch on alternately as polarity of input signal changes, thus maintaining direction of integrator current. V-f converter cannot distinguish between signal polarities of the same magnitude and so generates the same frequency for both.

Converters simplify design of frequency multiplier

by Michael K. McBeath
Burroughs Corp., Goleta, Calif.

By using a programmable digital-to-analog converter in combination with frequency-to-voltage and voltage-to-frequency converters, this circuit can multiply an input frequency by any number. Because it needs neither combinational logic nor a high-speed counter, it is more flexible than competing designs, uses fewer parts, and is simpler to build.

As shown in the figure, the V-f converter, a Teledyne 9400, transforms the input frequency into a corresponding voltage. An inexpensive device, the converter requires only a few external components for setting its upper operating frequency as high as 100 kilohertz.

Next the signal is applied to the reference port of the DAC-03 d-a converter, where it is amplified by the frequency-multiplying factor programmed into the converter by thumbwheel switches or a microprocessor. The d-a converter's output is the product of the analog input voltage and the digital gain factor.

R_3 sets the gain of the 741 op amp to any value, providing trim adjustment or a convenient way to scale the d-a converter's output to a much higher or lower voltage for the final stage, a 9400 converter that operates in the voltage-to-frequency mode. The 741 and R_3 can also be used to set circuit gain to noninteger values. The V-f device then converts the input voltage into a proportionally higher or lower frequency. □

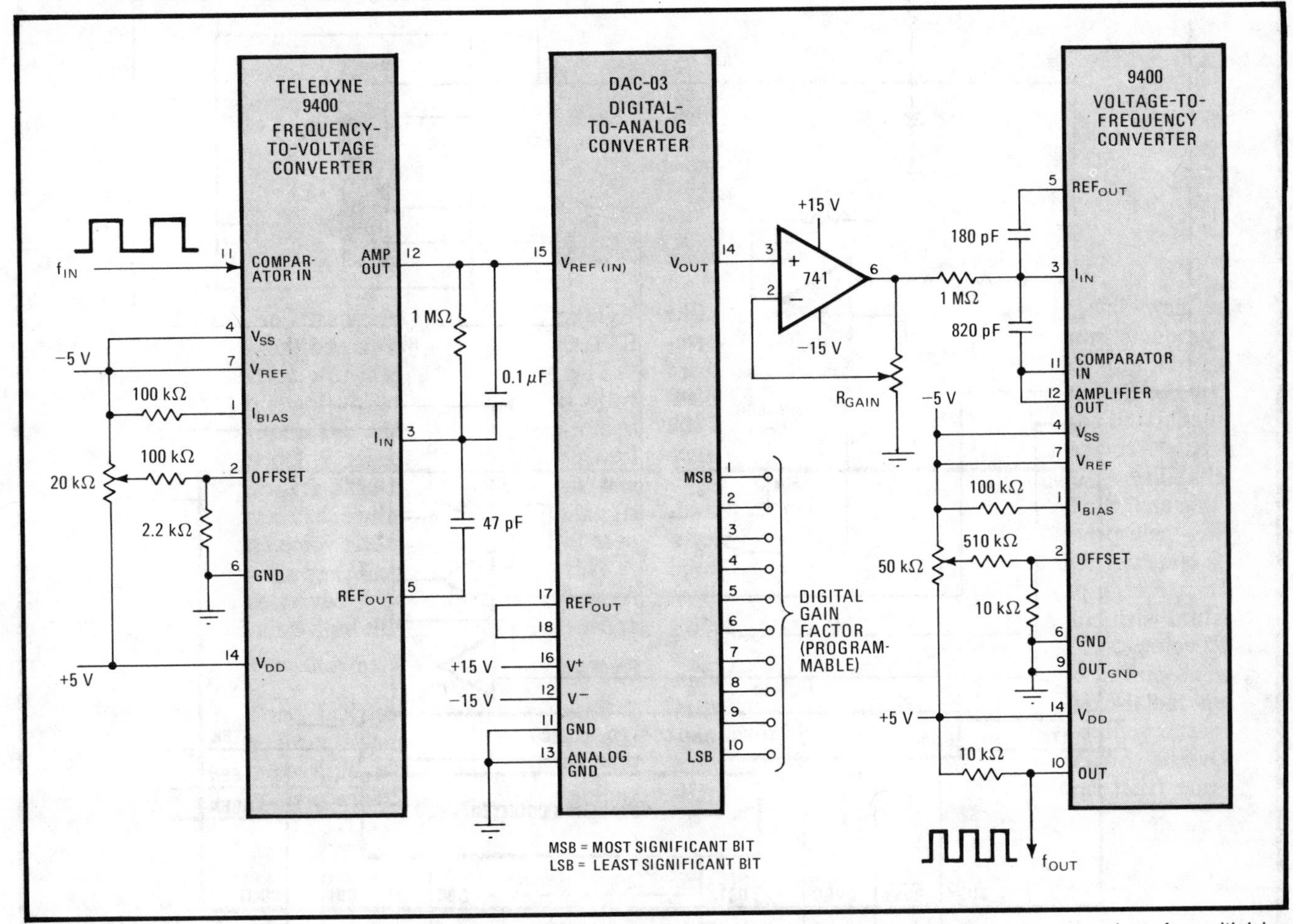

All numbers. Circuit uses frequency-to-voltage-to-frequency conversion, with intermediate stage of gain between conversions, for multiplying input frequency by any number. Digital-to-analog converter is programmed digitally, by thumbwheel switches or microprocessor, for coarse selection of frequency-multiplying factor; 741 provides fine gain, enables choice of non-integer multiplication values.

Buffer improves converter's small-signal performance

by Will Ritmanich
Precision Monolithics Inc., Santa Clara, Calif.

If an operational amplifier is placed in the integration and error compensation loop of the popular AD7550 analog-to-digital converter, the converter can measure microvolt-level signals without errors being introduced by its own or the op amp's offset-voltage drifts. Adding this buffer also increases the impedance of the signal (input) port of the converter, allowing it to be driven from high source impedances.

To understand why an external buffer can improve converter performance, it is necessary to know how the converter processes the input signal. The AD7550 uses the quad-slope integrating method, an improved version of the often-used dual-slope method.

A shortcoming of the dual-slope type is that errors at the input of the integrator due to noise or comparator

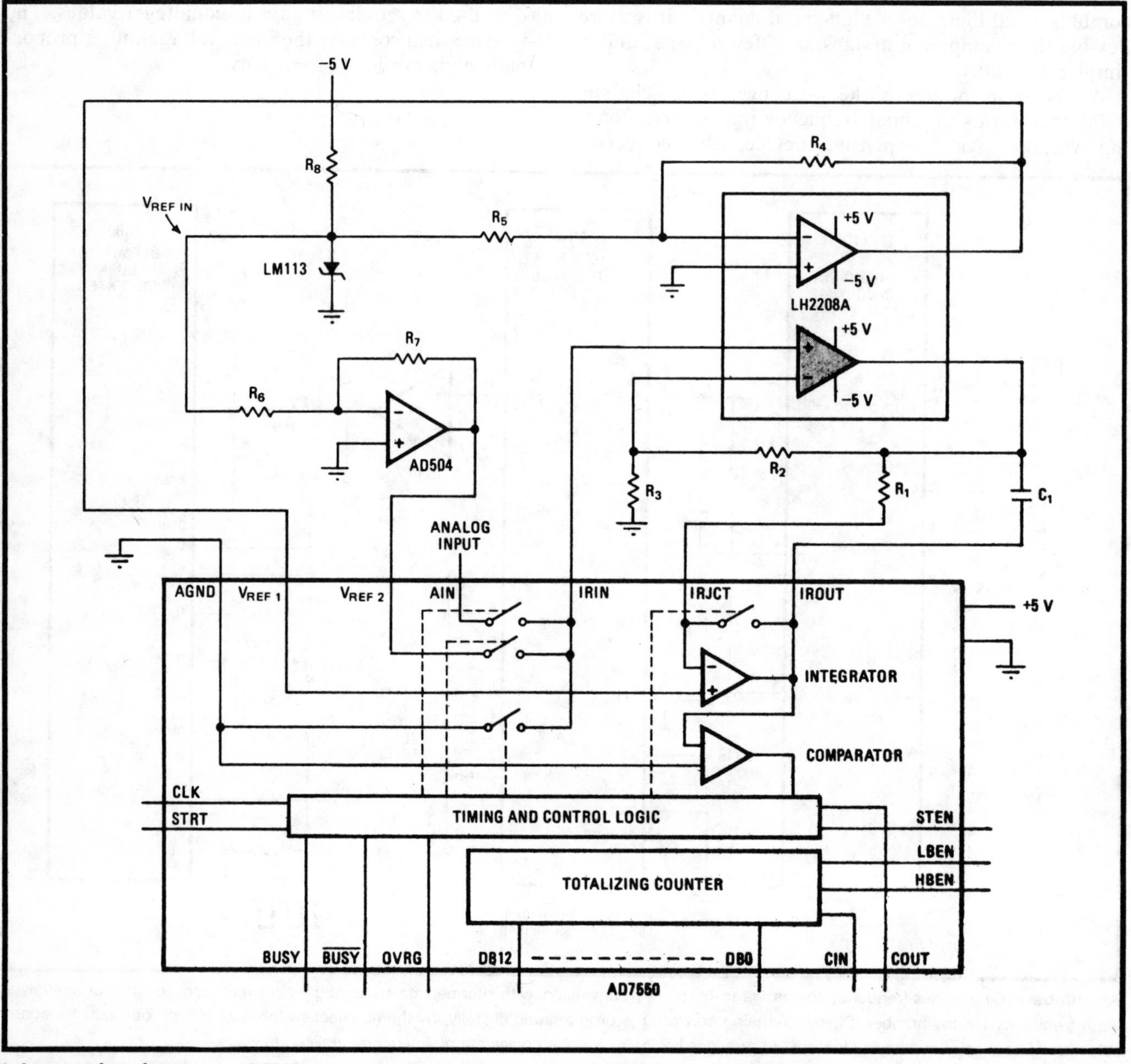

1. Improved performance. AD7550 accurately measures even microvolt-level analog voltages if op-amp buffer is placed in converter's integrator and error compensation loop. Converter views offsets as just another error that has been introduced in integrator loop.

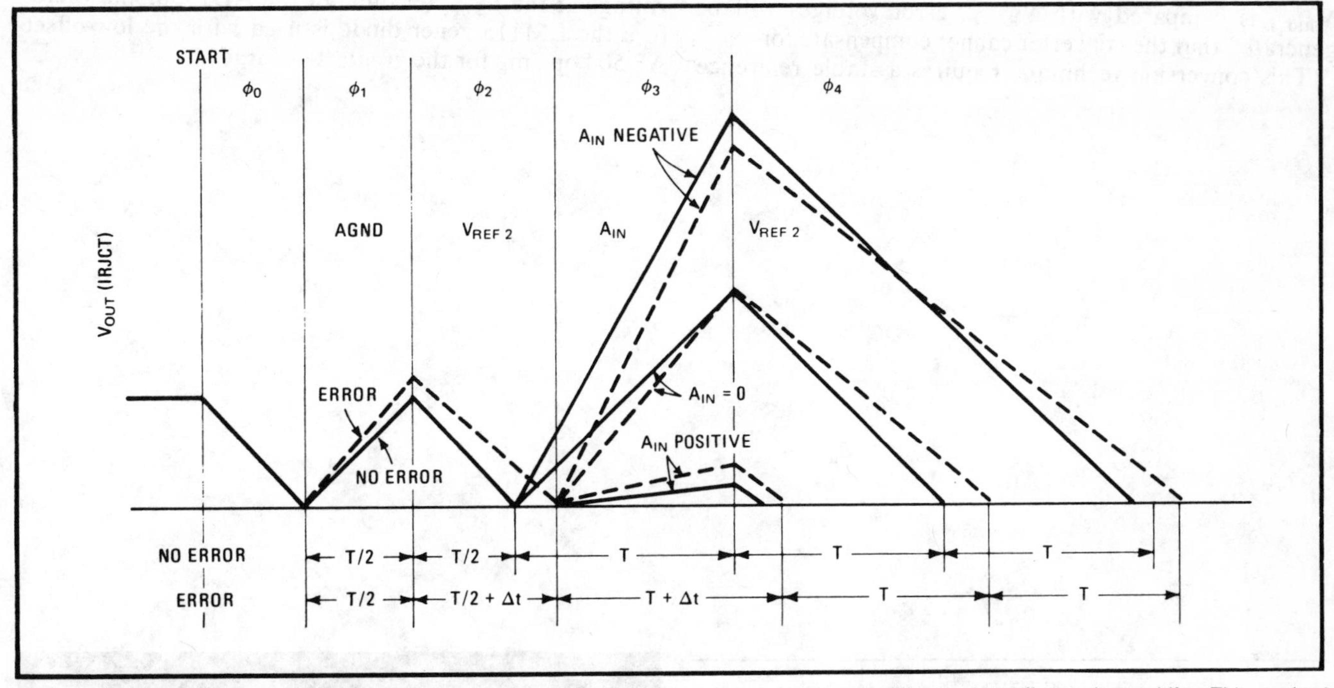

2. Four-phase integration. AD7550 uses quad-slope method to detect most errors, including op-amp offset voltage drifts. This method eliminates error by measuring comparator trip-time delay, Δt, converting Δt to a voltage and subtracting the voltage at end of cycle.

offset drift show up as errors in the digital output word. The quad-slope method detects all AD7550 offset-voltage errors, converts them to a digital equivalent during the integrating cycle, and subtracts them from the final count at the end of the cycle.

Normally, a ground signal, a reference voltage, or an analog input is connected through the converter's integrator input (IRIN) terminal to the integrator junction (IRJCT) port. As shown in Fig. 1, the buffer is connected between the terminal and the port, in series with integrating resistor R_1. The choice of signal is determined by internal logic-controlled switch settings.

Before a four-phase a-d conversion begins, a reference voltage, $V_{REF\,2}$, is applied to the inverting input of the integrator, resulting in a negative-going ramp at the integrator's output port. The ramp resets the output to zero. When the output voltage is equal to the comparator's trip voltage, phase ϕ_1 is initiated.

At this time, a digital counter starts and the integrator is connected to input AGND to determine whether the converter's negative supply voltage is at ground potential. If not, an error voltage is generated that is proportional to the difference.

The output of the integrator rises proportionally to the error voltage until the counter stops counting system clock pulses, at fixed interval T/2, (Fig. 2.) Then ϕ_2 is initiated. The accumulated voltage on the integrating capacitor, C_1, is proportional to AGND × T/2.

During ϕ_2, the integrator is connected to $V_{REF\,2}$ and any error voltage on this signal. The output of the integrator falls at a rate proportional to the error until the comparator trip point is reached. If there are no errors, the counter stops after time T; otherwise, the trip time is delayed by Δt.

ϕ_3 then begins, starting a second counter that ultimately displays the digital equivalent of the analog input voltage introduced during the phase. $V_{REF\,2}$ is again applied during ϕ_4, and the counter continues to run until the integrator falls below the comparator trip point. $V_{REF\,2}$ is normally equal to $2V_{REF\,1}$, and the converter detects any difference.

Because ϕ_1 through ϕ_3 occupy a total period of 2T if there are no errors, any errors detected during these phases must show up as a delay in trip time by Δt. Since the charge gained during ϕ_3 by the integrating capacitor is proportional to $A_{IN\,av} \times T$, and the charge lost will be equal to $V_{REF\,2} \times \Delta t$ during ϕ_4, then the number of counts (N) relative to the normalized (full-count) time is equal to:

$$\Delta t / t = A_{IN\,av}/V_{REF\,2} = N$$

and is displayed directly. Thus any offset error in the buffer will not affect system accuracy, because it will be viewed as an additional (series-introduced) error existing during ϕ_1 through ϕ_4. At the same time, the amplifier can still be used to boost small-signal input voltages.

The gain of the amplifier may be determined by resistors R_2 and R_3 and is equal to:

$$G = \frac{(R_2+R_3)}{R_3}$$

To maintain the proper relationship between $V_{REF\,1}$ and $V_{REF\,2}$, resistors R_2 through R_7 are employed in the circuit and are determined by the equations:

$$V_{REF\,1} = (R_4/R_5)V_{REF\,in}$$

and:

$$V_{REF\,2} = \frac{R_7}{R_6}\,\frac{R_2+R_3}{R_3}V_{REF\,in} = 2V_{REF\,1}$$

$V_{REF\,1}$ and $V_{REF\,2}/2$ should be within 6% of each other at all times. Otherwise, in ϕ_2 and ϕ_4, during which time

$V_{REF\,1}$ is compared with $V_{REF\,2}$, error voltages will be generated that the converter cannot compensate for.

This conversion technique requires a stable reference voltage at the V_{REF2} terminal of the AD7550, and therefore the LM113 zener diode is used with the low-offset AD504 op amp for the greatest accuracy. □

Programming a microcomputer for d-a conversion

by Richard T. Wang

Department of Zoology, University of Texas, Austin, Texas

If the speed of conversion is not an important consideration, a microcomputer can be used for digital-to-analog number conversion with the aid of this simple pulse-width modulation program. A resistance-capacitance filter and a latch, which may be connected to any one pin of the microcomputer's output ports, are the only items of microprocessor hardware required. Almost any strip-chart recorder can be used to record the filtered output voltage. Although intended for the plotting of bar graphs, histograms, and spectrograms from data stored in memory, the program needs only slight modification to be able to handle real-time data.

The program in the table was written for the Zilog Z-80 microcomputer system. All data for the graphs to be plotted is stored as an array in memory. The first two bytes of the array contain the number of data elements stored in the area labeled ARRAY. These bytes are loaded into registers B and C, which keep track of the number of bytes in ARRAY remaining to be sampled. The elements of ARRAY are loaded into register A and passed to a timing loop one at a time, where each will eventually be converted into an analog voltage.

The timing loop uses the alternate (primed) registers. Register B′ is loaded with a number (N) determined by the relationship N = PS/E, where E is the number of elements per inch recorded for a given chart speed (S) and output pulse rate (P) to the RC filter. B′ contains the number 192—the number of pulses per element (N) needed to ensure that 50 elements per in. are recorded for a given chart speed of 30 seconds/in., and a pulse rate of 320/s.

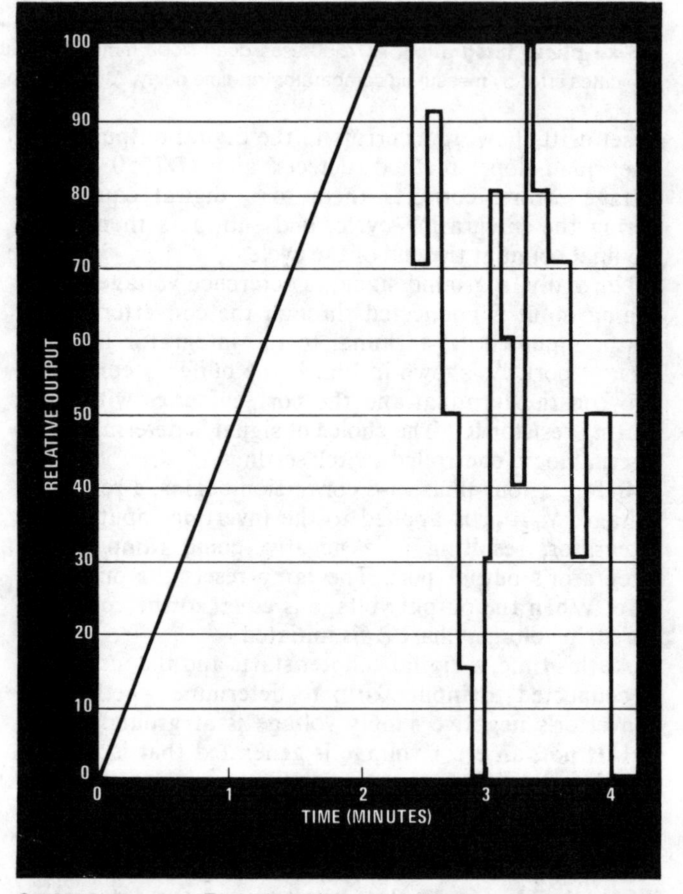

System response. Typical graph is accurately reproduced by Heath EU-205-11 strip-chart recorder. System linearity is confirmed with a ramp-test pattern, as shown on left. Right half shows a sample histogram. Total recording time is 4 minutes.

Register E′ is loaded with a selected value of 201, a number that is one greater than the largest number that may be plotted. E′ is decremented through a loop consisting of six instructions. A NOP instruction is included for timing compensation. When E′ equals the contents of A, the state of the output is changed from a 0 to a 1 and begins to charge the capacitor in the RC filter. This continues until E′ decrements to zero. At this time the output returns to 0. E′ then returns to 201, and the process repeats until 192 pulses are generated for that element. The next element then undergoes an identical process until the entire array has been processed. The output pulse width (W_i) is related to the digital numbers (D_i) for a system clock at 2.5 megahertz by $W_i = 15.2\ D_i$ where W_i is given in microseconds.

The critical timing in this program occurs between the START and TERM instructions. When the number to be converted is 1 (the smallest number that can be encoded), this instruction sequence takes 38 clock periods, or T states, exactly the same as that required by the START-TERM loop. The program is thus written so that multiples of 38T are added through the loop for numbers greater than 1, ensuring a linear relation of the output pulse width to the digital number stored.

With the output pulse rate of 320/s, an RC filter with a time constant equal to or greater than 3 milliseconds is sufficient to smooth the output voltage. To calibrate the system, the input to the chart recorder should be temporarily grounded, and its pen position should be set to zero. ARRAY is then filled with C8 (hexadecimal), the full-scale value, and the program is initiated. The recorder sensitivity control is then adjusted for a full-scale indication.

The figure shows a graph recorded on a Heath strip chart instrument. The ramp voltage is intended to display the system's linearity. □

Z-80 PULSE-WIDTH-MODULATION PROGRAM FOR DIGITAL-TO-ANALOG CONVERSION

```
 LOC   OBJ CODE   STMT SOURCE STATEMENT

                     1 *H D/A CONVERSION
                     2 ; SUBROUTINE DAC: D TO A CONVERSION BY PULSE
                     3 ; WIDTH MODULATION.  THE GRAPH TO BE DISPLAYED
                     4 ; IS SET UP IN ARRAY. THE FIRST TWO BYTES
                     5 ; INDICATE THE LENGTH OF ARRAY.   ANY NUMBER
                     6 ; GREATER THAN FULL SCALE (200 DECIMAL) WILL
                     7 ; BE DISPLAYED AS ZERO.
                     8 ;
                     9 ARRAY    EQU 3000H      ;ADDRESS OF ARRAY
2800                10          ORG 2800H
2800   210030       11 DAC      LD HL,ARRAY    ;ADDRESS OF ARRAY IN HL
2803   4E           12          LD C,(HL)      ;FIRST TWO BYTES
2804   23           13          INC HL         ;   INDICATE LENGTH
2805   46           14          LD B,(HL)      ;    OF ARRAY
2806   23           15 DSLOOP   INC HL         ;DISPLAY LOOP
2807   7E           16          LD A,(HL)      ;NUMBER TO BE DISPLAYED
2808   D9           17          EXX            ;GET ALTERNATE REGISTERS
2809   06C0         18          LD B,192       ;REFRESH CYCLES
280B   0E01         19          LD C,01        ;OUTPUT PORT ADDRESS
280D   1620         20          LD D,100000B   ;BIT 5 AS OUTPUT PIN
280F   1EC9         21 SUBDAC   LD E,201       ;SET E TO MAXIMUM + 1
2811   1D           22 LOOP     DEC E          ;E AS TIMING COUNTER
2812   280B         23          JR Z,TERM      ;TO TERMINATE PULSE
2814   BB           24          CP E           ;A EQUALS TO E?
2815   2803         25          JR Z,START     ;IF YES, START PULSE
2817   00           26          NOP            ;TIMING COMPENSATION
2818   18F7         27          JR LOOP        ;LOOP IF NOT EQUAL
281A   ED51         28 START    OUT (C),D      ;START OUTPUT PULSE
281C   C31128       29          JP LOOP        ;CONTINUE LOOP
281F   ED59         30 TERM     OUT (C),E      ;TERMINATE PULSE
2821   10EC         31          DJNZ SUBDAC    ;REFRESH UNTIL B ZERO
2823   D9           32          EXX            ;GET BACK MAIN REGISTERS
2824   0B           33          DEC BC         ;END OF ARRAY?
2825   79           34          LD A,C
2826   B0           35          OR B
2827   20DD         36          JR NZ,DSLOOP   ;IF NOT, SHOW NEXT NUMBER
2829   C9           37          RET
                    38          END
```

Line-frequency converter transforms 50 Hz into 60 Hz

by Juan E. Piquinela
Montevideo, Uruguay

Low-power equipment driven from the 60-hertz power line can usually be expected also to work properly at 50 Hz—that is, except for electric clocks and other time-keeping devices, to which many a traveler outside North America will attest. For such devices, a circuit that provides a multiplication ratio of 6:5 for generating a 60-Hz output from a 50-Hz input is required. Such a low-cost, low-power circuit is shown here.

The 555 timer, operating as an astable multivibrator at 300 Hz, provides the 4017 counter, A_1, with six count-pulses for every reset pulse from the 50-Hz line. The timer's period of oscillation—about 3 milliseconds—is not critical as long as six of its cycles are completed in less than 20 ms, the period of the 50-Hz line frequency.

On the sixth pulse, Q_6 of A_1 moves high and disables the timer through transistor Q_1 by shorting capacitor C. Thus, independent of the period set for the 555, its average frequency is $50 \times 6 = 300$ Hz. At the positive zero-crossing of the line voltage that occurs shortly after the sixth pulse, A_1 is reset through $R_1 - R_2$, C_2, and $D_1 - D_2$, and the process repeats.

Counter A_2 provides a divide-by-five function at 300 Hz, thereby generating an output frequency of 60 Hz. C-MOS drivers or transistors can provide increased current capability as required. □

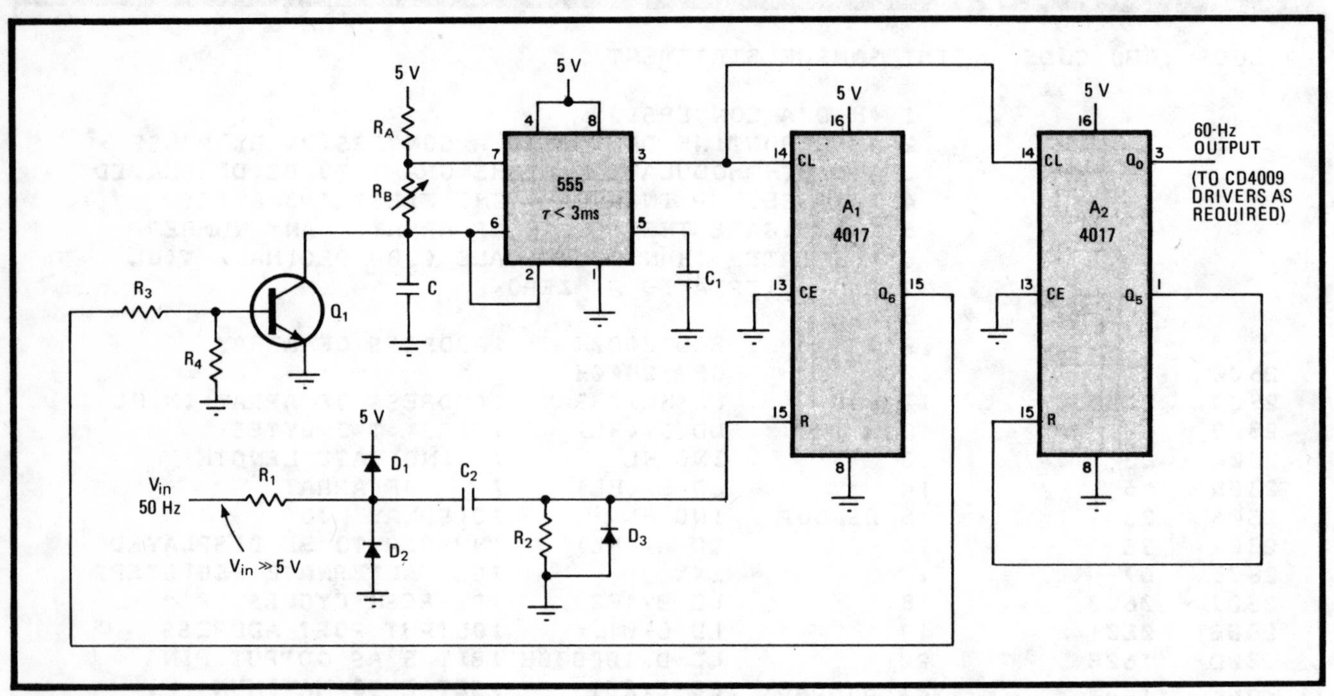

On time. Three-chip multiplier converts 50-Hz power-line frequency into 60 Hz for devices used in the U. S. A_1 generates six pulses for every 50-Hz cycle, forcing 555 timer to generate average frequency of 300 Hz. A_2 provides divide-by-five function on 555 waveform.

Chapter 9
COUNTERS

C-MOS counter-decoder pair sets divider's modulus

by Bradley Albing
Hickok Electrical Instrument Co., Cleveland, Ohio

The availability of one-chip complementary-MOS counters and decoders makes it possible to construct this low-power, variable-modulo divider with only two integrated circuits and a single-pole, multiple-position switch. Divider ratios of from 1 to 16 can be ordered.

A_1, a synchronous binary counter having a parallel-load feature, is stepped by an input signal of frequency f_{in}, as shown. Advancing from zero, its output is introduced to A_2, a 4-to-16 line decoder.

A_2 serves as a hexadecimal decoder, where each output port n (of m total ports) moves high sequentially after n cycles of f_{in}. Switch S_1 selects the desired port from which the output signal ($\overline{TC}$) is fed back to reset A_1. Thus it is seen that $f_{out} = f_{in}/n$.

It is easy to display the value of the divisor if a binary-coded-decimal–to–seven-segment decoder (that is, a 74C48) and a suitable seven-segment display are available. Here, the output of A_1 drives the decoder as well as the corresponding inputs of A_2. Note that A_1 may be reset to 1 or any other number as required.

A 10-part CD4028 decoder may be used in place of a 74C154 if a divisor of 10 or less is desired. In this case, an inverter will be needed between the TC line and the $\overline{PE}$ port of the 74C161 counter. □

Decimal divider. Binary counter (A_1) and one-of-sixteen decoder (A_2) form variable-modulo divider. Ratios are selected with multiposition switch. CD4028 can replace 74C154 if a divisor of 10 or less is required. C-MOS circuit draws low power—typically a few milliamperes.

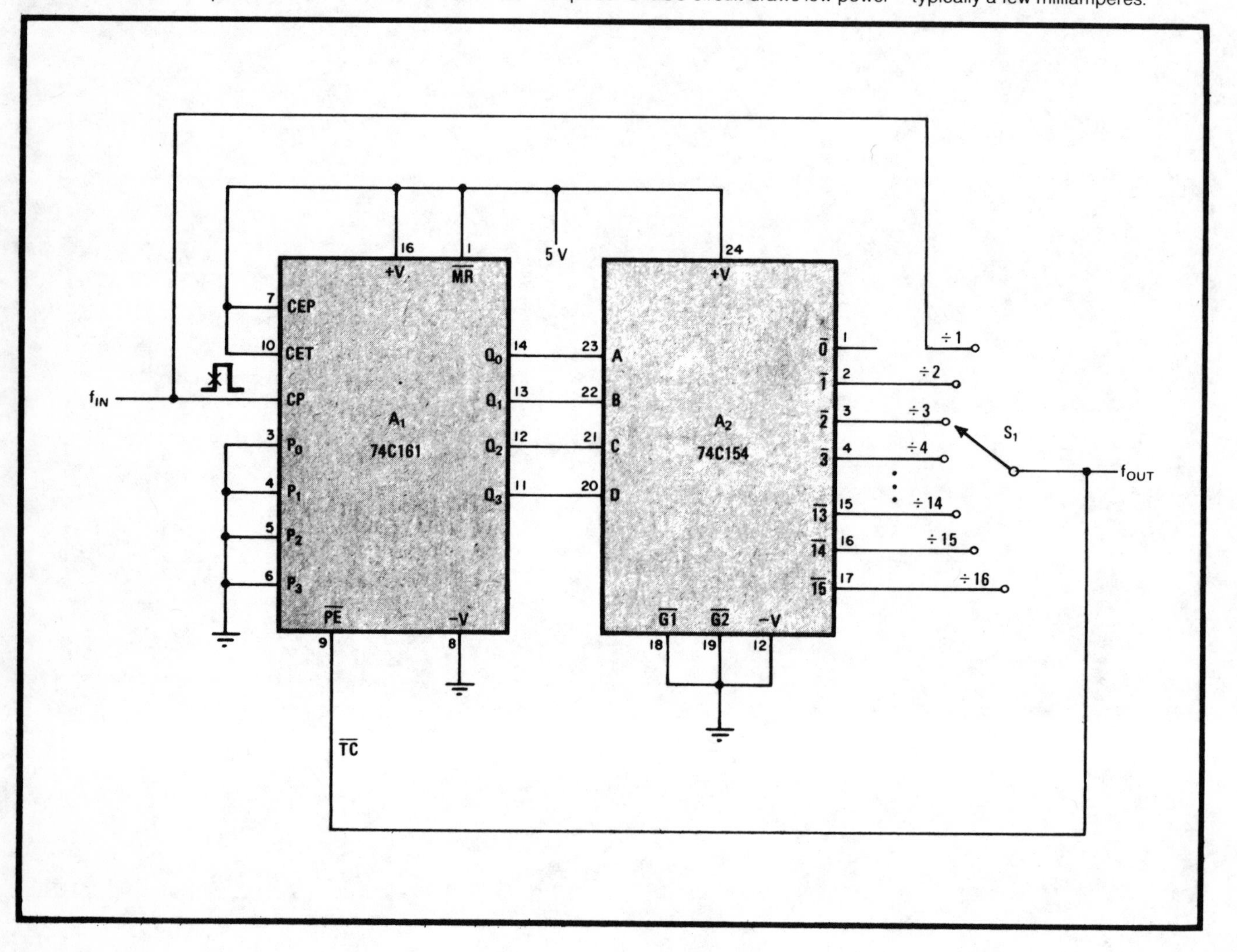

C-MOS counter sets divider's modulus

by Arie Shavit
Kiriat Tivon, Israel

The cost and power consumption of Albing's C-MOS variable-modulo divider[1] can be reduced even further with this circuit, which uses logic gates and four low-cost binary switches to replace one counter and the multiple-pole selector, respectively. Although the counter's modulus is set with the binary elements, thereby sacrificing the convenience of ordering up values in the familiar decimal form, the ease of interfacing the counter to microprocessor-based control systems is immensely enhanced. Divider ratios of from 1 to 16 can be selected.

The 40161 synchronous binary counter, A_1, which has parallel-load capability, is stepped by input frequency f_{in}, as shown in (a). Switches S_1–S_4 set the binary representation of $16-n$ at the parallel-load inputs P_0–P_3, where n is the desired divider ratio, as shown in (b).

Output pin TC of A_1 moves high after n cycles of f_{in}. Thus the output signal from gate G_1 is a pulse of short duration having a frequency of $f_{out} = f_{in}/n$. TC is then inverted by gate G_2 and used to reset the counter.

Gate G_3 comes into play if a modulus of 1 is set. Under these conditions, TC remains high and f_{in} serves to gate itself to the output. □

References
1. Bradley Albing, "C-MOS counter-decoder pair sets divider's modulus," p. 82

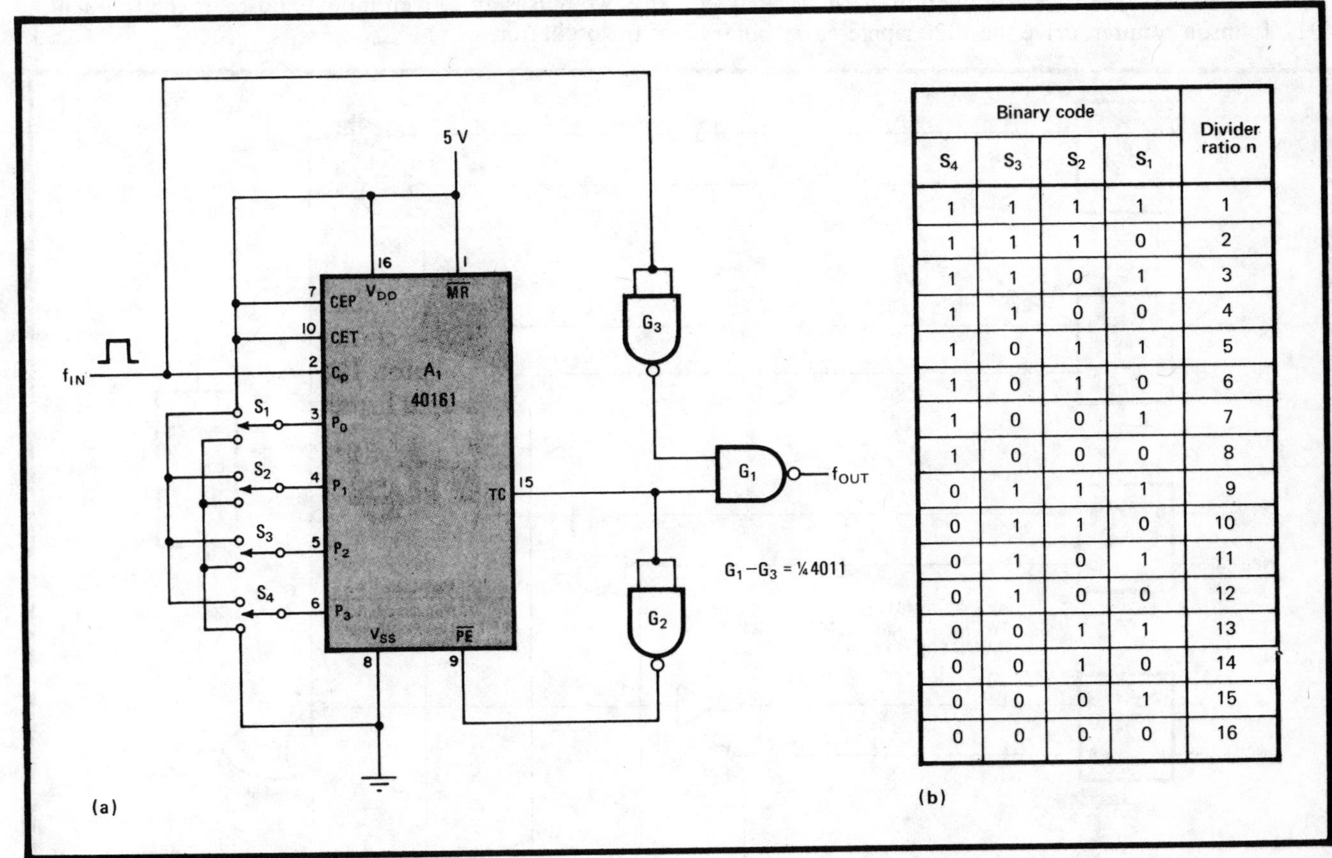

Binary code				Divider ratio n
S_4	S_3	S_2	S_1	
1	1	1	1	1
1	1	1	0	2
1	1	0	1	3
1	1	0	0	4
1	0	1	1	5
1	0	1	0	6
1	0	0	1	7
1	0	0	0	8
0	1	1	1	9
0	1	1	0	10
0	1	0	1	11
0	1	0	0	12
0	0	1	1	13
0	0	1	0	14
0	0	0	1	15
0	0	0	0	16

Binary breakup. Single counter and three gates simplify design of variable-modulo divider (a). Binary switches S_1–S_4 set counter to $16-n$, where n is desired divider ratio (b). Output of gate G_1 is a pulse with a frequency equal to $f_{out} = f_{in}/n$, for $1 \leq n \leq 16$.

Eight-port counter handles coinciding input pulses

by Gary Steinbaugh
Owens/Corning Fiberglas Corp., Technical Center, Granville, Ohio

In applications where an asynchronous counter is driven by more than one input source, there is always the possibility that pulses will arrive simultaneously, causing an incorrect count. This circuit overcomes the problem by latching all input pulses as they occur and multiplexing the latch outputs so that the pulses may be applied to the counter one at a time.

As shown in the figure, four 4013 dual latches and their associated gates, in conjunction with a clocked 4022 Johnson counter, drive the 4020 ripple-carry binary device that is used to accumulate a count. (In the interests of space, the identical circuitry of A_5–A_8 and their gates is not shown.) One eight-input NAND gate, the 4068, is used to clock the 4020.

A_1–A_4 will latch on the rising edge of any pulse that appears at their respective inputs. The 4022, driven by an oscillator, scans the contents of each flip-flop; if the output of the flip-flop scanned is at logic 1, a pulse appears at the output of the 4068 and the 4020 is advanced. At the same time, the flip-flop is reset.

The counter may be configured for any number of inputs. Note that the frequency of the oscillator driving the Johnson counter must be high enough to permit each flip-flop to be reset before it is set by a following input pulse; otherwise, the event will not be recorded. For an n-input counter, $f_{in} < nf_{osc}$, where f_{in} is the highest input frequency expected (the reciprocal of the time between any two pulses at a given input) and f_{osc} is the frequency of the oscillator. □

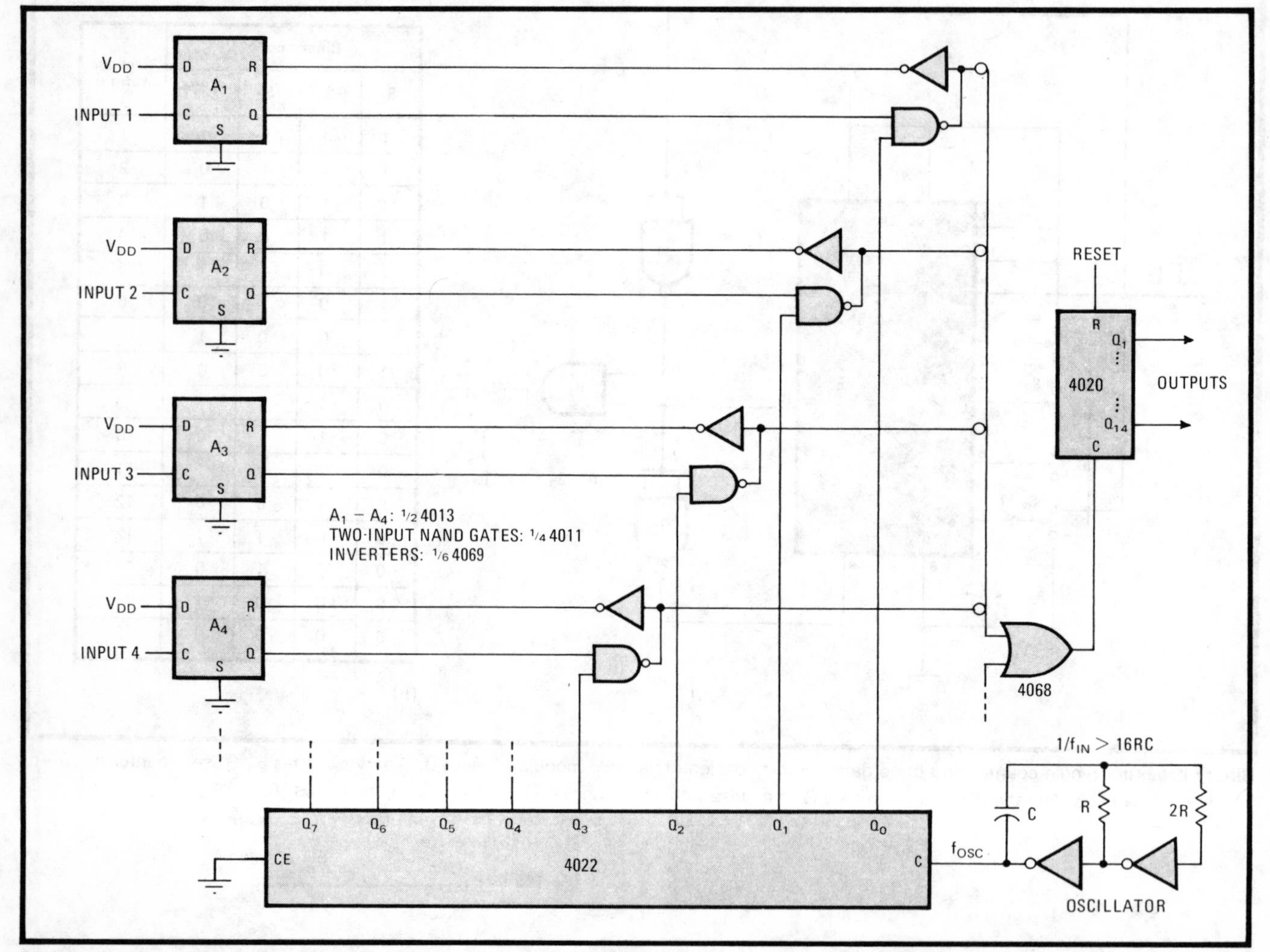

Serial stepping. This multiplexer circuit avoids the difficulties associated with multi-input counters that are called upon to handle simultaneously occurring pulses. The clocked 4022 Johnson counter, driven by a simple RC oscillator, scans each flip-flop so that input pulses previously latched are applied one by one to 4020 accumulator. The counter can be easily configured for any number of inputs.

Counter banks stagger radar's pulse rate

by Prakash Dandekar
Tata Electric Companies, Bombay, India

In many radar applications, the instantaneous pulse-repetition frequency must be varied in an orderly fashion to improve the read-out accuracy of the system's moving-target indicator. Considerable circuitry is usually required to achieve the so-called staggered operation, but as shown here, two sets of synchronous counters can be easily connected to control the prf over any range, while providing superior MTI performance.

Normally, designers resort to transmitting pulses at each of three selected periods only, in order to simplify circuitry. Specifically, a popular technique is to transmit a group of three 1-microsecond pulses spaced at 1, 1.1, and 1.2 milliseconds repeatedly. When this is done, however, the filtered output of the MTI is not uniform and so—aside from causing discontinuities in the curve of MTI filter output versus target velocity—this method creates blind velocity points, or ranges over which velocity cannot be determined accurately.

With this circuit, a perfectly smoothed response is achieved by increasing the number of staggered pulses per given time. Thus in this case, a group of 200 pulses, each having a time between pulses of (1,201 − M) micro-seconds, where M denotes the Mth pulse of 200, are generated.

As shown, 12-bit counters A_1–A_3, comprising the main counter chain, advance at a 1-megahertz rate. When the counter reaches its maximum, the carry output of A_3, serving as the synchronous output, is generated.

The same signal is used to preset the main counter to a 12-bit binary number, N, which is determined by the state of the offset counter A_4–A_6. Because A_4–A_6 is also clocked, this unit is incremented with every sync pulse, so during each cycle the main counter is initialized at a higher value than it was previously. Thus the repetition time is reduced by 1 μs on each pass.

Note that the offset counter is initialized at a minimum value of $B51_{16}$ (see A–D inputs of A_4–A_6) and advances to a maximum of $C18_{16}$ ($= 2^{12}$) before it is reset by logic gates G_1–G_3. Thus, the difference between the counter's maximum and minimum is 200 counts, meaning the instantaneous pulse-repetition rate will vary from 1,200 to 1,001 microseconds. The maximum and minimum values may be easily changed, however, so that any pulse-repetition frequency range can be set.

When the counter reaches 3,096, corresponding to a rate of 1,001 μs, A_4–A_6 is loaded with $B51_{16}$. The rate becomes 1,200 μs once more, and the cycle is repeated.□

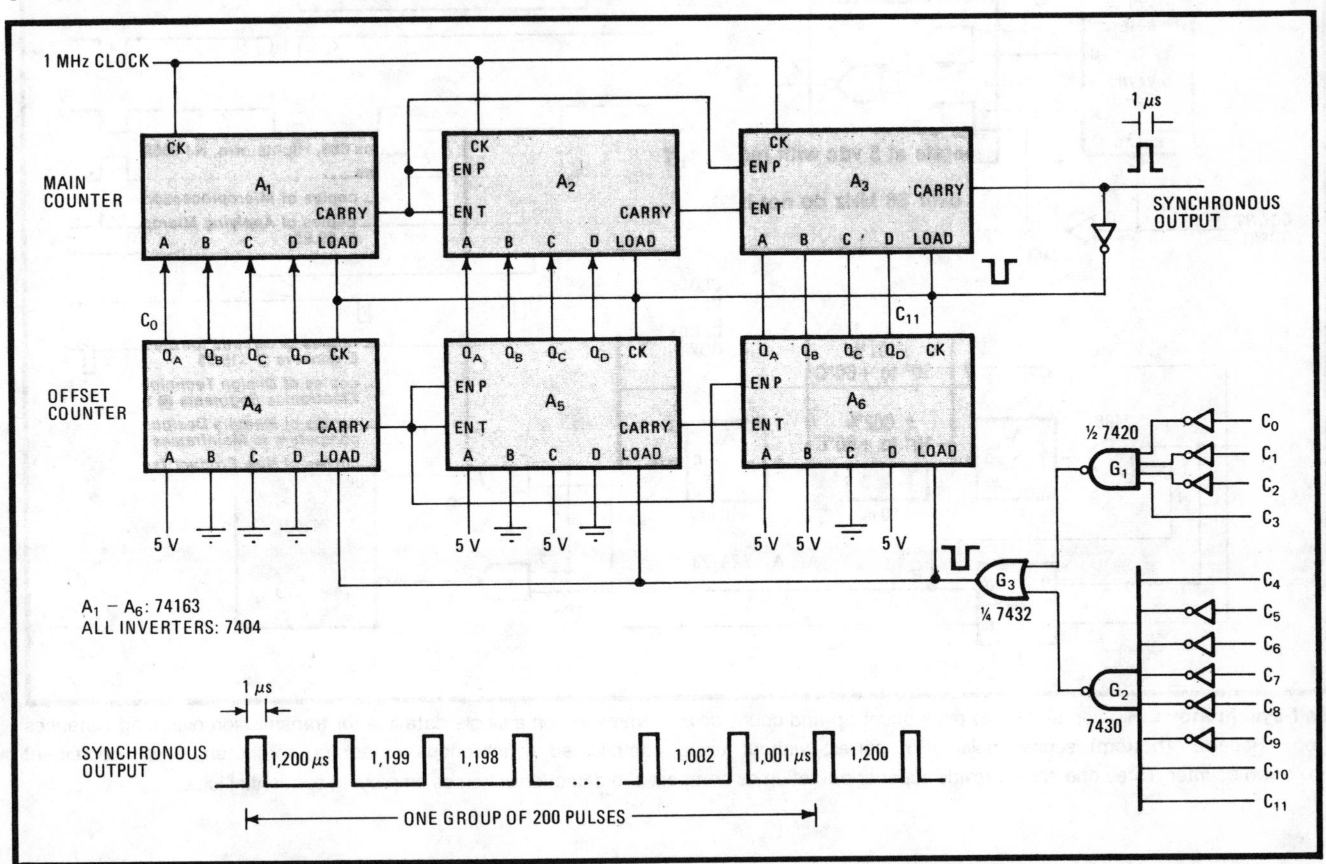

Smooth staggering. Two 12-bit counter chains generate a group of repeating N pulses spaced at (1,201 − M) μs, where M denotes the Mth pulse of N. for incremental staggering of the radar-pulse rate. Master clock sets absolute value of maximum pulse-repetition frequency.

Up/down counter processes data over single channel

by N. Bhaskara Rao
U. V. C. E., Department of Electrical Engineering, Bangalore, India

Two inputs are normally required for incrementing and decrementing an up/down counter—a count-up command line and a count-down line. But commands can be sent and received over a single channel if they are first multiplexed. Shown here is a transmitter-receiver pair for the one-channel system, which is particularly cost-effective over long distances.

At the sending end, it can be seen that $\bar{x} = QU + \bar{Q}D$, where x represents the transmitted signal, U is the count-up variable, and D is the count-down variable. It is assumed that the width of the U and D pulses is not more than a few hundred nanoseconds, and that the frequencies of U and D are relatively low.

When negative-going count-up pulses appear on the U line, the 7470 flip-flop is preset to Q = 1 and so $x = \bar{U}$ thereafter. Thus x moves from logic 1 to logic 0 on the trailing edge of the count-up command.

Similarly, a count-down command clears the 7470 and so x = D. Thus x moves from logic 0 to logic 1 on the trailing edge of the down command. The timing diagram details transmitter operation.

The receiver determines if variations in x are due to changes in U or D by using an indirect and inexpensive phase-locking scheme. Transitions in x are introduced at the input of the 7486 exclusive-OR gate. Feedback loop A_1, A_2, and A_4 acts to bring point L high, regardless of whether the transition is positive- or negative-going. One-shot A_1, having a set delay time, τ, slightly greater than the width of the U and D command pulses at the transmitter, then triggers one-shot A_2. By this time, x has stabilized, so that a logic 0 (U pulse) or a logic 1 (D pulse) is clocked into A_4, thus bringing point L low τ nanoseconds after it goes high.

Meanwhile, one-shot A_3 is fired, and the active contents of A_4 are transferred to either the clock-up or the clock-down port of the 74193. The receiver's timing diagram should be inspected to clarify operation. □

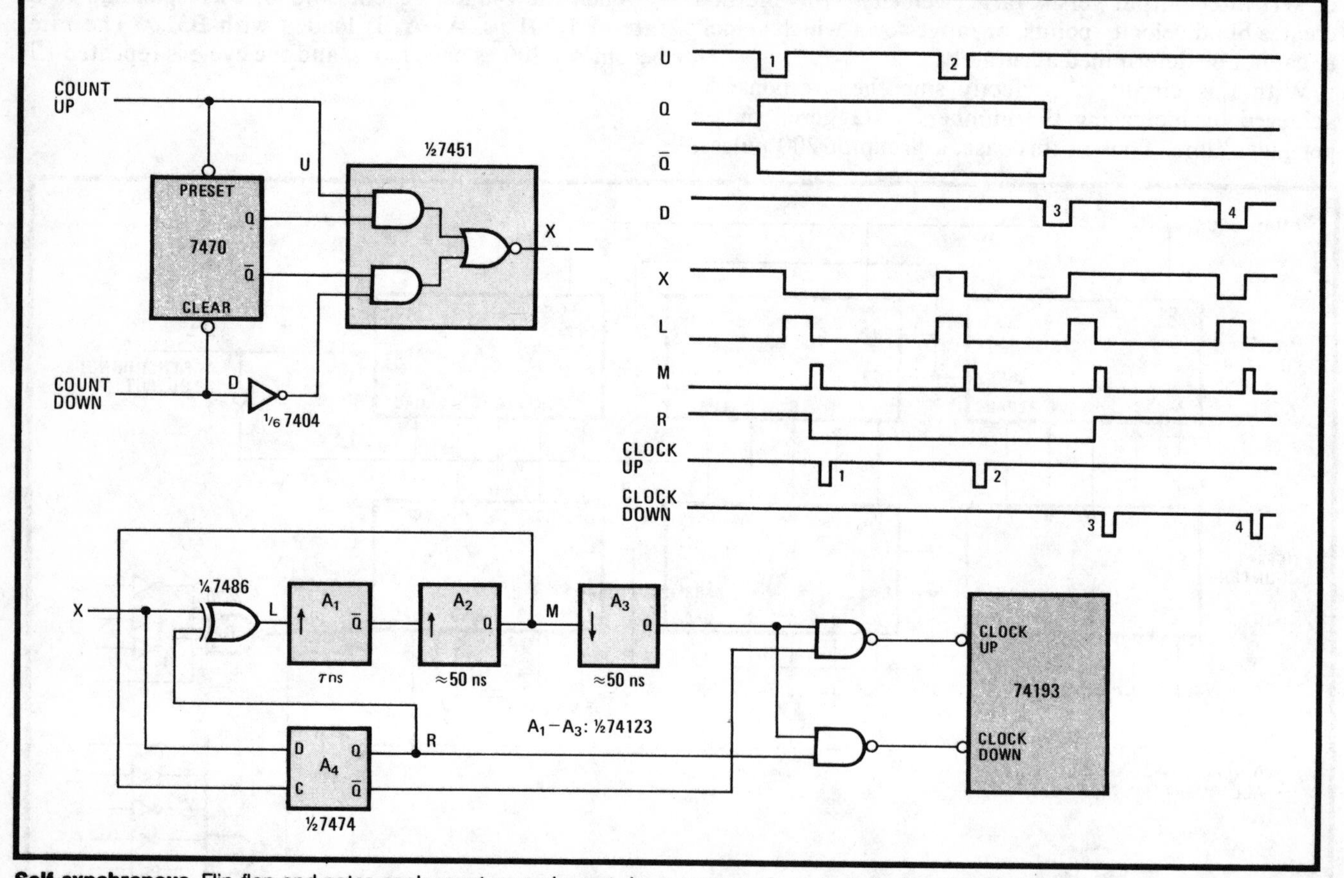

Self-synchronous. Flip-flop and gates pack count-up and count-down commands on a single data line for transmission over long distances (top). Receiver (bottom) senses polarity of pulses, without using complicated synchronous detectors, to increment or decrement up/down counter. Three one-shots provide superior operation as compared to detection schemes employing sequential logic.

Zero-sensing counter yields data's magnitude and sign

by Gary A. Frazier
Richardson, Texas

Upon sensing when an ordinary up-down counter is about to count down through zero, this circuit reverses the direction of the count to enable it to express a negative number by its magnitude and sign, instead of by the more usual but less convenient 2's complement. The circuit similarly represents positive integers, making it easy for any stored value to be handled directly by such data-system devices as digital-to-analog converters or microprocessors.

The unit shown in the figure has been found particularly useful in a digital-averaging application. Data is sampled during an interval in which the ADD/$\overline{\text{SUB}}$ line is toggled at some frequency, which in this case varies with time, as shown. Any data in phase with the ADD/$\overline{\text{SUB}}$ signal will accumulate in the n-stage counter (count-up mode), and digital noise will be averaged out (subtracted). The output of the counter is sampled at the end of the data-collection interval.

If the contents of the counter should ever decrease to zero, a borrow pulse is generated by the most significant counter, C_n, and toggles the flip-flop. In turn, the NAND gate connected to the clock-up port of C_1 is enabled. In this way, the output of the counter is mirrored about 0, as shown in the timing diagram, and is equal to the absolute value of the difference between the number of the add-to and the subtract-from counts. The sign bit is set each time the data actually drops below zero.

Note that when using the up-down counter, it is necessary to keep the counter cleared while the $\overline{\text{BORROW}}$ is low to avoid difficulties should the clock-up line suddenly become disabled while the clock-down line is activated. Otherwise, a decrement will take place. The frequency and symmetry of the ADD/$\overline{\text{SUB}}$ waveform is arbitrary so long as the counter does not overflow on any half cycle of a sampling interval. □

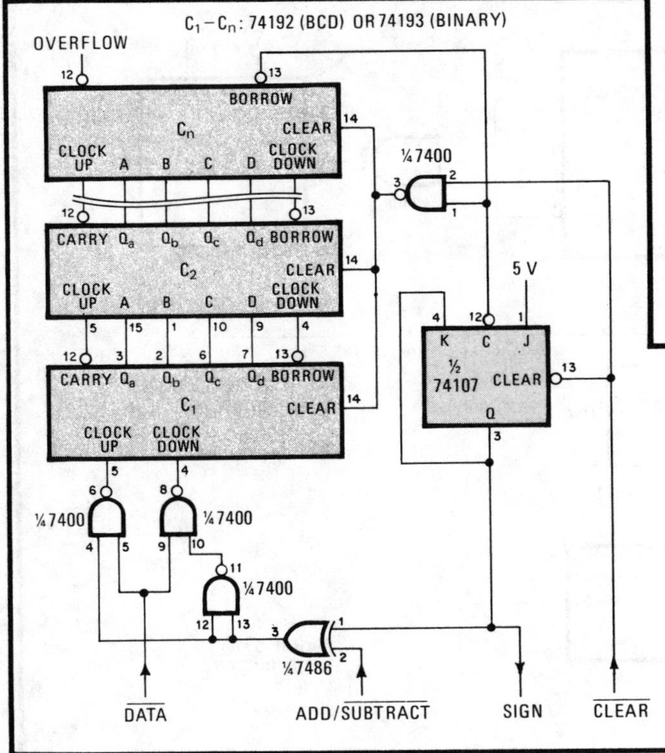

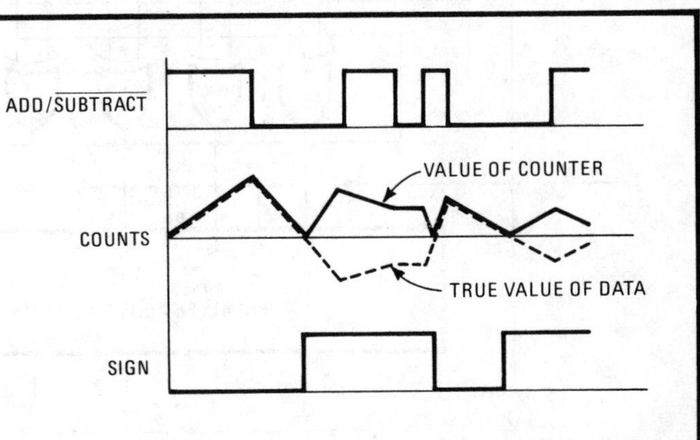

About face. Circuit inhibits up-down counters C_1–C_n from decrementing below zero, instead forcing them to count up and mirror the result of a negative-number addition in binary form. A negative-sign bit is also generated. Thus, all numbers are suitable for direct handling by a d-a converter or other data-system device.

Counter delivers data in signed or complemented form

by N. Bhaskara Rao
U.V.C.E., Department of Electrical Engineering, Bangalore, India

Adding one chip to the basic digital-averaging circuit proposed by Frazier [see previous article] forms a data counter whose outputs express a negative number not only by its magnitude and sign, but by its 2's complement as well. Expressing numbers in the latter form enhances the circuit's usefulness by allowing direct interfacing with computer circuits.

The circuit shown works differently from Frazier's, which generates a sign bit and reverses the direction of the count when the counter is about to move down through zero. Instead, the 74193 is allowed to go below zero, where its R output becomes 15 and the state of its borrow output changes, thereby clocking the 7470 flip-flop. The 7470's Q output, which is the sign bit, S, then moves high.

The sign bit and the R bits are applied to the 7483 4-bit full adder (not part of the original circuit) through the 7486 exclusive-OR gates. The 7483 and 7486 together form an add-or-subtract unit. As a result, the output of the adder, M, will be equal to R if the 74193 indicates it holds a positive count (S = 0), and will be equal to 16 − R if the 74193 has gone negative (S = 1).

If S = 0, the R bits, as seen at the output of the 74193, will represent the binary equivalent of the number. Any negative number will be represented by its 2's complement value and an enabled borrow bit. □

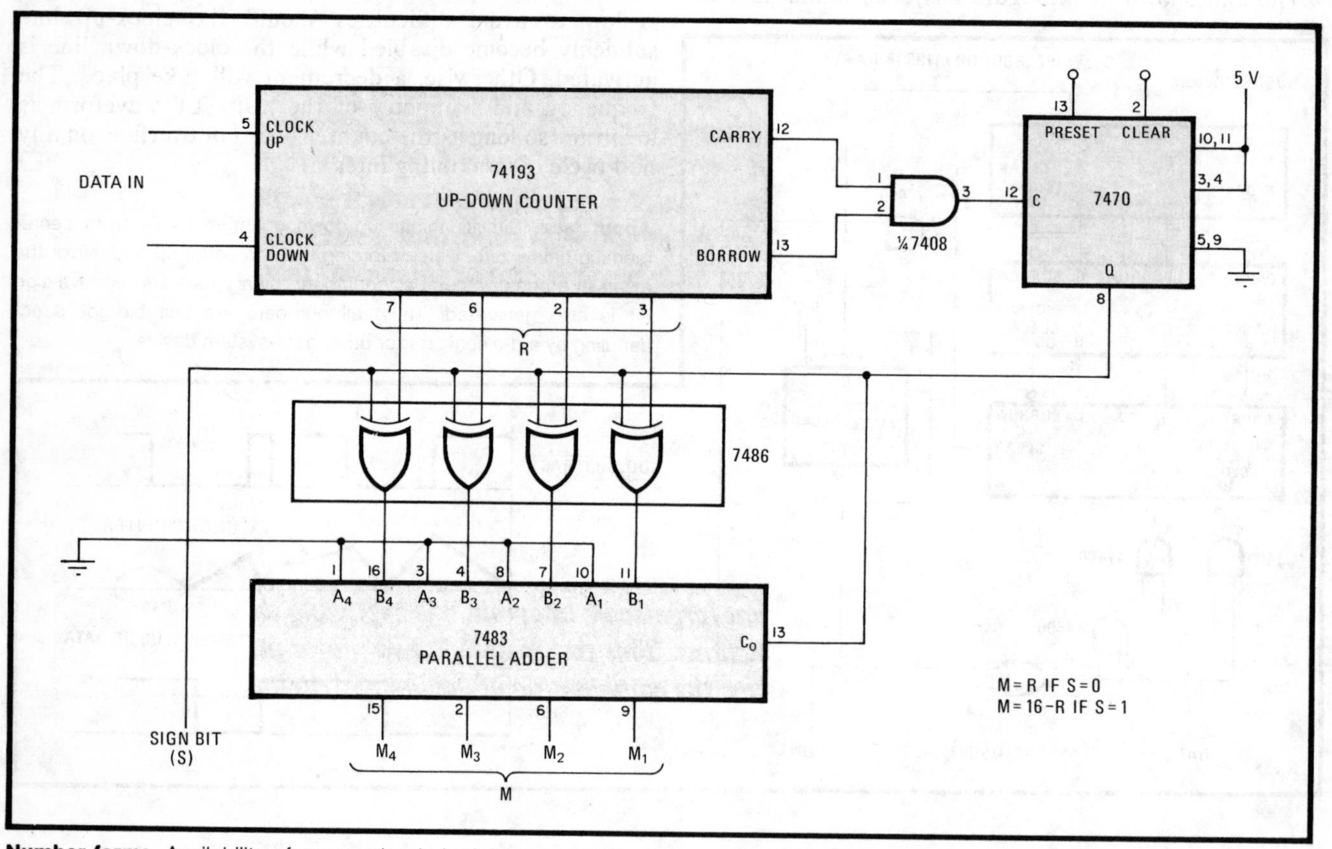

Number forms. Availability of any number in both binary equivalent or magnitude-sign forms enhance circuit's interfacing capability. Data generated by 74193, expressed as binary number (R bits), is converted into magnitude-sign form by add-or-subtract unit 7483–7486. R represents the 2's complement of any negative number when counter steps down through zero.

Prescaler and LSI chip form 135-MHz counter

by Gary McClellan
La Habra, Calif.

Combining a prescaler designed for very high frequencies with large-scale integrated circuits and a few other devices builds a multifunction frequency counter capable of working at 135 megahertz. The counter, which uses complementary-metal-oxide-semiconductor and emitter-coupled-logic chips, has many desirable qualities, including the ability to measure the period of a waveform, moderate power consumption (100 milliamperes, including displays), good sensitivity (16 millivolts at 135 MHz), and portability.

In the frequency mode (selected by switch S_1), signals at the input are limited by the resistance-capacitance network and two diodes (see figure) so as to prevent overloading of A_1, the National DS-8629N prescaler. A_1, amplifies the signal and divides it by 100, and then it is counted by A_2, the Intersil ICL-7216B.

The ICL-7216B contains a counter, a display multiplexer, a seven-segment decoder, and digit and segment circuits for driving A_3, the HP 5082-7441 display. The chip also provides timing for the system, including the necessary oscillator circuitry and frequency dividers to generate the gate, latch, and reset pulses for multiplexing the display and controlling the sampling interval (selected by S_2).

In the period mode, input signals are limited in order to protect Q_1, an impedance converter. A simple preamp and Schmitt trigger, A_3, converts the signal to appropriate levels for A_2.

The counter also has provision for an external oscillator input. When properly used with an external 10-MHz standard, it makes measurements with a high degree of accuracy. S_3 enables the measurement.

Calibration is easy. A signal of known frequency is connected to the frequency input (a 100-MHz signal is ideal), and the 15-picofarad capacitor, which is in parallel with crystal Y_1, is adjusted for a matching counter reading. Accuracy is not greatly affected by the supply

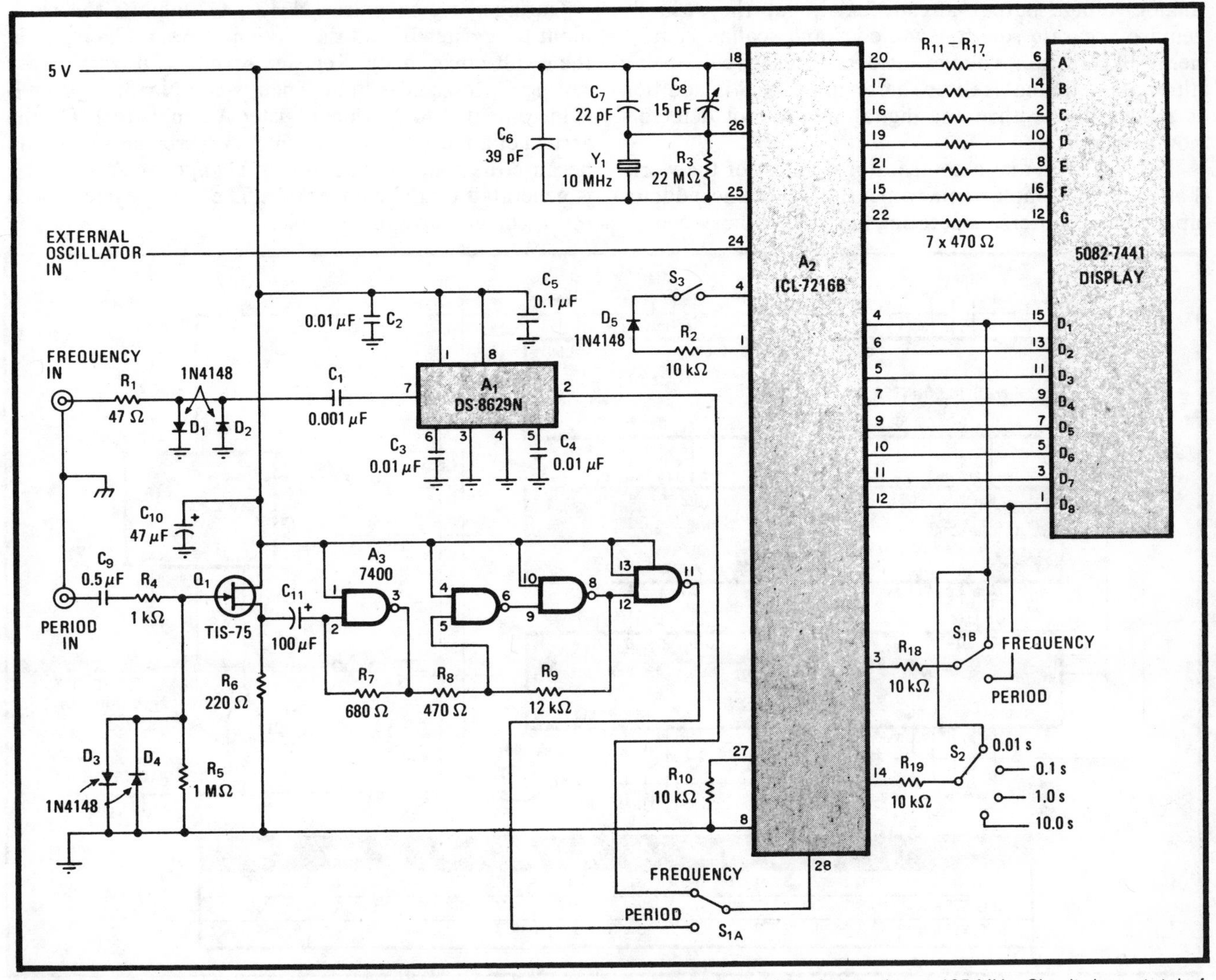

Counting high. Three-chip circuit uses prescaler and LSI chip in frequency counter capable of operating at 135 MHz. Circuit draws total of 100 milliamperes, has good sensitivity, is portable. Counter should be built on double-sided pc board for best performance.

voltage, as the counter holds to within two counts of the displayed frequency over 4.5 to 5.5 volts.

The counter is best built on a double-sided printed-circuit board. Parts layout is not critical, with the exception of the input leads, which should be positioned away from the display. Also, A_1 should have foil running on its underside, to act both as a heatsink and as shielding. □

Square-root counter calculates digitally

by S. H. Tsao
National Resarch Council, Ottawa, Canada

A square-root counter has many uses, notably the measurement of the root-mean-square output voltage of a transducer that has a square-law response. This counter finds the square root digitally, rounded off to the nearest bit, of any analog voltage that has been converted into a digital train of N pulses by a voltage-to-frequency converter. In other words, the circuit finds the square root of N.

Of course, square-root circuits that generate an analog-voltage output already exist, but they usually require accurate square-law diodes and scaling components like resistor voltage dividers. This circuit, being all-digital, eliminates those shortcomings and in addition is much simpler than the digital square-root detectors now available.

Two registers and a comparator do most of the work. Four 4029 complementary-metal-oxide-semiconductor up/down counters, operating in the binary-coded-decimal mode, form the two synchronously clocked registers, A and B, as shown in the figure. Register A clocks up or down, depending on the state of the C-MOS 4027 flip-flop. Register B advances by one count any time an input clock pulse is received and the carry-out signal ($\overline{C}_o$) of register A is active, or low. The negative-going transitions of the input signal clocks both the flip-flop and B, while positive transitions clock A.

Initially, the flip-flop is reset and both registers are cleared by means of the JAM lines. After the clock pulse ($N=1$), register B advances to state 1 and the flip flop is set, placing register A in the count-up mode. Register A then advances to 1 and $\overline{C}_o$ moves high, setting the J input of the flip-flop low.

The C-MOS 4063 comparator detects that the contents of registers A and B are now equal (as they were before, at initialization, and it brings the K input of the flip-flop high.

During the next pulse ($N=2$), G_1 inhibits the clock input to register B, and the flip-flop is reset, placing A in the count-down mode. Register A returns to zero, again setting $\overline{C}_o$ low and J high. Then, when $N=3$, register B is incremented to 2, while register A counts to 1. On the arrival of a pulse ($N=4$), counter A increments to 2, and because the contents of B do not change, an $A=B$ pulse is generated by the comparator. The entire cycle is then repeated, as shown in the table.

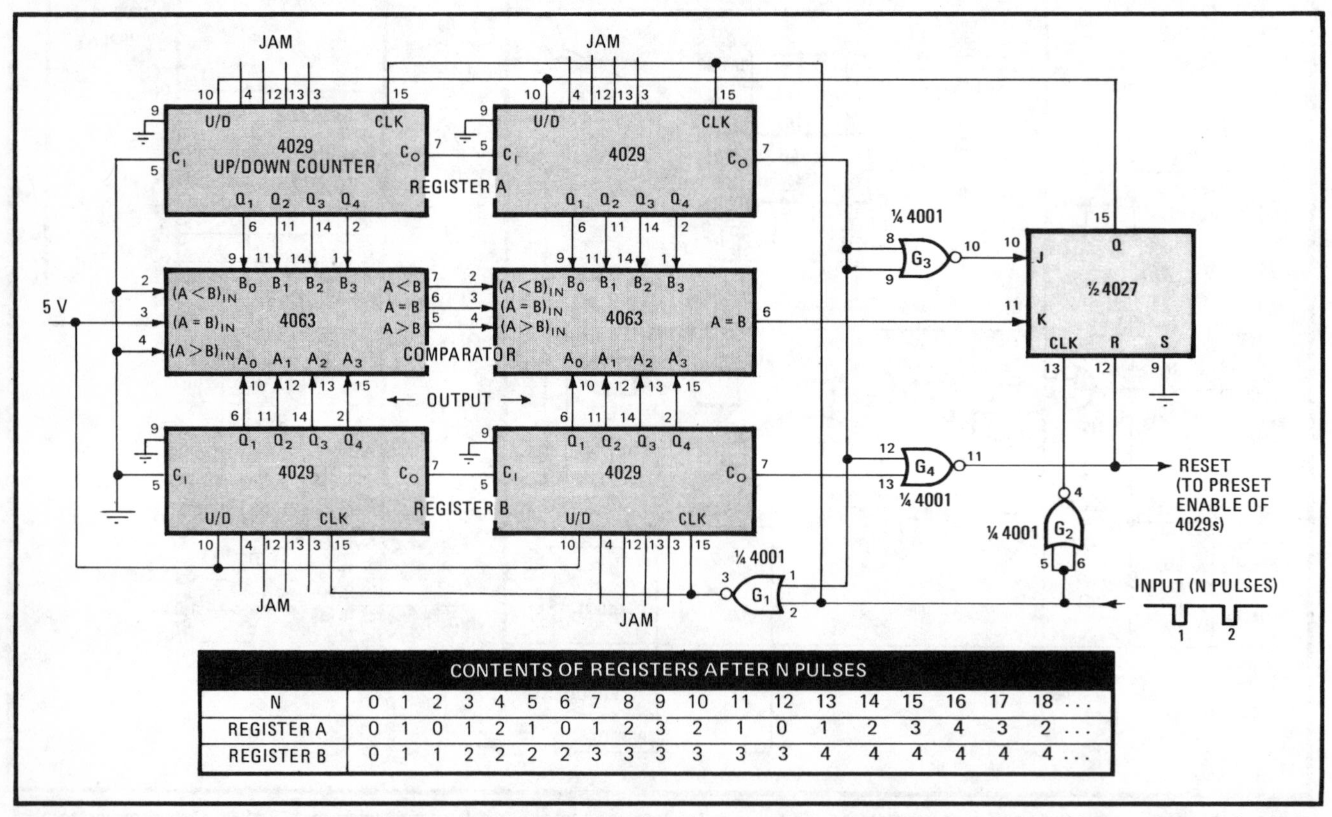

CONTENTS OF REGISTERS AFTER N PULSES																			
N	0	1	2	3	4	5	6	7	8	9	10	11	12	13	14	15	16	17	18 . . .
REGISTER A	0	1	0	1	2	1	0	1	2	3	2	1	0	1	2	3	4	3	2 . . .
REGISTER B	0	1	1	2	2	2	2	3	3	3	3	3	3	4	4	4	4	4	4 . . .

$N^{1/2}$. Register B in the circuit yields the square root of N, rounded off to the nearest bit, where N is the number of input pulses received after the circuit has been reset to zero. Digital circuit eliminates shortcomings of detectors that measure analog voltages directly.

Thus, register A starts an up count one cycle after it has reached 0, and initiates a down count one cycle after it has determined its contents are equal to B. As a result, the output of register $B = N^{1/2}$.

The reset line is activated by G_4 whenever both registers overflow, and this sets both registers to their initial values. It can also be activated regularly by an appropriate timebase for time-averaging measurements. For special applications, the registers may be preset to a nonzero number via their JAM lines. The registers may also operate in the binary mode without their square-root counting function being affected, while the counters may easily be extended beyond their 8-bit capacity, simply by being cascaded. □

Chapter 10
CURRENT SOURCES

Split current source damps reactive load oscillations

by Yishay Netzer
Haifa, Israel

A standard bilateral current source of the type shown in the first part of the figure (a) will often generate oscillations in circuit loads that are grounded and have an impedance (Z_L) that is not purely resistive. Inductive loads such as cathode-ray-tube deflection yokes and torque motors are best driven by the modified circuit shown in the second part of the figure (b). Adding the differential amplifier and feedback network to the circuit eliminates undesirable responses while ensuring that the output current will be virtually independent of the load impedance.

When the load is reactive, it may cause the circuit's step response to be underdamped and consequently unstable, with the result that the output current will become dependent on the load impedance. As shown in the equation in (a), the output current is dependent on the circuit transfer function, which is:

$$G(s) = K\frac{\omega_o^2}{s^2 + 2\zeta\omega_o s + w_o^2}$$

where ω_o is the natural undamped frequency of the circuit, K is a constant, and ζ is the damping factor.

The various parameters are determined by circuit constants R, R_S, and Z_L. Of particular importance is the fact that once determined by the circuit configuration, ζ cannot be modified, and that is why oscillations can result. Furthermore, the oscillations may be impossible to eliminate in the standard circuit because adding components may affect the output impedance, making any type of compensation impractical.

The circuit in (b) circumvents the problem by splitting the current source into two parts:

- A balanced difference amplifier (A_1), which converts the load current into a single-ended voltage feedback signal.
- A power amplifier (A_2), which, aside from assuming its original function, reduces the effect of Z_L upon I_L by making use of the feedback voltage.

Note that by adding the feedback network, resistor R_C has been introduced to the circuit, and therein lies the major advantage of this circuit. R_C can vary (compensate) ζ right down to its optimum value ($2^{1/2}$ in this case), without disturbing the proportional relation of V_{in} to I_L throughout the useful range of the circuit; that is, below ω_o. The basic transfer function of the circuit is not altered by the modified configuration, either.

The value of R_C is best found experimentally while observing the circuit response to a square-wave input. Note that the constant, K, in G(s) will be slightly smaller than its original value because of the feedback currents through R_C. □

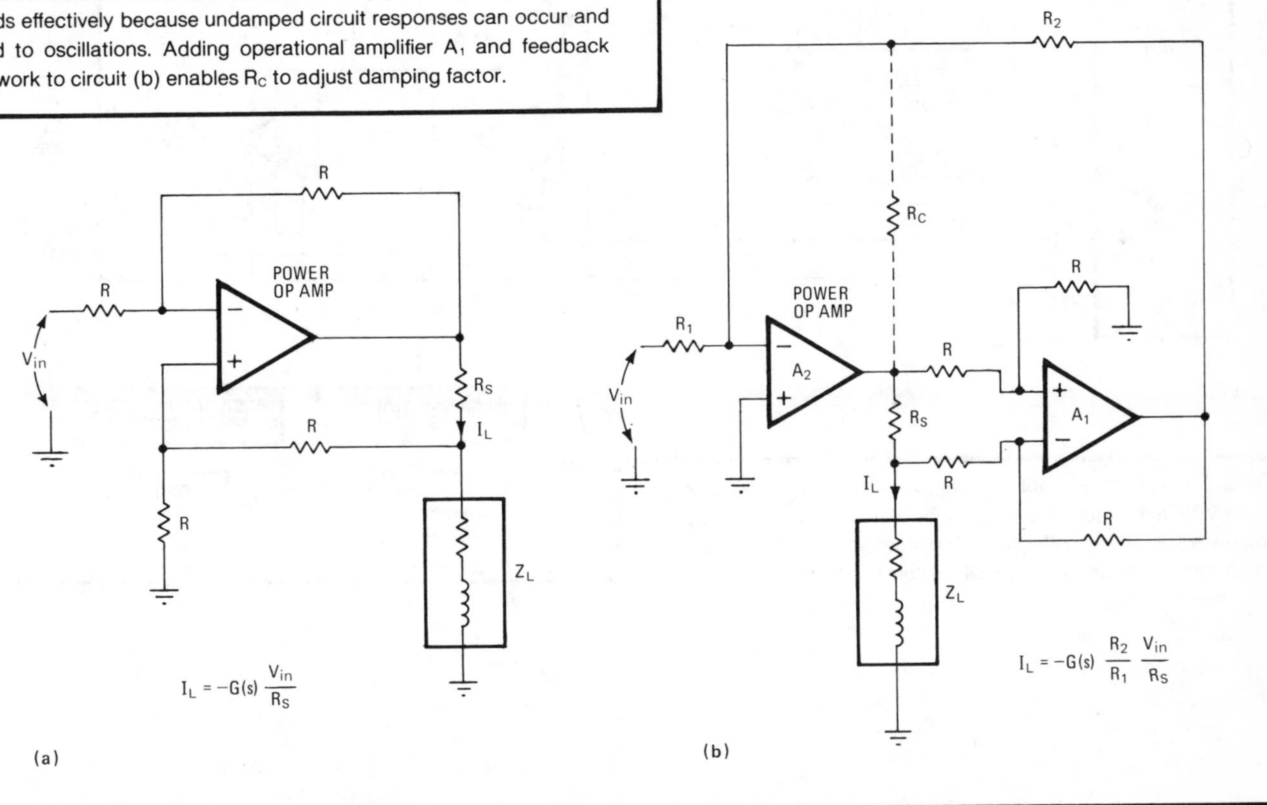

No oscillations. Standard current source (a) cannot drive inductive loads effectively because undamped circuit responses can occur and lead to oscillations. Adding operational amplifier A_1 and feedback network to circuit (b) enables R_C to adjust damping factor.

Floating current source drives automatized test fixture

by Richard M. Fisher
ADT Security Systems, Clifton, N. J.

This generator provides a programmable current to drive any load, making the unit ideal for production-line testing. Because the constant-current source floats—that is, is not connected to ground—it can drive loads energized either by positive or by negative potentials of as much as 90 volts.

The output current is resolved to 50 microamperes by the 10-bit input to a digital-to-analog converter (a). The maximum current that can be delivered to the load is slightly more than 50 milliamperes.

As shown, the 10-bit command input is transferred to the d-a device through optocouplers, thus isolating the DAC-10Z from ground paths under virtually all conditions. Note the 5-, +15-, and −15-V potentials for the generator are obtained from circuitry associated with the isolated secondary winding of the transformer in the power supply.

Operational amplifier A_1 inverts and scales the output of the d-a converter. The maximum output voltage from the converter is −9.99 volts and results in a full-scale output voltage of 5.115 V from A_1. A_2, in conjunction with R_1 (= 100 ohms), thus provides a full-scale output current of 51.150 mA.

The V-groove MOS field-effect transistor, Q_1, serves as a voltage-to-current converter. Q_1 performs the conversion at high accuracy, because the V-MOS device requires no gate current.

As for using the current source, implementation is easy with any energizing potential. If the device—the load—under test is driven by a positive voltage (b), it is necessary to connect the generator's positive output to the supply voltage. The negative port of the generator is brought to the load.

For negative potentials, the situation is similar, with source's positive terminal being connected to the load as shown. The negative port is connected to the supply voltage. □

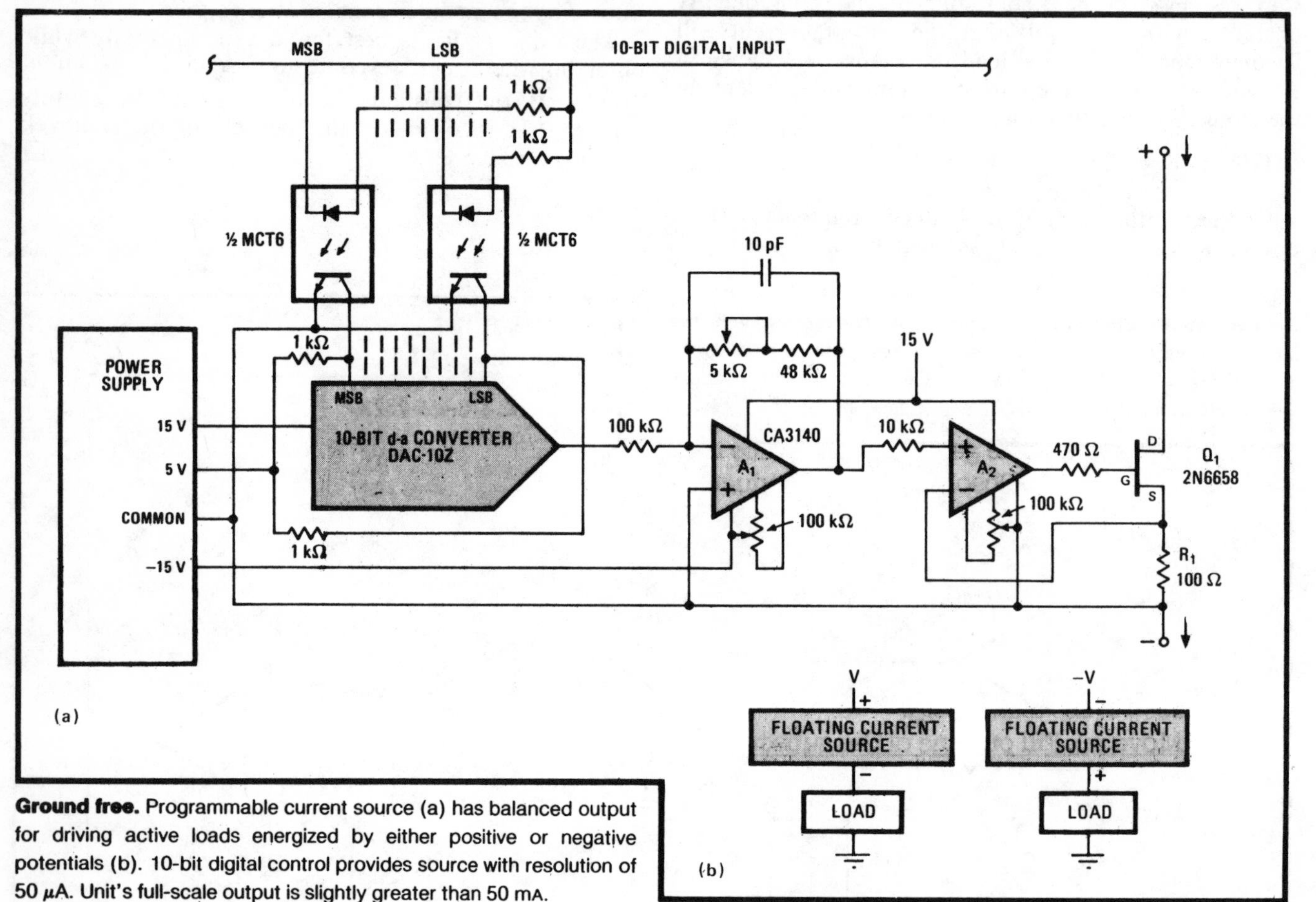

Ground free. Programmable current source (a) has balanced output for driving active loads energized by either positive or negative potentials (b). 10-bit digital control provides source with resolution of 50 μA. Unit's full-scale output is slightly greater than 50 mA.

Current source for I^2L saves energy

by Stephen H. Nussbaum
Data/Wave Development Inc., San Diego, Calif.

To capitalize on the low-power advantages of integrated injection logic (I^2L), a power source that also dissipates relatively small amounts of energy is required. This switched-mode supply provides programmable currents of up to 300 milliamperes at 2.3 volts to boards utilizing I^2L loads, with an overhead of only a few milliamperes needed for running the circuit.

The voltage-current characteristics of I^2L devices resemble those of the standard switching diode, whose operation is determined by the amount of driving current available. It is therefore necessary to drive these loads with a current source. Although a single high-value resistor in series with a voltage source would serve to deliver constant current, large amounts of power would be dissipated in the resistor. The difficulty is overcome with this circuit.

Q_1 and its associated components provide a reference current for the complementary-MOS quad analog switch, A_1, in the reference-resistance subcircuit. The R_2C_3 combination helps to stabilize the output against changes in input voltages.

A_1's switches are wired together such that its equivalent series resistance may be set to one of two values by a control signal. It is possible to order as many as five current levels with this switch if additional programming inputs are introduced.

A_1, with the aid of R_3, serves partly as a current-to-voltage converter, so that low-power oscillator A_2 sees the reference current as a representative voltage at its inverting input. This potential will cause Q_2 to switch on periodically. R_4 provides positive feedback for hysteresis, thus controlling the rate at which A_2 and Q_2 are switched—16 kilohertz, in this case. The 10 to 30 millivolts of hysteresis also appears at the output, but this poses no problem with I^2L loads.

L_1 and C_1 comprise the switcher's required storage elements, acting to release energy to the load through R_{sense} when Q_2 is off. R_{sense} is part of a feedback network used to set I_{out}.

Because the reference current and the output current at summed at the output node, A_2's input sees only the difference of these currents scaled to a voltage by their respective resistors, R_{sense} and R_{ref}. Thus the output current is set solely by the feedback loop. As a consequence of this arrangement, $I_{out} \approx I_{ref}/R_{sense}$. The efficiency of the supply is maximized by using a lower value of R_{sense}, a faster op amp for A_2, and a storage inductor (L_1) with as little dc resistance as possible.

With R_3 = 220 ohms and with I_{ref} = 0.60 mA, I_{out} = 220 mA if a logic 1 is applied to the control input. I_{out} = 300 mA for a logic 0. These values can be changed by suitable selection of I_{ref}, of course, but R_3 may also be varied. Note that:

$$R_{ref} = [(r_{on}/n)R_3]/[(r_{on}/n) + R_3]$$

where n = number of switches and r_{on} = on-state resistance of one switch in A_1, typically 600 Ω. □

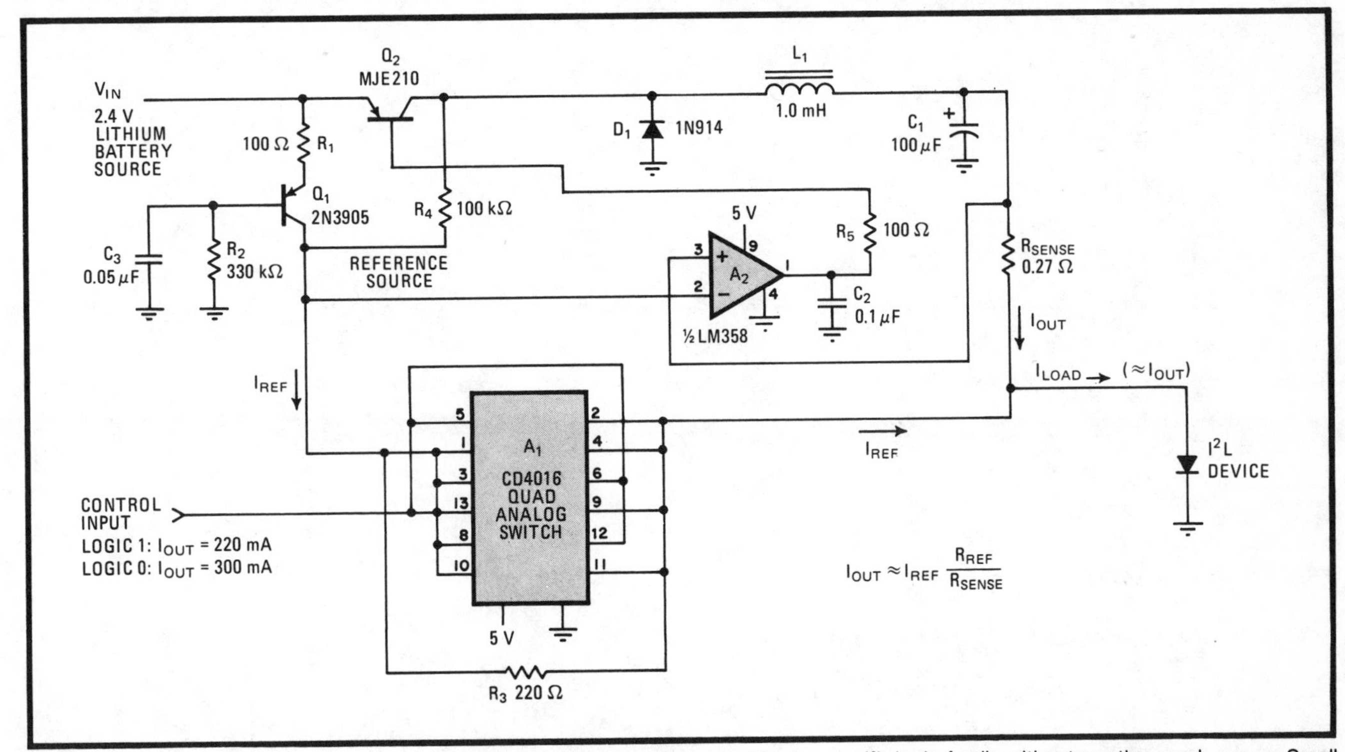

Injecting current. Switching source delivers constant current to members of low-power I^2L logic family without wasting much power. Small reference current, C-MOS switches, and low-power oscillator contribute to circuit efficiency. Two-level current source, which generates up to 300 milliamperes at 2.3 volts, can provide one of five current values if additional programming inputs are introduced at switch A_1.

Chapter 11
DETECTORS

Fast-attack detector optimizes ultrasonic receiver response

by Paul M. Gammell
Jet Propulsion Laboratory, Pasadena, Calif.

A radio-frequency amplifier with low output impedance greatly enhances the performance of an ultrasonic receiver. The low impedance, achieved by an open-collector output configuration, permits design of a detector having a fast rise time and somewhat slower fall time—characteristics important for high-accuracy distance-measuring applications. This inexpensive circuit is a wideband receiver with good overload-signal recovery. It has many pulse-echo uses, including nondestructive evaluation and depth finding, and performs well in biomedical ultrasonic applications.

As shown in the figure, a pulsed signal is simultaneously applied to a transducer and transistor amplifier Q_1. The transducer is one of many commercially available devices that will convert a typical 200-volt, 0.1-to-1-microsecond pulse into an ultrasonic (compressional) wave aimed at a distant target. The echo from the transducer, with an amplitude on the order of 1 to 10 microvolts, may return only microseconds later. Q_1 must therefore recover quickly from the large initial pulse in order to respond to the echo signal.

Components C_1, C_2, and D_1 through D_4 aid in isolating the amplifier circuits from the high-voltage pulse. Resistor R_1 improves the transient recovery by draining off any charge present on C_1 and C_2 that remains because D_3 and D_4 have not sufficiently bypassed the pulse to ground.

Q_1 is saturated by the initial pulse. Diodes D_1 and D_2 isolate the pulsed current from Q_1 (and the transducer) after the pulse drops below 0.7 v, thereby keeping the nonthermal noise contributions of the pulser out of the receiver. Q_1 drives the MC1350 radio-frequency/intermediate-frequency amplifier with a phase-split signal to increase the effective gain of the circuit and aids in isolating it from the pulser. To achieve a circuit gain of more than 50 decibels, a second MC1350 amplifier is added.

The output of the second amplifier drives a diode detector composed of D_5 and D_6 and the filter $R_{10}C_3$. Fullwave rectification is essential for optimum resolution, since the echo may be shifted 180° by the reflecting surface producing it.

A simplified equivalent circuit of the output stage of the MC1350 is shown at the bottom of the figure. A fast rise time for incoming signals is achieved by discharging C_3 through the 1-kilohm impedance of one of the output transistors. The longer fall time needed for a smooth

Pulse-echo receiver. Circuit (a) responds to wideband radio-frequency signals with good signal-handling capability. Time-dependent gain control is provided for sophisticated sonar applications. Choice of low-impedance rf amplifier (b) optimizes response.

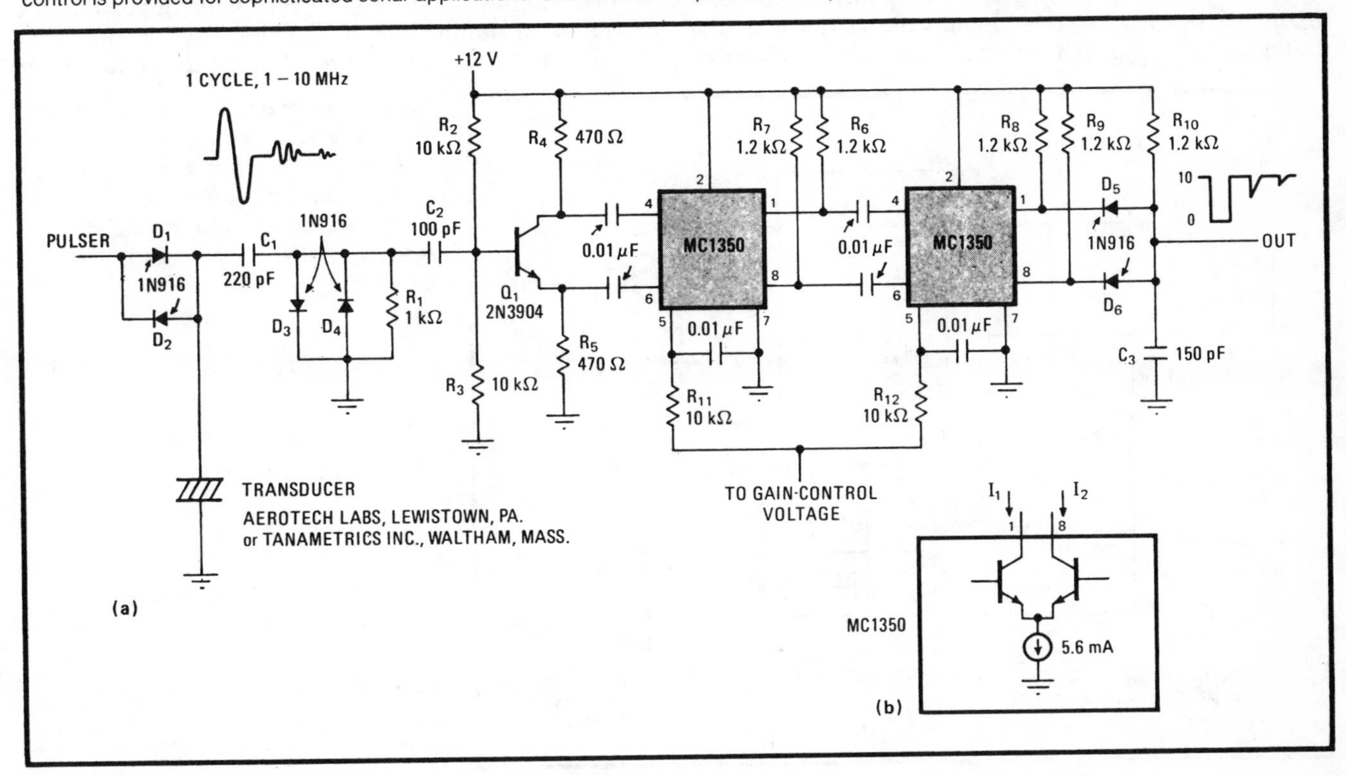

echo envelope is attained by recharging C_3 through R_{10}. In some applications, it is desirable to provide a control to vary the fall time by adjusting C_3 and to reject echoes of small amplitude by varying R_8, R_9, and R_{10}.

At small signal levels, an approximate square-law response is provided by the logarithmic characteristic of the 1N916 diodes. The sum of currents I_1 and I_2 in the MC1350 is 5.6 milliamperes typical, as specified by the manufacturer. The currents determine the operating point of the diodes. With the circuit values shown, a quiescent current of approximately 0.8 mA flows through each of the diodes D_5 and D_6.

Although integrated circuits are widely available that perform low-level detection, the desired detector characteristics can be more easily achieved with the circuit described. Furthermore, a stable (oscilloscope) baseline and a large dynamic range are easier to attain if the detector is driven by a reasonably high rf voltage source. The amplifiers in this circuit make this possible by providing adequate predetection amplification.

The signal at the output has a well-defined leading edge suitable for determining time differences between the pulse and its echo with an oscilloscope. A 10-v offset occurs at the output. It may be removed with a base-line restorer circuit consisting of a decoupling capacitor, a resistor tied to 12 v, and a germanium diode to clamp the baseline to about 0.3 v. Another approach uses a differential amplifier.

For sophisticated systems, a time-dependent gain control can be added to the rf amplifiers. A 0-to-12-v ramp voltage, the amplitude of which depends on the range and anticipated attenuation of the echo, can be applied to pin 5 of the devices through R_{11} and R_{12}, which control the distribution of stage gain. The ramp voltage should be positive and decrease with time to provide a gain that increases with time. If pin 5 is grounded, the amplifier operates at full gain. Although the ramp is usually synchronized with the transmit pulse, it may be synchronized with other sources, such as the surface echo from an attenuating target. □

Biasing the diode improves a-m detector performance

by Antonio L. Eguizabal
Vancouver, British Columbia, Canada

The sensitivity, dynamic range, and linearity of a standard amplitude-modulation diode detector improve if the diode is biased into its conducting region. Sensitivity and dynamic range increase because the incoming radio-frequency signal does not encounter the barrier-potential voltage of the diode before the onset of rectification. Linearity improves because the biasing voltage shifts the operating point of the diode into the linear portion of its characteristic curve.

As shown in the figure, signals are applied to D_1, a Schottky diode that can be used at frequencies approaching 1 gigahertz. It has a low barrier potential (0.35 volt), about one half that of conventional silicon diodes, and allows linear operation at a lower biasing voltage than is possible with conventional diodes.

A 5.6-v zener and the voltage divider composed of the 10-kilohm resistor and D_1 send 350 mv across D_1, making it conduct. The voltage drop across D_1 is in effect eliminated for rf signals, making possible the detection of millivolt-level signals. The demodulated signal appears across the resistance-capacitance filter

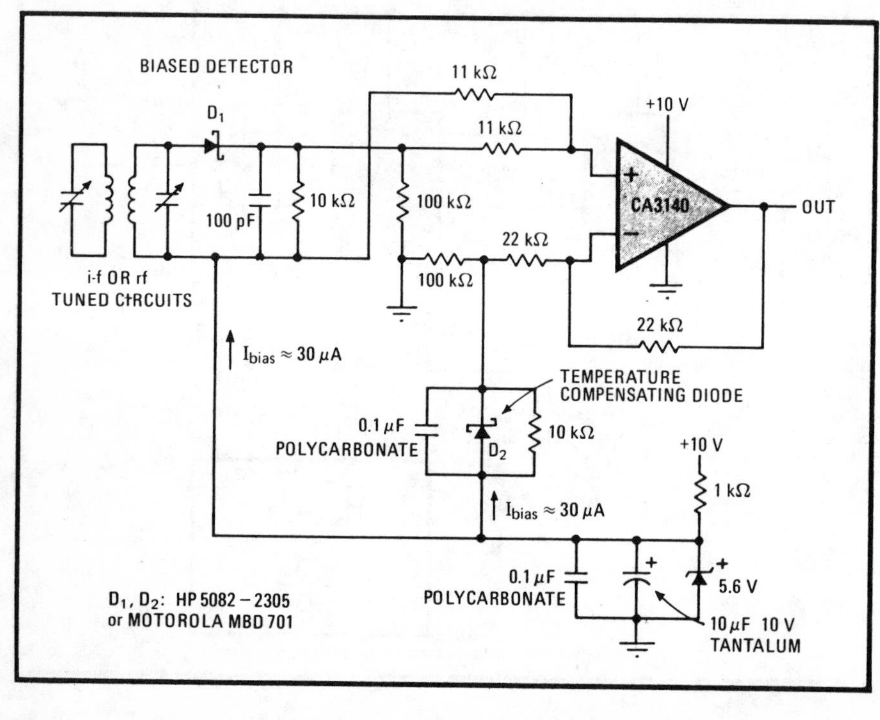

Tweaked detector. Even the old reliable a-m detector can be improved. Increased sensitivity, linearity, and dynamic range are obtained by dc biasing diode D_1 into conduction, to eliminate voltage drop as seen by rf driver. D_2 provides temperature compensation.

composed of the 100-picofarad capacitor and the 10-kilohm resistor.

Most diode detectors are used in conjunction with automatic-gain-control circuits to produce a constant audio output over a wide dynamic range. To ensure minimum response time of the diode detector, which is required when driving the agc, direct coupling between the two is necessary. Temperature compensation is therefore needed to eliminate drift caused by small temperature-dependent offset voltages.

A second Schottky diode and an operational amplifier, the CA3140, provide this function. Diode D_2 is biased similarly to D_1, with the result that voltage changes caused by temperature variations are almost identical across both devices. The CA3140 operates as a unity-gain wideband amplifier. It has a high common-mode rejection ratio and draws low input-biasing currents, qualities required in a good buffer amplifier.

The noninverting port of the op amp accepts the demodulated signal from the RC filter. The temperature-variable voltage drop across D_1 tends to increase the output voltage, but this is countered by an identical voltage across D_2 at the inverting port. Thus the net voltage change at the output as a function of temperature is approximately zero. Note that the op amp is operating with a single-ended power supply, which assures that D_2 is used in a temperature-compensating capacity only and does not upset the operation of the rf detection circuit.

This detection circuit can respond to a 10-mv rf input signal, and its dynamic range is 10 to 15 decibels greater than that of conventional diode detectors. In addition, the linearity of the circuit, especially at low levels, is noticeably better. □

Maximum voltage detector needs no a-d conversion

by Ronald Lumia
University of Virginia, Charlottsville, Va.

A detector that determines which of a set of analog voltages has the greatest positive value is useful for pattern-recognition systems and other classifying schemes. Only the relative magnitude of the input voltages is important in these applications, so that costly analog-to-digital converters are not needed.

As shown in the figure, a sample set of voltages is introduced at the noninverting port of a bank of 711 dual differential voltage comparators, which initially switch each amplifier high. All inverting inputs are driven by any monotonically increasing waveform, such as a ramp voltage.

Each comparator switches into the low state as the ramp voltage exceeds the particular sample voltage connected to its input, until only one comparator remains in the high state. The signal at this op amp input has the greatest amplitude in the sample set.

The combinational logic at the 7430 NAND gate array then generates a pulse to the 74174 device, clocking the lone logic 1 signal from the op amp input into the D input of its associated flip-flop and presenting it to the processing device. By this time, the ramp voltage has returned to its minimum value, and the next sample set may be again introduced.

As shown, the range over which the sample-set voltages may be detected lies between zero and 5 volts. This range may be changed by suitable adjustment of the ramp and supply voltages to the comparators. The use of complementary-metal-oxide-semiconductor logic circuits is advised when the operating voltage to the comparators exceeds 5 volts. □

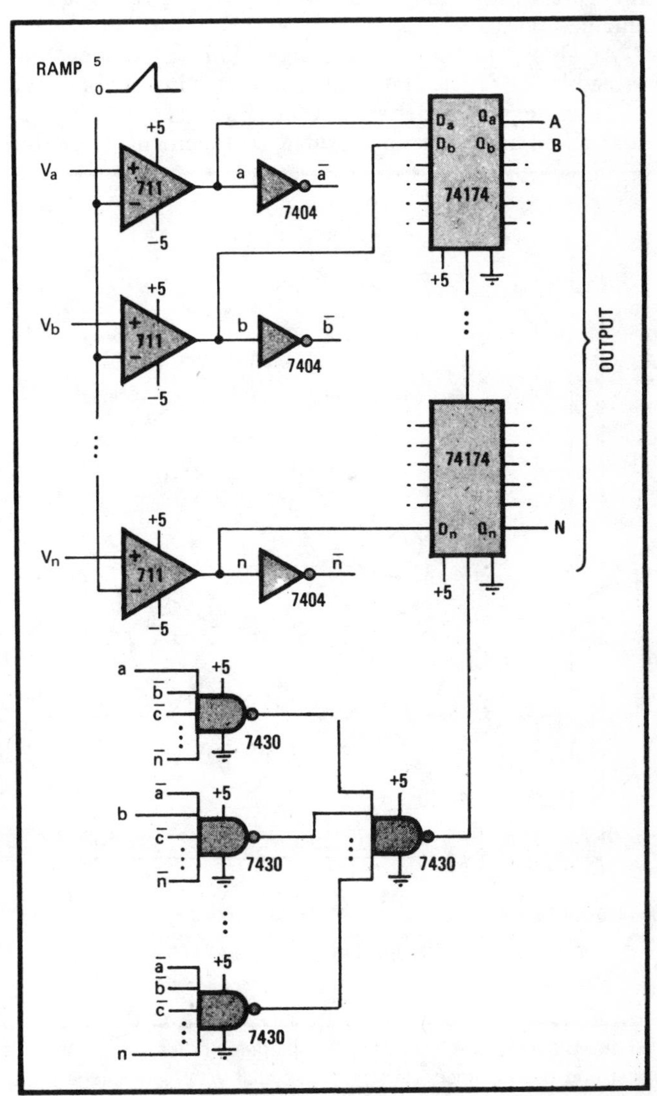

Maximum-voltage detector. Relative maximum of sample set $V_1 - V_n$ is determined by ramp generator and logic circuit. Sample voltages drive all comparators high if values lie above the minimum ramp voltage. Logic detects lone op amp remaining high during the ramp sweep and clocks that state to its D flip-flop.

Optoisolators slash cost of three-phase detector

by G. Olivier and G. E. April
Concordia University and Montreal Polytechnic Institute, Canada

Optically coupled isolators replace transformers in a zero-voltage detector for synchronizing the firing of a thyristor in three-phase control applications, making this circuit cheaper, less bulky, and simpler than most competing designs. Moreover, the optoisolators eliminate the need for a low-pass filter, required in standard detectors for eliminating spurious zero-crossings caused by the thyristor's switching transients. They also provide high-voltage isolation and present much lower capacitive coupling to the circuit than a standard transformer, in fact presenting about as low a coupling as double-shielded types.

As shown in the figure, a light-emitting diode (contained in the GE H11A1 optoisolator) is inserted in each of three legs of a delta network. Each LED is wired to four standard diodes in a bridge arrangement, to enable it to respond to both polarities of the power-line input.

During most of the cycle, all phototransistors are on. At times when the voltage between any two lines is within 0.7 volt of zero, however, no current will flow through the LED connected across those lines. Therefore its corresponding phototransistor will be off, causing pin 2 of the 74LS221 one-shot to fire and a phase-identification pulse (P) to be generated twice every cycle.

In the case illustrated, the phototransistors are wired so that a pulse will be generated at the output each time the input voltage, as measured across ϕ_a and ϕ_b, passes through zero. Note that the one-shot should be adjusted so that the trailing edge of the output pulse corresponds to the actual zero-crossing point.

Identification pulses are also generated for all three phases collectively and these can be accessed, if required, at the zero-voltage pulse output, Z. These pulses occur three times as often as P.

Because at least one LED is conducting at any one time, no transient will normally be generated, so no low-pass filter is needed. Furthermore, the phototransistor's slow response of a few microseconds acts to suppress any transients that might occur near the zero-voltage points, thereby increasing the circuit's noise immunity. □

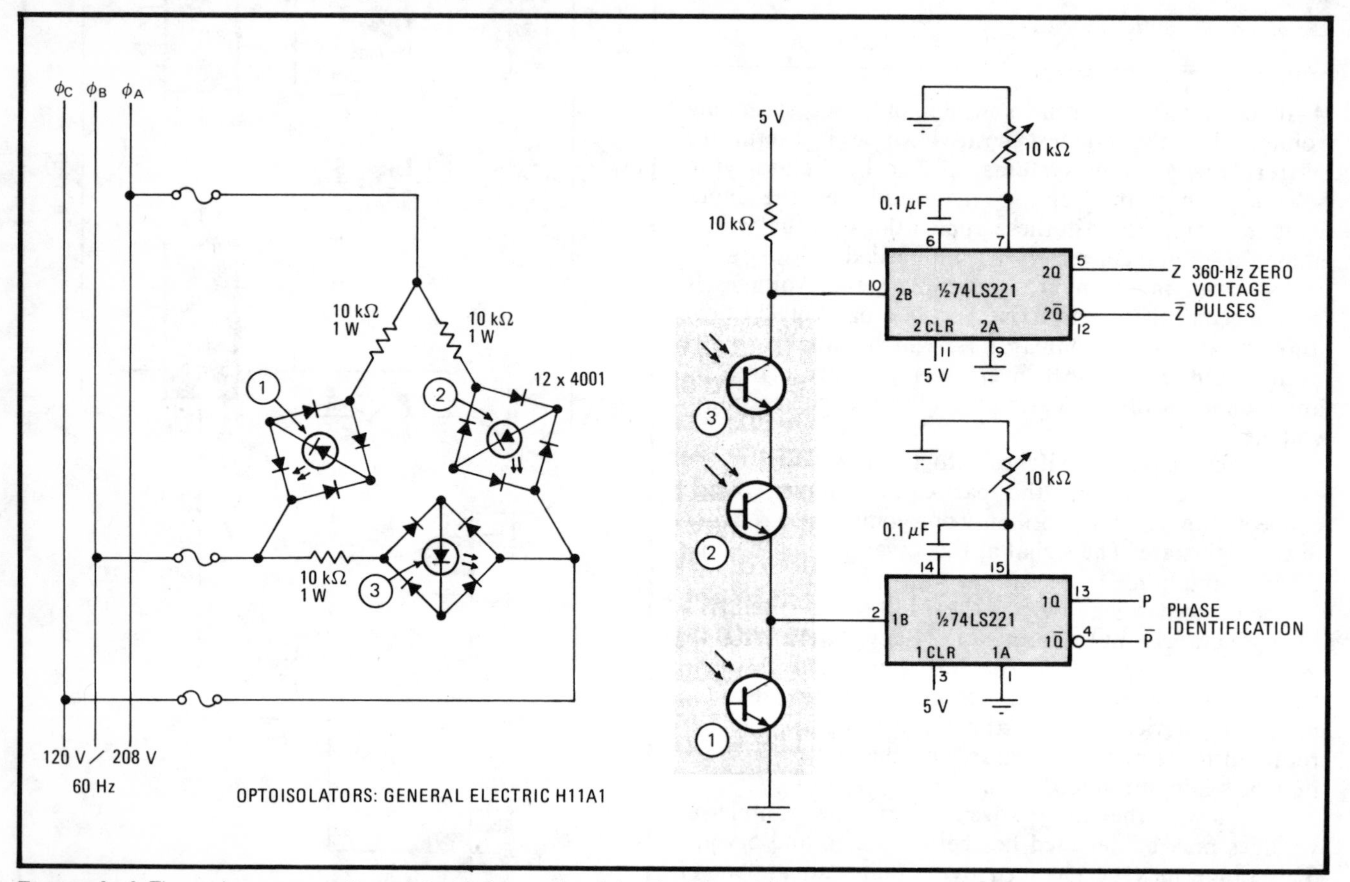

Economical. Three-phase, zero-voltage detector for synchronizing the firing of thyristors uses optoisolators in place of transformers to cut cost and bulk. Optocouplers provide circuit with high-voltage isolation and lower capacitive coupling than transformers.

Peak detector recovers narrow pulses accurately

by Jerome Leiner
Loral Electronic Systems, Yonkers, N. Y.

This peak detector can accurately process input data pulses as narrow as 50 nanoseconds and as high as 3 volts. The recovered voltage is always within 1% of the input signal's true value.

In the circuit shown, emitter-coupled logic generates a −0.2- to −3-volt signal for input into amplifier A_1. Assuming a pulse with a 50-nanosecond width and a rise and fall time of 5 ns, that leaves storage capacitor C_4 only 45 ns in which to charge. The LH0024 op amp used for A_1 has wide bandwidth and a high slew rate to accommodate the fast charging required.

Q_1 acts as a buffer to prevent C_4 from discharging through R_7 between system reset pulses. The voltage at Q_1 appears at Q_2 and is fed back to A_1, to be compared with E_{in}. When E_{out} reaches E_{in}, D_2 becomes back-biased and the stored charge is held until C_4 is intentionally discharged by the reset signal. D_2 remains back-biased during discharge.

A_1 is normally used as an amplifer, and so it will be driven into negative saturation whenever the input signal drops below the output level. D_1 prevents this by clamping the amplifier output.

C_1 and R_2 provide A_1 with input- and feedback-signal stabilization. C_5 compensates for A_1's input capacitance. Note that if C_1R_2 were placed at the output of A_1, a larger charging current would be required for a given input signal. Because this current is usually limited, A_1's effective slew rate would be reduced.

The peak detector is optimized by shorting D_2 and then adjusting C_1, R_2, and C_5 for minimum overshoot and ringing on a series of fast data pulses. □

Fast and precise. Using one op amp, one transistor, and two field-effect transistors, peak detector recovers data pulses having amplitudes of up to 3 volts and widths as narrow as 50 nanoseconds. Output voltage is within 1% of the input data's true value under all signal conditions.

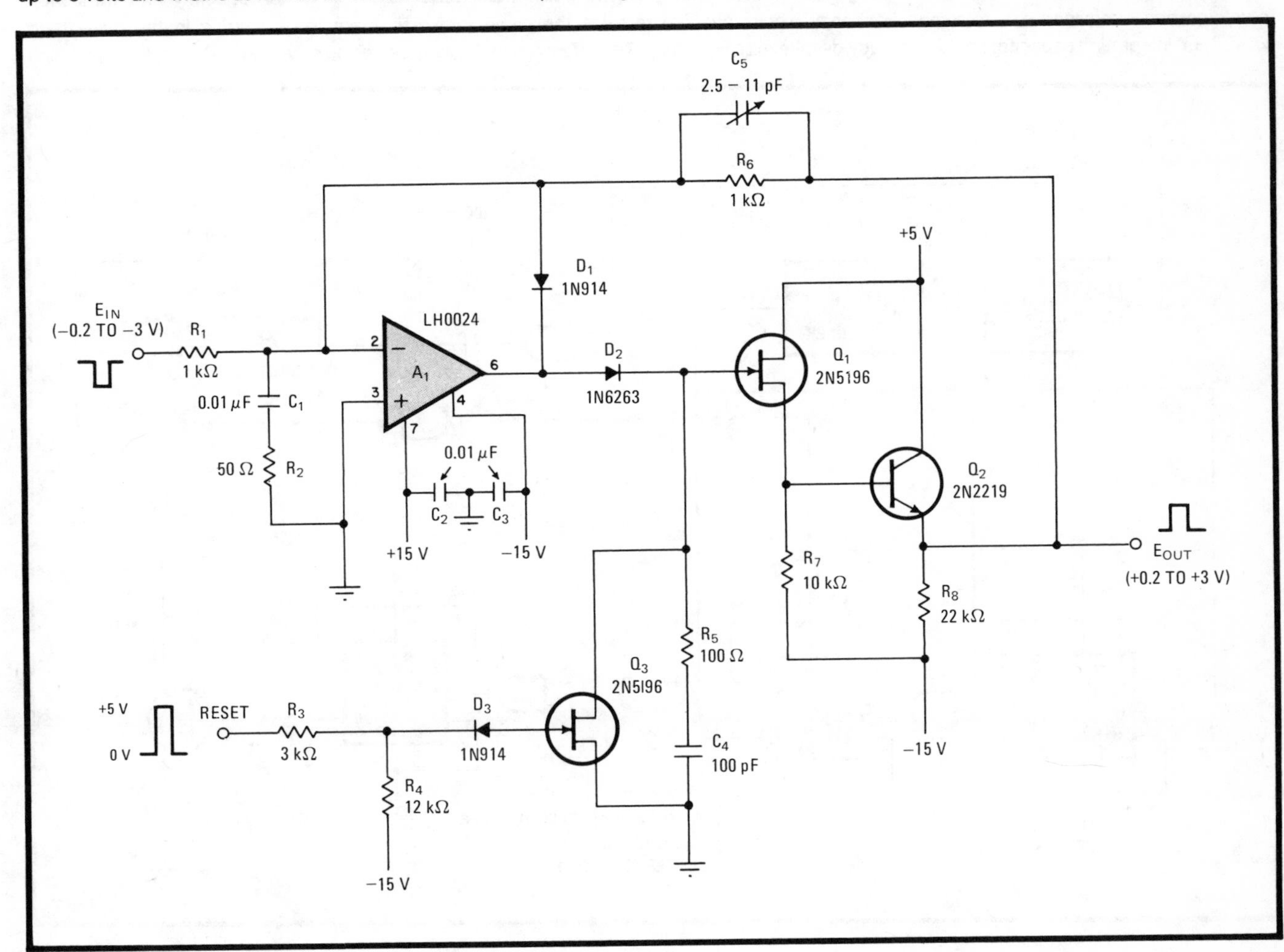

Missing-pulse detector handles variable frequencies

by Joe Lyle and Jerry Titsworth
Bendix Corp., Aircraft Brake and Strut Division, South Bend, Ind.

Virtually all missing-pulse detectors require an input signal of fixed frequency in order to operate satisfactorily. They malfunction when the input frequency varies because their circuits employ detection networks that have a fixed time constant. Through the implementation of inexpensive voltage-to-frequency and frequency-to-voltage converters to derive an average, or reference, frequency that tracks the input signal, this circuit can pinpoint missing pulses without being affected by input frequency variations.

As shown in the figure, A_1 and A_2 establish the reference frequency, f_{ref}, using input frequency f_{in}. Missing pulses do not change the reference because of the integrating capacitors within the converters. Meanwhile, the three NAND gates comprising the one-shot produce pulses of 10 microseconds in duration, with a frequency determined by the input signal.

The chip labeled A_4 is clocked by f_{ref} and A_3 through a NAND gate. As long as the input train is continuous, the Q output of A_4 is low. If a missing pulse is detected, however, the one-shot will not generate a pulse to the reset pin of A_3, and the Q output of A_3 (which is also clocked by f_{ref}) will go high to clock A_4. A_4 and f_{ref} will then switch A_5's Q output to high.

This turns on transistor Q_1 and the pilot lamp glows. Switch S_1 is used to reset the circuit after a missing pulse has been detected. Note that circuit operation remains independent of the input frequency, since the arrival of f_{ref} and the 10-μs pulse at A_3 is synchronized to f_{in}.

The circuit should be calibrated by setting A_1 for an output voltage of 10 when a 10-kilohertz input signal is applied. Similarly, A_2 should be set to generate a 10-kHz signal for a 10-V input. □

Synchronous. The circuit detects the missing pulse independently of the pulse train frequency. Voltage-to-frequency and frequency-to-voltage converters derive a reference frequency whose average remains the same for small anomalies occurring in the pulse train: converters' integrating capacitors hold f_{ref} steady despite missing pulses. The reference in this way serves as a synchronous clock.

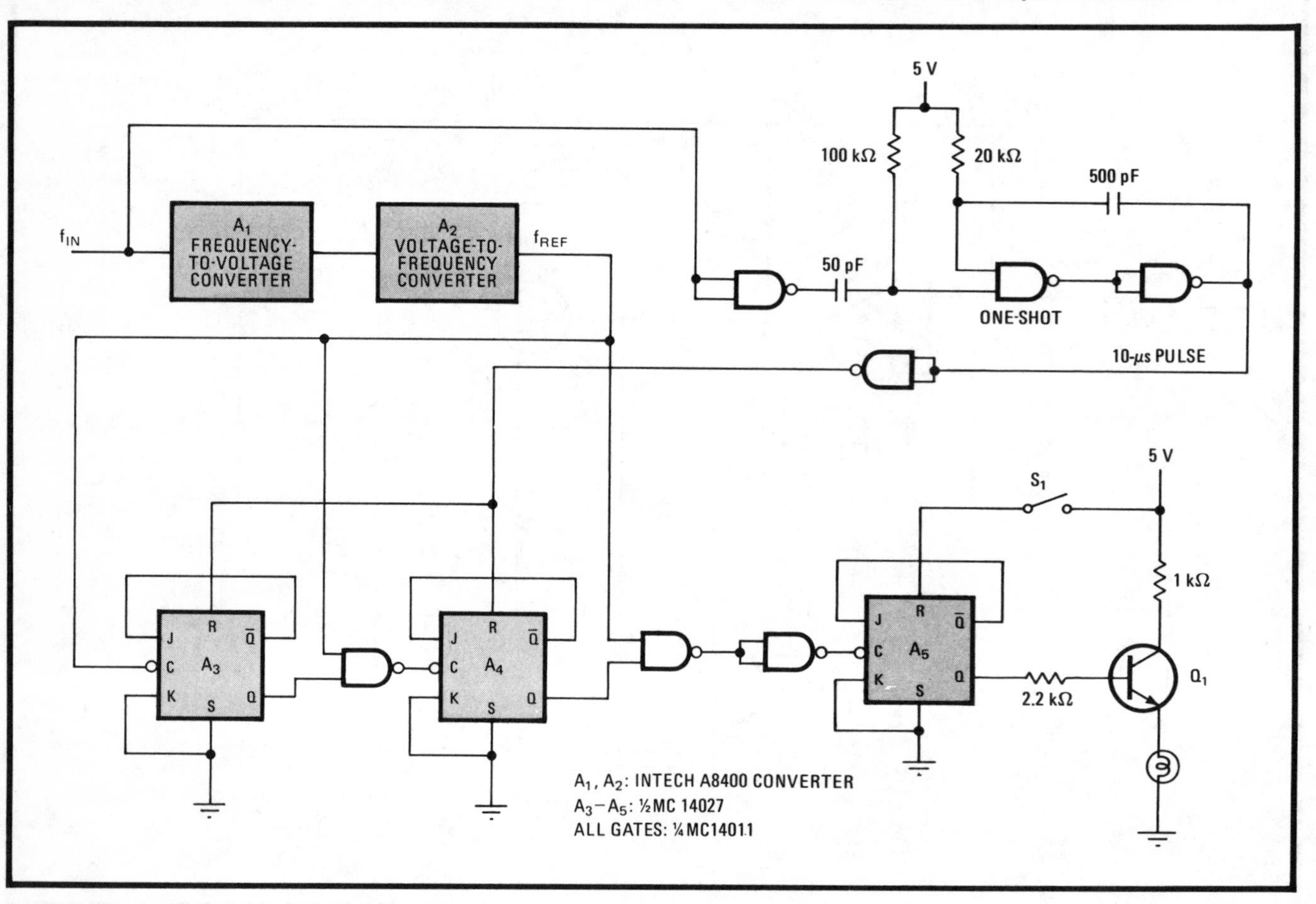

Digital peak detector finds 4-bit highs and lows

by N. Bhaskara Rao
U. V. C. E., Department of Electrical Engineering, Bangalore, India

This circuit finds the maximum and minimum value of a 4-bit data signal over any given time interval. Here, a latch-comparator feedback scheme using standard TTL operates as a real-time memory bank to determine the peaks and valleys of the input signal.

A negative-going start pulse latches the incoming data, D, in A_1 and A_2 at the beginning of the sample period. Thus initially, $D_o = N_o = M_o$, where N and M are the stored maximum and minimum data values, respectively.

When the system clock first moves high, gate G_3 is enabled and N_o appears at the output of the data selector, A_3. N_o is then compared to the present value of the input data at A_4. If D should exceed N_o, pin 5 of A_4 will go high and A_1 will therefore latch the input data.

Similarly, when the system clock first moves low, gate G_4 is enabled and A_4 compares M_o to the present data value. If D is less than M_o, A_2 will latch the input data. This process is repeated during each system clock until the end of the sample period, which may be terminated by the user in any of several ways, depending on the application. The maximum and minimum values of D appear at the output of A_1 and A_2, respectively.

No difficulties have been encountered with the generation of signal spikes or transients from pins 5 and 7 of A_4 during A_3's switching periods, and no problems will occur as long as the data-input lines are settled during those times. □

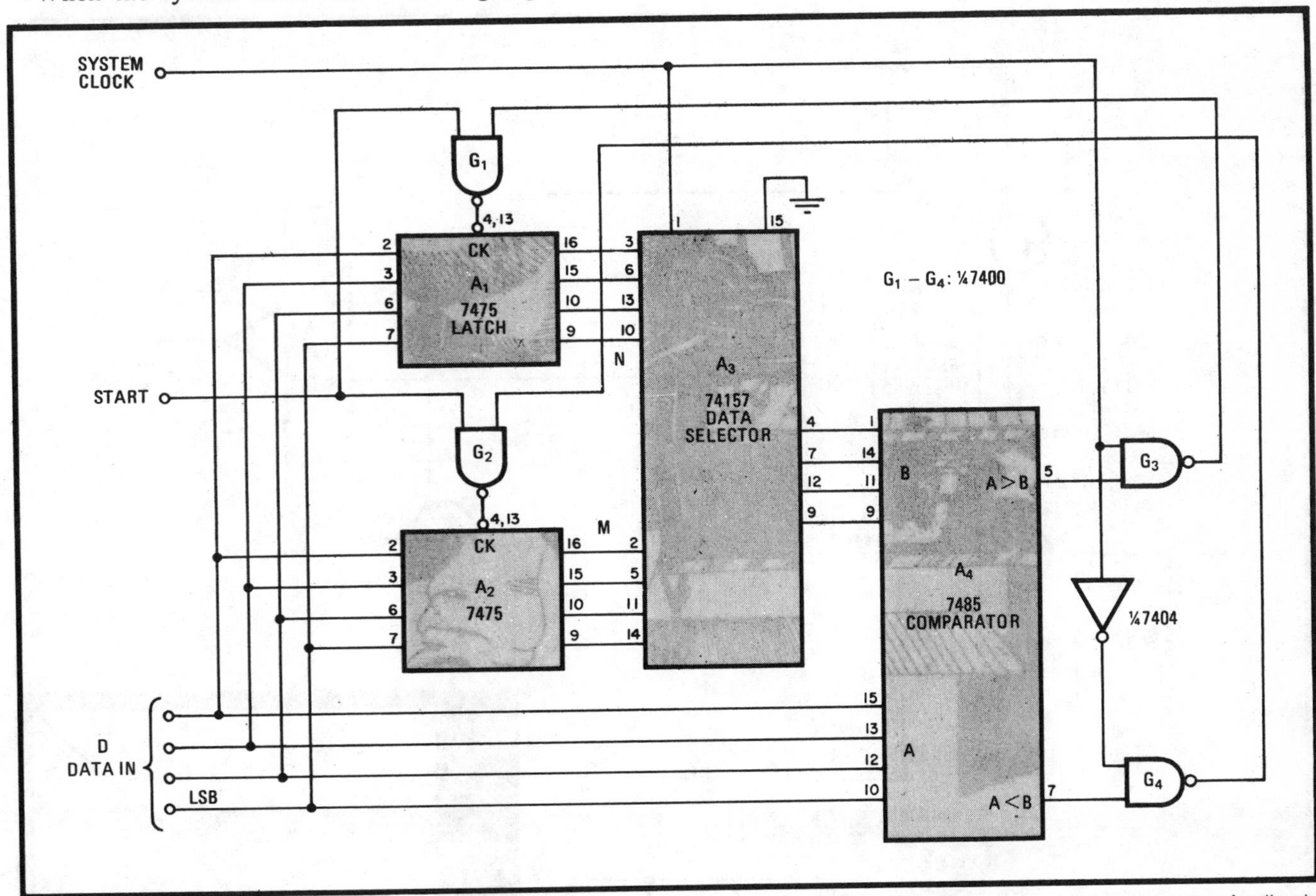

Peaks and valleys. Maximum/minimum detector for 4-bit digital data uses standard TTL elements in latch-comparator feedback arrangement. Circuit performs continual comparison of previously stored maximum and minimum data versus present value of input data, and relatches latter when necessary. Updated high and low values appear at output of A_1 and A_2, respectively.

Wideband peak detector recovers short pulses

by Saul Malkiel
Advanced Technology Systems, Roselle, N. J.

Using a Schottky-barrier diode for detection and a wideband operational amplifier, this peak detector recovers data pulses as narrow as 10 nanoseconds in the range of 0.1 to 1.3 volts. The circuit's linearity as a percentage of the full scale output is 4%.

The input signal, whose rise time is assumed to be a minimum of 10 ns, is applied to one side of the differential source follower, A_1-A_3, via a 50-ohm coaxial cable. A blocking capacitor removes any baseline shifts. The output of this JFET follower is then applied to the wideband amplifier, A_2. The amplifier has a gain-bandwidth product of 1 gigahertz.

The output of the amplifier switches on diode detector D_1 so that a charging current can be delivered to storage capacitor C_s. When there is overshoot, source follower A_3 is turned on more heavily than A_1 so that A_2 may be driven negative. D_2 then comes into play, acting to limit the amplifier's negative excursion. The excursion, coupled through D_1's shunt capacitance, reduces the output voltage by 30 millivolts. This pullback effect is one cause of error in small-signal detection.

The storage capacitor is a silver-mica type, chosen for its low loss and its high stability. Its value is 22 picofarads, a tradeoff between A_2's hold-drift (100 microvolts per microsecond) and pullback characteristics.

The detected voltage is now applied to A_4, which serves as a buffer and provides a low-output impedance for driving the external circuitry. The LF356 operational amplifier used has wideband characteristics, notably a settling time of approximately 200 ns, after which its

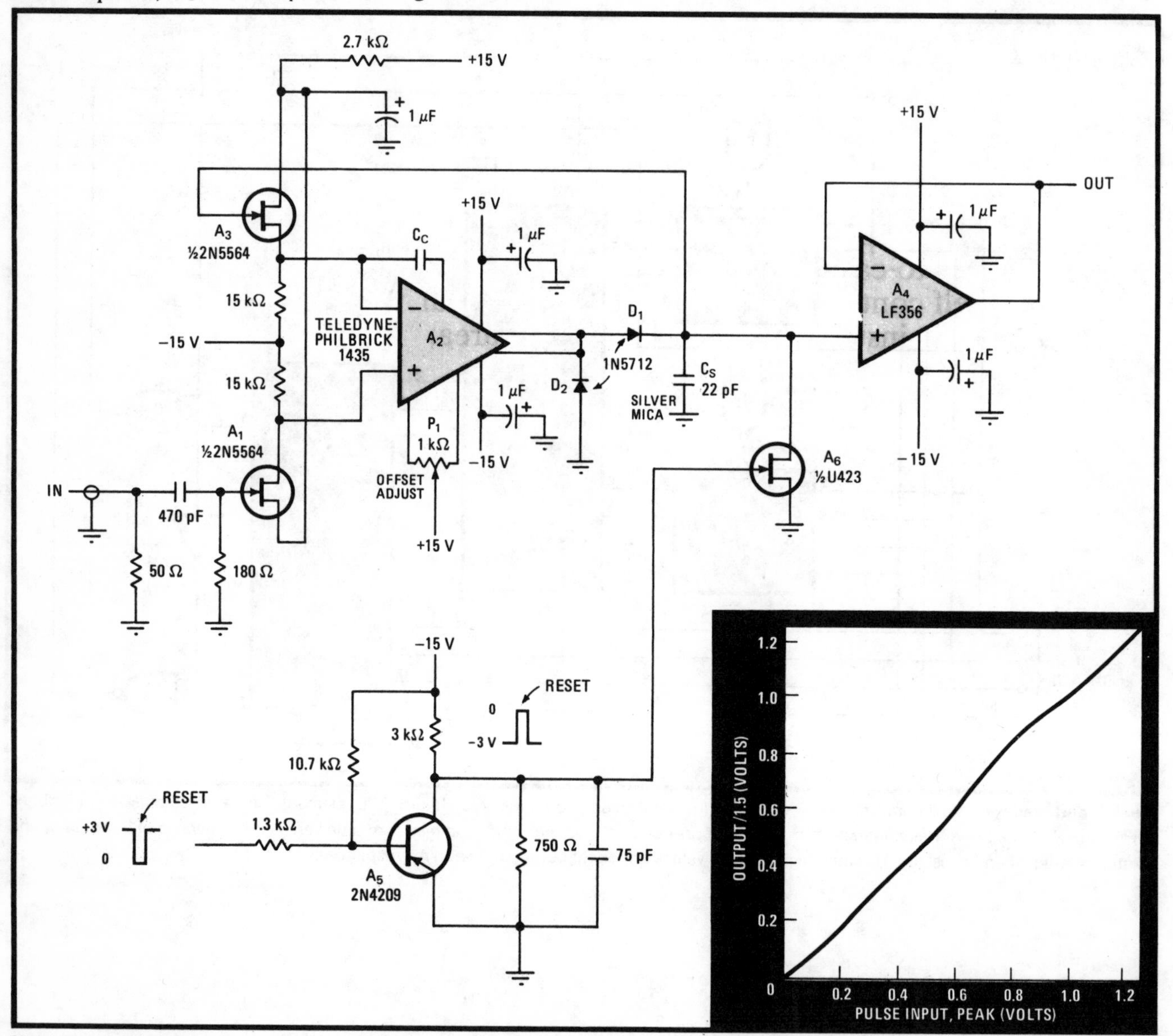

Narrow capture. Wideband amplifier and fast diode detect pulses having widths as small as 10 ns. Output-to-input voltage linearity of circuit is 4%, and linearity is virtually independent of pulse width. Response is illustrated in curve at bottom right.

output becomes valid.

A pulsed command can then be applied to the 2N4209 transistor, A_5, to reset the circuit. The reset command is then transferred to A_6, the U423, and the stored voltage on C_s discharges exponentially toward zero. In order to minimize the discharge time, the output of A_2 is biased at 10 mV dc under no-signal conditions. Potentiometer P_1 is used to adjust the required offset. A total time of about 2 microseconds is required to reset the circuit. □

C-MOS twin oscillator forms micropower metal detector

by Mark E. Anglin
Novar Electronics, Barberton, Ohio

A battery-powered metal detector can be built with the four exclusive-OR gates contained in the 4030 complementary-metal-oxide-semiconductor integrated circuit. The gates are wired as a twin-oscillator circuit, and a search coil serves as the inductance element in one of the oscillators. When the coil is brought near metal, the resultant change in its effective inductance changes the oscillator's frequency.

Gates A_1 and A_2 in the figure are the active elements in the two simple oscillators, which are tuned to the fundamental frequencies of 160 and 161 kilohertz, respectively. A_1 serves as a variable oscillator containing the search coil, and A_2 oscillates at a constant frequency.

The pulses produced by each oscillator are mixed in A_3, and its output contains sum and difference frequencies at 1 and 321 kHz. The 321-kHz signal is filtered out easily by the 10-kHz low-pass filter at A_4, leaving the 1-kHz signal to be amplified for the crystal headset connected at the output. The headset has a high impedance (2,000 ohms) and therefore will not impose a big load on A_4.

A change in the output frequency indicates a frequency change in the variable oscillator due to the mutual-coupling effect between a metal and the search coil. The device's sensitivity, determined largely by the dimensions of the search coil, is sufficient to detect coin-sized objects a foot away.

This device's effectiveness derives from the twin-oscillator approach, because it is not feasible to directly vary a single oscillator operating at 1 kHz. An oscillator operating in this range requires high values of L and C, and these elements would load down the gate and consequently reduce circuit sensitivity. In addition, the cost of high-value inductors and capacitors is great. □

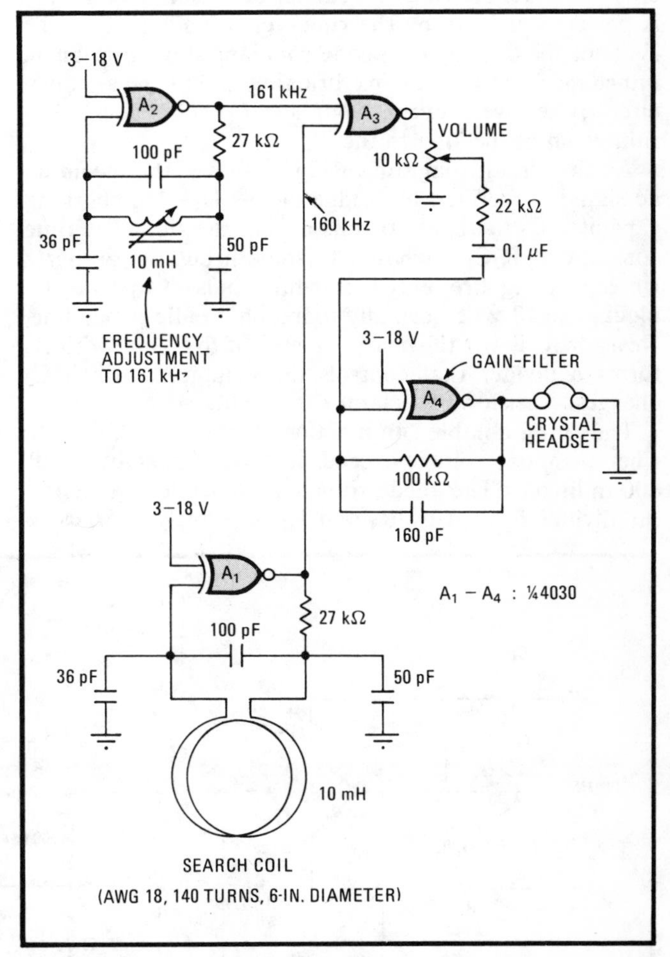

Metal detector. Two oscillators and a search coil form a simple metal detector. Objects near search coil change A_1's frequency of oscillation and the 1-kHz output note produced by the mixing of oscillators A_1 and A_2. A_4 amplifies and filters audio signal.

Telephone-ring detector eliminates relay

by Joe Gwinn
Baltimore, Md.

A dial-up telephone-line interface to a computer can detect telephone ringing signals with a circuit that includes a varistor, a programable unijunction transistor, and a photon coupler. This circuit replaces the more conventional ring detector that consists of a capacitor-isolated full-wave bridge driving the coil of a small relay.

Like the relay, the arrangement described here isolates the phone line from the computer logic and is immune to noise interference. Nor does it suffer from the inherent mechanical disadvantages of relays. Indeed, this circuit has advantages of its own. It emits one and only one pulse per ring —it cannot be teased. Its line side is powered entirely by the ring, yet it loads neither the ring nor the line. By telephone-company standards, an ac impedance of 47 kilohms bridging a line is an open circuit; however, check your local phone company's rulings on connecting to the line.

As the circuit diagram in Fig. 1 shows, the incoming ac signal is rectified by bridge diodes $D_1 - D_4$, charging capacitor C_2 to about 10 volts during the ring. The time constant of R_2C_2 is chosen to smooth out the 20-hertz ripple, leaving a roughly rectangular pulse. Capacitor C_1 blocks the 48 v dc normally found on an idle phone line. Resistor R_1 limits the charging current to C_2 and with R_2 forms a divider that controls the voltage to which C_2 charges. Varistor VR_1 clamps transients.

The programable unijunction transistor, Q_2, fires when its anode voltage exceeds its gate voltage by about 600 millivolts. The anode voltage is controlled by C_3 and the divider formed by R_4 and R_5. C_3 slowly charges to half the voltage across C_2 and reaches 4 v or so by the end of the ring. The gate voltage is controlled by R_3C_4 and follows the voltage on C_2 closely. When the ring ends, the anode voltage is 4 v and decaying slowly, while the gate voltage is 10 v and decaying quickly (Fig. 2). In about 100 milliseconds, the gate voltage catches up with the anode voltage; Q_2 then fires, dumping C_3's charge through current-limiting resistor R_6 and the light-emitting diode of the photon coupler.

When the LED conducts the current pulse from C_3's discharge, it turns on transistor Q_1 in the photon coupler. Transistor Q_3 is then turned on, producing an output voltage pulse that is adequate to drive C-MOS circuitry. If transistor-transistor logic is to be driven, the pulse will require some sharpening by a buffer transistor and a Schmitt trigger such as the 7413.

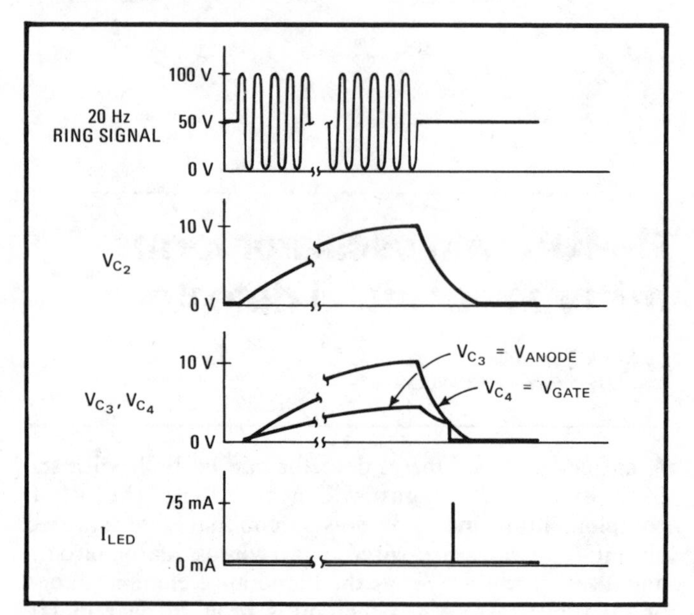

2. One pulse per ring. Waveforms of circuit in Fig. 1 show programable unijunction transistor firing when gate voltage drops below anode voltage, dumping C_3's charge into LED.

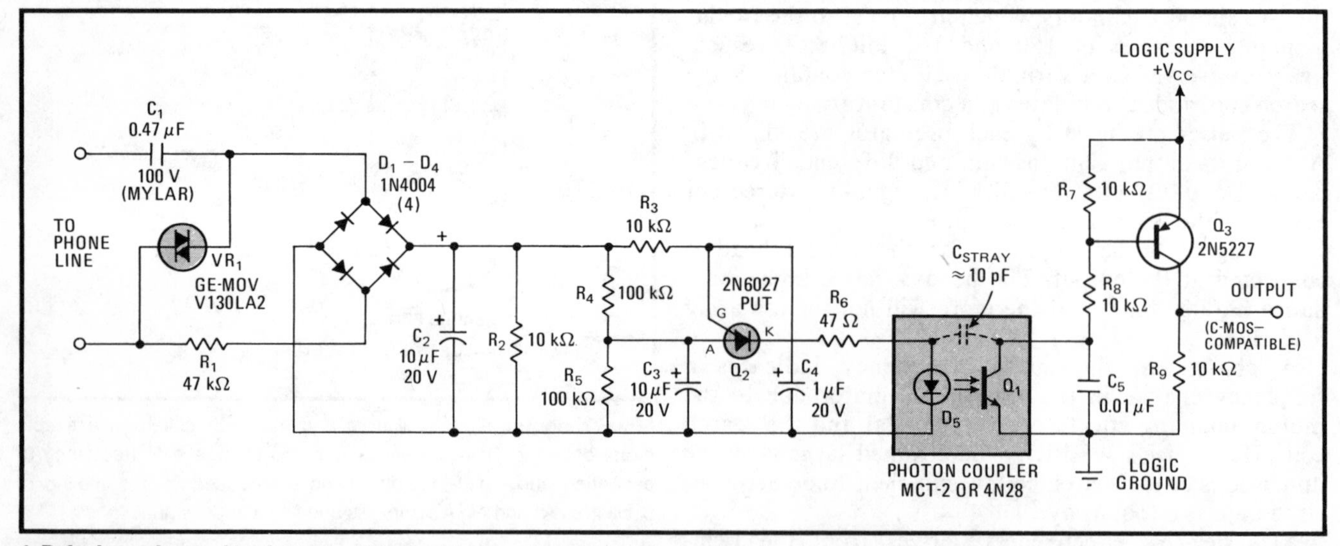

1. Relayless phone-ring detector. This circuit produces a single output pulse when a ringing signal comes in on the telephone line. It is useful in a computer-to-phone-line interface because it provides isolation, is immune to line noise, and is reliable. The programable unijunction transistor, fired as capacitors discharge after ringing stops, pulses the photon coupler to produce output.

Capacitor C_5 is included to provide noise immunity. There is enough common-mode noise on telephone lines to falsely switch Q_3 even after it has passed through the stray capacitance of the coupler. C_5 forms a voltage divider with the stray capacitance, reducing the noise to insignificance. □

Versatile phase detector produces unambiguous output

by L. E. S. Amon and B. Lohrey
University of Otago, Department of Physics, Dunedin, New Zealand

A dual monostable multivibrator and integrator network forms a detector that not only measures phase difference between two signals throughout the entire 360° range, but also produces an unambiguous output signal for various phase-advance and -retard conditions by generating a voltage and slope output combination that is unique for every angle.

As shown in (a) of the figure, a positive zero crossing of reference signal A triggers the comparator C_1 and

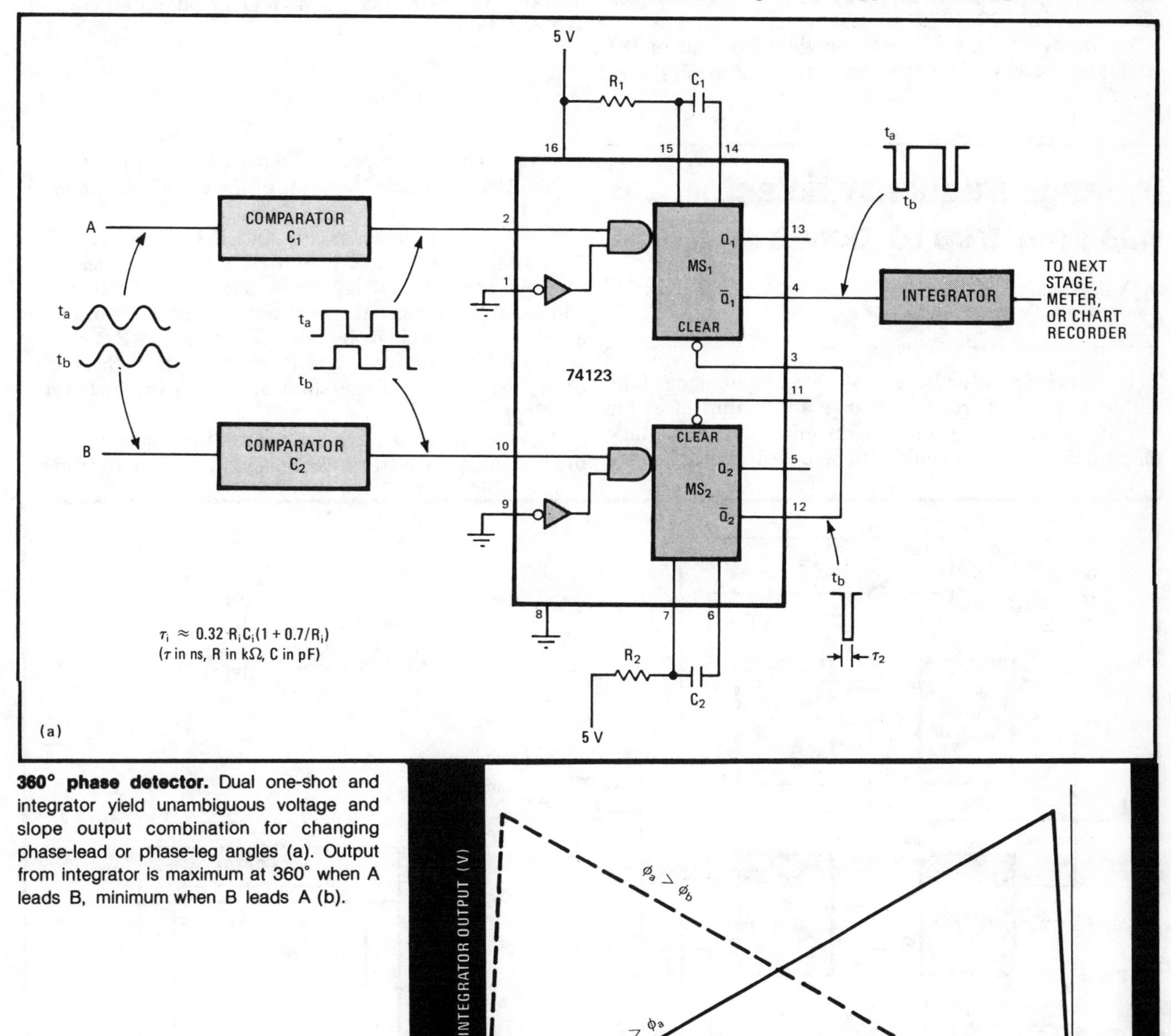

360° phase detector. Dual one-shot and integrator yield unambiguous voltage and slope output combination for changing phase-lead or phase-leg angles (a). Output from integrator is maximum at 360° when A leads B, minimum when B leads A (b).

one-shot MS_1 at time t_a, where MS_1 is one half of the 74123 device. This one-shot, in the retriggerable mode and set so that its pulse width τ_1 is greater than T_a, the period of the reference signal, stays on until comparator C_2 and one-shot MS_2 are fired by signal B at time t_b. The narrow pulse produced by MS_2 resets MS_1.

The phase of B with respect to A may be related to the duty cycle of the output signal from MS_1. The duty cycle may be expressed by:

$$W = \frac{t_b - t_a}{T_a}$$

The output signal may be converted to a dc voltage by the integrator network connected to the output of MS_1. Alternatively, the phase may be measured digitally [see *Electronics*, Dec. 20, 1973, p. 119].

As shown in (b), an increase in the dc output of the integrator occurs when the phase angle of B increases with respect to A. The output decreases when the phase angle of B decreases with respect to A. Thus the frequency relation of B to A may be determined from the slope of the integrator's output if the two signals are of similar but not identical frequency.

When the two signals are not harmonically related, their phase relationship will change with time. The circuit is therefore useful as a phase-modulation detector for applications in communications receivers.

Further versatility can be achieved by placing frequency dividers at the input ports of the phase detector to achieve a full-scale output at $N \times 360°$. The sensitivity can be increased, on the other hand, if both of the input frequencies are multiplied by N. This reduces the range of phases that yield an identical output voltage to 360°/N. □

In-range frequency detector has jitter-free response

by A. J. Nicoll
Instromedix Inc., Beaverton, Ore.

This simple circuit will detect when an input signal falls within a specified frequency range and is thus ideal for use as an out-of-tolerance alarm or as a rudimentary phase-locked loop. It could also be called unusual, since it uses hysteresis to provide separate lock and capture ranges that eliminate the jitter of the circuit's logic-level output.

The diagram shown in (a) and the hysteresis curve shown in (b) help make the circuit's operation clear. A_1 and A_2 are two retriggerable one-shots. Their pulse widths, and therefore their maximum frequency of operation, are controlled by R_1–R_4. Whether R_1 or R_2 controls the width of A_1 and R_3 or R_4 controls the width of A_2 depends upon the state of the A_3 or A_4 D-type flip-flops.

Assume R_1 and R_4 are the controlling elements as an input signal of arbitrary frequency, f_{in}, arrives to trigger

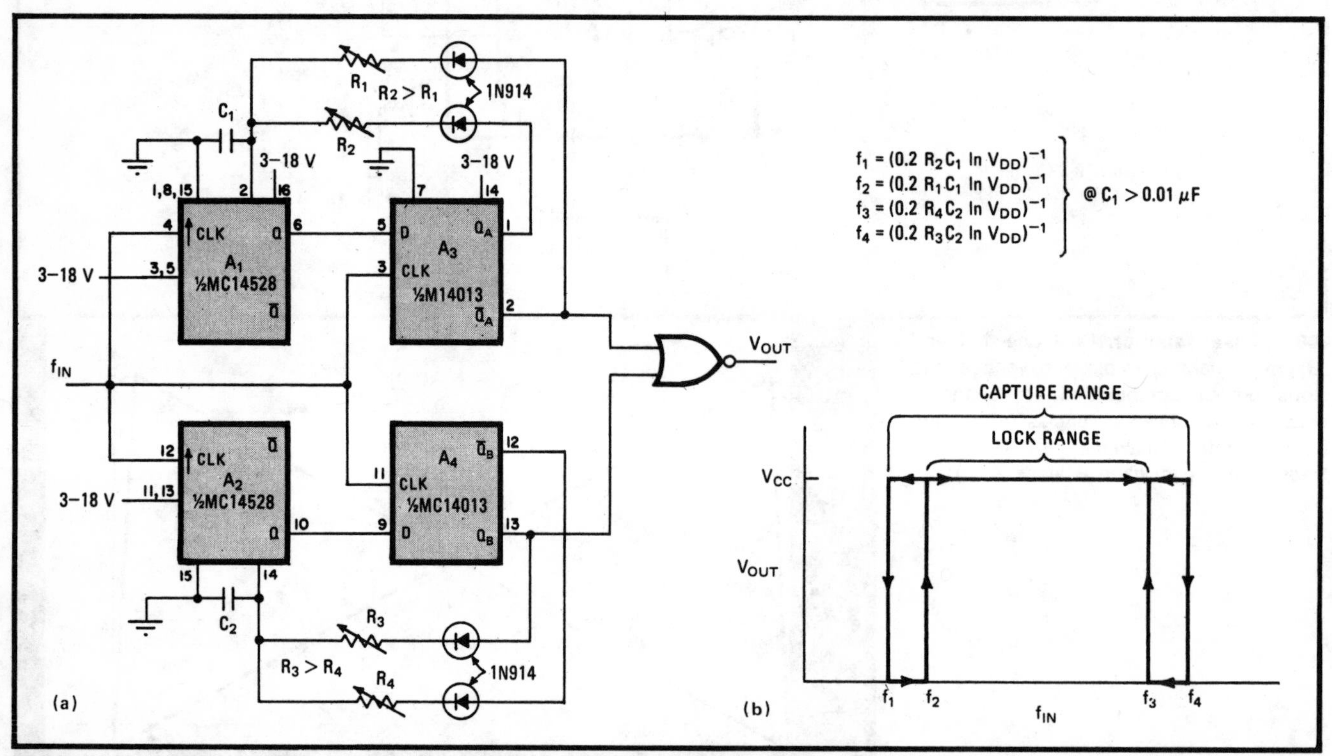

Within limits. Circuit (a) detects whether input signal is within user-set frequency range f_2–f_3 (b). Flip-flops enable selectable hysteresis so that circuit, once locked, will not change state until f_{in} moves below f_1 or moves above f_4. Lock and capture ranges are controlled by R_1–R_4. Hysteresis eliminates jitter that would normally occur at output if f_{in} were near f_2's or f_3's edges.

both one-shots simultaneously. The positive transition of f_{in} then fires A_1 and A_2, as shown. The next positive-going transition will trigger both A_1 and A_2 again while clocking the previous output states, which were generated before retriggering, into A_3 and A_4.

If this second transition occurs before either one-shot has returned to its time-out state, a logic 1 will be clocked into its respective flip-flop, changing the state of that flip-flop. Once the flip-flop moves from a 0 to a 1, the pulse width of the one-shot will be controlled by one of the two timing elements, R_2 and R_3.

The curve (b) shows more clearly how the lock and capture ranges are controlled by R_1–R_4, where f_1–f_4 are equal to the reciprocals of the pulse widths determined by C_1–R_1 or –R_2 and C_2–R_3 or –R_4. A_3 will move high when f_{in} rises above f_2, and it will not move back to its initial state until f_{in} falls below f_1. Similarly, A_4 will change from a 0 to a 1 when f_{in} rises above f_4, and it will change back to a 0 only when f_{in} falls below f_3. The amount of hysteresis acting upon f_1–f_2 and f_3–f_4 can be chosen by simply selecting the appropriate resistance values for R_1–R_4.

The NOR gate output moves high when f_{in} is within the set limits of f_2–f_3. It will not move low again until the input frequency falls below f_1 or above f_4. If desired, an OR gate can be used instead of a NOR gate, since both the Q and $\overline{Q}$ outputs are available in D-type flip-flops. □

Chapter 12
DISCRIMINATORS

Two 555 timers build pulse-height discriminator

by R. Karni and T. Assis
I.A.E.C. Nuclear Research Center, Negev, Israel

When making pulse and noise measurements, it is sometimes necessary to count the number of transients or pulses of a given amplitude. By configuring two 555 timers as adjustable-threshold monostable multivibrators with the output of one inhibiting the other, the resulting pulse-amplitude discriminator generates a clean square-pulse output only when it receives an input pulse of predetermined magnitude.

As shown in the figure, each 555 is connected as a basic monostable. In this configuration, a negative transition below a voltage level of $V_{CC}/3$ at the trigger input (pin 2) generates an output pulse at pin 3 of duration 1.1(RC). The voltage-dividing network at pin 2 permits an adjustable bias from $V_{CC}/3$ to almost V_{CC}. Therefore, since a V_{CC} of 15 v has been chosen, triggering can be made to occur on negative pulses from a minimum of nearly 0 volt on up to a maximum of almost 10 volts.

Both timers have trigger inputs biased in the same fashion, and both receive the same input pulse through decoupling capacitors C_1 and C_2. But the output of timer A is connected through an inverting buffer-transistor to the reset input (pin 4) of timer B; consequently, whenever monostable A is triggered, B is inhibited. It is this arrangement that permits the pulse discrimination.

If the threshold of timer A is set higher than that of B, only pulses having magnitudes between the two thresholds will produce a pulse at the output. Pulses of a magnitude less than the threshold of B will trigger neither monostable, and those of a magnitude greater than the threshold of A would trigger both—but the inhibiting action of A on B will allow no output pulse to be produced.

In using the discriminator, the control of B sets the threshold of the incoming pulse, while the control of A is set higher than B, to determine the "window" or the difference of the two thresholds.

With the components shown, and a 15-v supply, the pulse threshold is adjustable from 0 to about 10 v, and the window can be varied from a maximum of just under 10 v (when the threshold of B is set to minimum), down to zero, when B is set to its maximum. If signals of greater amplitude are to be encountered, suitable dividers may be added to avoid transitions below ground at the pin-2 trigger inputs of the timers.

The output, which can drive up to 200 milliamperes of transistor-transistor-logic loads, may be connected to a counter or monitoring device. □

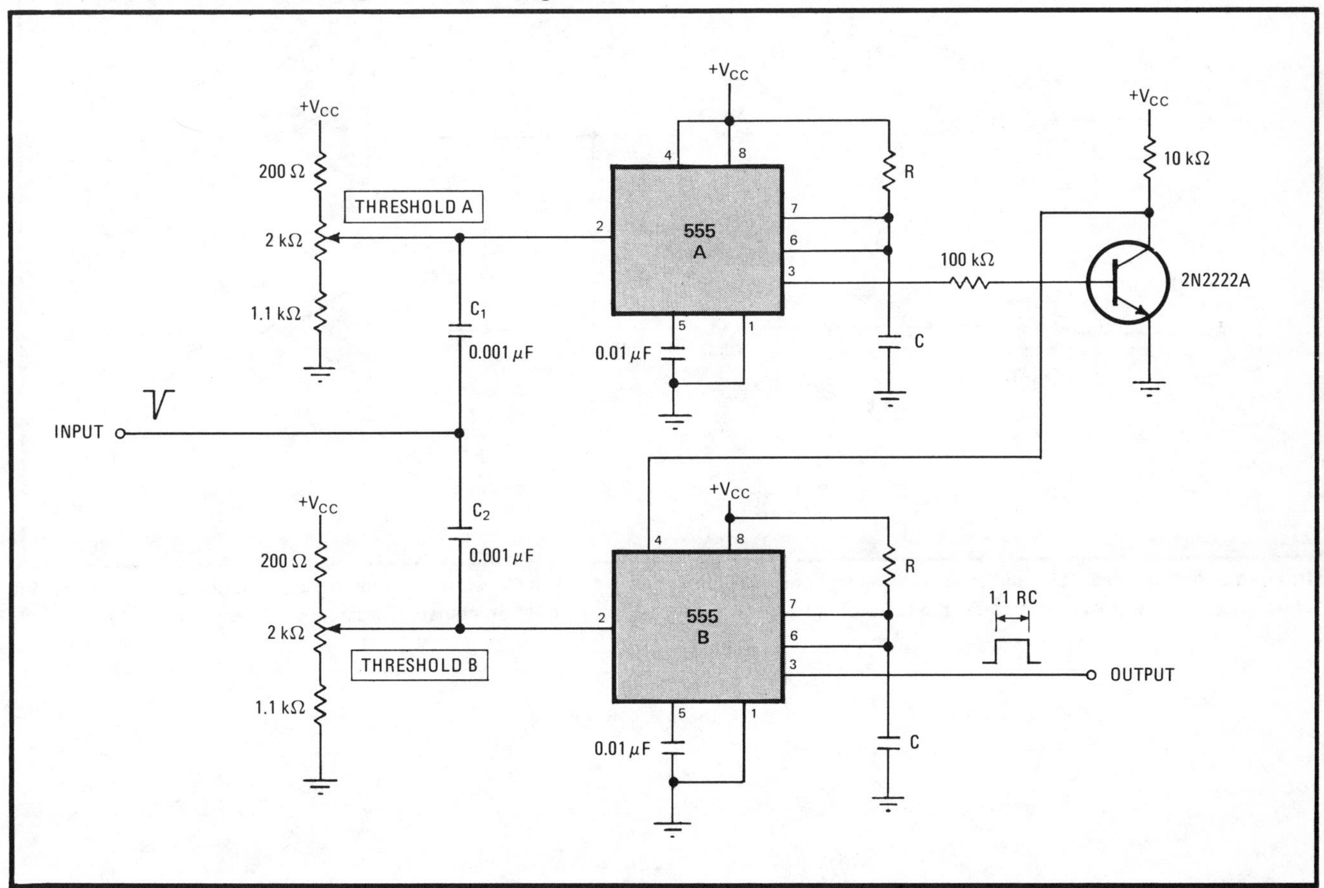

Pulse-amplitude window. Two 555 timers, hooked up as monostables with differing thresholds, select pulses by height. Timer A inhibits B so that an output occurs only when pulse level is within window set by two controls. TTL output-pulse duration is 1.1(RC).

Glitchless TTL arbiter selects first of two inputs

by Yukihiro Mikami
Ottawa, Canada

Failure to differentiate between independently timed asynchronous signals only a few nanoseconds apart can cause glitches in standard transistor-transistor-logic circuits that spell disaster to system operation. Once the glitch is detected, it must be eliminated, so that an arbiter circuit like the one described here is essential in applications like dynamic memory controllers.

As shown in the figure, the 74S02 cross-coupled NOR latch will respond to a signal on request line A or B. The signal arriving first will appear on the corresponding output line. With simultaneous inputs, however, a single-pulse glitch, a sinusoidal oscillation, a metastable-state response (0.8-to-2.1-volt unassigned or guard-band region), or a combination of such responses may occur at each output for an indefinite period before the gate decides to switch into its desired state.

An RC network and Schmitt-trigger buffer eliminate these undesirable responses from the circuit outputs while minimizing the decision time. The amplitude of the glitch or oscillation is essentially filtered or damped, as the case may be, by the RC combination of the 330-ohm resistors and the 22-picofarad capacitor.

The resistors also combine with the 820-Ω resistor at the input of either Schmitt trigger to form a voltage divider. This divider effectively raises the positive-going switching threshold at this point by 0.6 to 2.1 V. Thus there is no spurious response, even for a 3-V glitch, a 1.5-V peak-to-peak oscillation, or a metastable-state response at the NOR gate output. The circuit responds with an output when the transient dies away, as the latch may then decide which signal came first. In the event of the arrival of two truly simultaneous signals, the gate would render an arbitrary decision.

Nonsimultaneous signals appearing at the appropriate output of the NOR gate will pass through the deglitching circuit relatively unaffected. Typical propagation delay of the arbiter is 20 nanoseconds for nonsimultaneous inputs and 25 ns for simultaneous input signals. □

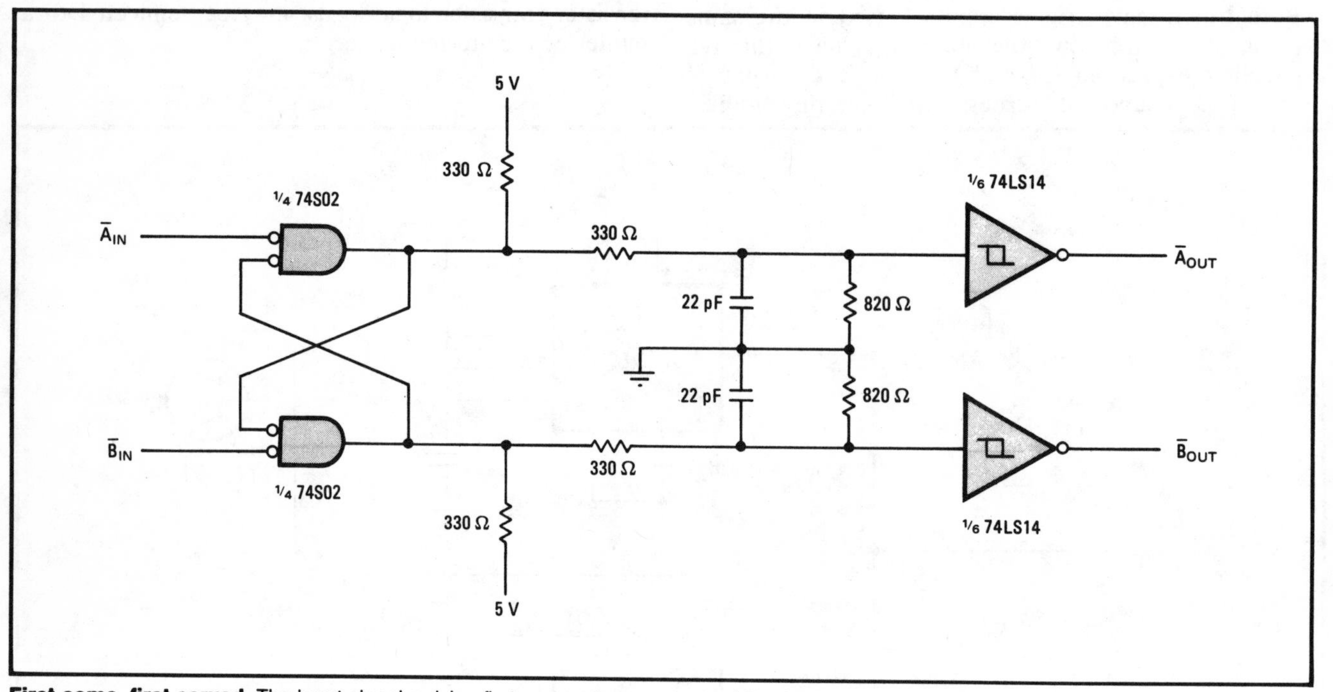

First come, first served. The input signal arriving first passes to an output, while simultaneously arriving signals produce an eventual output but no glitch. The resistor network raises the threshold of the Schmitt trigger, and the RC combination reduces glitch amplitude.

Frequency discriminator has ultra-sharp response

by S. J. Collocott, *CSIRO Division of Applied Physics, National Measurement Laboratory, Sydney, Australia*

Most rudimentary circuits for discriminating between two frequencies or two bands of frequencies sacrifice selectivity to simplicity. But this simple circuit, which uses just a frequency-to-voltage converter and a couple of general-purpose comparators, can differentiate between two frequencies separated by only a few hertz.

In this application, the circuit rejects all frequencies below 2.1 kilohertz, while passing others, although it is a simple matter to modify the discriminator to handle signals at any frequency. Input signals are introduced into the LM311 comparator (A_1), which operates as a zero-crossing detector. Its output is then applied to one input of a dual NAND gate and A_2, the LM2917 frequency-to-voltage converter.

The converter, which drives the noninverting input of comparator A_3, generates an output of one volt for each kilohertz applied at its input. Thus, when f_{in} is less than 2.1 kHz, the output of the converter is less than 2.1 volts, and A_3 (whose noninverting input is biased at 2.1 V by diodes D_1–D_3) is low. Therefore, output gate A_4 is disabled. If f_{in} moves above 2.1 kHz, A_3 will go high and enable A_4, thereby permitting f_{in} to appear at the output.

The sharpness of the cutoff, which is determined by the transfer function of A_3, is approximately 1 Hz. The response time of the circuit is adjusted by C_3 and R_5C_4. These components act to control the integration time at the output, ensuring that a steady dc voltage is attained after a nominal number of periods of f_{in}. If a fast response time is desired, R_5 and C_4 should be deleted.

The circuit is made to handle signals at any frequency by applying a variable control voltage at pin 3 of A_3, in lieu of the D_1–D_3 and R_6 combination. And the discriminator can be used in other modes, to reject high frequencies, for example, or as a bandpass discriminator.

The discriminator that rejects high frequencies may be realized by simply reversing the inputs to A_3. For bandpass applications, A_3 is replaced by a dual comparator, where the low- and high-cutoff frequencies are set by control voltages on the inverting and noninverting inputs of the comparators, respectively. A_4 must then be replaced with a triple-input NAND gate. □

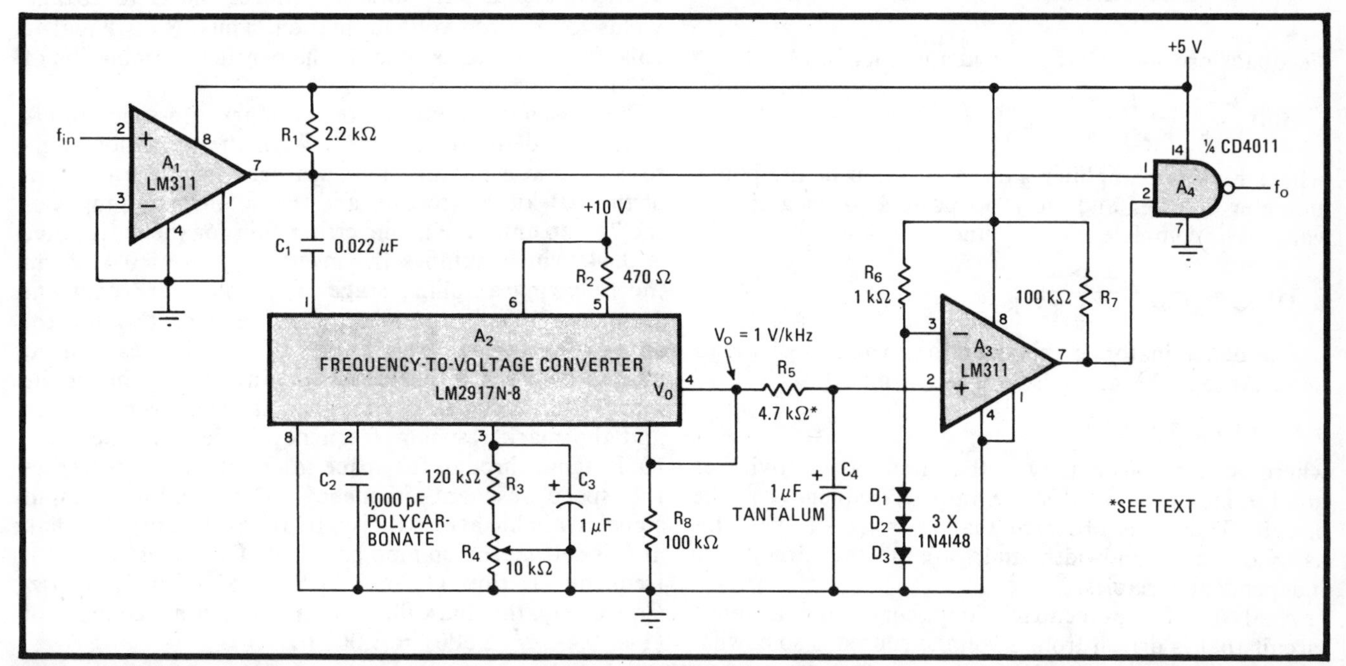

Cycle cutoff. Frequency-to-voltage converter and comparators combine simplicity and selectivity in this frequency discriminator. Transfer function of LM311 determines sharpness of cutoff, in this case being 1 hertz. Circuit can be made to handle signals at any frequency if variable voltage is introduced at pin 3 of A3. Discriminator, configured in high-pass mode, can easily be modified for low-pass or bandpass duties.

Resistor-controlled LC network drives tunable discriminator

by John W. Newman
U. S. Army Electronics Materiel Readiness Activity, Warrenton, Va.

A single potentiometer can adjust the fixed-tuned circuits that determine the mark-and-space frequencies in an audio-frequency-shift-keyed discriminator. This can be accomplished if the potentiometer controls the feedback current that passes through the inductor of each LC combination. Such calibrated single-control tuning is an advantage when reception of any one pair of several widely used shifts is necessary, because the mark-and-space filters do not have to be individually and repeatedly set by a frequency counter or by some other instrument.

A LaPlace analysis of a current-driven tuned circuit will show the dependence of the resonant frequency on the amount of feedback. The tuned circuit in Fig. 1 has a transfer function that is:

$$A(s) = \frac{s + R_L}{s^2 + R_L s + 1}$$

Feedback provided by the second amplifier is:

$$B(s) = \frac{K}{s^2 + R_L s + 1}$$

where K is the amplifier gain, a function of the potentiometer setting, and may be positive or negative in value. The complete transfer function becomes:

$$H(s) = \frac{s + R_L}{s^2 + R_L s + 1 - K}$$

The denominator of this equation, which is of major importance in this analysis, is of the form:

$$s^2 + (AQ)s + \omega^2$$

where A is a constant, Q is the circuit's selectivity or quality factor, and ω is the radian frequency of the circuit. Thus it is observed that $\omega = (1-K)^{1/2}$. This assumes that bandwidth and gain of the circuit are independent variables.

Analysis of the feedback loop containing a tuned circuit that is driven from a voltage source is somewhat more complicated, but the results are similar. For the actual voltage-driver circuit in Fig. 2, the transfer function is approximately:

$$H(s) = \frac{(s + R_L)}{R_4[s^2 + s(R_L + {}^1/R_4 + {}^1/R_5) + 1 \pm K]}$$

where the radian frequency term is the same, but the value of Q depends largely on resistors R_1 and R_2, and the value of K is dependent on R_3.

Determination of R_1, R_2, and R_3 is most important for circuit optimization of Q and transient response. After limiting of the 2-to-3-kilohertz input signal by the first operational amplifier, the 14-volt output signal must be reduced by one half by the voltage divider consisting of R_1 and R_2. This will prevent overdrive of subsequent stages containing two identical tuned circuits with equivalent impedance Z_x. In addition, the dc output of the circuit is a function of the relationship of the mark-and-space frequency to the frequency of each tuned circuit (and thus Z_x, R_1, and R_2).

An unloaded (no-feedback) Q of about 100 is to be expected at 2,500 Hz from each resonant circuit, providing a Z_x of 138,200 ohms. It is reasonable to set a loaded Q of 25, providing a bandwidth of 100 Hz. Resistor R_2 is selected for a Q of 50 so that the parallel equivalent of Z_x, R_1, and R_2 reduces the Q to 25 and yields the desired voltage division. Thus, R_2 is equal to 138,200 Ω, and R_1 is equal to the parallel combination of Z_x and R_2, or 69,100 Ω.

The resonant frequency of the mark-and-space filters is directly determined by R_3. With the potentiometer's resistance at a minimum as measured from the junction of the 101-kilohm resistor and the noninverting input of the 741 op amp, the noninverting gain for the mark filter is 0.44, which nullifies the inverting gain of 0.44 from the following amplifier stage. Thus the mark resonant frequency remains at 2,500 Hz. Feedback through the space resonant circuit is zero, and it is also resonant at 2,500 Hz. When R_3 increases, the inverting gain for the mark filter becomes greater than the noninverting gain and the mark resonant frequency increases. The feedback signal through the space resonant circuit decreases the space resonant frequency. The maximum input signal available across R_3 is 0.09 times the output signal; at this setting the op amp gain is 4. The lowest resonant frequency is thus $(1 - 0.36)^{1/2}$ (2,500) = 2,000 Hz. Conversely, the mark filter has a maximum frequency of $(1 + 0.44)^{1/2}$ (2,500) = 3,000 Hz.

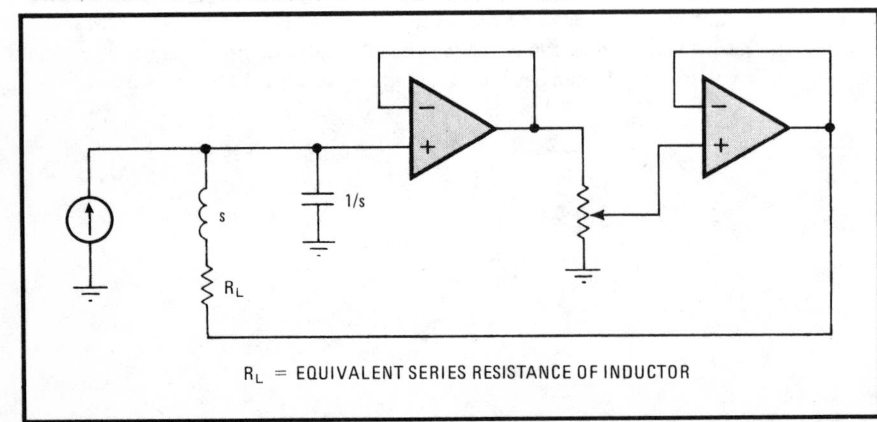

1. Current analysis. Resonant frequency of tuned circuit is affected not only by L and C values but also by magnitude of feedback current through inductor. Potentiometer may control gain of amplifier and thus resonant frequency. Circuit is simpler to analyze but yields results similar to voltage-driven discriminator network described in text.

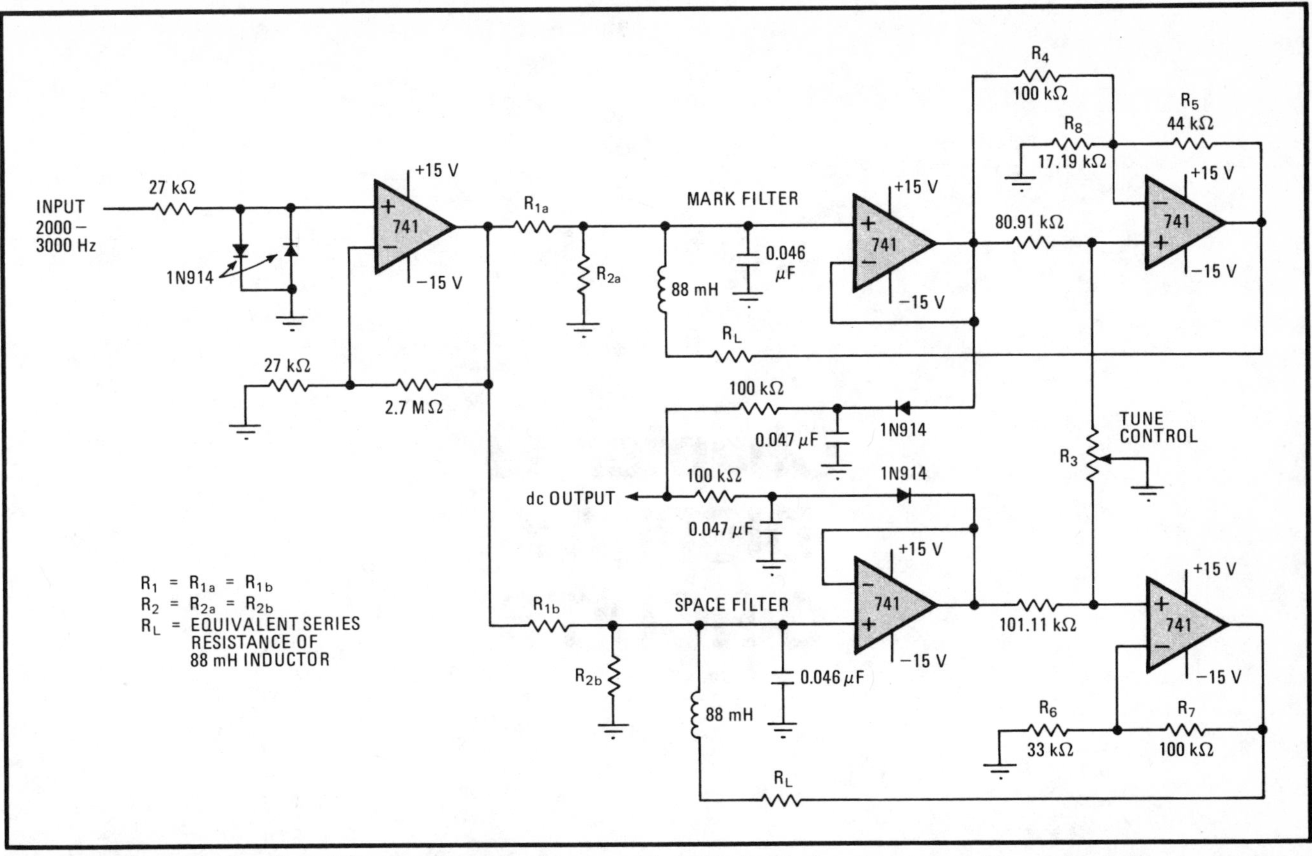

2. A discriminating network. Resistor-tuned filters provide one-control adjustment of mark-and-space frequencies. Shifts are continuously adjustable from zero to 1 kilohertz at a center frequency of 2,500 hertz and are linearly proportional to potentiometer setting.

RESISTOR-TUNED DISCRIMINATOR		
POT ROTATION	FREQUENCY (Hz)	
(%)	MARK	SPACE
0	2,500	2,500
10	2,549	2,451
20	2,598	2,401
30	2,647	2,352
40	2,697	2,302
50	2,747	2,252
60	2,797	2,202
70	2,847	2,152
80	2,897	2,102
90	2,949	2,051
100	3,000	2,000

The table shows the relationship of the potentiometer setting to the mark-and-set frequency pairs. The resonant frequency of the mark filters should be trimmed to a center frequency of 2,500 Hz by R_8. The space filter's lower limit should be trimmed by R_6 or R_7; the mark filter's upper limit should be set by R_4 or R_5.

The dc output is derived from intermediate op amps in conjunction with half-wave rectifier networks. The output voltage will always be positive for received mark frequencies and negative for space frequencies, permitting a suitable source for transistors that will drive radio teleprinter relays and similar equipment. Rejection of off-frequency mark-or-space signals is excellent. Mark-and-space frequency pairs can be within 100 Hz of each other while still providing good circuit performance. □

Chapter 13
DISPLAY CIRCUITS

LEDs track signal level in visual data monitor

by Michael O. Paiva
Teledyne Semiconductor, Mountain View, Calif.

In this circuit, a matrix of light-emitting diodes is combined with an analog-to-digital converter and multiplexing logic to form a dual-setpoint meter. The LEDs are used to track changes in the input data and display the setpoints. The unit may be considered a variation of the analog panel meter; it will find many uses in industrial control applications where it is necessary to observe quick changes while determining whether data is within a preset range. By adding a dual comparator to the circuit, an alarm can be sounded when the input level goes outside the range.

The signal to be tracked, V_{in}, and the upper and lower setpoint voltages are applied to an eight-channel multiplexer, A_1. The a-d converter, A_2, samples channel 1 first. After the conversion, pin 23 of A_2 goes high and advances the binary counter, A_3. A_3 then addresses the second channel of the multiplexer, and so on, until each channel is scanned in sequence.

Bits 2 through 8 of the a-d converter, representing the binary equivalent of the voltage sampled, drive A_4–A_7, which are wired as two one-of-eight decoders. Thus each voltage is converted into a control signal that drives one diode in the eight-by-eight-diode matrix.

The setpoint voltages are sampled twice during each scan cycle (each is connected to two input channels), while the input signal, V_{in}, is connected to four channels and so is scanned four times per cycle. Thus the brightness of the LED corresponding to V_{in} is twice that of the setpoint LEDs, making it easy to differentiate between the three signals.

The circuit's worst-case response time is two scan cycles. A_2 has a conversion time of 1 millisecond, so that the display will require 2 ms to follow a change in V_{in} and 4 ms to follow any change in the setpoint potentials. Because of this high refresh rate, no flicker will be observed on the LED display. □

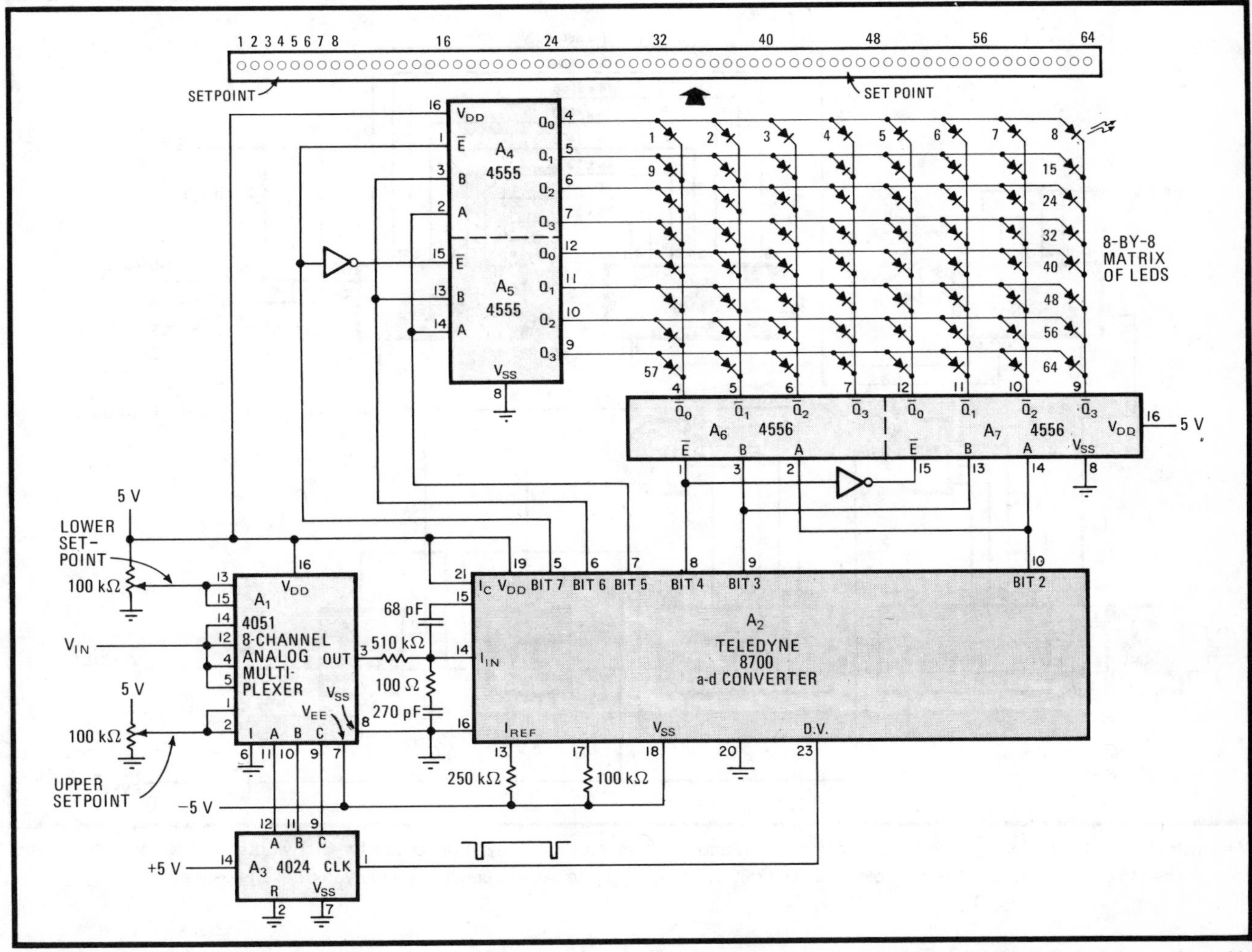

Visualizing voltage. Eight-by-eight-LED matrix and C-MOS logic form analog dual-setpoint meter. Setpoint potentials and data voltage V_{in} are each introduced into multiplexer and scanned in sequence, then converted into control signals that light the appropriate LEDs in the matrix. Setpoint LEDs are half the brightness of the LED representing V_{in} because they are sampled at half the rate.

Standard ICs provide raster scan interface

by Serge Poplavsky
University of New South Wales, Kensington, Australia

One master oscillator and a few counters, flip-flops, and gates make a baud-rate clock with a sync generator that produces horizontal and vertical pulses for noninterlaced raster scanning of a cathode-ray tube. In it, standard integrated circuits are used to generate rates up to 38.4 kilobauds (the upper limit of most generators is only 9,600 bauds) and sync pulses for producing 312 lines at a 50-hertz refreshing rate (or 260 lines for a 60-Hz rate). Thus the circuit is a low-cost solution to building an important part of any data terminal.

All clock rates are derived from one crystal-controlled oscillator, A_1, which uses the 7209 and a crystal cut for 7.9872 megahertz. The master clock frequency is then divided by 13 by A_2, the 74LS161 presettable counter. A_2 generates a frequency 16 times 38.4 kilobauds, suitable for interfacing with universal asynchronous receiver-transmitters or similar modems.

This frequency is further divided by A_3, the 14040 binary counter. Thus, rates extending from 16×300 bauds to 16×19.2 kilobauds will appear at its output.

S_1, which is used in A_4, the 74C151 one-of-eight-line multiplexer, selects the baud rate desired. When all switches are open, a rate of 300 bauds is selected. Closing switch a, the least significant bit (i.e., binary number 1), selects a baud rate of 600, and so on.

The horizontal sync pulses for the cathode-ray tube are derived from A_5 and a four-input NAND gate, which generates a frequency of 15,600 Hz, each pulse lasting 4 microseconds. The vertical sync frequency, either 50 or 60 Hz, is obtained from A_6–A_7 and a four-input NAND gate.

Dividing the horizontal sync pulses by 312 or 260 (for 50 or 60 Hz, respectively), the binary counter, A_6, which in this case is wired as a divide-by-312 device, drives two flip-flops (A_7). A_7 resets the counter and generates vertical sync pulses, each 256 μs long. □

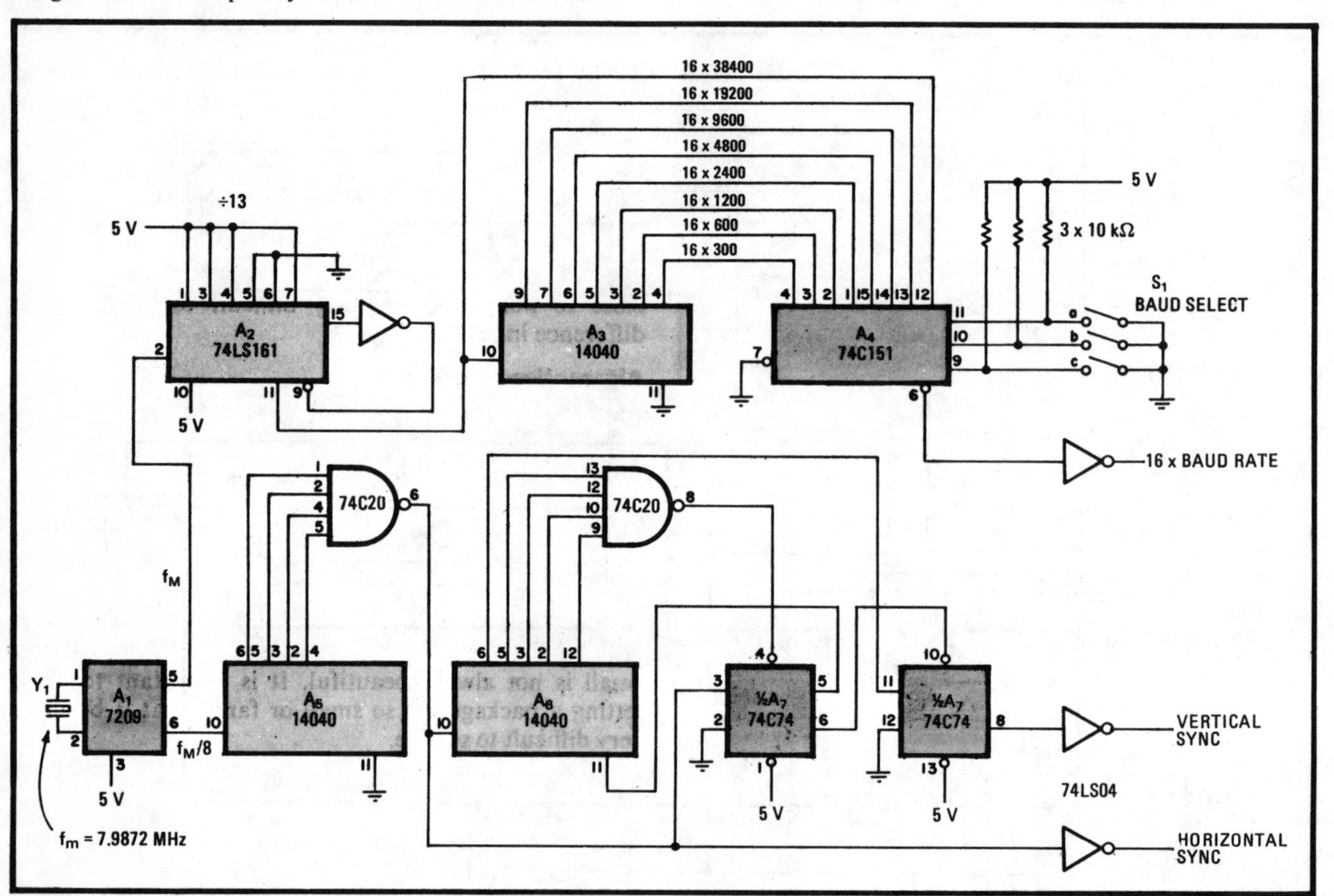

Versatile. Using standard integrated circuits, combination baud-rate generator and sync generator for CRT works up to 38.4 kilobauds and can be wired to generate sync pulses suitable for a 50- or 60-Hz refreshing rate. All frequencies are derived from one oscillator.

Time-shared counters simplify multiplexed display

by Darryl Morris
Northeast Electronics, Concord, N. H.

Although multiplexed display circuits reduce the number of components otherwise required for decoding on a per-digit basis, additional hardware is then needed to select and multiplex various lines to the display. But if a display is driven by a frequency counter, as is often the case, the counter itself can be made to perform the multiplexing with only minimal extra circuitry.

Multiplexing is done by using a master clock having several times the frequency of the normal clock, depending on the number of digits to be multiplexed, and by time-sharing the counters between the count and display mode. In the count mode, the $\overline{\text{LOAD}}$ and enable-P (EP) inputs of the counters shown are high and A_1–A_5 function as a conventional cascaded counter circuit under control of the enable-T (ET) input of A_1. The counter circuit advances one count for each clock period during which the count control line is high.

During the display mode, the control line and $\overline{\text{LOAD}}$ input of A_1–A_5 move low. The counters now accept data at their preload inputs, P_A–P_D. Because the preload inputs are connected to each preceding set of a counter's outputs, A_1–A_5 operates as a 4-bit-wide recirculating shift register when clocked. Thus, the contents of each counter is rotated past the seven-segment decoder (A_6) during its display interval, and the appropriate digit in the display is strobed by the mode controller, A_7 and A_8.

This technique offers the best saving in chip count when the count rate is slow or numbers are to be displayed only after the counted event has terminated. □

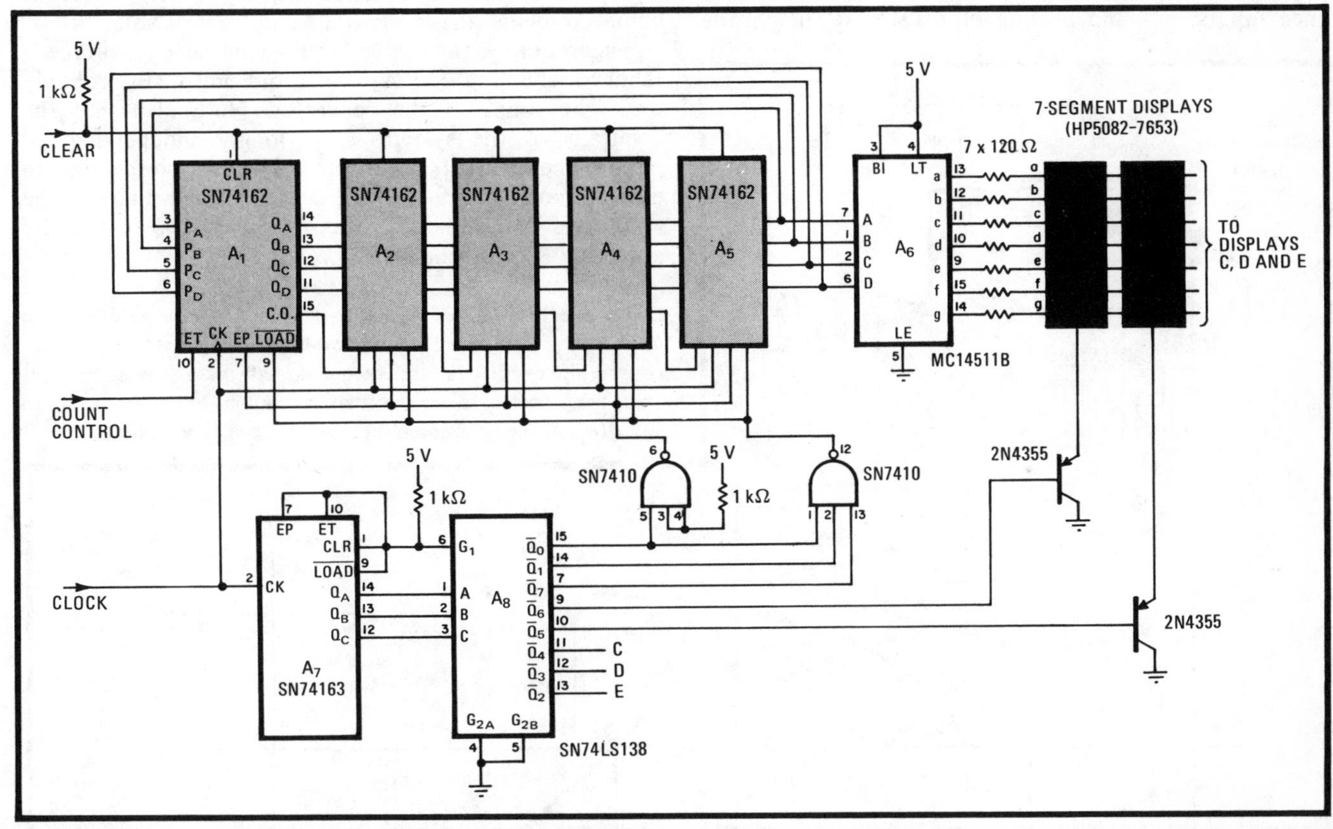

Time-shared. Counter circuit switches between count and display modes without selector devices. Counter operates as 4-bit-wide recirculating shift register. Master clock frequency is assumed to be several times that used for the counting circuits.

One-chip DVM displays two-input logarithmic ratio

by David Watson
Intersil Inc., Reading, Berks., England

The popular ICL7106 series of analog-to-digital converters that serve so widely nowadays as one-chip digital voltmeters can be easily converted to display the logarithm of the ratio between two input voltages, making them useful for chemical densitometry, colorimetry, and audio-level measurements. Only slight wiring modifications at the device's input and integrating ports are required.

Shown in (a) is the new configuration. The modifications from the standard a-d converter connection include the addition of a resistive divider, R_1–R_2, at the reference inputs, and the placing of resistor R_p in parallel with the device's integrating capacitor.

As shown with the aid of the timing diagram in (b), the time constant of the integrating network is given by $\tau = C_{int}R_p$, with the asymptotic endpoint voltage of the integration voltage being $V_{as} = R_p (V_1 - V_2)/R_{int}$, where V_1 and V_2 are the input voltages to be measured. The final integrator voltage therefore becomes $V_{int} = R_p (V_1 - V_2)(1 - e^{-T/\tau})/R_{int}$, where T is the fixed integration period.

During the deintegration portion of the cycle, the exponential decay moves toward the total voltage, V_{tot}, which equals $V_{int} + V_{ref} (R_p/R_{int})$. But $V_{ref} = kV_2$, where k is set by the resistive divider, so that $V_{tot} = R_p (V_1 - V_2)(1 - e^{-T/\tau})/R_{int} + R_p k V_2/R_{int}$. The integrator voltage actually crosses zero when the exponential waveform reaches $V_{final} = V_{ref}R_p/R_{int} = R_p kV_2/R_{int}$.

As seen, the time needed to reach the zero crossing is given by $T_{DEINT} = \tau \ln(V_{tot}/V_{final})$. Making $k = (1 - e^{-T/\tau})$ and $\tau = T/2.3$, it is realized that $T_{DEINT} = T \log_{10}(V_1/V_2)$. For this condition, $k = 0.9$, which is achieved by making $R_1 = 1\ M\Omega$ and $R_2 = 9\ M\Omega$.

Theoretically, the system's full-scale output voltage is reached when $\log_{10}(V_1/V_2) = 2$, but noise will probably limit the range of the converter. Note also that the accuracy of the system is no longer independent of passive component variations. The simplest way to ensure that $k = 0.9$ is to use a pretrimmed divider. The system is calibrated by making $V_1 = 10V_2$ and by adjusting R_p until the display reads 1.000. □

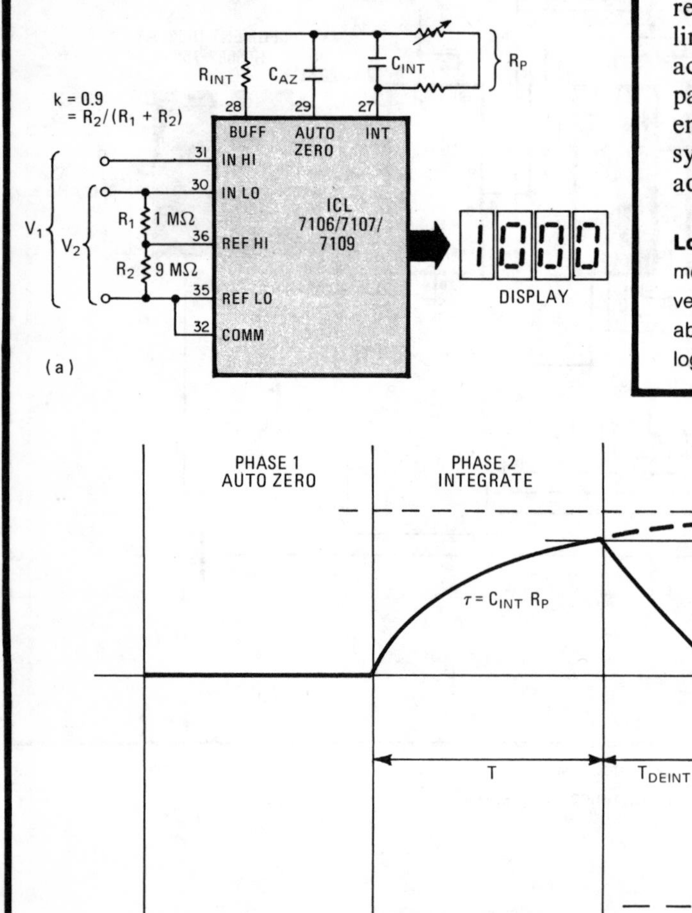

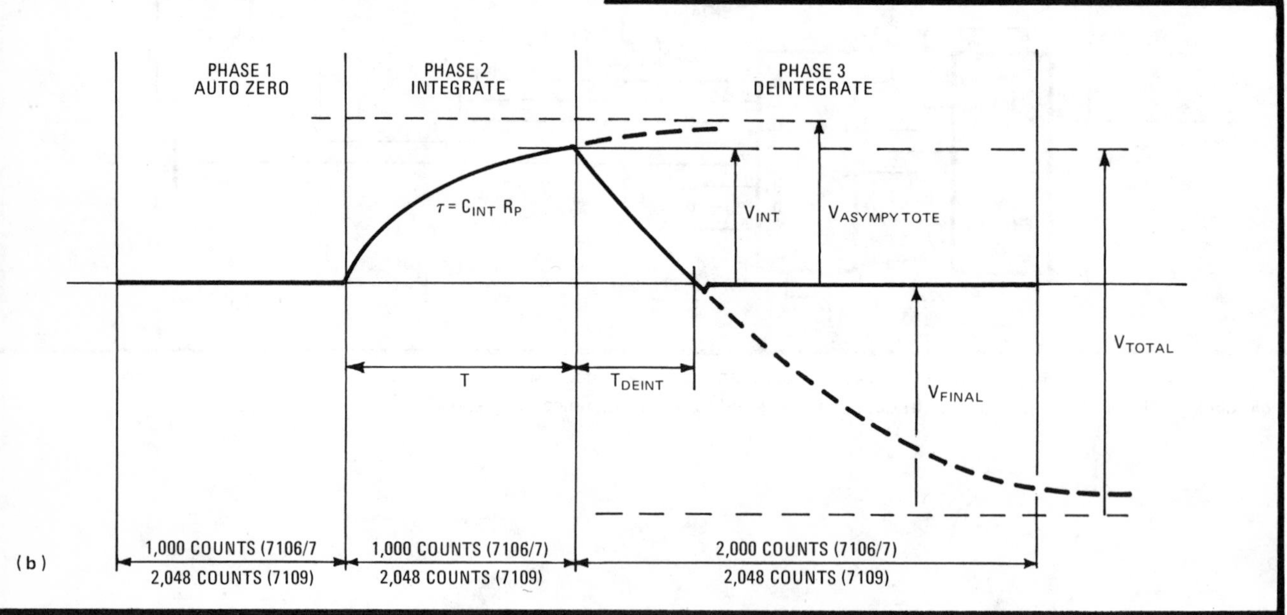

Log converter. ICL7106 analog-to-digital converter may be used to measure the logarithmic ratio of two input voltages. Modifying converter's input circuit (a) and integrating network and selecting suitable time constants ensure that its output is proportional to $\log_{10}(V_1/V_2)$. Timing diagram (b) clarifies circuit operation.

555 timer IC freezes digital panel meter display

by Howard M. Berlin
Wilmington, Delaware

Connecting a 555 timer as an astable multivibrator produces a sample-and-hold circuit that will increase the display time of many digital panel meters. This increase is an advantage in situations where the displayed value changes rapidly, making it difficult to determine an average reading. For example, a sensor monitoring pressure changes near a source of mechanical vibration can produce readings on a 4½-digit DPM that vary as much as ±10 digits in a quarter second, making a visual approximation virtually impossible.

When used with a DPM with an external-hold input such as the 4½-digit Fairchild 540921, the 555's signal overrides the internally controlled sampling period at pin N of the meter (see figure). The external-hold port is level-sensitive, so it is desirable to sample the test signal for short times. The output of the timer is a pulse train with a duty cycle of $d = (R_1 + R_2)/(R_1 + 2R_2)$, which can approach 100%, and a frequency of $f = 1.443/C(R_1 + 2R_2)$ hertz.

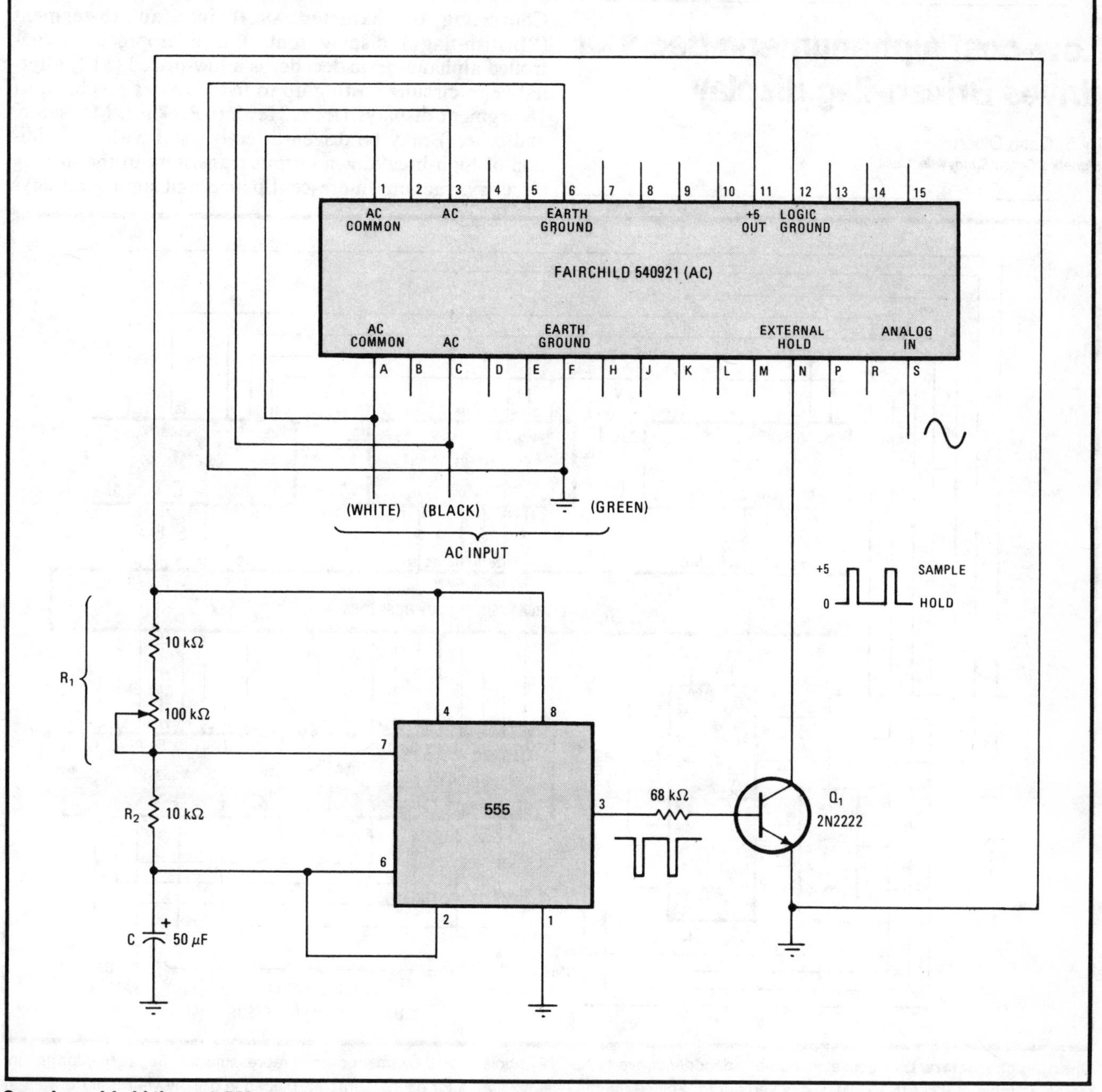

Sample-and-hold timer. A 555 extends the display period to relax hyperactive digital panel meters. The display can be frozen for 0.7 to 4.2 seconds, and updated every 1 to 4.5 seconds. Power for the circuit can be obtained from the DPM's 5-volt supply or a battery source.

The 555 output is inverted by transistor Q_1 so that a logic 1 is periodically presented to the external-hold input to sample the input analog voltage at pin S for a time that is small compared to the total sampling time. When the collector of the transistor is at logic 0, the sample is held and displayed for the number of seconds determined by the ratio d/f. With the values shown, the signal can be displayed for times ranging from 0.7 to 4.2 seconds. Updating is possible every 1 to 4.5 s.

Power for the timer is obtained from the meter, and the current drain is only 4 milliamperes. The circuit has been used for controlling several DPMs simultaneously, with additional transistor circuits connected to pin 3 of the timer. □

Low-cost alphanumeric decoder drives British-flag display

by S. Cash Olsen
Signetics Corp., Sunnyvale, Calif.

Converting 64-character ASCII into an 18-segment ("British-flag") display font, this microprocessor-controlled alphanumeric decoder is a low-priced ($12) alternative to circuits costing up to five times as much. Most 18-segment displays (from Hewlett-Packard, Monsanto, and others) may be driven directly. And with the addition of high-breakdown output transistors to the driving circuitry, vacuum fluorescent panels and similar displays

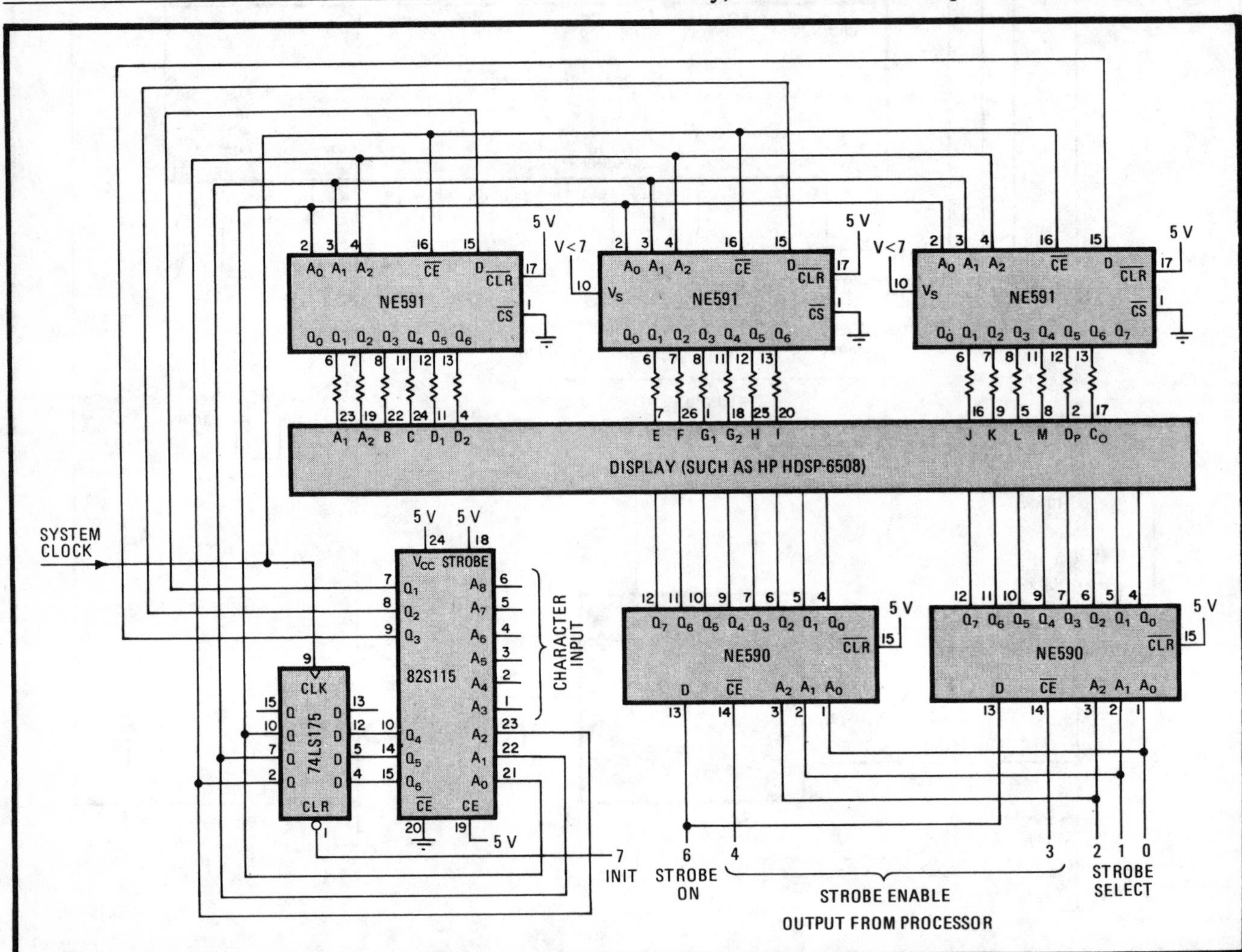

Charting characters. Low-cost alphanumeric decoder converts ASCII symbols into 18-segment display representation. Segment information, stored as table in 82S115 PROM, is clocked out in 3-bit segments over six states for each character, placed in display via NE591 drivers at any location by NE590 strobe latches. PROM character-generation table outlines method utilized to create symbols.

READ ONLY MEMORY CHARACTER GENERATION

Address	Q_6	Q_5	Q_4	Q_3	Q_2	Q_1	Symbol
000_8	0	0	1	X	X	X	commercial @
001	0	1	0	1	1	0	
002	0	1	1	1	1	0	
003	1	0	0	1	1	0	
004	1	0	1	0	0	0	
005	1	1	0	1	0	0	
006	1	1	1	1	0	0	
007	1	1	1	X	X	X	
010	0	0	1	X	X	X	letter A
011	0	1	0	1	1	0	
012	0	1	1	1	1	0	
013	1	0	0	1	1	0	
014	1	0	1	1	1	0	
015	1	1	0	0	0	0	
016	1	1	1	0	0	0	
017	1	1	1	X	X	X	
⋮			⋮				
770	0	0	1	X	X	X	punctuation ?
771	0	1	0	1	0	0	
772	0	1	1	1	0	0	
773	1	0	0	1	0	1	
774	1	0	1	0	0	0	
775	1	1	0	0	0	0	
776	1	1	1	0	0	0	
777	1	1	1	X	X	X	

Key

A_1	E	J
A_2	F	K
B	G_1	L
C	G_2	M
D_1	H	D_P
D_2	I	C_O

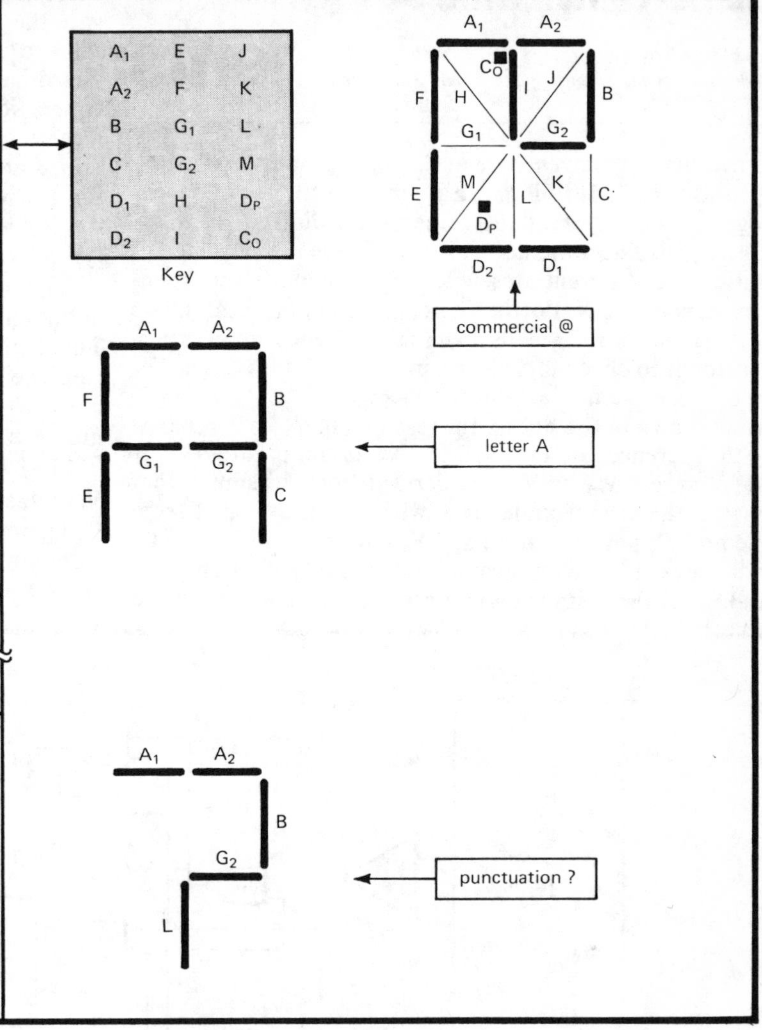

requiring high voltage may be accommodated, also.

In general operation (see figure), the microprocessor coordinates character selection, strobe-timing, and overall control duties with the aid of the NE590 strobe drivers, the NE591 peripheral display drivers, and the 74LS175 quad latch. When suitably addressed, the 82S115 512-word-by-8-bit PROM, which stores all the ASCII characters, delivers a logic-state table corresponding to the character selected via the clocked 74LS175 and the NE591s.

The PROM functions both as a character-request look-up table and as a state machine, with the quad flip-flop holding the current machine state. Bit 7 of the processor initializes the state to zero at the beginning of a character-decode cycle.

Logic signals corresponding to the character desired are then applied to pins A_3–A_8 of the PROM, and the device is clocked through seven states (see table) so that the desired segments are excited. The display is then strobed and the character thus placed in any desired location via command from pins 0 to 6 of the processor via the strobe latches, each latch of which is enabled separately. This process is repeated for up to 64 characters, the maximum that may be placed on the display at any given instant. Thereafter, as in all multiplexed displays, only one character is enabled at any time. All characters will appear to be displayed continuously, however, because of the high scanning rate.

As seen in the table, during each clocked state the PROM generates 3 bits of segment information. Six such states define the character produced. Thus only three display segments switch during each NE591 latching period, substantially reducing load transients and large load-current variations, which tend to cause difficulty in circuits of this kind. Only six of each NE591's eight outputs are used, to reduce power dissipation. Note that each device handles 6 of the total of 18 display segments for each character.

As the circuit is digital, neither layout nor component values are critical. The clock frequency, typically less than 5 MHz, should have a minimum pulse width (t_w) of 100 ns, however, in order to ensure proper display and strobe latching. □

Dynamic logic probe displays five states

by Mihai Antonescu
Federal Institute of Technology, Lausanne, Switzerland

Providing the convenience of the logic probe proposed by Prasad and Muralidharan[1], which utilizes a seven-segment display readout rather than discrete LEDs or lamps, this five-state detector also senses the presence of pulses and differentiates between the logic 0 and open-circuit conditions. Furthermore, provision has been made to bias the probe via its external-reference inputs, permitting it to check logic levels in circuits built with either TTL or complementary-MOS devices.

As shown in the figure, the test signal, V_{in}, is compared with reference voltages V_{ref1} to V_{ref4} at comparators A_1 to A_3. Voltages V_{ref1} to V_{ref2} are derived from the supply that powers the circuit under test, with V_{ref3} (developed from the probe's power source) and V_{ref4} at approximately 0.05 volt and 0.1 V below ground, respectively. Switches S_{1a} and S_{1b} at the voltage divider are used to set the required logic-level references at A_1 and A_2 in order to check either TTL or C-MOS circuits.

Gates G_1–G_{13}, comprising the combinational-logic detector, determine V_{in}'s relation to V_{ref1}–V_{ref4} and activate the appropriate segments of the display (see table). A_1 and A_2 are used to check for the logic 1, 0, and guard-band conditions. A_3 is used to detect the no-connection, or open-circuit, condition, which can be differentiated from the logic 0 state because V_{ref4} is maintained at 0.1 V below ground. Note that the logic 0 state for TTL will typically be 0.4 V and is hardly ever below 0.2 V, whereas the logic 0 state for C-MOS will be typically 0.01 V.

Switch S_2 must be depressed to catch any expected input pulses having a width down to 15 nanoseconds. This action resets the 555 one-shot, enabling it to override the displayed symbol with the letter P when the pulse arrives. If a train of pulses having a repetition rate greater than about 0.2 seconds (the time constant of the 555) is detected, the P will be displayed indefinitely. With faster one-shots, pulses of 5 ns can be snared.

The probe can be powered by any dc source having a minimum voltage of 19 V. Resistor R should be selected to pass about 30 mA to the probe circuit. □

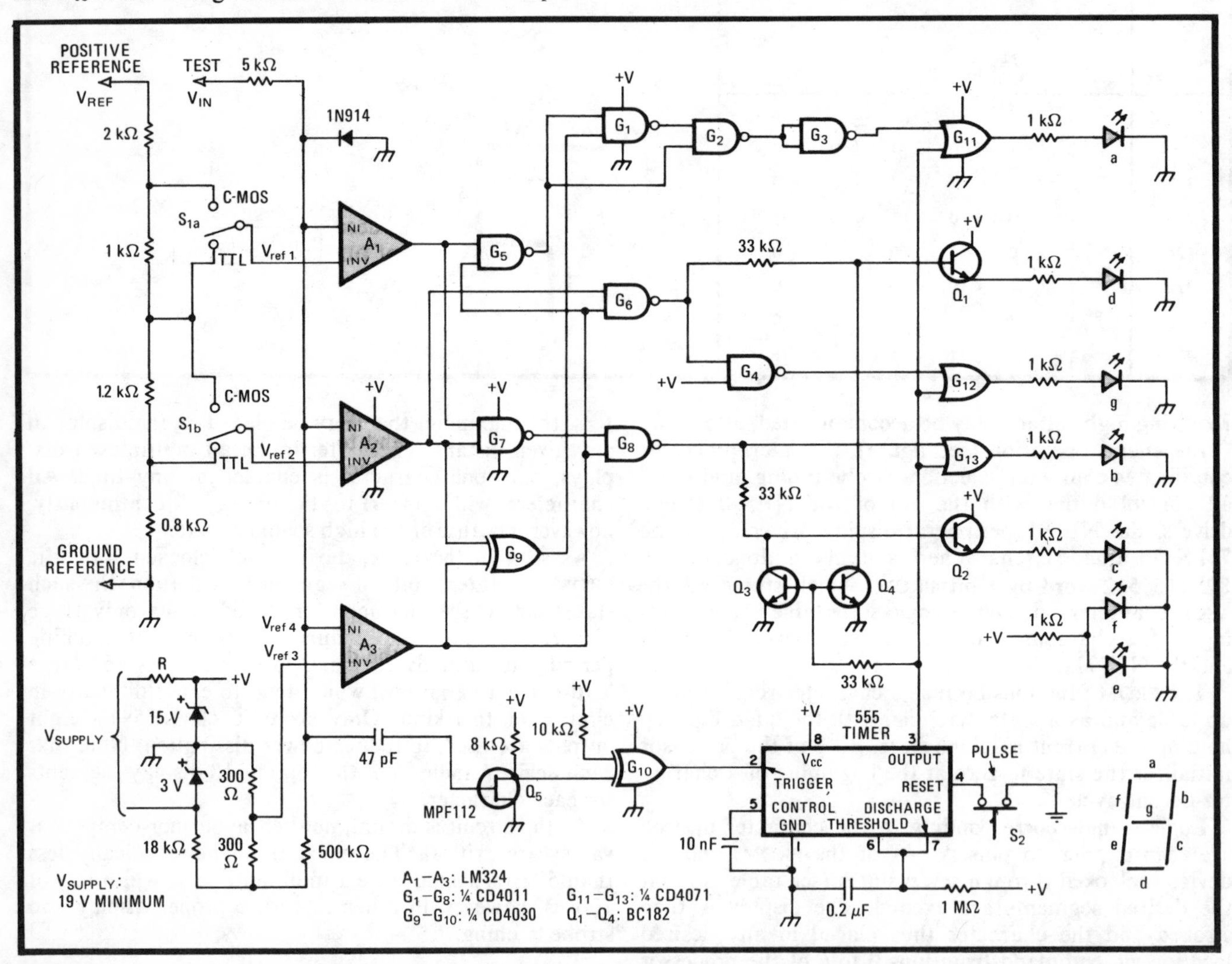

All-state. Logic probe having seven-segment display detects logic 1 and guard-band conditions, can differentiate between logic 0 state and open circuit, and senses a pulse train. Unit can check logic levels in circuits built with the two most popular families, C-MOS and TTL.

RESPONSE OF FIVE-STATE LOGIC PROBE												
Test voltage		Display	Comparator			Display segments						
C-MOS*	TTL	Letter	A_1	A_2	A_3	a	b	c	d	g	e	f
$V_{in} < 40\%$	$V_{in} < 0.8$	L	L	L	H				on		on permanently	
$40\% < V_{in} < 60\%$	$0.8 < V_{in} < 2$	F	L	H	H	on				on		
$V_{in} > 60\%$	$V_{in} > 2$	H	H	H	H		on	on		on		
NC	NC	O	L	L	L	on	on	on	on			
pulse	pulse	P	—	—	—	on	on			on	*% of V_{ref}	

References

1. S. Jayasimha Prasad and M. R. Muralidharan, "Logic tester has unambiguous display," *Electronics*, March 3, 1977, p. 117.

LSI counter simplifies display for a-m/fm radio

by Gary McClellan

Beckman Instruments Inc., Fullerton, Calif.

The design of a display providing a direct readout of any frequency tuned by an a-m/fm radio is made simple with this circuit, which uses a large-scale integrated counter-driver to determine the frequency of the receiver's local oscillator. The counter is unique in that it contains circuitry that subtracts the receiver's intermediate frequency from the local oscillator frequency in order that the true channel frequency may be found. The

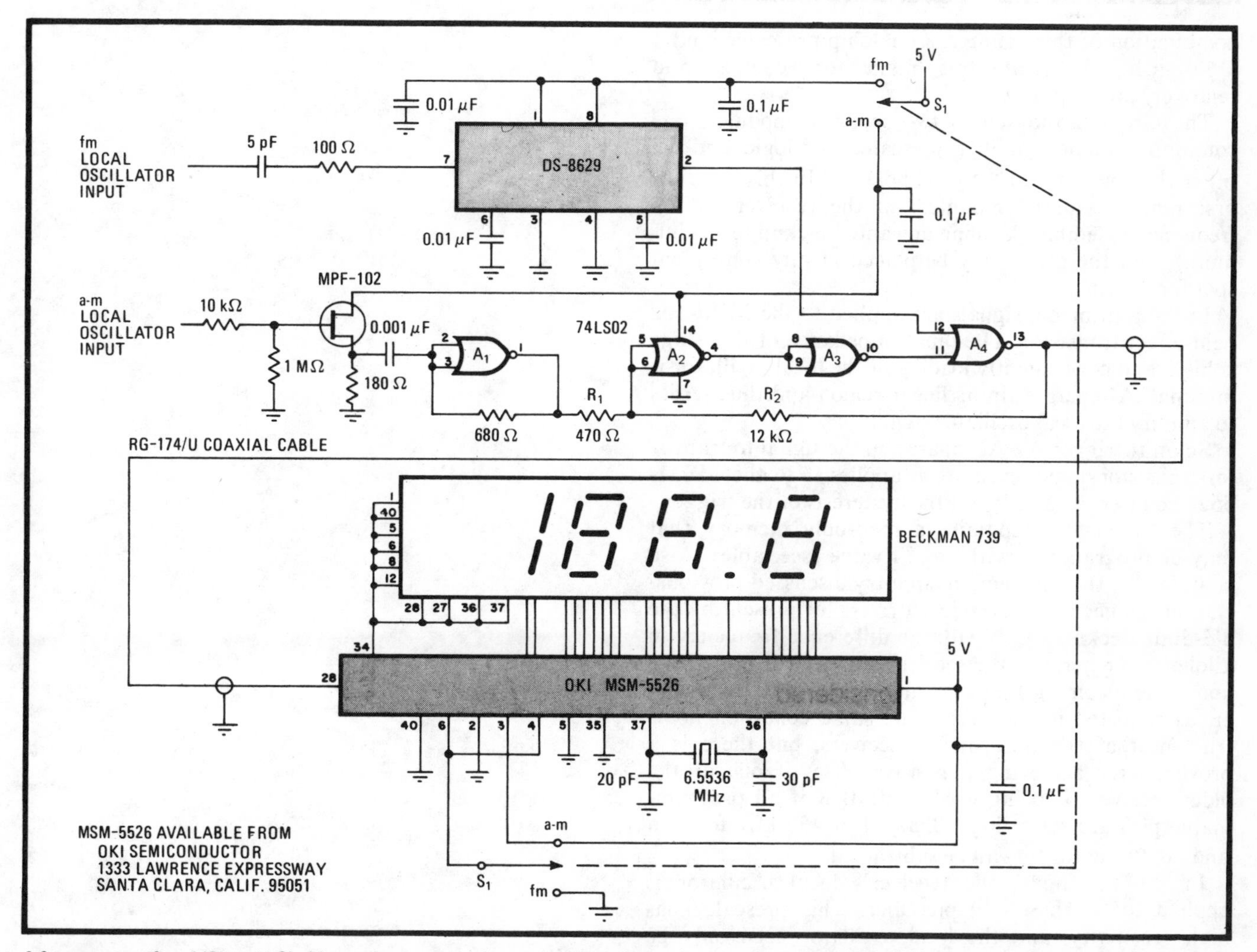

I-f compensation. MSI and LSI chips reduce cost and complexity of display for a-m/fm radio. MSM-5526 counter has circuitry for subtracting receiver's i-f frequency (see table) from radio's local oscillator input so that the true channel frequency may be displayed.

MSM-5526 INTERMEDIATE-FREQUENCY OFFSET					
Display mode	Input-pin state				Offset (a-m in kHz, fm in MHz)
	2	3	4	5	
a-m (pin 6 high)	H	H	H	X	−452.5
	L	H	H	X	−454.5
	H	L	H	X	−456.5
	L	L	H	X	−465.5
	H	H	L	X	−467.5
	L	H	L	X	−469.5
fm (pin 6 low)	H	H	H	H	10.68
	L	H	H	H	10.71
	H	L	H	H	10.75
	L	L	H	H	10.79
	H	H	L	H	10.82
	L	H	L	H	−10.58
	H	L	L	H	−10.60
	L	L	L	H	−10.61
	H	H	H	L	−10.62
	L	H	H	L	−10.63
	H	L	H	L	−10.65
	L	L	H	L	−10.66
	H	H	L	L	−10.69
	L	H	L	L	−10.70
	H	L	L	L	−10.72
	L	L	L	L	−10.73

combination of this counter, a one-chip prescaler, and a 3½-digit liquid-crystal display makes for a compact and relatively low-cost unit.

The circuit is housed in two separate modules, one containing the preamplifier, prescaler, and logic, and the other the counter and LCD components. In this way, the first module can be mounted on the receiver's radio frequency assembly (keeping unwanted pickup to a minimum), and the other may be placed at any convenient spot for viewing.

In the a-m mode, signals are applied to the MPF-102 field-effect transistor. The input impedance of this stage is high, and consequently loading of the local oscillator is minimal. A_1 operates in its linear region and thus serves to amplify the local oscillator signal.

Schmitt trigger A_2–A_4 squares up the signal to transistor-transistor–logic levels, then applies it to the MSM-5526 counter. R_1 and R_2 set the hysteresis of the trigger.

The MSM-5526 contains a read-only memory that may be programmed with any i-f value (see table). Also contained is the subtraction circuitry discussed previously, and the necessary decoders/drivers for presenting the 3½-digit Beckman LCD with the difference frequency in kilohertz. Generally, the local oscillator will always lie above the incoming frequency in the modern a-m receiver, as reflected in the table. The same condition holds true most of the time in fm receivers, but there is a provision for achieving a positive offset if one of the older receivers is being used. Note that if all programmable pins are set at logic 0, an i-f of 455 kHz for a-m and 10.7 MHz for fm will be subtracted.

In the fm mode, the receiver's local oscillator is applied to the DS-8629 prescaler. This prescaler has high sensitivity, and the local oscillator need only be capable of supplying a minimum of 12 mV at 100 MHz.

The DS-8629 divides the incoming frequency by 100. Then the signal is gated through to the counter via A_4. In this configuration, an i-f equal to 1/100 the fm receiver's nominal value (10.7 MHz) is subtracted from the input frequency to the counter, and the result is displayed in megahertz.

A number of practical considerations must be taken into account when building this display. Specifically, the presence of the prescaler will introduce a typical shift of 500 Hz in the read-out frequency. The error may be eliminated entirely by simple adjustment of the 30-picofarad air-variable trimming capacitor, located at pin 36 of the counter.

The first module should be shielded from the receiver's tuner if noise in the fm mode is to be held to a minimum. Housing the module in an aluminum enclosure will suffice in most cases. And although the liquid-crystal display will tend to generate less noise than many light-emitting-diode displays now available, shielding it may also be necessary in extreme cases.

Both modules should be coupled via a coaxial cable. Otherwise a broadband hiss may be heard when the unit is placed in the a-m mode. □

Moving-dot indicator tracks bipolar signals

by Ted Davis
Riverton, Ill.

Although bar- or dot-display chips are a simple means of indicating the instantaneous value of a signal, they respond only to unipolar levels, a definite drawback in processing audio-frequency signals with asymmetrical (bipolar) inputs. If reduced resolution is acceptable, one solution is to offset the audio voltage to the display chip. In this way it will be centered at half scale to allow for positive and negative signal excursions. Such a method is implemented in the scheme shown here.

The circuit is configured to detect signal changes in 6-decibel steps, making it useful for audio-level monitoring. Other steps may be ordered by rewiring the output circuit appropriately. The unit may also be used as a bin-sorter or percent-change indicator for ac inputs or, with removal of capacitor C_1 and consolidation of resistors R_4 and R_5, dc inputs.

Operational amplifier A_1 applies a reference voltage to the inverting input of A_2 so that it and the LM3914 bar/dot display may be offset by the desired amount. The value of the reference voltage, which is derived from the LM3914, is $V_r = 1.25[-2R_9/(R_8+R_9)+1]$ assuming that $R_6 = R_7$ and the reactance of C_1 is negligible. The offset signal thus applied to the signal input (pin 5) of the LM3914 is $V_r k$, where $k = R_3/R_4$.

Assuming also that $R_5 = R_3 - R_4$, the offset voltage can be made to vary linearly from $-1.25k$ to $+1.25k$ and be centered at any value simply by adjusting R_8 and R_9. To set the value at the mid-level digital output of the LM3914 dot or bar display, for example, R_8 and/or R_9 is varied so that Q_5 trips and, through the 74LS47 BCD-to-seven-segment decoder/driver, dims light-emitting diode 1. The user should then back off on the setting until Q_5 goes high again and then move the corresponding potentiometer halfway towards the position that would dim the LED once more.

Superimposed on the reference signal will be the component added by the audio signal, which at the

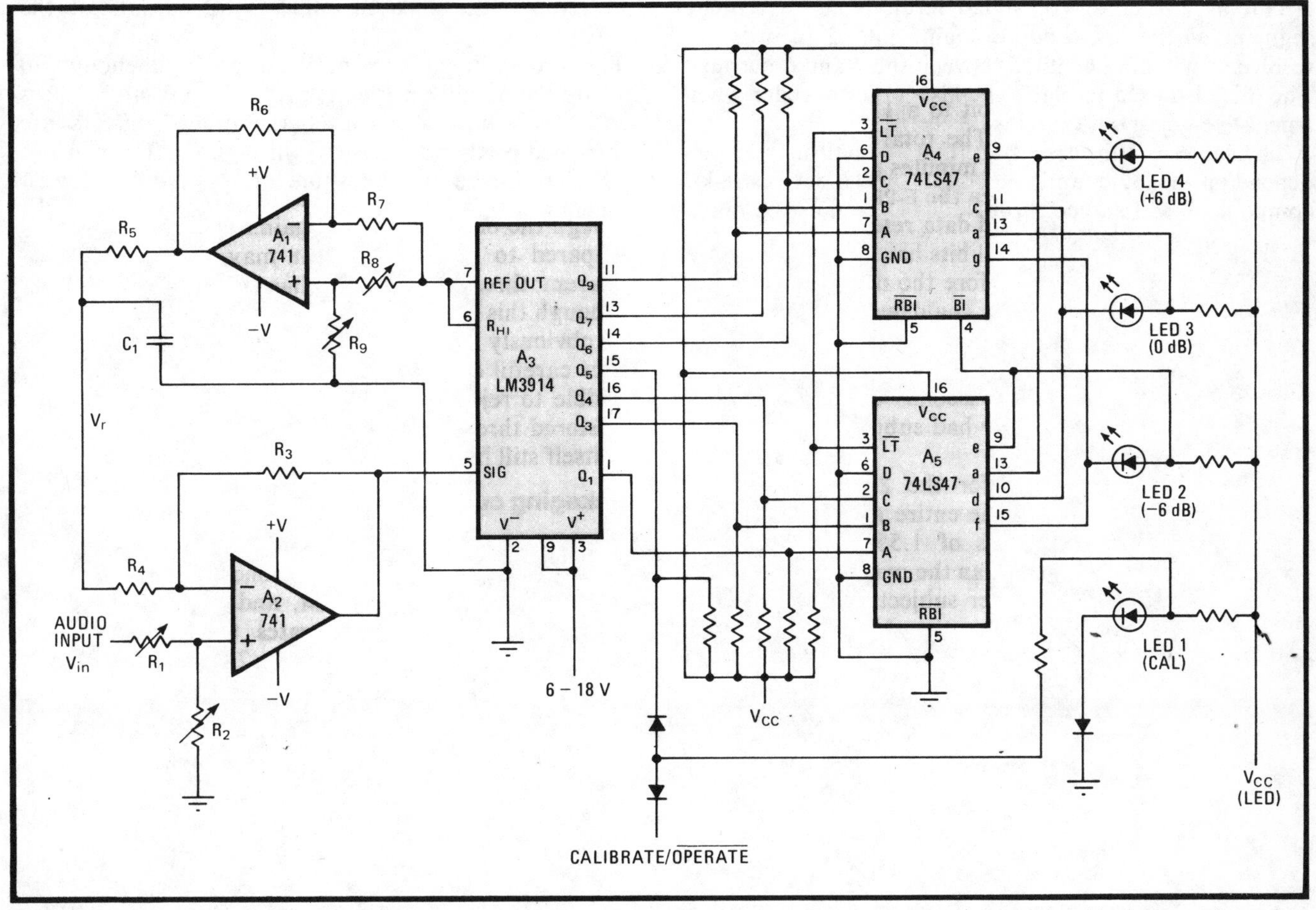

Plus and minus. Input of bar- or dot-display chip LM3914 is biased at user-set dc level so that it will respond to bipolar excursions of ac signals. Three LEDs serve as moving-dot indicator with a resolution of 6 dB. Truth table outlines circuit operation.

TRUTH TABLE: SIGNAL-LEVEL INDICATOR

INPUT V_{in}	A_4 D	C	B	A	$\overline{RBI}$	$\overline{BI}$	a	c	e	g	A_5 D	C	B	A	$\overline{RBI}$	a	d	e	f	LED 2	3	4	
below 10%	0	1	1	1	1	1	0	0	1	1	0	1	1	1	0	0	1	1	1			λ	(+6 dB)
10% to 20%	X	X	X	X	X	0	1	1	1	1	0	1	1	0	0	1	0	0	0		λ		(0 dB)
20% to 40%	0	1	1	1	1	1	0	0	1	1	0	1	0	0	0	1	1	1	0	λ			(−6 dB)
40% to 60%	0	1	1	1	1	1	0	0	1	1	0	0	0	0	0	1	1	1	1	ALL OFF			(underrange)
60% to 70%	0	0	1	1	1	1	0	0	1	0	0	0	0	0	0	1	1	1	1	λ			(−6 dB)
70% to 90%	0	0	0	1	1	1	1	0	1	1	0	0	0	0	0	1	1	1	1		λ		(0 dB)
above 90%	0	0	0	0	1	1	0	0	0	1	0	0	0	0	0	1	1	1	1			λ	(+6 dB)

X = don't care   Input voltage V_{in} normalized to full scale at pin 5 of LM3914

output of A_2 is equal to $V_{in}R_2(k+1)/(R_1+R_2)$. Thus positive and negative excursions of the ac signal will be detected by the LM3914. The scale factor is adjusted by applying the user-standard audio level to the input and adjusting R_1 and/or R_2 until the 0-dB LED just lights up.

The truth table outlines the overall operation of the circuit as a function of signal level. Note that the e segment of the low-order 6 shunts LED 2 in order to resolve a switching conflict between the 4 and 6 outputs. The 6 is also used to blank the high-order decoder when a negative-going 0-dB level is detected.

The values of the current-limiting and pull-up resistors depend on the logic family utilized; for TTL devices, 1-kΩ components will suffice throughout. Care must be taken to ensure that the voltages developed at the e output satisfy the noise-margin requirements of the BI input of A_4; that is, the total sink current at e must not raise the voltage above the maximum logic 0 level and the drop across LED 2 in series with the sink transistor must exceed the minimum logic 1 level. R_8+R_9, in parallel with R_7, set the sink current of the outputs of the LM3914.

The programmed current must be high enough to saturate the output transistors given the pull-up resistors used. The values of most of the other resistors are determined by the values of R_7 through R_9. The value of C_1 is determined by the value of R_4 and the lowest frequency of V_{in}. □

Chapter 14
ENCODERS AND DECODERS

Priority encoder simplifies clock-to-computer interfaces

by John P. Oliver
Astronomical Time Mechanisms, Gainesville, Fla.

Although a computer can easily be programed to keep track of time internally, in many cases it is better to allow the processor to read an external clock. This is especially so when the clock must be kept running during downtime and program loading or debugging. Using a 9318 priority-encoding integrated circuit to interface the National MM5314 digital clock chip to a computer permits easy loading of time data into memory under interrupt control, and the entire system requires only a handful of parts.

The complete clock and interface unit is shown in Fig. 1. The MM5314, powered from the ac line, derives its time base from the 60-hertz line frequency as well. The 100-kilohm resistor is connected to the secondary of the power transformer, to supply the power-line frequency to the 50/60-Hz input of the clock chip. The outputs of the seven-segment digits are multiplexed at a rate of about 1 kilohertz, determined by the RC time constant, and the sequential order is from unit seconds to tens of hours.

The 9318 encodes the address of the highest-priority data line (ordered from D_7 to D_0) having a logical-1 input, into a 3-bit binary code at outputs $\overline{A}_0$ through $\overline{A}_2$. Thus, the encoder is used to indicate which of the seven-segment-output digits is being scanned by the multiplexer, thereby telling the computer the time in a serial format—first in seconds, then tens of seconds, then minutes, and so on. The digits are identified by the address lines, and information for each digit is conveyed by five of the seven segments—a, b, e, f, and g. Segments c and d are not sent to the computer.

Figure 2 details how the omission of the c- and d-segment signals still permits unambiguous identifica-

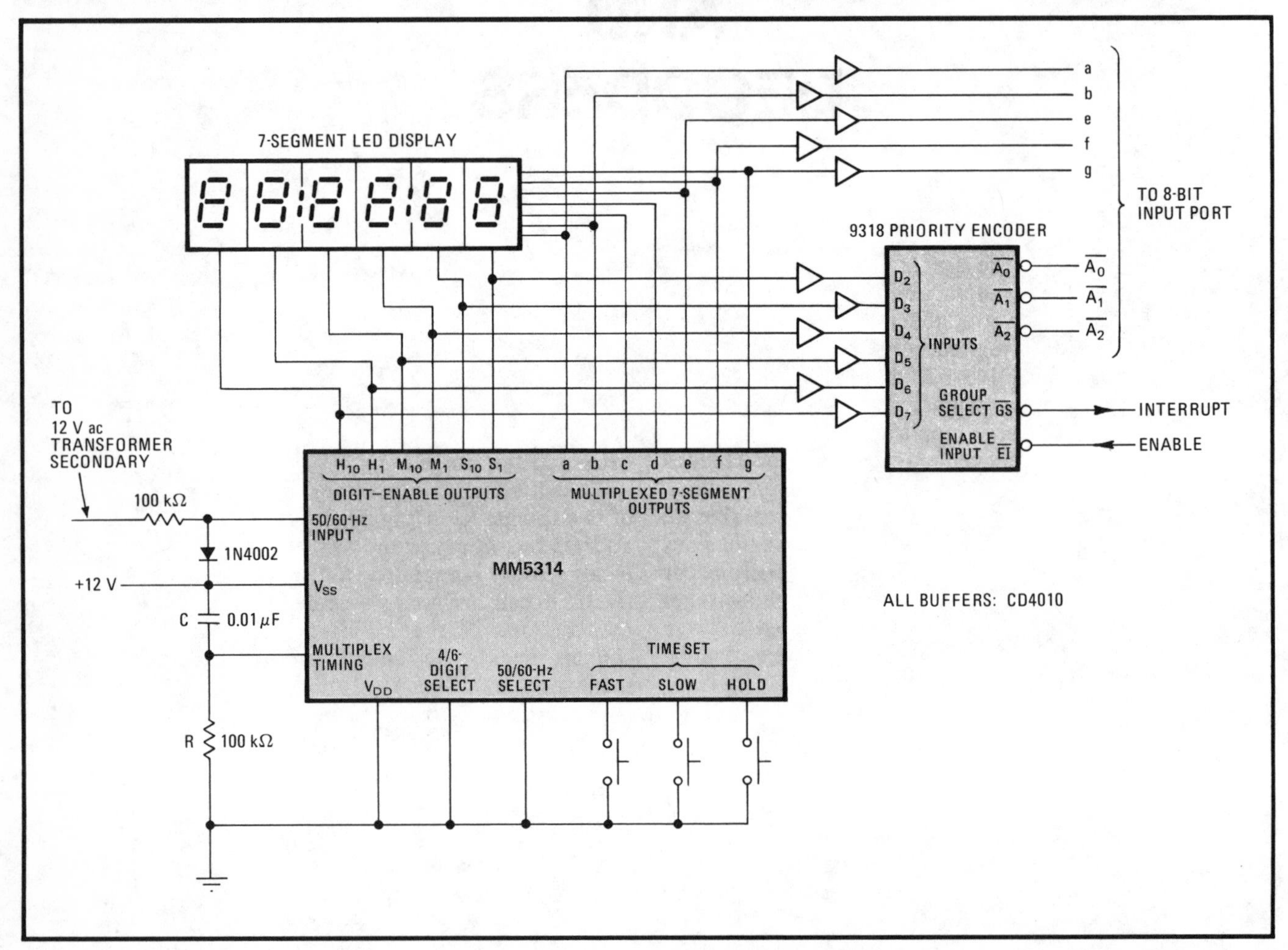

1. Punch in. Complete clock-to-computer interface requires few parts and includes seven-segment LED display. Note that no digit- and segment-drive transistors are required, as the MM5314 chip can power most small displays to reasonable brightness.

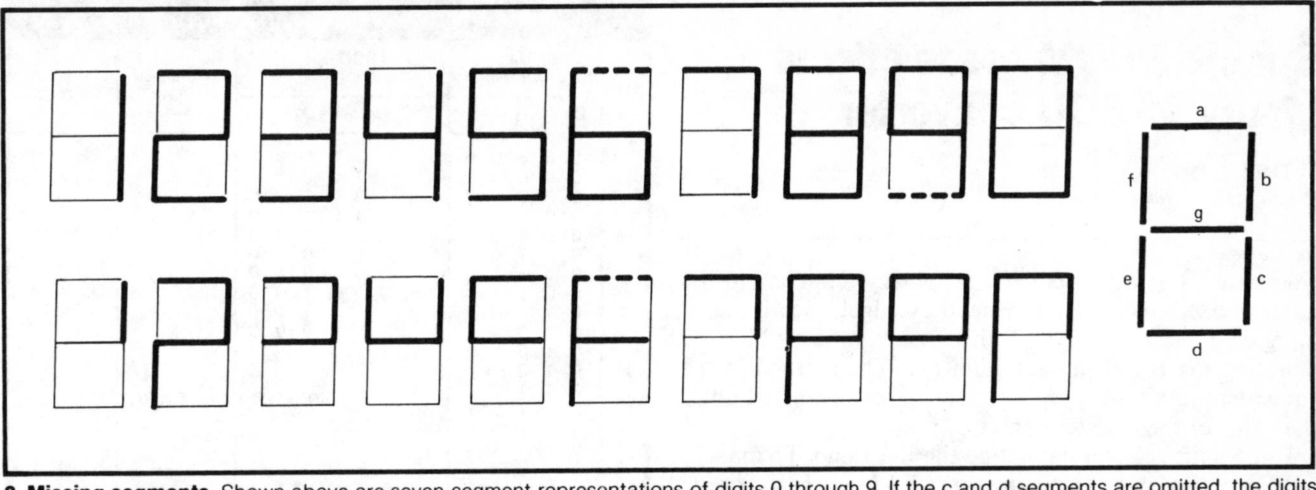

2. Missing segments. Shown above are seven-segment representations of digits 0 through 9. If the c and d segments are omitted, the digits retain their integrity, as seen in the lower figures. Some clock chips will illuminate the "tail" on the digits 6 and 9.

3. Outputs. The upper eight signals go to the input port of the computer, while the remaining two, the group-select ($\overline{GS}$) and enable-input ($\overline{EI}$) require an input and output line, respectively.

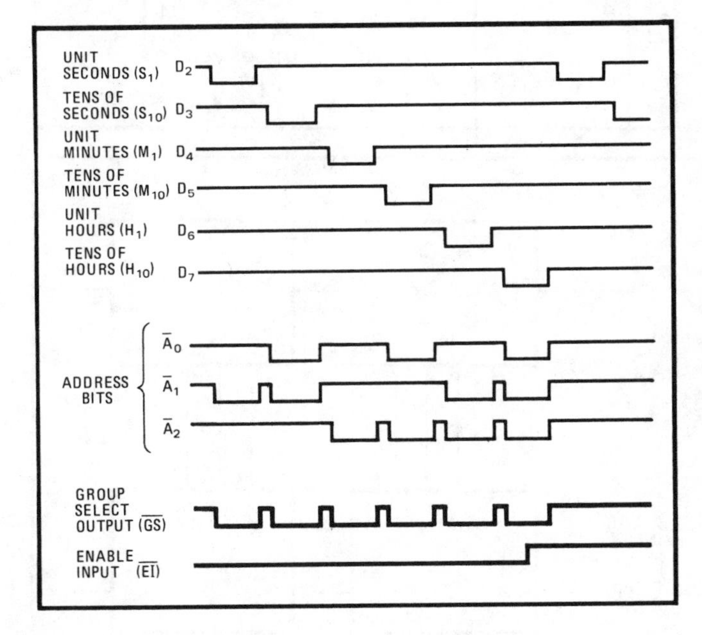

tion of the digits. Actually, the minimum number of lines required for decimal encoding is four—and clock chips having binary-coded-decimal outputs, such as the MM5311, MM5312, and MM5313, will reduce to seven the number of lines to the computer's input port.

The waveforms at the outputs are shown in Fig. 3. Note that all the signals are negative logic. The group-select output ($\overline{GS}$) of the encoder indicates that priority inputs are present by going low, as it is a negated output. The $\overline{GS}$ output, putting out a pulse as each digit of the display is scanned, can provide an interrupt to the computer. These interrupts can be enabled and disabled within the computer programing or, alternatively, an output line from the computer can be sent to the enable input ($\overline{EI}$) of the priority encoder. Holding the $\overline{EI}$ high inhibits the $\overline{GS}$ output upon completion of the next digit-segment-output, as implied by the waveforms.

If the latter method of enabling and disabling the interrupts is chosen, the user should beware of possible spurious $\overline{GS}$ output spikes as the $\overline{EI}$ is switched by the computer. □

Special PROM mode effects binary-to-BCD converter

by D. M. Brockman
The Boeing Co., Seattle, Wash.

Computer software converts a binary number to its binary-coded-decimal equivalent by algebraically summing each bit, starting with the most significant bit, and doubling the result of each addition. The method is the equivalent of a binary shift register where a shift doubles the value of the register contents.

But a shift register cannot do the job alone. To make a hardware binary-to-BCD converter also requires a storage register and a programmable read-only memory that is operated in an unusual mode. The circuit is especially advantageous when processing large binary numbers, since PROMs may be simply cascaded for additional digits. The circuit speed is typically a few megahertz.

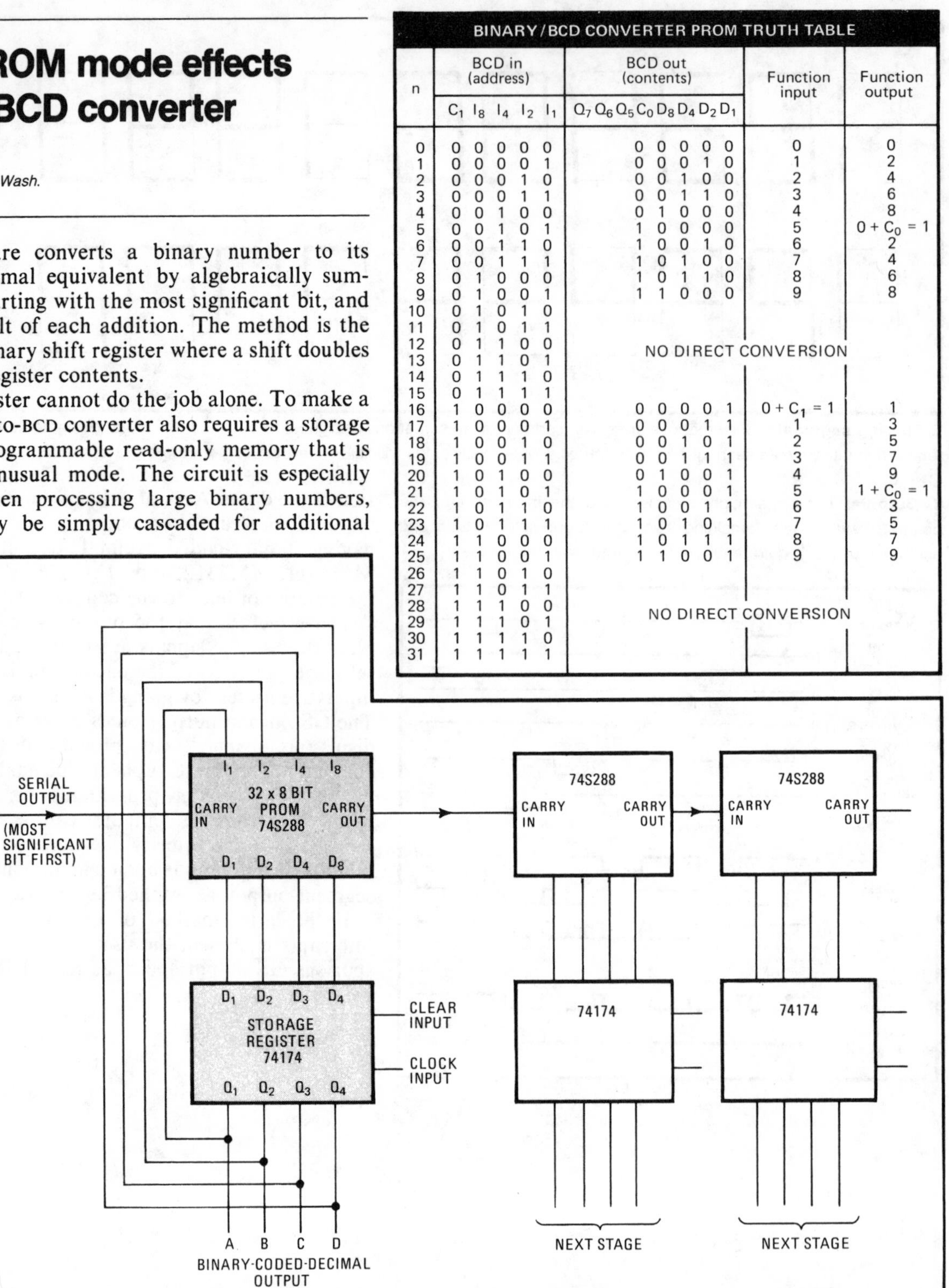

BINARY/BCD CONVERTER PROM TRUTH TABLE

n	BCD in (address)	BCD out (contents)	Function input	Function output
	C_1 I_8 I_4 I_2 I_1	Q_7 Q_6 Q_5 C_0 D_8 D_4 D_2 D_1		
0	0 0 0 0 0	0 0 0 0 0	0	0
1	0 0 0 0 1	0 0 0 1 0	1	2
2	0 0 0 1 0	0 0 1 0 0	2	4
3	0 0 0 1 1	0 0 1 1 0	3	6
4	0 0 1 0 0	0 1 0 0 0	4	8
5	0 0 1 0 1	1 0 0 0 0	5	$0 + C_0 = 1$
6	0 0 1 1 0	1 0 0 1 0	6	2
7	0 0 1 1 1	1 0 1 0 0	7	4
8	0 1 0 0 0	1 0 1 1 0	8	6
9	0 1 0 0 1	1 1 0 0 0	9	8
10	0 1 0 1 0	NO DIRECT CONVERSION		
11	0 1 0 1 1			
12	0 1 1 0 0			
13	0 1 1 0 1			
14	0 1 1 1 0			
15	0 1 1 1 1			
16	1 0 0 0 0	0 0 0 0 1	$0 + C_1 = 1$	1
17	1 0 0 0 1	0 0 0 1 1	1	3
18	1 0 0 1 0	0 0 1 0 1	2	5
19	1 0 0 1 1	0 0 1 1 1	3	7
20	1 0 1 0 0	0 1 0 0 1	4	9
21	1 0 1 0 1	1 0 0 0 1	5	$1 + C_0 = 1$
22	1 0 1 1 0	1 0 0 1 1	6	3
23	1 0 1 1 1	1 0 1 0 1	7	5
24	1 1 0 0 0	1 0 1 1 1	8	7
25	1 1 0 0 1	1 1 0 0 1	9	9
26	1 1 0 1 0	NO DIRECT CONVERSION		
27	1 1 0 1 1			
28	1 1 1 0 0			
29	1 1 1 0 1			
30	1 1 1 1 0			
31	1 1 1 1 1			

BCD converter. Storage register in feedback loop with programmable read-only memory allows continuous updating of the memory's contents. Multidigit conversion, if necessary, is relatively simple because cascade stages are connected by one lead only.

The circuit of the converter is shown in the figure. After loading the 74165 shift register, which accepts either parallel or serial data, an n-bit binary word is serially clocked into the carry input of the 74S288 256-bit PROM. This 32-by-8-bit memory has input and output stages that accept or generate BCD data.

The unusual features in the PROM's use are that its contents are updated in each clock cycle and that the updating forms the circuit's desired equation:

$$\text{BCD output} = 2(\text{BCD input}) + \text{carry input}$$

The PROM's truth table shows the BCD output for a

given BCD input plus carry input. Any BCD sum exceeding 9 will generate a carry output, thus permitting multidigit conversion if the carry output is brought to the carry input of the next stage of the circuit. If the most significant bit of the n-bit number is shifted out first to the PROM, then after n clock pulses, the resulting BCD-equivalent number will be at the output of all the 74174 storage registers.

The output of each storage register is the input for its PROM, and its content is that of its PROM during the previous clock period (ignoring the carry out). The circuit can convert a 20-bit number to its BCD equivalent at speeds of at least a few megahertz. □

RC-discharge clock makes a-d encoder logarithmic

by V. Ramprakash
Electronic Systems Research, Madurai, India

This analog-to-digital converter produces a 12-bit digital output that represents the natural logarithm of an input audio signal in the 1-to-1,000-millivolt (60-decibel) range. It is useful for monitoring slow changes in many natural processes. The circuit also may be adapted for use as a combination voice compander and encoder in a digital communications system.

Operation is based on the principle that the discharge rate of a voltage stored across a resistance-capacitance network is proportional to the logarithm of the ratio of the instantaneous to the initially applied voltage. When the voltage across the RC network is used to control the gating time of a counter, the counter's output is a binary-coded decimal number equal to ln V_i, where V_i is the input voltage.

The voltage across a discharging capacitor, V_c, in an RC network is given by:

$$V_c = V_r e^{-t/RC}$$

where V_r is the initial voltage. This equation, when transposed, becomes:

$$t = -RC \ln(V_c/V_r)$$

Let t_i represent the time it takes the capacitor to discharge from V_r to the input voltage V_i. If during this time a down counter is gated while being clocked at frequency f_c, the number of counts reached will be:

$$n_d = f_c t_i = -f_c RC \ln(V_i/V_r)$$

More generally, if the down counter is initially at a count of n_i, then the net count n_n after time t_i will be:

$$n_n = n_i - n_d = n_i + f_c RC \ln(V_i/V_r)$$

Letting V_r = 1,000 mv and n_i = 690, we arrive at:

$$n_n = 690 - 100 \ln 1000 + 100 \ln V_i = 100 \ln V_i$$

which can be reduced to $n_n = \ln V_i$ with appropriate scaling in a practical circuit.

The circuit shown in the figure implements the derived equation. When the output of the 555 timing clock is high, the 74192 up-down counter is loaded with the number 690 as shown in the table. At the same time, capacitor C_T is charged to 3.3 volts. R_1 facilitates the scaling of this voltage so that the potential as seen at the noninverting input of the comparator is 1 v.

When the output of the timing clock moves low, the 74192 begins to count down at a 10-kilohertz rate and C_T begins to discharge. As the voltage at the noninverting input drops below the sample voltage, V_i, the

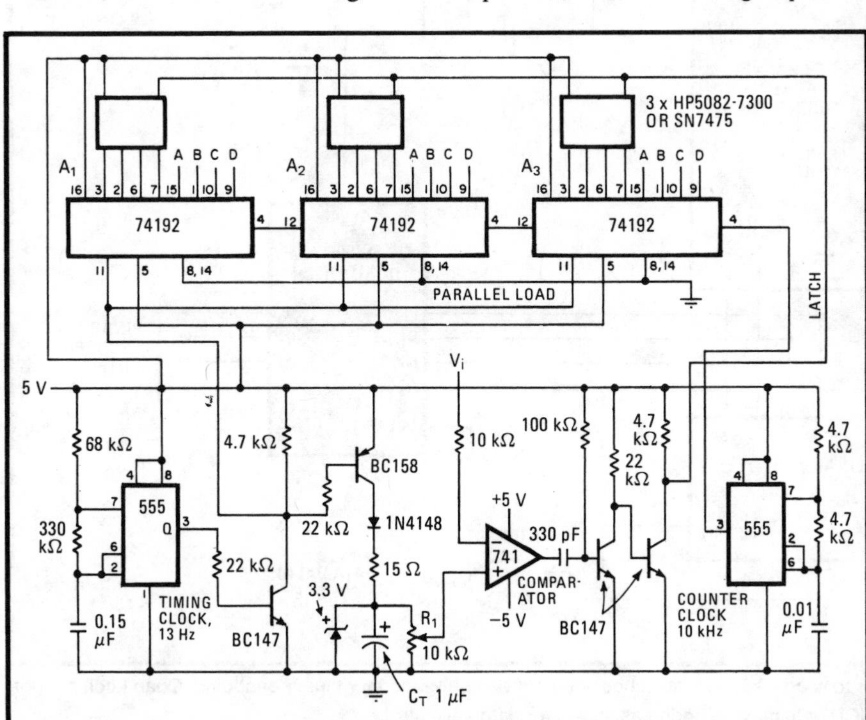

LOADING OF 74192 UP-DOWN COUNTER

COUNTER	NUMBER	CODING			
		A	B	C	D
A_1	6	1	1	0	0
A_2	9	1	0	0	1
A_3	0	0	0	0	0

Natural processing. Circuit is logarithmic a-d converter and digital encoder in one. Voltage across C_T, which decays exponentially at start of each sampling cycle, controls gating time of 74192 counters, ensuring logarithmic response. Counters are preset to 690 before each encoding to eliminate constant-coefficient terms inherent in circuit's transfer function, so that output from counter is n = ln V_i.

comparator output moves low and generates a latch pulse for the BCD-to-seven-segment displays, or 4-bit latches, as required. Thus the contents of the counter are stored in either the display or the latches. The sequence is then repeated. Note that a decimal point is located in the most significant display (corresponding to counter A_1) so that the natural logarithm of a 1,000-mV input signal will be correctly displayed as 6.90.

The low-frequency clock limits the input signal sampling rate to 13 hertz. However if the clock frequency is increased to 5 kHz or so, and the clock counter is replaced by one that can run at a few megahertz, the circuit will serve as an excellent speech encoder. □

Multiplexer scans keyboard for reliable binary encoding

by Merritt E. Keppel,
Richmond, Va.

The myriad keypad-coded requirements in point-of-sale and other data-entry applications require a reliable, bounceless binary encoder. This inherently bounceless complementary-MOS circuit encodes up to 16 inputs and latches each BCD number for stable output.

As shown in the figure, the keypad is scanned by an RCA CD4067 16-channel analog multiplexer. Each of the inputs has a 1-megohm pull-up resistor. When a key is depressed, the appropriate channel is driven low. A clock source, which can be derived from the device using the keypad, drives a CD4029 4-bit binary counter. Scanning occurs as the counter addresses the multiplexer.

When the counter has addressed an input to the multiplexer that is low, the common-out pin of the multiplexer goes low for one clock cycle. A differentiating network changes this pulse into a negative spike used to strobe the counter data (the binary representation of the particular key input) into the CD4042 quad latch. The binary word remains in the latch until another key is depressed, providing a stable output.

The circuit is inherently bounceless, since a noisy input will latch itself on its first negative transition and remain stable until another input is selected. The scan rate is not critical, but the clock frequency should be greater than 200 hertz for a normal key-entry rate.

In some cases, the processing equipment to which the circuit is connected will require a data-valid strobe. This can be accomplished by gating the clock with the carry-out pin of the counter. A two-input NOR gate, such as in a CD4001 quad package, provides a data-valid strobe one half a clock pulse before the beginning of every scan cycle. □

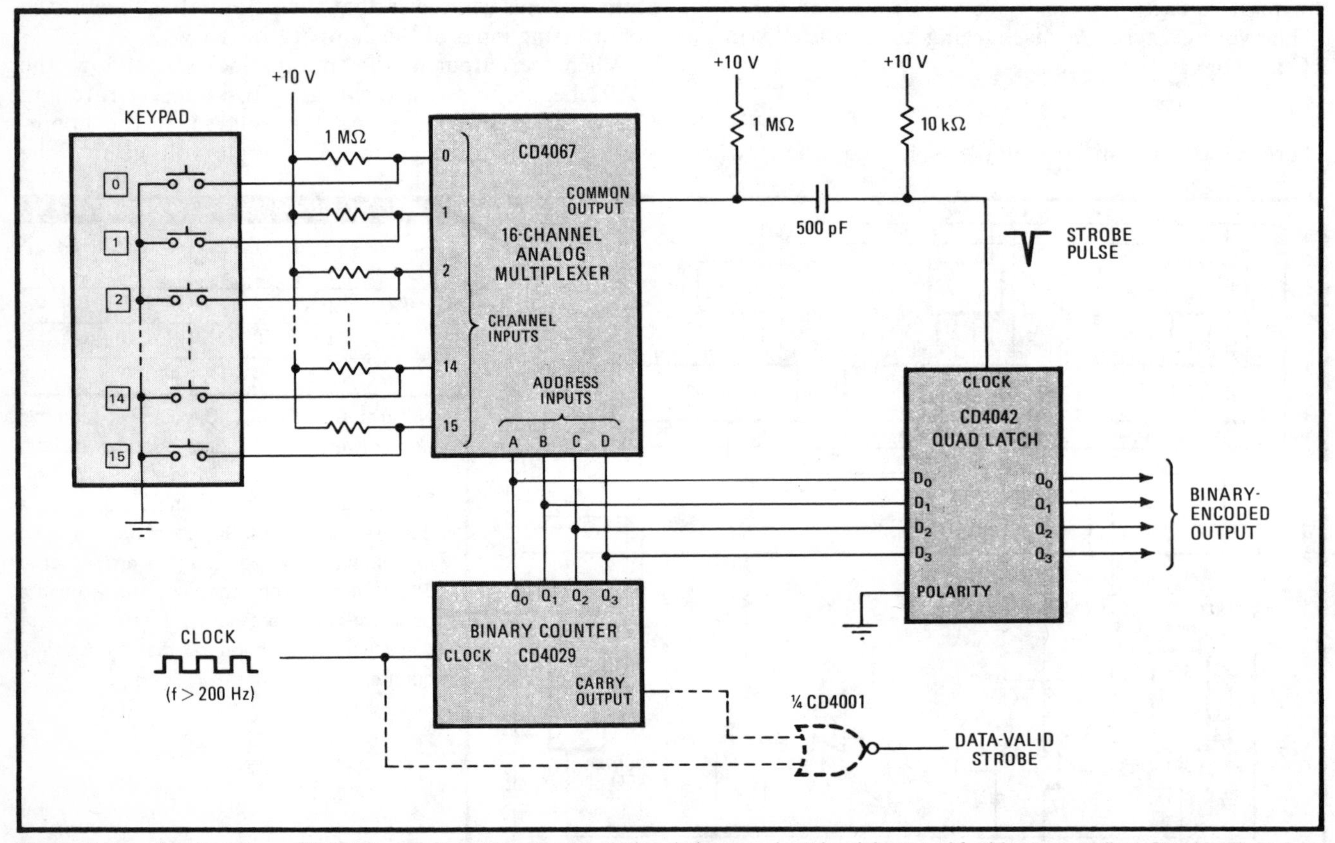

Keypad encoder. By using a CD4067 multiplexer to scan a keypad, switches need not be debounced for binary encoding. Quad latch at right holds digit encoded until next digit is entered. A NOR gate may be added as shown for data-valid strobe.

4-by-4–matrix chip encodes larger arrays

by James H. Nixon
Southwest Research Institute, San Antonio, Texas

With this circuit idea, the standard encoder for a 4-by-4–matrix keyboard can be made to handle arrays as large as 4 by 8. In fact, arrays as large as 4 by 10 can be readily accommodated if a 5-by-4-line encoder, such as Motorola's 74C923, is used. Interfacing with a microprocessor is easy and the generation of interrupts is not required.

As seen in the circuit example of a 3-by-8-line encoder, the 74C922 scans keyboard columns 1 through 4 (pins 7, 8, 10, and 11) at a rate set by its internal oscillator components and C_1. Each line is grounded in sequence until a key closure forces one of the row sense lines low, which in turn drives the data-available line high and halts the scanning process until the key is released. Meanwhile the appropriate input of the key encoder corresponding to the key pressed is brought high and the results of the row-column detection appear at the analog-to-digital outputs of the 74C922.

Each time the column 4 line is scanned, pin 7 of the key encoder moves high and clocks the 4013 flip-flop, F_1, causing the complementary-MOS 40257 data selector to switch between rows 1 to 4 and rows 5 to 8. The output from F_2 and A_6 thereby indicates which of the two row sets was accessed. The key closure line indicates if the data has been previously read and thus prevents redundant entries to the processor. □

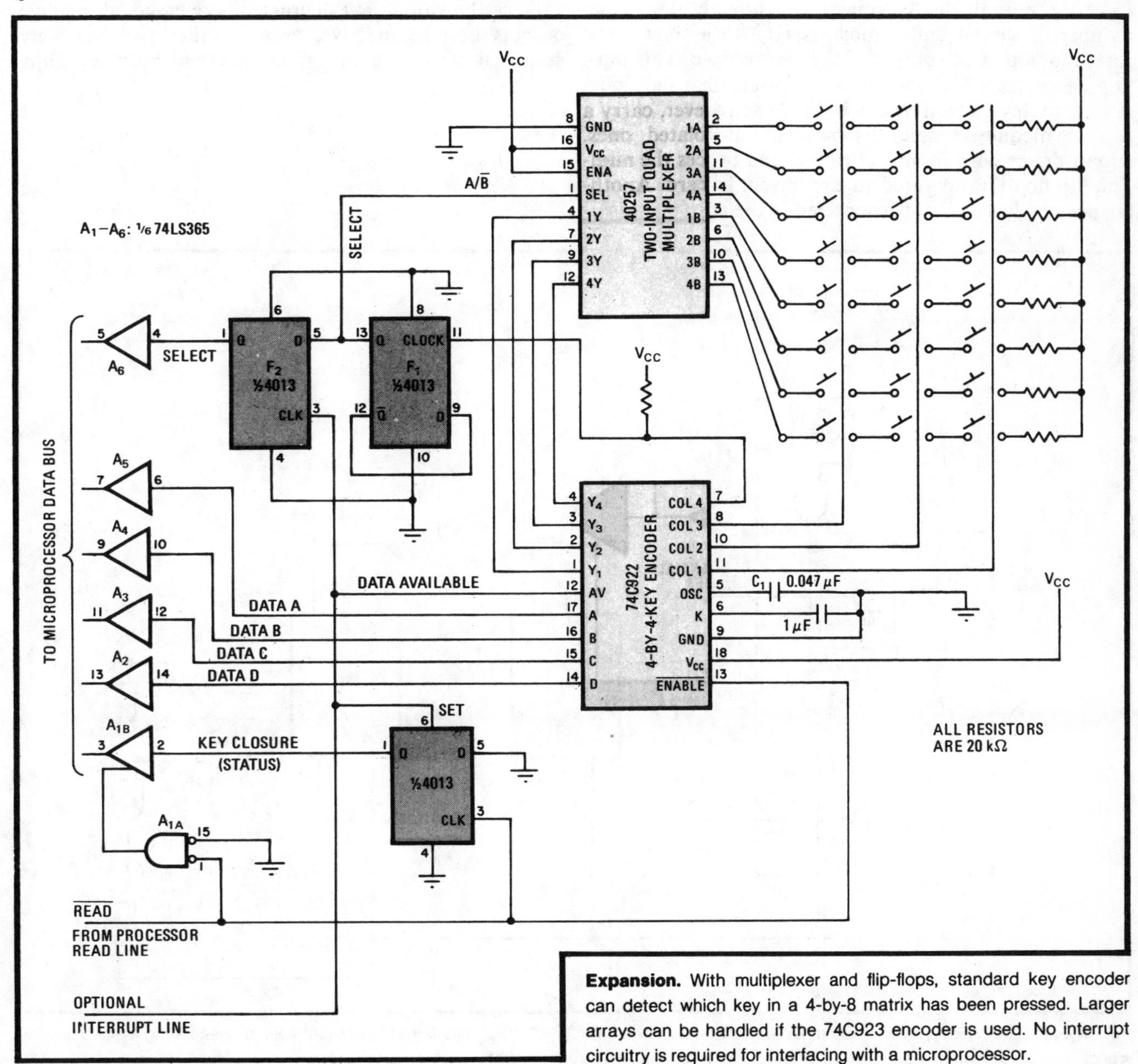

Expansion. With multiplexer and flip-flops, standard key encoder can detect which key in a 4-by-8 matrix has been pressed. Larger arrays can be handled if the 74C923 encoder is used. No interrupt circuitry is required for interfacing with a microprocessor.

Modular switch array includes priority encoder

by Thomas L. Sterling
Sigma Consultants Inc., Virginia Beach, Va.

The output ports of this momentary-contact switch array respond to the first command received, and the circuit locks out all subsequent commands, providing a time-sequence priority scheme often needed in industrial systems. The low-cost circuit prevents simultaneous switch depressions from spoiling system operation, and the modularized design technique employed makes it fairly simple to implement.

The structure for this switch array is shown in the figure. A number of single-pole, double-throw switches (SW_n) are individually interfaced through the same number of switch buffer modules (B_n) to a single set-reset flip-flop at the output of the last module. This flip-flop generates a lock-out signal to ensure that only one module at any given time can be in the active state.

Each module contains its own SR flip-flop, a gated output driver with inhibit circuitry, and two AND gates. The flip-flop is configured to circumvent switch contact bounce and has its inputs connected to the normally open (NO) and normally closed (NC) contacts of each switch. The inverting output is combined with the M_i input signal through an AND gate to derive the M_o output. The noninverting output of the flip-flop is applied to the gated output driver. Each module is cascaded by connecting the M_o and A_o ports to the M_i and A_i ports of the next buffer.

Depressing any switch drives the $\overline{P}_i$ output of its associated module low. The M_o and A_o ports of the last buffer module in the chain move low at this time, permitting generation of the LCK signal.

The inhibit gate in each module will prevent the output gate from going low, irrespective of the state of the input flip-flop, if the buffer module output, $\overline{P}_i$, is inactive. If the buffer module is active, the inhibit gate will be inactive, independently of the state of the LCK output. The lockout flip-flop is not reset until the M_o signal at the last flip-flop moves high again. This occurs when the switch is released.

Once the switch first depressed is released, all module outputs become inactive, even if other switches were activated after a particular module had been set. Only

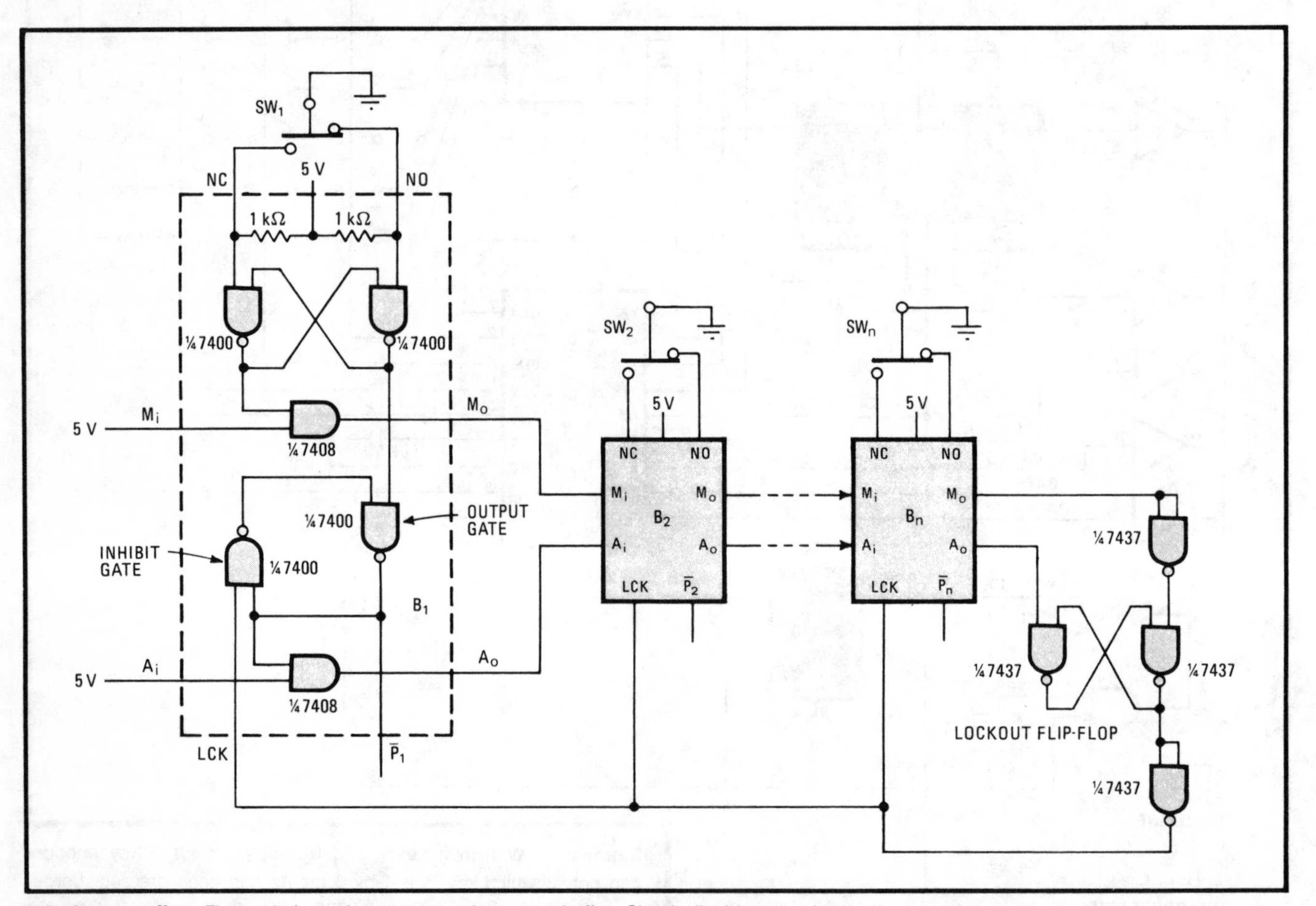

Priority encoding. First switch to close captures its output buffer. Circuit disables all other buffer output lines by generating a lockout (LCK) signal. Release of switch or switches automatically resets circuit. The truth table of the switch array is illustrated.

TRUTH TABLE FOR SWITCH ARRAY BUFFER

Time	SW_1	$SW_2 \ldots$	$SW_i \ldots$	$SW_R \ldots$	SW_{n-1}	SW_n	$\bar{P}_1$	$\bar{P}_2 \ldots$	$\bar{P}_i \ldots$	$\bar{P}_k \ldots$	$\bar{P}_{n-1}$	$\bar{P}_n$	M_{o_n}	A_{o_n}	LCK
t_0	0	0	0	0	0	0	1	1	1	1	1	1	1	1	0
t_1	0	0	1	0	0	0	1	1	0	1	1	1	0	0	1
t_2	0	0	1	1	0	0	1	1	0	1	1	1	0	0	1
t_3	0	0	0	1	0	0	1	1	1	1	1	1	0	1	1
t_4	0	0	0	0	0	0	1	1	1	1	1	1	1	1	0
t_5	0	0	0	1	0	0	1	1	1	0	1	1	0	0	1
t_6	0	0	0	0	0	0	1	1	1	1	1	1	1	1	0

after all switches are released can one of the buffers become active again. The operation of the switch array is shown in the illustration.

This design can accommodate up to 30 switches. This limit on their number is set by the driving capability of the inverter driver of the lockout flip-flop. □

Simplified priority encoder has low parts count

by Tomasz R. Tański
Warsaw, Poland

The number of chips in the priority encoder circuit first described by Sterling [in previous article] can be drastically reduced by replacing the modularized gate arrays by D flip-flops and a wired-OR gate. As in the original circuit, the output ports of this modified momentary-contact switch array responds to the first command received and locks out all subsequent commands, so as to provide a time-sequence priority scheme that is useful in many industrial systems. But this circuit is simpler to build and test, because the interconnections between elements are minimized.

Depressing any switch, SW_n, shown in the figure sets the corresponding D input of its flip-flop, D_n, high. The switch signal also quickly propagates through the 7416 inverters, which have their open-collector outputs wired together to form an n-input OR gate, and fires the 7437 flip-flop (G). A few microseconds later, the rising edge of the resultant output from G_1 stores the signal generated by SW_n into D_n before the D line, at logic 1 for a time measured in milliseconds, can return to ground.

The strobing signal (LCK) will stay high until all switches return to their normally closed (NC) position. Thus altering the output state of any flip-flop is impossi-

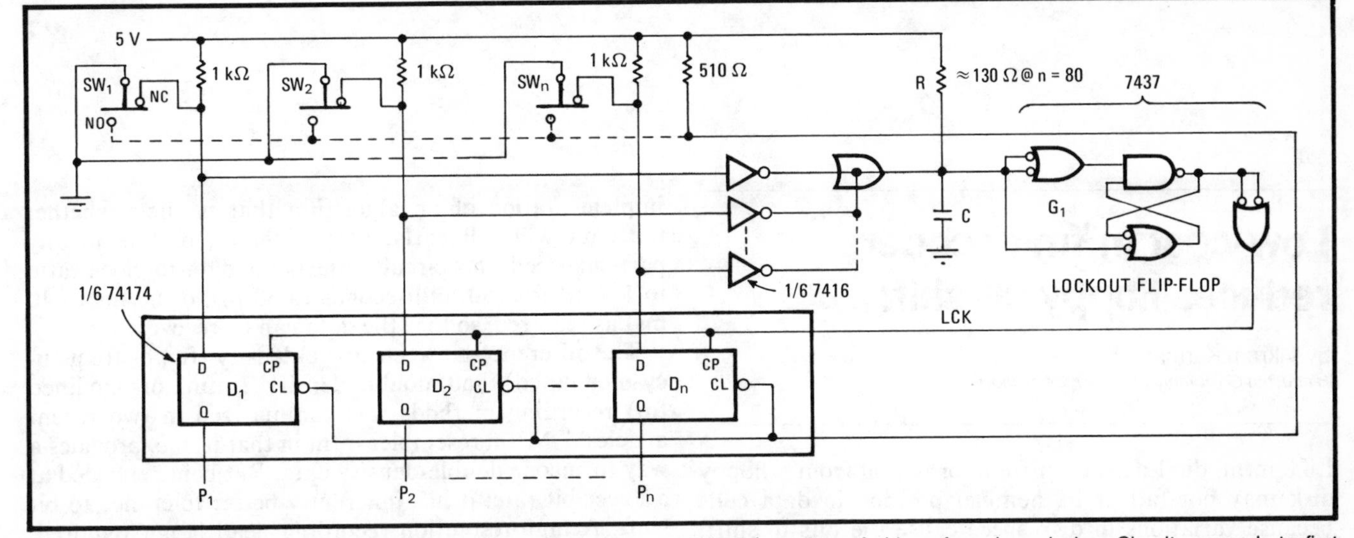

Simplification. Momentary-contact priority encoder uses about one fifth the number of chips of previous design. Circuit responds to first command received by switches and locks out all subsequent commands, thus forming a first-come, first-served switch.

ble, because all other switch commands are locked out.

Resistor R serves in a dual capacity. Its primary function is as load resistor for the open-collector inverters forming the wired-OR gate. Its value is selected so that the maximum current drawn is limited to the full-on collector current of one gate, independent of how many gates are activated. Secondly, R, in combination with C, provides effective switch debouncing. For optimum debouncing performance, the value of C should be selected (with the aid of a scope) to provide the trailing-edge delay required for signals from G_1.

Not considering the lock-out flip-flop, only one sixth of a 7416 and 74174 device are needed per switch, compared with the 1½ integrated circuits required in the original circuit. Only one 7437 is required, even for a large number of switches. Each input lead of the 7437 can accommodate an 80-input, wired-OR gate. The LCK signal can drive up to 180 flip-flops. □

Low-cost m²fm decoder reduces floppy bit-shift

by Vikram Karmarkar
Hindustan Computers Ltd., New Delhi, India

Like mfm, dual-density m²fm information from a floppy disk may not fall at its nominal position in data cells because variations in disk speed cause the bits to shift, thus reducing the cells' data margin and causing errors in the system's decoding circuitry. Using a hardware implementation of an algorithm that predicts whether the data will fall early, at the center, or late in any particular cell, this circuit adjusts the data-to-clock ratio to 1:1 or 3:2 (50 milliseconds to 50 ms, or 60 ms to 40 ms) as required, so that the data can be recovered.

The difference between single-density (fm, or frequency-modulated) and double-density (mfm, or modified fm) recording methods was summarized in two recent articles[1,2]. M²fm resembles mfm in that it, too, provides a way to encode double-density data. But being encoded at a lower bit rate, it has inherently better tolerance to bit shift. A high-resolution recording head is not required, as in mfm. Further, fm systems can be upgraded to m²fm encoders without the need to change disk drives.

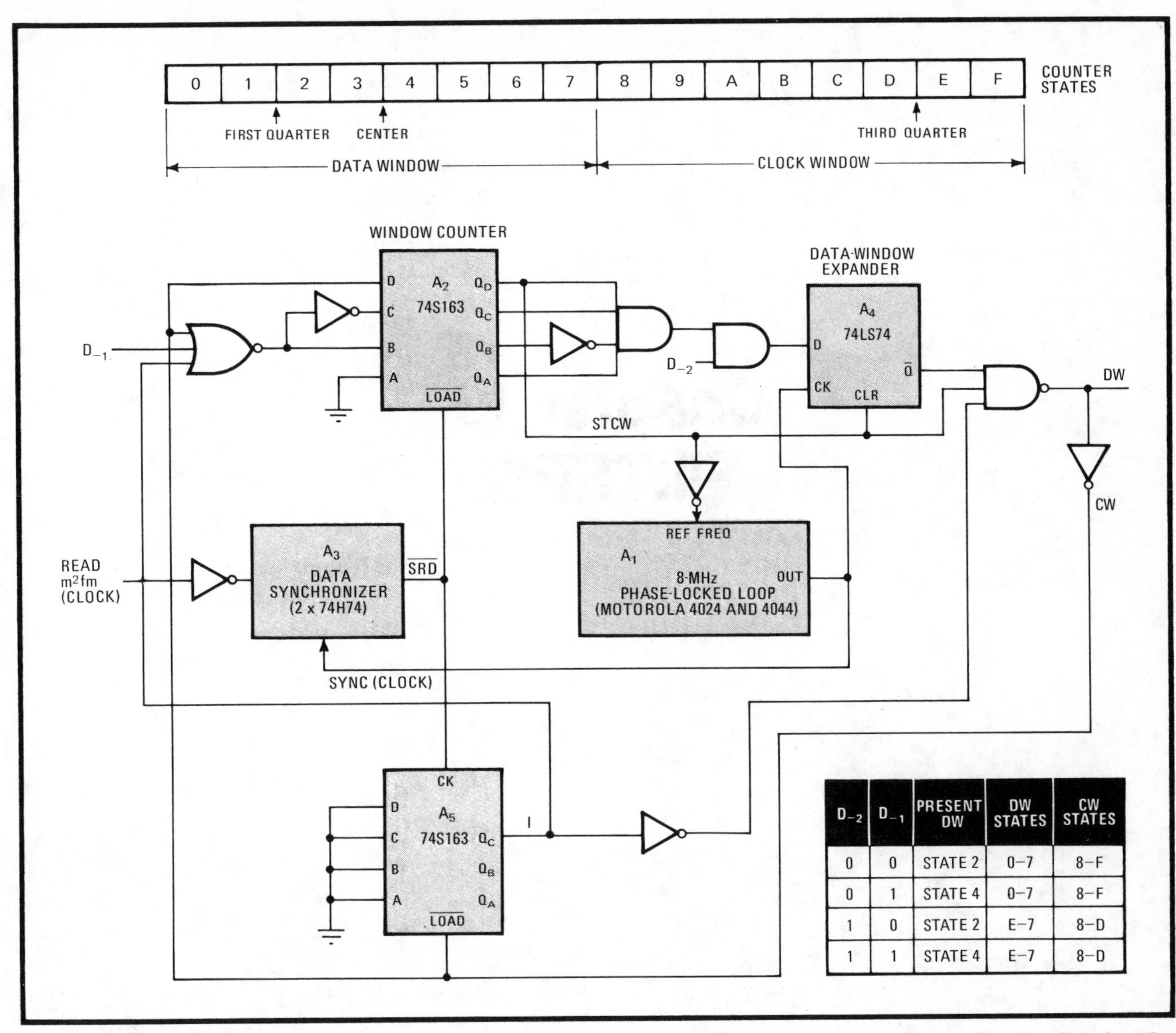

D_{-2}	D_{-1}	PRESENT DW	DW STATES	CW STATES
0	0	STATE 2	0–7	8–F
0	1	STATE 4	0–7	8–F
1	0	STATE 2	E–7	8–D
1	1	STATE 4	E–7	8–D

Guess when. Decoder for double-density floppy determines where to place receiving window for recovering m^2fm data, based on flux transitions of two previous data bits that can cause resultant shifts in position. Circuit adjusts data-to-clock ratio at 1:1 or 3:2 for each cell as required to recover data, compensates for bits removed ±250 ms from their ideal position.

The typical m^2fm cell has 16 states (see figure, top), 8 of which comprise the data window and the remaining 8 the clock window. The algorithm for decoding the data:

- Applies each system clock pulse to the center of the clock window.
- Initializes the data window at the first-quarter point in the cycle at the arrival of a data pulse if the previous data bit, D_{-1}, equals 0. The data window is centered if $D_{-1} = 1$.
- Initializes the data window at the third-quarter point in the clock window for the current data bit if the second previous data bit, D_{-2}, equals 0.

An acid test for this algorithm is the handling of the mark bytes found in the data stream, since these bytes do not fit into the m^2fm clock pattern. The hardware used to decode the data will process the data, the ID address, and selected data markers with no problem, while maintaining the system margin.

The circuit used to achieve the decoding is shown in the main part of the figure. A_1 generates a center frequency of 8 megahertz for the modulo-16 window counter, A_2, thereby providing cells of 125 ms in width. States 0 through 7 constitute the nominal (50-to-50–ms) data window, and states 8 to F the clock window. Note the PPL can be replaced by a fixed-frequency (crystal) oscillator with no degradation in circuit performance if little additional drive current is required.

Any clock pulse appearing at the read m^2fm input presets A_2 to state 2 or 4, respectively, as governed by the truth table. A_4 converts the data-to-clock ratio to 60:40 when necessary. Thus it is seen that the circuit can handle a bit tolerance of ±250 ms in position. Inversion counter A_5 places the clock and data pulses into their corresponding windows. □

References

1. Curt Terwilliger, "Pattern generator simulates double-density disk data." p. 168

2. John G. Posa, "Peripheral chips shift microprocessor systems into high gear," *Electronics*, Aug. 16, 1979, p. 93.

Chapter 15
FILTERS

Tunable equalizers set amplitude and delay

by P. V. Ananda Mohan
Indian Telephone Industries Ltd., Bangalore, India

Equalizing networks providing constant amplitude and/or delay over a wide range of frequencies are easily realized by utilizing the feed-forward and feed-back techniques of these tunable circuits. More specifically, a parallel-T arrangement of resistors and capacitors as shown in (a) makes it possible to select the equalizing delay with a single potentiometer. Equalization and selection of amplitude can be attained by adding a single operational amplifier stage of variable gain (b) to the basic circuit. If the parallel-T is made tunable (c), the equalizer's center frequency can be adjusted with very little difficulty.

The circuit that is illustrated in (a) is so configured that its transfer function is that of an all-pass network having a roll-off dependent on circuit Q, or:

$$\frac{e_{out}}{e_{in}} = \frac{s^2 - s(\omega_o/Q) + \omega_o^2}{s^2 + s(\omega_o/Q) + \omega_o^2}$$

where $\omega_o = 1/RC$ and $Q = 1/4 + R'/2R$. The amount of delay is selected with potentiometer R′, as the amount of phase shift introduced by the RC network is:

$$\theta = t - 2\tan^{-1}[\omega\omega_o/Q(\omega_o^2 - \omega^2)]$$

where θ will vary little about ω_o, provided ω is sufficiently removed from ω_o.

If the op amp circuit (b) is placed between ports A and B in (a), the equalizer's gain at mid-frequency ω_o becomes $G = (2R'm)/R - 1$, and so the gain may be set by varying R′ and/or m. Note that the delay is still a function of R′ and that G will not vary significantly over a wide range of frequencies.

The center frequency of the equalizer may be adjusted if the parallel-T network shown in (c) replaces the network in (a) enclosed between points 1 and 3. In this case, $\omega_n = \omega_o/(1-K)^{1/2}$ and $Q = (1/4 + R/2R)/(1-K^2)^{1/2}$, where K is the fraction of the total resistance of P_2, as measured from its lower end. □

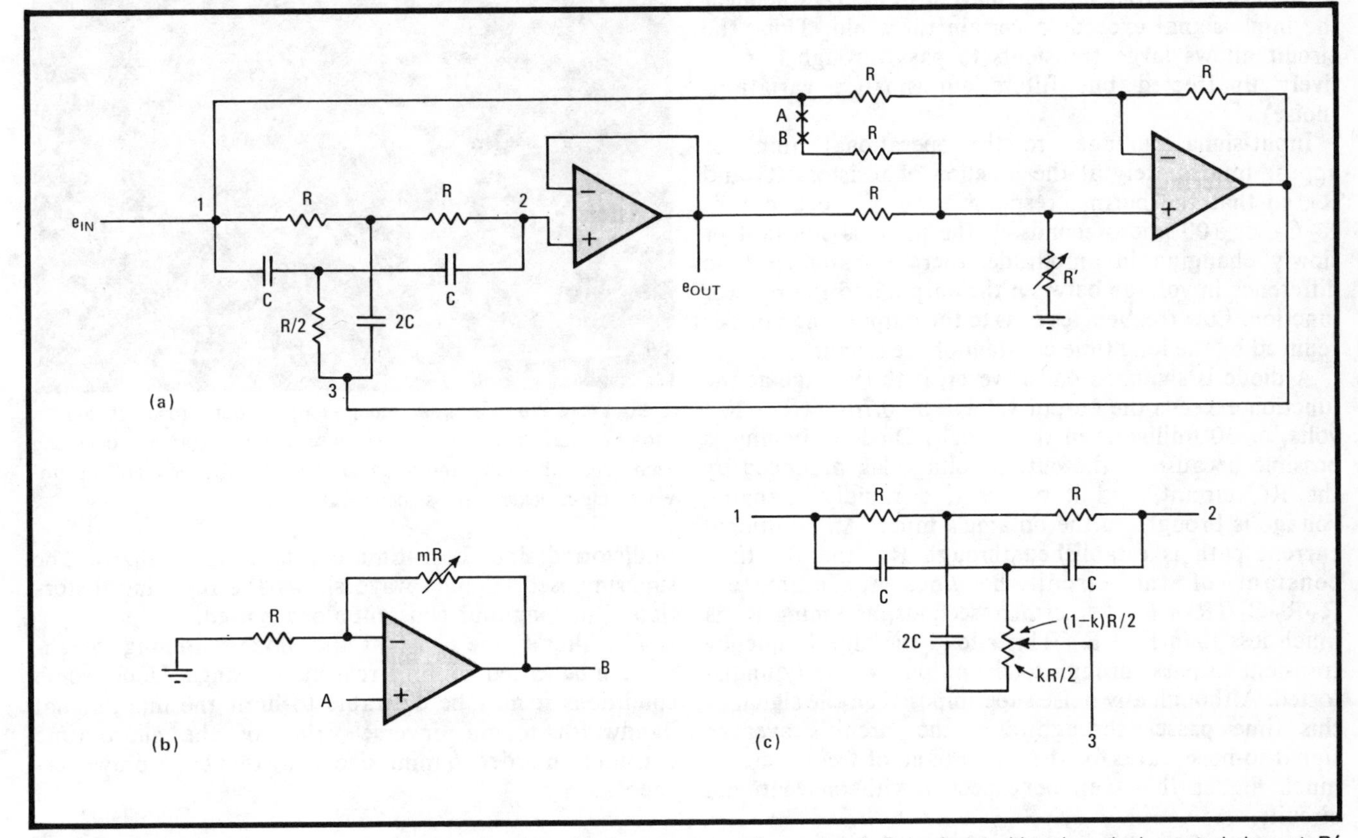

Selection. Parallel-T RC network simplifies design of tunable two-stage equalizer network. Delay is set with only a single control element, R′ (a). Amplitude equalization or adjustment in network's center frequency is attained by adding op amp (b) and tunable RC (c), respectively.

Noise-reducing filter switches time constants

by Martin V. Thomas
Boston University Medical Center, Boston, Mass.

To reduce high-frequency noise in a signal waveform without significantly distorting the signal is often beyond the capacity of conventional low-pass filters. For this purpose, a piecewise-linear filter is far more effective, especially for complex waveforms such as square waves and sawtooth signals.

The circuit shown in Fig. 1 achieves this improvement in signal-to-noise performance and has been used for the precise determination of input-signal amplitudes in the presence of noise. It makes use of the fact that although signal amplitude varies significantly with time, the variation of the root-mean-square value of the superimposed noise with time is smaller and relatively constant. The filter normally has a comparatively long time constant, T_1, but switches to a shorter time constant, T_2, whenever the input signal exceeds a certain threshold. Thus, the circuit allows large transients to pass through it relatively unaffected but filters out smaller variations (noise).

Input-signal voltages to the operational amplifier appear immediately at the junction of resistors R_1 and R_2, so that the normal response time of the circuit is $R_2 \cdot C_2$, or 100 microseconds. If the input is constant or slowly changing in amplitude, there is essentially no difference in voltage between the output and the resistor junction. Low frequencies pass to the output, and noise is reduced by the long time constant of the circuit.

A diode is switched on, however, if the voltage at the junction exceeds the output voltage by $0.7 \cdot R_2/(R_1+R_2)$ volts, or 30 millivolts in this circuit. Diode switching is possible because of the output voltage lag produced by the RC circuit, and it occurs if a rapidly changing voltage is brought to the op amp's input. An additional current path is established through R_3, and the time constant of the circuit becomes approximately $R_3 \cdot R_2 \cdot C_2/(R_1+R_2)$, or 2 microseconds, assuming R_3 is much less than R_1+R_2. This allows the high-frequency transient to pass through to the output, virtually undistorted. Although any noise superimposed on the signal at this time passes through also, the circuit's average signal-to-noise ratio for the entire band of frequencies is much higher than can be expected with conventional circuits.

Outputs of the filter with a 1-volt, 5-kilohertz square wave input are shown in Fig. 2. The square wave is undistorted since both time constants are utilized. The superimposed triangle wave shows the resulting distortion if the long time constant alone is used.

The filter's time constant and diode switching threshold can be varied within a reasonable range. Under some conditions it may be desirable to limit the input signal bandwidth to the inverse value of the short time constant, in order to minimize distortion caused by overshoot. □

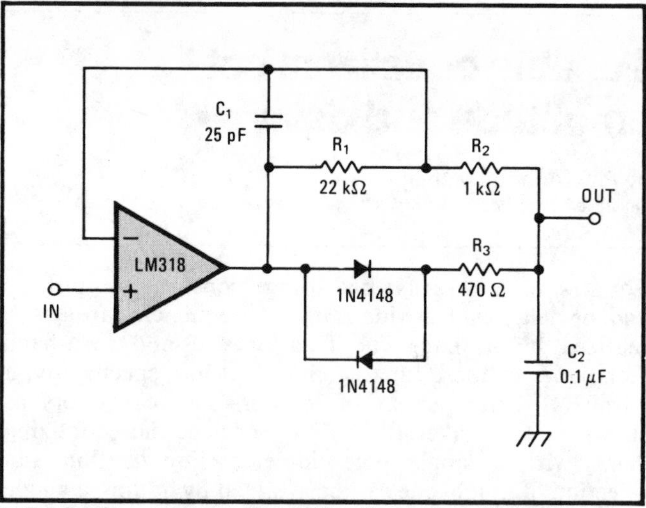

1. Dual-value response-time filter. For best performance, C_2 is made relatively large. R_3 maintains stability by limiting charge current, C_1 prevents oscillation in feedback loop.

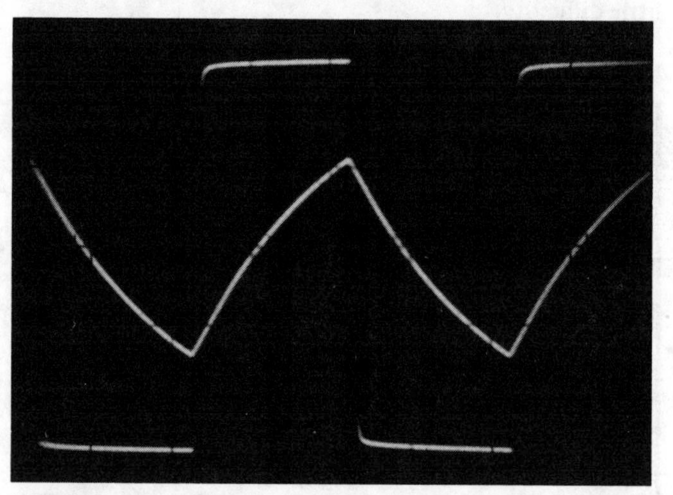

2. Square-wave response. Square-wave output is result of circuit's use of a short as well as a long time constant. If filter uses only long time constant, overfiltered triangular waveform results. Full square-wave output yields higher signal-to-noise ratio.

Frequency-controlled gate makes high-Q filter

by Noel A. Sivertson
Denver, Colo.

A bandpass filter in conjunction with a frequency-controlled gate provides a highly selective filter at low cost. Its high Q is especially useful for single-frequency discrimination, where conventional filters falter.

The circuit utilizes the response of the bandpass filter itself to pass desired signals, and it absolutely rejects all other frequencies. As indicated by the figure, input signals encounter A_1, a standard bandpass filter. Resistors R_1-R_4 and capacitors C_1 and C_2 determine the filter center frequency and bandwidth, and their values are selected accordingly. The output of A_1 is then presented both to A_2, which serves as a peak-rectifier detector, and to the input of the CD4016 analog transmission gate. The output of A_2 is a dc signal having an amplitude that is proportional to the filter's transmission coefficient—that is, the filter's center frequency produces the largest dc output. This output represents the envelope of the filter response.

A_2's output is compared to a user-determined reference voltage at A_3, and when it exceeds this reference voltage, A_3 signals the transmission gate to transfer its input signal to the output. Thus comparator A_3 is in effect a bandwidth adjustment control.

Depending on the quality of the band-pass filter, the ultimate bandwidth of the circuit could be as sharp as a few cycles. To retain this sharpness, however, it is required that the signal be of a reasonably constant amplitude at the input to the circuit.

This circuit has also been used in an amplifier-squelching device and to control the range of a sweep oscillator. □

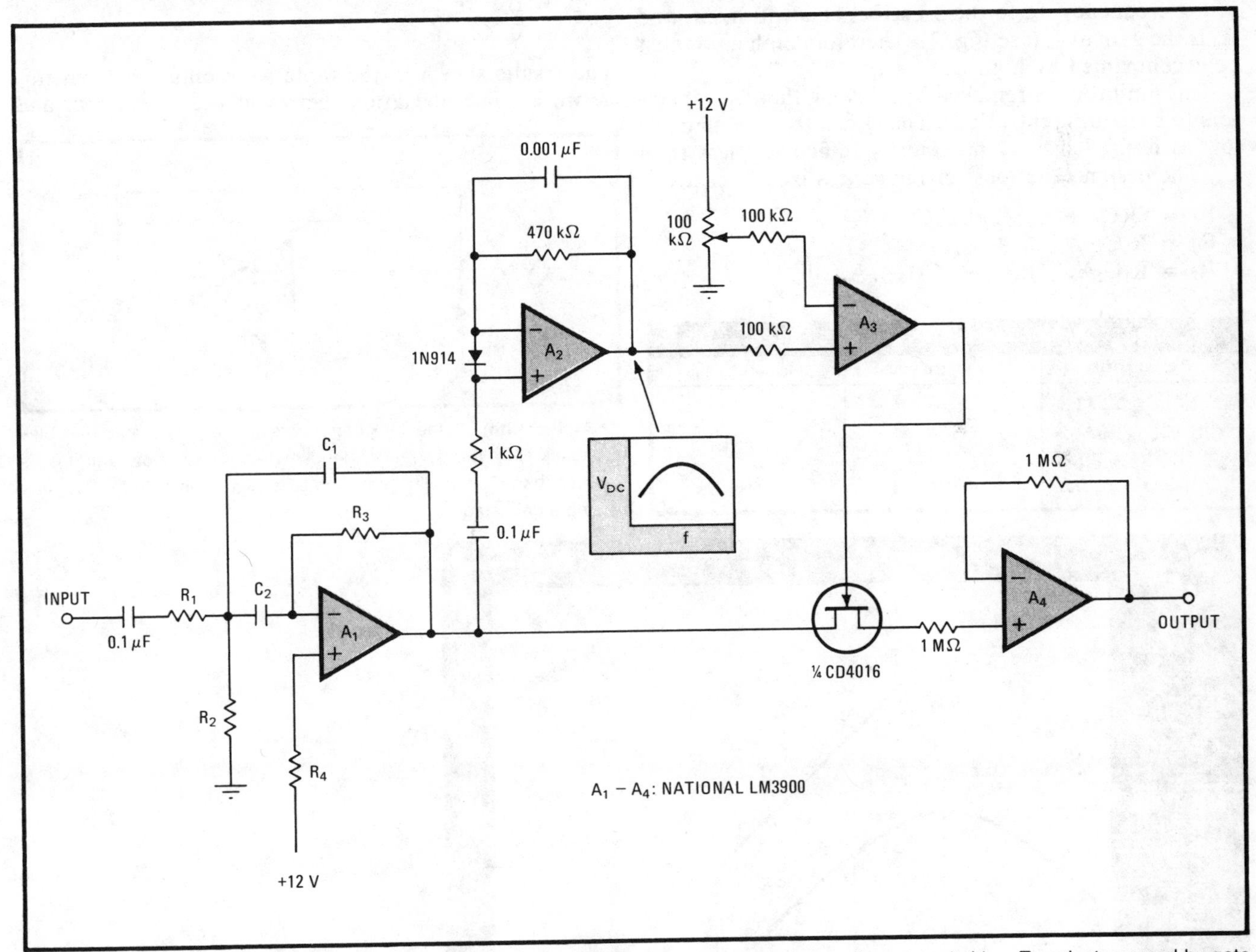

Gated filter. Circuit uses pass-band response of ordinary filter in generating voltage reference for gate switching. Transients caused by gate switching in the audio range are not detrimental to circuit performance but may cause concern at higher frequencies.

Multiple-feedback filter has low Q and high gain

by Gregory O. Moberg
Eastman Kodak Co., Rochester, N.Y.

The multiple-feedback operational-amplifier circuit found in many applications manuals for operational amplifiers readily generalizes into an active filter with both high gain and low Q. This versatile design saves one or more gain stages and is useful in any filtering application.

The multiple-feedback bandpass filter is shown in Fig. 1. To simplify its design equations, C_1 and C_2 are usually made equal. The price paid for this simplification, however, is that the Q of the filter can not be made less than $(A_o/2)^{1/2}$, where Q is the ratio of the filter's center frequency, f_o, to the 3-decibel bandwidth, Δf, and A_o is the gain at f_o (see Fig. 2). Therefore high gain must be accompanied by high Q.

This limitation is removed by allowing the two capacitors to have different values. The Q can then be made as low as desired just by making C_2 sufficiently less than C_1. The design equations for the generalized filter are:

$$R_3 = Q(C_1 + C_2)/2\pi f_o C_1 C_2$$
$$R_1 = R_3 C_1/A_o(C_1 + C_2)$$
$$R_2 = R_1 C_2 A_o/[Q^2(C_1 + C_2) - A_o C_2]$$

PERFORMANCE OF DESIGN-EXAMPLE FILTER

Characteristic	Calculated value	Measured value
A_o	104	101
f_o	101 Hz	102 Hz
Q	3.0	2.8

To design a filter that has a given A_o, f_o, and Q, the steps are as follows:

1. Assume C_1 = 1 microfarad.
2. Choose a standard value for C_2 equal to or less than $C_1Q^2/(A_o - Q^2)$.
3. Calculate R_1, R_2, and R_3.
4. Multiply all the resistors and divide all the capacitors by a scaling factor that produces convenient impedance levels. For example, it may be convenient to have the scaled value of R_1 much greater than the source resistance of whatever is driving the filter.
5. Make sure that the closed-loop gain of the filter, A_{CL}, is 20 dB below the open-loop gain of the op amp, A_{OL}.

As an example, a filter was constructed with a National Semiconductor LM307 op amp and the following components:

R_1 = 9.1 kilohms
R_2 = 11 kilohms
R_3 = 1 megohm
C_1 = 0.1 μF
C_2 = 0.005 μF

The results shown in the table were obtained from this network. The deviation between the measured and

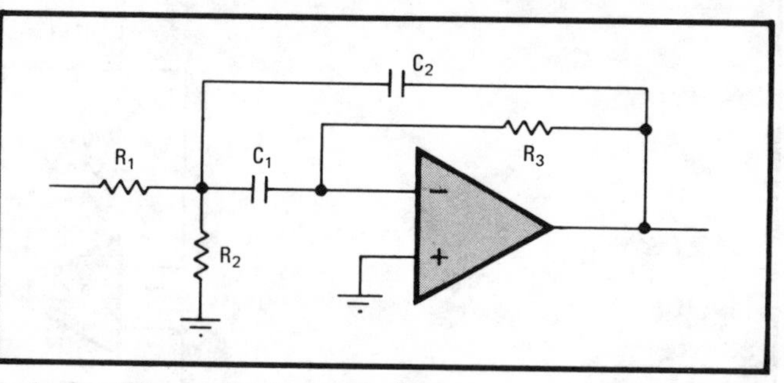

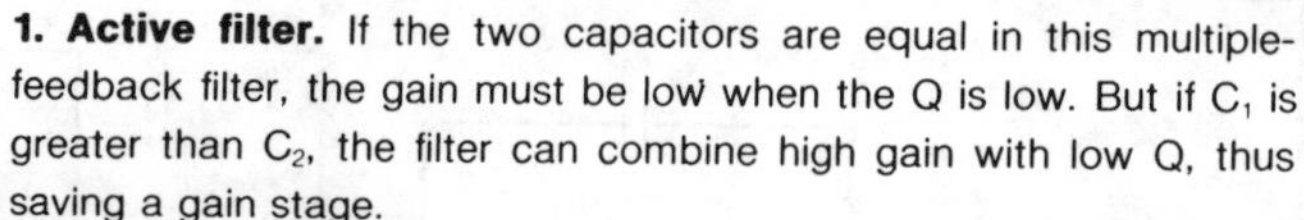
1. Active filter. If the two capacitors are equal in this multiple-feedback filter, the gain must be low when the Q is low. But if C_1 is greater than C_2, the filter can combine high gain with low Q, thus saving a gain stage.

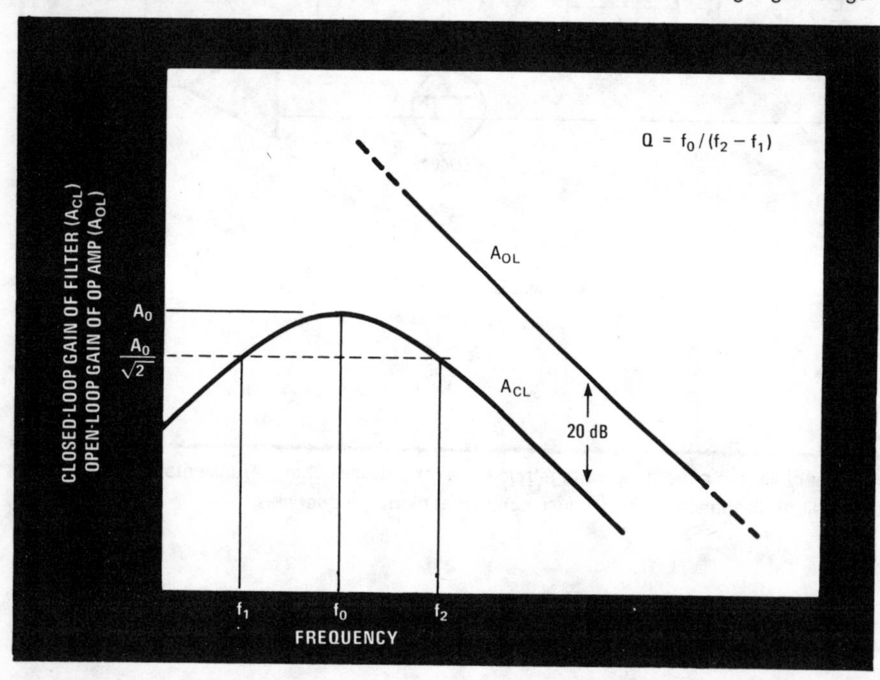

2. Frequency response. For given A_o and f_o, the Q of the filter in Fig. 1 can be set by the choice of capacitor values. The filter should have a closed-loop gain that is 20 dB below the open-loop gain of the op amp. This will ensure ideal behavior of the op amp.

calculated performance is well within the range expected from the 5% components used.

It is important that the difference between the closed-loop gain of the filter and the open-loop gain of the op amp be at least a factor of 10, in order to ensure ideal behavior of the op amp. A sufficient condition for this is that $A_o f_o(1 + 1/2Q)$ be equal to or less than 0.1 BW, where Q is equal to or greater than unity, and BW is the op-amp unity-gain bandwidth. For low gain or low f_o designs, the LM307 op amp (BW = 800 kilohertz) will suffice. For higher gains or higher f_o, an op amp such as the LM 318 (BW = 15 megahertz) should be used. □

Wien bridge and op amp select notch filter's bandwidth

by Dominique Fellot
Thomson-CSF, Gentilly, France

The band over which a notch filter provides rejection of unwanted frequencies can be selected with this circuit, which uses a Wien bridge plus an operational amplifier with fixed gain. Such a circuit represents one of the simplest configurations for easily adjusting the selectivity of the filter, which has a notch depth of nearly 60 decibels, independent of component precision.

Circuit operation is most easily described with the transfer function shown in the figure, which is:

$$\frac{V_o}{V_i} = \frac{1 - R^2C^2\omega^2}{1 - R^2C^2\omega^2 + j3(1-k)RC\omega} = \frac{1 - x^2}{1 - x^2 + j3(1-k)x} \quad (1)$$

where $x = \omega/\omega_c = f/f_c = RC\omega$, ω is the frequency of interest, f_c is the center frequency of the notch filter, and k is the percentage of the output voltage from A_1 that is introduced at the noninverting input of A_2. This transfer function has a transmission zero at $f_c = 1/2\pi RC$, which is the center frequency of the notch.

The amount of phase shift provided by the Wien bridge (A_1 and the RC components), from Eq. 1, is:

$$\tan\phi = [-3(1-k)x]/(1-x^2) \quad (2)$$

The width of the rejected band at the −3-dB points can be easily expressed with respect to k by setting $\phi = 45°$, so that Eq. 2 becomes:

$$|\tan\phi| = 1 = [3(1-k)x]/(1-x^2)$$

or:

$$k = (x^2 + 3x + 1)/3x, \text{ for } x < 1 \quad (3)$$

where $x = f_r/f_c$ and f_r is defined as the difference in frequency as measured at the −3-dB points. Thus, for example, if $f_c = 10$ kilohertz and the desired x = 0.9 (or f_r = 9 kHz), then k must be set at 0.93.

It will be noted that although the transfer function for the popular twin-T variety of notch filter (not shown) is almost identical to that in Eq. 1 (the constant 3 is replaced by the number 4), in practice, the twin-T is not very easily adjusted. This is because a greater number of components must be trimmed, and more careful adjustments made, to achieve the desired degree of selectivity and notch depth required. □

Selectable stopper. Notch filter, which operates at up to 200 kHz, uses modified Wien bridge to select bandwidth over which frequencies are rejected. RC components determine filter's center frequency. P_1 selects notch bandwidth. Notch depth is fixed at about 60 dB.

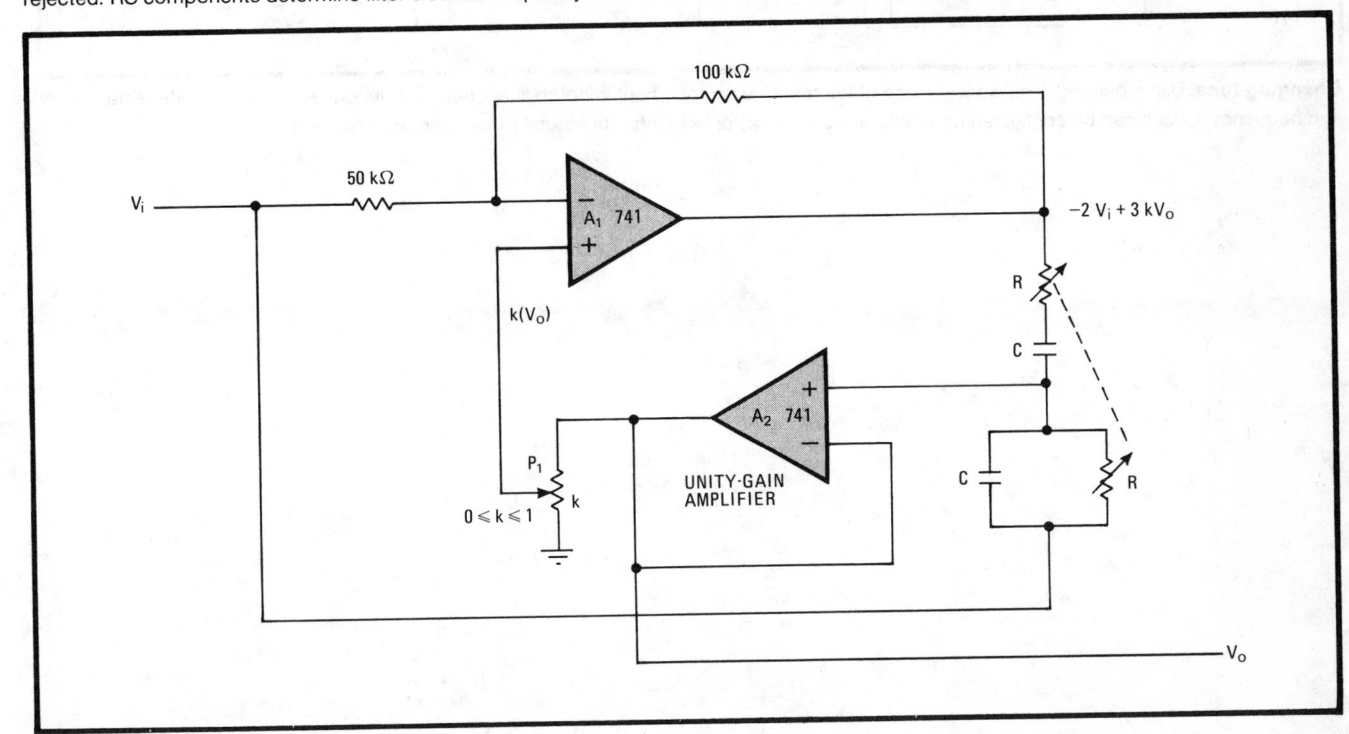

Bridged-T selects filter's notch frequency and bandwidth

by P. V. Ananda Mohan
Indian Telephone Industries Ltd., Bangalore, India

If a bridged-T network is used in place of the Wien bridge in the notch filter proposed by Fellot,[1] both the bandwidth and the frequency may be independently adjusted. The bridged-T approach has been explored previously[2] as an extension of some work carried out on parallel-T notch filters, and, as illustrated here, the technique offers an excellent way of building units that are simple and versatile.

R_N and R_Q comprise the balancing arms of the bridged-T network (note A_1 is a unity-gain buffer) as seen in (a). In this configuration, the circuit's transfer function is:

$$e_o/e_i = [ns^2 + \omega_o^2]/[s^2 + 3(1\text{-}q)s\omega_o + \omega_o^2]$$

where n and q are selected by R_N and R_Q, respectively, and $\omega_o = 1/RC$. Note that $0 \leq n, q \leq 1$, and that the frequency of the notch is

$$\omega_n = \omega_o/n^{1/2}$$

Therefore for this circuit ω_n will always be equal to or greater than ω_o.

The bandwidth is adjusted with R_Q, and Qs greater than 1,000 will be realized when high-gain operational amplifiers are used. In general, Qs will be higher than can be achieved with parallel-T networks. The notch depth is at least 50 decibels throughout the operating range. R_N and R_Q must only be at least 10 times smaller than R to achieve the stated filter characteristics.

By modifying the circuit slightly, as in (b), the transfer function becomes:

$$e_o/e_i = [s^2 + n\omega_o^2]/[s^2 + 3(1\text{-}q)s\omega_o + \omega_o^2]$$

and the notch frequency ω_n is made tunable for frequencies below ω_o, so that $\omega_n = \omega_o n^{1/2}$. □

References
1. Dominique Fellot, "Wien bridge and op amp select notch filter's bandwidth," p. 145
2. "An Active RC Bridged-T Notch Filter," *Proc. IEEE*, August 1977, p. 208.

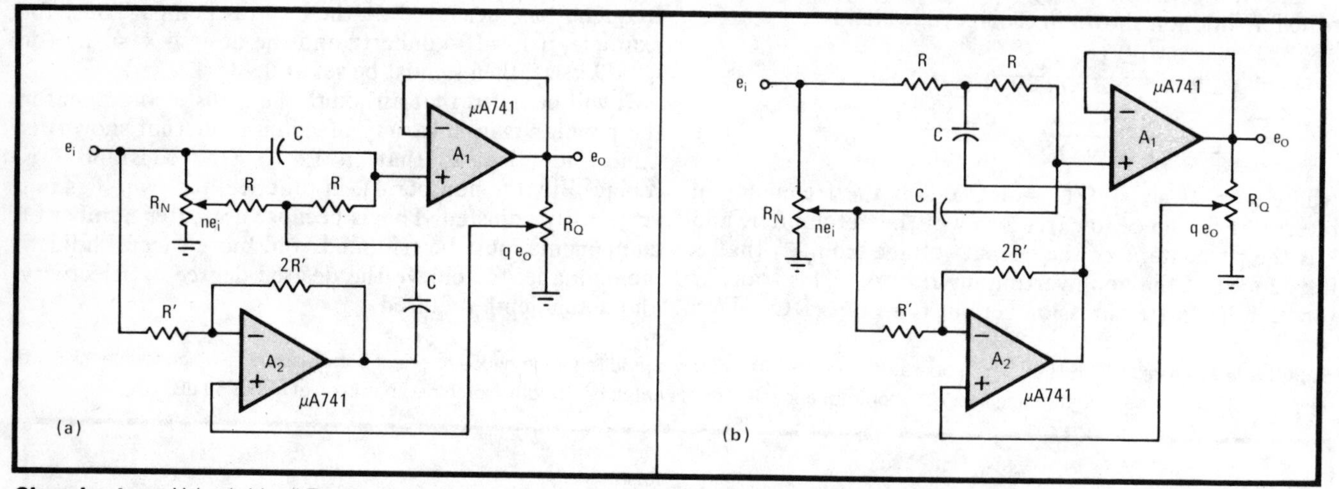

Changing tune. Using bridged-T network in place of Wien-bridge arrangement in notch filter enables independent control of filter's bandwidth and frequency. Circuit can be configured for tuning filter above (a) or below (b) its natural radian frequency $\omega_o = 1/RC$.

One-chip gyrator simplifies active filter

by Kamil Kraus
Rokycany, Czechoslovakia

Now that gyrators, or impedance inverters, are available on a single integrated circuit, active filters with both high input impedance and few component-sensitivity problems can be easily built in a small area with a minimum of parts. This one uses only one other active component—a dual operational amplifier—to provide low-pass, high-pass, bandpass, or band-reject response at reasonable cost over the dc range to 10 kilohertz.

The low-cost TL-083 op amp has been selected for use in the circuit because of its virtually negligible input-offset voltage and input-bias current. These characteristics are required to achieve a high input impedance over the range of interest and to realize the optimum response of the filter.

The Signetics TCA 580 gyrator simulates the relatively large inductor needed for the required LC (passive) network. The inductance across pins 6 and 11, and thus in parallel with capacitor C, is $L = C_0R_0^2$. The resonant (center) frequency of the LC combination, in turn, is $f_0 = 1/2\pi\ (LC)^{1/2}$, with its quality factor $Q = R(C/L)^{1/2}$. Thus the filter, which is shown configured on the bandpass mode, can be made to work at any frequency and Q, once its components are suitably selected.

The circuit can be transformed into a low-pass filter if its output is applied to an integrator whose time constant is RC. Similarly, it will function as a high-pass filter if its output is applied to a differentiator whose time constant is RC. For band-reject operation, C must be placed in series with the simulated inductor. □

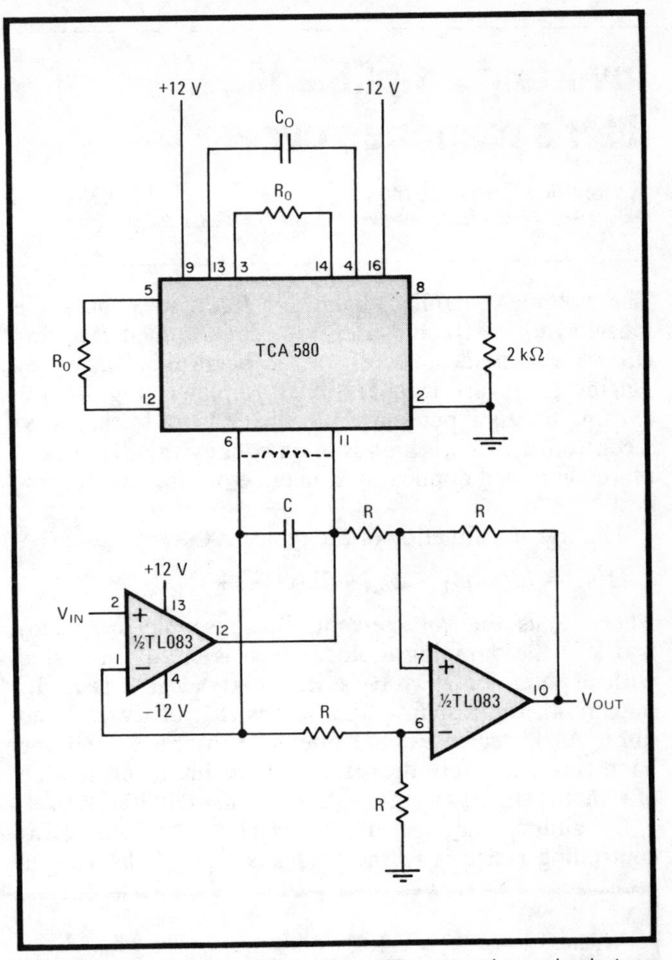

Optimal. Gyrator and operational amplifier comprise a simple two-chip bandpass filter with high impedance and state-of-the-art stability. Filter configuration remains the same for high-pass and low-pass operation; only an external differentiator or the integrator, respectively, need be added. For the band-reject mode, C is placed in series with a simulated inductor.

Inverting amplifier flips filter's response curve

by Henrique Sarmento Malvar
Department of Electrical Engineering, University of Brazilia, Brazil

The voltage-controlled bandpass filter so popular in music synthesizers and useful in remote-tuned receivers can be also made to work in the band-reject mode by placing an inverting operational amplifier stage in the existing filter's input/output feedback loop. In that way, a controlled notch, which is often equally valuable in the aforementioned applications, may be put together at low cost.

The transfer function of the typical VCF is given by:

$$H(s) = (s/\omega_o)/[(s/\omega_o)^2 + 2k(s/\omega_o) + 1]$$

where ω_o is the voltage-controlled resonant frequency and k is the damping factor, which is usually adjusted with a potentiometer. It is seen that when k is at its maximum, the response approaches that of a wideband filter. As k decreases, the filter's Q increases, and the bandwidth therefore decreases. At the limit, for a value of k that is slightly negative, the system oscillates at ω_o.

By adding the operational amplifier and its gain-controlling resistors to the feedback loop of the VCF as shown in (a), the output voltage generated is:

$$V_o(s) = -V_o(s)H(s) - V_i(s)$$

This expression leads to the transfer function:

$$H'(s) = \frac{V_o(s)}{V_i(s)} = \frac{(s/\omega_o)^2 + 2k(s/\omega_o) + 1}{(s/\omega_o)^2 + (2k+1)(s/\omega_o) + 1}$$

This function has two zeros and two correspondingly equal poles. But although the absolute value of both pairs is the same, for $k > -0.25$ the poles will be more damped than the zeros, and thus the filter's frequency response will be mainly determined by the zeros. Therefore, the system will operate as a band-reject filter (b).

The deepest null is attained at $k = 0$, and a theoretically infinite attenuation is thereby achieved at $\omega = \omega_o$. Thus the filter can be tuned to null the fundamental frequency of any synthesized signal, leaving only its harmonics. If the frequency control is simultaneously fed with a low-frequency sine or triangle wave, the so-called phaser sound used for special effects is obtained.

As k is increased toward infinity, the null becomes less sharp and the filter offers almost no attenuation at any frequency, thereby behaving as a quasi–all-pass network. Note that as k increases beyond $k = 10$, the filter response approaches that of $k = -0.25$. Clearly, k should not be less than -0.25, because the poles of the function will again become prominent and the system will once more behave as a bandpass filter. At $k < -0.50$, the system will oscillate. □

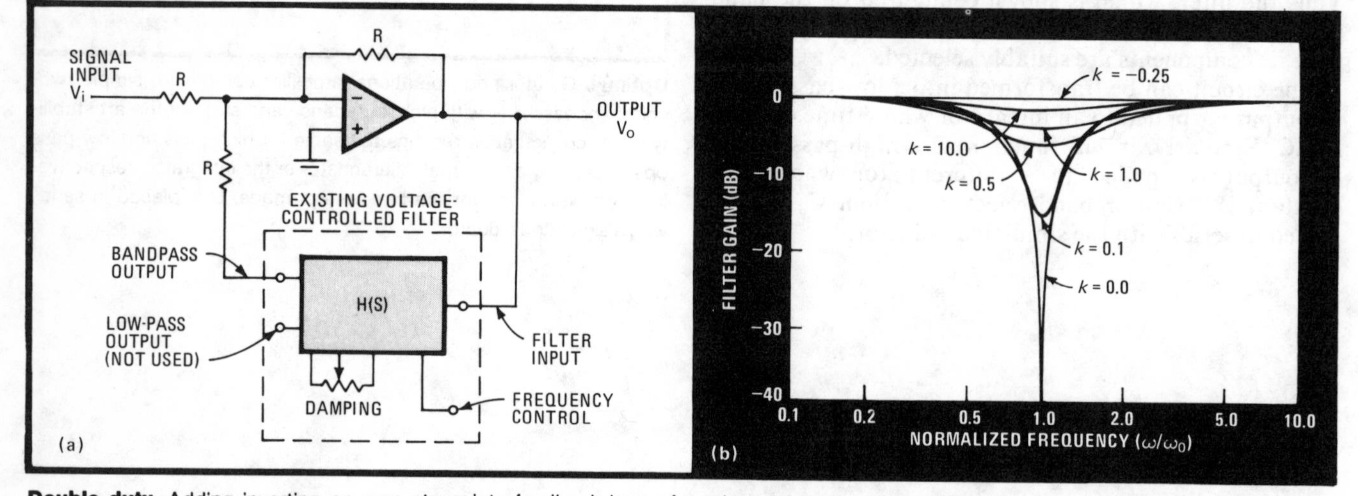

Double duty. Adding inverting op-amp stage into feedback loop of music synthesizer voltage-controlled bandpass filter (a) adapts it for band-rejection duties. Notch depth (b), selected by filter's damping potentiometer, may be adjusted for a maximum value of −60 dB.

Dual-slope filters optimize speaker's crossover response

by P. Antoniazzi and A. Hennigan
SGS-ATES Electronic Components, Milan, Italy

The crossover response, and thus the overall performance, of a two-way high-fidelity loudspeaker system can be significantly improved with these high- and low-pass networks. Staggering two cascaded RC filters in the woofer channel yields a slope of 6 dB/octave near the cutoff frequency, f_c, and a notably steeper 12 dB/octave beyond f_c. When combined with the complementary (inverted response) output of the tweeter section, optimum crossover characteristics are achieved at low cost and without audio-frequency discontinuities at f_c.

In general, many simple low-pass networks can provide a 6-dB/octave response at frequencies approaching f_c from the low side. When a single-pole filter is used, however, as is still done on occasion, the maximum roll-off beyond f_c can never be greater than 6 dB/octave. Unfortunately, the typical loudspeaker does not have a linear enough response to handle high-level signals (degraded by only 6 dB/octave) at its high-frequency limits, and so distortion results.

With second-order filters (12 dB/octave), a loss of audio usually occurs at the crossover point. This phenomenon is caused by the +90° phase shift of the low-pass network, which when combined with the −90° output of the system's high-pass filter tends to cancel the audio output.

Using a third-order Butterworth filter solves both of

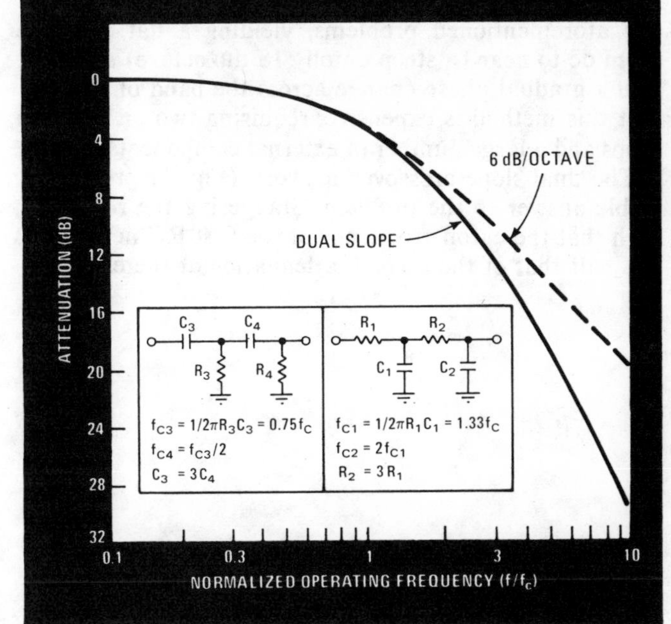

2. Response. Dual-slope filter, using staggered RC networks, virtually eliminates drop-off in audio output at the cutoff frequency of hi-fi speakers, while providing a roll-off of greater than the usual 6 dB/octave. Equations for woofer and tweeter sections summarize design.

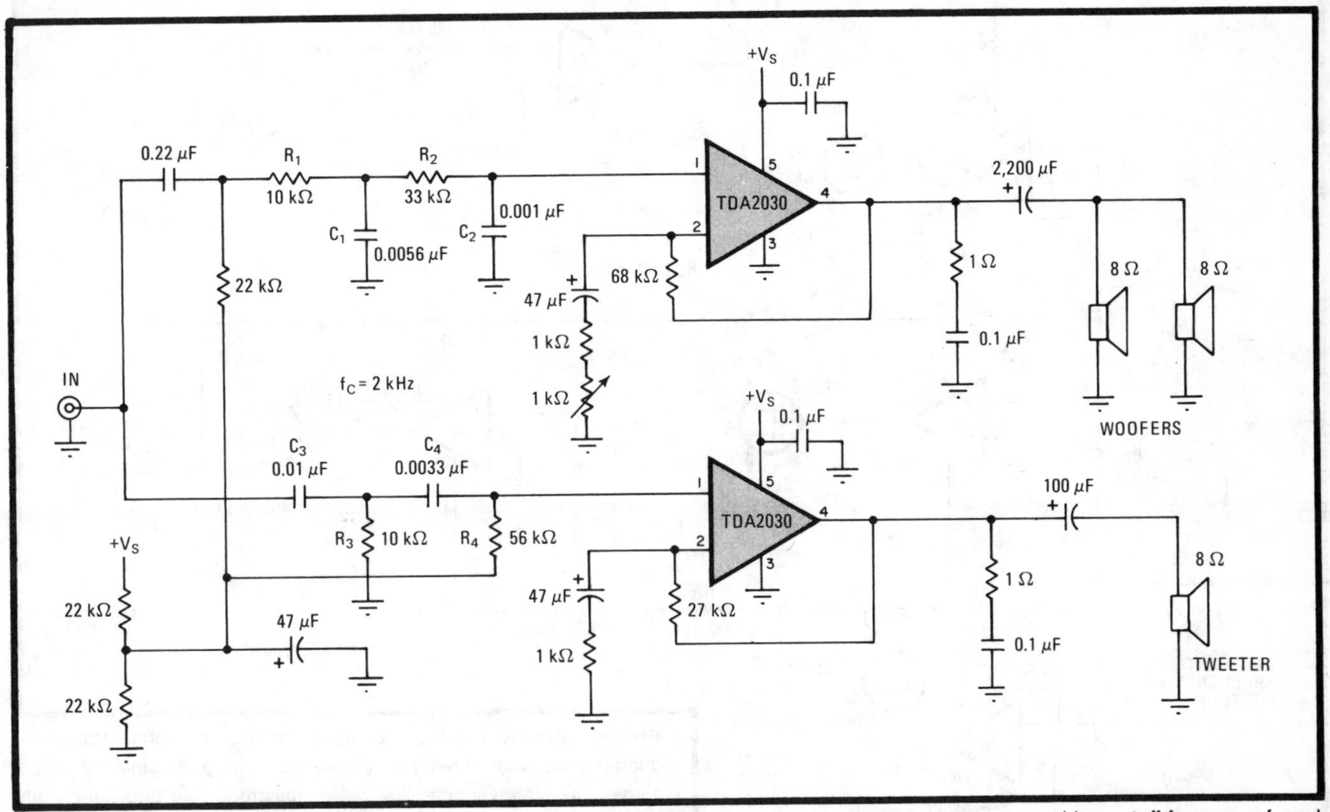

1. Distortionless. Staggered low-cost, low-pass filters in woofer channel achieve slope of 6 dB/octave approaching cutoff frequency f_c and 12 dB/octave above f_c without introducing quadrature phase shift and accompanying distortion produced by loudspeakers. When combined with complementary output of high-pass section, system achieves crossover characteristic devoid of audio discontinuities at f_c.

the aforementioned problems, yielding a flat response from dc to near f_c, steep cutoff (18 dB/octave) above f_c, and a gradual phase change across the band of interest. But this method is expensive, requiring two or three op amps and a large number of external components.

The dual-slope crossover network (Fig. 1) provides a viable answer to the problem. Staggering the responses such that the cutoff frequency of the first RC network is one half that of the second, attenuation at the crossover frequency will be 3 dB as in other systems, but the phase shift at f_c will be 60°; thus the cancelation problem typical of second-order filters is avoided. This circuit is ideally suited to active loudspeaker systems.

The plotted response of the woofer section is shown in the curve, which is complete with the required design equations. Corresponding equations for the tweeter are also included. □

Voltage-controlled integrator sets filter's bandwidth

by Henrique Sarmento Malvar
Department of Electrical Engineering, University of Brazilia, Brazil

Because it uses operational transconductance amplifiers as electrically tuned integrators for selecting a state-variable filter's center frequency and bandwidth virtually independently of one another, this circuit is ideal for use in music and speech synthesizers that require individual voltage-controlled setting of these parameters.

As shown, input signals to be processed are applied to transconductance amplifier A_1, whose bias current is

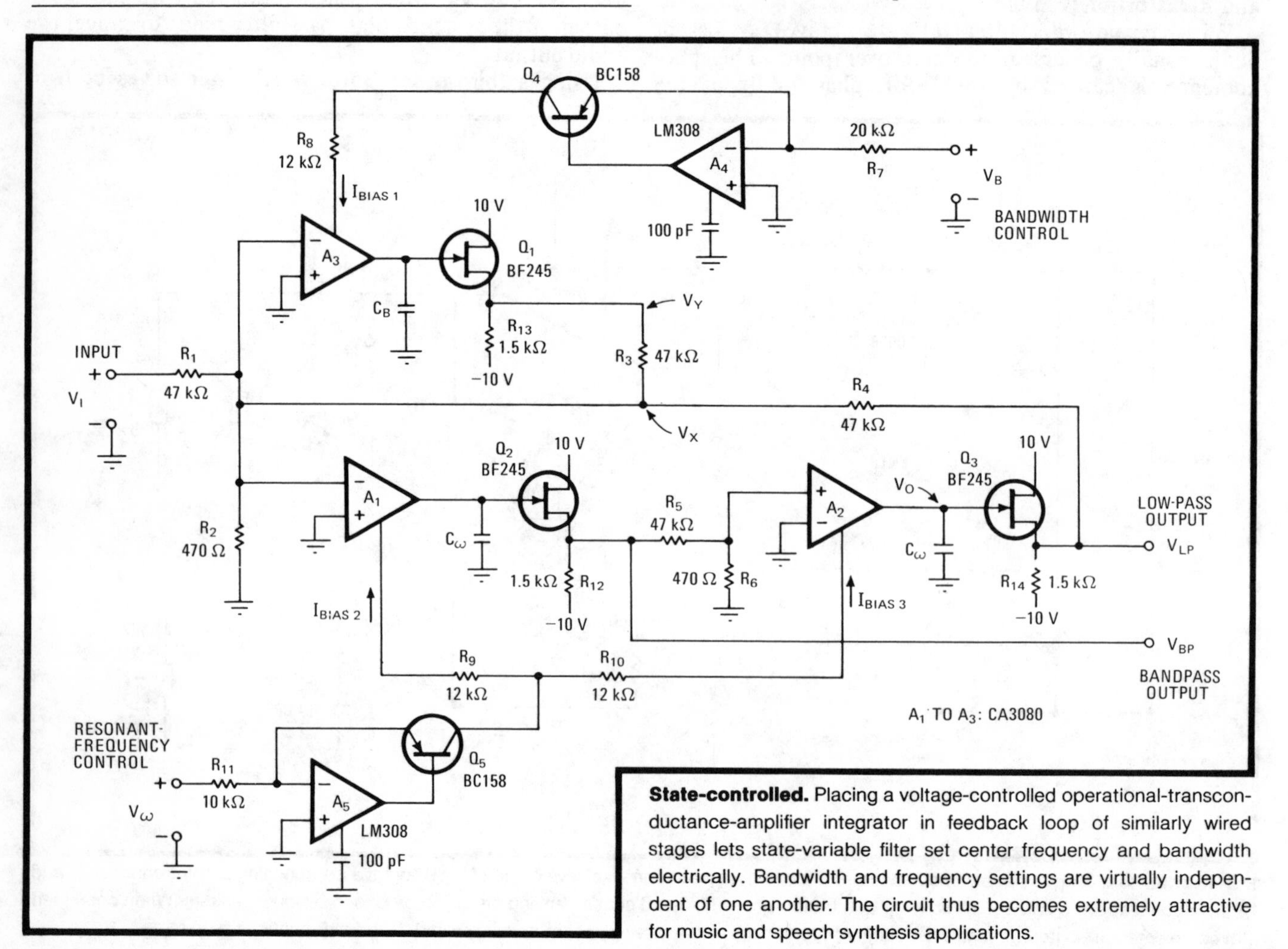

State-controlled. Placing a voltage-controlled operational-transconductance-amplifier integrator in feedback loop of similarly wired stages lets state-variable filter set center frequency and bandwidth electrically. Bandwidth and frequency settings are virtually independent of one another. The circuit thus becomes extremely attractive for music and speech synthesis applications.

derived from voltage-to-current driver A_5Q_5. The bias current, in turn, is set by control voltage V_ω, which ultimately determines the filter's center frequency. In general, $I_{bias} = V_\omega/2R_{11}$.

Placing the capacitor C_ω across the output of the operational transconductance amplifiers A_1 and A_2 converts the two-stage network into a noninverting integrator, the transfer function becoming $V_o(s)/V_i(s) = k/s$, where k is the 3080's transconductance factor k'/C, and transconductance k' as given in its data sheets is $k' = 19.2\ I_{bias}$. Substituting for I_{bias}, $V_o(s)/V_i(s) = 19.2\ V_\omega/sRC_\omega$.

Placing integrator A_3 in the feedback loop of A_1 and A_2 equips the basic filter with the capability to control bandwidth. As seen from inspection of the circuit:

$$V_x(s) = R_1||R_2||R_3||R_4\left[\frac{V_1(s)}{R_1} + \frac{V_y(s)}{R_3} + \frac{V_{LP}(s)}{R_4}\right]$$

where the low-pass function is given by:

$V_{LP}(s) = -(k_2/s)^2V_x(s)R_6/(R_5+R_6)$

and where $V_y(s) = -k_1V_x(s)/s$. Constants k_1 and k_2 are given by:

$k_1 = 19.2\ V_B/R_7C_B$

$k_2 = 19.2\ V_\omega/2R_{11}C_\omega$

and the bandpass function, $V_{BP}(s)$, equals $-k_2V_x(s)/s$.

Assuming $R_1 = R_3 = R_4 = R_5 = R'$ and $R_2 = R_6 = R''$, with $R''/R' = a<<1$, it follows that $R_1||R_2||R_3||R_4 = R_2$. Substituting the appropriate quantities back into all the above equations, it is seen that:

$$V_{LP}(s)/V_i(s) = -1/[(s/\omega_o)^2+(B/\omega_o)(s/\omega_o)+1]$$

$$V_{BP}(s)/V_i(s) = -(s/\omega_o)/[(s/\omega_o)^2+(B/\omega_o)(s/\omega_o)+1]$$

where the resonant frequency and the bandwidth are given by:

$$\omega_o(\text{rad/s}) = ak_2 = 19.2aV_\omega/2R_{11}C_\omega$$

$$B(\text{rad/s}) = ak_1 = 19.2aV_B/R_7C_B$$

Parameter a should be near 0.01, to ensure the best compromise between distortion and signal-to-noise ratio. The bias current of the operational transconductance amplifier should not be more than 500 microamperes. □

Chapter 16
FREQUENCY SYNTHESIZERS

Frequency multiplier uses combinational logic

by R. J. Patel
Tata Institute of Fundamental Research, Bombay, India

Relying on a technique that uses digital logic rather than high-speed system clocks or nonlinear generators to perform frequency multiplication, these circuits derive a square wave with an output frequency of up to four times that of the input signal. Extremely easy to understand and implement, the general method used in synthesizing these combinational-logic circuits is useful for achieving practical high-order frequency multiplication of up to eight times.

Since frequency-doubler circuits are relatively simple and well-known configurations exist, the logic technique is shown in Fig. 1 for a frequency tripler. For the logic circuit to perform tripling, the waveform at Z must traverse three full cycles, or six half-cycles (represented by states 101010), during the time of one input cycle (represented by 111000) at V. Thus the circuit must detect six different logic states, and so a minimum of three input variables, V, W, and X, is required.

Note, however, that the input signal at port V is the

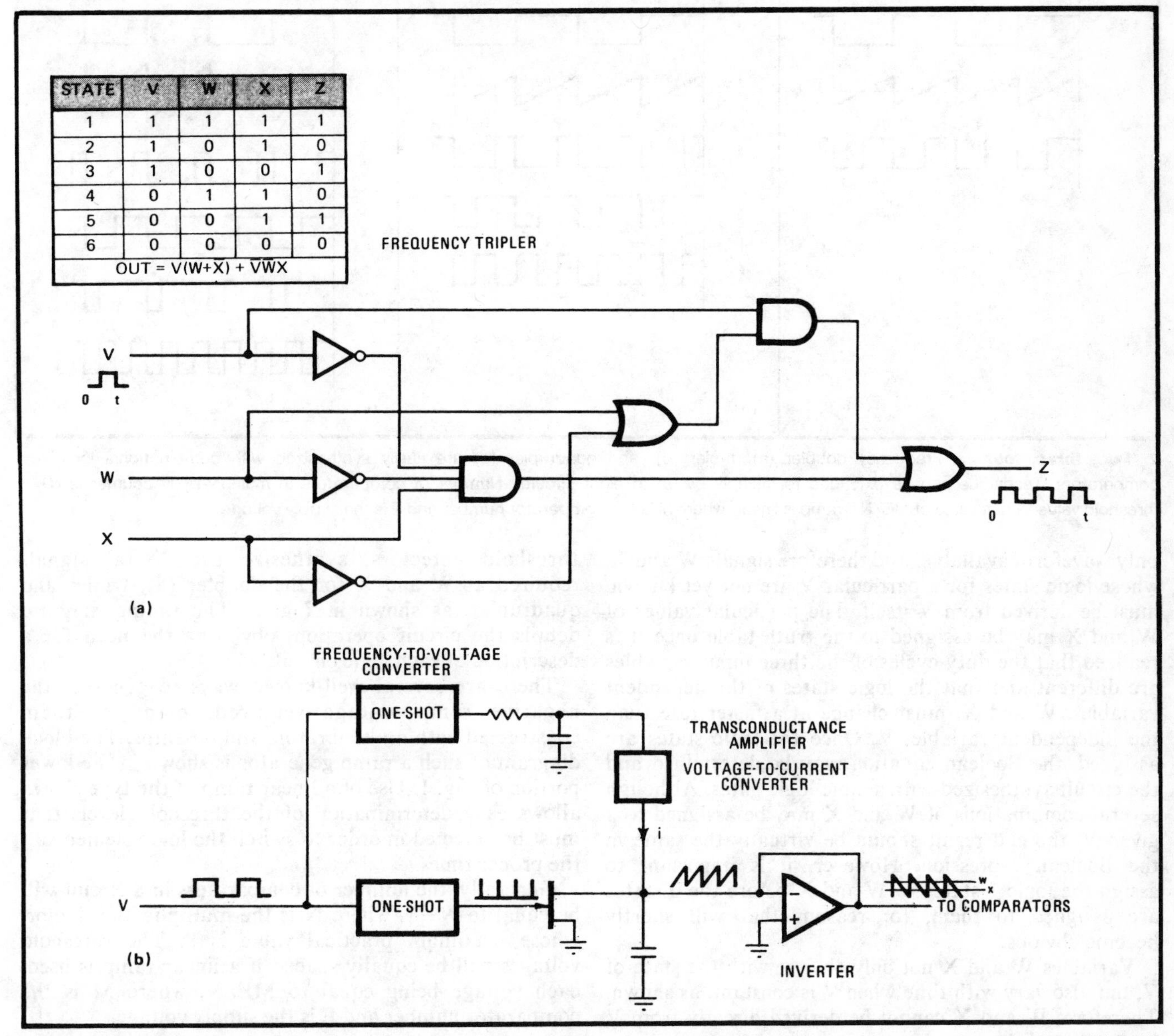

STATE	V	W	X	Z
1	1	1	1	1
2	1	0	1	0
3	1	0	0	1
4	0	1	1	0
5	0	0	1	1
6	0	0	0	0
OUT = $V(W+X) + \bar{V}\bar{W}X$				

1. Multiply by 3. Digital frequency multiplier is an alternative to multipliers using high-speed clocks and nonlinear generators. States W and X are derived from V, although transformation cannot be done digitally (a). Ramp and comparators can generate the required digital voltages from V, however (b). Use of linear ramp allows easy determination of the threshold levels that must be detected to switch logic elements.

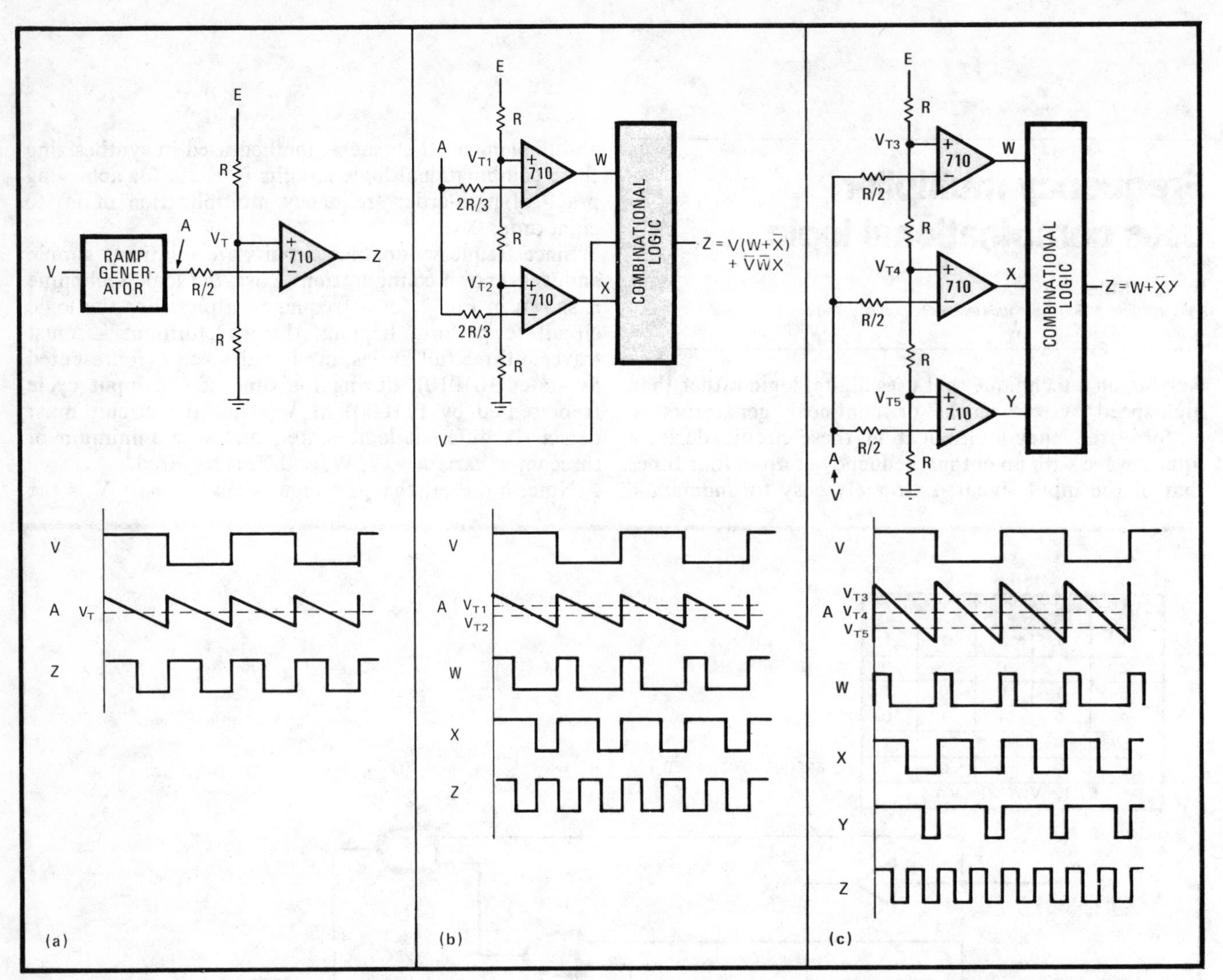

2. Two, three, four.... Frequency doubler (a), tripler (b), and quadrupler (c) are easily synthesized with combinational logic and comparators. Technique can be extended to multiply by up to eight circuits. Number of comparators in multiply-by-N counter is N − 1; threshold value V_T has value of ME/N if ramp is linear, where M is the comparator number and E is the supply voltage.

only waveform available, and therefore signals W and X, whose logic states for a particular V are not yet known, must be derived from V itself. The particular values of W and X may be assigned to the truth table once it is realized that the duty cycles of the three input variables are different and that the logic states of the dependent variables, W and X, must change at a faster rate than the independent variable, V. Once the logic states are assigned, the Boolean equation may be determined and the circuit synthesized with simple logic gates. Although several combinations of W and X may be assigned to a given V, the end result should be virtually the same in the Boolean expression. However, it is important to assign the logic 1 states to W and X before the 0 states are assigned to them, for reasons that will shortly become obvious.

Variables W and X not only change with the state of V, but also vary with time when V is constant, as shown. Therefore, W and X cannot be derived directly from V in the digital domain. However, a negative-going ramp voltage whose sweep rate is equal to twice the input frequency can, with the aid of operational-amplifier threshold detectors, synthesize the digital signals required at W and X for the doubler (a), tripler and quadrupler, as shown in Fig. 2. The timing diagram details the circuit operation, obviating the need for a description of each logic circuit.

There are several well-known ways to generate the negative ramp voltage required, many of them constructed with multivibrators and op amps. The block diagram of such a ramp generator is shown in the lower portion of Fig. 1. Use of a linear ramp of the type shown allows easy determination of the threshold levels that must be detected in order to switch the logic elements at the proper times.

Generally, the number of comparators in a circuit will be equal to N − 1, where N is the multiplication factor, whose maximum practical value is 8. The threshold voltages will be equally spaced if a linear ramp is used, each voltage being equal to ME/N, where M is the comparator number and E is the supply voltage. □

Programmable multiplier needs no combinational logic

by Noel Boutin
University of Sherbrooke, Electrical Engineering Department, Quebec, Canada

The frequency multiplier circuit proposed by R. J. Patel [see preceding article] can be vastly simplified for applications where the duty-cycle of the output waveshape must or may be kept low. Specifically, no combinational logic is needed for a frequency multiplication of N, and only one comparator is used in place of the N − 1 comparators required in Patel's circuit. A programmable voltage reference, which can be just a potentiometer, a digital-to-analog converter, or a keyboard [see chapter 28, p. 249], can be used to set the multiplication ratio desired.

The modified circuit is shown in Fig. 1. Two one-shots are fired on the positive edge of each input pulse, f_{in}. The input signal may have almost any duty cycle.

Emanating from the transconductance amplifier is a sawtooth waveform, which has a period determined by M_2. The one-shot drives one input of an OR gate that switches gate G_1 on at regular intervals. Connected to the other input of the OR gate is the output of a comparator, which turns on G_1 and resets the ramp whenever the value of the sawtooth amplitude exceeds the value set by the programmable reference voltage at the comparator's inverting port. This operation is shown in the timing diagram of Fig. 2 for multiplication ratios of 2, 3, and 4, respectively. Thus frequency multiplication is achieved by the feedback signal from the comparator, not by the use of combinational logic, which derived the output wave from the input signal in Patel's circuit. The number, N, of equally spaced spikes that appears at the output of the comparator during each period of the input signal is given by $N = V_{max}/V_{ref}$, where V_{max} is the maximum voltage that would be reached if the feedback signal were not present.

If a symmetrical output (square wave) is required, a flip-flop operating a divide-by-2 counter can be added to the output of the comparator. It will also be necessary to adjust the multiplication ratio to twice the desired value (2N) in order to recover a square-wave output frequency of Nf_{in}. □

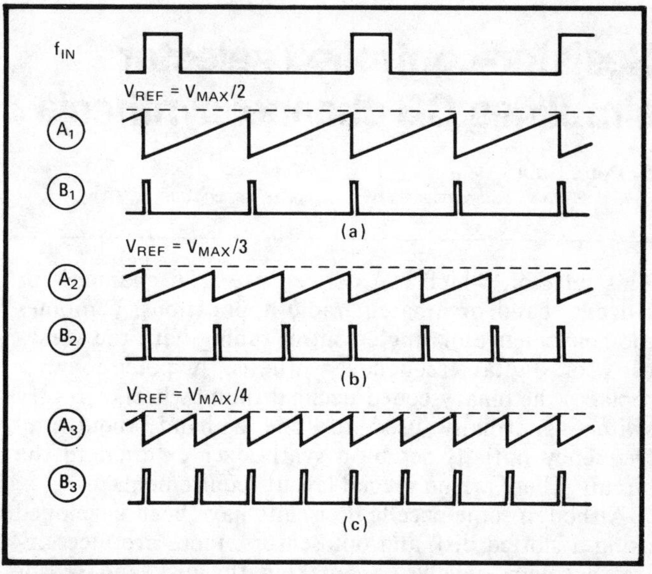

2. Two, three, four . . . Frequency doubler (a), tripler (b), and quadrupler (c) are easily synthesized by adjusting the ratio V_{max}/V_{ref} with the programmable reference source. Output signals are pulses, but may be converted to square waves by adding a flip-flop operating as divide-by-2 counter to comparator output.

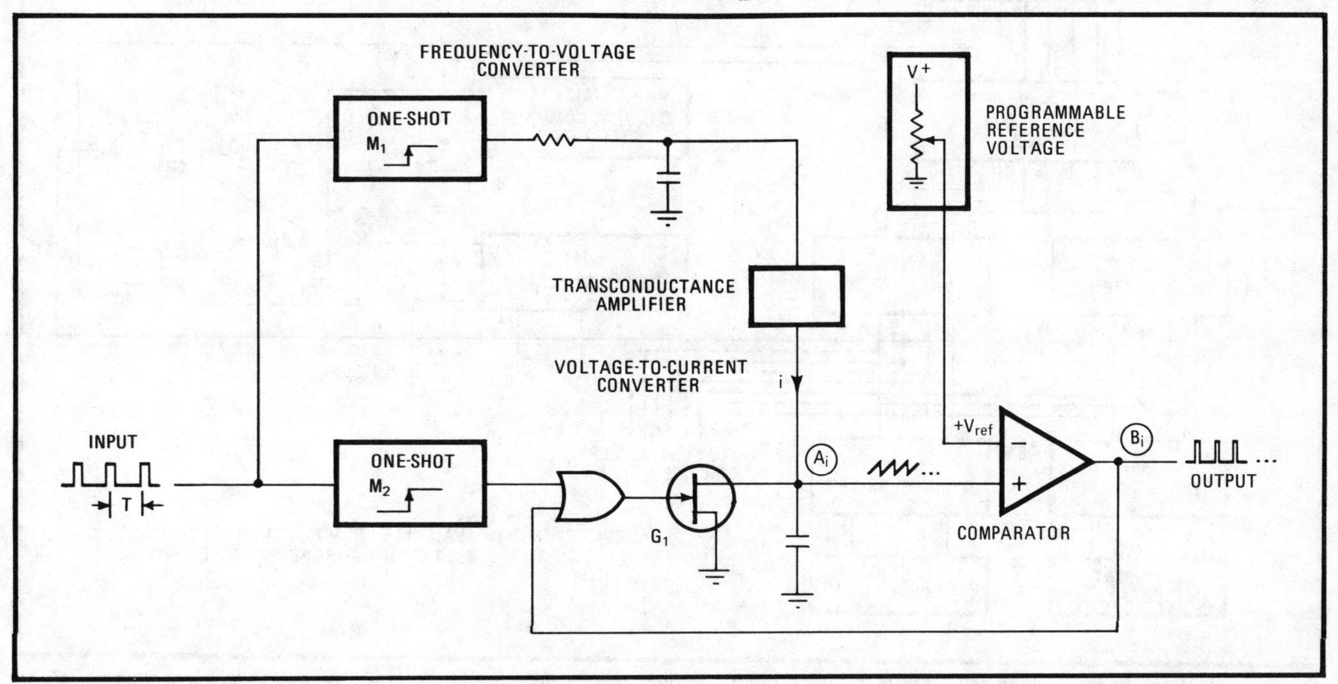

1. Simplified. Circuit can achieve frequency multiplication without using logic or more than one comparator (see text). Multiplying signal is derived from input wave and comparator, which periodically resets sawtooth each time the ramp itself exceeds a present reference voltage.

Resistor-controlled selector simplifies CB channel synthesis

by Peter Saul
Ferranti Electronic Components Division, Lancashire, England

This selector, which can derive up to 256 channels for citizens' band or amateur-radio applications, combines the convenience of single-control tuning with the versatility of digital frequency synthesis. A potentiometer replaces the binary-coded-decimal thumbwheel or rotary switches normally used to tune a band about any frequency initially set by a synthesizer external to the circuit. There are no special layout requirements.

Although some excellent circuits have been developed using a slotted disk and optosensors, there are mechanical problems involved in making the disk and sensor mountings, and the systems are generally too complex and costly. In addition, these systems do not return to the set frequency after a power-down. This circuit overcomes those drawbacks.

The system uses a potentiometer with 10 turns or more, an analog-to-digital converter, and straightforward logic designs as shown in the figure. A good choice for the analog-to-digital converter is the Ferranti ZN425E, which has an on-chip counter.

A voltage derived by P_1 (in the tuner circuit) from the on-chip voltage, V_{REF}, is applied to the noninverting input of the ZN424P low-noise, high-gain amplifier. The voltage must be converted to its BCD equivalent by the a-d converter.

The conversion begins as the ZN425 starts counting in binary after the initialization from the convert-command oscillator. The counter is advanced by the clock oscillator, which is gated through the 7400 array. As the counter advances, a linear ramp voltage is generated at its output at pin 14. This signal is introduced to the noninverting port of the ZN424. When the ramp voltage exceeds the voltage set by P_1, the output of the op amp falls, resetting the flip-flop in the 7400 and terminating the count. This process is repeated 70 times per second.

The 7475 latches capture the data and drive the 7447 decoder and drivers so that the channel number, from 0 to 63, may be displayed. The 7-bit word also drives the synthesizer. For 256 channels, two additional 74185 converters are needed, as are an additional 7475, 7447, and seven-segment display. The actual frequency of the channel is determined by the design of the oscillator in the synthesizer.

The circuit returns to its initially set channel after power-off or power failure, because the same comparison voltage appears at the noninverting input of the ZN424P. Temperature and voltage stability are good,

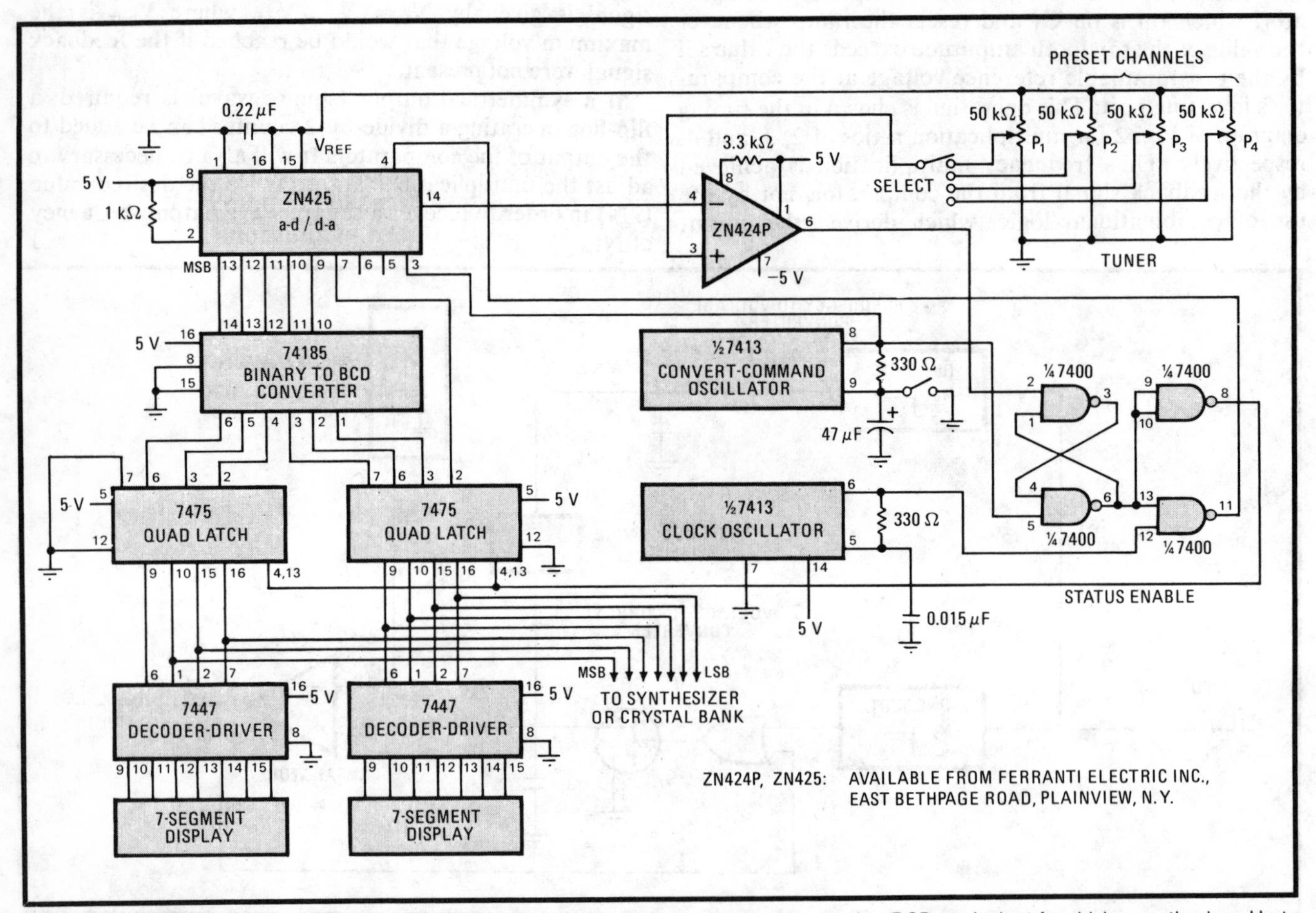

Single-control synthesis. Voltage derived from V_{REF} by 10-turn potentiometers is converted to BCD equivalent for driving synthesizer. Up to 64 channels are generated by 74185 and ZN425. Additional BCD converters, latches, and displays are needed for a 256-channel synthesizer.

provided good-quality potentiometers are used. Under extreme combinations of temperature and voltage changes, a one-channel shift has been observed, but that is not a serious problem. In addition, it is extremely difficult to set the frequency between channels. Power drain from the circuit is 350 milliamperes, but it may be reduced considerably if low-power transistor-transistor logic replaces the standard devices. □

Ring counter synthesizes sinusoidal waveforms

by Timothy D. Jordan
Texas A & M University, College Station, Texas

A digital circuit composed of only two counters and a weighted resistor network is as good at producing sine and cosine waveforms as many quadrature oscillator networks. Because matched components are not used, design considerations are radically simplified.

Use of the digital technique eliminates many components. The upper frequency limit of the oscillator is 250 kilohertz, and it is not affected by the frequency limitations of operational amplifiers, because no op amps are used. Tweaking the oscillator is not necessary, because no special circuitry is needed. And the sine and cosine waveforms are equal in magnitude at every frequency, because no integrating or differentiating circuits are used. It is even possible to transform the circuit into a digital-to-sine-wave converter with little modification, if the counters' parallel input ports are used to accept binary signals.

As shown in the figure, two cascaded 4018 complementary-metal-oxide-semiconductor integrated circuits wired as a single ring counter are driven by the master clock. The 4018s divide the input frequency by 12. The digital clock advances the ring counter by one count on the positive clock transition, and each output port moves from the high to low state sequentially.

The resulting current through the weighted resistor network at the counter's output produces a 12-step approximation of a sine wave. The output stages of the second 4018 produce a cosine wave, since it is delayed three clock periods, or one quarter of a cycle, with respect to the first counter.

The first appreciable harmonics to appear at the output are the 11th and 13th, and they may be filtered out with a passive resistance-capacitance filter. Identical filters should be used for each counter so that the phase shift introduced is equal for both output waveforms. The input frequency may be as high as 3 megahertz; above 1 MHz, no filtering is necessary. □

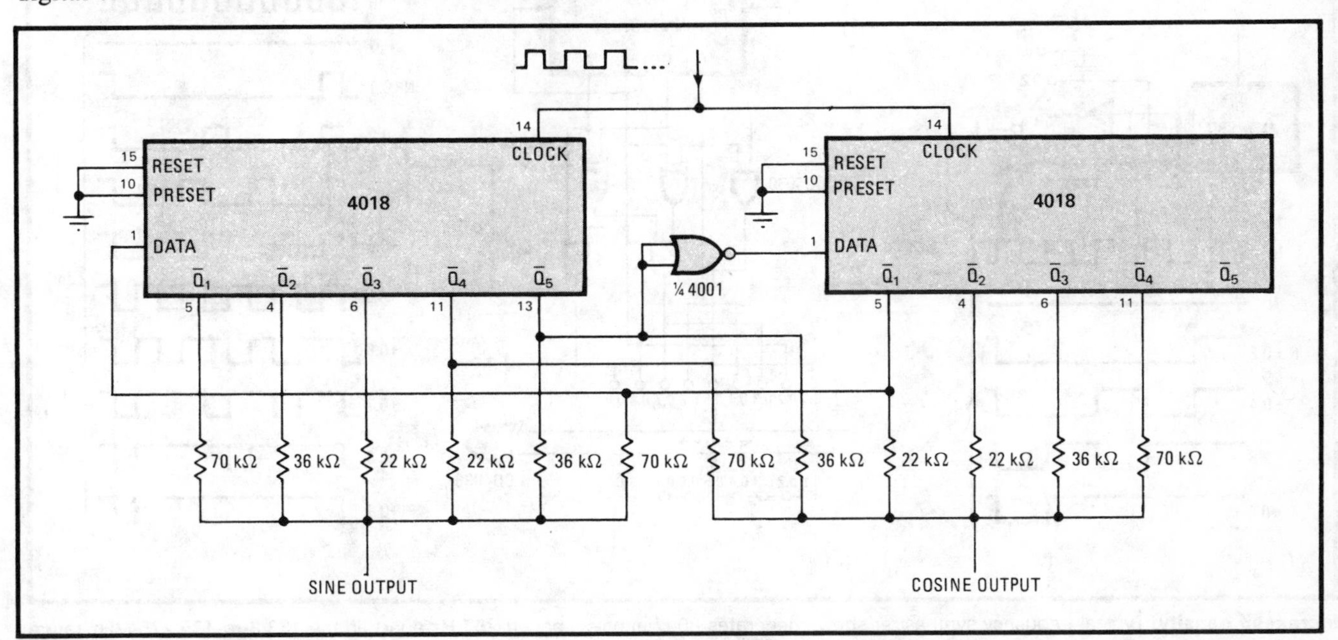

Digitized waves. Frequency dividers and weighted resistor network generate sine and cosine waveforms. The 4018 counters divide input frequency by 12. First counter provides 12-bit approximation of sine function. Second device lags by 90°, producing cosine waveform.

Digital interpolator extends number of synthesizer channels

by Jo Becker
University of Stuttgart, Institute of Biomedicine, Stuttgart, Germany

The extensive modifications needed to increase a commercial or already built frequency synthesizer's channel-generating capability over a given range can be a designer's nightmare. But the addition of only a counter and a few gates will multiply the synthesizer's channel capacity 5 to 10 times—and does so almost without touching the basic circuit.

The technique used for channel multiplication may be

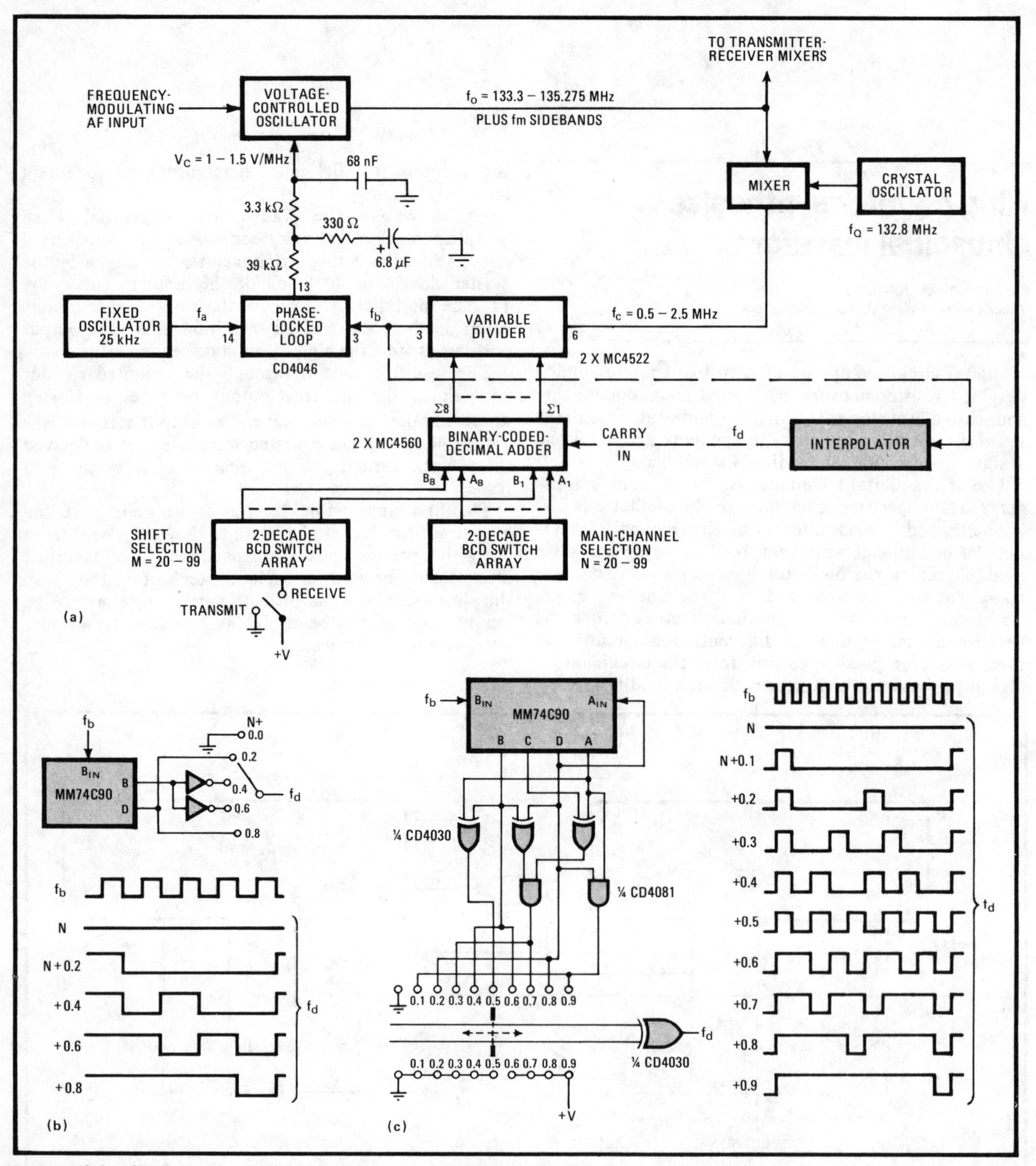

Increased density. Typical frequency synthesizer shown generates 80 channels, spaced 25 kHz apart, in the 133.3-to-135.275-MHz range (a). Number of channels in the given band may be increased by 5 by adding only one bi-quinary counter and two inverters (b). Channel multiplication can be extended to 10 if a few more gates are added (c). N values are selected with a rotary switch and an exclusive-OR gate.

understood with the aid of a block diagram of a typical frequency-modulated synthesizer (a). Here, the voltage-controlled oscillator generates 80 carrier frequencies, or channels, in the 133.3-to-135.275-megahertz band, where the channel separation is 25 kilohertz. The channel spacing is determined by the value of f_a. The output frequency is:

$$f_o = f_q + Nf_a$$

where f_q is generated by a local crystal oscillator and N, an integer, is selected by a binary-coded-decimal switch array and a suitable BCD adder, plus a variable frequency divider and a standard feedback circuit that uses a phase-locked loop. An additional BCD switch array control makes it possible for the synthesizer to generate a frequency, f_o, such that:

$$f_o = f_q + (N+M)f_a$$

and this option is useful in cases where the synthesizer is employed in a receiver circuit to generate separate transmitting and receiving frequencies.

To enable the VCO to oscillate at frequencies between the channels, noninteger values of N or M must be generated, and therefore it is necessary to connect the interpolator (b) to the basic synthesizer circuit. The interpolator circuit simply uses a MM74C90 counter and two inverters to convert the divide-down signal, f_b, into a bi-quinary code in order to modify the BCD output of the MC4560 adder, the device that is primarily programmed by the switch array.

When the bi-quinary patterns, represented by f_d, are introduced to the carry-in port of the 4560, the average value of N produced per given time interval is selectable throughout the range N+0.0 to N+0.8 in increments of 0.2, enabling 400 channels with 5-kHz spacing to be generated by the VCO. The instantaneous value of N changes, of course, and therefore the phase-locked-loop filter shown at the output of the 4046 (a) is required to ensure that there will be minimal ripple in the control voltage, V_c. V_c will thus be smoothed sufficiently to prevent spurious signals from being generated between adjacent 5-kHz frequencies.

A second circuit (c) uses the interpolation technique to increase the number of channels to 800, spaced 2.5 kHz apart. □

Chapter 17
FUNCTION GENERATORS

Timer generates trapezoid for musical synthesizers

by Roland Bitsch
Anzefahr, West Germany

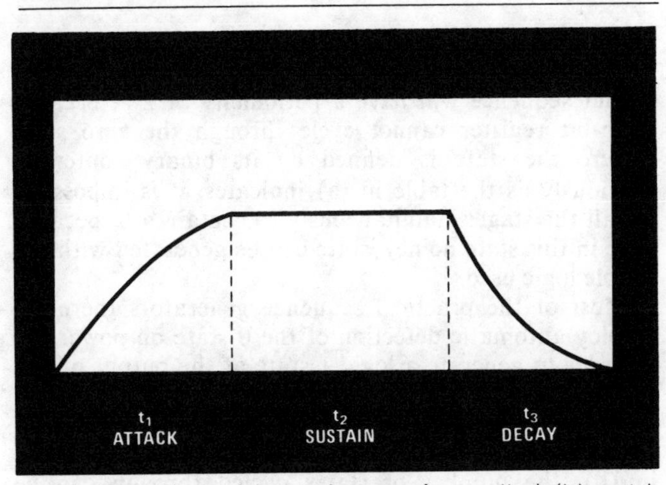

1. Envelope. Shape of the envelope waveform—attack (t_1), sustain (t_2), and decay (t_3)—in addition to timbre (harmonic content) distinguish the sounds of musical instruments. Woodwinds have fast attack, quick decay; plucked strings are the opposite.

To build a trapezoidal-waveform generator of the kind used as an envelope in electronic music synthesizers, only a 555 timer integrated circuit and a few discrete components are needed. The circuit can vary the waveform's attack, sustain, and decay times over a wide range and puts out a signal with a maximum amplitude of 8 volts.

Figure 1 shows the three time increments characteristic of this envelope generator. Since its attack and decay signals are functions of charging capacitors, its waveform exhibits exponential rise and fall times. But linear ramps may be obtained by charging each capacitor with a constant current—a bipolar or field-effect transistor inserted in its charging circuit does the job.

Figure 2 is the generator circuit. The input is a 0-to-2-v positive pulse, which is compatible with most keyboard-type voltages. The RC network differentiates the pulse, and transistor Q_1 inverts it, to provide the necessary negative-going trigger for the timer.

Triggering the timer sets the flip-flop within the chip, and the discharge pin (7) goes high. Diodes D_1 and D_2, now reverse-biased, permit timing capacitors C_1 and C_2 to charge through sustain control R_1 and attack control R_2, respectively. C_2 ceases to charge upon reaching the breakdown voltage of the zener diode (8.6 v), and this event ends the attack period, t_1.

C_1 continues to charge until the voltage at the threshold pin (6) of the timer, determined by the R_4-R_5 summing junction, reaches approximately 10 v. The flip-

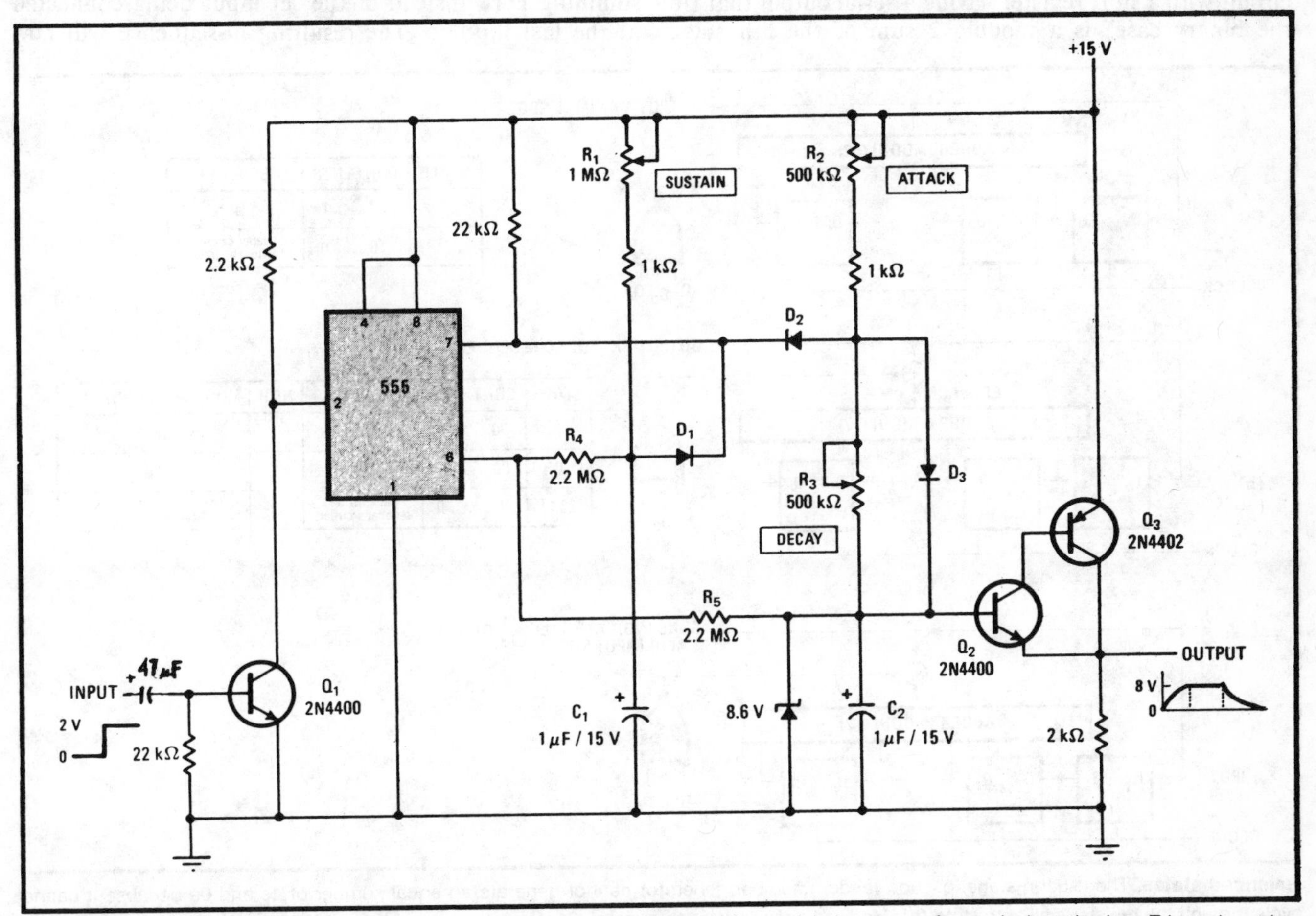

2. Waveform generator. Timer circuit puts out envelope waveform suitable for modulating tones of a musical synthesizer. Trigger input is a 2-V positive pulse; output has a maximum amplitude of 8 V and variable attack, sustain, and decay times of up to several seconds each.

flop within the chip is then reset, and this event ends the sustain period, t_2.

Reset of the internal flip-flop drives the discharge pin (7) low, allowing C_1 and C_2 to discharge through forward-biased diodes D_1 and D_2. Capacitor C_2, discharging through potentiometer R_3, determines the decay time period, t_3. Note that R_3 is shorted by D_3 when C_2 is charging during the attack time.

For linear ramps, the constant current sources must be placed in the R_2 and R_3 charging legs. However, the audible difference between linear and exponential attacks and decays of short duration is not so great. □

Feedback extends sequence of random-number generator

by J. T. Harvey
AWA Research Laboratory, New South Wales, Australia

Altering the periodicity of its sequence from 2^{n-1} to 2^n makes a pseudorandom number generator more useful in such applications as frame synchronization, numbering, or identification in a digital communications system. The modification has little effect on the pseudorandomness of the number sequence, and it can be implemented with just a simple shift register and a slightly changed feedback loop.

The usual pseudorandom, or maximal-sequence, generator is shown in (a). It combines a simple feedback circuit with a shift register having a serial output that (in the binary case) is a modulo 2 sum of the nth stage output and one or more of the previous stage outputs. No matter what feedback taps are chosen, however, the output sequence will have a periodicity of 2^{n-1}, because an n-bit register cannot cycle through the zero state (where the state is defined by its binary contents). Obviously as the table in (a) indicates, it is impossible for all the stages simultaneously to contain a 0, because once in this state no new state can be generated with the simple logic used.

Most of the practical sequence generators therefore employ automatic detection of the 0 state on power up, in order to generate a logic 1 state at the output of one register, using a NOR gate as shown. After initialization, though, the random generator will cycle with a period of 2^{n-1}, because the 0 state never recurs. However, as shown in (b), the number of states cycled through can be increased to 2^n if the last input to the NOR gate is connected to the Q_{n-1} stage, instead of the output of the Q_n stage, and if the output of the NOR is connected to a summing port, instead of the set input being connected at the last flip-flop. The resulting m-sequence will now

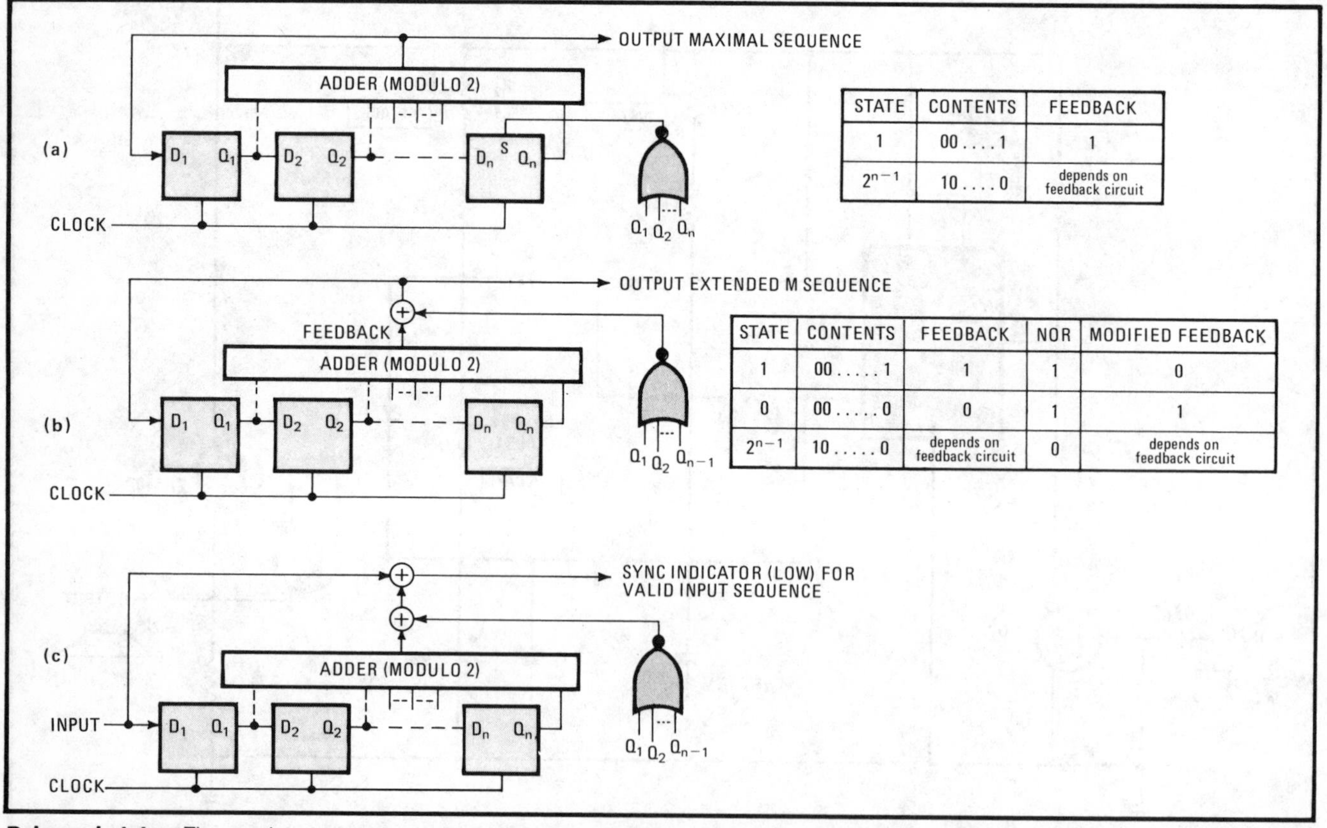

STATE	CONTENTS	FEEDBACK
1	00....1	1
2^{n-1}	10....0	depends on feedback circuit

STATE	CONTENTS	FEEDBACK	NOR	MODIFIED FEEDBACK
1	00.....1	1	1	0
0	00.....0	0	1	1
2^{n-1}	10.....0	depends on feedback circuit	0	depends on feedback circuit

Balanced states. The usual n-stage, pseudo-random number generator cannot generate an equal number of 1s and 0s because it cannot cycle through the register's all-zero state (a). Modified circuit cycles through all 2^n states (instead of the usual 2^{n-1} states) and detects 0s (b). Circuit can be used in feed-forward mode to detect repeating sequences (c).

contain an equal number of 1s and 0s.

The logic needed to implement the summing port is small. In most cases, it need consist of a half-adder or a few resistors only. Note from the table in (b) that the all-zero state is detected and used to generate a logic 1.

The same basic circuit can be used in the feed-forward mode to detect a sequence of n bits, too, as (c) proves. This method for checking the presence of a given sequence is of value when the error probability is low. During high-error bursts, however, the circuit's efficiency is reduced because isolated, incorrect bits generate m+1 parity errors when there are m taps driving a half (modulo 1) adder.

Note that the sequence-recognizer circuit in (c) does not require additional logic to prohibit the acceptance of the (false) all-zero data state. A standard sequence checker will give a valid output when it is fed continually by 0s, because it would predict the next input to be a 0, once the register is filled. Thus, an additional n-input NOR gate and an exclusive-OR gate must be added to a standard feed-forward checker to reject an all-zero sequence within the data stream. □

Smoke-detector chip generates long time delays

by J. Brian Dance
North Worcestershire College, Worcs., England

Long, repeatable time delays are attained with this circuit, which uses a single low-power complementary-metal-oxide-semiconductor chip usually found in smoke-detection systems. A small number of additional passive elements are used, though only three of them—the timing components and a load resistor—are actually required for operation.

Unlike more popular timers such as the 555, the Supertex SD1A can provide long time delays because it does not load down the timing components. The input impedance of its comparator input, to which the timing components are connected, is 10^{13} ohms, enabling timing resistors of up to 10^{12} Ω to be used. To avoid problems associated with capacitor leakage, however, resistor values of up to 2.5×10^{10} Ω and a 10-microfarad non-electrolytic capacitor have been used to yield time constants of 10 hours or more.

When the start switch is opened, capacitor C_t charges through resistor R_t (see figure). The timing period ends when the potential at the junction of R_t and C_t has fallen to half the supply voltage. The period is approximately $0.69R_tC_t$. This time can be varied by placing a potentiometer (50 kilohms to 50 megohms) between pin 2 (pin

A smoking timer. Special-purpose chip, for smoke detectors, can serve as one-shot, providing longer delays than popular timers, such as the 555. SD1A's input does not load timing network, enables setting of time constants to more than 10 hours. Versatile chip also has provisions for flashing LED to check chip operation during long time intervals and for sounding alarm (R_L) if battery voltage is low.

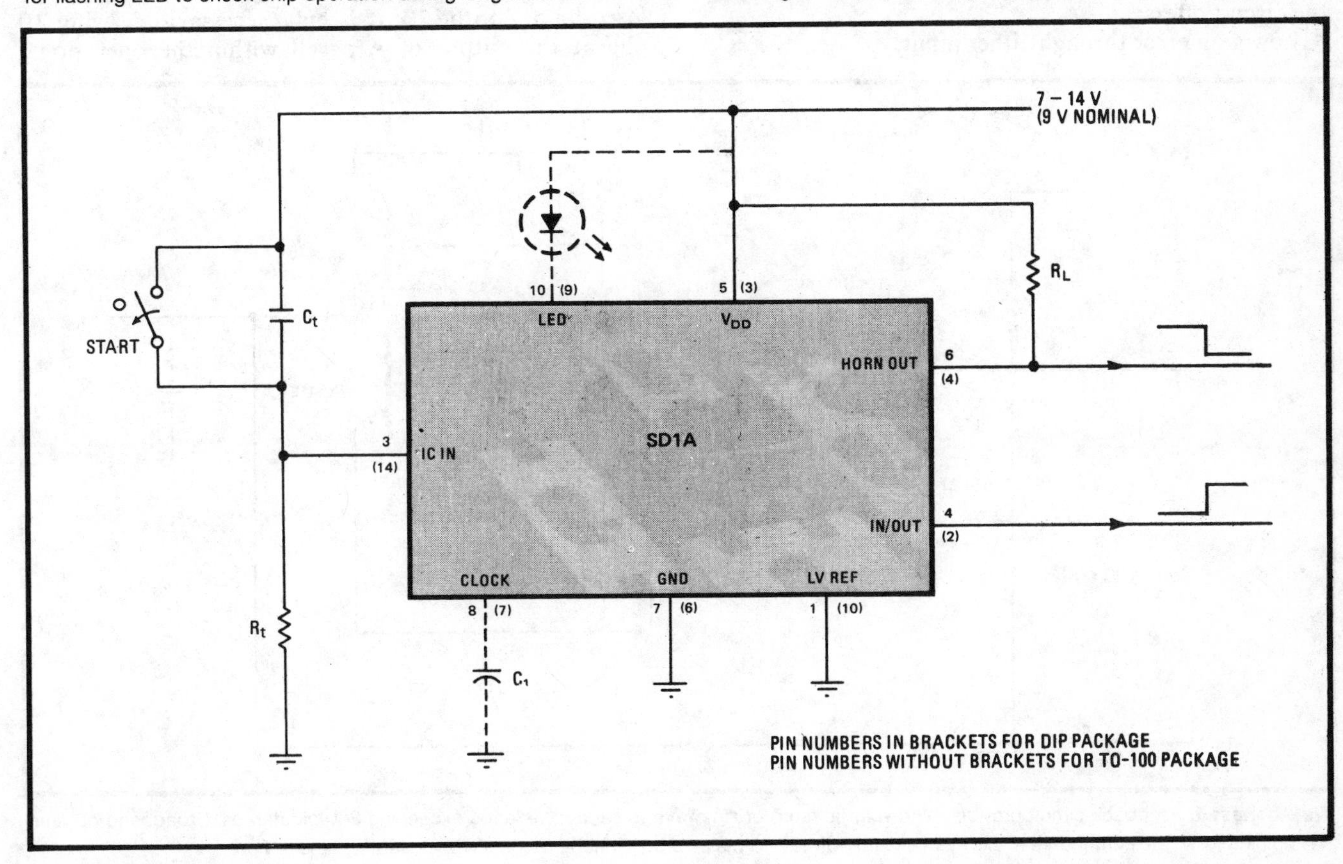

12 of the dual in-line package) and the supply.

At the end of the timing period, the voltage at the input/output port rises to $V_{DD} - 0.5$ volts. The horn output falls to near zero, serving as a current sink for the load, R_L, which may be an alarm or just a resistor. The horn output can sink at least 300 milliamperes.

A light-emitting diode connected to the appropriate port of this versatile chip can be made to flash every 40 seconds during the timed period if a 10-microfarad capacitor is connected to the clock pin. The flashing provides a useful indication that the circuit is operating during long timing periods. The frequency of the flashing is controlled by the value of C_1 or can be adjusted by connecting a potentiometer between the clock pin and V_{DD} for decreasing the period or between the clock pin and ground for increasing the period.

The SD1A also includes a circuit for sounding the alarm every 40 seconds if the supply voltage is low (below 7 V). Here the low-voltage reference pin, LV REF, has been grounded so as to disable this feature.

The cost for the SD1A is $2.50 in the TO-100 package version, and $2 in the DIP. The manufacturer does not yet have authorized distributors for the device, however, and thus at this time the SD1A can only be purchased in lots of 50 or more directly from Supertex, 1225 Bordeau Dr., Sunnyvale, Calif. 94086. □

Serrodyne amplifier generates wideband linear ramp

by Roy Viducic

Loral Electronic Systems, Yonkers, N. Y.

Producing an extremely linear 200-V ramp or peak-to-peak sine wave over a frequency range of 50 kilohertz, this circuit is ideally suited for modulating the helix of a traveling-wave tube (serrodyning) or generating large voltages for circuit synchronizers. Because of the inherent symmetry of the circuit design, it can produce both positive-going and negative-going ramp waveforms.

In addition to the advantages previously mentioned, the circuit offers:

- Low gain error through either input.
- Fast response for both ramp polarities—the output's flyback-settling time is 1 microsecond.
- Wide dc-offset capabilities.

To achieve this performance, an operational amplifier with a high slew rate and broadband response is used in a simple feedback arrangement.

The low-level ramp or sine wave signal is applied to either input of A_1. Note that the polarity of the ramp at the output is reversed if the input signal is applied to the opposite port of A_1. The broadband characteristics of A_1 provide sufficient output, even at a closed-loop gain of 20. This is the minimum loop gain required to obtain the linearity, precision, and dynamic range that were initially specified for this application.

After inversion and further amplification by Q_1, the signal passes through complementary-transistor pair Q_2–Q_3 to the output. Because Q_1 is designed for a closed-loop gain of 20, it is only necessary to swing 10 volts at the output of A_1, well within the operational

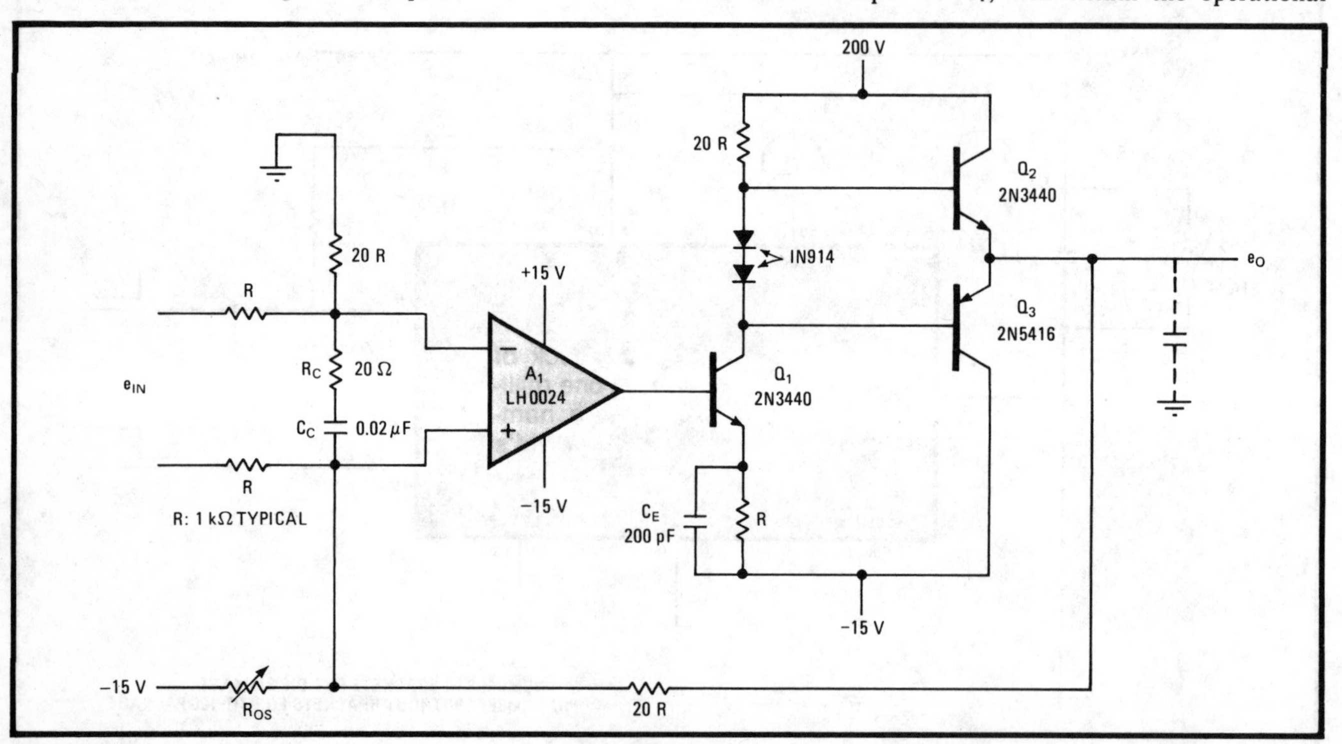

Well-behaved. Feedback circuit provides high-voltage ramp or sinewave to capacitive loads exceeding 200 picofarads. Broadband op amp having high slew rate contributes to excellent linearity of waveforms over 50 kilohertz range. Flyback settling time is 1 μs.

amplifier's capabilities in its linear region.

Capacitor C_E compensates for the total base-to-emitter capacitance at Q_2 and Q_3, which act to reduce circuit speed. Network R_cC_c stabilizes the entire loop. Potentiometer R_{os} is used to adjust the output offset.

Considerable heat will be generated in Q_1–Q_3 because of high collector voltage. In spite of this, the power dissipated by them is well within the ratings of their TO-5 cases. Therefore, the transistors do not require external heat sinks. □

Diode-shunted op amp converts triangular input into sine wave

by Giovanni Righini and Franca Marsiglia
Florence, Italy

Finding applications in low-frequency function generators, linear voltage-controlled oscillators, and, in general, wherever it is necessary to have an output with a constant amplitude that is independent of frequency, this converter transforms a bipolar triangular wave into a sinusoidal output having a total harmonic distortion of less than 1%. The circuit uses a single operational amplifier and several diodes placed across its input and output ports to synthesize a transfer function that decreases the amplifier gain as output voltage increases. Thus it generates a piecewise-linear approximation of a sine wave for corresponding changes in triangular-signal amplitude.

The principle upon which the converter operates is simple, as shown in (a). If the input voltage to the op

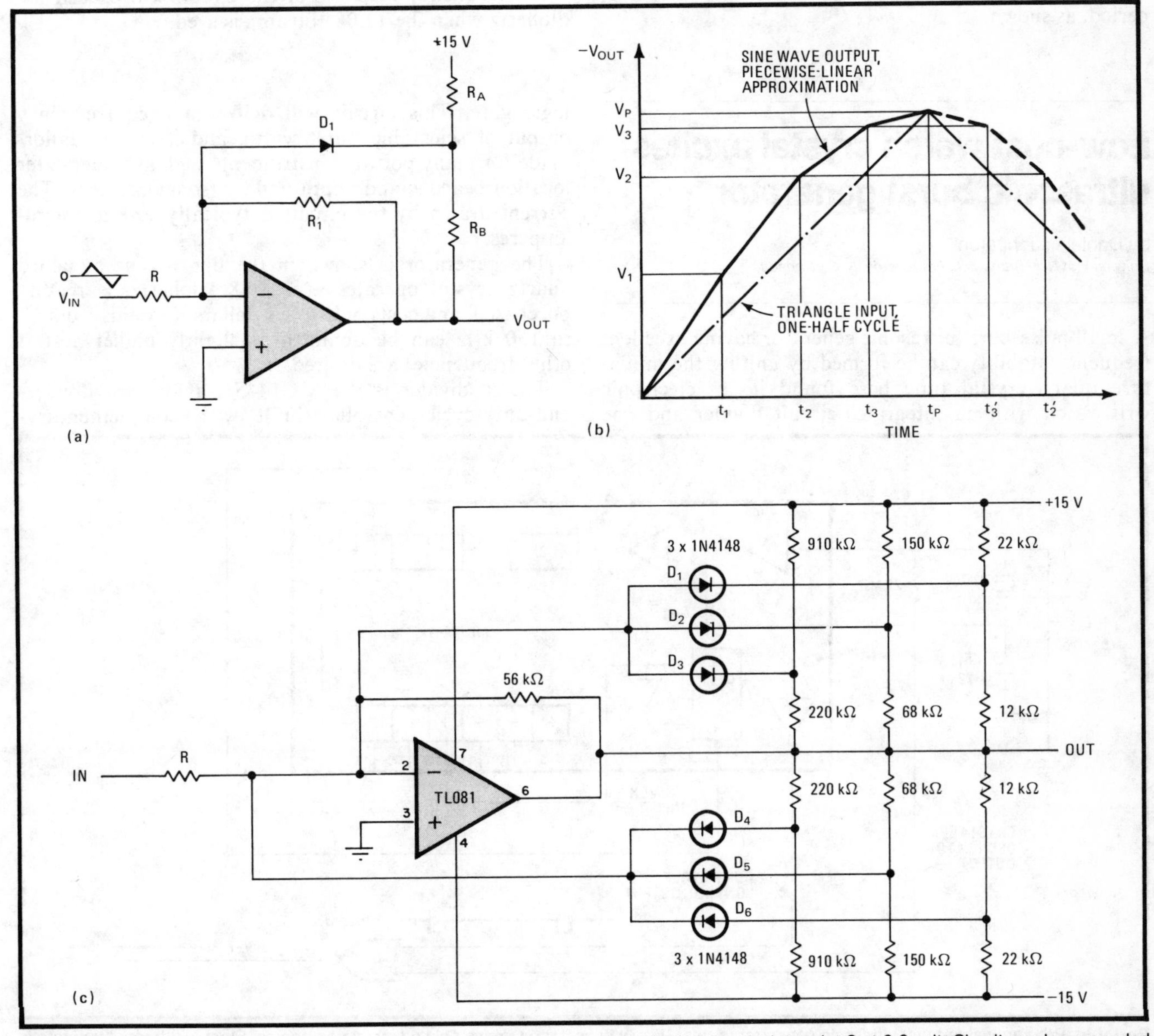

Segmented sines. Diode and resistors (a) reduce op amp gain when output signal falls, placing point C at 0.6 volt. Circuit can be expanded to generate piecewise-linear approximation of half a sine wave (b) from triangle wave when other diode-resistor networks are added (c) to reduce gain further with increasing input voltage. Adding D_4–D_6 networks permits generation of other half cycle of sine wave.

amp is such that the output voltage is positive, diode D_1 does not conduct and the amplifier gain will be R_1/R. But when a positive input voltage from the triangular input is applied to the inverting port, the output will go negative and D_1 will conduct when point C reaches a threshold value of -0.6 volt.

In effect, if the internal impedance of the diode is neglected so long as it is conducting, resistor R_b is in parallel with R_1, and the amplifier gain is reduced to $R_1R_b/(R_1+R_b)R$. The change in gain occurs at point V_1 on the transfer function (b).

By placing additional resistor-diode networks that include D_2 and D_3 at the output of the op amp (c) and by selecting resistor values appropriately, the voltage at which conduction occurs for each diode can be set. So, for each progressive rise of the triangular voltage the transfer function can be extended to points V_2 and V_3. A good approximation to a half sinusoid can be obtained by using four segments for every quarter of a (sine wave) period, as shown.

To process the negative-polarity portion of the triangular wave (which is required to generate the other half of the sine wave), resistor-diode networks including D_4, D_5, and D_6 are placed in the circuit. Note that the diodes will conduct in sequence as the op amp's output voltage exceeds the predetermined positive values.

The component values have been selected to produce three segments corresponding to end-point values of $|V_1|=4.35$, $|V_2|=7.7$, $|V_3|=9.15$. V_P is limited by the maximum voltage of the input signal. Only resistors having a tolerance of 5% or below should be used in this circuit, to keep distortion to a minimum.

The gain of the op amp is most conveniently controlled by varying R and can be selected to handle a wide range of input voltages—from a few tens of millivolts to tens of volts. R should be about 1.8 kilohms for input voltages of 1 V peak to peak and should be trimmed to minimize sine-wave distortion at the output.

The frequency response of the circuit will exceed 30 kilohertz when the TL081 op amp is used. □

Low-cost watch crystal excites ultrasonic burst generator

by Daniel F. Johnston
University of New Brunswick, Fredericton, N. B., Canada

A small pulse-burst ultrasonic generator having excellent frequency stability can be formed by uniting the miniature quartz-crystal time base found in an electronic wristwatch with an integrated-circuit divider and one logic gate. This circuit will deliver a fixed frequency output of selectable burst width, and it is thus tailor-made for many portable instruments such as underwater location beacons and depth-finding (sonar) devices. The current drawn by the circuit is typically several microamperes.

The generator is shown in the figure. The standard quartz crystal operates at 32.768 kilohertz, is readily purchased, and costs only a few dollars. Crystals from 17 to 150 kHz can be obtained at slightly higher cost if other frequencies are desired.

The IC divider is the MC14451, a low-cost divider-and-duty-cycle-controller built with complementary-

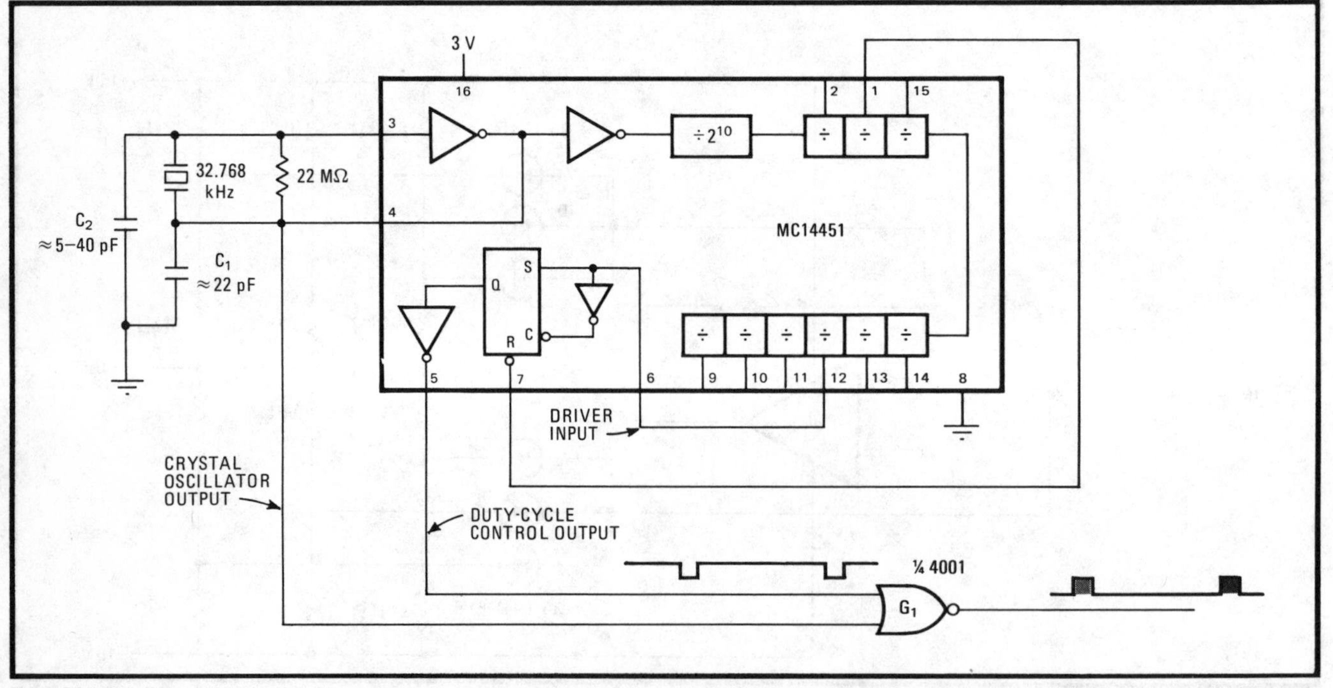

Portable. Programmable-burst ultrasonic generator is small, is low in cost, and draws only a few microamperes. Burst-width repetition rate is selected by connecting the appropriate buffered output of the tapped binary divider in the MC14451 to pin 6. Burst width, adjustable from about 31 milliseconds to more than 1 second, is selected by connecting a second buffered output to pin 7 of the device.

metal-oxide-semiconductor technology that may be powered by a source of from 1.3 to 3.0 volts. The device contains an 19-stage binary divider (with taps available at any register port from 2^{11} to 2^{19}, inclusive), and a buffered flip-flop circuit for duty-cycle, or burst-width control.

The crystal is placed in a conventional oscillator circuit, as shown, with one inverter of the MC14451 serving as the active positive-feedback element. C_1 and C_2 in the oscillator are trimmed to achieve the required accuracy and are on the order of 22 picofarads for C_1 and 5 to 40 pF for C_2. The output of the oscillator is simultaneously fed to the MC14451 and G_1.

The crystal oscillator signal appearing at the output of G_1 is gated by the duty-cycle control output of the MC14451. To select the burst-width repetition rate, the appropriate buffered output of the divider must be connected to the driver input of the duty-cycle flip-flop, pin 6. Another buffered output, whose output period corresponds to twice the required burst width, must be connected to the duty cycle reset port, pin 7. The duty-cycle control output will toggle as required, switching low for the interval specified by the reset line at a rate controlled by the driver-port signal. The output from G_1 will therefore be a burst of a constant frequency.

Current consumption of the circuit is only 5 μA when a 32-kHz crystal is used and only twice that for a 65-kHz crystal. Thus the circuit can be powered by a small-capacity battery. □

RAM and d-a converter form complex-waveform generator

by William A. Palm and G. A. Williamson
Magnetic Peripherals Inc., Minneapolis, Minn.

A random-access memory and digital-to-analog converter are at the heart of a programmable waveform generator that will produce almost any wave, however complex, at low frequency. It is only necessary to determine the digital equivalent of the desired wave and store that in a 256-word-by-8-bit RAM, for later transformation into an analog output voltage by the d-a converter.

The maximum repetition rate of the signals produced

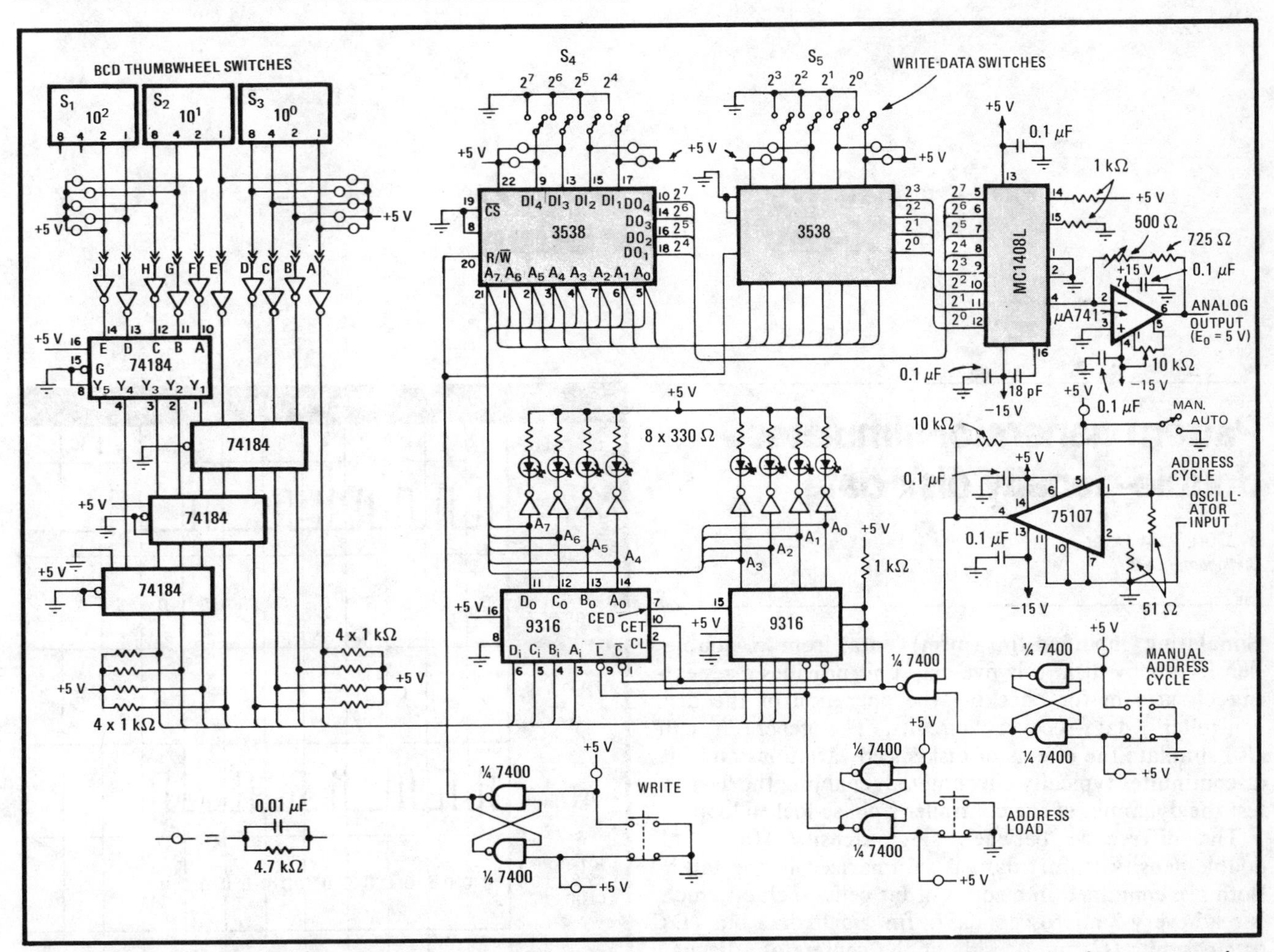

Simple storage of complex waveshapes. Programmable generator is capable of producing almost any complex low-frequency waveform. Once the digital equivalent numbers of the waveform are determined, they are stored in the 3538 random-access memories. The MC1408 digital-to-analog converter transforms the random-access memory's contents into the desired analog waveform.

is only 7.8 kilohertz, being limited by the RAM's access speed and the number of locations; but more often than not, a complex wave is required only in very-low-speed applications. If higher speed is desired, the use of emitter-coupled-logic RAMs will increase the repetition rate to nearly 400 kHz, on the assumption that a high-frequency clock is available.

As shown in the figure, binary-coded-decimal thumbwheel switches $S_1 - S_3$ are used to address one of 256 RAM locations, and S_4 and S_5 select one of 256 waveform amplitudes (over the range 0 – 255 in base 2). Thus each location can be stored with an amplitude that can be resolved to 1 bit in 256.

When each location of the RAM is stepped (at a maximum rate equal to the system clock frequency divided by the number of locations), its output at any given instant is in essence converted to one point in a 256-by-256 matrix. The RAM is formed by two 256-word-by-4-bit Fairchild 3538 devices.

Four 74184 BCD-to-binary counters, two 9316 4-bit counters, a 75107 line receiver, and several gates and switches make up the binary address counter. The 75107 allows flexibility in choice of the clock that addresses the RAM—almost any driving signal will do, whether a sine wave, ramp, pulse, or square wave. The address counter either may be single-stepped manually to the desired location (needed when loading the RAM) or cycled continuously when the desired waveform is generated.

Two momentary-contact switches perform the actual address-loading and data-writing operations, and the address at any given time is displayed by eight light-emitting diodes. The maximum clock frequency permitted is limited to 2 megahertz by the slow response of the RAM, and therefore the repetition rate of the output waveform is 2 MHz/256 = 7.8 kHz.

Storing 256 numbers is a tedious task, especially if it must be done often. In cases where a given waveform is needed frequently, consideration should be given to storing it permanently in a nonvolatile read-only memory. The ROM could be substituted for the RAM in this circuit.

When cycled, the output of each RAM location is converted into a current by the MC1408 d-a converter. This is then transformed into an analog voltage by the 741 operational amplifier. □

Pattern generator simulates double-density disk data

by Curt Terwilliger
Burlingame, Calif.

Simulating modified-fm (mfm) data from a double-density floppy disk, this five-chip unit provides a repeating bit stream for checking the operation of the disk controller's data-recovery circuits. The generator can also simulate the effects of disk-speed variations and bit discontinuites typically encountered, enabling the user to test the dynamics of the controller's phase-locked loop.

The difference between single-density (fm) and double-density (mfm) data is summarized in the table. Both are contained in a series of bit cells, each of which is read every 2 microseconds. In fm, single data bits (D) or clock bits (C) may reside at the centers of adjacent cells. Note that in mfm, however, no clock cells are present and the data is compressed; here, n data bits,

COMPARING FLOPPY-DISK DATA FORMATS

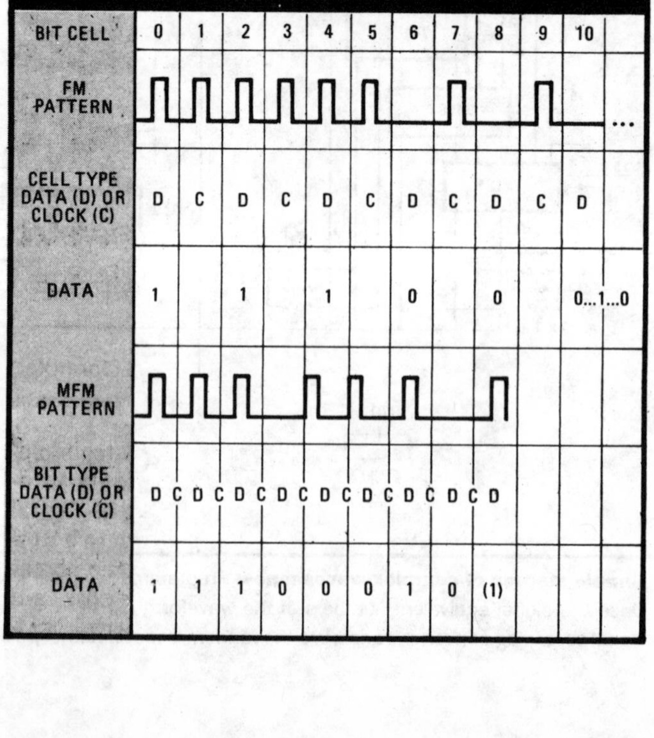

BIT CELL	0	1	2	3	4	5	6	7	8	9	10
FM PATTERN											...
CELL TYPE DATA (D) OR CLOCK (C)	D	C	D	C	D	C	D	C	D	C	D
DATA	1		1		1		0		0		0...1...0
MFM PATTERN											
BIT TYPE DATA (D) OR CLOCK (C)	D C	D C	D C	D C	D C	D C	D C	D C	D		
DATA	1	1	1	0	0	0	1	0	(1)		

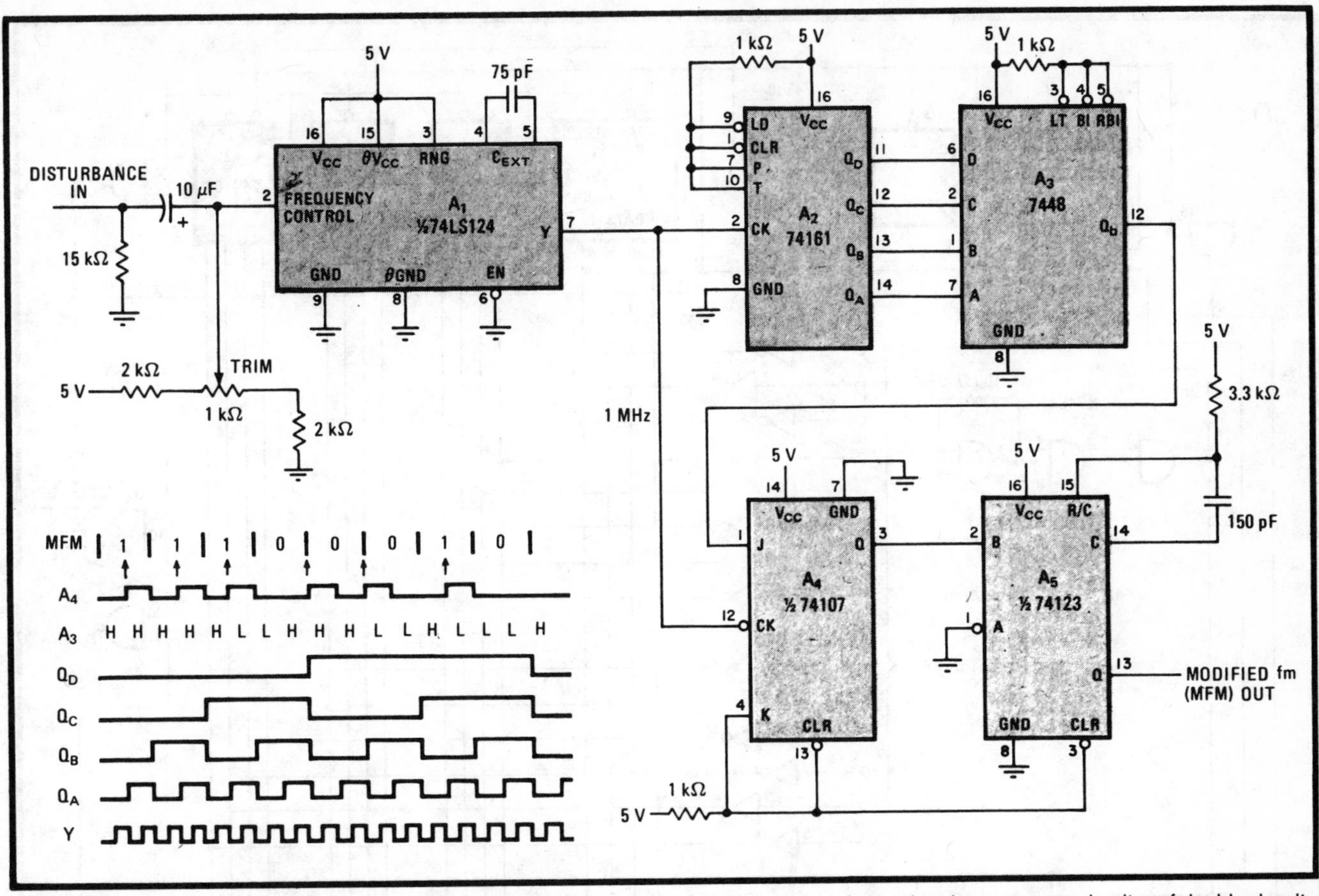

Mimicking mfm. Five-chip generator provides preprogrammed, modified-fm data stream for testing data-recovery circuitry of double-density floppy disk. Bit rate can be externally controlled to mimic disk-speed variation in order to check dynamic range of disk-controller's PLL.

which would be contained in 2n cells in fm, fit into n cells. The circuit provides the compressing operation by generating the clock pulse at the leading edge of appropriate cells, so that only the D bits are considered by the system's data-recovery circuits for every cell read.

The circuitry required for data compression is straightforward and in this case provides a pattern in which all three of the bit-to-bit intervals normally encountered in mfm appear (that is, 2, 3, and 4 microseconds). One bit stream that provides these intervals is 11100010, and it is selected for use here because it is easily generated with the use of readily available binary counters.

As shown in the figure, A_1, the 74LS124 programmable oscillator, clocks binary counter A_2 at a 1-megahertz rate. The output of A_2 is then applied to the BCD-to-seven-segment decoder, A_3, whose port b output appears as shown in the timing diagram. Note that any arbitrary pattern could be produced if A_3 were replaced by a programmable logic array (PLA) or a programmable read-only memory. The 7448 used in this circuit is in essence an inexpensive PLA.

When port b of A_3 is high and a clock pulse arrives, flip-flop A_4 is placed in the toggle mode and a pulse of 250 nanoseconds is produced at the ouput of one-shot A_5 for every other clock. As a consequence, A_2 counts to 16 for each pass through the 8-bit pattern. Although pulses are generated at the start of cell bits 4 and 5 with this method (see table), the mfm recovery circuits sample data during the middle portion of the cell and so will detect a logic 0.

When modulated at the disturbance input, A_1 can be made to deviate from its 1-MHz nominal frequency. Thus variations in disk speed can be simulated. □

Synchronous counters provide programmable pulse delays

by R. E. S. Abdel-Aal
Sunderland Polytechnic, Sunderland, England

In this circuit, cascade counters are digitally programmed to provide pulse delays of 50 nanoseconds to 3.25 milliseconds, accurate to within 50 ns. Selected by a 15-bit binary number, N, the delays can be ordered in 100-ns steps.

Two chains of synchronous counters, driven by a 20-megahertz clock, delay the input pulse's leading and trailing edge separately. Input pulses, which are asyn-

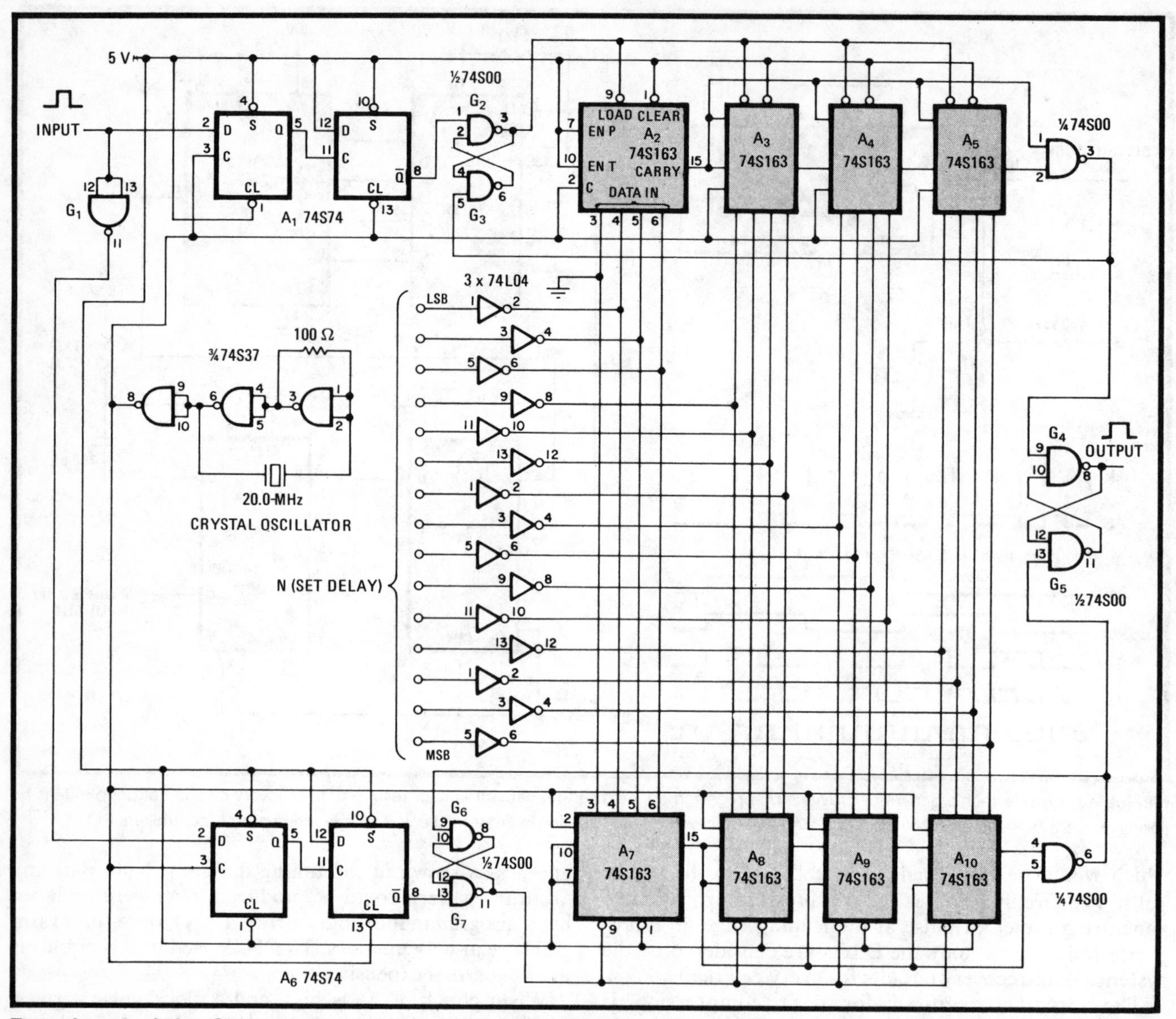

Two-edge retardation. Counter chains, of which one starts counting on arrival of the positive edge of an input pulse, the other on its negative edge, use a 20-MHz clock to provide repeatable delays of 50 ns to 3.25 ms on both edges of signal. Delays produced are accurate to within 50 ns. Amount of delay is selected with 15-bit word (M). Width of input signal to be processed must be at least 60 ns.

chronous, are first applied to a dual D flip-flop, A_1, as shown. A_1 generates a single negative-going clock pulse when triggered by the 20-MHz clock. The pulse is used to set the bistable latch, G_2–G_3, thus enabling counters A_2–A_5 to delay its leading edge.

A_2–A_5 are wired to perform a fast look-ahead operation for the high-speed multistage counting required. To eliminate glitches that might upset the counter, the carry outputs of both A_1 and A_5 are brought to a NAND gate and then to G_2–G_3. This ensures the latch will be reset and the desired N value loaded into the counter only after the previous delay period is ended, despite the differential delays that exist in the signal path.

The input pulse is inverted by G_1 and applied to flip-flop A_6 for the counters that provide delay on the trailing edge of the pulse, A_7–A_{10}. G_6–G_7 and A_7–A_{10} perform the function identical to G_2–G_3 and A_2–A_5.

In actual operation, both counter chains are programmed by the set-delay lines, which are connected to their data-in ports. The delay time is given by $T = 50(1 + 2N)$ ns. At the end of the delay period, N is loaded into both counter chains. Note that the circuit is wired such that the N inputs must be active high, ensuring that maximum delay will correspond to a value of N that, when read as a number in standard binary form, is maximum.

Upon the arrival of the leading edge of the input pulse, A_2–A_5 count up, starting from the number loaded. Similarly, A_7–A_{10} count up on the trailing edge of the pulse. The carry pulse generated by A_5 signifies the end of the delay period. This signal is united with the carry pulse generated by the trailing-edge counter A_{10}, at flip-flop G_4–G_5. The signal at the output of G_4–G_5 thus has the same width as the input pulse, but is delayed by a period of time proportional to the number N. □

Up-down ramp quickens servo system response

by R. E. Kelly
Southwest Research Institute, San Antonio, Texas

Servo systems often require a ramp waveform to control both the acceleration and deceleration rates of heavy loads, but all too often system complexity and reaction time are increased because only a positive-going, fixed-slope ramp is available for performing the dual function. Using a dual-polarity ramp generator that allows individual selection of both the up- and down-ramp rates, such as that shown in the figure, ensures optimum servo system response at low cost.

Command-input voltage derived from the servo system's joy-stick (control) position is applied to operational amplifiers A_1 and A_2, as shown. Diodes D_1 and D_2 provide a 0.7-volt drop of the input signal before it is introduced to A_1.

When the control-input voltage is above or below zero by more than 0.7 v, the system load must be accelerated to a position corresponding to that control voltage. A_1 detects that the input voltage is other than zero and switches phototransistor A_3 on, which in turn switches A_4. Meanwhile, A_2 will switch from a positive voltage to −12 v if the control voltage is equal to or greater than the ramp's output voltage. A_3 provides a high-impedance input for A_1 and an output signal suitable for driving transistor-transistor or similar logic. A_4 is a solid-state, single-pole, double-throw relay that initiates an up-ramp waveform whose slope is partially controlled by the value of R_1.

The positive-going ramp emanating from A_5 will move the system load toward its desired position, and as that occurs, the servo system's feedback voltage will act to reduce the control-input voltage. When the input voltage moves within 0.7 v of ground, A_1 moves low and A_3 turns off. A_4 now initiates a down-ramp waveform to decelerate the system to a stop at a rate determined partially by R_2.

The ramp generator, A_5, is an op-amp integrator that produces waveforms whose rate is proportional to the input voltage, e_i. This voltage, in turn, is set by R_1 or R_2. The time-dependent voltage output from the ramp generator is approximately equal to the voltage across C. C is charged by a steady current $I = e_i/R_3$, so that:

$$e_o = e_i t/R_3C \qquad (1)$$

because $e_o = e_c = It/C$.

Equation 1 is useful for setting the values of R_3 and C needed for a given ramp slope. For example, assume that the maximum positive-going or negative-going ramp rate must be 20 v/second and that e_i is 12 v maximum. Then, from Eq. 1, $20 = 12/R_3C$ and therefore $R_3C = 0.6$. Assuming a nominal value of 0.33 microfarad for C, R_3 is then equal to 1.81 megohms.

Using the component values shown in the circuit allows a linear ramp to be generated over a 10:1 frequency range. Note that once the component values for the maximum ramp rate are determined, the actual ramp rate may be selected by setting R_1 and R_2 for the corresponding value of e_i. □

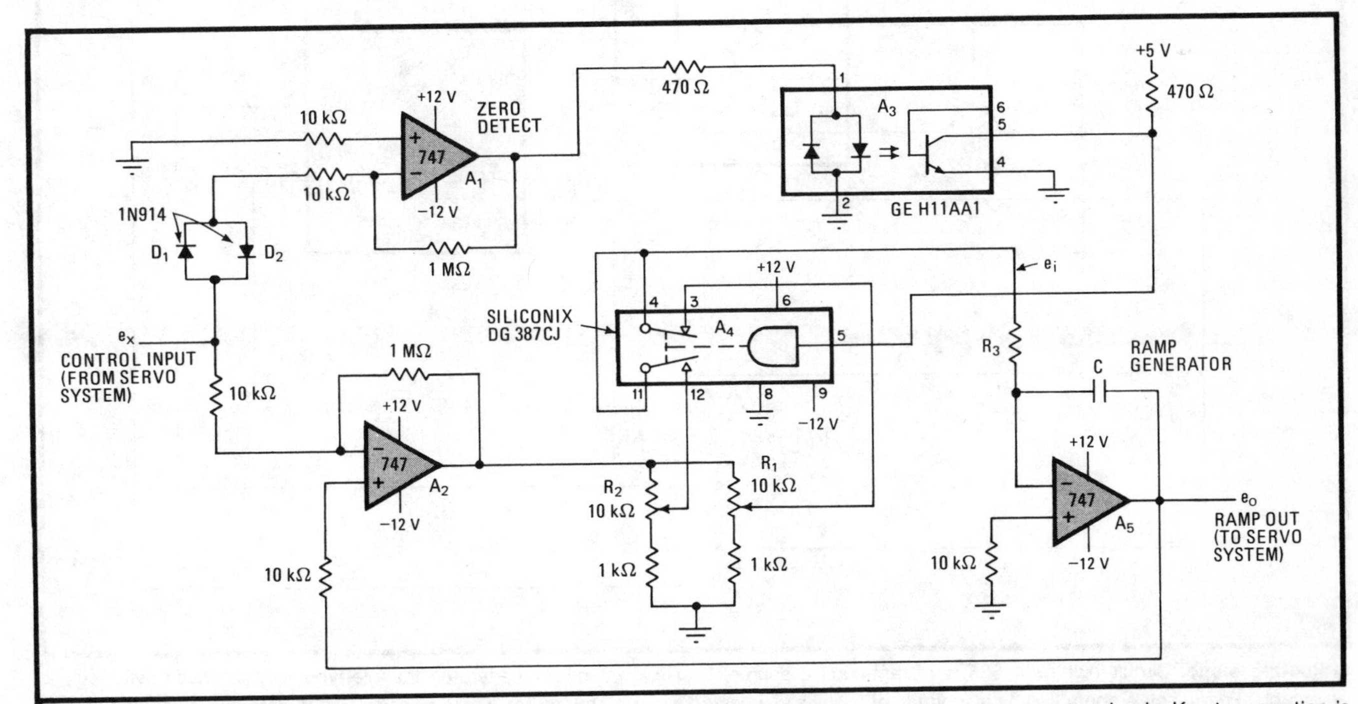

Placement. Up-down ramp generator with selectable-slope capability can efficiently position heavy servo-system loads. Key to operation is solid-state relay A_4, which switches from up ramp to down ramp to decelerate load when it zeros in on terminating position.

Timer generates sawtooth with switchable symmetry

by Roberto Tovar-Medina
Institute of Applied Mathematics, University of Mexico

With some inexpensive components, a 555 timer operating in the astable mode forms a circuit that generates triangular waves of selectable symmetry. The cost of the unit is below $6.

The triangular waves are generated by charging a capacitor with a constant current, I, and discharging the capacitor through a current mirror that sinks I. The symmetry is controlled by selecting the rate at which the capacitor is charged and discharged.

When capacitor C is virtually discharged and the switch, S, is at position 1 as shown, pin 7 of the 555 (the discharge port) is high, and Q_1 and Q_2 are off. Thus the capacitor is charged at a rate of It/C, where t is time, until the voltage at pin 6 (threshold port) of the 555 exceeds two thirds of the supply voltage, V_{cc}.

Pin 7 then moves low, Q_1 and Q_2 turn on, diode D_1 becomes back-biased, and C discharges through Q_1, Q_2, and Q_3 at a rate of It/C. When the voltage at pin 6 falls below a third of the supply voltage, pin 7 moves high again and the process repeats at a frequency given by $f = 3I/2CV_{cc}$. At low frequencies (below 1 kilohertz), D_1 may be omitted, because C can charge through the base-collector junction of Q_3. At high frequencies, D_1 must be included to avoid the discontinuities in the triangular waveform that are caused by switching.

A fast charge time is attained if the output of the 555 (pin 3) is connected to pin 6 through diode D_2. This connection is achieved by placing S in position 2. The discharge time is the same as before. Connecting the discharge port directly to the threshold pin yields a normal charge time and a fast discharge time. S must be placed in position 3 to achieve this symmetry. In either case, the operating frequency is exactly twice what it was previously, or $f = 3I/CV_{cc}$.

Given the component values shown, the circuit oscillates at about 1 kHz. It can be made to work at frequencies up to 30 kHz, however. □

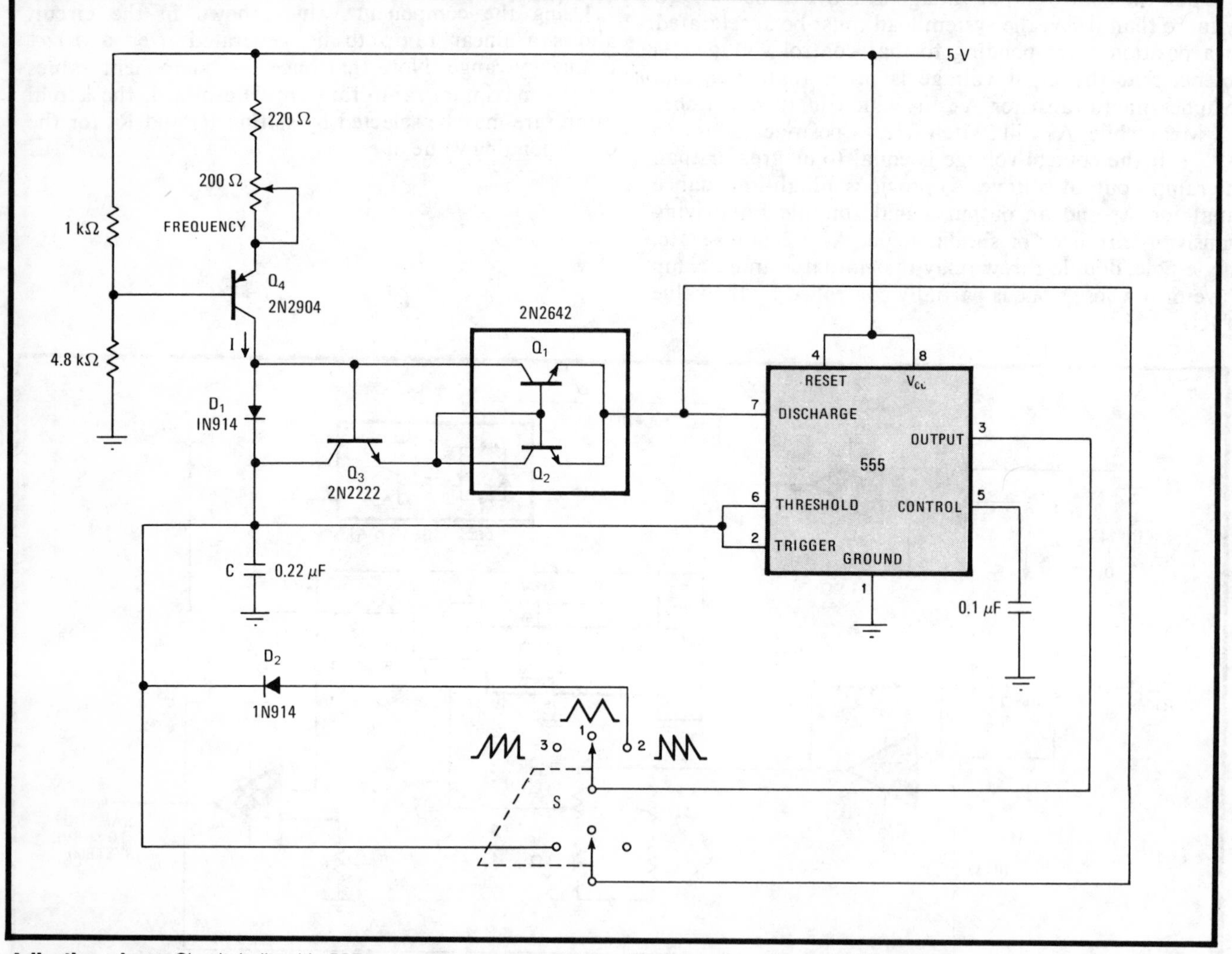

Adjusting slope. Circuit built with 555, constant-current source, and current mirror generates triangle-type waves whose symmetry is selectable. Frequency is controlled by magnitude of I, which is adjusted with potentiometer. Three-position switch determines rate at which C is charged or discharged, so generator can produce standard or fast-rising sawtooth waves or waves that are truly triangular.

Logic-function generator needs no power supply

by P. R. K. Chetty, *Department of Electrical Engineering, California Institute of Technology, Pasadena*

When the same logic and control signals that drive this generator are also used to power it, 14 functions of two input variables are generated without the benefit of a bona fide supply. Consequently, dc power is conserved, and the physical size of the generator is also reduced. It is only necessary that each input signal be able to provide a minimum average current of 15 milliamperes at 5 volts to energize the generator.

Given input variables X and Y, the generator (see figure) will derive most of the popular logic functions of both when ordered by control inputs P_0–P_3. A logic 0 and logic 1 output can also be generated.

The X and Y signals are applied to one or more of three exclusive-OR gates and directly combined with three of the P lines, as shown. Each of the exclusive-OR gate outputs is combined in various ways with the aid of one AND and one OR gate so that the general logic function appearing at the output of the fourth exclusive-OR gate is denoted by the Boolean expression shown. By suitable choice of the logic value of the P_i terms (see table), the desired logic function will be generated.

Although discrete transistors, resistors, and diodes are used, the circuit can easily be condensed into a four-chip device using a transistor array, such as the CA3081, resistor arrays (Beckman 898-3) and a diode package (LM3039). In either case, the circuit will be compatible with TTL, although the unit will accept and generate a wide range of voltages, corresponding to the magnitude of the driving signals. □

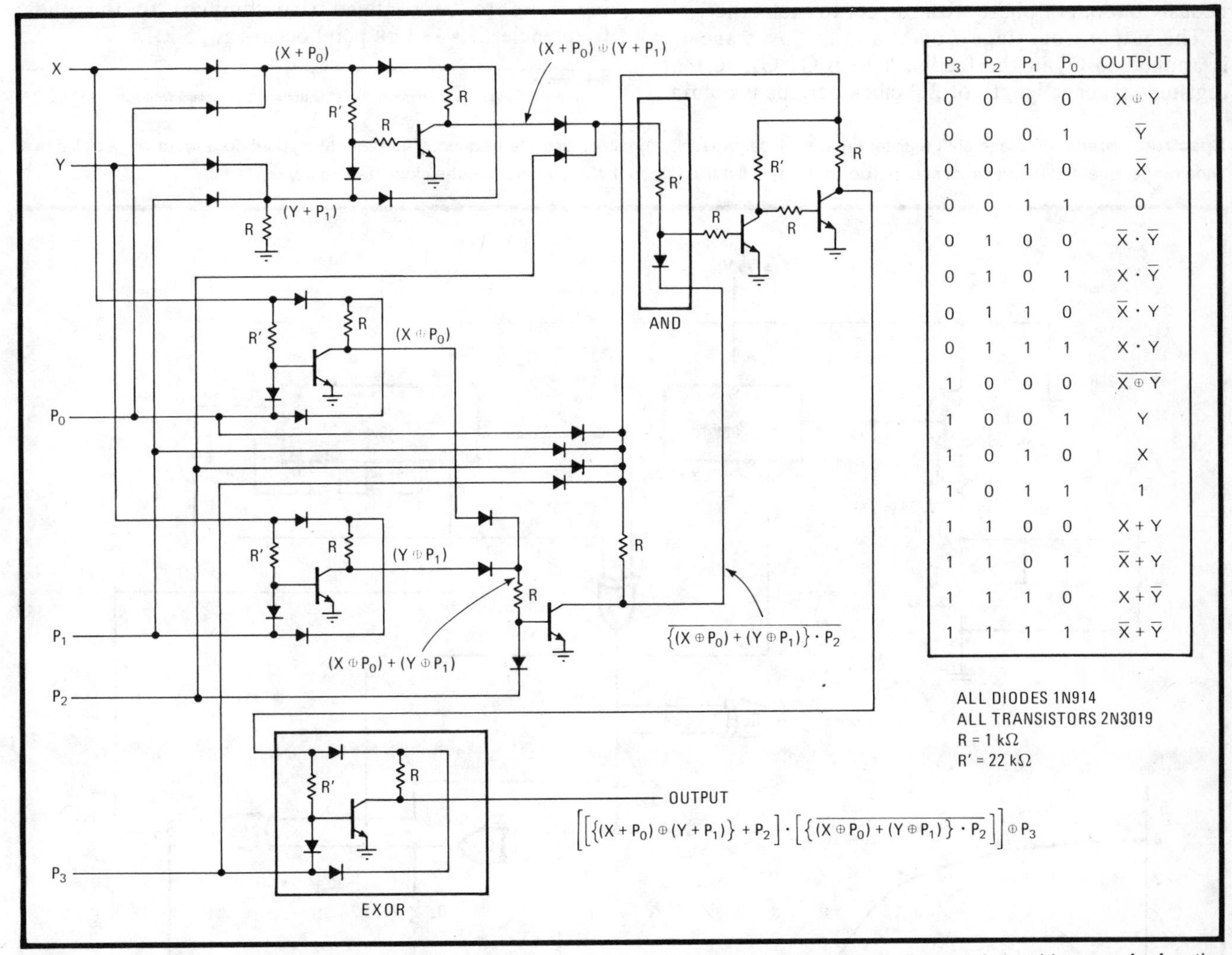

P_3	P_2	P_1	P_0	OUTPUT
0	0	0	0	$X \oplus Y$
0	0	0	1	$\overline{Y}$
0	0	1	0	$\overline{X}$
0	0	1	1	0
0	1	0	0	$\overline{X} \cdot \overline{Y}$
0	1	0	1	$X \cdot \overline{Y}$
0	1	1	0	$\overline{X} \cdot Y$
0	1	1	1	$X \cdot Y$
1	0	0	0	$\overline{X \oplus Y}$
1	0	0	1	Y
1	0	1	0	X
1	0	1	1	1
1	1	0	0	$X + Y$
1	1	0	1	$\overline{X} + Y$
1	1	1	0	$X + \overline{Y}$
1	1	1	1	$\overline{X} + \overline{Y}$

Signal power. X-Y driving logic, and control signals that select 1 of 16 possible outputs (see table) simultaneously provide power for function generator. All inputs should be capable of providing at least 15 mA to circuit, which uses wired-AND, OR, and EXOR gates throughout.

Maximum-length shift register generates white noise

by Henrique Sarmento Malvar
Department of Electrical Engineering, University of Brasilia, Brazil

Using a circuit based on a maximum length sequence generator,[1] this simple unit inexpensively provides a source of white noise over a range of up to 200 kHz. It is far superior to generators that use a reverse-biased base-to-emitter transistor junction, which provides quasi-white noise over a very limited portion of the spectrum. Using two integrated circuits comprising a 25-stage shift register, it can be built for less than $6.

A_1 and A_2 form the n-stage shift register driven by clock G_1–G_2, with A_1 an 18-stage device and A_2 being eight stages in length. A_1 and A_2 are driven simultaneously but out of phase with respect to each other.

The output from stage 7 of A_1 and the last stage of A_2 is applied to G_3 in the feedback loop G_3–G_4, so that a register sequence length of 2^{n-1} clock periods is obtained. Note that G_4 provides signal inversion, so that on power up (the all-zero output state of A_1 and A_2), the noise generator will be self-starting.

It can be shown that the spectrum of the signal at the output of A_2 will contain several discrete frequencies, separated by $f_c/(2^{n-1})$, where f_c is the clock frequency, in this case 200 kHz. Because n is large, the separations between the discrete frequencies become so close (here, it will be 0.006 Hz with a sequence period of 150 seconds), that the spectrum may be considered continuous. So although the noise is pseudorandom because of the method used to produce it, the difference in the spectral properties of the noise as compared with the ideal is minimal.

As for the amplitude of the output envelope, it will vary with frequency as $(x^{-1} \sin x)^2$, where $x = f/f_c$. Here, the −3-dB point will occur at $f = 0.45 f_c$, as shown in the curve at the lower left of the figure.

Q_1 serves as a buffer. The network $R_1R_2C_2$ is a low-pass filter that has been added for an application requiring noise to be confined (bandlimited) to the audio frequencies. Its −3 dB point occurs at 25 kHz. □

References
1. I. H. Witten and P. H. C. Madams, "The Chatterbox-2," Wireless World, Jan. 1979, p. 77.

Spectrum spread. 25-stage shift register creates closely spaced signals of discrete frequency for generating pseudorandom white noise over wide range. Spectral response of source (bottom left) is flat from dc to 0.45 f_c, where f_c is the clock frequency.

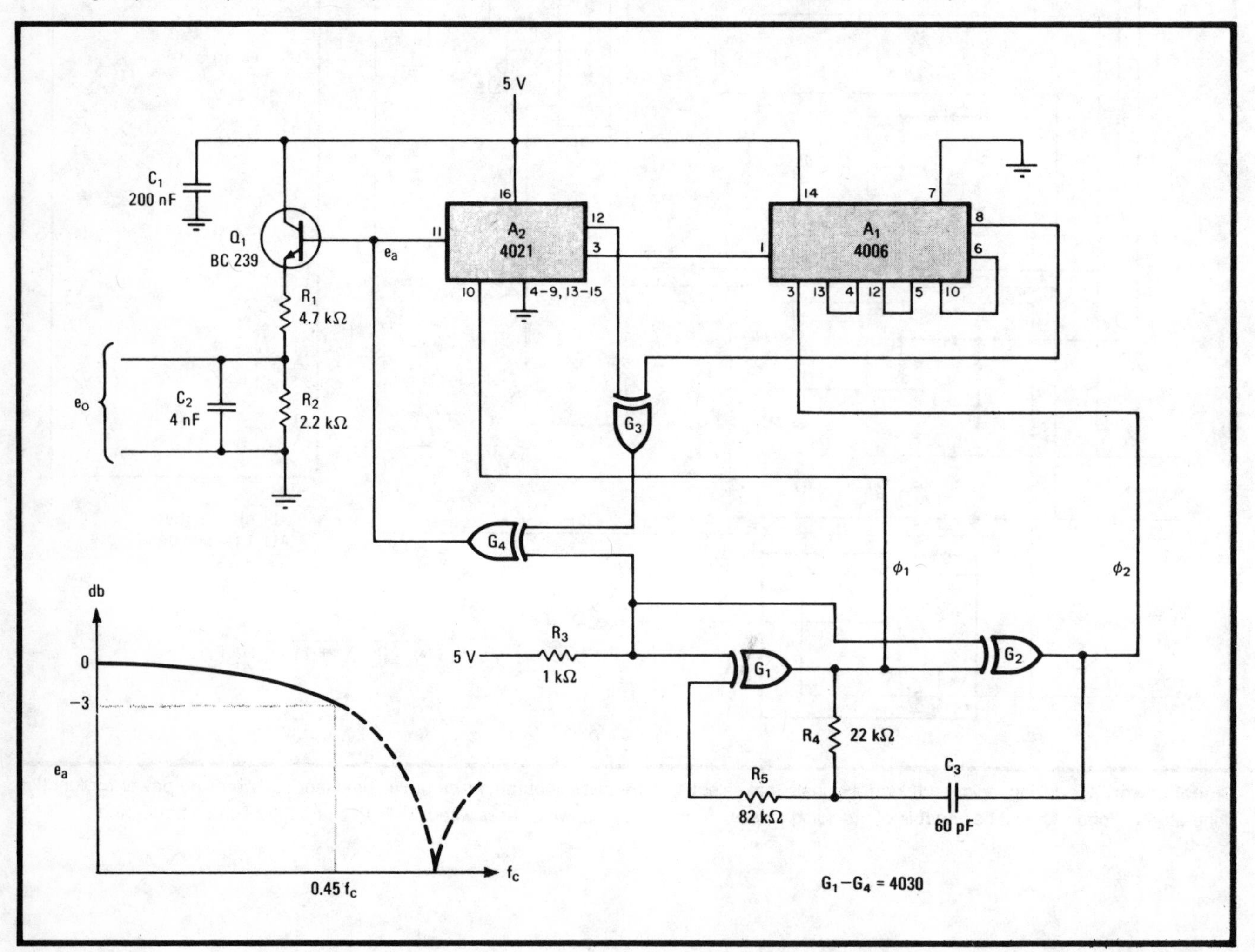

Pseudorandom generator has programmable sequence length

by Ajit Pal
Indian Statistical Institute, Calcutta, India

Providing a pseudorandom binary sequence of order i in the range of 2 to 16, this generator will find many uses in fault-detection and speech-scrambling equipment. Any sequence having a maximum length of $2^{16}-1$ can be generated. If the sequence can be selected electronically, instead of mechanically by means of a manual-switching arrangement as shown, the unit will be extremely useful in automatic-test environments.

The pseudo-random sequence is produced with the aid of a 16-bit shift register and appropriate circuitry for providing a feedback signal to the register's first stage. A_1–A_4 are the 4-bit registers that comprise the 16-bit stage, wired to shift bits from left to right on every system clock. A_5–A_7, connected at the register's outputs, and A_8 are exclusive-OR gates used to generate the feedback signal, which is determined by switches S_i^F. The switch positions are set in accordance with the primitive polynomial of the binary sequence to be generated. Note that the settings of the switches in the figure correspond to a sequence of length $2^{15}-1$, or an equivalent primitive polynomial of $x^{15}+x+1$.

A_9–A_{13} detect the all-zero condition of A_1–A_4 and ensure that the register will not be locked in that state on power-up or during normal operation. The mode control input otherwise allows A_1–A_4 to be set at any point in the sequence as determined by the S_i^I switches. □

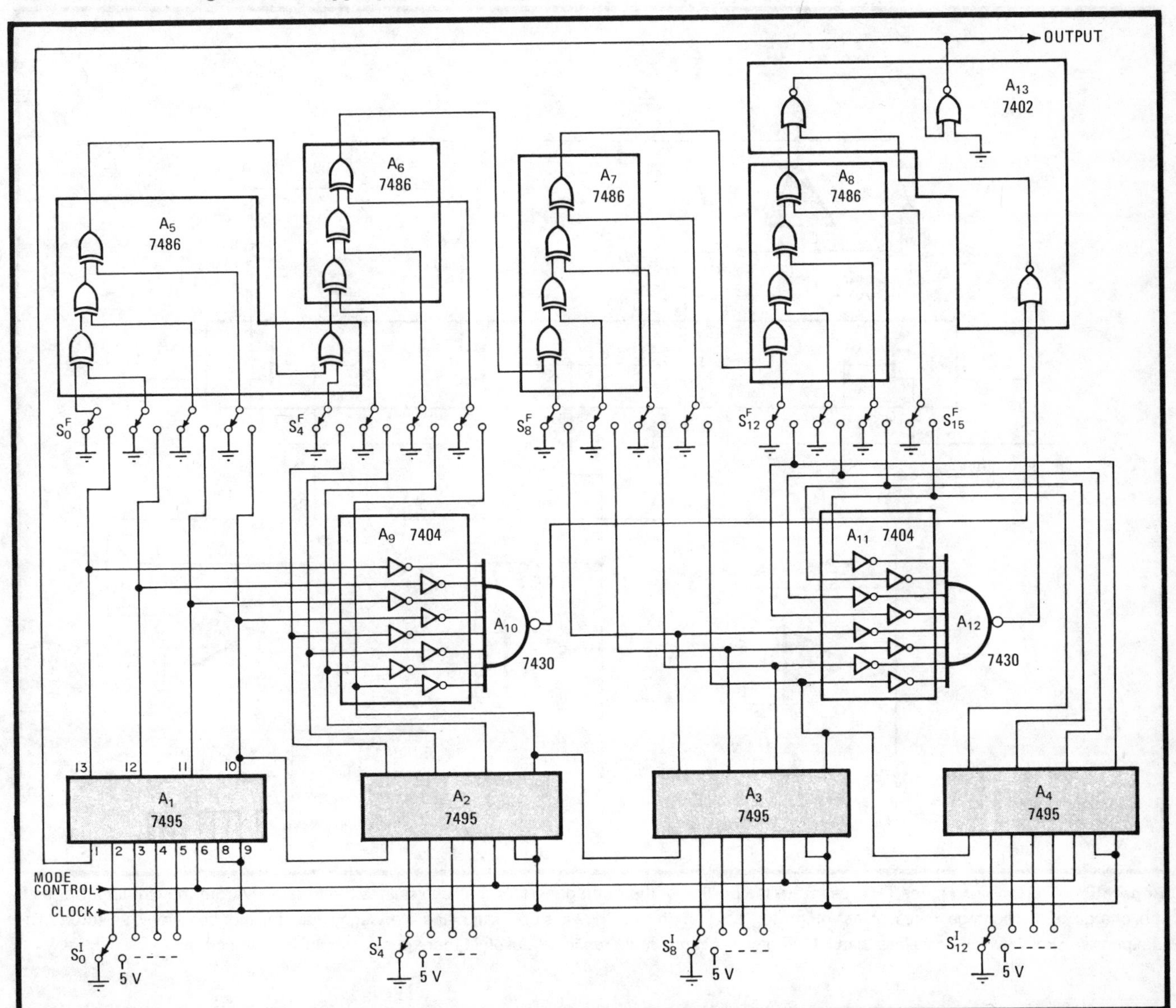

Selection. Switches S_i^F and A_5–A_8 derive suitable feedback signal so that shift register A_1–A_4 can generate a pseudorandom binary sequence of selectable length. A_9–A_{13} detect register's all-zero state and prevent register lock-up by generating logic 1 bit to A_1–A_4 input during power-initialization period. Switches S_i^I initialize registers at any point in a sequence that may extend to $2^{16}-1$ bits.

Ramp generator has separate slope and frequency controls

by Henrique Sarmento Malvar
Department of Electrical Engineering, University of Brazilia, Brazil

Isolating with four analog switches the frequency-determining portion of the circuit from that controlling the charging and discharging of its RC integrator, this ramp generator achieves independent selection of slope ratio and repetition rate. Such a unit is useful in a music synthesizer, where timbre must be changed without affecting a note's fundamental frequency.

Analog gates T_1 and T_2 are initially switched on, and therefore V_c is applied via operational amplifier A_1 to the integrator built around A_2 (see figure). Thus, $-V_c$ appears at the inverting input of A_2, and its positive-going output reaches voltage V_H in $T_1 = 2V_HC(R_1+R_2)/V_c$ seconds, where $V_H = V_{cc}R_5/R_6$.

At this time, A_3 switches on and A_4 goes off. T_1 and T_2 are thus disabled, and T_3 and T_4 are brought high so that $+V_c$ is applied to the integrator. The output at A_2 thus falls linearly toward $-V_H$, where time $T_2 = 2V_HC(R_3+R_4)/V_c$.

The frequency of the ramp is given by:

$$f = 1/(T_1+T_2)$$
$$= R_6V_c/[2CR_5V_{cc}(R_1+R_2+R_3+R_4)] = kV_c$$

where k is a constant (in the approximate range of 1 kHz/V) that can be adjusted with potentiometer P_1. Because R_1+R_3 is a constant, it is seen that an adjustment in potentiometer P_2 will affect the slope ratio, but not the frequency. With the values shown, the slope ratio can be selected from 1/11 to 11. The slope ratio is given by $T_1/T_2 = (R_1+R_2)/(R_3+R_4)$. □

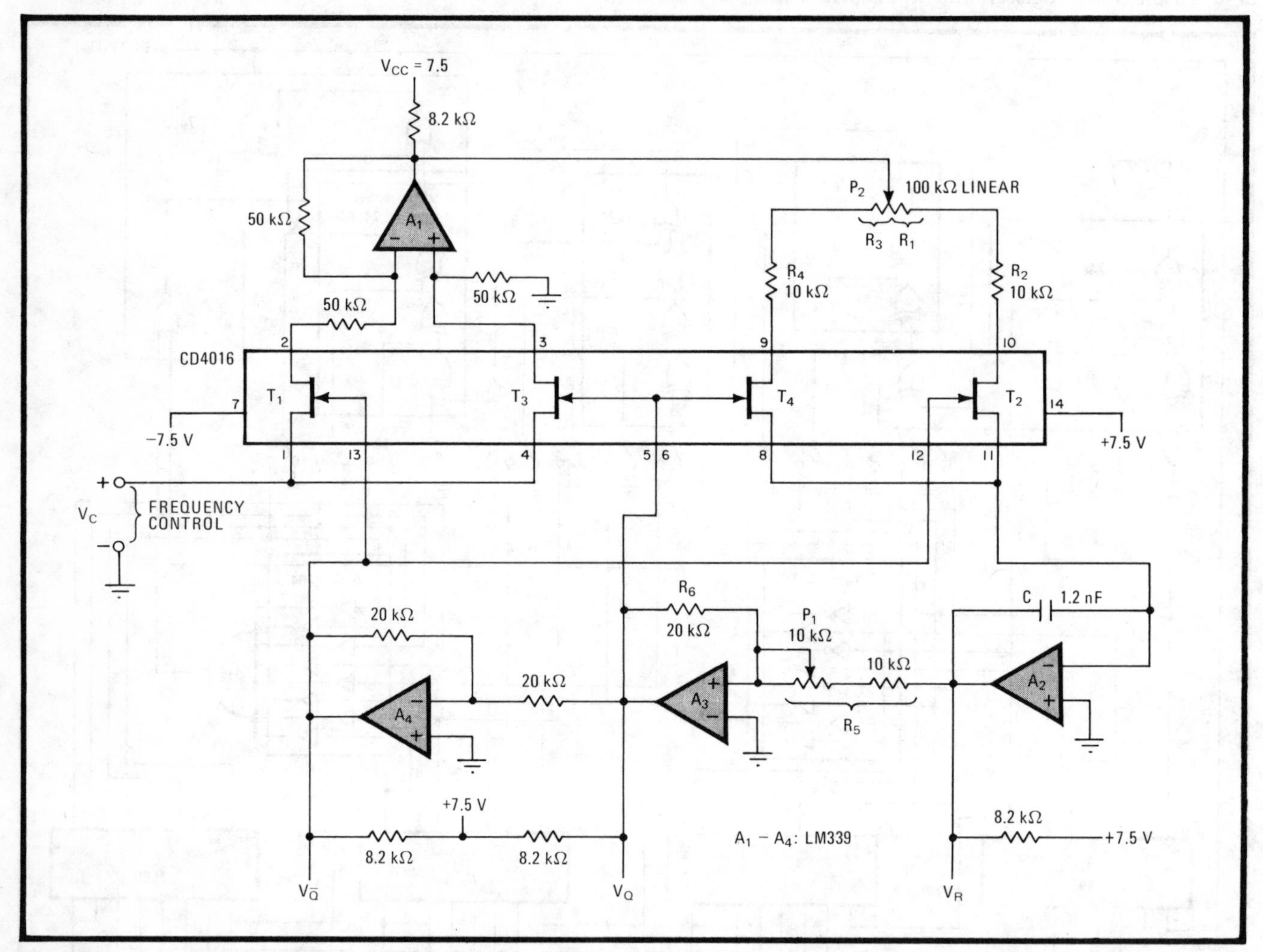

Separation. Transmission gates T_1–T_4 separate the portion of the ramp generator that determines the frequency from the circuitry that sets the charge and discharge times of its integrator, so that the up/down slope ratio and frequency can be independently selected. The inexpensive circuit, which costs less than $10 and works in the audio range, is a useful timbre control in music synthesizers.

Chapter 18
INSTRUMENT CIRCUITS

Instantaneous-frequency meter measures biomedical variables

by T. G. Barnett and J. Millar
Department of Physiology, The London Hospital Medical College, England

Providing immediate feedback on respiration, heart rate, and other biomedical parameters, this device generates an output proportional to instantaneous frequency (1/interpulse interval) at 100 hertz and less. The meter is simple and inexpensive and uses readily available components.

As shown in the diagram, an input spike generated by the variable being measured fires monostable A_1, whose timing components R_eC_e are selected to provide an output pulse of 500 microseconds. A_1's output initializes A_2, an 8-bit digital-to-analog converter, which begins to count up at a rate determined by clock A_3.

Thus A_2 generates a precision ramp whose slope is controlled by the clock frequency $f = 1.44/(R_a+R_b)C_a$. After passing through buffer A_4, the ramp is introduced to a transmission gate, A_5, which passes the ramp to a sample-and-hold amplifier A_6 until A_1 times out. The value of the ramp is stored at this time in capacitor C_H. One-shot A_7, triggered by the falling edge of A_1, then resets the d-a converter to zero, initiating generation of a new ramp.

One problem with an earlier version of this circuit lay in handling too long a pulse interval, in which case A_2 would reach a maximum and reset, giving an erroneously small output voltage. To remedy this situation, comparator A_8 has been added. When the ramp rises above the user-selected reference voltage, the comparator goes high and turns oscillator A_3 off. Thus the maximum ramp voltage can be set to any level.

If a voltage proportional to the pulse interval is required, it may be tapped from the output of the sample-and-hold directly. For securing a voltage proportional to 1/pulse interval, the output may be taken from analog divider A_9, which generates an output equal to $10z/x$, where x is the input voltage and z is a reference (in this case, 0.1 volt).

Although the frequency response of the circuit is affected by the choice of monostable pulse widths and by the oscillation frequency, the factor limiting the response is determined by the minimum voltage A_9 can detect. The voltage will be nominally 100 millivolts, yielding a maximum countable frequency of 100 Hz. More expensive analog dividers may improve on this figure. □

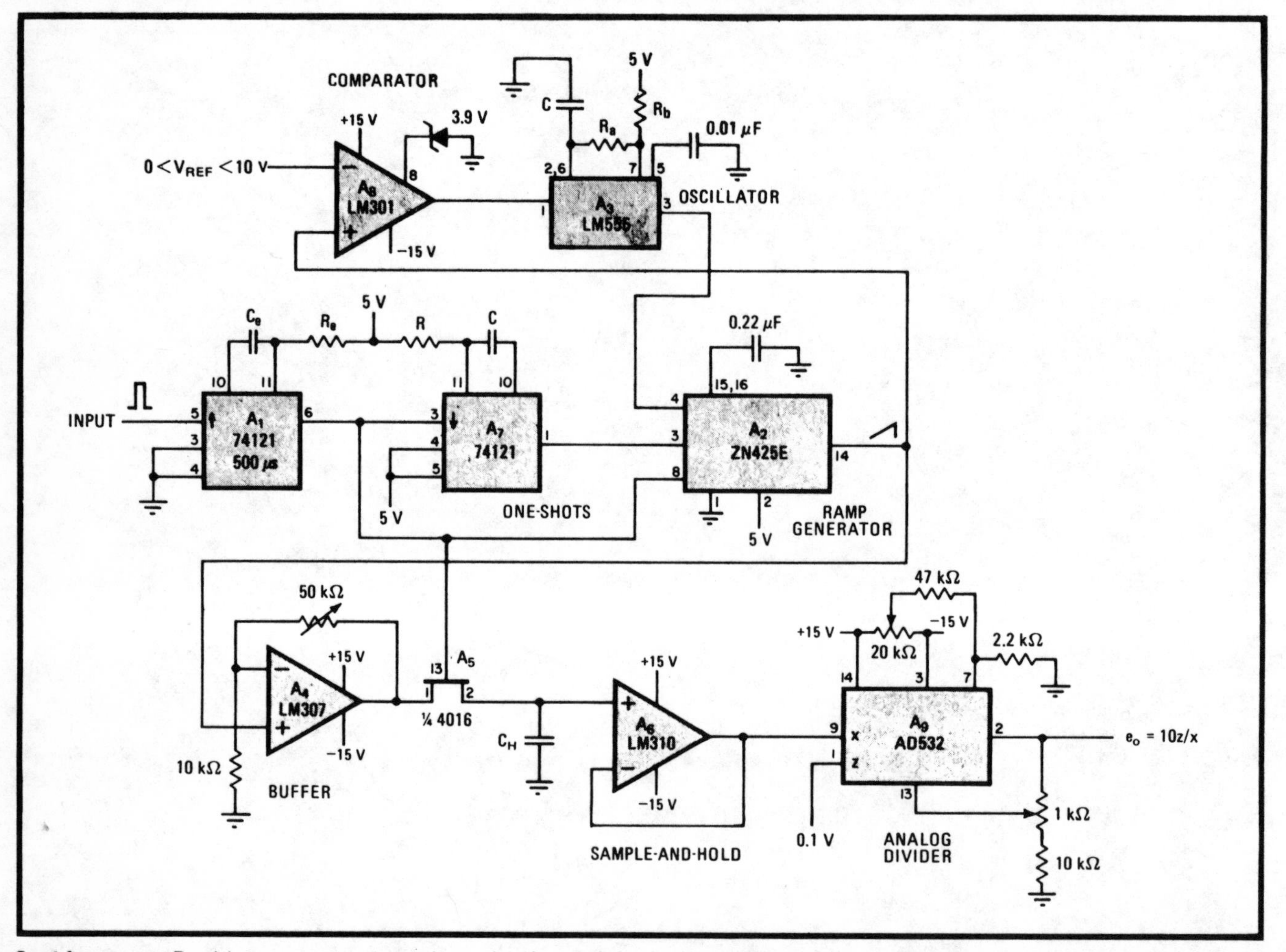

Spot frequency. Precision ramp generator and associated control circuitry generate analog voltage both directly and in inverse proportion to instantaneous input frequency. Providing maximum countable frequency of 100 Hz, meter is useful in biomedical applications.

Timer and converter generate slow ramp for chart recorders

by David Wingate
London Hospital Medical College, London, England

A long sweep time increases the effectiveness of an X-Y chart recorder in monitoring slow system changes. An astable multivibrator operating at low frequency and a versatile digital-to-analog converter can be used to generate an extremely linear, low-frequency ramp for the X axis, and at low cost, too. Although many good chart-recorder systems have the ability to hold down the distance the pen traverses to 0.02 centimeter per second, this circuit can limit it even further—in the neighborhood of 1 centimeter per hour, if needed.

In this circuit, the rate of increase in the ramp voltage produced by the converter is controlled by the frequency of the timer. As shown in the figure, the clock for the circuit is the Ferranti ZN1034E precision timer, operated in the astable mode. This device is somewhat unusual in that the pulse width of the oscillations produced is controlled by R_1C_1 but the frequency of oscillation is controlled by combination $R_TC_TR_{cal}$. The second combination is part of the astable network, detecting the fall of the output pulse and retriggering the timer, and thus controlling its basic repetition rate.

R_T and C_T are held constant, with R_{cal} the variable resistor. The table in the figure shows the relation of R_{cal} to the pulse frequency. The pulse width is not critical for timebase applications, but it can be varied if required. The width equals $0.6\ R_1C_1$ seconds, where R is expressed in kilohms and C is expressed in picofarads.

After passing through a buffer stage, the clock drives the binary counter in the Ferranti ZN425E d-a converter, which is wired to serve as a ramp generator. The rate of increase in the magnitude of the ramp voltage is a function of the clock frequency; since the output rises incrementally, a staircase approximation of a ramp is generated. After passing through a buffer, the ripples in the staircase waveform are smoothed by a standard low-pass filter tuned to the clock frequency.

By increasing the capacity of the converter from an 8- to a 12-bit device, it is possible to generate a ramp voltage having even greater resolution. The circuit has served well in an application involving prolonged recording of biomedical and biological data, but, many other nonmeasurement applications are possible. □

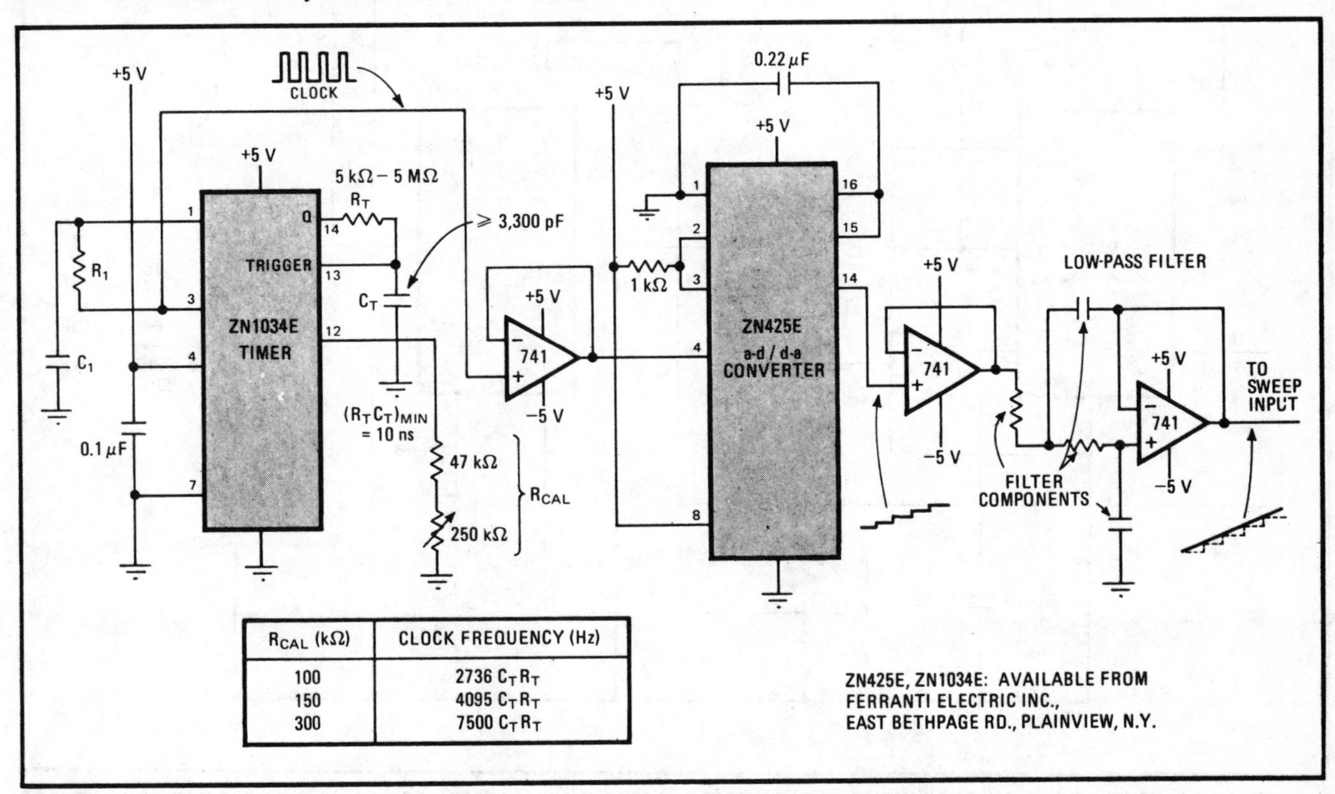

R_{CAL} (kΩ)	CLOCK FREQUENCY (Hz)
100	2736 C_TR_T
150	4095 C_TR_T
300	7500 C_TR_T

Slow time base. Low ramp frequency for X input of chart recorder, permitting long-period monitoring assignments of various systems, is generated by d-a converter and timer. Slope of ramp voltage with respect to time is determined by clock frequency.

Monolithic timers form transducer-to-recorder interface

by T. George Barnett
Laindon, Essex, England

Capacitive transducers often require an expensive capacitance bridge to transform sensed capacitance variations into a voltage for presentation on a chart recorder or oscilloscope. A circuit using two monolithic timers can provide both a capacitance-to-voltage interface and a simple and accurate method for measuring the transducer capacitances.

As shown in the figure, the transducer serves as a capacitive frequency-determining element for the 555 timer. This makes it possible to measure transducer capacitances indirectly, while isolating the transducer from the scope or chart recorder to minimize the loading effect. The LM555CN timing device is connected in the astable mode, its free-running frequency set by R_x, R_y, and C_x. The transducer, typically in the range of 0.001 to 100 microfarads, is element C_x in the timing network.

As the transducer capacitance varies in response to the physical parameter being measured, the output frequency of the 555 varies linearly. The ratio of R_x to R_y sets the duty cycle, which depends on the frequency range desired.

The output of the 555 is presented to the LM322N timer. This circuit, wired as a monostable multivibrator, and combined with the one-pole resistance capacitance filter, forms a frequency-to-voltage converter. The dc output voltage varies linearly with the input frequency, and has a slope of 0.1 volt per kilohertz. The linearity is within 0.2% over the output voltage range of 0 to 1 v.

A 1-kilohm potentiometer connected to pin 7 of the 322 adjusts the output pulse width, serving to calibrate the system to a specified voltage at 10 kilohertz or some other frequency. To ensure linearity, the collector of the output transistor, pin 12, is fed to pin 4 (V_{REF}), so that the amplitude of the pulse at pin 1, the emitter of the output transistor, is constant. The period of the one-shot should be much less than the period of the astable multivibrator for best results. □

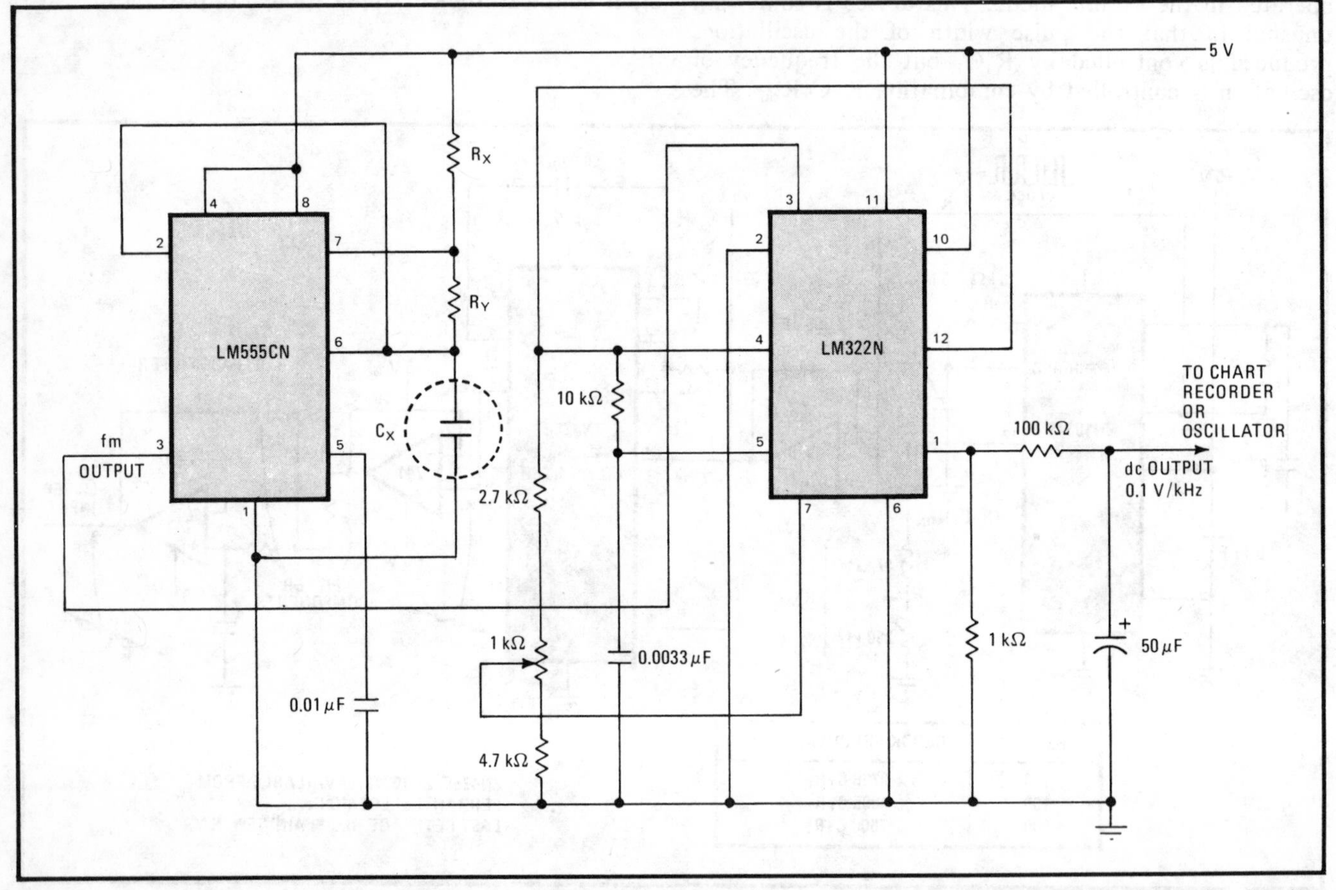

Transducer-to-recorder interface. Two timers determine transducer capacitance, perform capacitance-to-voltage conversion for chart recorder, while isolating transducer from output-circuit loading. Transducer placed in timing network of 555 astable multivibrator determines its frequency. LM322 one-shot, which should have a much shorter period than the multivibrator, transforms frequency into voltage.

Compact industrial ammeter measures 10-ampere peaks

by Paul Galluzzi
Beverly, Mass.

A self-contained sample-and-hold amplifier and a digital panel meter make this device small, rugged, and suitable for measuring peak currents in industrial applications. Although not inexpensive (it can be built for about $65 in small quantities), it is an easy to build, accurate, and reliable peak-reading ammeter.

The 0–10-ampere pulses to be measured are converted to a voltage by instrumentation amplifier A_1, such that the input to peak detector A_2 swings 1 volt for each ampere at the circuit's input. A_2's output voltage then drives digital voltmeter M_1.

The timing pulses for A_2's sampling cycle are derived from a 10-Hz clock signal by gates G_1–G_6 and binary-coded-decimal counter A_3. The decoded clock pulse introduced at gate G_2, along with the delayed clock pulse from G_3 and G_4, form the read pulse for sampling the output of the peak detector for a 50-millisecond interval, and holding that reading for the remainder of the one-second cycle.

During that time, clock inputs A and B reset A_2 to zero. Flutter caused by the 50-ms sampling pulse cannot be detected visually. □

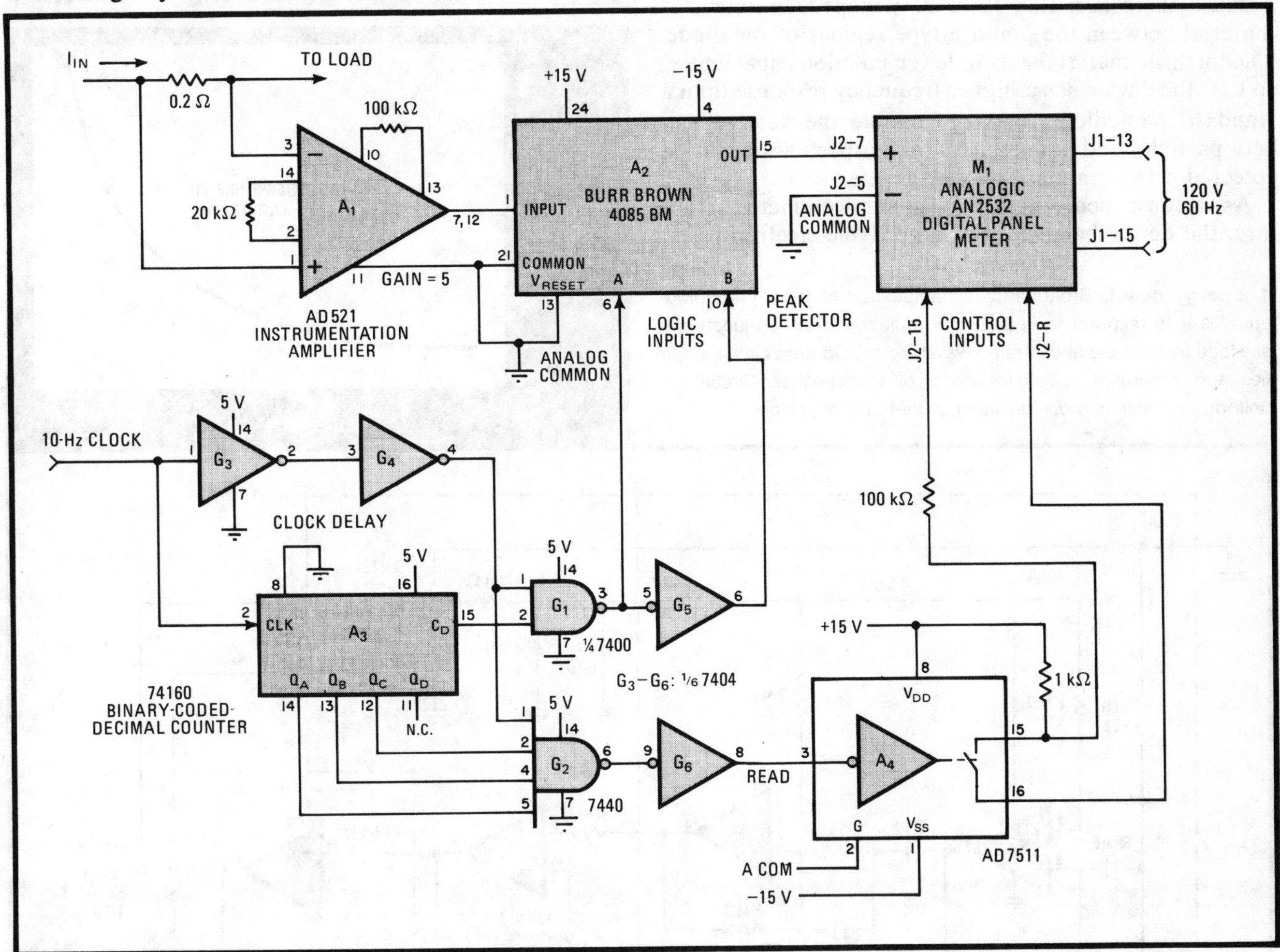

Rugged and reliable. Ammeter using hybrid sample-and-hold amplifier, digital panel meter, and standard logic components is ideally suited for measuring pulses of peak current in hostile environments. Readout is in peak amperes, accurate to 0.1% over −25° to 85°C.

Photodiode and op amps form wideband radiation monitor

by Grzegorz Hahn
Institute of Nuclear Research, Swierk, Poland

A sensitive radiation monitor may be simply constructed with a large-area photodiode and a quad operational amplifier. Replacing the glass window of the diode with Mylar foil will shield it from light and infrared energy, enabling it to respond to such nuclear radiation as alpha and beta particles and gamma rays.

The general circuit is shown in Fig. 1. The HP-5082-4203 device is a p-i-n photodiode, called that because there is a thin layer of undoped, or intrinsic, material between the p and n type regions of the diode. The intrinsic material acts to lower junction capacitance, so that the device has a higher frequency response than a standard photodiode, making possible the detection of beta particles and gamma rays (alpha particles could be detected with a standard photodiode).

As a consequence of the p-i-n semiconductor structure, the device bandwidth is large. Hole-electron pairs, and thus charge (Q), can be accumulated across the photodiode by all forms of ionizing radiation. When the junction is shielded from visible and infrared light, the photodiode output is a function of the nuclear-type radiation only.

The junction charge generated by the ionizing radiation is $Q=\Delta E/\epsilon$, where ϵ is the ionizing constant of silicon (3.66 electronvolts at 300 K), and ΔE is the energy stored across the active region of the photodiode. The output voltage from integrator A_1 is thus:

$$V = (Q/C_f)\,(1-e^{-t/R_fC_f})$$

where t is measured from the instant ΔE appears across the junction and R_fC_f is the time constant of the integrating network. A_2 is a quasi-Gaussian filter that shapes the pulse in order to inprove the signal-to-noise ratio of the small output signal at A_1. A_3 generates a

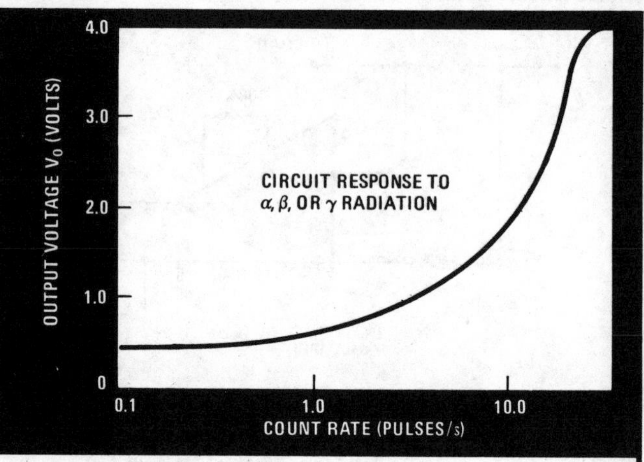

1. Energy count. Broadband characteristics of p-i-n photodiode enables it to respond to α, β, and γ radiation when pn junction is shielded from visible and infrared wavelengths. Op-amp circuit amplifies and integrates pulses for meter or loudspeaker. Circuit has uniform response to radiation, independent of energy class.

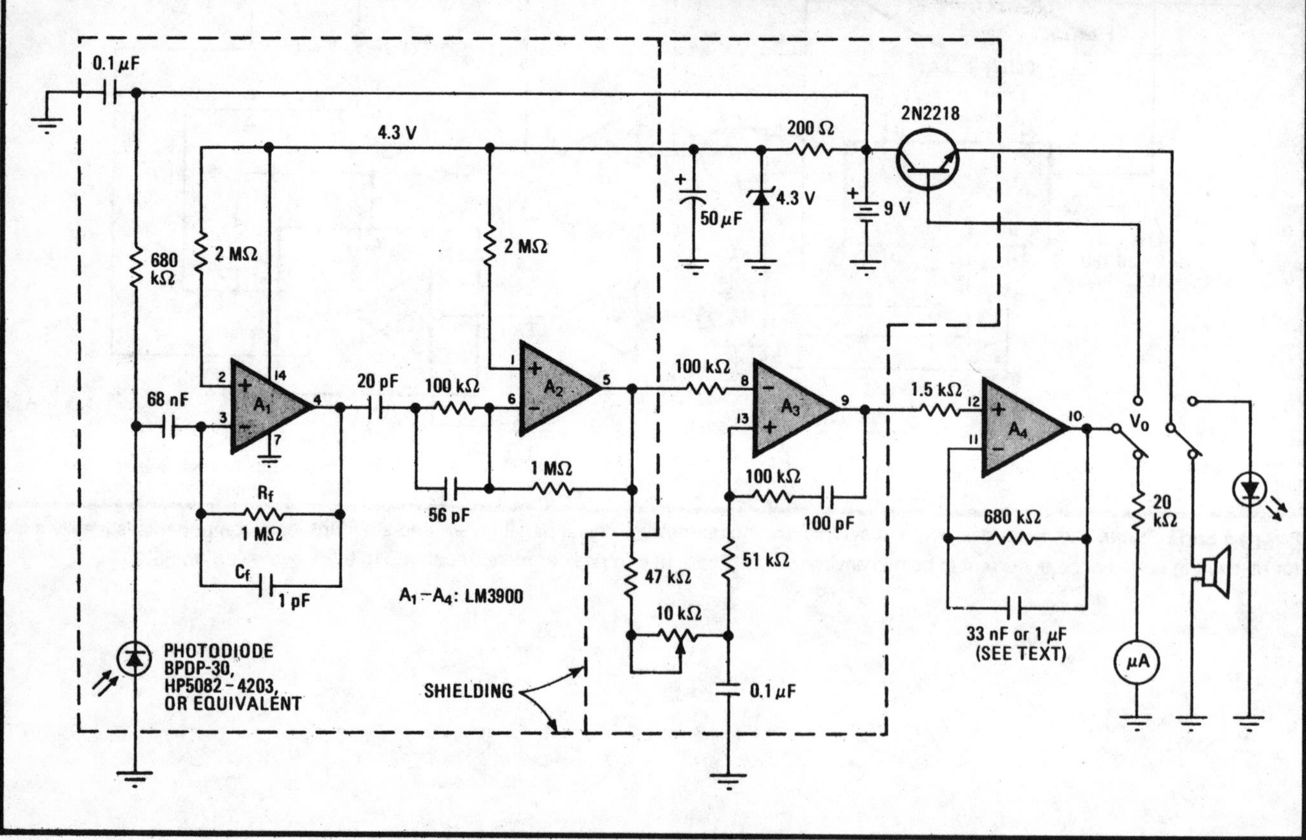

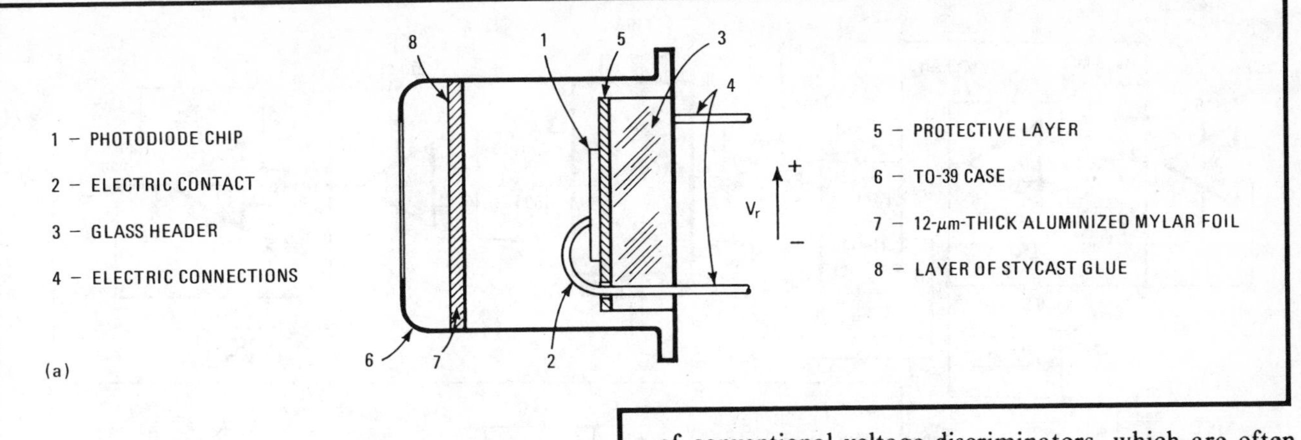

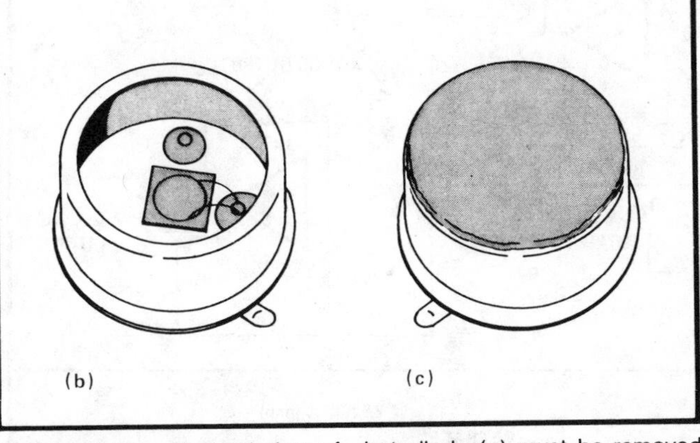

2. Adaptation. Glass window of photodiode (a) must be removed and replaced by opaque material to shield pn junction from light. Top of photodiode case is first cut out (b) by turning lathe, then layer of aluminized Mylar foil is secured in place (c) with Stycast glue.

rectangular pulse with a width proportional to the input amplitude for every signal that exceeds a threshold set by the user.

The threshold control is used if a radiation alarm circuit is desired (indicating maximum limits have been exceeded) or if it is necessary to discriminate against hash in high-noise environments. It should be noted that A_3 is in principle a current discriminator and that its temperature-drift coefficient is probably lower than that of conventional voltage discriminators, which are often used in radiation monitors.

A_4 integrates the output of A_3 in order to drive a microammeter. A 1-microfarad capacitor is used in the integrating network. A lower value, say, 33 nanofarads, will make it possible to drive a small loudspeaker (50-hertz output signal) or light-emitting diode.

Figure 2 illustrates the procedure used to replace the photodiode's glass window with a 10-micrometer-thick aluminized Mylar foil. The photodiode shown here is the BPDP-30, a European make, but most photodiodes made in the U. S. are very similar.

As shown in Fig. 2b, the top of the TO-39 case containing the window must be cut away with a turning lathe. Care should be taken not to touch the pn junction within. Sharp edges are then filed smooth with care. The new window is then secured to the edges of the device by means of black Stycast glue (available from Emerson and Cuming Inc., Canton, Mass.). Figure 2c is a view of the completed diode.

The circuit response is seen in the upper part of Fig. 1. Note that this device is a radiation monitor, as opposed to a radiation meter, and so cannot distinguish between α, β, and γ radiation. Because of the photodiode's wide bandwidth, each energy class generates the same output voltage for a given radiation intensity. Thus the monitor is intended for use where an *a priori* knowledge exists of the type of energy to be encountered. □

Op amps and counter form low-cost transistor curve tracer

by Forrest P. Clay Jr., Clarence E. Rash, and James M. Walden
Old Dominion University, Department of Physics, Norfolk, Va.

For a curve tracer, this relatively simple circuit is unusually inexpensive. Used to test small-signal bipolar transistors as well as junction diodes, it generates the waveforms needed to display or plot their characteristic curves on an oscilloscope or X-Y plotter, interfacing directly with either. Operational amplifiers, one transistor, and a single binary counter are the only active devices needed.

Central to the circuit is a current generator made up of an op amp driven by the counter. It supplies eight levels of base current in sequence to the transistor under test. Op amps A_1 and A_2, with the aid of the R_1-R_2-C_1 timing network, initially produce both square and triangular waves at test points A and C (TPA and TPC), respectively. S_1 selects the waveform frequency—either

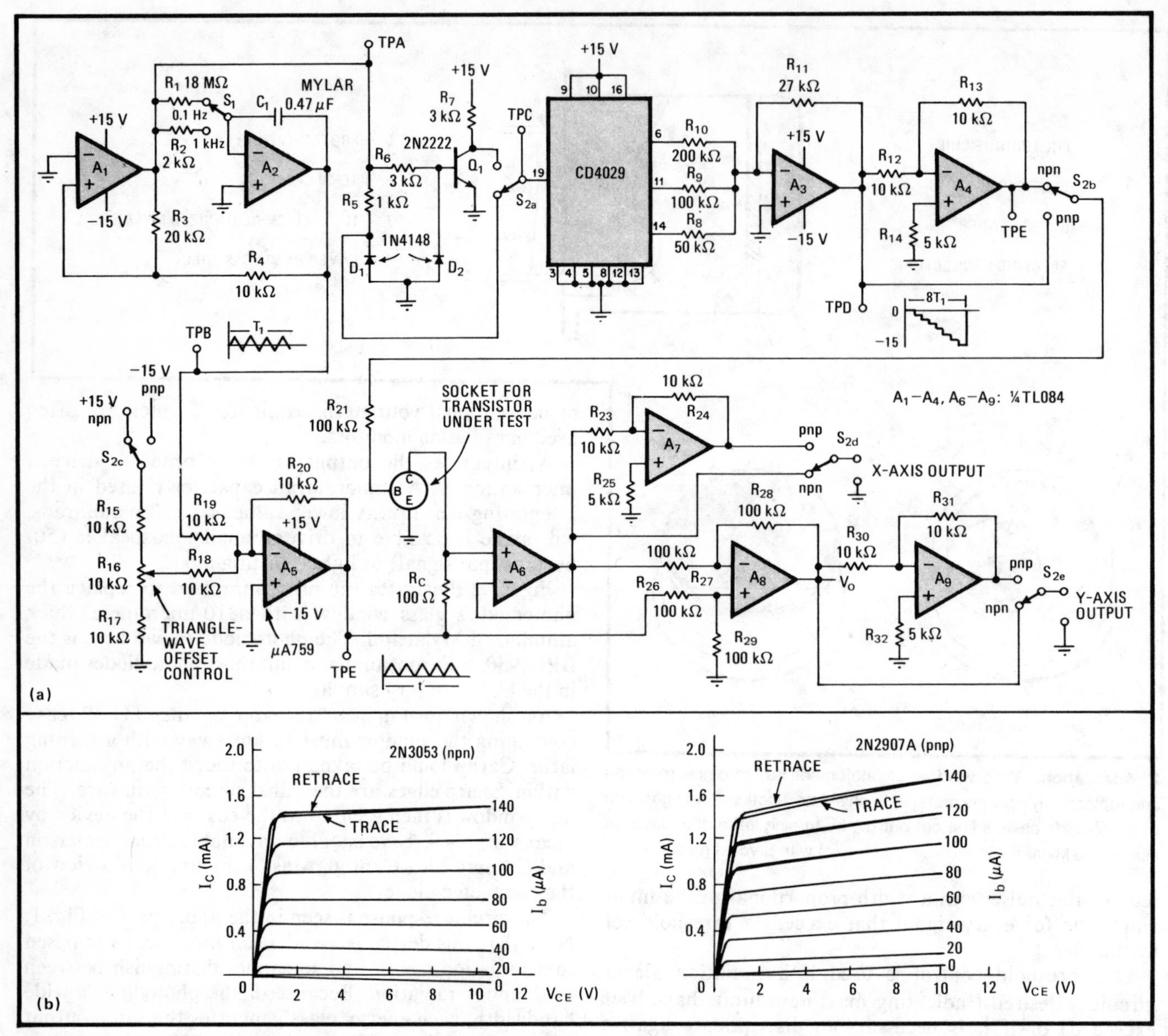

Current family. Tracer produces set of eight curves of collector current vs collector voltage from npn or pnp transistor under test. Diodes may also be checked in circuit's pnp-transistor mode (a). Representative curves are plotted using X-Y recorder (b). Note temperature effects seen on the retrace portion of curves for higher values of I_b and I_c.

1 kilohertz for output onto an oscilloscope or 0.1 hertz for plotting with an X-Y recorder.

The waveform at TPA is then shaped by D_1–R_5 or D_2–R_6 into a clock pulse suitable for the 4029 binary counter. The signals emanating from the Q_a, Q_b, and Q_c ports of the counter, when fed into a binary-weighted summation network (R_8–R_{11}), produce an eight-step staircase waveform at the output of A_3 or A_4, depending upon whether a pnp or npn transistor is under test. The actual base current value is determined by appropriate selection of R_{21}. The collector current can be calculated from $I_c = V_o/R_c$.

Both the collector-biasing voltage for the transistor under test and the linear-deflecting voltage for the X-axis output to the scope are derived from the triangle wave. The first voltage is obtained by using S_{2c} and R_{16}, which permit the proper dc component to be added to the triangle signal.

Note that the Y axis is stepped at one eighth of the rate at which the X axis is scanned. Thus, if the sampling rate is 1 kHz, each of the eight current levels is swept at a rate 125 Hz, well above the rate at which flicker is detectable on a scope.

The circuit is easy to use. Simply place ganged switch S_2 into whichever position is correct for the type of transistor being measured (npn or pnp); place the transistor into the test socket; and apply circuit power. To test a diode, insert its anode and cathode leads into the emitter and collector sockets, respectively, and put S_2 in the pnp mode.

Figure 1b shows two representative families of curves the circuit produced on an X-Y recorder for the two types of bipolar transistor. □

High-accuracy calibrator uses no precision components

by Walter Allen
The Audichron Co., Atlanta, Ga.

A calibrating voltage source with an output of 1 to 10 volts in 1-V steps can achieve an accuracy of within 0.1% even though all the resistors in the circuit have ±10% tolerances. The precision is obtained by digitally chopping the reference voltage with a decade counter and then integrating the varying duty cycles back into voltage steps.

As the schematic shows, the 10-V reference source is built around a feedback-stabilized operational amplifier that is one section of an MC3403 quad op amp. This reference voltage is the V_{DD} supply for a CD4017 decade counter and powers all the digital chips in the circuit as well. The CD4017 is a Johnson counter that produces 10 decoded outputs (0 – 9) that are pulsed sequentially.

Half of the CD4001 quad, two-input NOR gate is connected as an astable multivibrator that, going to the clock input, drives the counter at approximately 200 kilohertz. The other half of the CD4001 is configured as a set-reset flip-flop.

The flip-flop is set by the counter's decoded output 0 and reset by one of its 1 – 9 decoded outputs, depending on the position of the voltage step-selector switch. The flip-flop's output is therefore a rectangular pulse train of approximately 20 kHz that has a duty cycle varying from 0% to 100% by 10% steps.

The signal is then integrated by the RC low-pass network and buffered by another section of the MC3403 quad op amp for output. An adjustable divider and additional op-amp follower provide stable voltages between steps.

The MC3403 quad op amp is used for its high current capability—up to 30 milliamperes—and also for its ability to swing down to within a few millivolts of the negative supply rail.

Potentiometer R_1 sets the 10-V reference voltage. When this is precisely adjusted, the 1-V switch-select-

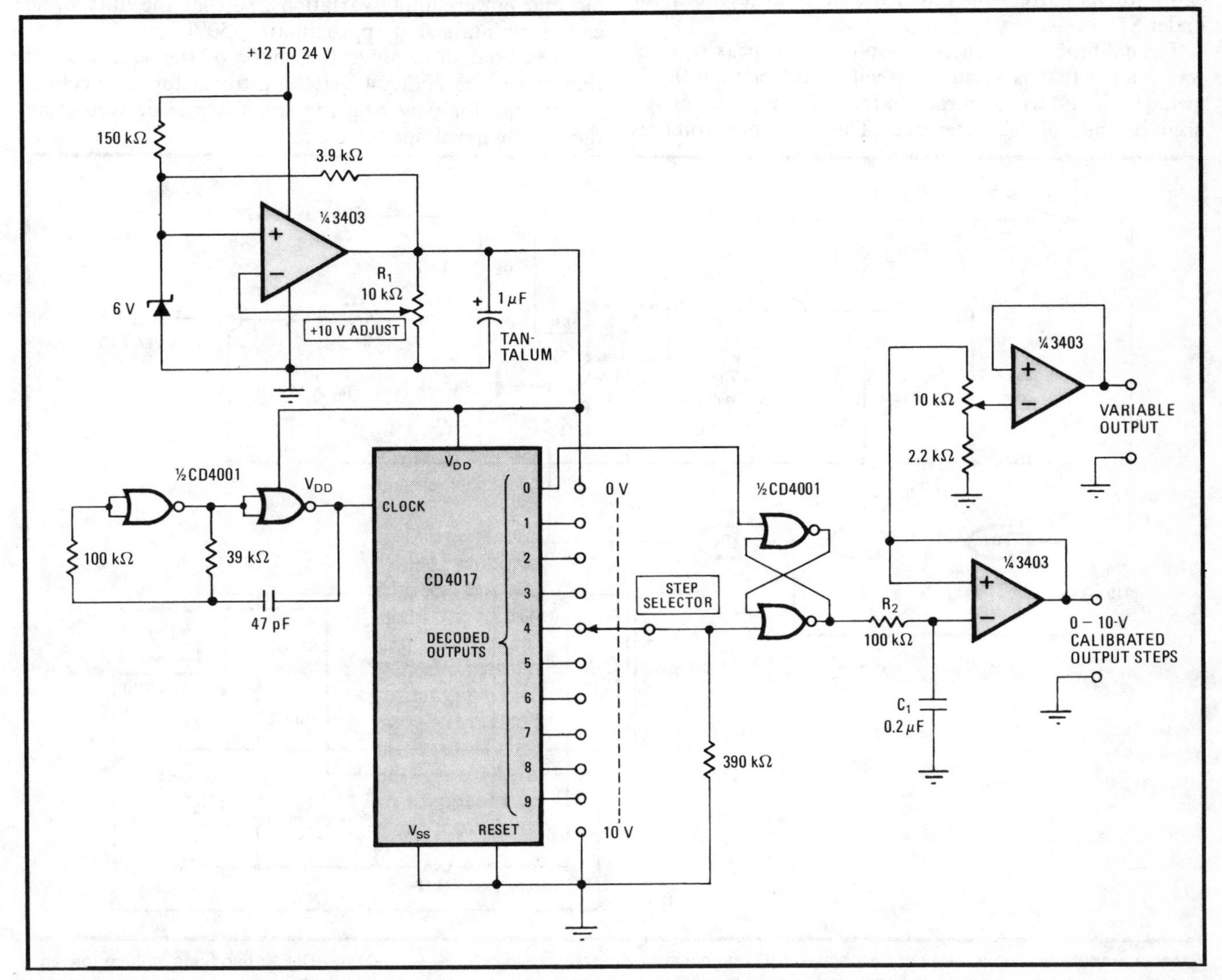

Digital division—analog accuracy. Changing the duty cycle of the pulse train is the key to accurate voltage division. Frequency and duty cycle of clock at left do not affect output, since decoded counter outputs accurately determine duty cycle.

able output steps are accurate to within 0.1% with any loads up to 20 mA. The adjustable output tracks between steps with high linearity, and the 20-kHz ripple component in the output is less than 1 mv.

The capacitor in the integrating network should be a high-quality, low-leakage unit, though its value is not critical since it determines only the step response time. With the value of 0.2 microfarad shown, the response time is about 100 milliseconds. A 1.0-μF tantalum capacitor would give better filtering characteristics, but its response time is on the order of 0.5 s. □

High-accuracy calibrator uses band-gap voltage reference

by Henno Normet
Diversified Electronics, Leesburg, Fla.

The Analog Devices' AD581J voltage reference can be used to build a low-cost and extremely accurate voltage calibrator for oscilloscopes that either do not have one built in or have one of inadequate accuracy. If this calibrator is battery-operated, the unit can be built for under $15.

The calibrator generates a 1-volt peak-to-peak square-wave signal that is accurate to within better than 0.5%, owing its long-term accuracy to the band-gap technology used in the voltage reference. The reference voltages produced by the band-gap method are more temperature-stable than that produced by a zener diode, because the method makes use of the inherently constant potential that exists between adjacent electron energy levels in the semiconductor material of the integrated circuit itself. Here, the potential across selected energy bands is used to derive a 10-v reference that will vary no more than ±13 millivolts over the temperature range of 0°C to 70°C (for the 581L, the variation would only be 2.5 mv).

In this circuit, the output of a 1-kilohertz square wave is scaled to 1 v with the aid of the reference as shown in (a). An astable multivibrator, G_1–G_2, is used as the square-wave oscillator. R_2 compensates for input-threshold and power-supply variations, so that the duty cycle can be maintained at approximately 50%.

G_3 is used to improve the shape of the square wave that drives the 2N3904 switching transistor. The collector voltage for powering the transistor is derived from the band-gap voltage reference.

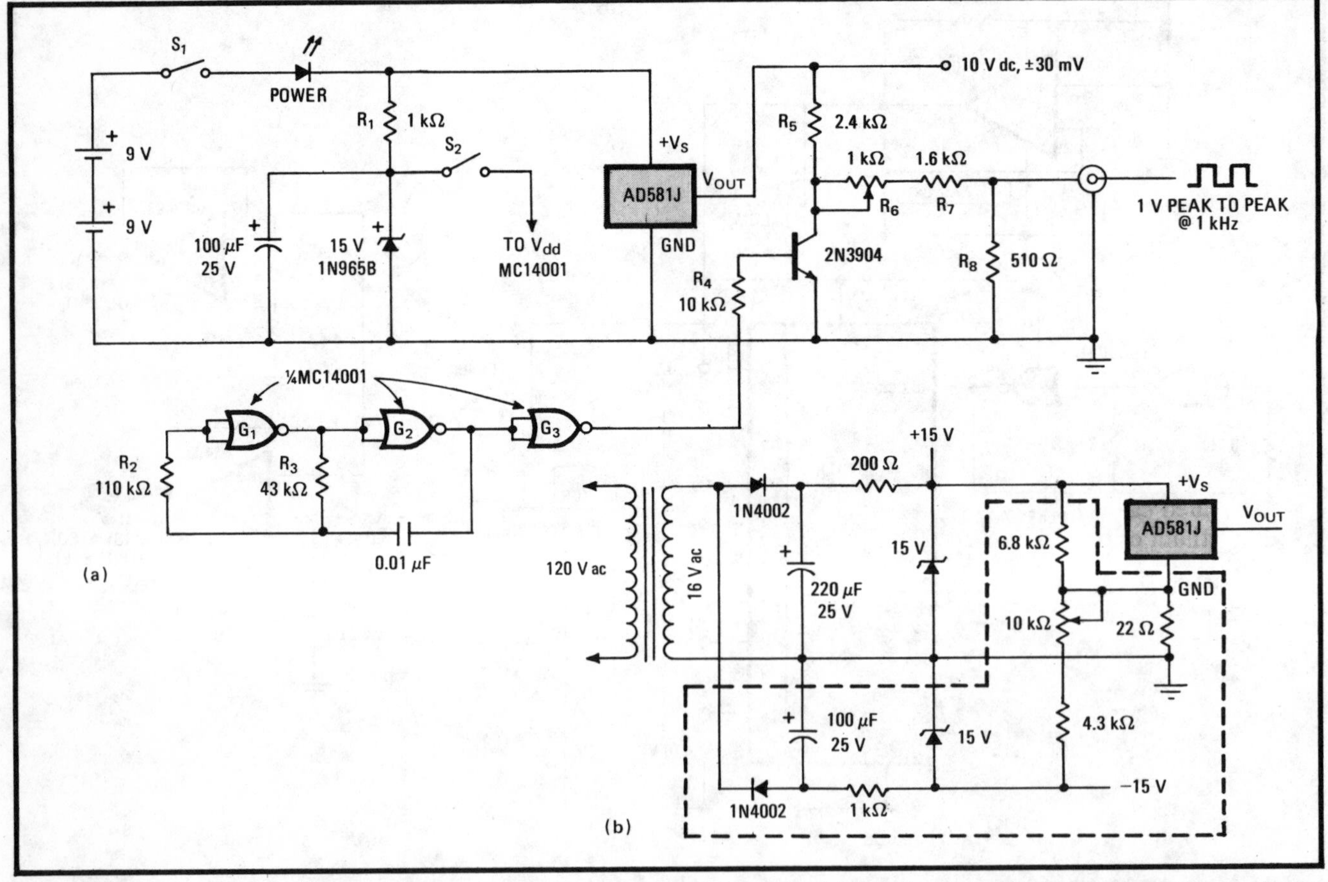

Precise amplitude. Scope calibrator generates 1-V square wave, accurate to within 0.5%, with aid of AD581J band-gap voltage reference (a). Unit may be powered by two 9-V batteries. Current drain is 6 mA. Ten-volt reference voltage is available at output jack for voltmeter calibration, etc. If more than occasional use is contemplated, the unit should be powered from the ac mains (b).

To eliminate the errors due to the transistor's offset voltage, R_6 is used to adjust the output for a collector swing of 0.005 to 1.005 V so that the difference voltage of 1 V peak to peak will appear at the output.

A dc digital voltmeter can be used for accurate adjustment of the output voltage, contributing to the ease with which the circuit can be calibrated. Accurate peak-reading ac voltmeters are not readily available, and root-mean-square voltmeters will not yield accurate results if the output is not perfectly symmetrical (having a duty cycle of 50%).

To calibrate the circuit, it is necessary to open S_2, which disables the 1-kHz oscillator by removing the supply voltage to G_1–G_3. R_6 is then adjusted for an output voltage of 1.005 V. Then S_2 is closed, the input of G_1 is grounded and the output voltage is measured again. The difference between these two readings should be exactly 1 V. The two-step procedure should be repeated as necessary; R_6 should be adjusted for a voltage slightly removed from the 1.005 V originally set, then G_1's input grounded, and so on, until a difference voltage of 1.000 V is obtained.

Two 9-V batteries will provide many hours of operation. Battery drain is approximately 6 milliamperes. Typical units will work well down to a supply voltage of about 12 V.

If more than occasional use is anticipated, the unit should be powered from the 120-V ac line, as shown in (b). A bipolar (15-V) supply is derived from the power-line voltage, and a fine-trim circuit added as shown inside the dotted line, so that the set accuracy of the 581J's 10-volt output may be improved. □

Chapter 19
INVERTERS

Autocorrecting driver rights pulse polarity

by Shlomo Talmor
Hartman Systems, Huntington Station, N. Y.

This circuit provides positive-going output pulses for corresponding input signals of either polarity without the need for any manual intervention (that is, polarity switches). Utilizing a simple RC integrator, the unit automatically propagates positive-going signals having a duty cycle of less than 50% through to the output and inverts signals that have a duty cycle of greater than 50%, or are negative-going. The circuit is particularly useful in instrumentation and test-facility applications, where the polarity of the signal emanating from a port is often both positive- and negative-going at different stages of a complicated test sequence.

As shown in the figure, input pulses having a width and rate in the range of 1 microsecond to 1 millisecond are applied through inverter A_{1a} to integrator R_1C_1, thereby developing a dc voltage across the capacitor. Thus, operational amplifier A_2 goes high for positive-going input pulses having a duty cycle of less than 50%. A_{1b} and A_{1d}, along with NAND gates A_{3a} to A_{3c}, therefore propagate the input signal through to the output without an inversion.

If the duty cycle is greater than 50%, or if the incoming pulses are negative-going, the voltage developed on the integrating capacitor will be below 5 volts, which is the potential applied to the inverting input, and A_2 goes low. A_{1b}, A_{1d}, and A_{3a} to A_{3c} then act to invert any negative-going wave at the input to a positive one at the output and vice versa. In the case where the duty cycle is 50%, inverter A_{1c} provides the necessary hysteresis to A_2 for proper switching.

Light-emitting diodes D_1 and D_2, with A_{1f} and A_{1g}, provide visual indication of pulse polarity. Both the red and green diodes will light softly when the input signal has a duty cycle approximating 50%.

The range of the pulse width handled can be extended, at the expense of a lower rate, by increasing R_1. The response time of the circuit will of course be lowered. The supply voltage can vary from 5 to 15 volts, but if it is less than 7.5 V, a 15-kilohm resistor should be placed in series with pin 6 of A_2 in order to properly drive the rest of the circuit. □

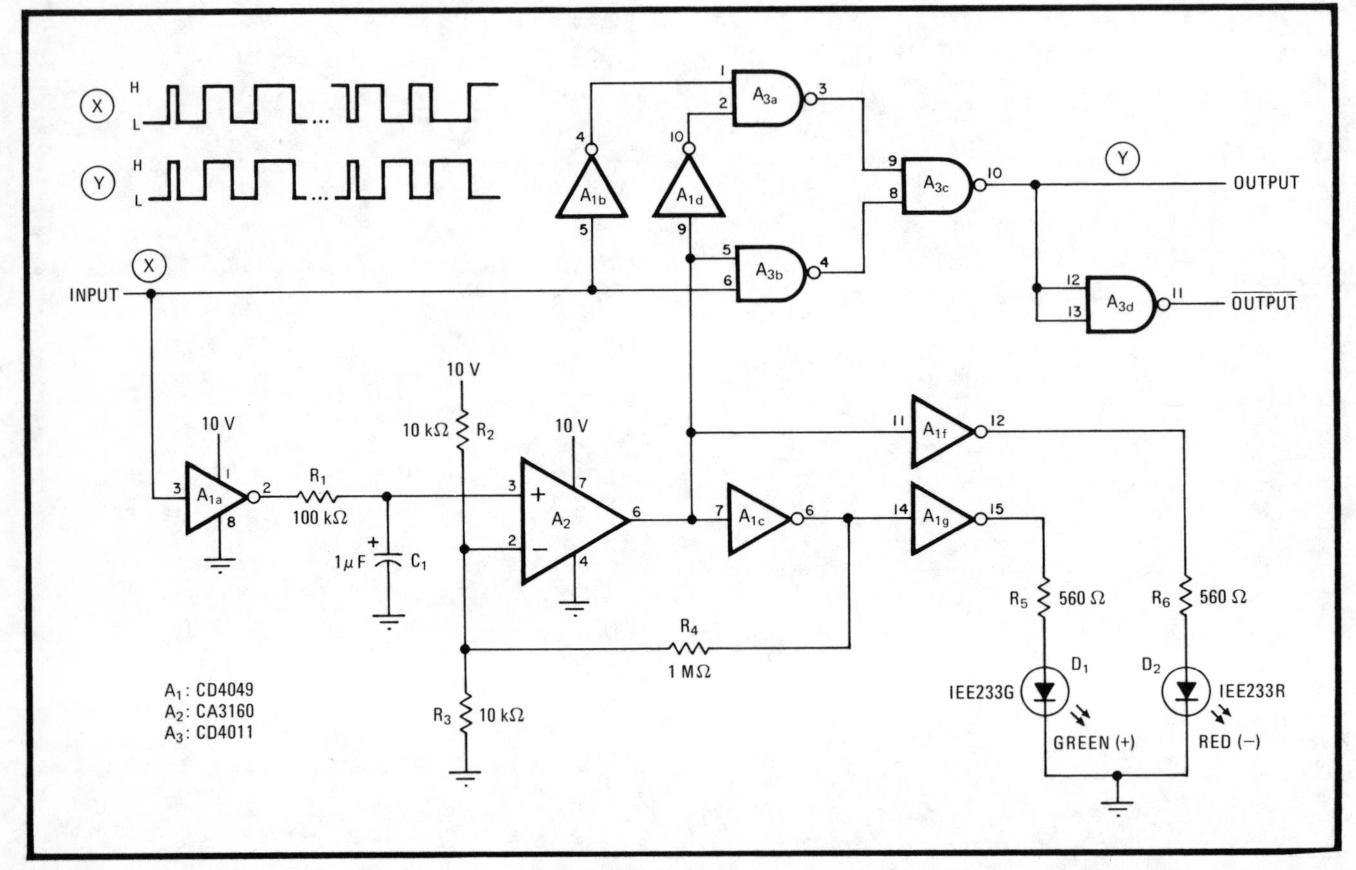

Switch. Three-chip driver accommodates both positive- and negative-going input signals, converting them both into positive-going output. Circuit propagates positive-going signals having a duty cycle less than 50% through to the output while inverting all other signals.

Bipolar current mirror scales, inverts signals

by Henry E. Santana
Hewlett Packard Co., Loveland, Colo.

A pair of operational amplifiers and a few resistors build this precision current mirror. Though simple and low in cost, the circuit excels the usual designs because it not only offers true bipolar operation but also can scale and/or invert any ac or dc input signal.

Input currents are applied to op amp A_1, which is biased by V_{ref}. If I_{in} is generated by a constant current source, V_{ref} may be brought to zero. Otherwise, it should be set to some arbitrary value to maintain circuit bias.

A current-to-voltage converter at the input and a voltage-to-current converter at the output comprise the current mirror. As a consequence of the configuration, the voltage appearing at the output of A will thus be:

$$V_{A1} = V_{ref} + I_{in}R_1$$

for $R_1 >> R_2$ and R_L. The voltage applied to the output circuit is therefore:

$$V_2 - V_1 = I_{in}R_1$$

Writing the nodal equations for V_L, V_3, and V_4 yields these results:

$$I_L = -V_L(1/R_2 + 1/bR_2) + V_3(1/bR_2) + V_2(1/R_2)$$
$$V_3 = A_2(s)\,(V_L - V_4) = (GB/s)(V_L - V_4)$$
$$V_4 = V_1[a/(1+a)] + V_3[1/(1+a)]$$

where GB is A_2's gain-bandwidth product. Substituting V_3 and V_4 into the equation for I_L, it is seen that $I_L = (R_1/R_2)I_{in}$, given that $a = b$ and $s << GB/(1+b)$.

The output impedance can be set, within limits, by selection of aR_2 and bR_2. The output impedance is:

$$Z_o(s) = \left[\left(\frac{a}{1+a}\right)\left(S+\frac{GB}{1+b}\right)R_2\right] \div \left[S+\left(\frac{a-b}{1+a}\right)\left(\frac{GB}{1+b}\right)\right]$$

Since a must equal b for the circuit to work, this equation simplifies to $Z_o(s) = [1/(1+a)]\{1 + [1/1+a)]GB/s\}\ R_2$, and no other assumptions about resistor ratios are made.

In addition to its use as a scaled current mirror, the circuit will find other not-so-obvious applications. Such an example is its use as a capacitance multiplier (b). □

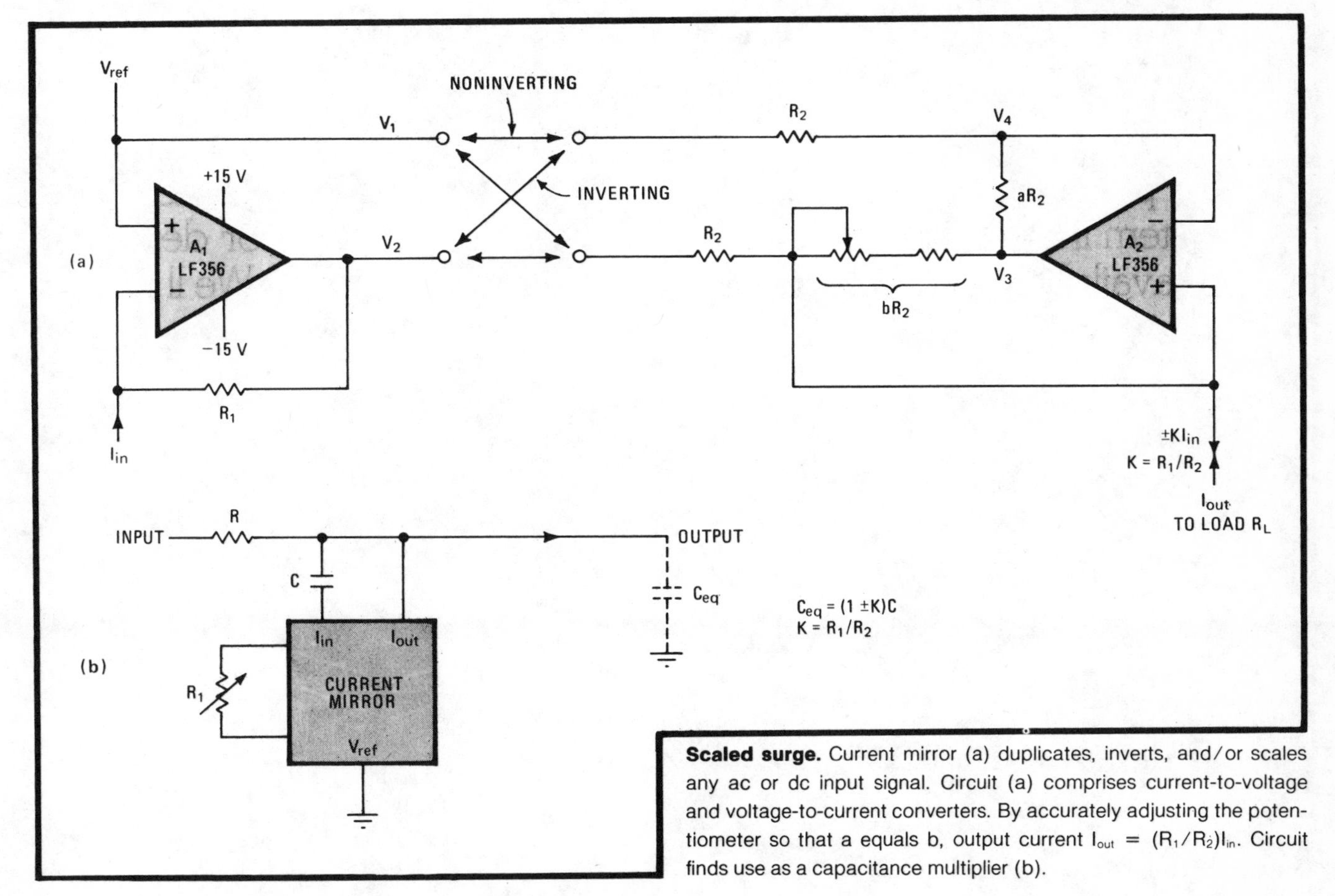

Scaled surge. Current mirror (a) duplicates, inverts, and/or scales any ac or dc input signal. Circuit (a) comprises current-to-voltage and voltage-to-current converters. By accurately adjusting the potentiometer so that a equals b, output current $I_{out} = (R_1/R_2)I_{in}$. Circuit finds use as a capacitance multiplier (b).

Chapter 20
LIMITERS

Double-ended clamp circuit has ideal characteristics

by Keith Wilson
Herga Design and Development Ltd., London, England

The quad operational amplifier and the analog multiplexer in this dual-threshold clamp circuit give it ideal characteristics—in particular, razor-sharp clamping. Two op amps in the array serve as comparators, determining the relationship of the input signal amplitude to the high and low reference levels, and the multiplexer passes one of the three signals, depending on the op amps' decision. The absence of feedback networks simplifies design and optimizes performance and ensures that component values and layout will not be critical.

The ideal characteristics of the clamper are made possible mainly by the use of op amps. They permit independent adjustment of upper and lower clamp points and simplify the solution of any temperature-compensation problems. In addition, the op amps will detect millivolt-level differences between the input signal and the set thresholds, so that the multiplexer will accept signals within the thresholds and pass them undisturbed to the output but will block signals exceeding the thresholds.

As shown in the figure, an input signal is introduced to the noninverting ports of two op amps making up one half of the LM348 device. The high and low reference levels are applied to the inverting input ports. The op amps, operating as comparators, run at open-loop gain and are powered from a single-ended supply. Both drive the CD4052 differential four-channel multiplexer.

Each comparator may be in one of two states at any given time. Therefore, in combination, both comparators may assume one of four possible states. The control signals from the comparators are applied to the A and B inputs of the multiplexer, which routes the input signal, the upper threshold voltage, or the lower threshold voltage to the output amplifier. Thus a hard clamping action takes place. Between the limits of the clamping levels, circuit response is linear, as can be determined from the truth table at the bottom of the figure.

Changes in the channel resistance of the multiplexer switches are caused by supply voltage and temperature variations, but are due chiefly to the amplitude variations of the input signal. The purpose of the output amplifier circuit is to compensate for the amplitude distortion introduced by the multiplexer switches.

Without compensation, the signal encounters a voltage divider composed of the switch resistance and the 10-kilohm resistor at the input to the output amplifier (added to improve switch linearity). By using another 10-kΩ resistor and the remaining switch in the 4052, the gain of the op amp can be made to increase proportionately to switch resistance.

Supply voltage changes amounting to several volts cause negligible change at the output of the op amp. Temperature variations create errors amounting to only a few millivolts. □

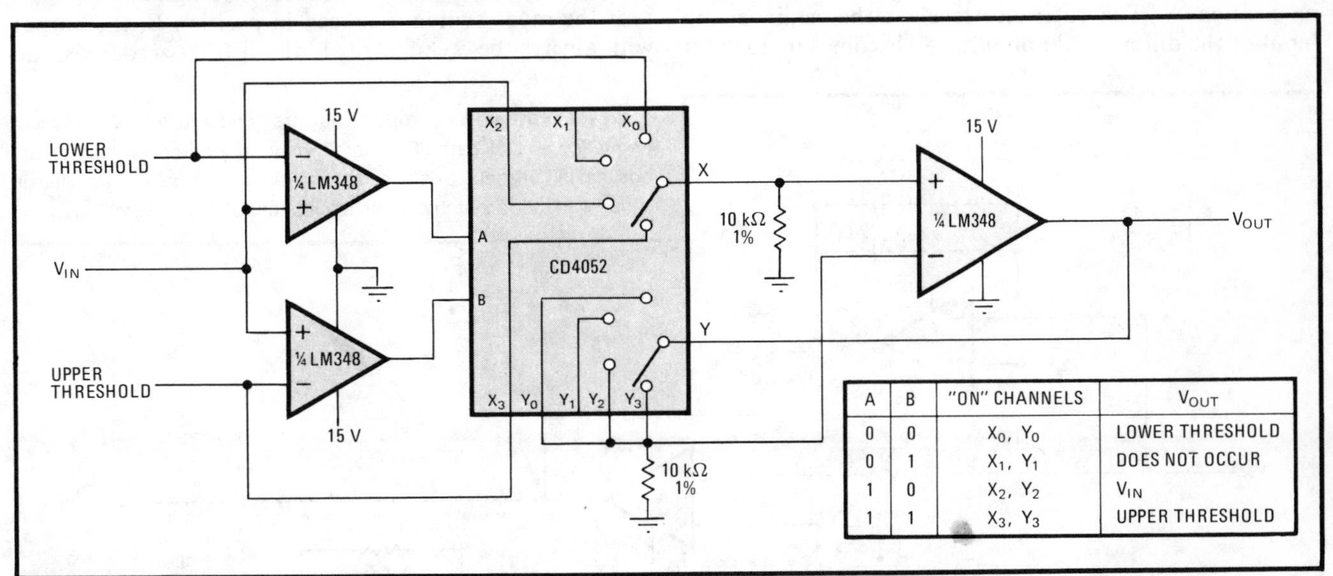

A	B	"ON" CHANNELS	V_{OUT}
0	0	X_0, Y_0	LOWER THRESHOLD
0	1	X_1, Y_1	DOES NOT OCCUR
1	0	X_2, Y_2	V_{IN}
1	1	X_3, Y_3	UPPER THRESHOLD

Dual-threshold clamper State of comparators driving analog multiplexer ports A and B determines which of three inputs appears at V. Circuit response is linear between threshold limits. Truth table describes circuit function.

Automatic clamp controls symmetrical wave-form offset

by George O. Wright
Washington, D.C.

It is often necessary to shift the dc level or offset of a signal. Driving transistor-transistor-logic circuits, for example, requires positive-only signals, but sometimes circuits may need negative-only signals, or symmetry of a wave form about a given dc level may be required. An inverting operational amplifier, an integrator, and a half-wave rectifier can be used to form such an offset control simply and inexpensively.

At the flick of a switch, this circuit clamps the input level at almost any desired value (Fig. 1). Operation of the circuit is based on the principle that the dc value of any periodic wave form can be found by integration. Since the input is dc-coupled, the integrator output V_{dc} will be equal to the dc component of the signal plus any initial dc offset. As a consequence of its integrating operation, the circuit can generate an output symmetrical about the zero axis.

As shown in Fig. 2, the input signal, which may be a sine, triangle, square or ramp wave form, is summed with the inverted output of the integrator A_1 at the inverting input of op amp A_2. The input wave form at this point is displaced downward by an amount equal to its integrated voltage value V_{dc}.

To clamp the output at 0 v, the mode switch is placed in position A. The reference level at the noninverting input of the differential amplifier A_2 becomes minus half its initial peak output swing because of the signal's subsequent peak detection at the output of A_3. Operation of amplifiers A_1 and A_2 assure that the initial signal to A_3 is symmetrical about 0. The polarity of the output voltage depends on the orientation of the diode—for the direction shown, it is positive-going.

To clamp the output at some other value than 0 v, the mode switch must again be placed in position A, and a dc amplifier must also be inserted in the loop between the rectifier output and the noninverting input of op amp A_2 to control the value of the reference signal.

If the mode switch is placed in position B, the output will always be symmetrical about the zero axis, no

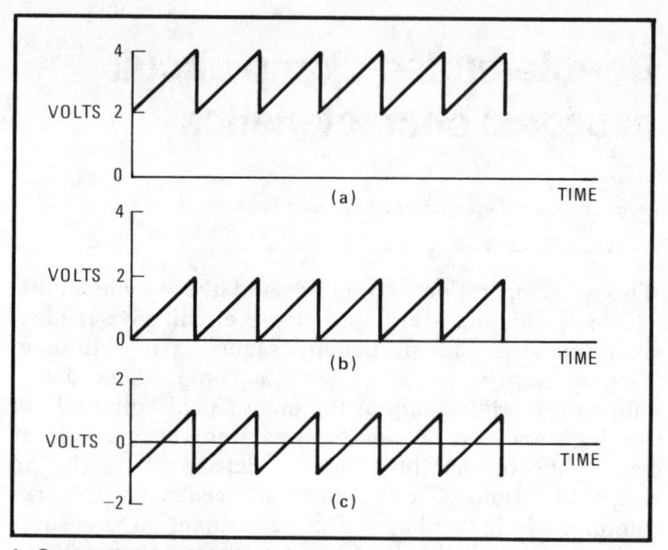

1. Symmetry or clamping desired? Input ramp at (a) is clamped to zero (b) or is symmetrical about zero (c). Circuit may be modified so that clamping to nearly any voltage level is possible.

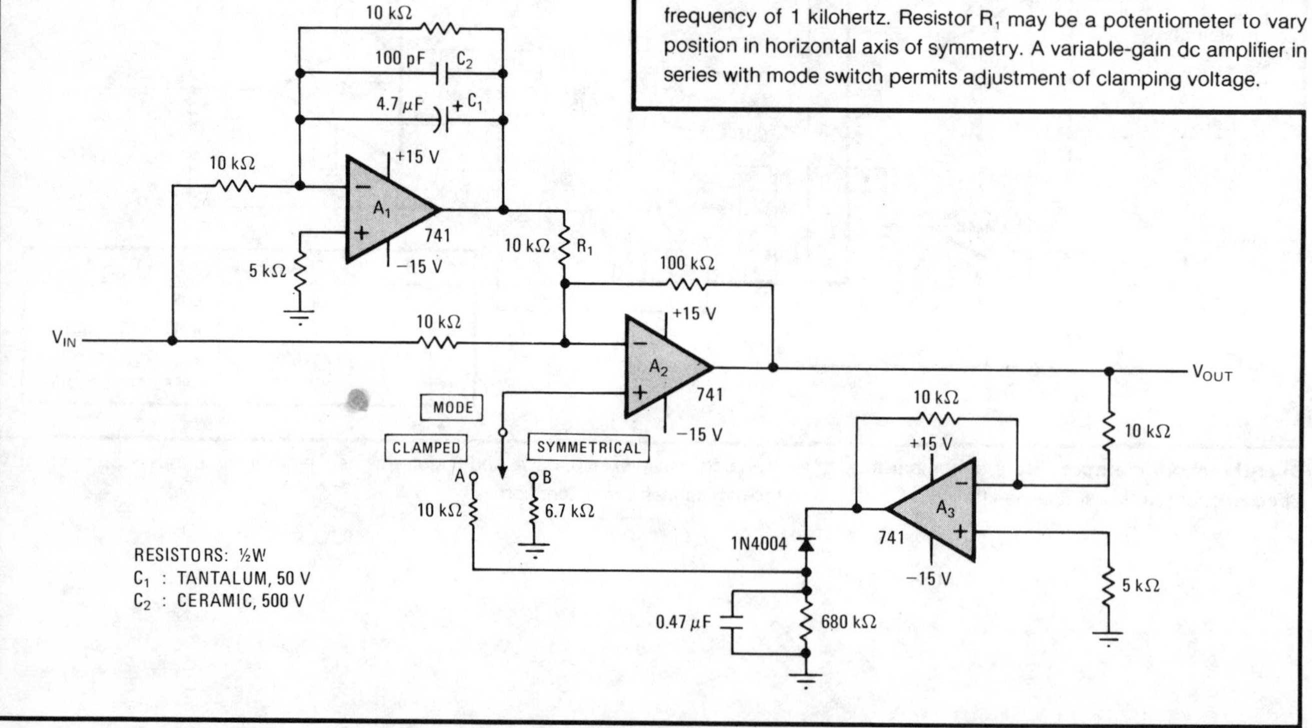

2. Level adjuster. Components are chosen for a minimum frequency of 1 kilohertz. Resistor R_1 may be a potentiometer to vary position in horizontal axis of symmetry. A variable-gain dc amplifier in series with mode switch permits adjustment of clamping voltage.

matter what the original input level is. This operation is roughly equivalent to removing the integrator from the circuit and placing a capacitor in series with the input lead to block the dc component of the signal.

Component values in the integrator circuit shown will effectively process signals of 1 kilohertz and higher, up to the frequency limit of the op amp. In general, the circuit will process almost any signal that is symmetrical about a horizontal axis. The exception is pulse trains of single polarity: their short duty cycle makes a symmetrical output impossible because the dc component is less than half the peak value. In this case, R_1 may be adjusted to provide the desired output level, which can be varied over a small range. □

Adjustable limiter controls telephone-line signal power

by Tom Hilleary
HMS Corp., Shawnee Mission, Kansas

To minimize crosstalk and other unwanted line radiation on telephone lines, it is often necessary to limit the input power of the data or voice signals, and to do so without amplitude distortion. This circuit does so by using two shift registers to monitor the relation of a set reference to the input signal amplitude (which is directly proportional to power at constant line impedance) and then divert some input current from the line transformer's

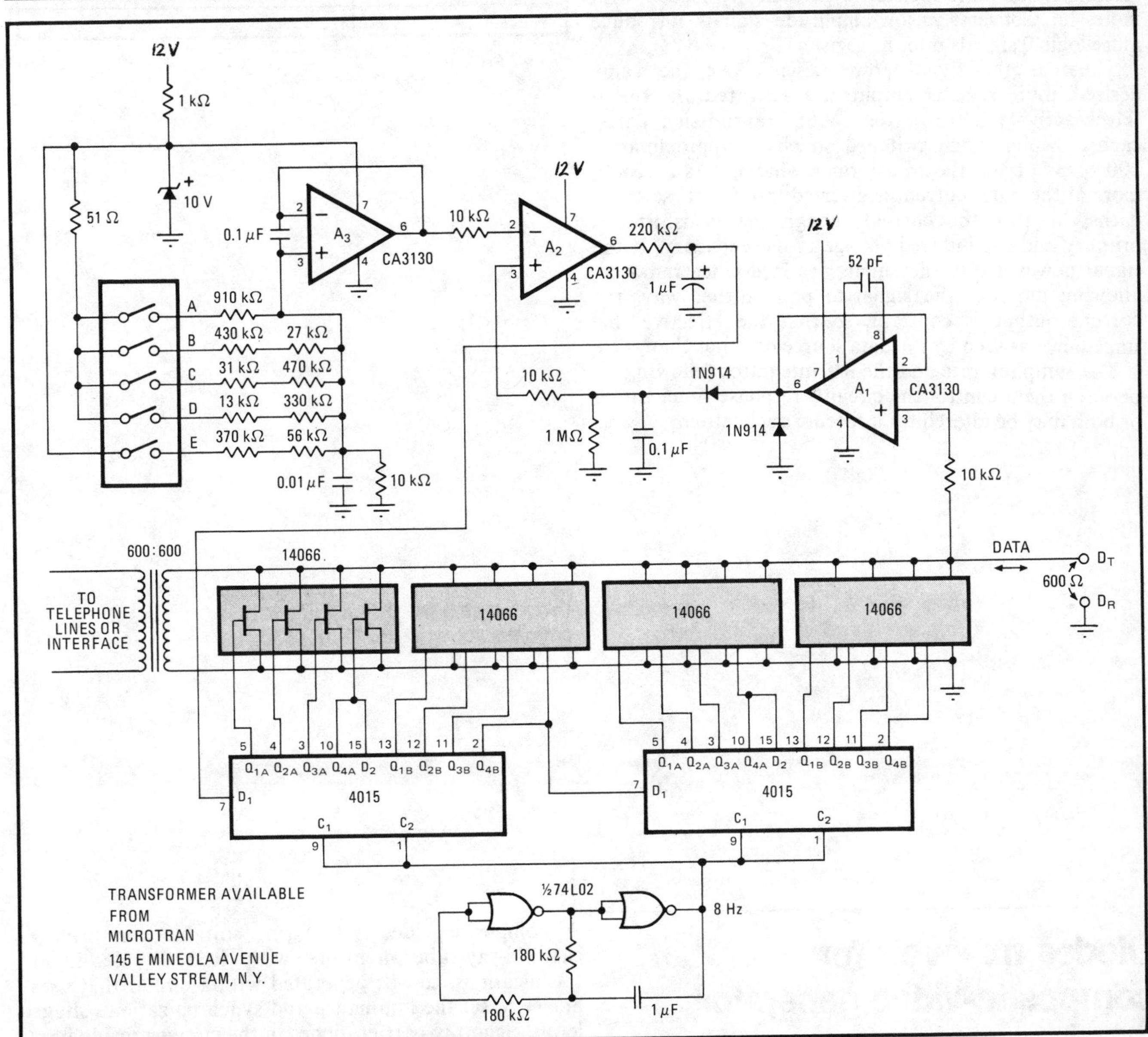

Digital limiter. Circuit provides 0 to −12 dBm of signal attenuation. Relative amplitude of constant-current data signal to set reference determines number of logic 1s generated by A_2 into 4015 registers over a given time. Signal above threshold loads more 1s into register, turning on more transmission gates as register is clocked and thus diverting the input current away from the primary winding.

primary winding whenever necessary. The limiter is effective when used with any constant-current data source. It provides 0 to −12 dBm of limiting and is accurately adjustable to 1 dB.

As shown in the figure, data or voice signals having a maximum amplitude of 1.4 volts at 0 dBm are applied to operational amplifier A_1. The output of A_1 in turn feeds the noninverting input of comparator A_2. The desired limiting (reference) level is set by five switches, operated in accordance with the table. A_3 is a voltage follower and passes the reference voltage to the inverting input of A_2.

A_2 compares the peak value of the input signal to the reference and when the reference voltage is exceeded generates a logic 1 for the data input of two 8-bit 4015 shift registers. The 4015s are configured as a single 16-bit device that is clocked at an arbitrary 8-hertz rate. If the input signal is mostly larger in amplitude than the reference, the shift register is periodically loaded with more 1s. Conversely, low-amplitude signals introduce more logic 0 signals into the register.

Thus, as the signal power rises above the value desired, more register outputs are activated and themselves activate more of the 14066 transmission gates, each of which when switched on places approximately 100 ohms across the transformer winding. As a result, more of the input current is diverted through these shunt paths, limiting the current though the transformer's primary and the induced voltage in its secondary. If the signal power drops, the shunts are in essence removed, allowing more of the signal to pass. Either way, the current output is constant, so that the effective line impedance as seen by the data source does not change.

The sampling rate and the RC integrator following A_2 between them control the circuit's response time. Either or both may be altered for a specific application. □

DIGITAL LIMITER

D_T PEAK LEVEL (dBm)	CLOSE SWITCHES				
	A	B	C	D	E
0	■	■	■	■	■
−1	■		■	■	■
−2			■	■	■
−3	■			■	■
−4	■	■			■
−5		■			■
−6		■	■		
−7	■				■
−8	■		■		
−9	■	■			
−10				■	
−11			■		
−12		■			

Diodes fix levels for composite-video generator

by Robert H. Lacy
Applied Automation Inc., Bartlesville, Okla.

A composite video waveform, suitable for driving cathode-ray-tube monitors with RS170 video-input specifications, can be generated with a circuit that uses diodes to set the luminance and synchronization voltage levels. Schottky barrier diodes in the circuit provide fast, clean pulses free of overshoot and ringing, with rise and fall times of about 12 nanoseconds.

As shown in the composite video waveform example of

Fig. 1, the circuit fixes four discrete voltage levels—sync, blanking, black, and white. Since the design is intended for displaying digital alphanumeric data on the CRT, only the two extremes of the gray scale are supplied—black and white. However, the peak level of the video signal could be clamped to 2.5 volts, for example, to provide a gray output as well.

If the display of raster lines on the screen is not desired, control of the blanking level may be omitted; however, in a high-contrast CRT, for example, the lines should remain visible, as they counteract reflections off the CRT glass. Thus blanking is needed to eliminate the retraace that appears at higher brightness settings.

The schematic is shown in Fig. 2. The three inputs—sync, blanking, and video—are transistor-transistor-logic-compatible. Schottky-TTL inverters buffer the inputs for fast rise and fall times.

The most positive voltage, producing a white luminance level from a logic-1 video input, is established at approximately 3.4 volts by the R_1-R_2 voltage divider. The black level, corresponding to a logic-0 video input, is obtained from the 1.9-V total forward voltage drop of the two junction diodes and just one Schottky diode. The blanking and sync levels are determined by the other diode combinations, as shown in the figure.

The reverse-recovery time of the three-diode string in the video input is very fast, thanks to the Schottky diode characteristics. The rise time of the video pulses is determined mainly by the time constant of R_1 with the stray capacitance, indicated here by C_{stray}.

With such quick rise and fall times, the circuit is more than adequate for CRT displays of 80 characters per row. If only 40 characters per row are required, conventional silicon switching diodes such as the 1N914 may be substituted for the Schottky devices, though with a sacrifice in edge definition.

A series-complementary emitter-follower pair, made up of transistors Q_1, Q_2, and associated resistors R_3 through R_5, provides a high-impedance buffer for driving coaxial-cable lines. The 68-ohm resistor in series with the output increases the output impedance of the Q_2 follower, preventing oscillation in the event an unterminated cable is connected. □

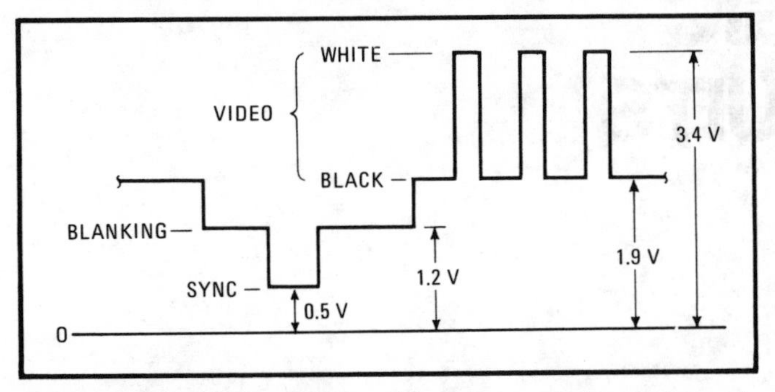

1. Composite video waveform. Four voltages shown are compatible with CRT monitors having RS170 specifications. Waveform is typical of digital input to CRT, with only white and black displayed.

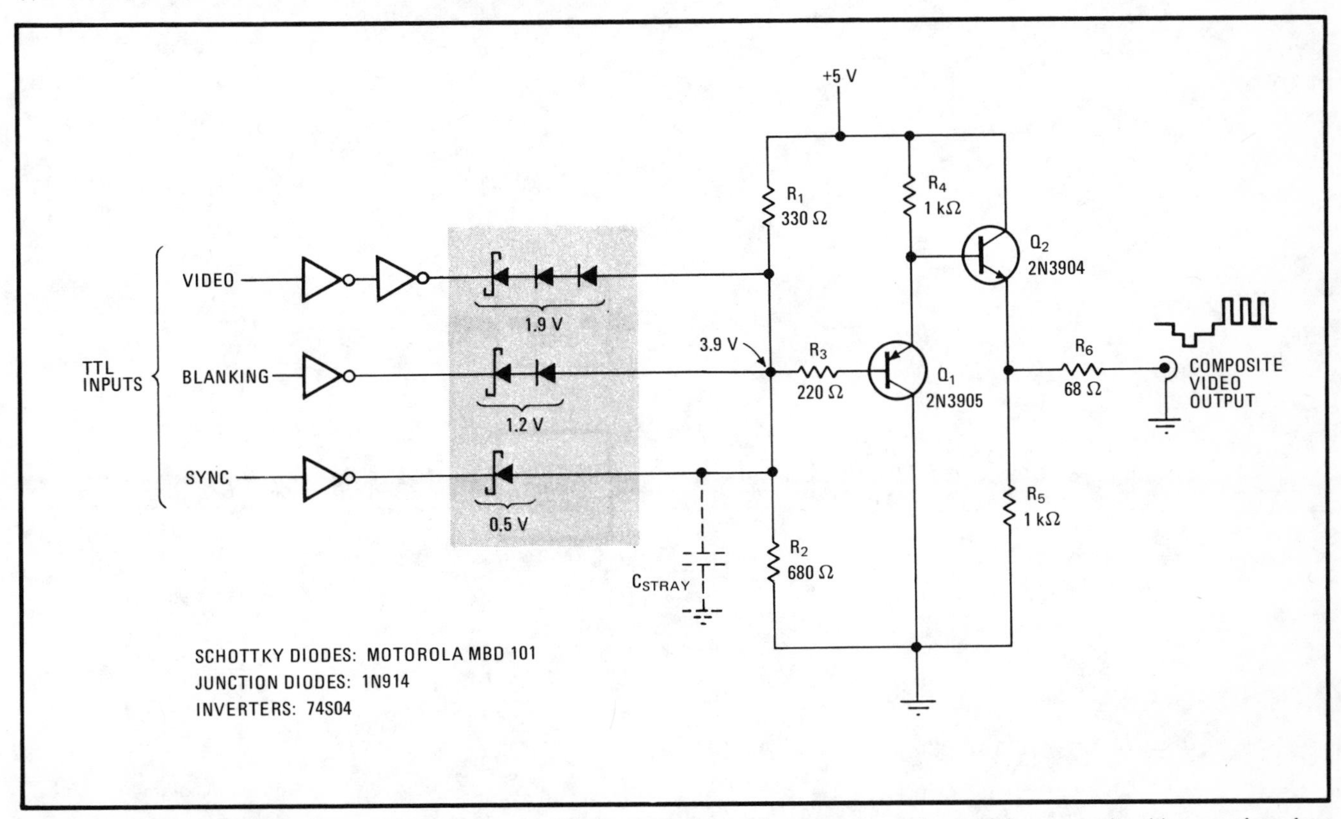

2. Video voltage-level generator. Voltage drops across diodes determine luminance and sync levels of the composite video waveform from TTL input signals. Blanking input is needed to eliminate retrace, if raster lines are brought up. Q_1 and Q_2 followers buffer the signal for driving a standard 75-Ω coaxial line, and output is short-circuit-proof.

Chapter 21
LOGIC CIRCUITS

Digital logic multiplies pulse widths

by N. Bhaskara Rao
U.V.C.E., Department of Electrical Engineering, Bangalore, India

Using logic elements to multiply the width of incoming pulses by a value selected by the user, this circuit is simple to build and provides a higher accuracy-to-cost ratio than its analog counterpart. It should therefore find numerous uses in synchronous systems, and although its prime function is to provide a multiplication factor of greater than unity, it can generate smaller values as well.

The figure will help make circuit operation clear. The multiplication factor is selected by presetting two 74S192 down counters, A_1 and A_2. Initially, counter A_1 is set to a value, M; A_2 is preset to a second value, N; A_3 is zero; and the Q output of flip-flop A_4 is high.

An incoming pulse of width T_1 (the signal to be multiplied) switches on the 7408 AND gate and enables A_3 to count to a number determined by f_1, which is derived from an external clock having input frequency f. Thus, at the end of the pulse, the counter contains the number f_1T_1. Note that $f_1 = f/M$.

When the trailing edge of the pulse arrives, A_5 is triggered and presets A_6 with the number contained in A_3. Meanwhile, A_4 is cleared (Q=0) by A_5, and the OR gate is thereby activated so that counter A_6 can initiate counting from its preset value. Note that A_6 is driven by A_2, the divide-by-N down counter, and that $f_2 = f/N$.

The time taken for A_6 to reach zero from its preset value is thus $T_2 = (f_1/f_2)T_1 = (N/M)T_1$. At this time, the output from A_6's borrow port clears A_3 and presets the flip-flop. Therefore the time between the flip-flop's move to logic 0 (at the trailing edge of the input pulse) and the time its Q output moves high again is T_2.

The output signal is not derived until the input pulse's trailing edge arrives. The multiplication factor, N/M, can thus be set to any value greater or less than unity because the conversion is carried out after a delay. Needless to say, f should be much greater than T_1 for accurate pulse-width multiplication. □

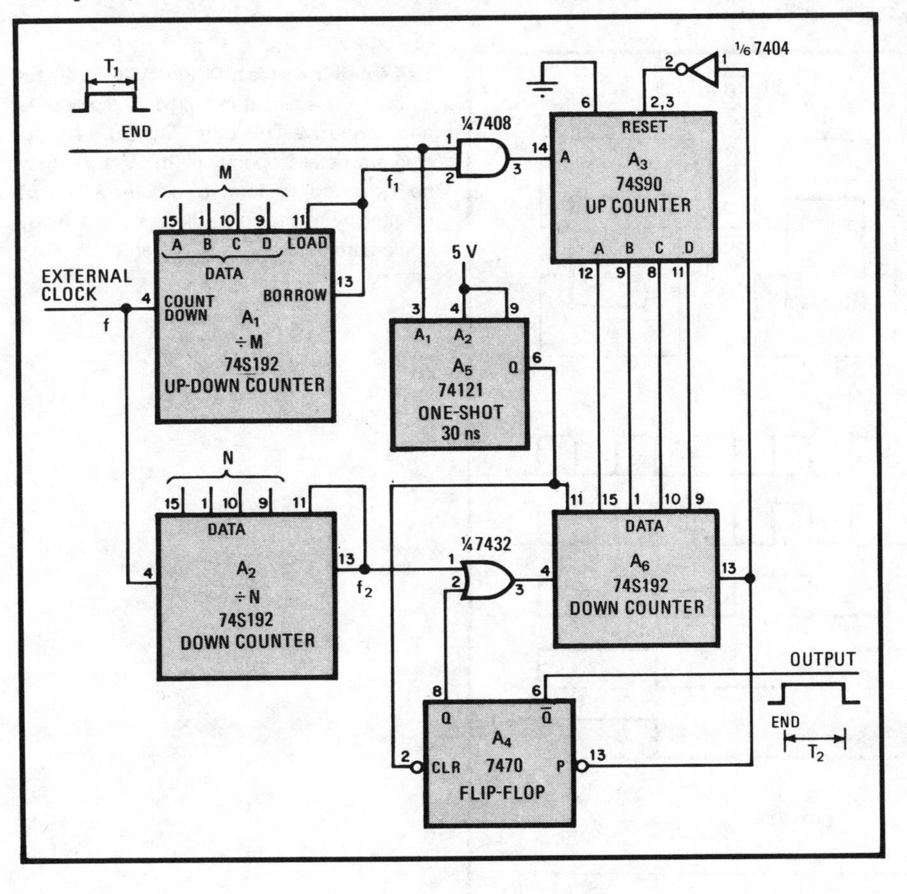

Width multiplier. Circuit has no analog elements. Multiplication factor is determined by ratio of N to M, set by user. A_6 is preset to f_1T_1; stepped to zero at f_2 rate, it signals flip-flop. Time between state changes of flip-flop is $T_2 = (N/M)T_1$.

Exclusive-OR gate and flip-flops make half-integer divider

by Tung-sun Tung
University of Illinois, Urbana, Ill.

A simple and inexpensive divide-by-(N − ½) counter can be built with D flip-flops and an exclusive-OR gate. The circuit uses very few parts, largely because, unlike many such dividers, it does not rely on a collection of monostable multivibrators. Nor are there difficult timing considerations, which would also complicate the circuit and act to increase its parts count.

Various logic circuits need to be driven by what may be termed a half-integer clock signal that is derived from a master clock. For instance, the MM-57100 video-game integrated circuit is driven by a 1.0227-megahertz signal, which the chip's manufacturer (National Semiconductor Corp.) recommends be derived from the television set's 3.579545-MHz color-burst crystal oscillator. This application requires a divide-by-3½ counter, which may be built by modifying a divide-by-4 circuit as shown in Fig. 1a.

Flip-flops A_1 and A_2 form the divide-by-4 counter. The master clock, C_{IN}, drives A_1 through the exclusive-OR gate, G_1. However, the output state of G_1 is additionally controlled by A_3.

As a result, A_3 converts the divider to a divide-by-3½ counter. A_3 is situated in the feedback loop such that it enables generation of a short pulse at C_O for every 3½ counts of C_{IN}. The pulse is generated if C_{IN} either falls to logic 0 or rises to logic 1, which depends on the state of Q_C. Its shortness is due to the delay between the initial 0-to-1 transition of C_O (caused by C_{IN}) and the change of state of Q_C—as Q_C changes state, it disables gate G_1, causing C_O to fall almost immediately after it has reached logic 1. The timing diagram given in Fig. 1b details circuit operation.

This method may be extended to the general case of the divide-by-(N − ½) counter, as shown in Fig. 1c, simply by substituting a single synchronous or asynchronous counter for the flip-flops A_i that would otherwise be required. □

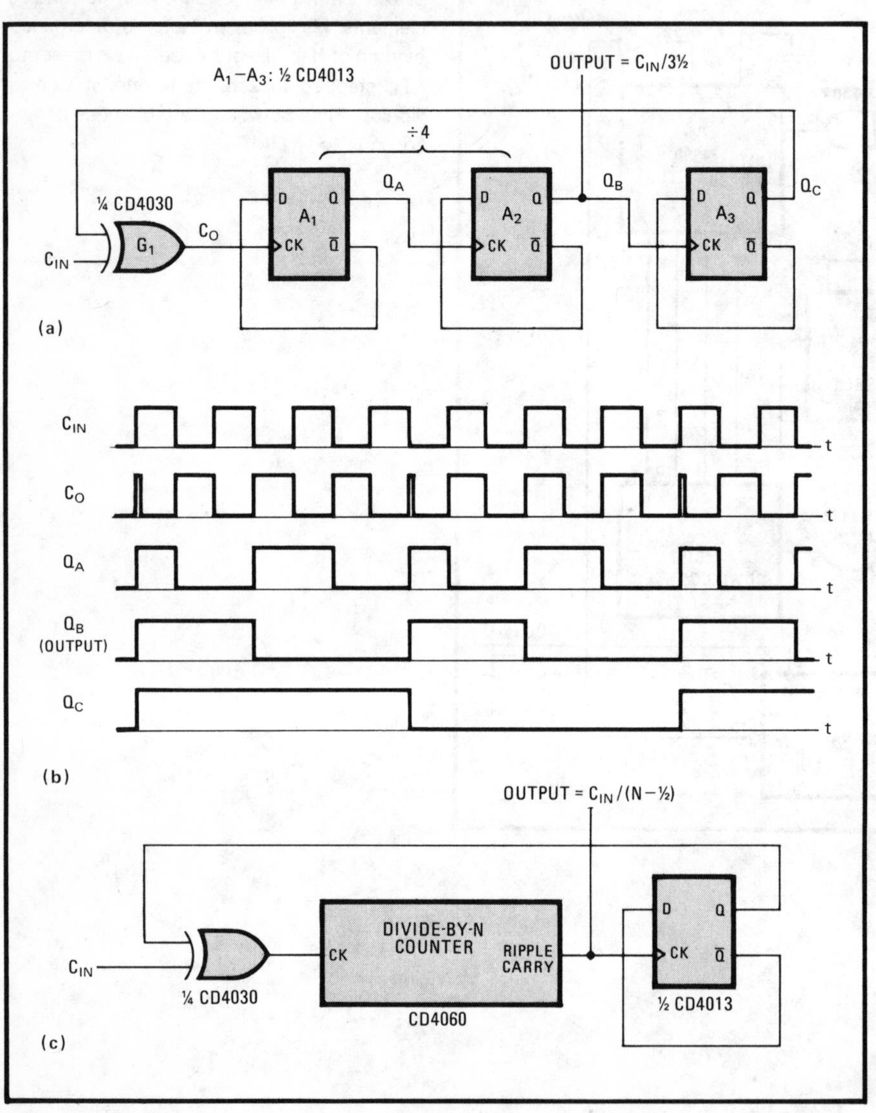

Half-integer divider. Divide-by-3½ counter is divide-by-4 circuit modified by flip-flop A_3 and exclusive-OR gate G_1 (a). Timing diagram details operation (b). Method may be extended to form a divide-by-(N − ½) counter by replacing flip-flops A_1 and A_2 by ripple-carry, divide-by-N counter (c).

Full adders simplify design of majority-vote logic

by Zhahai Stewart
Penfield Engineering, Boulder, Colo.

A majority-vote circuit—one with an output state determined by the logic states of the majority of its input variables—can be implemented by 4-bit binary full adders to yield lower chip counts or cost advantages than with standard designs using programmable read-only memories. The adders are easily combined to synthesize circuits used for decision-making or time-and-event-counter applications.

The circuits described use the 7483 full adder, but other adders that provide a similar logic function can be used. Each bit adder in the 7483 (Fig. 1a) has inputs A_i, B_i, and C_i; the outputs are S_i and C_{i+1}. Each adder is cascaded internally by connecting the C_i ports together.

The characteristics of the device are such that there will be a logic 1 output from the sum (S_i) terminal of each adder if one input is high or all three inputs are high. The carry output (C_{i+1}) of each stage will assume the high state if any two inputs or all inputs are high. This is the basic design tool used in the design of any n-bit majority gate.

Only one adder is needed to form majority gates with as many as five inputs. For example, a three-input device can be built by introducing the input variables to A_3, A_4, and B_4 of the 7483, while tying all other inputs high to form the "trivial case" solution. In reality, only one quarter of the adder is used to full advantage. The output, taken from the carry-out port of the adder, will be high if any two of the three inputs are high.

Figure 1b shows an implementation of a more important case, the five-input gate. Output M is high if any three of the five inputs are high. The seven-input circuit (Fig. 1c) is a simple extension of the five-input case using the design rules outlined above, but it cannot be built using a single 7483 because its internal-carry connections are not accessible. While the adders cannot be directly cascaded, a majority gate of any size may be easily built as design experience is gained.

When the number of inputs is small, a PROM can generate the required gate function for slightly greater cost and design effort. However, when the number of input variables is greater than 10 or so, the use of adders will provide clear-cut advantages. An 11-input gate constructed from three adders is easier to design than with a PROM. (Fig. 2a).

When the number of inputs becomes very large, it is

1. Binary-adder majority gate. A single 7483 full adder (a) or similar device can be wired to produce up to a five-input majority-vote circuit. Five-input (b) and seven-input (c) gates are built using adders' truth table as the primary design tool.

2. No need for PROMs. Implementation with adders offer advantages over PROM designs when number of inputs exceed 10 (a). Tree networks can replace adders for additional ease in wiring when the number of input variables is higher (b).

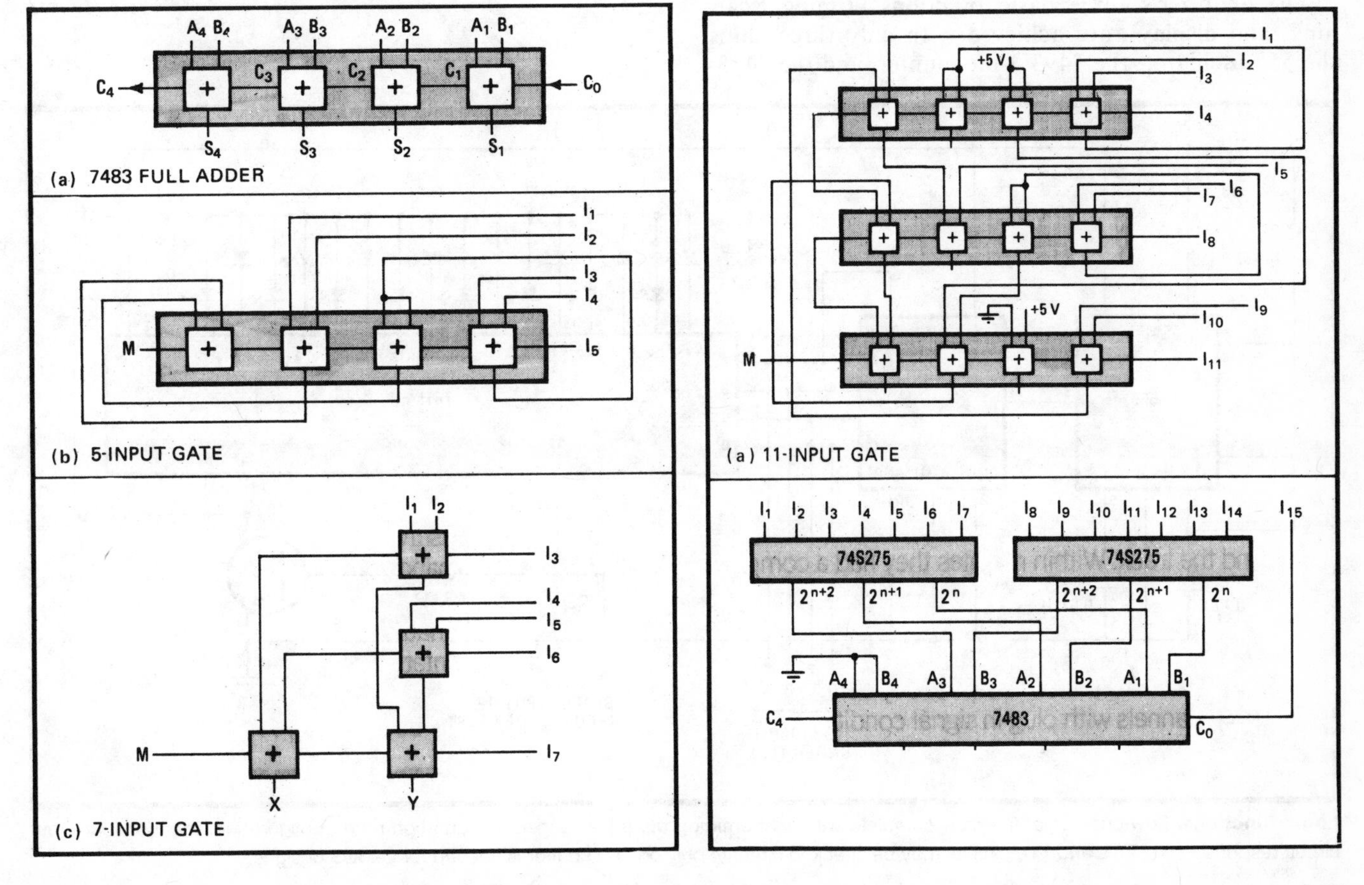

advantageous to search for devices with characteristics similar to the adder and use them to reduce wiring and design headaches—and, in many cases, to minimize chip count. In Fig. 2b, a 15-input gate is implemented using two 74S275 "Wallace trees" and a full adder; the 74S275 provides the function of four adders without the external wiring. When connected to the external adder, space and cost are minimized.

If the carry bits of the Wallace tree are wired low, each tree output will yield the binary sum of the number of input variables that are high. Alternatively, the X, Y, and M outputs of two 7-bit devices (example in Fig. 1) may provide the binary sum to the external adder, where Y is the least significant bit. The output of both trees are then added directly by the 7483, with the result that M will be high if eight or more inputs are high. □

Three-chip logic analyzer maps four-input truth table

by C. F. Haridge
University of Ottawa, Ontario, Canada

Providing an extremely simple and low-cost alternative to the use of an oscilloscope, this logic analyzer will determine the truth table of circuits with as many as four inputs. The state of the circuit for a single monitored output is displayed by a four-by-four array of light-emitting diodes arranged in a Karnaugh-map configuration. Resistor-, diode-, and transistor-transistor-logic circuits can be checked directly, and only one input/output buffer is required to check complementary-MOS designs.

The analyzer's three basic functions—timing, scanning, and display—are achieved with only three chips: the 555 oscillator, the 7493 4-bit counter, and the 74154 4-to-16-line decoder. The 555, running at a minimum frequency of 480 hertz to eliminate display flicker, clocks the 7493 through its 16 states continuously. As a result, a binary sequence of 0–15 periodically drives the four inputs of the circuit under test. These logic signals are also applied to the decoder chip. Consequently if the instantaneous output of the circuit point under test is high for any given set of input variables A–D, the LED corresponding to the 4-bit output number of the 7493 will light up.

The analyzer may be easily expanded to test circuits having more than four inputs by adding the appropriate number of counters, decoders, and LEDs. The clock frequency must also be increased to minimize flickering in the display.

A higher clock frequency will reduce the on-time of each LED, however. In order to compensate for this reduced brightness, resistor R_4 must be made proportionally smaller. □

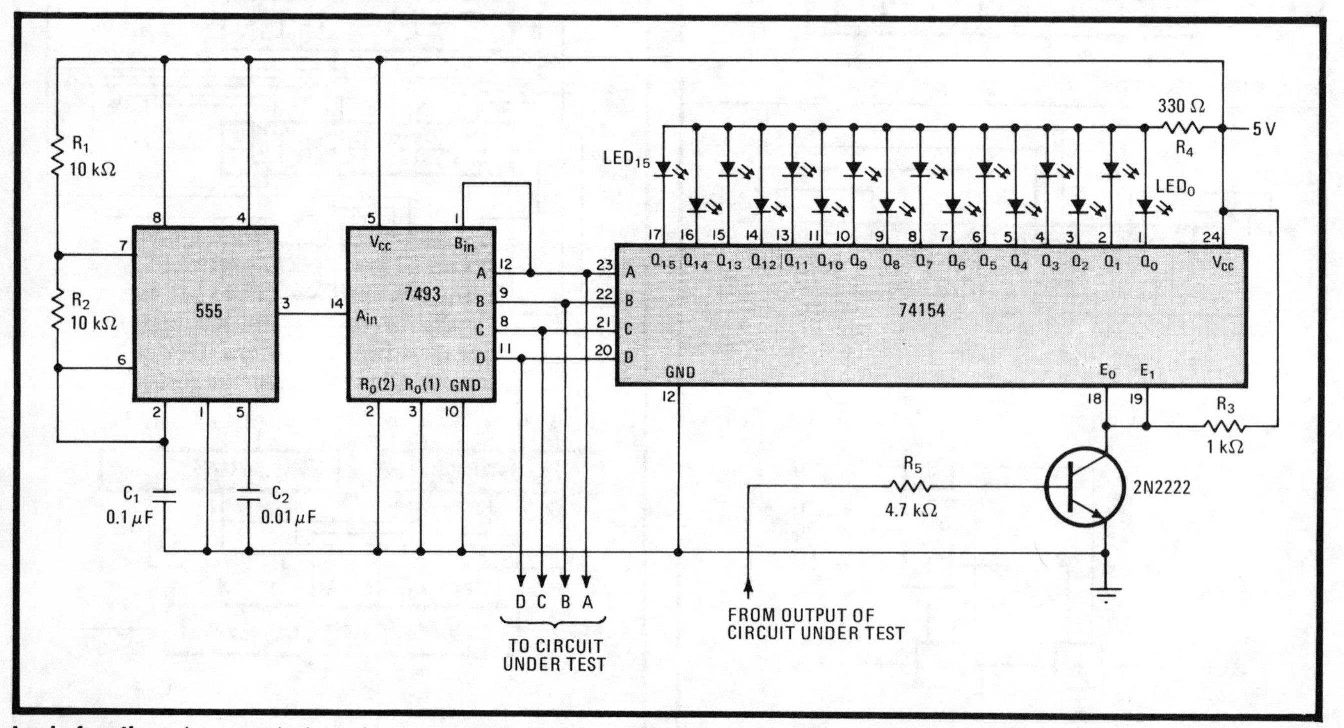

Logic functions. Low-cost logic analyzer, complete with light-emitting diodes arranged in Karnaugh-map configuration, monitor four-input circuit response. RTL, DTL and TTL circuits may be checked directly; only one I/O buffer is needed for C-MOS designs.

Gates convert J-K flip-flop to edge-triggered set/reset

by David L. DeFord
Colorado Video Inc., Boulder, Colo.

Set/reset flip-flops with edge-triggered inputs are not readily available in the common logic families, but can be built with a few standard gates and a J-K flip-flop. Both positive- and negative-transition triggering are easily configured with the use of NOR and NAND gating, respectively.

In the schematic below, when the flip-flop is triggered by an active transition at the set (S or $\overline{S}$) input, that input is disabled and the reset (R or $\overline{R}$) input is enabled. The flip-flop can then change state only upon an active transition at the reset input, and once this occurs, the set input is again enabled, completing the cycle.

As with all logic circuits, certain minimum pulse widths and timing restrictions must be observed. The input signal must remain stable following an active transition until the new outputs of the flip-flop have settled on the gate inputs. Using standard transistor-transistor-logic devices, the minimum pulse width for the negative-edge-triggered inputs is 65 nanoseconds in the worst case, but typically about 32 ns. In the positive-edge-triggered configuration the worst-case value is 77 ns, but typically 40 ns.

If low-power Schottky TTL is used, the worst-case minimum pulse width is reduced to 45 ns for both circuits and a typical value might be about 25 ns.

Since specifications vary between manufacturers, the minimum pulse widths may be calculated from the actual propagation delays to logic-1 (t_{pd1}) and logic-0 (t_{pd0}) levels of the individual gates and flip-flops. The formulas for the negative- and positive-edge flip-flop are:

$$\downarrow\ t_{w(min)} = t_{pd1}(\text{NAND}) + t_{pd0}(\text{NAND}) + t_{pd0}(\text{FF})$$
$$\uparrow\ t_{w(min)} = t_{pd1}(\text{NOR}) + t_{pd0}(\text{NAND}) + t_{pd1}(\text{FF})$$

Failure to observe the minimum pulse widths can result in extra pulse inputs to the J-K flip-flop, returning it prematurely to its original state. In addition, the set and reset pulses should be separated by at least a minimum pulse-width spacing, since neither of the inputs has priority, and erroneous triggering may otherwise result. □

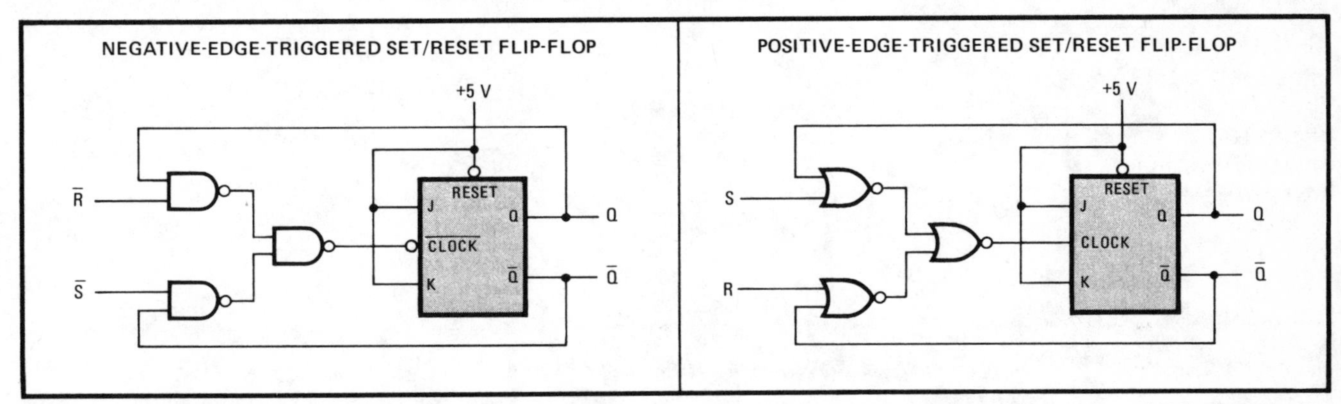

Getting the edge. Negative- or positive-edge-triggered R/S flip-flops can be made by adding NAND or NOR gates to a J-K flip-flop. The flip-flop at left must be negative-edge triggered, like the 7473, while the unit at right must be positive-edge triggered, like the 7470.

Chapter 22
MEMORY CIRCUITS

Saturable core transformers harden latch memories

by Gordon E. Bloom
IRT Corp., San Diego, Calif.

In certain critical applications, the loss of data in latch memories caused by supply and command-line failures or transients can be disastrous to system operation. However, at a slight sacrifice in switching speed, a memory-hardening circuit that uses saturable transformers can make the memory latch nonvolatile and insensitive to false command signals.

As shown in the figure, the subminiature saturable transformers are placed in the feedback paths connecting two NOR gates, which occupy half of a 7433 quad package. The transformers are incapable of an immediate state change, and because they stay magnetically biased without a voltage supply, they provide a reference state to which the memory latch returns when the power is reapplied.

The output voltage of each transformer is a function of its magnetization state (set by a gate output) and a power-up pulse, which attempts to examine this state. Each transformer is biased at all times—one in positive saturation (low-resistance core condition) and the other in negative saturation (high-resistance core condition). If power is lost, the magnetization state of each core is unchanged and remains so indefinitely. When power returns, a voltage is induced into the primary of the transformer that is in negative saturation. This voltage then drives the associated gate, restoring the original conditions at the latch outputs.

Transistor-transistor-logic gates are used. Resistor values have been chosen to ensure that primary-to-secondary coupling of the transformers produce currents that are sufficient to bias the core properly on its hysteresis curve, yet that do not overdrive the gates. In addition, the total flux capacity of the cores permits adequate switching time, while placing no undue restiction on the normal set and reset timing relationships during normal operation.

When a bit is initially stored in this circuit, the set or reset pulse must have a minimum width of 35 microseconds, and pulses must be separated by a minimum of 65 μs. Should power fail during operation, the power-supply fall time must be much less than 35 μs; otherwise, the latch may lose its contents. The same rise-time constraints accompany power-up.

The circuit may be implemented with other logic families, but choice of component values must take into consideration the family's impedance and switching characteristics. □

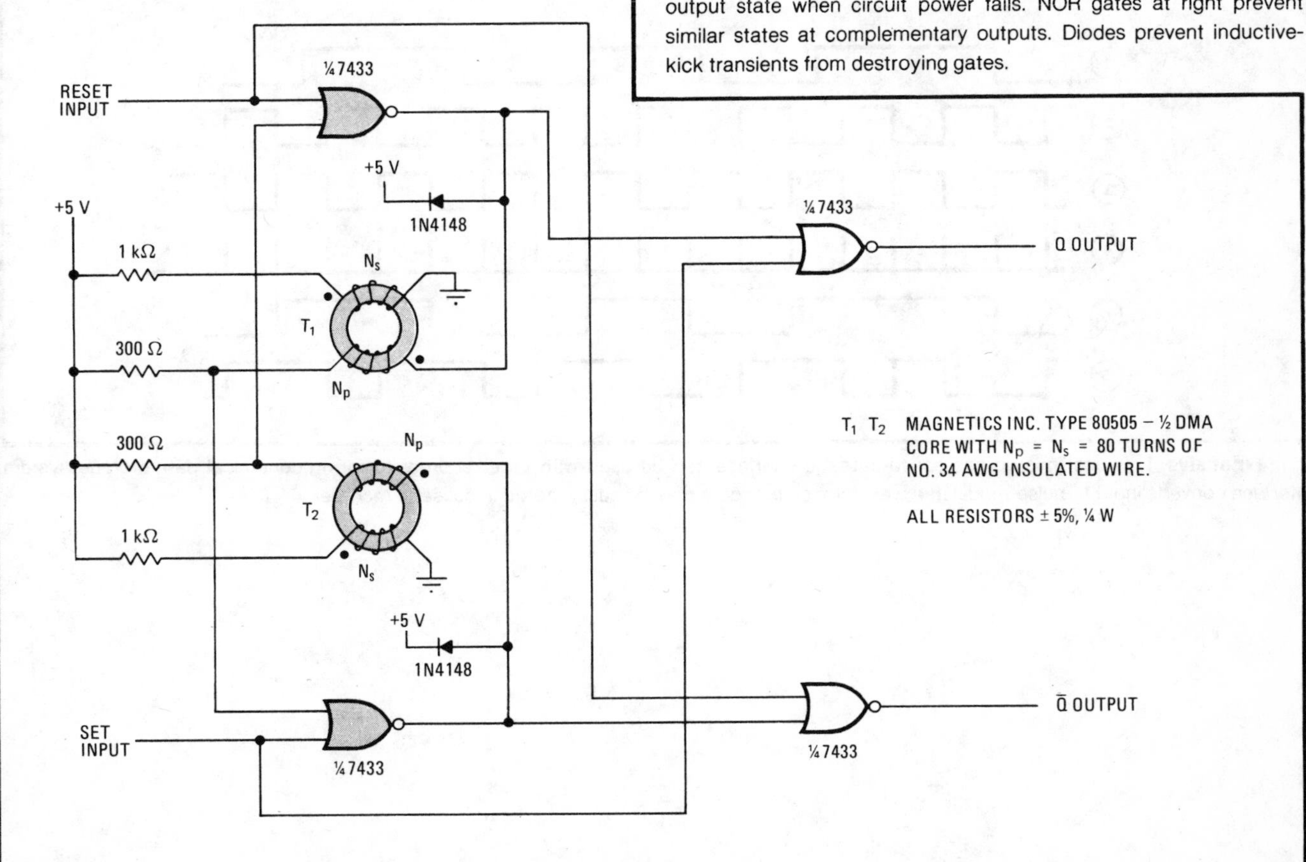

Nonvolatile memory. Saturable transformers remember previous output state when circuit power fails. NOR gates at right prevent similar states at complementary outputs. Diodes prevent inductive-kick transients from destroying gates.

Processor-to-cassette interface helps slash data-storage cost

by Pawel Mikulski
Finlux Television, Lohja, Finland

It doesn't pay to buy an expensive mass-storage device to store data handled by an inexpensive microprocessor-based data system. Storing data on a cassette tape recorder or reading data from one is a viable alternative, however, and these low-cost interfaces will provide an economical solution to the data-storage and -retrieval problem for read/write speeds of up to 4,000 bauds.

Microprocessor data is phase-modulated by the trans-

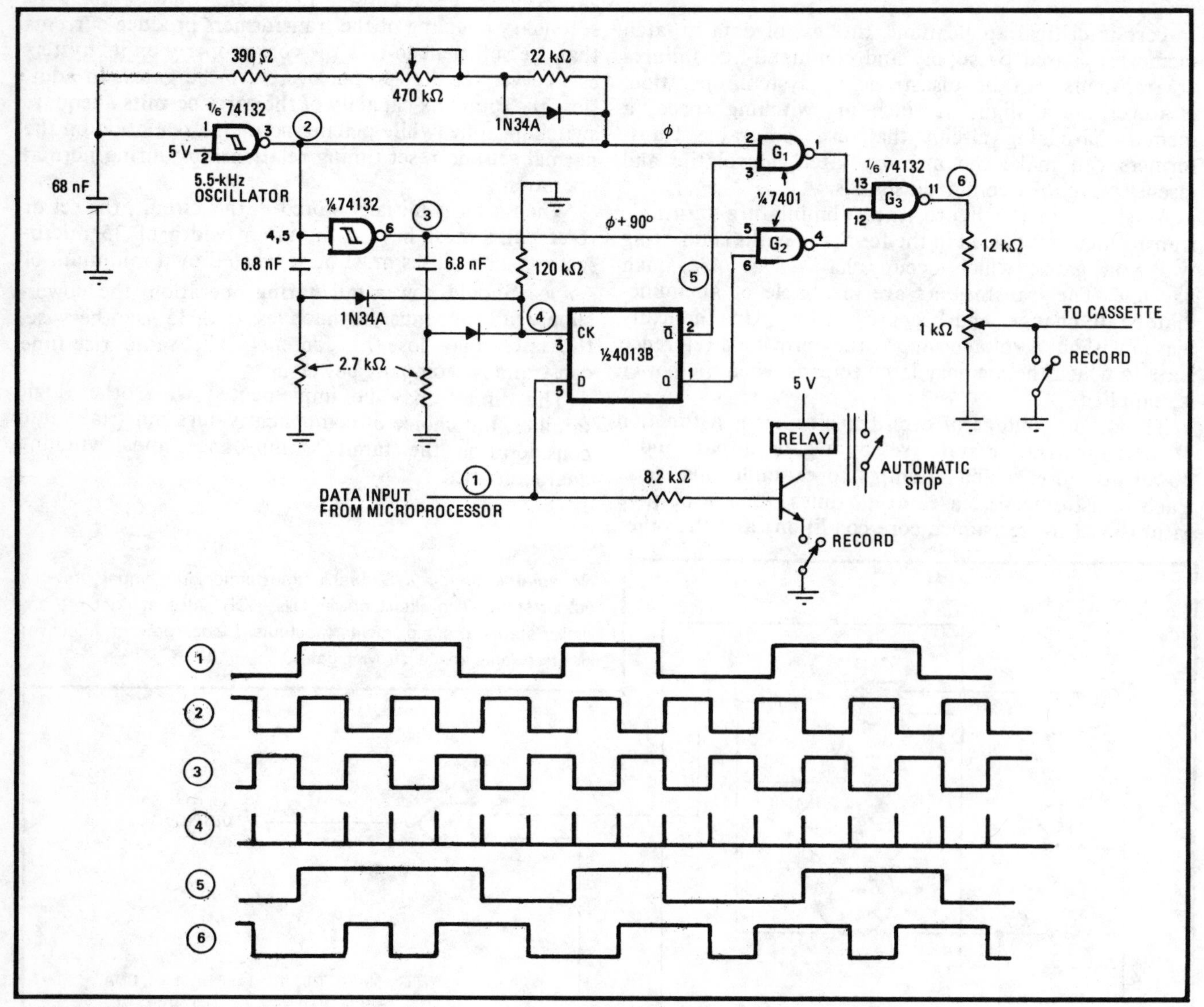

1. Inexpensive. Low-cost microprocessor-to-cassette interface can be used with tape recorder to form economical data-storage system. Interface converts input to pulse-modulated waveform so that data may be easily stored in cassette recorder.

mitter interface as shown in Fig. 1, in order that it may be stored in the recorder in a form that may be easily retrieved. The input signal drives G_1 and G_2 and, depending on the logic value (0 or 1), will determine if either phase ϕ or $\phi + 90°$ (both generated by a 5.5-kilohertz oscillator) appears at the output of G_3. This signal is then stored in the cassette. The transmitter timing diagram clarifies circuit operation.

Data played back to the receiver (Fig. 2) is applied first to A_1 and then to a Schmitt trigger/comparator (74132). A_2, a retriggerable one-shot, is fired on every rising and falling edge of the input signal and thus will stay high if the input signal pulses are separated by less than 130 microseconds. A_2 drives A_3, a D-type flip-flop wired as a T device, so that the output will be a replica of the data signal originally recorded. A_4 is a time-out one-shot, which moves high ($\overline{Q}$) if data input should cease for more than 300 milliseconds. Circuit operation can be clearly visualized with the aid of the receiver's timing diagram. □

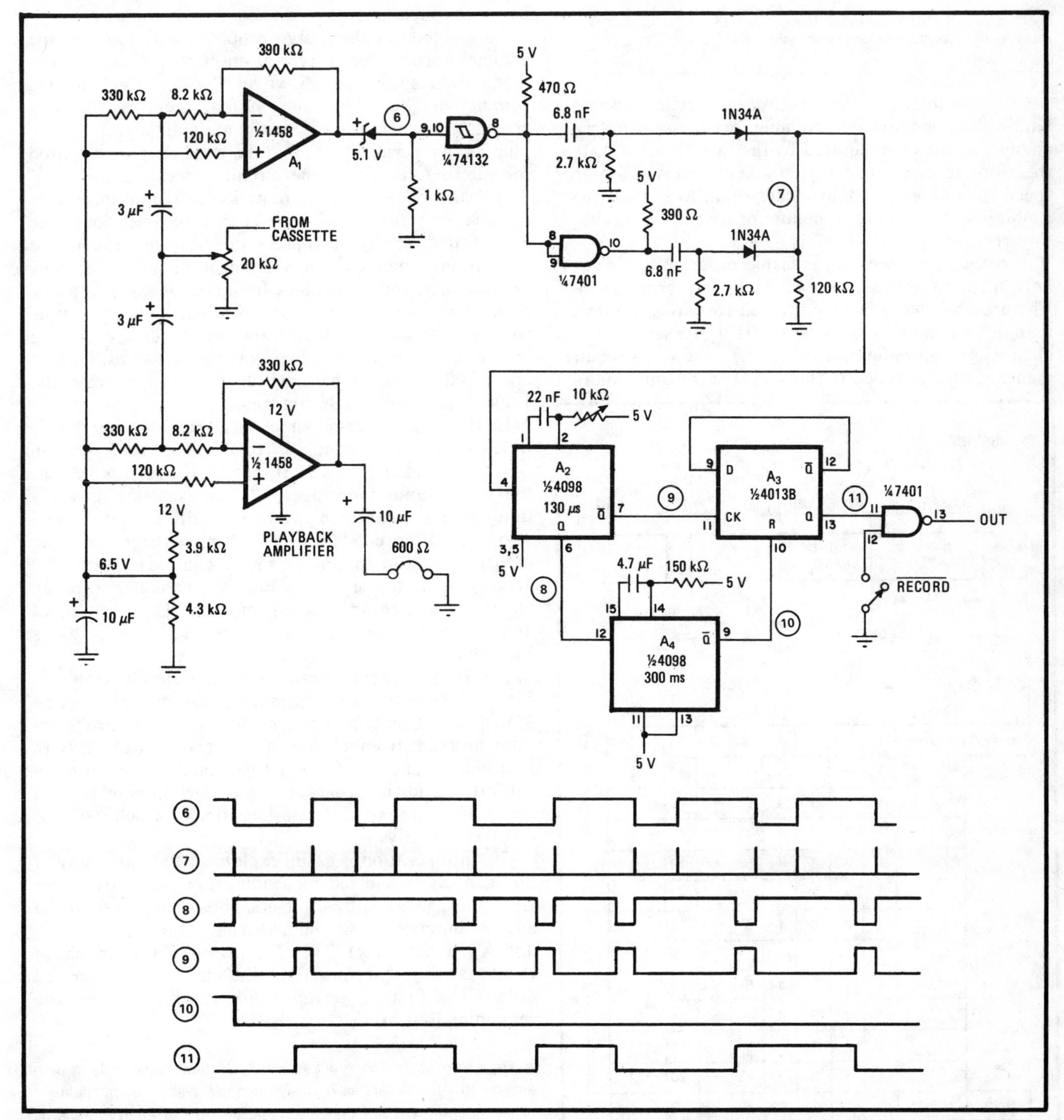

2. Retrieved. Data is recovered after passing through receiver interface using process essentially inverse to one used at transmitter. One-shot A_2 is used to convert input signal to two complementary 130-µs waveshapes. One waveform drives a second one-shot, A_4, which in turn resets A_3, while the other waveform drives its clock input. Timing diagram details operation. Output is a delayed version of the original signal.

EPROM achieves nonuniform data-channel sampling

by B. Bowles and T. U. Nelson
Chamber of Mines of South Africa, Johannesburg

In some multiplexed data-acquisition systems, adequate bandwidth is not always available for transmitting every channel at the rate required by the fastest channel. It is therefore necessary to sample some channels at a higher speed than others, each at a rate equal to at least twice its highest frequency. A nonuniform sampler circuit is thus required.

A circuit built with an erasable programmable read-only memory forms an effective nonuniform sampler. The erasable PROM is programmed so that a sequential scan of its memory locations results in the selection of a channel sequence determined by the bandwidth requirements of each channel. In this way, the transmission rate of any channel is made proportional to its bandwidth. Using the erasable PROM and the associated circuitry to generate the sequence is more practical than using the usual array of analog switches or the well-known shift-register–diode-matrix circuit, especially when the number of channels is large or the sequence is complex.

To understand the design problem, first consider the channel format for a typical multiplexed pulse-code-modulation system as shown in Fig. 1a. Each analog channel of this 20-channel system requires 10 clock periods—8 clock periods for quantizing a data word, plus 2 clock periods preceding the word that are required for internal timing in the circuit processing the data. Any 7 of the 20 channels are periodically multiplexed as a block, or subframe. Twelve such subframes form one main frame, as shown. Since each subframe, except the first, is preceded by a synchronization pulse 16 clock periods long, and if the clock frequency is assumed to be 50 kilohertz (period of 20 microseconds), the sync time will be 320 μs. The mainframe sync pulse occurring before subframe 1 is 32 clock periods, or 640 μs, long. It takes 200 μs to sample each channel or, equivalently, 1,400 μs to sample each subframe.

In this case, each channel has the following bandwidth: channels 1 through 4, 163 Hz; channels 5 through 14, 47 Hz; and channels 15 through 20, 16 Hz. Note that the total frame time, about 21 milliseconds, is three times greater than the highest bandwidth of any channel sampled, to prevent fold-over distortion (from sampling theory, 2 is the minimum ratio required). With this arrangement, the sampling rate of channel 1 through channel 4 will be 4 times that of channels 5 through 14 and will be 13 times that of channels 15 through 20, as shown in Fig. 1b.

The format can be realized by the circuit shown in Fig. 2 if the PROM is programmed, beginning at location 3, with the channel numbers shown in Fig. 1b. Programming proceeds from left to right, subframes 1 to 12. Both locations 1 and 2 of the PROM, each representing a subframe, and the programming of certain nondata bits are subject to special considerations, which will be discussed shortly.

Circuit operation is easily explained. Basically, any of 20 channels is selected by sending the channel number from the 256-word-by-8-bit erasable PROM (A_1) to the address inputs of an 8- and a 16-channel multiplexer (A_2 and A_3, respectively). The contents of each channel are then passed to the output of the circuit in prescribed order. The first 16 channels are handled by A_3, the remaining four by A_2.

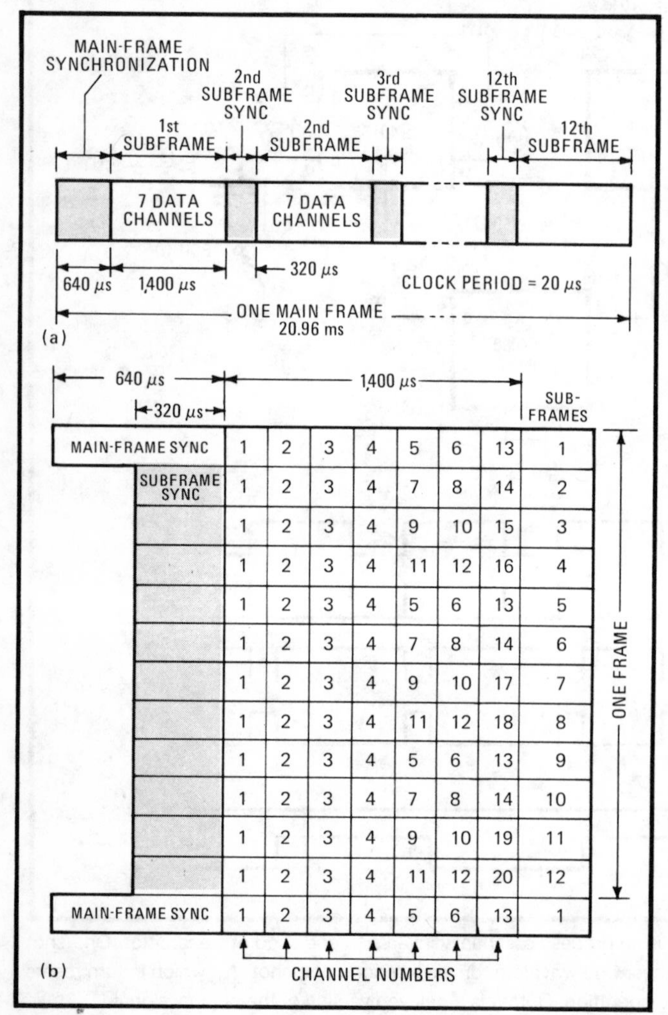

								SUB-FRAMES
MAIN-FRAME SYNC	1	2	3	4	5	6	13	1
SUBFRAME SYNC	1	2	3	4	7	8	14	2
	1	2	3	4	9	10	15	3
	1	2	3	4	11	12	16	4
	1	2	3	4	5	6	13	5
	1	2	3	4	7	8	14	6
	1	2	3	4	9	10	17	7
	1	2	3	4	11	12	18	8
	1	2	3	4	5	6	13	9
	1	2	3	4	7	8	14	10
	1	2	3	4	9	10	19	11
	1	2	3	4	11	12	20	12
MAIN-FRAME SYNC	1	2	3	4	5	6	13	1

1. Data format. Typical requirements for a multiplexed PCM system are shown. Any 7 channels of this 20-channel system are multiplexed in 12 subframes, such that the sampling rate of each channel is proportional to its bandwidth (a). Transmission format for this system (b) can be found once the channel bandwidths are known (see text). Data sampler can generate this or any other format desired.

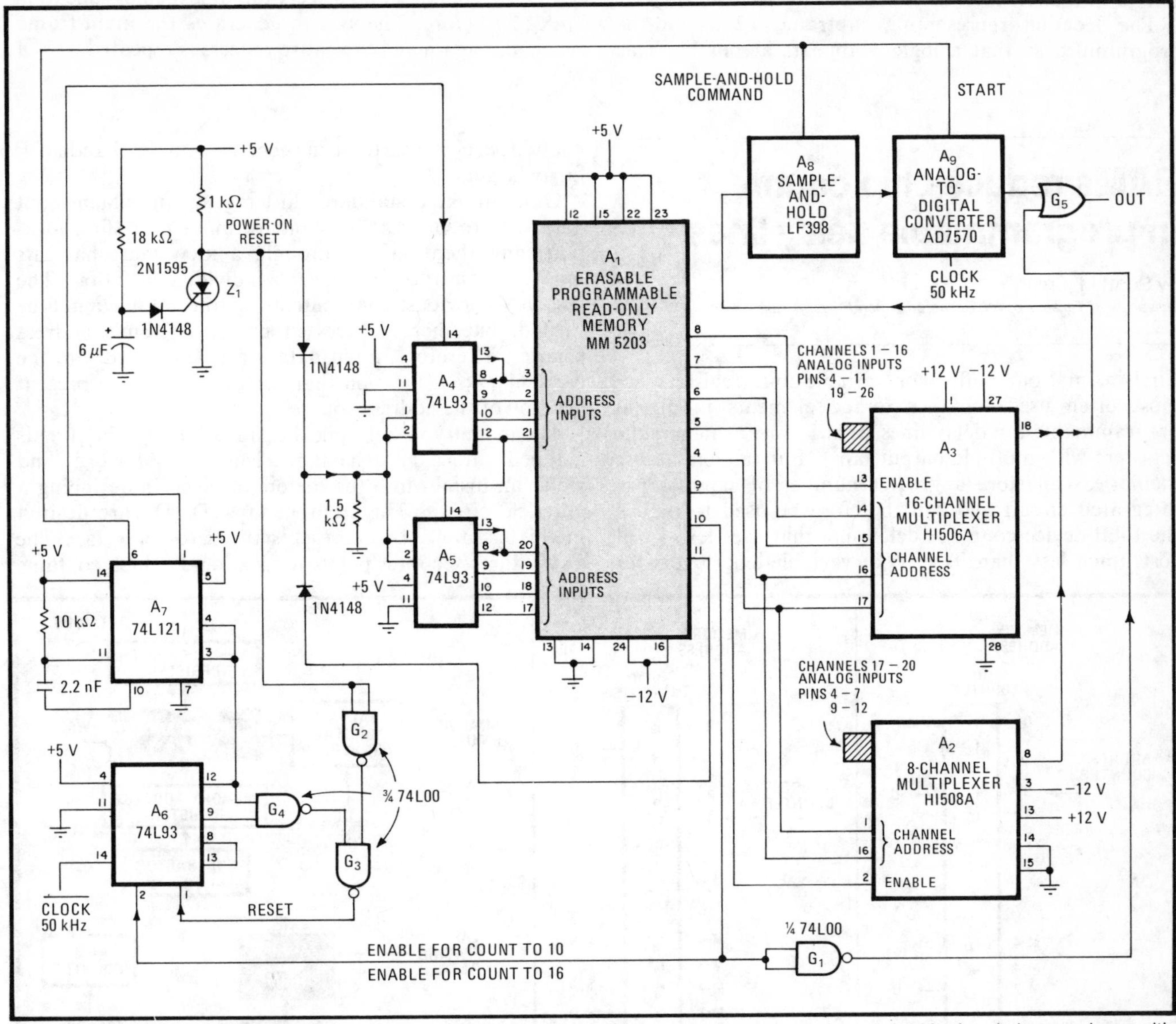

2. Data sampler. Circuit ensures that the rate at which all channels are sampled is proportional to the bandwidth of each, in accordance with the requirements of sampling theory. The erasable PROM can be programmed so that any desired channel sequence can be transmitted via the 16-channel multiplexer; the transmission rate for each channel will in this way be easily controlled.

On power up, thyristor Z_1 resets counters A_4 through A_6, thereby selecting the first PROM location. A_4 and A_5 will be placed in the standby mode and A_6 will be placed in the count-to-16 mode, because the PROM is programmed to generate a logic 0 at pin 10 of A_1. When Z_1's anode voltage drops to zero, A_6 begins to count, and this initiates the master sync cycle.

After 16 counts, the one-shot (A_7) fires and increments A_4, and therefore the second memory location of the PROM is selected. The second location is programmed so that its contents are identical to that of location 1, and consequently, 16 counts later, the third memory location is selected. This location contains the address of the first multiplexer channel.

Pin 10 of A_1 has moved high, thereby enabling G_5, and setting A_6 into its count-to-10 mode. Note that the multiplexers are addressed by 6 bits of A_1, not 8; the remaining 2 bits are required for the system sync-control circuit comprising G_1, G_5, and pin 10 of A_6.

Meanwhile the one-shot initiates the sample-and-hold command and resets A_9. The contents of channel 1 then appear at the input of A_9. At the second positive clock edge after the cessation of the pulse emanating from A_7, the most significant bit of data appears at the output of A_9 and thus at the circuit output. A_3 is now in the count-to-10 mode (caused by pin 11 being high). A_4 and A_5 are again incremented by A_7 after the monostable is triggered by pulse 10 of A_1. A_5 and A_6 select the next PROM address, which is 2 in this case. This process is repeated until all seven channels have been selected.

At the next memory location, the PROM must be programmed so that a 0 once again emanates from pin 10, to set A_6 into the count-to-16 mode once more and to generate a logic 1 at the system output by means of G_1 and G_5. Triggering A in this way sets the stage for the generation of a subframe sync pulse and the selection of the first channel in the second subframe after 16 pulses have been counted by A_1. The selection process continues until all the subframes have been scanned.

The location representing subframe 12 should be programmed so that a logic 1 appears at pin 11. This resets the counters and selects the first memory location in A_1, as before. The system generates the main-frame sync pulse again and the entire process is repeated. □

RAMs reduce chip count in programmable delay lines

by Scott M. Smith
University of Texas, Applied Research Laboratories, Austin, Texas

First-in, first-out buffers or variable-shift registers are most often used for the storage elements in digital programmable-tap delay lines (that is, one or more shift registers with multiple-output taps). But random-access memories can store a greater number of samples per integrated circuit and can therefore be used to reduce the total device count. A delay line that uses RAMs will cost much less than its FIFO or variable-length register counterparts if the total number of samples handled is fairly large.

Quite unlike a standard shift register, in which input data is introduced at its standard-input port (first location) and then shifted through, a RAM must have its input data introduced at each individual location. The reason is obvious: the contents of the RAM cannot be shifted, but merely accessed by the system's address counter. Therefore, input data must be entered into the particular RAM location that corresponds to the present location of the address counter.

The memory map in Fig. 1a shows how a delay line is mapped onto a RAM having a length of M words and yields an insight into the factors involved in designing a practical circuit. Three output taps, D_0–D_2, are desired in this example. D_0 represents the zero-delay tap. The RAM address counter points to location 3, which contains

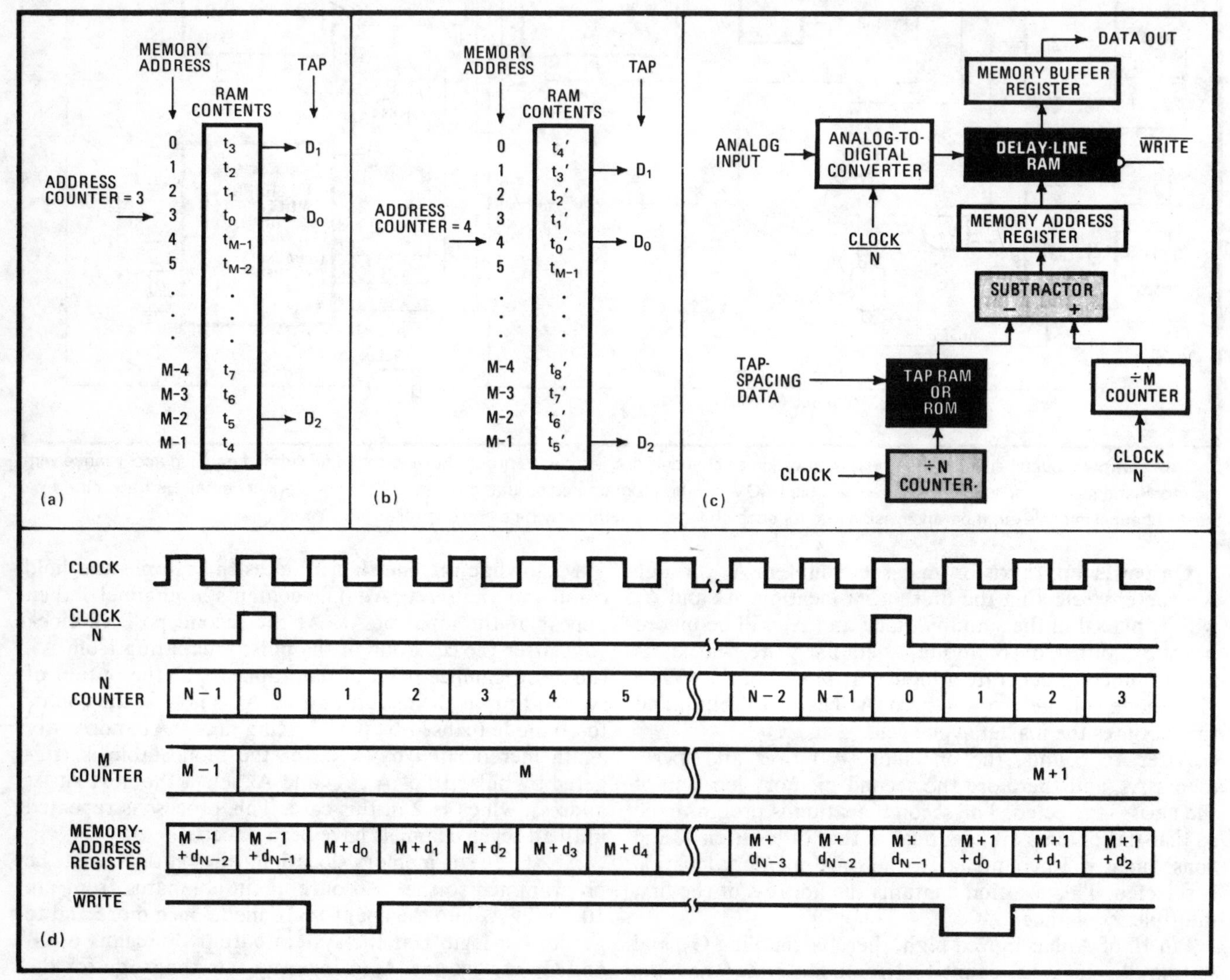

1. Super-long tapped delay. Memory map shows how an N-tap delay line is mapped onto an M-word RAM (a). Input data may be introduced into RAM by incrementing counter and placing sample there. Oldest data sample is destroyed and existing samples are redefined (b). Block diagram of system outlines procedure used to write data, examine output taps (c). Waveform diagram details timing constraints (d).

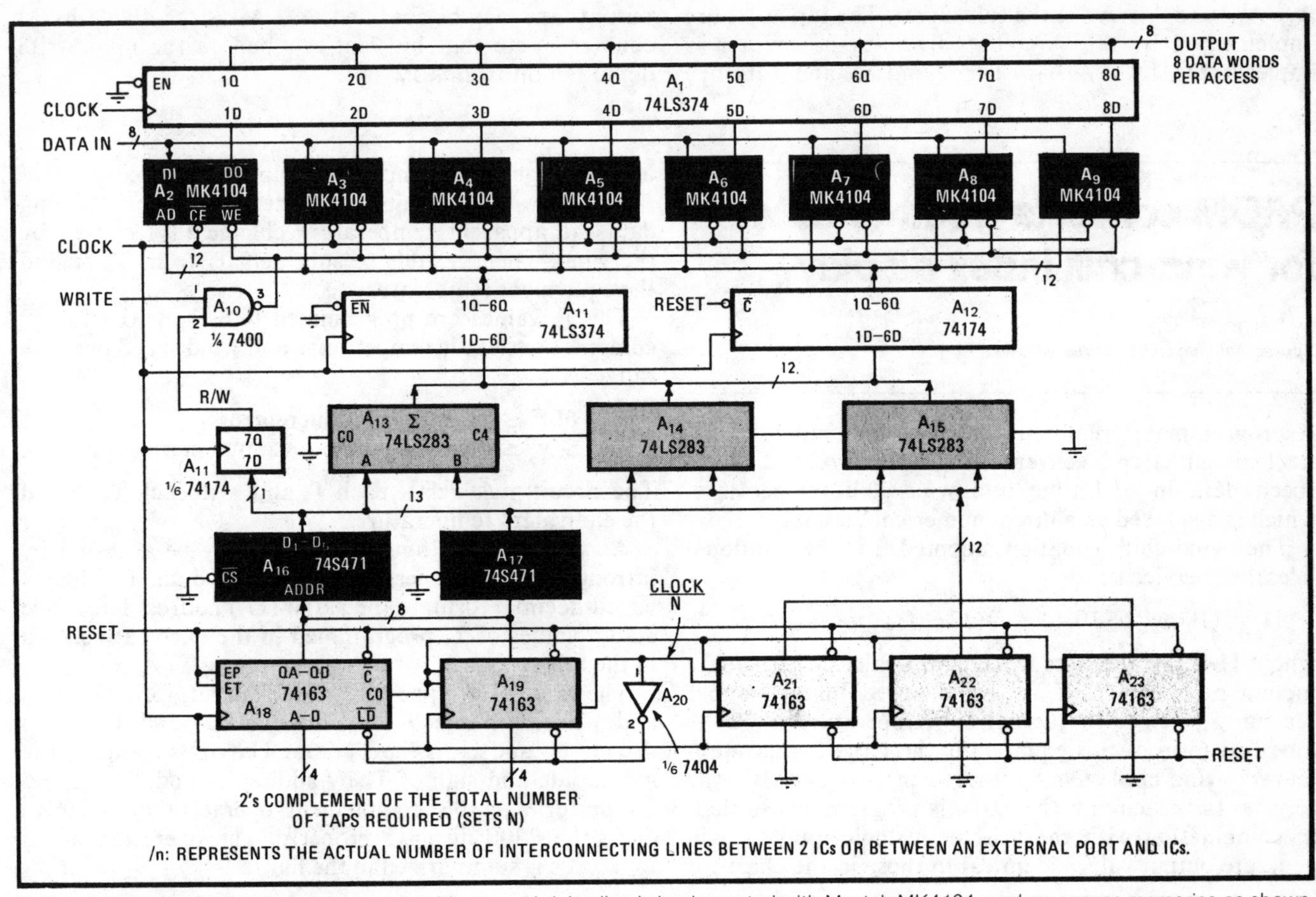

2. Great capacity. Eight-bit-wide programmable-tapped delay line is implemented with Mostek MK4104 random-access memories as shown. Each delay line is the equivalent of a 4,096-bit shift register. User may specify a total of 255 output taps with any desired spacing.

t_0, the most recent sample in the delay line. The next most recent sample is t_1, with t_{M-1} being the oldest sample. D_1 and D_2 are taps delayed three samples and five samples, respectively, with regard to D_0.

The value corresponding to memory address 3 (t_0= logic 1 or logic 0) would appear at D_0 if that tap were requested. Similarly, the sample at memory address 0 would be fetched if D_1 were to be requested, and the sample at address $M-2$ would be fetched if D_2 were requested.

Figure 1b shows how a new sample would be inserted into the delay line. The counter would be incremented, pointing to location 4 as shown, and the new sample would be written into the location, thus shifting the oldest data sample (t_{M-1}) out of RAM. The memory contents of RAM would otherwise be unchanged; each memory address would be simply redefined as being one sample older. If D_1 were queried, the sample at memory address 1 would be fetched; when D_2 were requested, the sample at location $(M-1)$ would be fetched.

The block diagram shown in Fig. 1c more clearly explains how data is written, and taps are specified and read. A divide-by M counter driven by a clock running at 1/N times the system clock frequency is required for pointing to the most recent sample (D_0). Also required is a tap RAM or ROM, which is programmed so that its output is equal to the distance in time (that is, the number of samples) each user-specified output tap is from the most recent sample, t_0. Thus, the spacing between taps is specified. A divide-by-N counter (where N is the number of taps) is needed to address each tap in a sequence that is selected by the user. Note that the N counter must move through one complete cycle for each increment in the M count. The subtractor determines the numerical difference between the tap distance and the zero-delay location and stores the result in the memory address register in order to access the memory address desired. The delay-line RAM is then accessed to obtain the data sample corresponding to the tap selected, or to write in a new data sample. Then the sample that has been read is stored in the memory buffer register, to be shifted out in serial form.

The timing considerations for the circuit are shown in Fig. 1d. As may be observed, provision should be made to ensure that the M counter advances before any new data (write) is stored in RAM, if necessary, to allow the oldest sample to be read before it is overwritten. There are no other major considerations. The taps may be accessed in any order and are selected by appropriate programming of the tap delay memory (tap ROM). The maximum shift rate (the frequency with which new samples are placed in memory) is $f_{s\ max} = 1/Nt_{c\ min}$, where $t_{c\ min}$ is the minimum cycle time of the system.

Figure 2 shows a design example that uses an 8-bit-wide programmable tapped-delay line. The RAM memories, each holding 4,096 1-bit words, form a 4,096-word-by-8-bit array. A_1 is the memory buffer register, A_2–A_9 is the RAM delay line, the memory address register is

A_{11}–A_{12}, and A_{13}–A_{15} is the subtractor. The tap ROMs are implemented by A_{16}–A_{17}. The divide-by-N counter is implemented by A_{18}–A_{20}. Two hundred and fifty-five output taps may be specified. A_{21}–A_{23} is the divide-by-M counter. Note that bit 7 of A_{11} buffers the read/write definition bit from ROM. □

PROM converts weather data for wind-chill index display

by Vernon R. Clark
Applied Automation Inc., Bartlesville, Okla.

A programmable read-only memory and four arithmetic/logic units can convert air-temperature and wind-speed data in real time into wind-chill temperature, which is displayed on a direct numerical readout.

The wind-chill equation adopted by the National Weather Service is:

$$H = (100w^{1/2} + 10.45 - w)(33 - T_a)$$

where H is the heat loss in kilogram-calories per square meters per hour, w is the wind speed in meters per second, and T_a is the actual air temperature in °C. A modified form of this equation is the basis for the well-known wind-chill temperature chart issued by the service. In this circuit, the PROM is programmed so that, in combination with the arithmetic/logic units, it will generate output values identical to those in the chart for a wide range of air temperatures and wind speeds.

Basically, the circuit determines from the incoming data the apparent temperature change (T_c) caused by the wind. Then, it adds or subtracts T_c from T_a to find the equivalent temperature (T_e).

The T_c values are programmed into the PROM for all combinations of air temperature and wind speed over the range:

$$-60\text{F} \le T_a \le 50\text{F (10 increments)}$$
$$0 \le w \le 46 \text{ miles per hour (2 mph increments).}$$

The circuit must relate each T_a and w to each T_c to find the equivalent temperature.

As shown in the figure, each T_c may be accessed by introducing air-temperature and wind data, in binary-coded-decimal form, to the PROM (U_1) address lines. The actual values of T_c programmed in the PROM are shown in the table.

The value of T_c appearing at the output, for a given T_a and w, is introduced to two ALUs, U_2 and U_3. Also driving U_2 and U_3 is the T_a data. The ALUs compute the magnitude and sign of T_e by adding T_a and T_c. U_4 and U_5 perform a 10's complement operation in order to drive the 7300 displays properly. The operation of all four ALUs is summarized in the figure.

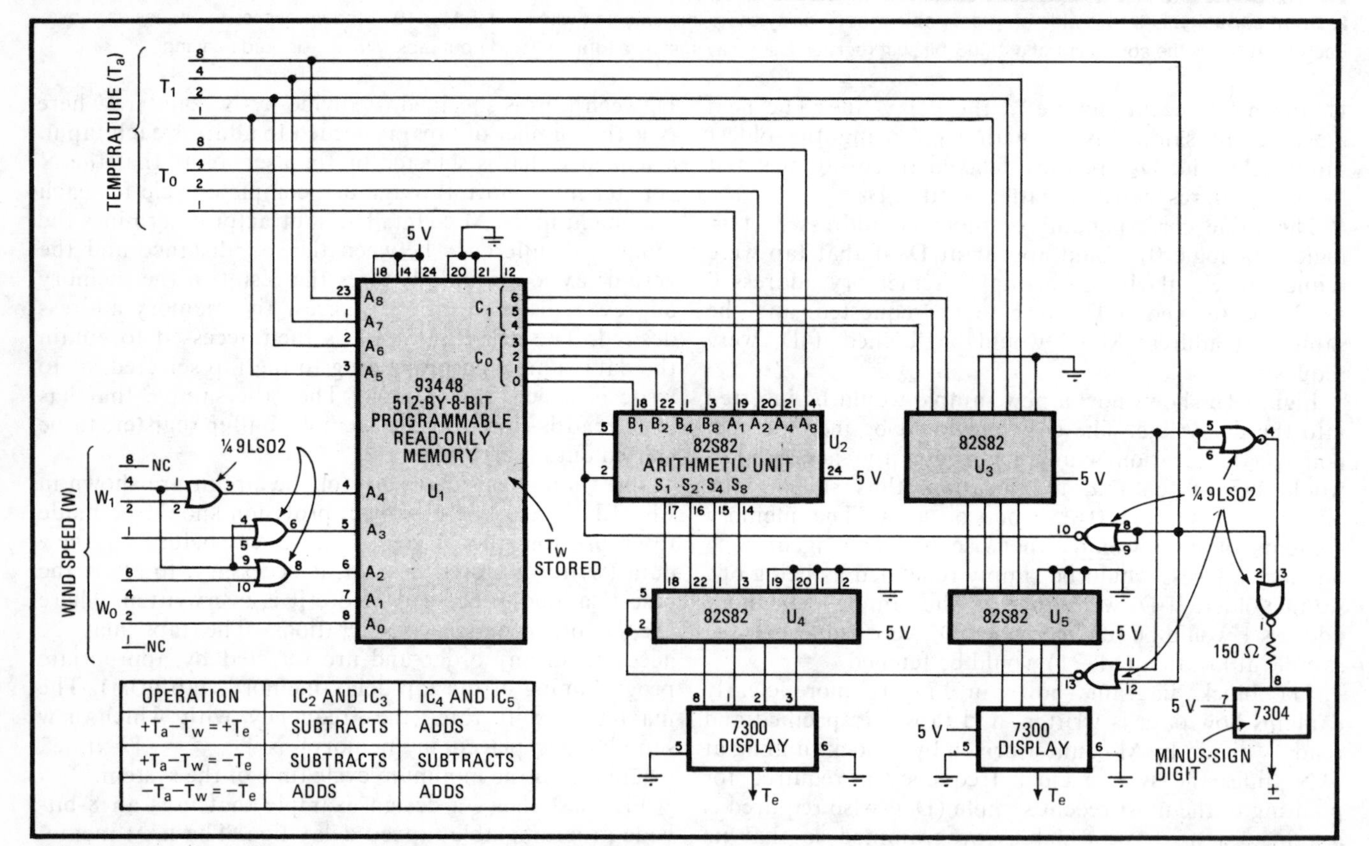

OPERATION	IC_2 AND IC_3	IC_4 AND IC_5
$+T_a - T_w = +T_e$	SUBTRACTS	ADDS
$+T_a - T_w = -T_e$	SUBTRACTS	SUBTRACTS
$-T_a - T_w = -T_e$	ADDS	ADDS

Cold solution. Circuit determines and displays wind-chill temperature (T_e). Air temperature (T_a) and wind-speed data (w) address PROM lines to access apparent temperature change (T_c) brought about by given w at T_a. Arithmetic/logic units U_2 and U_3 add T_a and T_c to find T_e. U_4 and U_5 perform a 10's complement operation for the digital display units, for which they serve as an interface.

PROM CONTENTS – WIND-CHILL INDICATOR

LOC		LOC		LOC		LOC		LOC	
0000R	0002	0070R	2628	00E0R	0000	0150R	4750	01C0R	0004
0002R	0409	0072R	3030	00E2R	0000	0152R	5355	01C2R	0714
0004R	1500	0074R	3100	00E4R	0000	0154R	5700	01C4R	2500
0006R	0000	0076R	0000	00E6R	0000	0156R	0000	01C6R	0000
0008R	2125	0078R	3233	00E8R	0000	0158R	5960	01C8R	3543
000AR	2933	007AR	3434	00EAR	0000	015AR	6162	01CAR	5055
000CR	3700	007CR	3637	00ECR	0000	015CR	6365	01CCR	6000
000ER	0000	007ER	3738	00EER	0000	015ER	6667	01CER	0000
0010R	3941	0080R	0001	00F0R	0000	0160R	0003	01D0R	6468
0012R	4344	0082R	0205	00F2R	0000	0162R	0511	01D2R	7275
0014R	4600	0084R	0900	00F4R	0000	0164R	2000	01D4R	7800
0016R	0000	0086R	0000	00F6R	0000	0166R	0000	01D6R	0000
0018R	4849	0088R	1215	00F8R	0000	0168R	2834	01D8R	8082
001AR	5051	008AR	1719	00FAR	0000	016AR	4044	01DAR	8485
001CR	5354	008CR	2100	00FCR	0000	016CR	4800	01DCR	8890
001ER	5556	008ER	0000	00FER	0000	016ER	0000	01DER	9294
0020R	0002	0090R	2223	0100R	0002	0170R	5255	01E0R	0000
0022R	0407	0092R	2425	0102R	0409	0172R	5759	01E2R	0000
0024R	1300	0094R	2600	0104R	1500	0174R	6200	01E4R	0000
0026R	0000	0096R	0000	0106R	0000	0176R	0000	01E6R	0000
0028R	1923	0098R	2728	0108R	2125	0178R	6466	01E8R	0000
002AR	2730	009AR	2929	010AR	2933	017AR	6768	01EAR	0000
002CR	3300	009CR	3031	010CR	3700	017CR	7071	01ECR	0000
002ER	0000	009ER	3132	010ER	0000	017ER	7273	01EER	0000
0030R	3537	00A0R	0000	0110R	3941	0180R	0004	01F0R	0000
0032R	3840	00A2R	0203	0112R	4344	0182R	0612	01F2R	0000
0034R	4200	00A4R	0700	0114R	4600	0184R	2200	01F4R	0000
0036R	0000	00A6R	0000	0116R	0000	0186R	0000	01F6R	0000
0038R	4343	00A8R	1012	0118R	4849	0188R	3036	01F8R	0000
003AR	4445	00AAR	1415	011AR	5051	018AR	4247	01FAR	0000
003CR	4749	00ACR	1700	011CR	5354	018CR	5200	01FCR	0000
003ER	5051	00AER	0000	011ER	5556	018ER	0000	01FER	0000
0040R	0002	00B0R	1818	0120R	0002	0190R	5660	0200R	
0042R	0407	00B2R	1920	0122R	0409	0192R	6365		
0044R	1200	00B4R	2100	0124R	1700	0194R	6700		
0046R	0000	00B6R	0000	0126R	0000	0196R	0000		
0048R	1620	00B8R	2222	0128R	2329	0198R	6971		
004AR	2426	00BAR	2323	012AR	3337	019AR	7374		
004CR	2800	00BCR	2425	012CR	4100	019CR	7677		
004ER	0000	00BER	2526	012ER	0000	019ER	7879		
0050R	3032	00C0R	0000	0130R	4346	01A0R	0004		
0052R	3436	00C2R	0000	0132R	4850	01A2R	0613		
0054R	3700	00C4R	0000	0134R	5200	01A4R	2300		
0056R	0000	00C6R	0000	0136R	0000	01A6R	0000		
0058R	3839	00C8R	0000	0138R	5354	01A8R	3340		
005AR	4040	00CAR	0000	013AR	5657	01AAR	4652		
005CR	4142	00CCR	0000	013CR	5960	01ACR	5600		
005ER	4243	00CER	0000	013ER	6162	01AER	0000		
0060R	0002	00D0R	0000	0140R	0002	01B0R	6064		
0062R	0306	00D2R	0000	0142R	0510	01B2R	6770		
0064R	1000	00D4R	0000	0144R	1800	01B4R	7300		
0066R	0000	00D6R	0000	0146R	0000	01B6R	0000		
0068R	1418	00D8R	0000	0148R	2632	01B8R	7577		
006AR	2022	00DAR	0000	014AR	3640	01BAR	7980		
006CR	2400	00DCR	0000	014CR	4400	01BCR	8283		
006ER	0000	00DER	0000	014ER	0000	01BER	8485		

Wind speed frequently varies over a wide range in a short time. This may cause rapid flickering of the display and make it hard to determine the average wind-chill temperature. One answer to this problem is to sample the input data periodically. Another is to use average-value sensor circuits for smoothing the data. □

Chapter 23
MICROPROCESSORS

Shift registers act as control interface for microprocessor

by Felix J. Sawicki
University of Texas, Austin, Texas

Parallel-in, serial-out shift registers such as the 74165 provide a simple, low-cost solution to the problem of interfacing a variety of control elements including push button switches, sense switches, and other single-bit monitor points to a microprocessor. The interface is easy to expand and uses a minimum of hardware, since microprocessor software performs most of the decision-making functions. This technique also minimizes the bit manipulations necessary to determine the state of the monitor points.

The figure shows a shift-register interface connecting a microprocessor to a keyboard, sense switches, and remote sensors. The interface is assigned an address in memory so that it can be easily accessed for read and write operations. Since the microprocessor already uses lines A_0 - A_9, the interface is assigned line A_{10} (hex address 200), which is logically ANDed with the valid-address line for enabling.

The read/write line ($\overline{R}$/W) line of the processor is connected to the mode-control line (SHIFT/LOAD) of each of the shift registers, which are strung together by connecting their shift-in and shift-out lines, as shown. This technique transforms the many parallel monitor points into a more manageable serial sequence of bits.

When a write operation is performed by the microprocessor at its interface address (A_{10}), the information is transferred from the switches and keys to the shift registers in parallel format. A series of read operations are then started at the interface address.

In a read operation, the $\overline{R}$/W line from the microprocessor goes low, causing the shift registers to work in the

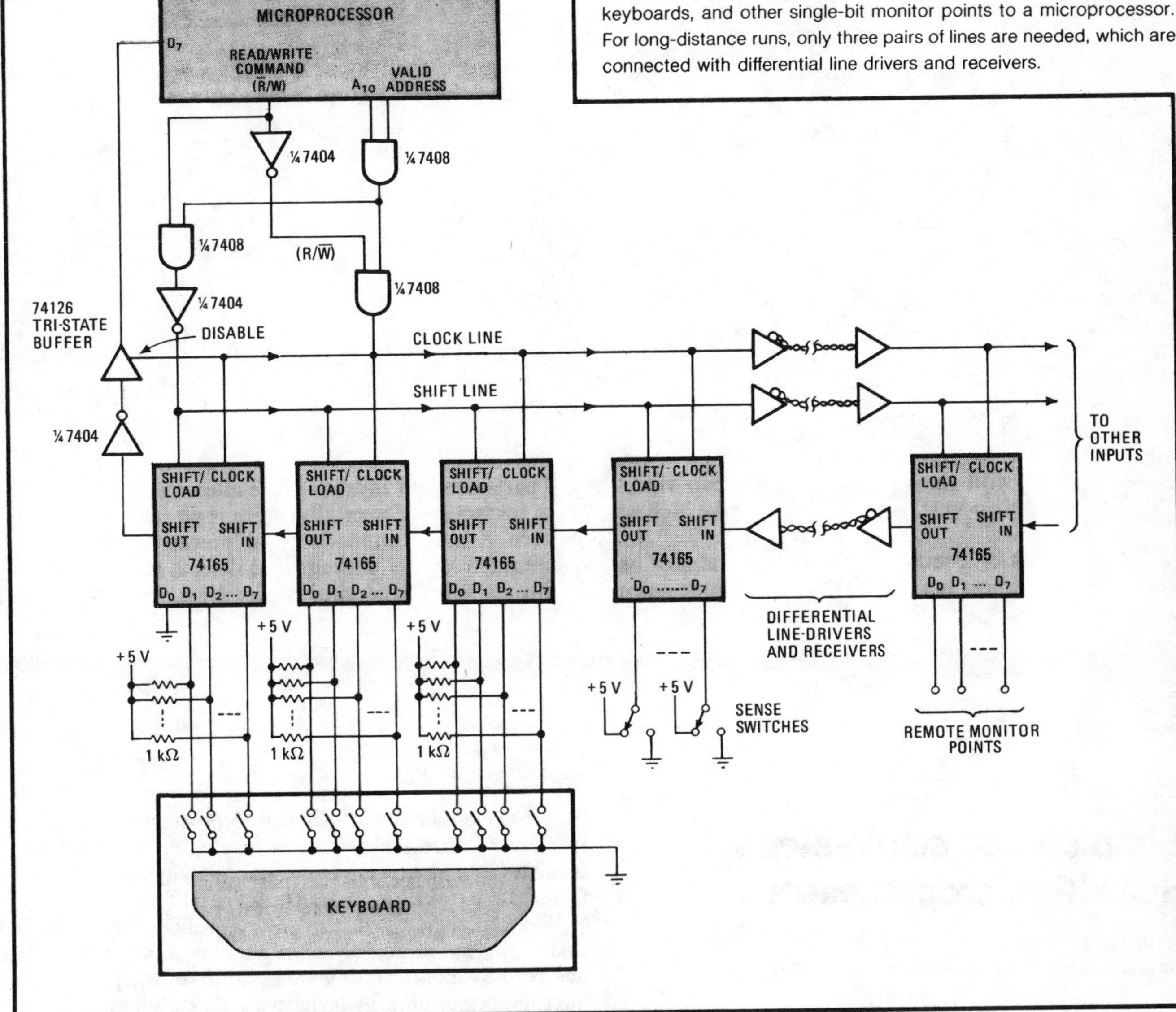

Microprocessor interface. 8-bit parallel-in, serial-out shift registers, such as the 74165, provide an easy method of interfacing switches, keyboards, and other single-bit monitor points to a microprocessor. For long-distance runs, only three pairs of lines are needed, which are connected with differential line drivers and receivers.

serial mode. The read operation shifts the contents of all the 74165s one location, allowing the microprocessor to examine a new bit. The first parallel input of the first shift register is not wired to a switch since the first shift operation occurs before the microprocessor samples the input bit. The input bit is connected to the most-significant-bit input of the microprocessor, D_7.

Since the bit being examined is the MSB of the word being loaded, the read operation sets the condition code of the microprocessor to indicate whether the input word is negative (MSB = 1) or nonnegative (MSB = 0). A branch on the condition code will then allow immediate action to be taken by the processor, based on the state of the switch just tested. Thus the input routine of the microprocessor can determine most efficiently whether the bit that is being read is a 1 or a 0.

It is very simple to expand this circuit to examine more switches by adding additional shift registers. If a group of switches is located at a distant point, only three pairs of differential line drivers and receivers (which provide noise immunity) are required. Two pairs are used for the read/write and clock lines controlling the sampling and shifting, and the other carries the incoming data.

The software needed for this interface will vary according to the type of devices being monitored and the application for which they are used. For the application in the figure, the main functions are: keeping track of which point is being examined; switch debouncing (for momentary contacts); and differentiating between newly and previously closed switches (for keyboards).

The problem of key bouncing can be circumvented by waiting a long enough time for the keys to settle before rescanning. This is usually about 20 to 50 milliseconds. If the initial scan of the keys is done while a key is bouncing, it doesn't matter if the key is being read or not, as it will have settled by the next scan.

A microprocessor register or memory location should be used as a counter to keep track of which switch or monitor point is being examined. □

Simple logic single-steps SC/MP microprocessor

by Richard Gersthofer
Industrial Automation and Research, Vancouver, B. C.

Single-cycle and single-instruction processing capability in a microprocessor system is invaluable for hardware troubleshooting as well as program debugging. An easy-to-build controller provides this capability.

When the mode switch of the controller is set to single cycle, all the stages of instruction processing may be observed by monitoring the logic states of the pins on the microprocessor chip. Alternatively, in the single-instruction mode, only the effects of each instruction may be observed.

Designed for use with the SC/MP central processing unit, the controller requires no initialization. As shown in the figure, the circuit has HALT and CONTINUE push buttons, in addition to switching between single-cycle, single-instruction, and continuous modes. Program execution may be halted with the push button or under software control by pulsing the halt flag line.

In the single-cycle mode, the continue input line, CONT, of the CPU is held at logic 1. This enables normal execution of the program stored in external memory. The input/output cycle extend input, NHOLD, is kept at logic 0. Pressing the CONTINUE button toggles the first flip-flop, FF_1, driving the NHOLD high, and permitting program execution. After executing one cycle, the CPU resets FF_1 by pulsing the address strobe output line, NADS, sending the NHOLD low again. The cycle time requires only a few microseconds.

In the single-instruction mode, the NHOLD line is held at logic 1 while the CONT line is kept at logic 0. Pressing the continue button toggles the second flip-flop, FF_2, driving the CONT line high, and the CPU continues execution. After completion of one instruction, the processor resets FF_2 by pulsing its NADS line. CONT goes low until the CONTINUE button is again pressed. □

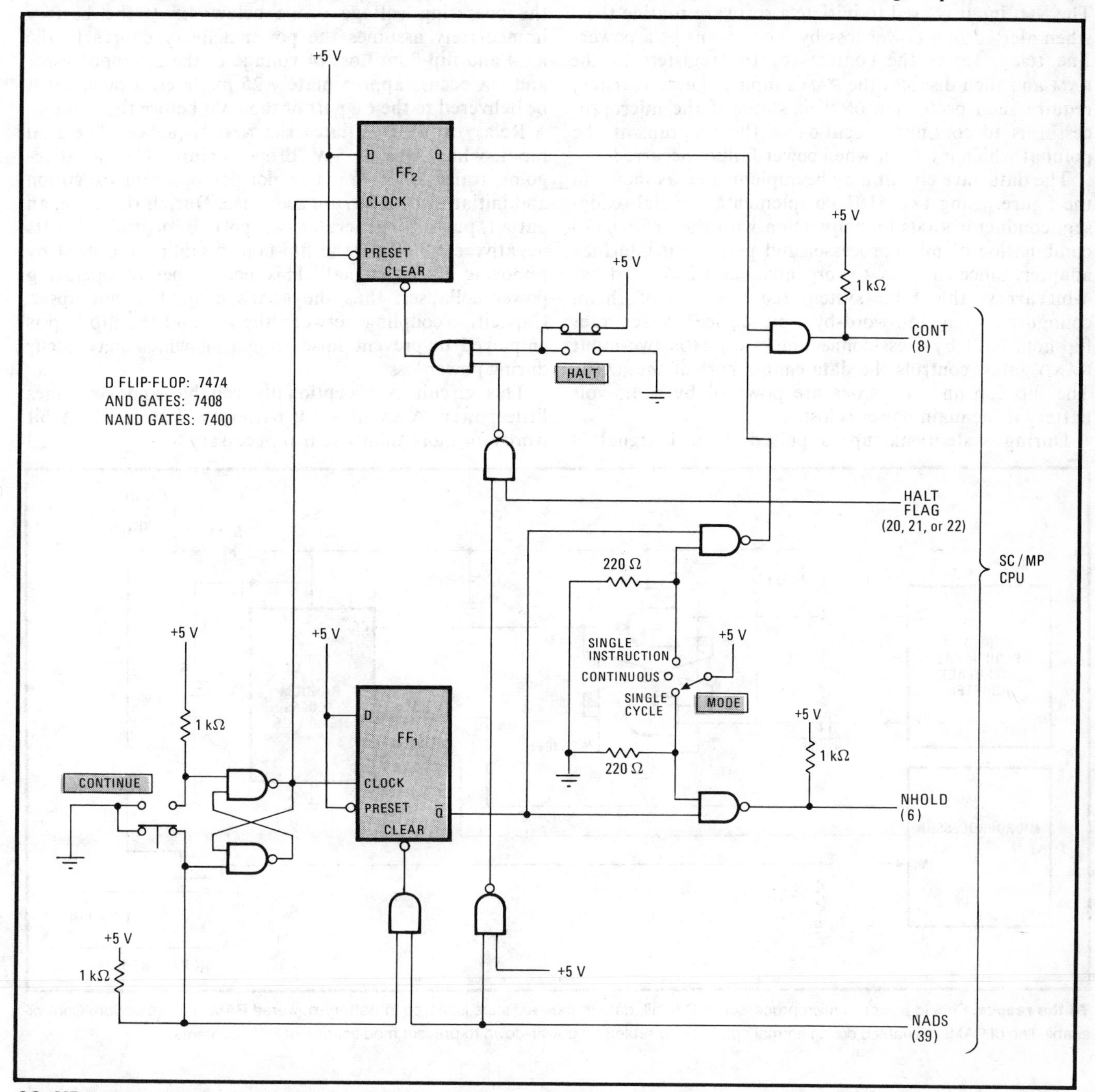

SC/MP system controller. Single-stepping through a program helps locate hardware or software errors. Circuit runs the microprocessor in single-cycle, single-instruction, or continuous mode, controlled by two push buttons. Halt may also be facilitated under software control, by having the SC/MP central processing unit pulse its halt flag line.

Nonmaskable interrupt saves processor register contents

by Ivars P. Breikss
Honeywell Inc., Test Instruments Division, Denver, Colo.

Linking a battery-powered random-access memory to the nonmaskable interrupt input available on many microprocessors will save the contents of the memory registers in a microprocessor system during power loss. The NMI input is used to initiate a software routine that, when alerted to a power loss by such means as a power-line relay, stores the contents of the registers in the RAM and then disables the RAM's inputs. These registers require such protection of their status if the microprocessor is to continue execution of the program at the point at which it left off when power failure occurred.

The data-save circuit may be implemented as shown in the figure, using two 5101 complementary-metal-oxide-semiconductor RAMs in conjunction with the 6800/6820 combination of microprocessor and peripheral interface adapter. Since each RAM is organized as a 256-word-by-4-bit array, this 8-bit system requires two of them, configured as a 256-word-by-8-bit device. A set-reset flip-flop, built by cross-connecting two C-MOS two-input NAND gates, controls the data-enable port of the RAMs. The flip-flop and the RAMs are powered by a 4.5-volt battery if the main power is lost.

During system startup, a pulsed logic 1 signal is generated at the A output of the PIA through software control. The negative-going edge of this pulse sets the flip-flop and drives the control-enable lead (CE) of the RAMs. This allows desired system parameters, which may have taken hours to determine initially, to be stored in the RAM for protection from power failure.

During a power-down cycle, the address and data lines of the RAM will usually assume random logic states for several milliseconds. This condition is likely to destroy or modify the contents of the RAM unless a logic 0 is applied to the CE input at least a few microseconds before the main power is lost.

A loss of line voltage causes the relay to open; once the operating voltage drops below 4.5 V the battery immediately assumes the power-delivery chores to the RAM and flip-flop. Loss of voltage to the microprocessor and PIA occurs approximately 25 ms later; a pulse must be delivered to the CE port of the RAM before that time.

Relay dropout initiates the NMI sequence. The NMI input, which was at 5 V, drops rapidly. The negative-going transition terminates normal program execution and initiates the interrupt sequence. During this time, an output pulse is generated at port B or the PIA. Its negative edge clears the flip-flop, disabling the RAM by removing its CE signal. This occurs before operating power collapses; thus the RAM's content is not upset. Capacitive coupling between the PIA and the flip-flop is employed to prevent false triggering which may occur during power loss.

This circuit is exceptionally reliable and consumes little power. A small 4.5-V battery will store 256 8-bit words for more than a year if necessary. □

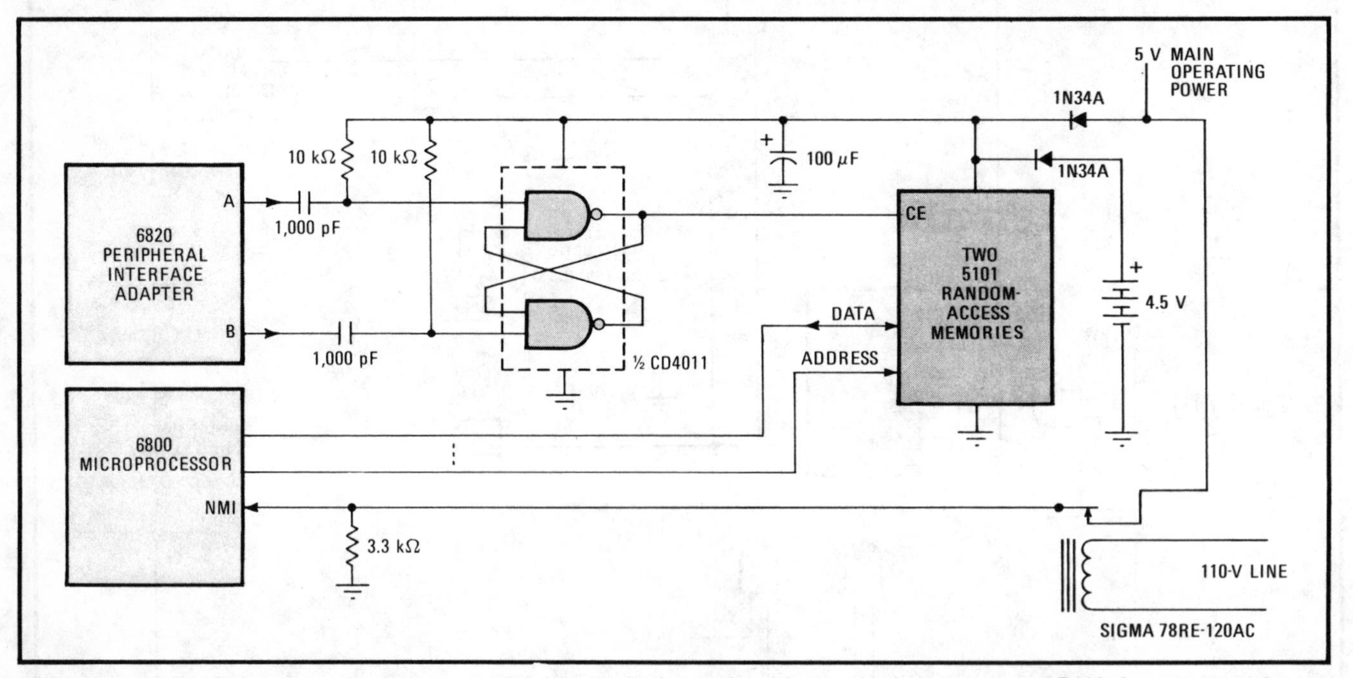

To the rescue. Should power to microprocessor or PIA fail, data in their registers is stored in battery-powered RAMs for protection. Control-enable line of RAMs is enabled during normal operation, disabled on power-down to prevent modification of RAM contents.

Scanner finds interrupts with highest and lowest priorities

by N. Bhaskara Rao
Department of Electrical Engineering, UVCE, Bangalore, India

In many multitask microprocessor applications, it is necessary to access not only the interrupt request of the highest priority, but the low-priority one as well. This scanning unit finds both for 14 priority-interrupt lines.

The scanning cycle is initiated when the output of counter A_1 reaches 1111, whereupon G_3 moves low and presets A_2. Thus, A_1 begins to count down.

A_1's output addresses A_3, a 16-to-1-line multiplexer, to determine the priority lines requesting a program interrupt. Connected to the multiplexer's data inputs are 14 priority lines, arranged from the least significant (E_1) to the most significant (E_{14}).

When the counter steps through the location corresponding to the first priority input at logic 1, A_3's output (W) moves low and sets A_6. Consequently, A_7 latches the counter's contents, which are the binary equivalent of the active priority input. This turns out to be the only priority input latched, because any succeeding output pulses from A_3 during the scanning cycle cannot effect the already active A_6.

Note that A_5 latches the highest-priority line, too, but that is unimportant because the low-priority input is determined in the latter part of the scanning cycle and it is assumed that the system logic strobes A_5 and A_7 at the end of the cycle.

When A_1 reaches zero, G_4 moves low and clears A_2 and A_4. A_1 then starts to count up. As the counter steps through the first active priority input, A_3's output again moves low and A_5 latches the contents of the counter. Note that priority inputs E_0 and E_{15} are uncommitted so as to prevent a simultaneous generation of preset and clear signals to A_4 and A_6 under conditions where only E_0 or E_{15} would be active.

As A_1 reaches 15, A_2 is set and A_6 is cleared. The scanning cycle is then repeated. □

Line list. Scanning interface determines which of 14 interrupt requests to microprocessor have the highest and lowest priorities at any given time. Highest-priority line number as seen by 16-line multiplexer (bottom, left) is latched as bidirectional counter A_1 steps from 15 to 0; low-priority number is latched as counter moves back to 15. Both numbers are strobed by system at termination of scanning cycle.

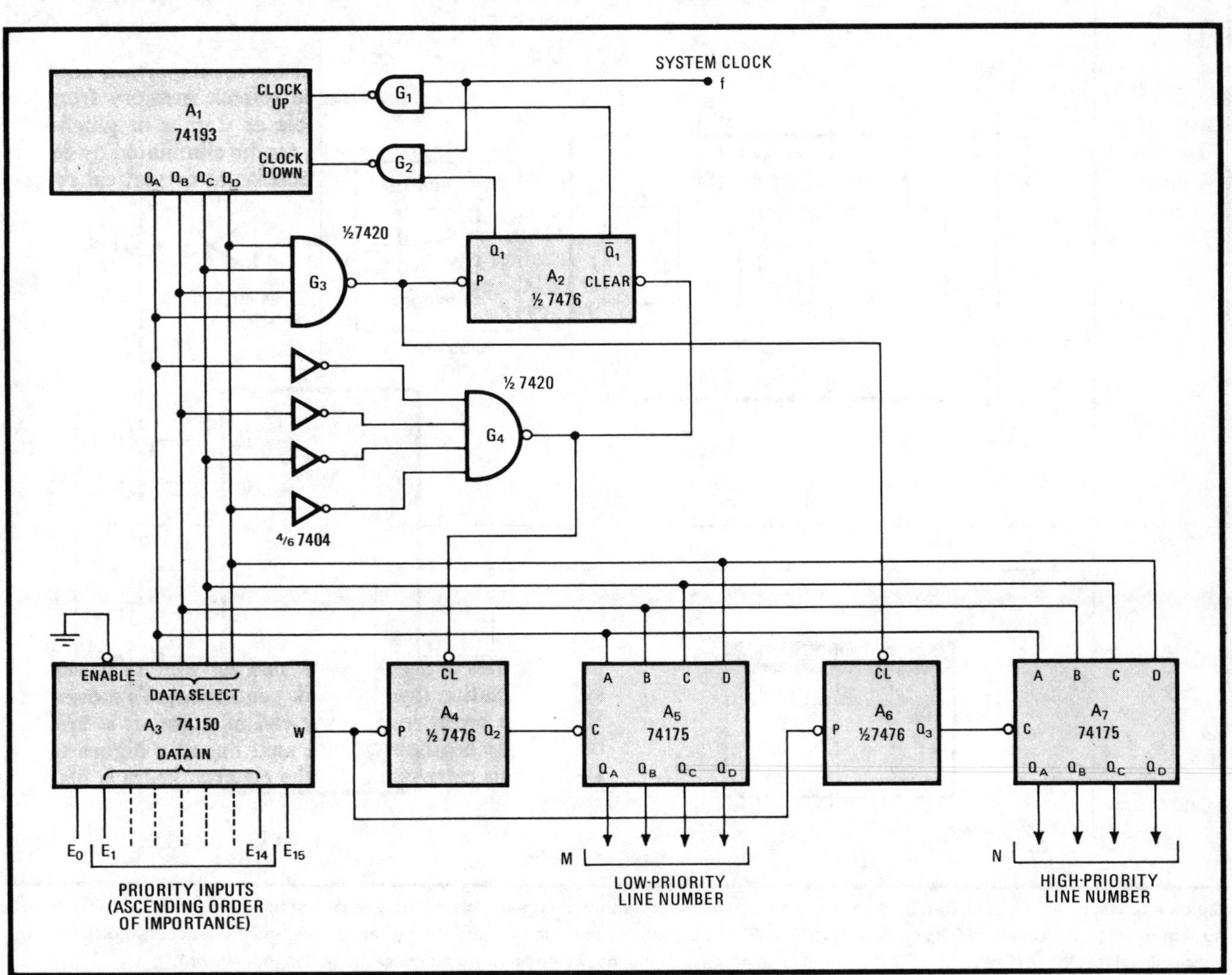

PROM adds bootstrap loader to Intellec-8 development system

by Bernard Boulé and Simon Gagné,
Laval University, Department of Electrical Engineering, Quebec, Canada

This three-byte bootstrap loader program will enable users of Intel's popular microcomputer-development package, Intellec-8, to immediately and automatically access the system's monitor, or executive-control routines, on power-up. The bootstrap is stored in a programmable read-only memory external to the system.

In normal operation, the Intellec's 8080 microprocessor is reset on starting up, thus clearing the system's program counter. Program execution then proceeds from memory location 0, but the monitor is located at starting address 3800H. Therefore, to enter the monitor, a jump instruction (programmed as C3 00 98) must be written into the first three locations of the system's random-access memories starting at location 0 after each power-up. Manual programming is a bothersome task, requiring that the memory-access port be activated and that each address be entered, loaded, and then incremented as the contents of each (C3, 00, 98) are set and loaded into memory.

An easier way to enter the monitor is to program the jump instruction into a 742188 PROM and dump its contents directly into memory during power-up, as shown in the figure. Although only 3 of the 32-word-by-8-bit device's locations are used, the low cost of the PROM and the convenience afforded by the modifications overshadow the waste of the 29 unused locations. The only other consideration with this circuit is to ensure that the system's RAMs will not be disturbed in any way by the PROM.

The PROM is programmed with the data shown in the table. With S_1 in the program position, the PROM's output lines (D_0–D_7) are disabled and the PROM can be loaded. Pull-up resistor R_1 ensures that the RAM memory enable (RME) line is active, so that any program loaded into RAM at address 0 can be run without interference stemming from programming the PROM.

The actual programming of the 742188 is simple and

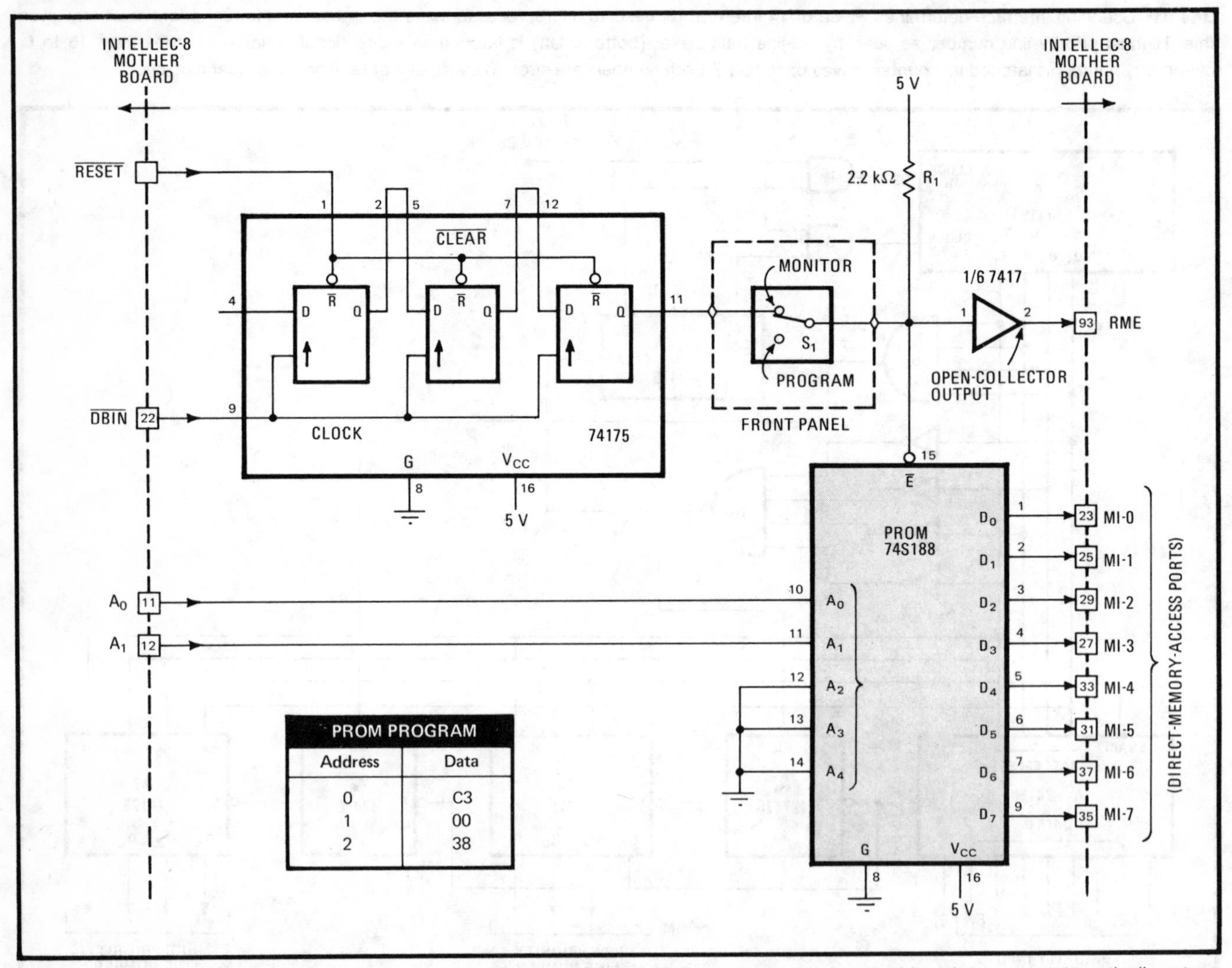

PROM PROGRAM	
Address	Data
0	C3
1	00
2	38

Quick access. Intellec-8 bootstrap loader is programmed into a programmable read-only memory, enabling the user to automatically enter a system monitor at location 3800H on power-up. The PROM's contents are dumped into the system's direct-memory-access ports on three successive data clocks (DBIN). The PROM operation in no way affects the system's random-access memories (not shown).

is done only once. A_0 and A_1 address the desired location, and after the normal supply voltage on pin 16 is brought to 10 volts, a 10-v, 65-milliampere current emanating from a constant-current source is applied to the output lines that are to be programmed to logic 1. The procedure is repeated for all locations.

S_1 is then placed in the monitor position. Immediately after a system reset, the three flip-flops in the three-stage 74S175 shift register and the RME line are reset. At the same time, the output of the PROM is enabled. Upon the arrival of the first three normally occurring system, or data-byte, clock pulses (DBIN), lines A_0–A_1 are incremented and the contents of the PROM are read into system memory. On the third pulse, the output of the shift register goes high, releasing the RME line and disabling the PROM, which remains inactive until a new reset cycle occurs. □

Gates replace PROM in Intellec-8 bootstrap loader

by Simon Gagné and Bernard Boulé
Université Laval, Department of Electrical Engineering, Quebec, Canada

In what may be the most cost-effective solution yet for implementing a bootstrap loader for Intel's popular Intellec-8 development system, logic gates are used to replace the jump-to-monitor routine stored in a programmable read-only memory (PROM).[1] This simple method of automatically accessing the system's monitor, or executive-control routines, is possible because the set of 8-bit logic signals required at the data bus to access

(a)

Instruction	Op code	D_7	D_6	D_5	D_4	D_3	D_2	D_1	D_0
JMP 3800H	C3	1	1	0	0	0	0	1	1
	00	0	0	0	0	0	0	0	0
	38	0	0	1	1	1	0	0	0
(AFTER JMP)	FF	1	1	1	1	1	1	1	1

(b)

Condition	Q_1	Q_2	Q_3
RESET	0	0	0
AFTER FIRST $\overline{DBIN}$	1	0	0
AFTER SECOND $\overline{DBIN}$	1	1	0
AFTER THIRD $\overline{DBIN}$	1	1	1

1. Synthesis. Table outlines logic sequence required on data bus for achieving jump operation in order to enter Intellec-8's executive program (a). Outputs of clocked three-stage shift register of basic circuit generate signals (b), which can be used to synthesize D_0–D_1 and D_6–D_7 signals if function $\overline{Q}_1 + Q_3$ is formed with logic gates. Q_2 output is by itself identical to D_3, D_4, and D_5. Q_3 output is identical to D_2.

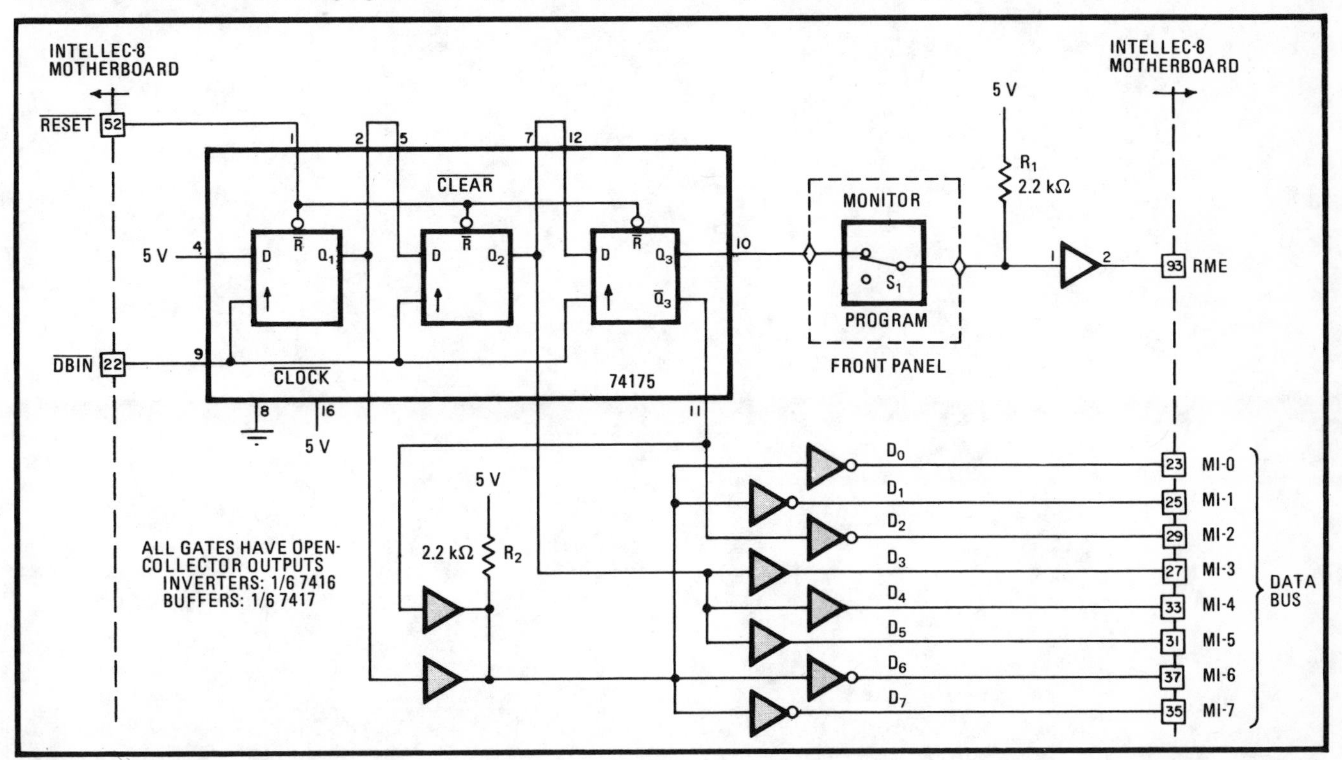

2. Implementation. Circuit for accessing system's monitor uses open-collector logic gates, which replace programmable read-only memory in original configuration. Circuit is armed by placing front-panel switch S_1 in monitor position before each power-up of system.

the monitor on power-up can be easily generated.

In the Intellec-8, the monitor is located at address 3800 (hexadecimal). Therefore, a jump instruction is used to advance the program counter from location 0 to 3800H. To avoid the manual programming required after each power-up, the previous solution was to use a PROM programmed with the instruction (C3 00 38).

The logic signals required on data bus lines D_0–D_7 for achieving the three-byte jump instruction are shown in Fig. 1a. The output of each flip-flop in the three-stage 74175 shift register, previously used to enable the PROM, together with simple logic, can synthesize this sequence.

When the outputs of the shift register (Q_1, Q_2, and Q_3) alone are stepped with system clock $\overline{DBIN}$, they generate the signal pattern shown in Fig. 1b. Comparison of Fig. 1a and 1b reveals that the identical patterns of lines D_0–D_1 and D_6–D_7 can be generated by forming the logic function $\overline{Q}_1 + Q_3$. It is also seen that bits D_3, D_4, and D_5 are identical to the Q_2 output of the shift register and the D_2 bit is identical to the Q_3 output, so that the Q_2 and Q_3 signals can simply be connected to these corresponding data lines through noninverting buffers. To make the final circuit, which appears in Fig. 2, operational, switch S_1 need only be placed in the monitor position for automatic loading on power-up. □

References

1. "PROM adds bootstrap loader to Intellec-8 development system," p. 220

Chapter 24
MODULATORS & DEMODULATORS

Tracking filters demodulate two audio-band fm signals

by Stephen Barnes
Center for Bioengineering, University of Washington, Seattle

Because of the way in which they retrieve recorded data, many systems designed for monitoring biomedical functions have to demodulate two closely spaced fm carrier signals in the audio-frequency band. The original data signals could be recovered with low-pass filters, but they are an expensive solution since their cutoff must be sharp to prevent cross modulation, signal blocking, or undue limiting of the bandwidth needed by one or both signals. The low-cost solution shown here, however, uses a dividing phase-locked loop to demodulate one signal and to provide the clock signal for a tracking notch filter that recovers the other channel of data.

The advantages of the circuit may be seen for a typical monitoring case in which a 30-hertz electrocardiogram signal (having a frequency too low to be recorded directly on cassette tape) is placed on a 9-kilohertz carrier. This signal is applied to the LF356 amplifier along with a 0-to-6-kHz signal from an ultrasonic doppler flowmeter that provides data on blood circulation.

Block A, the dividing phase-locked loop module, oscillates at a free-running frequency equal to $8f_m$. It is here that the 9-kHz carrier is directly demodulated. Also included in block A is a 74C193 divide-by-8 counter in the PLL's feedback loop, which provides the driving signals for the CD4051 multiplexer in block B. This block contains the sampling filter which passes frequencies equal to ⅛ the sampling frequency, $8f_m$, and its harmonics. Thus the doppler data is notched out by the sampling filter. But signal $\vec{f}_m$ is subtracted from the original input signal $\vec{f}_o + \vec{f}_m$ by the differential amplifier in block C. Therefore, only signal f_o will appear at the output.

Resistor R_Q, which is in the PLL's feedback loop, controls the width of the notch, which is given by $B = 1/8\pi R_Q C_Q$, where B is the width defined by the filter's upper 3-decibel frequency minus its lower 3-dB frequency. Potentiometer R_B is used for balancing the differential amplifier by nulling the 9-kHz feedthrough signal.

When the circuit is in the locked state, the minimum attenuation of the fm signal will be approximately 33 dB for input signals ranging from 540 millivolts to 8 volts peak to peak. Below signal levels of 540 mV, feedthrough from the multiplexer will reduce the attenuation.

Because the filter is a sampling device, it is subject to aliasing if any input-frequency components approach the Nyquist limit of $4f_m$, so that precautions should be taken to prevent this. Spurious output components that are higher harmonics of the 9-kHz fm signal or the clock signal can be removed easily.

Phase jitter in the PLL should also be minimized, for it causes narrow noise sidebands centered about the 9-kHz fm signal and separated from it by a frequency equal to the loop bandwidth of the PLL. The amplitude of these noise sidebands in a properly adjusted circuit should be down at least 33 dB from the level of the fm signals at the input. □

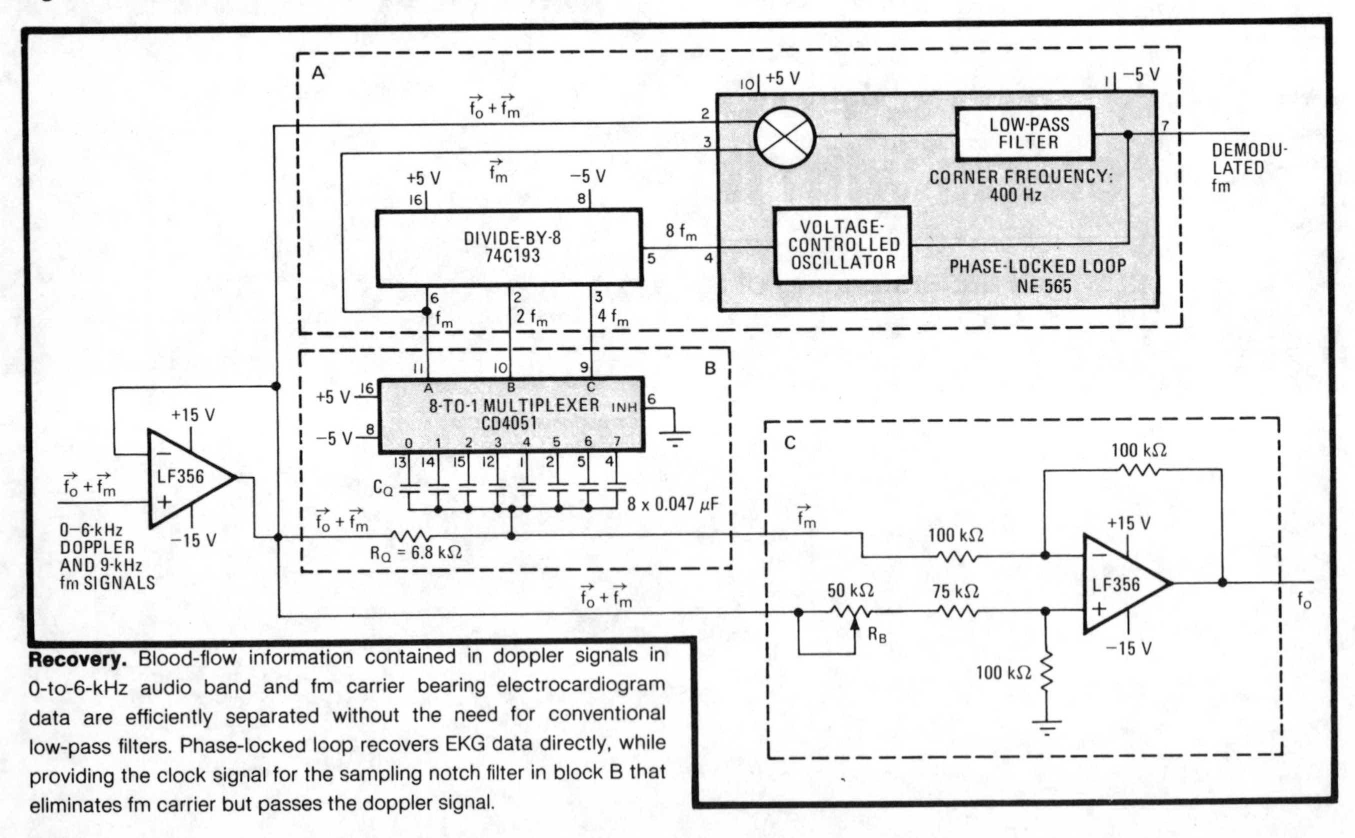

Recovery. Blood-flow information contained in doppler signals in 0-to-6-kHz audio band and fm carrier bearing electrocardiogram data are efficiently separated without the need for conventional low-pass filters. Phase-locked loop recovers EKG data directly, while providing the clock signal for the sampling notch filter in block B that eliminates fm carrier but passes the doppler signal.

PSK modulator resolves phase shifts to 22.5°

by Noel Boutin
University of Sherbrooke, Quebec, Canada

Requiring little more hardware than the circuit proposed by Chawdhury and Das[1] and eliminating the software entirely, this phase-shift-keyed modulator offers an even more versatile and less expensive solution to sending binary data over long distances. One of its great advantages is that carrier phase shifts can be resolved to 16 bits—22.5°.

Only three low-cost chips and a means for generating four-input binary data are required, as shown. The carrier signal is first divided by 16 and applied to the CD4015 eight-stage shift register. Because the register is clocked by the carrier at a rate 16 times that of the signal to be shifted, a discrete eight-phase version of the carrier appears at the output of the register, each shifted by 360/16 = 22.5° from its neighboring stage. These signals are then introduced to the CD4051 multiplexer.

The first 3 bits of each of the modulating data inputs, A–D, address the CD4051 also. Thus any desired phase shift from 0° to 157.5° may be selected (see large table). The eight remaining values, from 180° to 337.5°, may be selected with the aid of the D input, which at the output of the last stage of the register inverts the phase of the signals that have already been generated.

There may be instances where it is desirable to transmit fewer than 16 levels. The small table summarizes the A–D states required to achieve this. □

References
1. F. B. Chawdhury and J. Das, "8085 performs PSK modulation for data-line transmission," *Electronics*, Jan. 31, 1980, p. 108.

DATA				ϕ
D	C	B	A	
0	0	0	0	0°
0	0	0	1	22.5°
0	0	1	0	45°
0	0	1	1	67.5°
0	1	0	0	90°
0	1	0	1	112.5°
0	1	1	0	135°
0	1	1	1	157.5°
1	0	0	0	180°
1	0	0	1	202.5°
1	0	1	0	225°
1	0	1	1	247.5°
1	1	0	0	270°
1	1	0	1	292.5°
1	1	1	0	315°
1	1	1	1	337.5°

NUMBER OF PHASE LEVELS	A	B	C	D
2 ϕ	0	0	0	DATA
4 ϕ	0	0	DATA	DATA
8 ϕ	0	DATA	DATA	DATA
16 ϕ	DATA	DATA	DATA	DATA

Multiphase. Three-chip circuit performs PSK modulation on square-wave input, resolving carrier shifts to 22.5°. First three bits of modulating data inputs select shifts from 0° to 157.5°, with D input required for higher values. Truth tables summarize operation.

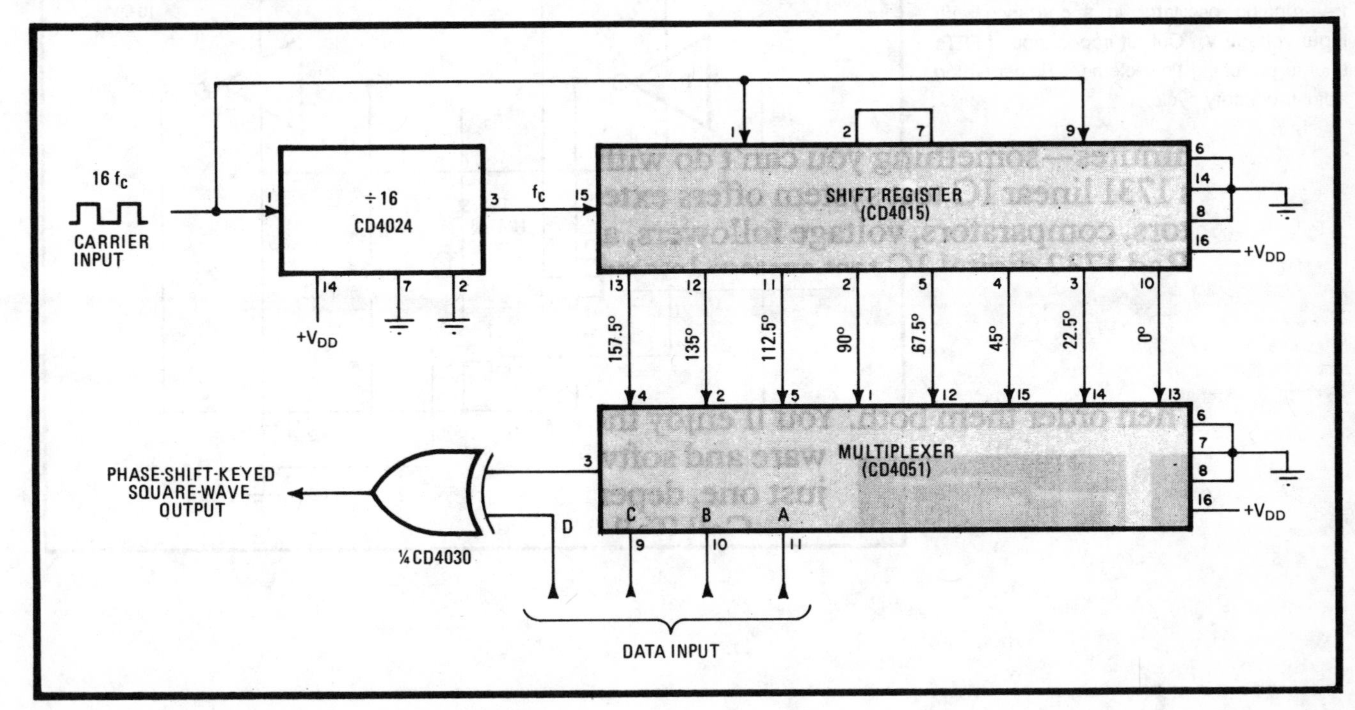

Single C-MOS IC forms pulse-width modulator

by Mark E. Anglin
Novar Electronics, Barberton, Calif.

A pulse-width modulator can be constructed with a single complementary-metal-oxide-semiconductor integrated circuit if the IC's field-effect transistors control the duty cycle of a free-running oscillator. The output resistance of the FET varies almost linearly with input voltage over portions of its characteristic curve, permitting the circuit to be used for applications in switching power supplies and analog conversion in data-communications systems.

This circuit uses a CD4007 dual complementary pair plus an inverter device, which comprises three n-channel and three p-channel enhancement-type MOS transistors. As shown in the figure, inverters I_1 and I_2, each formed by two gates in the 4007, themselves form an astable multivibrator in conjunction with resistor R and capacitor C. The frequency of oscillation of the multivibrator is given by:

$$f = 1/1.4\,RC$$

The 1-megohm resistor, left in the figure, limits the feedback current into I_1. This prevents the input circuit from burnout and the inverter from switching prematurely and affecting the desired frequency of operation.

The two remaining gates in the 4007, one n-channel and one p-channel gate, are connected across the output of I_1 and R. This network is designed to modify R and thus control switching times t_1 and t_2.

The oscillation (switching) times for this device may be expressed by two equations:

$$t_1 = RC \ln [(V_{dd} - V_{tr})V_{dd}]$$

and

$$t_2 = RC \ln (V_{tr}/V_{dd})$$

where t_1 is the on time and t_2 is the off time of the oscillator, V_{tr} is the threshold voltage at the gate's input, and V_{dd} is the supply voltage.

When the oscillator's output is high, diode D_1 may conduct. The p-channel output resistance will be in parallel with R, neglecting the diode's forward resistance. The p-channel gate's output resistance will decrease as V_c decreases.

Similarly, when the oscillator's output is low, D_2 may conduct, and the n-channel output is placed in parallel with R. The n-channel's resistance will decrease as V_c increases. The duty cycle of the oscillations (in other words, control of the on and off times) is variable from 1% to 99% of the operating frequency. The duty cycle is directly proportional to the amplitude of the control voltage V_c.

The component values in the figure yield an oscillation frequency of 1 kilohertz. The change in oscillation frequency as a function of duty cycle is minimal; although the basic frequency of oscillation is modified by the FETs across I_1, the output impedance of one always increases, and the other's decreases proportionately, for any value of control voltage V_c. Thus, the average resistance shunted across R during a cycle is constant. □

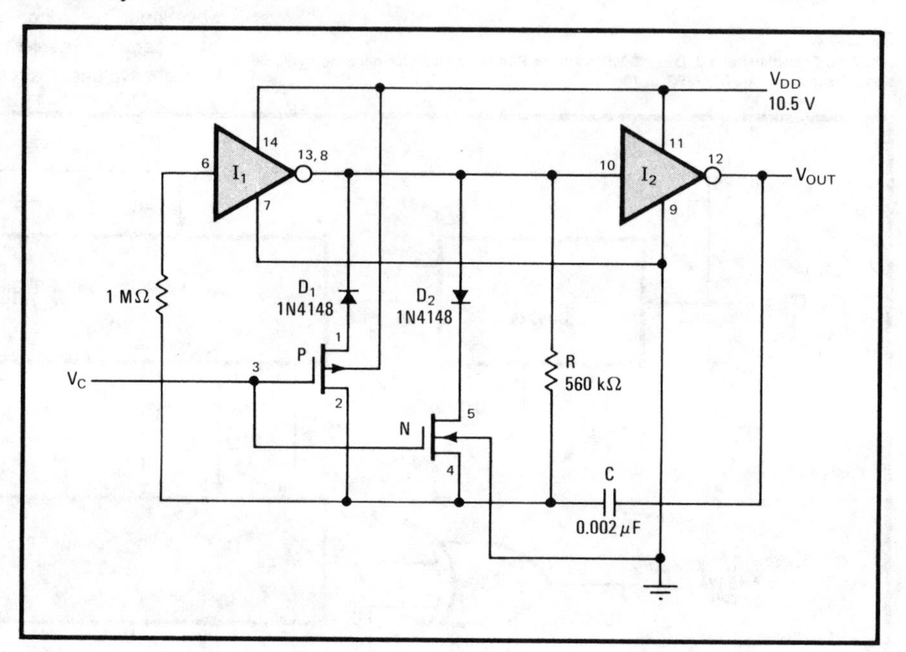

Single IC modulator. Pulse-width modulation is achieved by varying duty cycle of free-running oscillator in accordance with input voltage V_c. Output impedance of FETs shunts switching-time element R, permitting adjustable duty cycle.

Frequency modulator extends tape recorder's lf response

by W. B. Warren and W. L. Lively
TRW Subsea Petroleum Systems Inc., Houston, Texas

Tape recorders used in low-frequency data acquisition systems are, not surprisingly, expensive. But this simple frequency-modulator circuit extends to dc the effective low-frequency response of any inexpensive tape recorder. To do this, the modulator translates a low-frequency data signal to a frequency falling within the recorder's limits, thus ensuring that the data will be properly recorded.

The block diagram in Fig. 1a illustrates the operation of the circuit. Recording low-frequency data simply requires that the frequency of a voltage-controlled oscillator be varied by the amplitude of a low-frequency input signal. On playback, the recorded data can be recovered by a phase-locked-loop demodulator, which is formed by a phase detector, frequency discriminator, and the VCO that was used to record the original data.

The circuit is shown in Fig. 1b. In the record mode, data signals are introduced to amplifier A_1, which in turn drives the Exar XR2207 VCO. A 0-to-10-volt data signal will vary the VCO frequency from 10 to 5 kilohertz so that the corresponding frequencies can be recorded.

On playback, the recorded data is presented to the 2N2907 buffer, then introduced to one input port (pin 3) of the XR2208 multiplier, which serves as a phase detector. Driving the other input port at pin 5 is the VCO. The output signal from the phase detector is the demodulated waveform produced by the multiplication of the

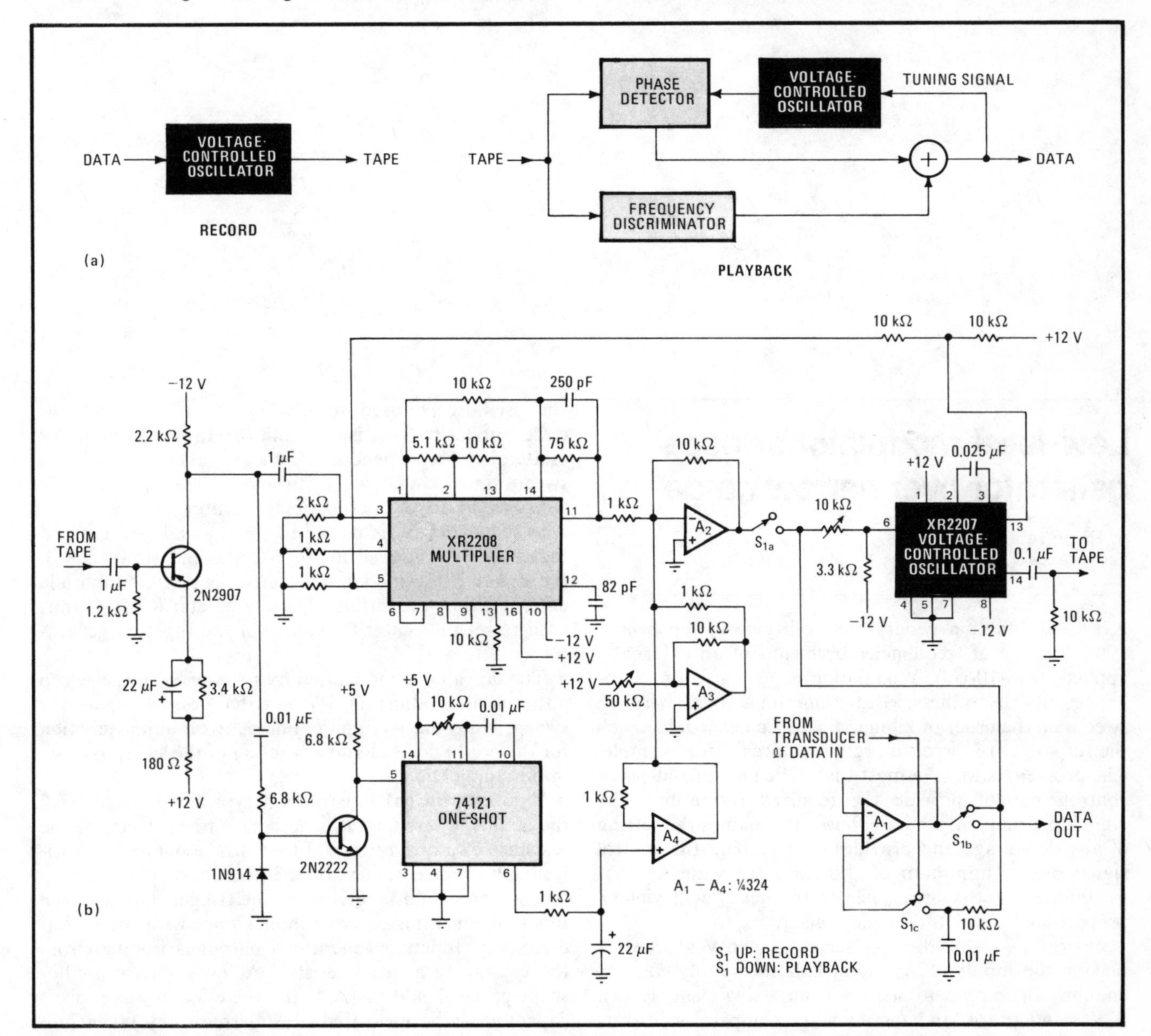

Data translation. Low-frequency data may be recorded on an inexpensive tape recorder if the signal is first modulated to a higher frequency. Signal is recovered from recorder by demodulation, using same VCO (a). Modulator uses four chips and two transistors (b).

playback signal times the output of the VCO.

The 74121 monostable multivibrator (one-shot) serves as a simple frequency discriminator, which drives the low-pass filter formed by the 1-kilohm resistor and the 22-micro-farad capacitor. The filter produces a voltage proportional to the frequency of the signal from the tape recorder. The addition of the discriminator to the circuit provides for rapid and reliable acquisition of the recorded data and reduces the dynamic range over which the phase-detector portion of the loop must operate.

The signals produced by the frequency discriminator and the phase detector are summed by A_2 through A_4 to generate the control voltage for the VCO. Therefore, the output signal, which is the control (tuning) voltage, is a dc voltage equal to that originally recorded. Using the same voltage-controlled oscillator for both record and playback modes ensures that VCO tuning nonlinearities have a negligible effect on system linearity.

The characteristics of the XR2207 are very stable with respect to supply-voltage and temperature variations. The primary sources of error between the record and playback signals will be those caused by tape stretch and the tape recorder's variations in speed. Both of these effects may be minimized by adjusting the potentiometer at pin 6 of the XR2207 to yield as large a frequency swing as possible for the range of signals being recorded. Better than 1% accuracy can be obtained with this circuit, even if an inexpensive tape recorder is used. □

Low-level modulator sweeps generator over narrow range

by Ralph Tenny
George Goode & Associates, Dallas, Texas

A typical function generator's ability to sweep over a 1000:1 range of frequencies by means of an externally applied 0-to-10-volt modulating signal certainly enhances its usefulness. But sometimes narrow-range sweeps on the order of kilohertz are also needed, to check the response of a precision resonant circuit, for example. The problem is that, in most cases, the unit's front-panel controls cannot provide the required resolution. The one-chip circuit shown here, however, enables the setting of any dc voltage and provides for sweeping the control signal over a minimum of ±0.1% of its value so that modulation of the preset center frequency will yield a proportionally small frequency variation.

Operational amplifier A_1 serves as a 6-V source for biasing the inputs of A_2–A_4 at half the supply voltage, enabling the circuit to operate from a single supply (a). A_2, an integrator, and A_3, a voltage comparator operating with heavy feedback, generate the 100-hertz triangle wave needed to sweep the generator and the x input of the oscilloscope used to display the response of the circuit under test. A_4 is a simplified Howland Pump[1], or bilateral current generator, which takes part of the sweep signal and uses it to modulate the preset dc voltage that drives the function generator.

When switch S_1 is placed in the manual position and R_3's arm is positioned at its extreme end (toward R_2), the signal at the modulation output is dc, its amplitude determined by the setting of potentiometer R_2. R_2 is thus used to set the center frequency of the function generator.

The dc value is modulated by placing S_1 in the sweep position and adjusting R_3 for the desired frequency sweep. Note that R_3 approximates a summing junction for the preset dc level and a fraction of the sweep voltage in this application.

The setup in (b) illustrates a typical application for the circuit, whereupon it is necessary to characterize the response of a quartz crystal that has resonant and antiresonant frequencies less than 3 kHz apart. The frequency counter should be driven by the trigger output of the function generator to avoid interference with the crystal drive. The function generator's output is isolated from the crystal by a large resistor. A low-capacity oscilloscope probe should be used, and the effect of the probe's capacity on the measured crystal frequency taken into consideration. A manual control switch allows the operator to measure the resonant and antiresonant frequencies

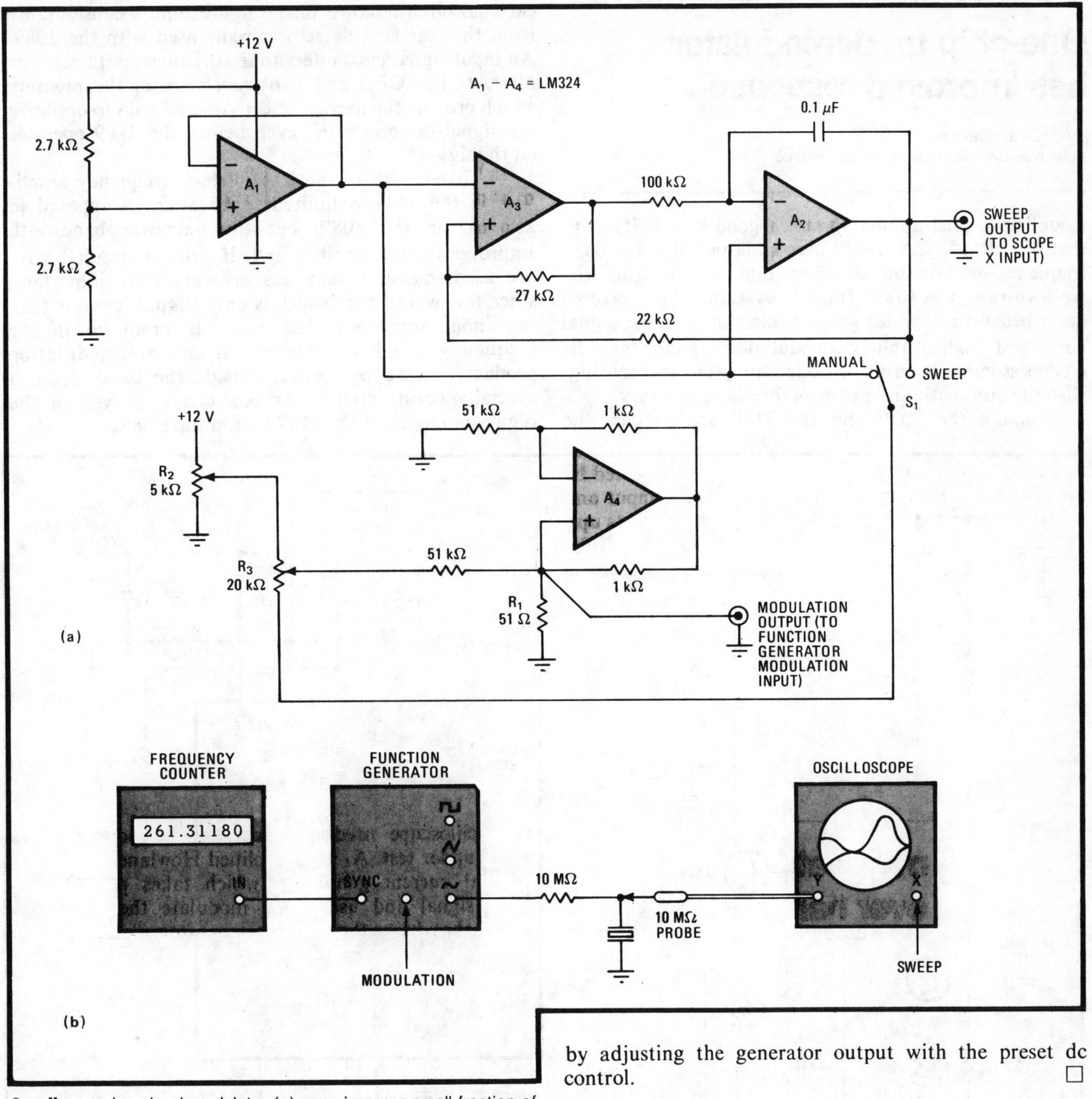

by adjusting the generator output with the preset dc control. □

Small scan. Low-level modulator (a) superimposes small fraction of 10-V triangle wave on preset dc voltage so that externally driven function generator can be swept over very narrow ranges not normally within the resolving power of unit's front panel controls. In typical application (b), response of crystal and isolation of its resonant and antiresonant frequencies are displayed and recorded.

References
1. Applications Manual for Computing Amplifiers, III.6, p. 66, George A. Philbrick Researches Inc., 1966.

One-chip fm demodulator has improved response

by J. Brian Dance
North Worcestershire College, Worcs., England

Coupling a preamplifier that has good selectivity with the new RCA CA3189E demodulator chip builds a frequency-modulation detector that outperforms the well-known CA3089 fm/i-f system. The circuit described here provides greater rejection of input-signal noise and high amplitude-modulation signals than its predecessor, while ensuring better audio-channel muting. Also the automatic-gain-control threshold can vary.

Although the 3089 and the 3189 are similar, the external circuit shown in the figure differs considerably from the standard detector circuit used with the 3089. An input signal encounters four 10.7-megahertz ceramic filters ($CF_1 - CF_4$) and two transistors in the preamp, which provide the needed selectivity and gain to optimize the signal-to-noise ratio, even before the 3189 operates on the signal.

The bandwidth of the intermediate-frequency amplifiers in the 3189 is limited to 15 MHz, as opposed to 25 MHz in the 3089; but the narrower bandwidth improves circuit stability and, if printed-circuit boards are used, makes layout less critical. More important, since the overall bandwidth is only slightly greater than the input frequency, less noise is produced in the frequency band of interest from intermodulation products caused by signals outside the band. Also, a specially constructed zener diode is employed in the regulator circuit of the 3189 to minimize noise.

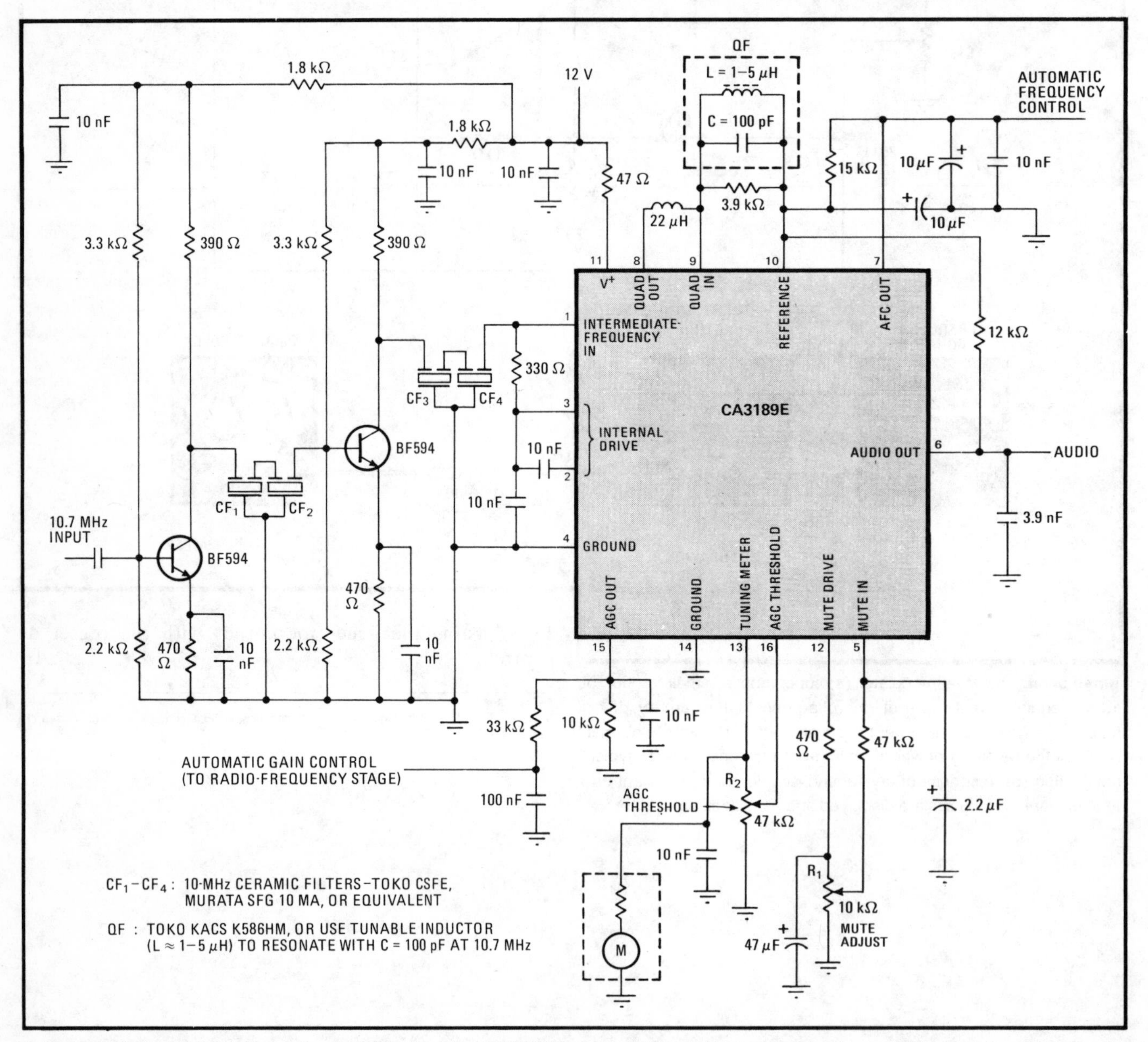

1. Improved fm detector. A good preamplifier and the CA3189 fm/i-f system provide optimum response. The circuit has greater noise immunity and better protection from a-m signal overload than detectors now used, and the agc threshold is selectable.

Two muting circuits are used. For noise between stations, part of the voltage change appearing at the mute-drive port at pin 12 (which is driven by the noise from previous input stages) is fed to the mute-control input at pin 5 through a voltage divider that includes potentiometer R_1. The muting threshold can thus be selected.

Although this arrangement is satisfactory for inter-station noise, additional components are needed to combat noise produced by tuning through a signal; otherwise, a sudden change in the output dc level will produce the familiar, low-frequency "thump" noise. An integrating circuit is formed by adding the 47- and 2.2-microfarad capacitors between pins 5 and 12 to reduce this noise, but even this step does not eliminate it if the integrating-circuit time constant is too small for fast tuning. However, the deviation-muting circuit formed by placing a resistor between pins 7 and 10 ensures the noise will be eliminated, provided the deviation is less than ±40 kilohertz. (The deviation is controlled by selecting a resistor of 15 kΩ.) Thus any voltage change caused by noise is reduced at the output, pin 6.

The agc threshold is determined by R_2, which sets the voltage fed to pin 16 from pin 13. The threshold point can be selected from 0.2 to over 200 millivolts. A 40-decibel agc range can be easily obtained if the very-high-frequency tuner driving the 3189 uses dual-gate metal-oxide-semiconductor field-effect-transistor stages or similar stages having wide dynamic range.

The signal-to-noise ratio is 50 dB for a 3-microvolt input signal. If the first transistor stage in the preamp is omitted, a s/n value of 20 dB can be obtained for the same input level using only two ceramic filters. For a deviation of ±75 kHz, the a-m signal rejection is 60 dB for input signal amplitudes greater than 500 μV, and the limiting sensitivity of the 3189 is typically 12 μV at 3 dB. The tuning meter has an approximately logarithmic response over an input signal range of 10 μV to 100 mV.

The quadrature tuned circuit between pins 9 and 10 determines the percentage of audio harmonic distortion—typically 0.3%. This figure can be reduced to 0.1% if the network is a double-tuned circuit.

The 12-kilohm load resistor sets the audio output level. The 3.9-nanofarad capacitor provides the 50-microsecond de-emphasis required for reception in Region 1 (Europe), and a 5.6-nF capacitor is suitable for the 75-μs de-emphasis required in Region 2 (U. S.). This capacitor should be omitted when the signal is fed to a stereo decoder circuit. □

ICs slash component count in Costas loop demodulator

by Carl Andren
E-Systems Inc., St. Petersburg, Fla.

Just three integrated circuits can build a Costas phase-locked loop that will detect differential phase-shift-keyed modulation. The Costas loop is named after its inventor, who first detected it with a PLL by regenerating the carrier in a double-sideband suppressed-carrier signal. The loop allows tracking of the desired frequency in a high-noise environment while ignoring carrier phase

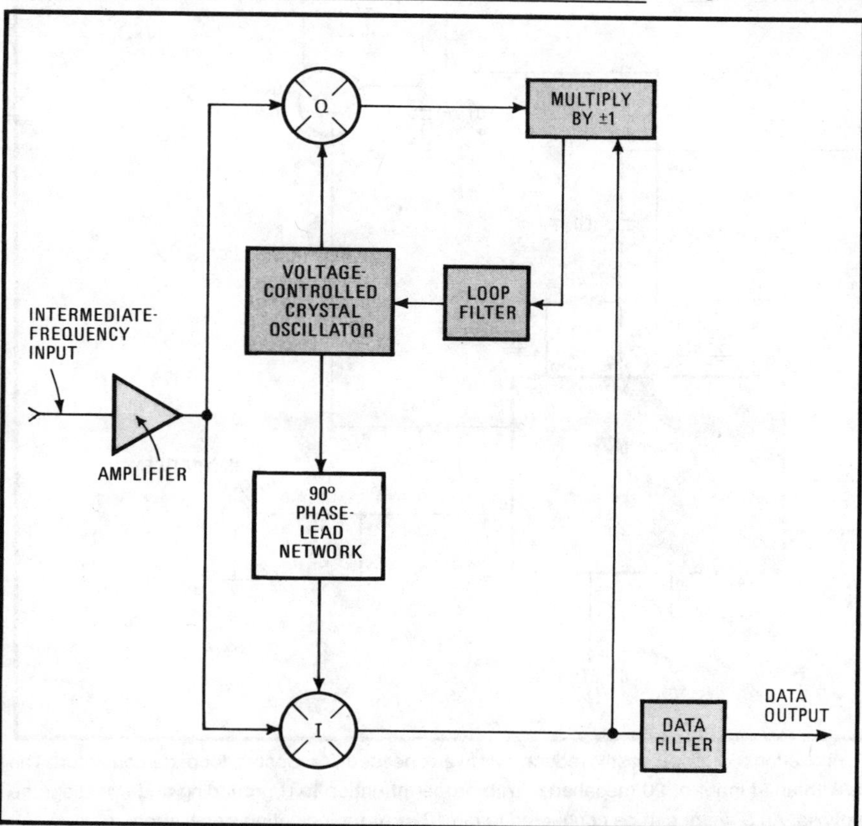

The Costas loop. An intermediate-frequency input signal is compared to the voltage-controlled crystal oscillator signal, and two quadrature signals are generated. The output from the Q (quadrature) channel is multiplied by ±1 depending on phase detected by the I (in-phase) channel. The feedback loop facilitates tracking of the carrier frequency in high-noise environments and maintains locking despite phase reversals of the carrier caused by modulation.

reversals caused by modulation.

The Costas loop (Fig. 1) has an input signal split into two channels. The in-phase channel (I) demodulates the data, and the quadrature channel (Q) tracks frequency and phase of the carrier.

The key to the loop's operation is the multiply-by-± 1 function, which inverts the phase of Q's output signal upon detection of a carrier phase reversal. This inversion is reflected in the feedback signal and maintains lock in the voltage-controlled crystal oscillator. The I channel, which detects the phase change, determines if the Q-channel output is to be inverted or multiplied by unity.

In the demodulating circuit of Fig. 2, the two CA3089 frequency-modulated intermediate-frequency systems and the MC1558 operational amplifier take over the Costas loop functions. This circuit was optimized for a data rate of 9.6 kilobits per second in a system with an intermediate-frequency bandwidth of 40 kilohertz.

The CA3089s replace more than 70 devices usually needed for the Costas loop. A_1 fills the Q-circuit function, while A_2 is wired to serve as the I circuit. The high-gain limiting amplifiers of A_1 and the monolithic two-pole crystal-filter in conjunction with the IN5462 varactor make up the voltage-controlled crystal oscillator. The loop filter consists of the RC network that drives the varactor. The MC1558 (A_3) serves the multiplier function. Data filtering is accomplished by capacitive loading (4,700 picofarads) at the output of A_2 in conjunction with the device's output impedance.

At the input of A_2, an i-f signal at 20 megahertz is

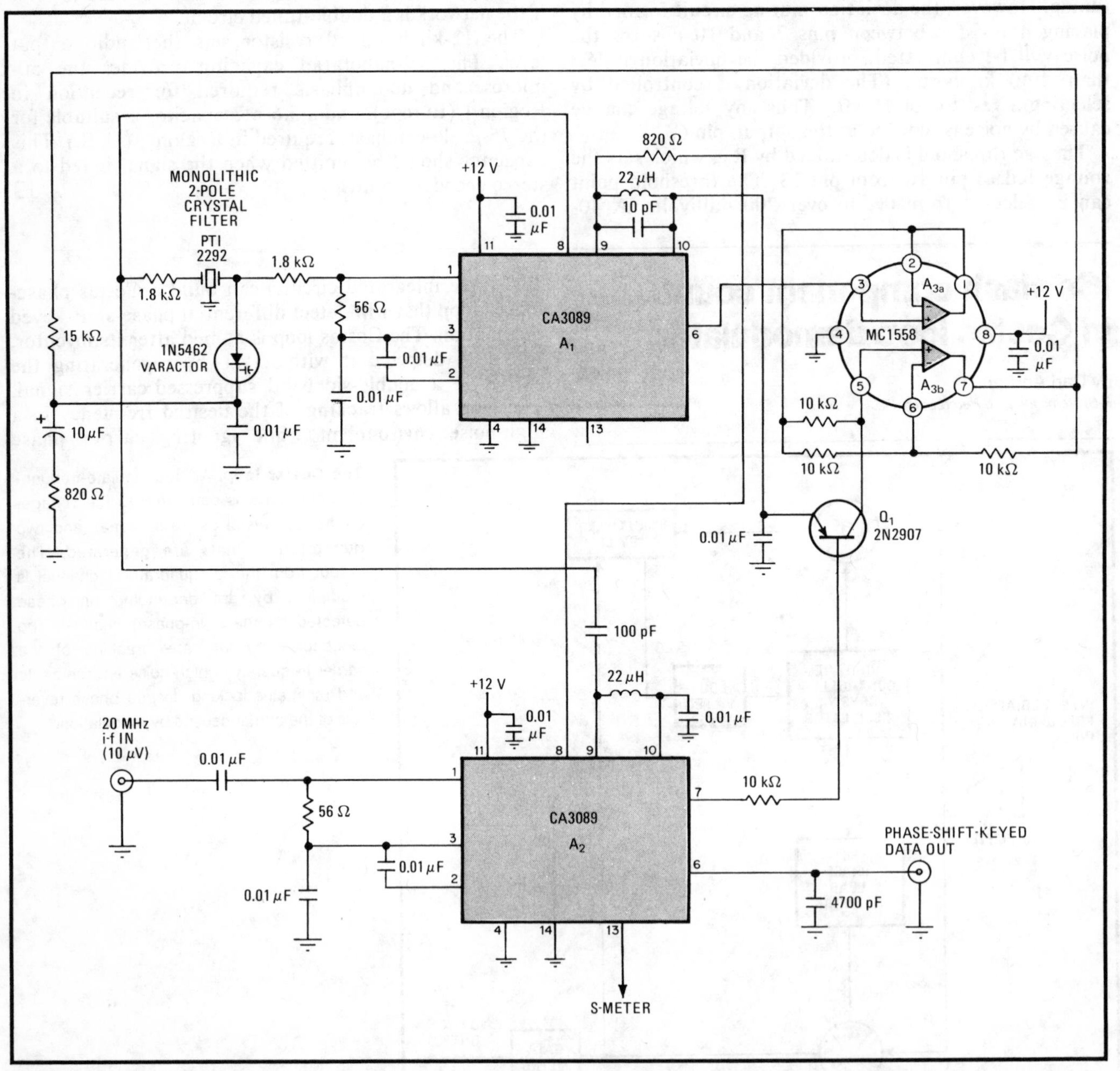

2. Differential-PSK demodulator. Two CA3089 phase-locked loops vastly reduce hardware needed for Costas loop demodulator. The circuit detects differential phase-shift-keyed signals with an i-f input of 20 megahertz. With proper attention to rf grounding and shielding, the detector can operate with signals as low as 10 microvolts. An S meter can be connected to pin 13 of A_2 for indicating signal strength.

amplified by the device's high-gain amplifiers, which provide three stages of amplification before presentation at the I-channel quadrature detector or mixer. A simple output buffer in this detector links it with the Q-channel quadrature detector in A_1, thus driving the mixers in both devices with virtually no phase difference.

The I-channel mixer is a balanced transconductance amplifier, biased relative to pin 10 of A_2. It is driven by the i-f input signal, as well as by the oscillator through a 90° phase-lead network. Its output drives two amplifiers in A_3; A_{3a} is in the feedback loop to the oscillator, and A_{3b} controls the multiply function through a transistor. The noninverting amp A_{3a} drives the op amp A_{3b}. The gain is -1 when transistor Q_1 is on and $+1$ when the transistor is off.

The multiplier's action causes a feedback voltage that varies the oscillator's frequency through the varactor in the tank circuit of the oscillator, and phase lock is readily accomplished. The phase shift of the oscillator's amplifier network is about 360° at 20 MHz, and the crystal filter element has no phase shift at its center frequency—thus allowing smooth operation near the lock frequency. The loop filter is designed to provide the correct loop damping and gain coefficients needed for proper operation. □

Chapter 25
MULTIPLEXER CIRCUITS

Double-balanced mixer has wide dynamic range

by Carl Andren, Eric Heinrich, and William Mosley
E-Systems Inc., St. Petersburg, Fla.

This double-balanced mixer is ideal for use in frequency-division-multiplexed systems and because of its extremely wide dynamic range will also find use as the baseband mixer in phase-shift-keyed demodulator circuits. Operating linearly on input signals extending from 5 microvolts to 5 volts at frequencies from dc to 1 megahertz. the circuit owes its wide range to a combination of factors, notably a balanced output-stage configuration, low offset voltages in its switching circuits, low local-oscillator feedthrough, and the low-noise output of the active devices used. The mixer has the additional advantages of a very low output impedance and extreme stability over a wide temperature range.

The mixer is shown in the figure. A_1, serving as a wideband buffer amplifier with selectable gain, routes input signal f_i toward A_2 through the CD4016 quad analog switches. The CD4016, which contains four transmission gates, is turned on through A_3 by a local oscillator signal (which equals, in this case, $2f_o$), alternately switching A_2 between its inverting and noninverting modes. This action varies the gain of the amplifier from +1 to −1, so that the amplifier performs a chopping (mixing) operation on the input signal.

The input signal, f_i, is thus translated into a frequency of $f_i \pm f_o$ at the output of A_2. Mixer balance is achieved by using the combination of a symmetrical driving source for the switches (Q and $\overline{Q}$ output of the 4013) and a symmetrical input circuit for amplifier A_2. Thus the local oscillator (carrier) will be effectively suppressed at the output—feedthrough will be 60 decibels below the amplitude of the f_i signal. Double balancing ensures that the input-signal feedthrough will also approach the −60-dB value.

An added benefit of A_2's balanced input circuit is that the switch-transient feedthrough is reduced. This is because the pulses introduced to both ports of the op amp are about equal, and because the differential input voltage is therefore near zero, the output of A_2 is approximately zero for these transient components. □

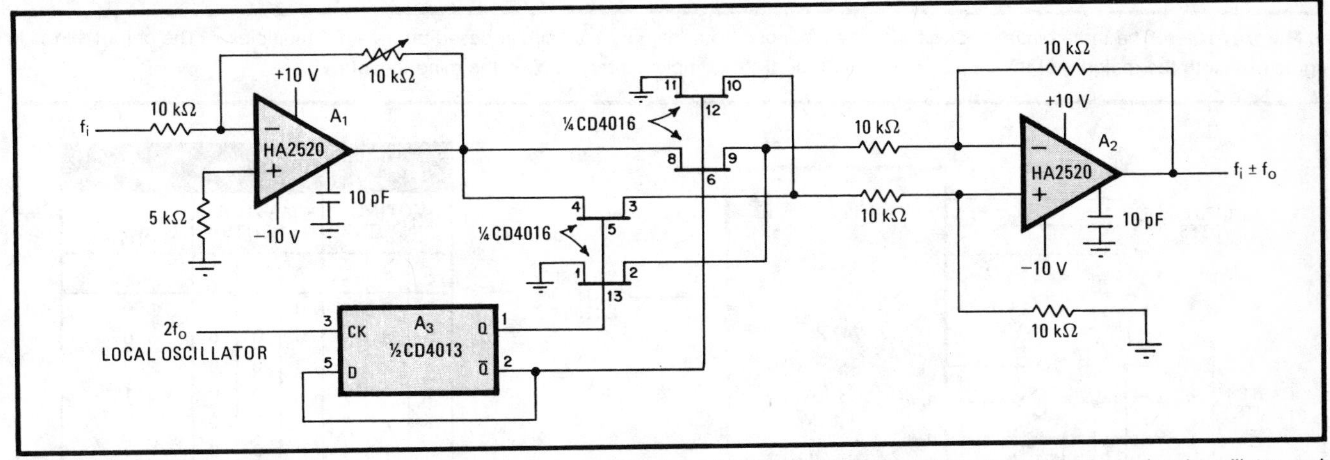

Dynamic. Mixer operates linearly over input-signal range extending from 5 μV to 5 V. Double-balanced circuit reduces local-oscillator and input-signal feedthrough to −60 dB below f_i. Switch-transient feedthrough is reduced by A_2's balanced input circuit.

Multiplexer chip forms majority-vote circuit

by Edwin P. Crabbe Jr.
GTE Automatic Electric Laboratories, Northlake, Ill.

□ An eight-channel digital multiplexer chip such as the 74151 can make an efficient three-input majority-vote circuit in which the output-logic level always agrees with the majority of the input levels. The uses of such a circuit range from the innocuous—decision making for electronic games—to the critical—reliability enhancement of triply redundant fail-safe systems.

As shown in the schematic of Fig. 1, the votes are registered at data-select lines A, B, and C of the multiplexer, and the data inputs, $D_0 - D_7$ are wired so as to produce an output at Y in accordance with the truth table.

With the addition of a few inverters, the system can be equipped with a master override, vesting any of the three voters with the power of veto. Figure 2 details the setup. A logic 1 at master-input A causes an output in agreement with the A vote, regardless of the votes of B and C.

Similarly, a logic 1 at master-input B effects an output equal to B's vote, and so on. With all the master inputs low, the output is again the majority of the inputs.

Systems with more than three voting inputs can only be practically implemented with a programable read-only memory, as majority-vote circuits cannot be cascaded. □

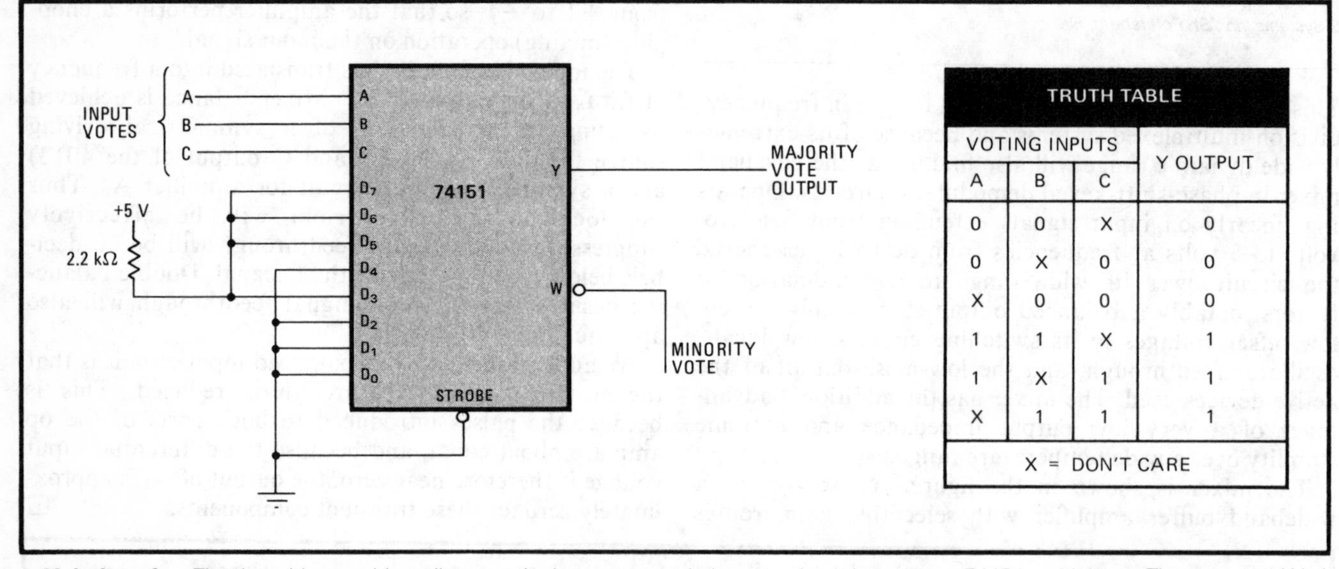

TRUTH TABLE

VOTING INPUTS			Y OUTPUT
A	B	C	
0	0	X	0
0	X	0	0
X	0	0	0
1	1	X	1
1	X	1	1
X	1	1	1

X = DON'T CARE

1. Majority rules. The three binary address lines are the inputs to a majority-vote circuit based on a 74151 multiplexer. The output at Y is in agreement with the majority of the inputs, and the output at W, complementary to Y, is the minority vote.

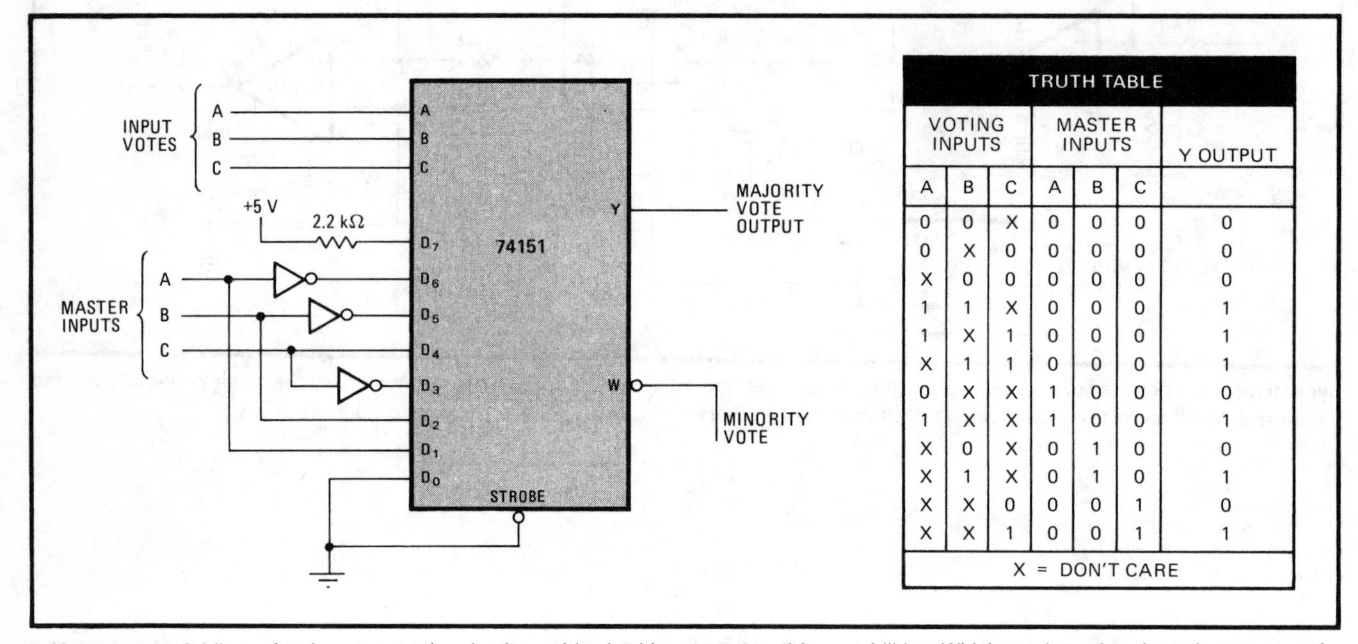

TRUTH TABLE

VOTING INPUTS			MASTER INPUTS			Y OUTPUT
A	B	C	A	B	C	
0	0	X	0	0	0	0
0	X	0	0	0	0	0
X	0	0	0	0	0	0
1	1	X	0	0	0	1
1	X	1	0	0	0	1
X	1	1	0	0	0	1
0	X	X	1	0	0	0
1	X	X	1	0	0	1
X	0	X	0	1	0	0
X	1	X	0	1	0	1
X	X	0	0	0	1	0
X	X	1	0	0	1	1

X = DON'T CARE

2. Veto power. Adding a few inverters to the circuit provides it with master-override capabilities. Whichever input is selected as master, the output always agrees with the vote of that input, regardless of the votes of the other two.

Intelligent multiplexer increases processor efficiency

by Edward Harriman
Boston, Mass.

The most efficient way to multiplex data into a computer is to initiate interrupts only when necessary. This circuit does just that. It polls the input lines and only in the event of an input change does it interrupt the processor, otherwise freeing it to attend to other tasks and therefore speeding overall system operation.

Shown in the figure, the circuit has worked successfully with the Motorola 6800 microprocessor. The hard-

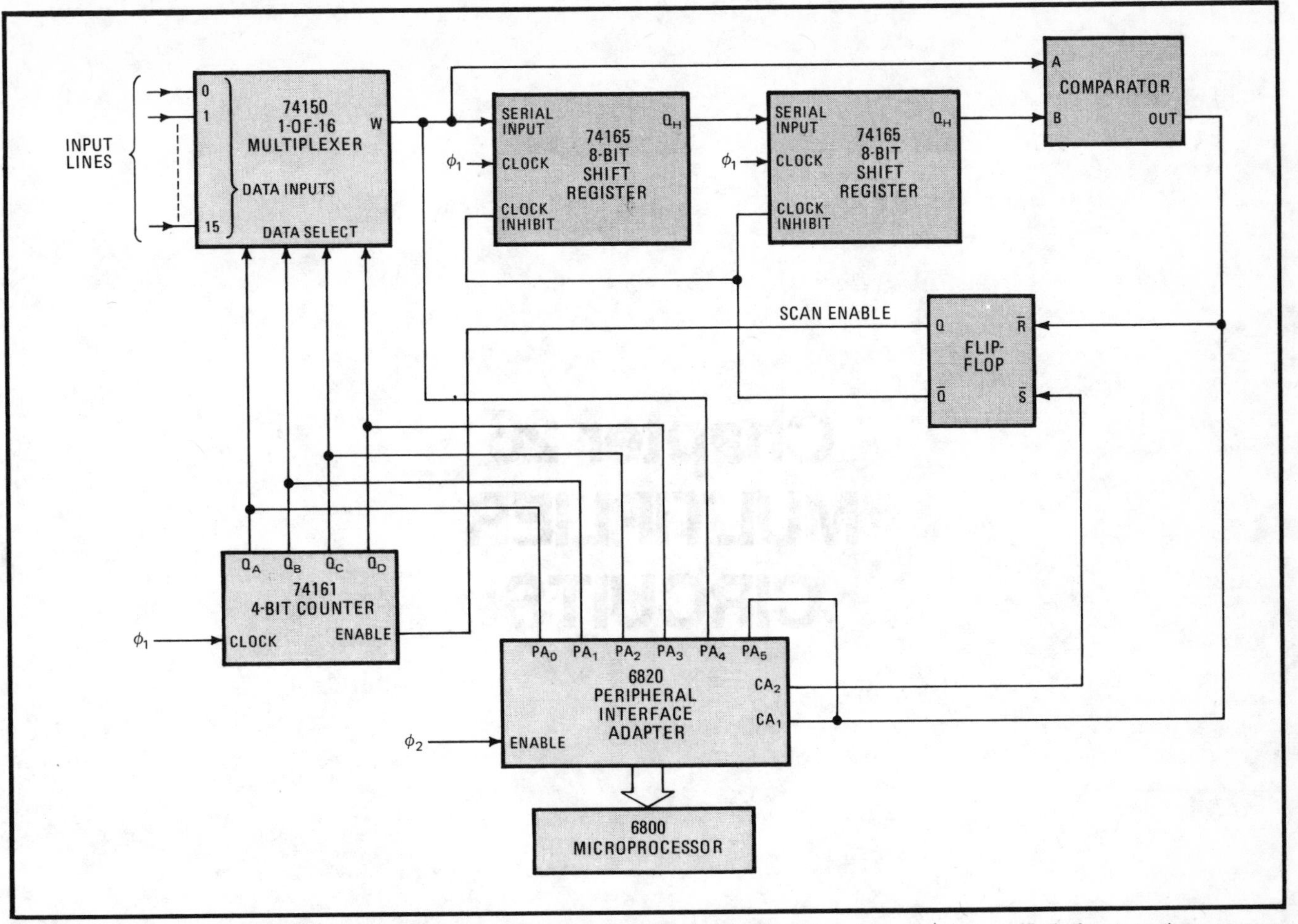

Smart controller. Sixteen lines are multiplexed to microprocessor, which performs input update only when notified of a state change on any line. The 6820 peripheral interface adapter is programmed to generate interrupt on negative transition of CA_1, which in turn generates negative-going CA_2 to advance counter by one. For eight input lines, an eight-input multiplexer and just one shift register would be used.

ware includes a 74150 1-of-16 multiplexer, a 74161 4-bit counter for encoding each line, and two 74165 8-bit shift registers for recording the previous state of each line. These elements, in conjunction with a standard comparator and $\overline{RS}$ flip-flop, control the actions of a 6820 peripheral interface adapter feeding the microprocessor.

To initialize the system, the microprocessor loads the 16-bit shift register (the two 74165 devices) by performing 16 read operations. The counter, which like most of the circuit is cycled by the system clock, selects each of the multiplexer's 16 input lines in turn. Assuming the system was initialized at line 0, on the 17th read operation the input of that line appears at the multiplexer's output and is fed to the comparator to be checked against the previous 0 line bit now stored in the last stage of the 16-bit shift register. If the comparator detects a difference in the two logic levels, it generates an interrupt to the processor through the peripheral interface adapter and also resets the flip-flop. This forces the scan-enable line low, disabling the counter.

Meanwhile, whether or not there has been a change from the previous state, the multiplexer's output is stored in the shift register. After 16 scans, the output of the last stage of the shift register is again compared to the present state of the input location 0. The loading and comparison operation takes place each scan for every input line.

If an interrupt is generated, the microprocessor, through prior programming, reads the location of change and the new data through the peripheral interface adapter. This operation also sets the flip-flop by generating a negative-going pulse from the control line CA_2. The pulse sends the flip-flop's Q output high for at least one cycle, enabling the 4-bit counter to advance one count, regardless of the state of the comparator that initiated the halt. The microprogram is written to observe if this second scan encounters a change in the state of the next location—a more efficient procedure than releasing the microprocessor immediately. This is done by testing line CA_1, the interrupt-status control line, which is connected to the comparator output. □

Chapter 26
MULTIPLIER CIRCUITS

Rate multiplier controls noninteger frequency divider

by Michael F. Black
Texas Instruments Inc., Dallas, Texas

Frequency dividers capable of dividing by integer and noninteger values can be built inexpensively from very few parts now that synchronous binary rate multipliers are available on single chips. To increase the resolution of the noninteger value, the rate multipliers are simply cascaded.

The ratio at which division is performed is set in an indirect manner by the 5497 rate multiplier. This number lies between two values preset in the 54161 synchronous counter, n and n−1. The circuit divides the input frequency by a ratio directly proportional to the time the counter spends in the n mode versus the time it spends in the n−1 mode.

The number of input pulses rate multiplier C_1 passes to synchronous counter C_2 is proportional to input address I. In this instance the counter is preloaded at either 14 or 15 by inputs A_{in} through D_{in}. C_1's output (pin 6) is connected to the counter's input A_{in}. Address I consequently controls the percentage of time the counter spends at divide values n = 2 and n = 3.

The rate multiplier's pulse-train output frequency is $f = f_{in}(I/M)$, where M is the size of the rate multiplier (in this case $2^6 = 64$). This particular circuit configuration results in $f_{out} = M(f_{in})/(nM - I)$. The actual divide ratio is $n' = n - (I/M)$.

The value of M determines the size of the available frequency step. The circuit as shown has been used to set f_{out} from 4 to 6 megahertz in steps of about 30 kilohertz; adding one more six-line rate multiplier would bring the step size down to about 400 hertz. Frequency steps in hundredths of a hertz can be easily obtained by cascading more multipliers.

This divider circuit will generate the exact number of clock pulses per second desired, but there will be some phase jitter, with $\Delta\phi = 360/n$. □

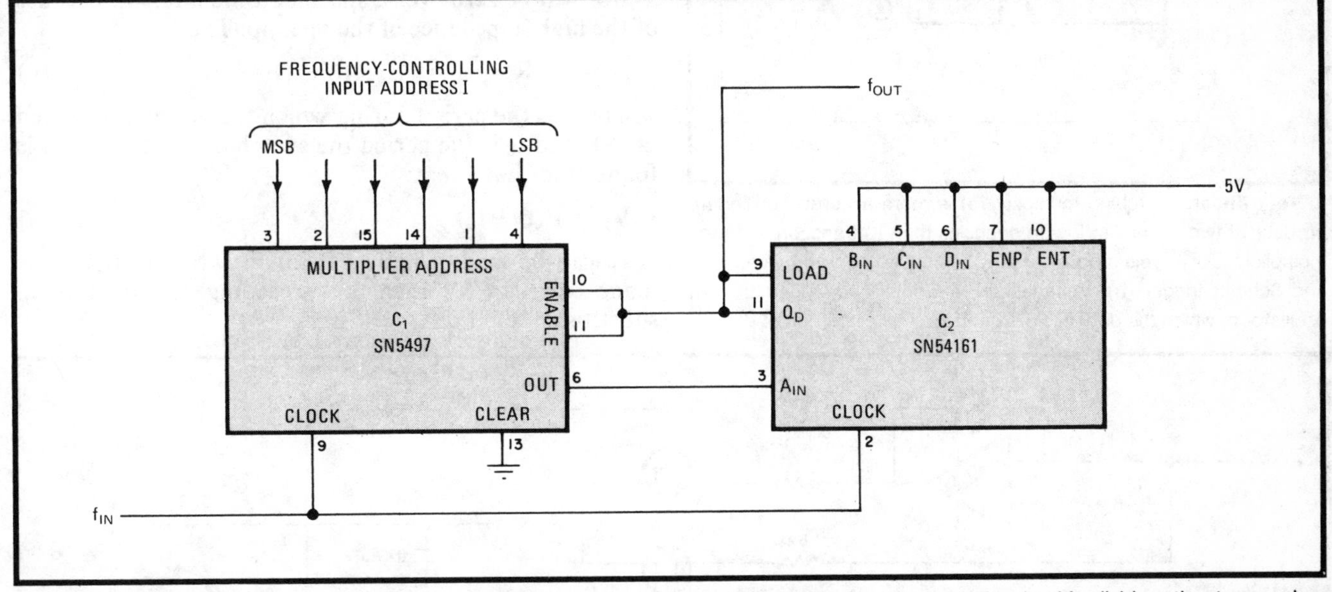

Continuous division. Synchronous frequency counter uses rate multiplier in two-chip circuit to program circuit's divide ratio at any value. Output frequency is proportional to the time spent between two preset divide values, n and n−1. Multipliers can be cascaded for step-size resolution all the way down to hundredths of a hertz. The amount of phase jitter at the output, in degrees, equals 360/n.

Switching multiplier is accurate at low frequencies

by Harold Anderson and Peter Hiscocks
Ryerson Polytechnic Institute, Toronto, Canada

When called upon to multiply low-frequency analog waveforms, switching-type multipliers are at least an order of magnitude more accurate than those that work on the principle of variable transconductance. The typical output error of a transconductance multiplier such as the MC1595 is 1%. But this circuit finds the product of two signals to within 0.05% of the true value and costs only $5.

In the basic switching multiplier (Fig. 1a), an analog signal of constant voltage V_x is applied to the circuit. The voltage at point V_a at any given instant is a function of the switch position, and the switch position in turn depends upon the control signal emanating from the duty-cycle modulator.

The average voltage at point A is:

$$V_a = V_x(t_{on} - t_{off}) \quad (1)$$

as shown in the graph, where t_{on} is the time during which the switch remains in contact with the $+V_x$ position and t_{off} is the time the switch remains in contact with the $-V_x$ position.

If the duty cycle, $t_{on}/(t_{on}+t_{off})$, can be made proportional to a second analog input voltage, V_y, then the following relation holds:

$$V_y = K(t_{on} - t_{off}) \quad (2)$$

where K is a constant. If Eqs. 1 and 2 are combined, the result is:

$$V_a = V_xV_y/K \quad (3)$$

A block diagram of a duty-cycle modulator which satisfies Eq. 2 can be constructed with an integrator network and Schmitt trigger (Fig. 1b).

The average current into the input port of the integrator will be zero over a specified time interval, because of the high impedance of the op amp. Thus:

$$\overline{i_1} = V_y/R = \overline{i_2} = (E/R)t_H - (E/R)t_L \quad (4)$$

where t_H is the period during which the switch is engaged at $+E$ and t_L is the period the switch dwells at $-E$. It is found from Eq. 4 that:

$$V_y = E(t_H - t_L) \quad (4a)$$

If t_H can be made equal to t_{on} of Fig. 1a and t_L can be made equal to t_{off}, then the preceding equation, when substituted into Eq. 1, yields:

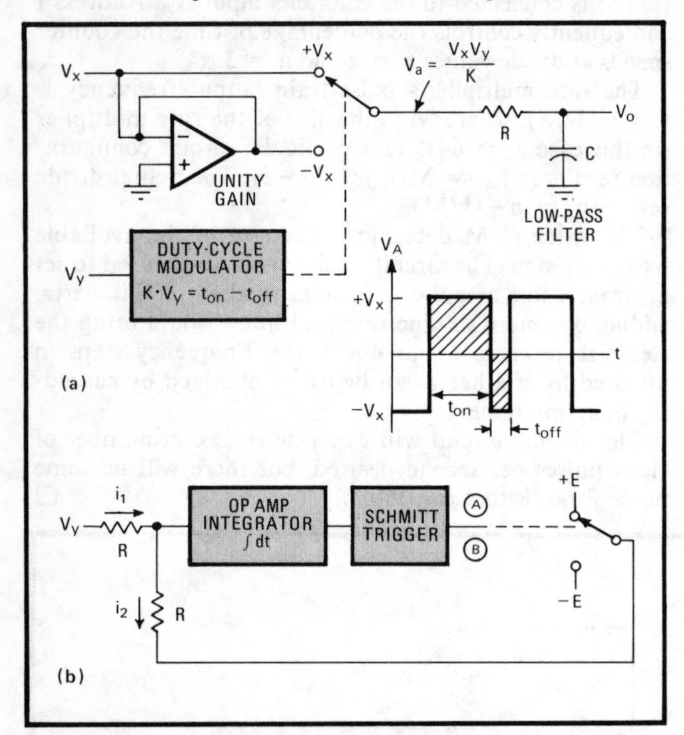

1. Very linear. Switching multiplier (a) is more accurate for finding product of two signals at low frequencies than IC transconductance multipliers. Duty-cycle modulator may be built with op-amp integrator and Schmitt trigger (b). Voltages $\pm E$ and integrator together form oscillator of which the duty cycle is controlled by amplitude of V_y.

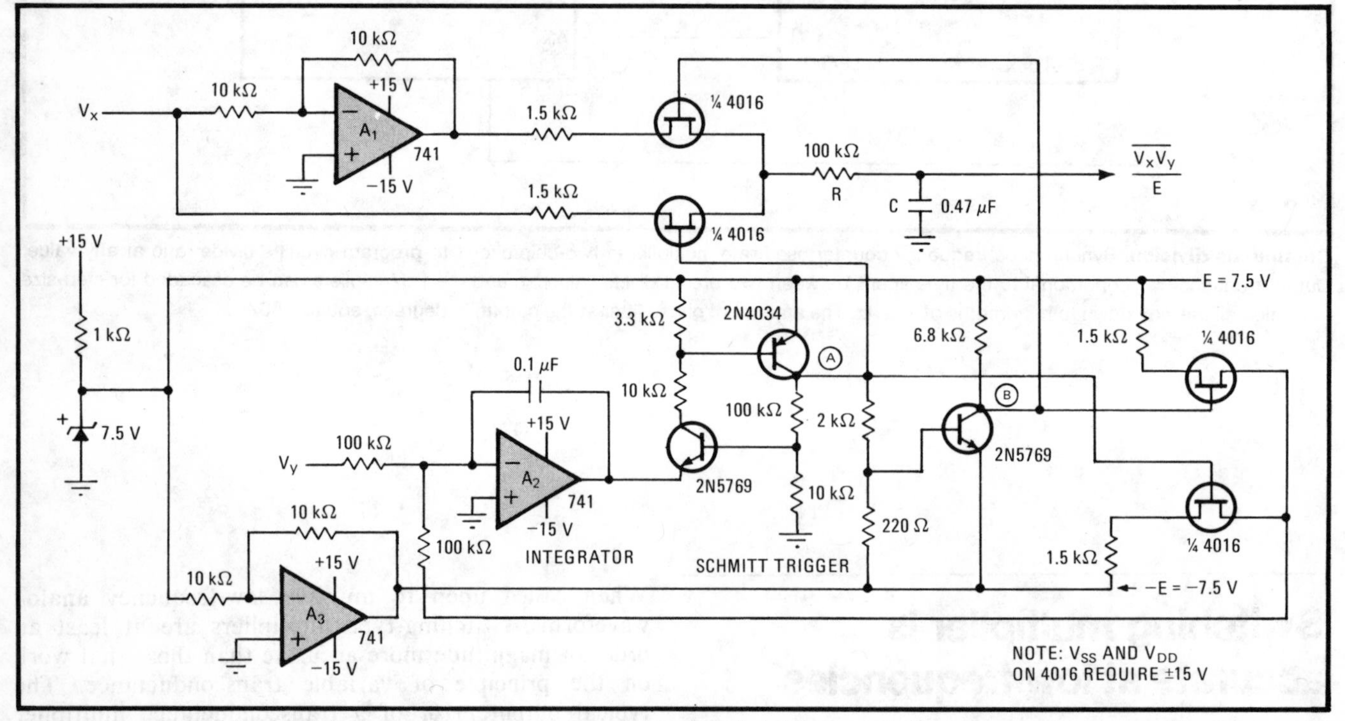

2. Multiplier. Building the circuit from the block diagram shown in Fig. 1 is straightforward. A_1 is inverting, unity-gain amplifier. A_2 serves as integrator. Three transistors form standard Schmitt trigger. Four transmission gates provide a practical switching circuit arrangement for $\pm E$, $\pm V_x$. Inverting operational amplifier A_3 ensures that $+E$ is equal in magnitude to $-E$.

$$V_a = V_xV_y/E \quad (5)$$

This condition is met by wiring the Schmitt trigger to engage not only the $\pm E$ ports but also the $\pm V_x$ ports of the circuit in Fig. 1a as well.

The block diagrams of Fig. 1 are therefore easily transformed into the practical circuit of Fig. 2. Note that voltages $\pm E$ are the feedback voltages to the summing integrator network at the input to A_2 necessary to ensure that $t_{on} - t_{off}$ is proportional to V_y. In essence, this part of the circuit is an oscillator, excited by V_y and driven by $+E$ or $-E$ feedback.

Also, although V_y controls the duty cycle, the basic oscillator frequency is virtually independent of it. (Note than when $V_y = 0$, $t_{on} = t_{off} + 0$).

The Schmitt trigger used is standard. The low-pass filter RC smooths out any transients that are caused by the switching process. Four complementary-metal-oxide-semiconductor transmission gates implement a practical switching circuit.

As suggested by Eq. 4a, this circuit does require that the magnitude of $-E$ tracks that of $+E$. Inverting op amp A_3 generates the mirror voltage required. □

Multiplier increases resolution of standard shaft encoders

by Frank Amthor
School of Optometry, University of Alabama, Birmingham

The resolution that can be attained by two-channel shaft encoders of the type used in speed controllers and optical-positioning devices may be increased by employing a digital frequency multiplier to derive a proportionally greater number of pulses from its TTL-compatible outputs. In this way, an up/down counter, which is normally driven by the encoder in these applications, can position the shaft more accurately and is more responsive to changes in speed and direction. Only two one-shot multivibrators and several logic gates are needed for the multiplier circuitry.

The circuit works well with a typical encoder such as the Digipot (manufactured by Sensor Technology Inc., Chatsworth, Calif.). In this case, the 128 square waves that are generated per channel for each shaft revolution (with output from the other channel in quadrature) are transformed to 512 bits per cycle.

When the shaft rotates in a clockwise direction, the output from port A of the encoder always leads the output from port B by 90°, and the logic will generate pulses only to the up input of the counter on both edges of both channel outputs. Thus, four pulses per square wave are generated. Rotation speed is limited by the duration time of the positive-edge–triggered one-shots, which should be kept to a few microseconds or less. Note that both the count-up and the count-down inputs of the counter are normally held high.

On the other hand, when the shaft's rotation is in a counterclockwise direction, the output of B leads that of A by 90°. In this case, four pulses per square wave are presented to the down input of the counter. □

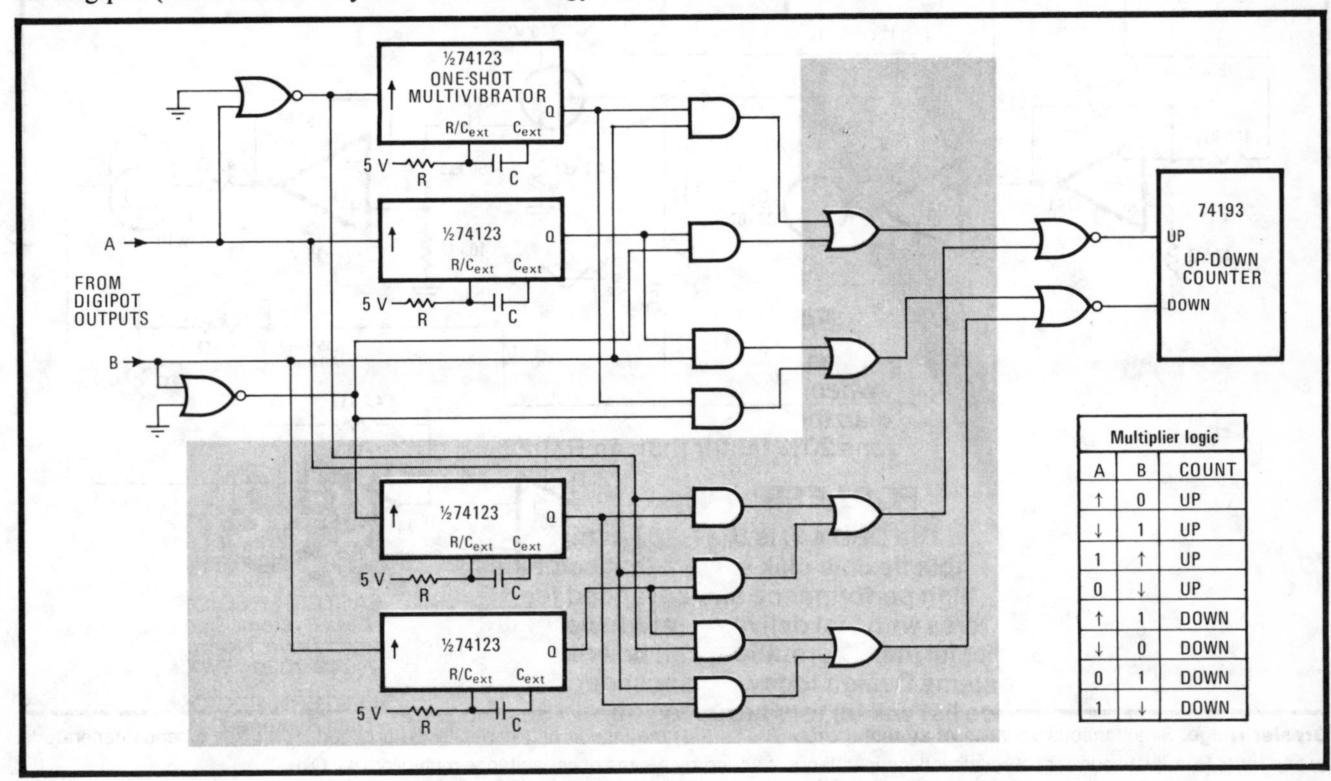

A	B	COUNT
↑	0	UP
↓	1	UP
1	↑	UP
0	↓	UP
↑	1	DOWN
↓	0	DOWN
0	1	DOWN
1	↓	DOWN

128 square-wave cycles (one revolution) from two-channel shaft encoder, for more resolution in speed controllers and optical positioning systems. Eliminating tinted area yields 256 bits per revolution.

Capacitance multiplier extends generator's sweep ratio

by A. D. Teckchandani
Eastern Electronics Ltd., Faridibad, India

Most low-frequency function generators can produce triangular waves over a frequency range of 100 to 1 by using the standard method—constant-current charging and discharging of a fixed-value capacitor. But the sweep ratio, or the ratio of the maximum to the minimum output frequency, can be extended to 500 or more by varying the magnitude of the current from the generators and the capacitance simultaneously. Using one potentiometer both to control the current source directly and to adjust the capacitance by means of a capacitance-multiplier circuit allows a greater frequency range, because the ratio of current to capacitance is varied over a wider range.

The frequency, or rate, at which a capacitor charges or discharges is directly proportional to the magnitude of the current from the generators and inversely proportional to the capacitance; that is, $f = Ki/C$, where K is a constant. This circuit expands the sweep ratio by ensuring that an increase in i is accompanied by a decrease in C, and vice versa.

The circuit shown below is so configured that the current sources formed by stages A_1, Q_1, and Q_2 aid in determining both the oscillation rate of a 555 timer, wired as a double-ended comparator (Schmitt trigger), and the amplitude of the current through timing capacitor C_o. The frequency of oscillation is also determined by stages Q_3 and A_2 and is equal to:

$$f = 1.5i/VC = i/3.6C$$

for V = 5.4, where:

V = supply voltage of 555
$C = C_o(1+|A_v|)$
$A_v = 4{,}700/(47+R_f)$ = voltage gain of A_2
R_f = on resistance of Q_2.

R_1 controls i, the current value being equal to $(10-V_1)/2R$. R_f is controlled by the voltage appearing at the gate of Q_3, which in turn is controlled by A_1 and ultimately R_1. Therefore, when there is an amplification in the A_2–Q_3 loop, the effective capacitance at the output of Q_1 increases by A_vC_o. (The phenomenon of output capacitance increasing with active-device gain was first observed in the Miller effect). Capacitance multiplication is thereby achieved by varying R_1.

The frequency range of the triangular waves that can be produced by this circuit varies from about 10 cycles to approximately 5,000 cycles. □

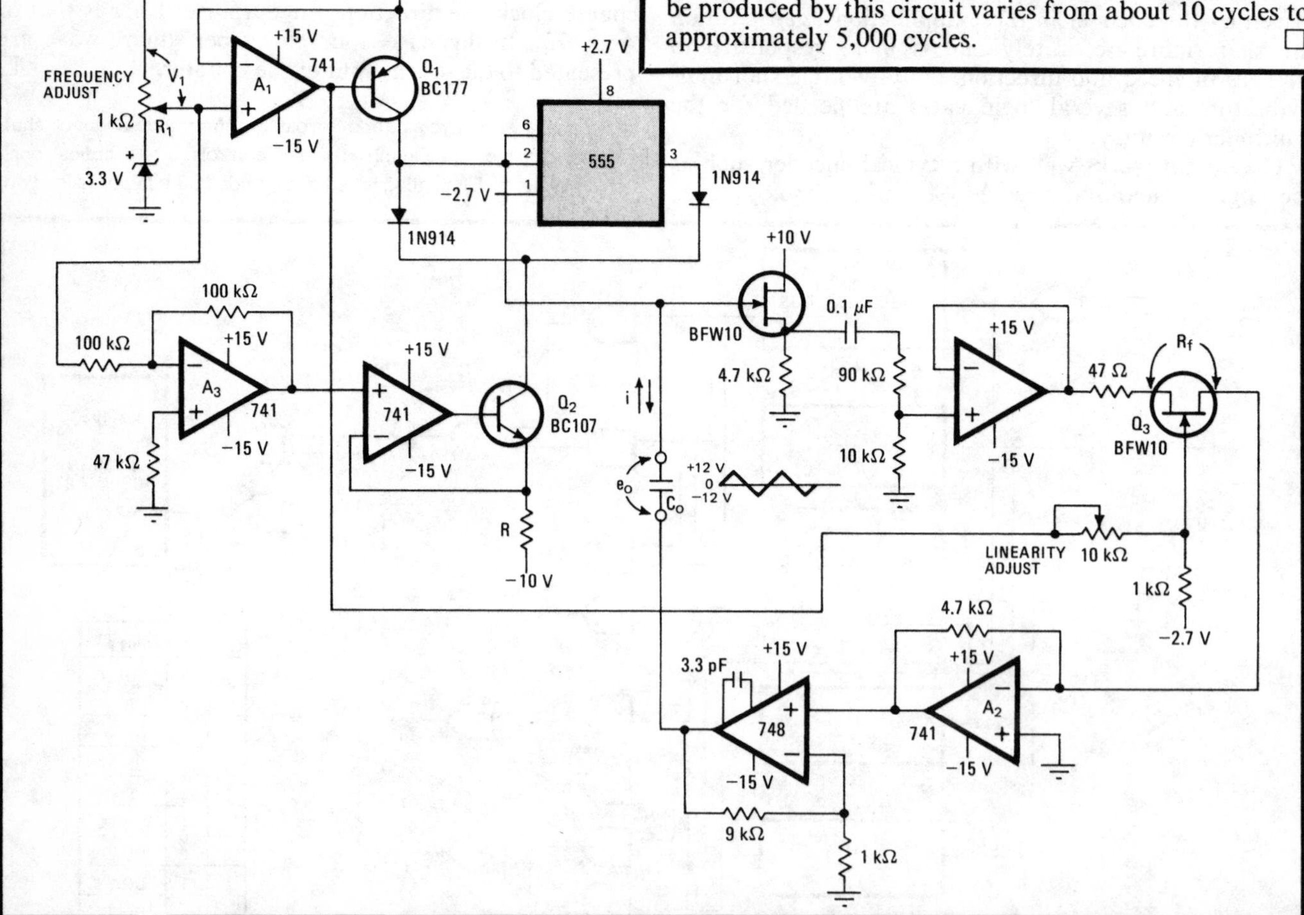

Greater range. Simultaneous variation of current-source (A_1, Q_1, Q_2) magnitude and capacitance charged by source extends generator's sweep ratio. R_1 adjusts current source directly; capacitance is varied by means of capacitance multiplier (A_2, Q_3).

Chapter 27
MULTIVIBRATORS

Dual one-shot keeps firmware on track

by Patrick L. McLaughlin
Teletech LaGuardia Inc., R&D Labs, Lafayette, Colo.

By noting the absence of pulses generated by status-reporting statements inserted in a running program, this missing-pulse detector reinitializes a microprocessor-based system when glitches on the power line or peripheral circuitry occur. The circuit provides more efficient system performance than a periodic reset timer and is much less expensive than installing line filters or isolators. Only one chip is required—a dual retriggerable monostable multivibrator.

Problems created by a power glitch—such as shuffling of information in the data registers and program jumps to undefined locations or to a location that gives rise to infinite loops—are conventionally solved by placing a timer in the system's reset line to initialize the system every 15 minutes or by using a brute-force power-line filter or even a dynamotor power isolator. A timer probably offers the best low-cost solution, but system speed is degraded by the unnecessary periodic interruptions.

A better solution is to provide a way for the program to report to the system hardware that it is running and on track. Using the 74123 dual one-shot, as shown in the figure, to monitor so-called report statements that are entered in the program's housekeeping loop automatically resets the microprocessor if and when the reports stop for longer than a specified period.

In general operation, both one-shots (one serving as the missing-pulse detector, the other as the output timer) trigger each other alternately in an astable, free-running mode, with R_1C_1 setting the report window, t_w, and R_2 setting the reset time, t_s. On power up, pin 12 of the 74123 is low and the processor is kept at rest until both one-shots time out. Then pin 12 is brought high, enabling the processor. If no report is made before time t_w, the cycle is repeated. An active-low series of report pulses made any time before t_w resets the missing-pulse detector (the output is Q_1), keeping pin 12 high and the processor running.

Usually, report statements are routinely entered before, after, or at both ends of the program's housekeeping loop and in most cases will be called frequently enough to fall within the t_w time window. In loops that may delay normal reporting, however, such as wait-for-data types, inclusion of additional report statements

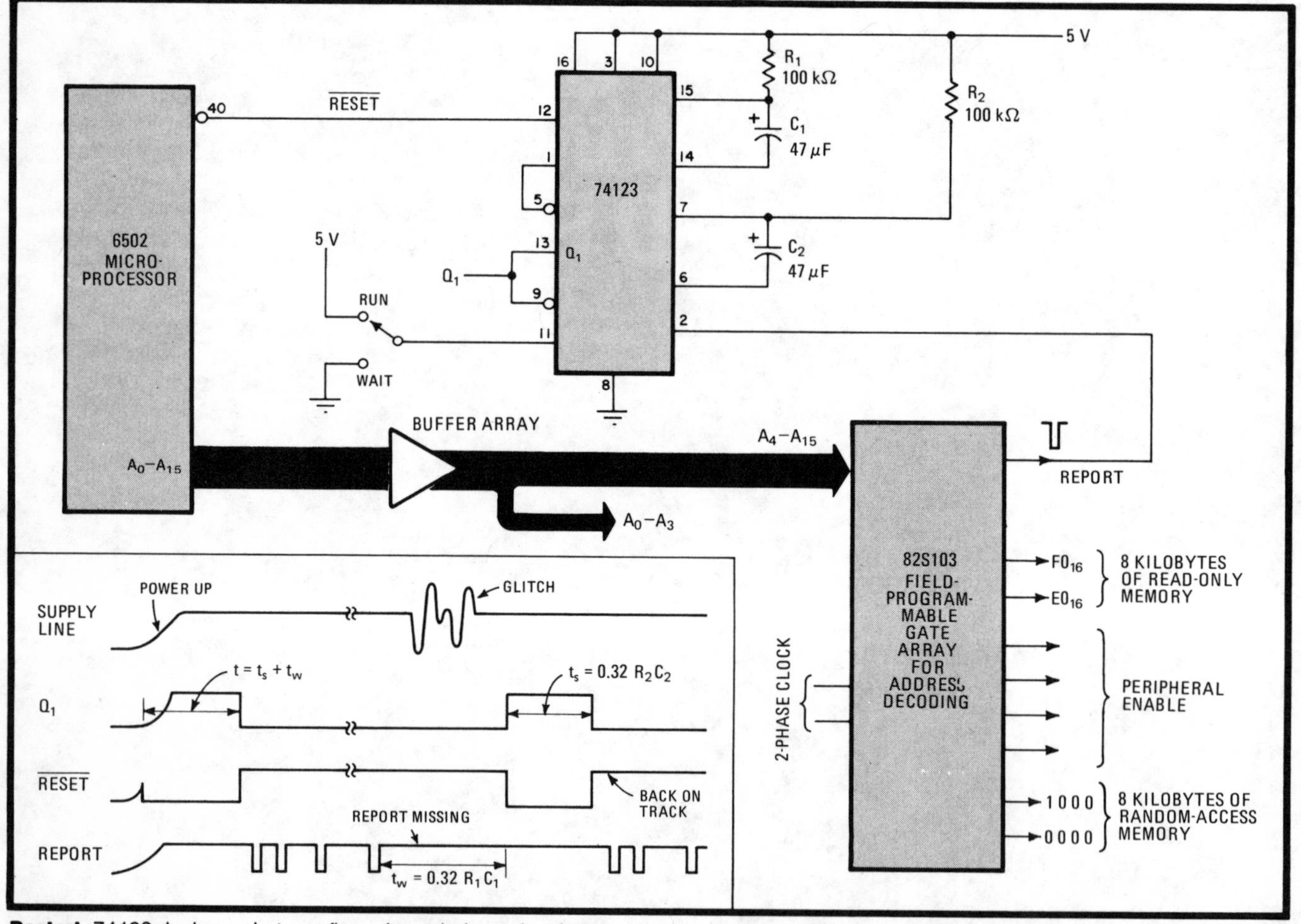

Restart. 74123 dual one-shot, configured as missing-pulse detector and output timer, detects absence of program report statements caused by power-line glitches in order to efficiently reinitialize microprocessor. Reports are entered as often as required in wait-for-data–type systems to ensure pulse rate falls within t_w window. Circuit accommodates static-type stop typical 8080/8085 wait instructions.

is advisable.

Note that if pin 11 is brought to ground, the one-shot at the output will be inhibited without resetting the processor. Thus, this circuit can accommodate static-type stops typical of the 8080/8085 wait instruction and is usable with slow-running programs and single-stepping arrangements.

The 74123 can be rewired to accept positive-going report pulses simply by introducing the report line to pin 1 of the chip and making pin 5 the output reset line. Pin 3 is then connected to 5 volts and pin 10 disconnected from the positive supply and connected to pin 4 instead. Finally, pin 2 is connected to pin 12. □

Differentiator and latch form synchronous one-shot

by Chacko C. Neroth
Amdahl Corp., Sunnyvale, Calif.

Many logic circuits require all operations to be synchronized with the system clock, including the firing of monostable multivibrators, even though the input signals to the one-shot are time-independent. However, a synchronous one-shot may be implemented using a D flip-flop and a differentiator network. In essence, the circuit substitutes a differentiator at the D input for a timing network in the one-shot in order to provide either immediate or time-delayed synchronous operation upon the arrival of a suitable trigger.

As shown in the figure, a positive-going input signal is applied to the D input of the 74C74 complementary-metal-oxide-semiconductor device through the resistance-capacitance network R_1C_1. The state of the flip-flop, which depends on the logic level at the D input during a clock pulse at the C input, assumes a 1 value at the arrival of the first clock. The output remains high until the voltage across the resistor has decayed below the 1 threshold of the D input (because of capacitor discharge), and the next clock occurs at the C input. Thus, assuming the clock period is high compared with the R_1C_1 period, the approximate pulse width, or on time, of the flip-flop is equal to 0.69 R_1C_1; the exact width of the pulse is an integral number of clock periods during which the D input is high.

The logic threshold of the C-MOS device at the D input is almost proportional to the supply voltage; therefore the pulse width output is relatively insensitive to supply voltage variations. For best results, the resistance-capacitance network should be selected to ensure that a decay in voltage across the resistor reaches the threshold level at the D input halfway between clock periods. Operation with negative-edge triggers is possible if R_1 is connected to the positive supply line instead of ground. □

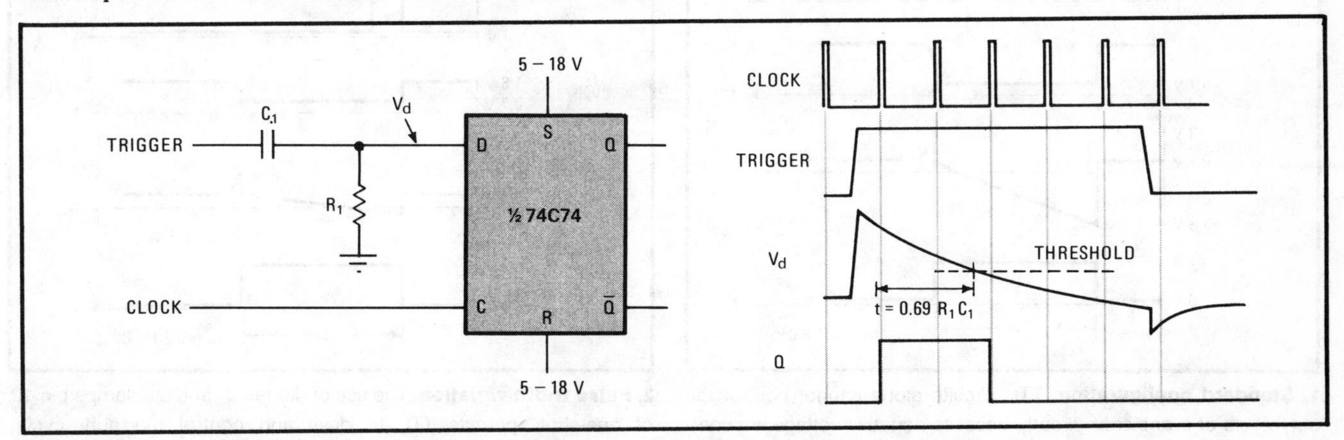

Monostable controller. Synchronous multivibrator is formed by differentiator and flip-flop, permitting initialization of clocked system by asynchronous data. Pulse width of multivibrator is determined by time constant of differentiator and clock rate.

One-shot multivibrator has programmable pulse width

by Stephen C. Armfield
MCI Inc., Fort Lauderdale, Fla.

The pulse width of a monostable multivibrator can be varied by digital control of its timing network. Using diode-modified gate circuits solves the interface problems inherent in driving the RC port with unipolar devices, while permitting the selection of resistors that shunt the timing capacitor to control its charging time.

As shown in the timing diagram in Fig. 1, a negative voltage is generated at pin 11 of a standard 74121 transistor-transistor-logic multivibrator although positive supply voltages are applied to the device. As a consequence of a triggering signal, the voltage at pin 10, which started at 5 volts, drops to 0.7 v. The voltage at pin 11 also drops by the same amount; since its initial voltage was only 0.7 v, however, its final voltage is −3.6 v. Thus, the timing (RC) network cannot be directly driven by standard TTL configurations.

With the use of diodes D_a and D_b, pin 10 can be clamped to about 1.6 volts without disturbing circuit operation, and the negative excursions at pin 11 will be restricted to a few tenths of a volt, as shown in Fig. 2. The 7405 open-collector gates can then be used in conjunction with isolating diodes D_1 through D_n to alter the charging rate of C. The alteration is accomplished by activating the desired digital inputs I_1 through I_n, which permit conduction through the isolating diodes, and consequently, shunting of resistance by resistors R_1 through R_n. The equivalent resistance is:

$$\frac{1}{R_T} = \frac{1}{R_1} + I_1\left(\frac{1}{R_1}\right) + I_2\left(\frac{1}{R_2}\right) + \ldots I_n\left(\frac{1}{R_n}\right)$$

where I_1 through I_n is equal to 1 or 0, corresponding to logic 1 or logic 0.

The current required by the clamping diodes D_a and D_b is 20 milliamperes or so and is supplied by a transistor internal to the multivibrator. If the increased power consumption can be tolerated, this programmable one-shot can be useful in many digital applications. □

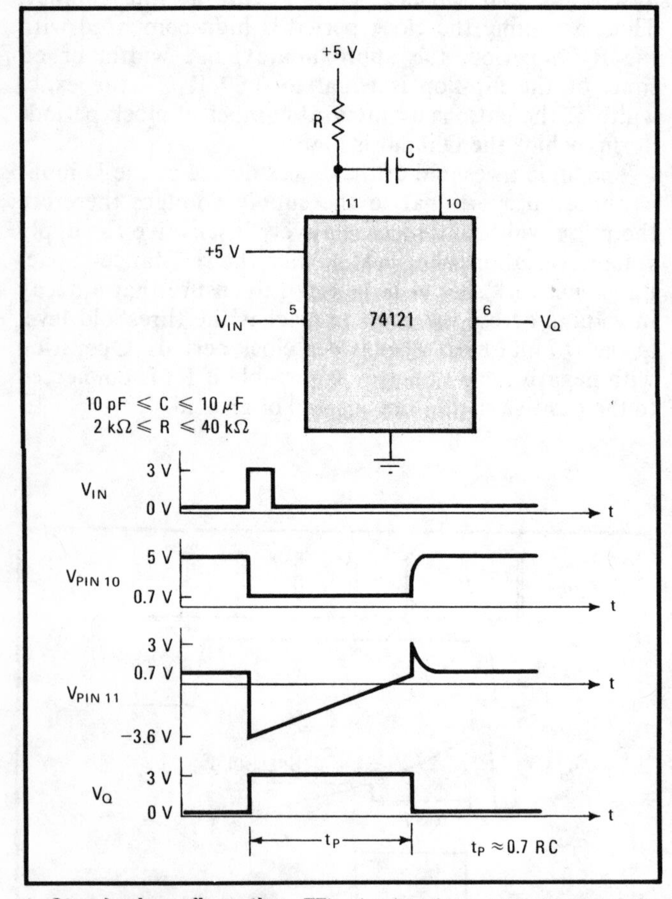

1. Standard configuration. TTL circuits alone cannot control the duty cycle of a one-shot directly because negative voltage is generated at timing port during normal operation. Reduction of this voltage to low level permits adjustment of pulse width.

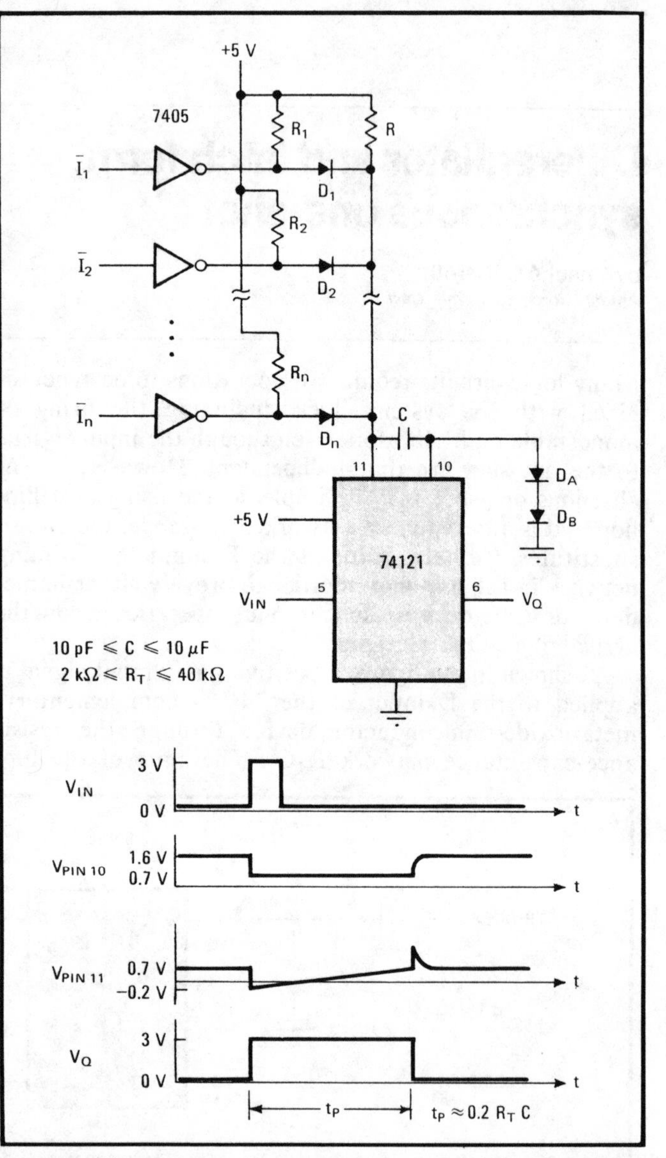

2. Pulse width variation. The use of diodes D_a and D_b clamps pin 10 of one-shot, permits TTL to drive and control the duty cycle. Actuating isolation diodes D_1 to D_n alters the charging rate of C, providing a choice of duty-cycle times.

Chapter 28
OPERATIONAL AMPLIFIERS

Current-compensated op amp improves OTA linearity

by Jacob Moskowitz
Raytheon Co., Portsmouth, R. I.

The control-input resistance of operational transconductance amplifiers which is nonlinear at low bias voltages, must be made linear if the device is to be used in certain voltage-controlled amplifier or automatic gain control applications. Here is a control scheme that provides a linear gain characteristic while preserving the full output-voltage swing of the OTA.

In OTAs such as the RCA CA3080, a forward-biased pn junction between the bias-current terminal and the chip substrate (negative supply) causes the nonlinearity of the control input resistance. Placing a resistor between the control voltage source and the bias terminal swamps nonlinear effects for large bias voltages, but fails to linearize the gain for biases of less than a volt or so.

A better solution, proposed by Walter Jung in the April 1975 Journal of the Audio Engineering Society, provides good linearity by means of a controlling op amp. This solution is practical for OTAs operating in an op-amp summing mode, but since the negative supply pin of the OTA is tied to the output of the controlling op amp, the output-voltage swing of the OTA is limited to values between the output voltage of the op amp and the positive supply voltage.

A linearizing scheme permitting full output voltage swing is shown in Fig. 1. In this version, a pnp transistor in the feedback path of the controlling op amp produces a current source linearly proportional to bias voltage V_C. This provides both gain-to-control-voltage linearity over a large control range (almost four decades), and minimum gain (in the linear portion) when V_C is zero. It is also much better suited for vca or agc applications than are four-quadrant multipliers, since control voltages below 0 will not produce negative outputs. Instead this circuit only turns off harder, which is most practical in audio work. In addition, the circuit of Fig. 1 requires no adjustment or trimming.

Figure 2 shows a similar circuit for use with negative control voltages. Since the control voltage is applied to the noninverting input of the op amp, the circuit exhibits a much higher input impedance than the circuit of Fig. 1, while the linearity characteristics are much the same. Both circuits offer a minimum linear gain at $V_C = 0$ and will only turn off harder for control voltages of the opposite polarity. Both allow full bipolar output voltage swing. □

1. Linearizing the OTA. A pnp transistor provides current feedback for controlling op amp to linearize operational transconductance amplifier under bias of less than 1 V. The higher the transistor's current gain, the better.

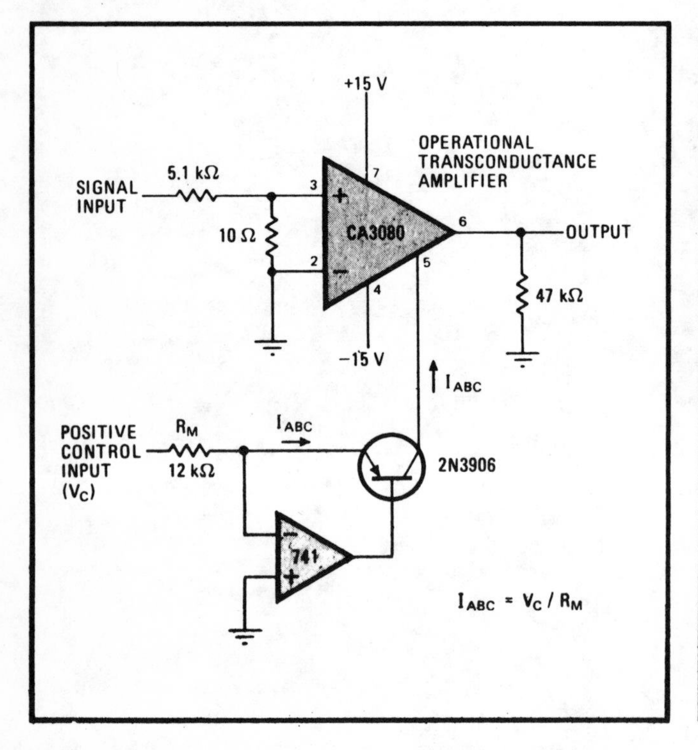

2. Negative OTA controller. For biasing OTAs with negative voltages, transistor/op-amp network is reconfigured as shown. Note controlling voltage is applied to noninverting input of op amp; input impedance is thus much greater than in circuit at left.

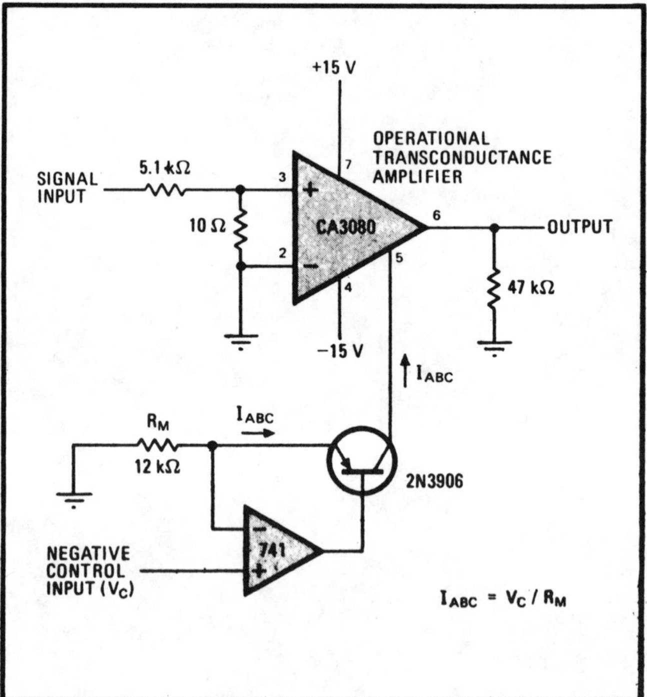

Keyboard programs the gain of an operational amplifier

by P. A. Benedetti
LAFBIC-CNR, Pisa, Italy

Placing a standard keyboard and a few precision resistors in the feedback loop of an operational amplifier produces a handy gain-programmable amp, useful for generating any one of several equally spaced voltages at the push of a button. Applications vary from testing components to controlling a computer program that employs an analog-input channel.

As the figure indicates, depressing 1 of 16 keys on the normally open contacts of the keyboard selects the value of the feedback resistor placed across the 558 operational amplifier A_1, to which a fixed input voltage V_{in} has been applied. A_1's gain varies with feedback resistance, of course, and so the output voltage also varies and assumes 1 of 17 equally spaced values (including 0), depending on which button has been depressed. The resistance values in the feedback loop have been selected so that the output voltage at A_2 is:

$$V_{out} = \left[\frac{\text{Key number depressed}}{16}\right] V_{in}$$

As might be expected, the programmable-gain principle applies to a keyboard of any size. Resistor precision must vary accordingly, however, becoming greater as the number of keys increase.

Key-bounce effects are not a problem except in some computer-based applications. A solution is to include double-testing of contact points in the software. □

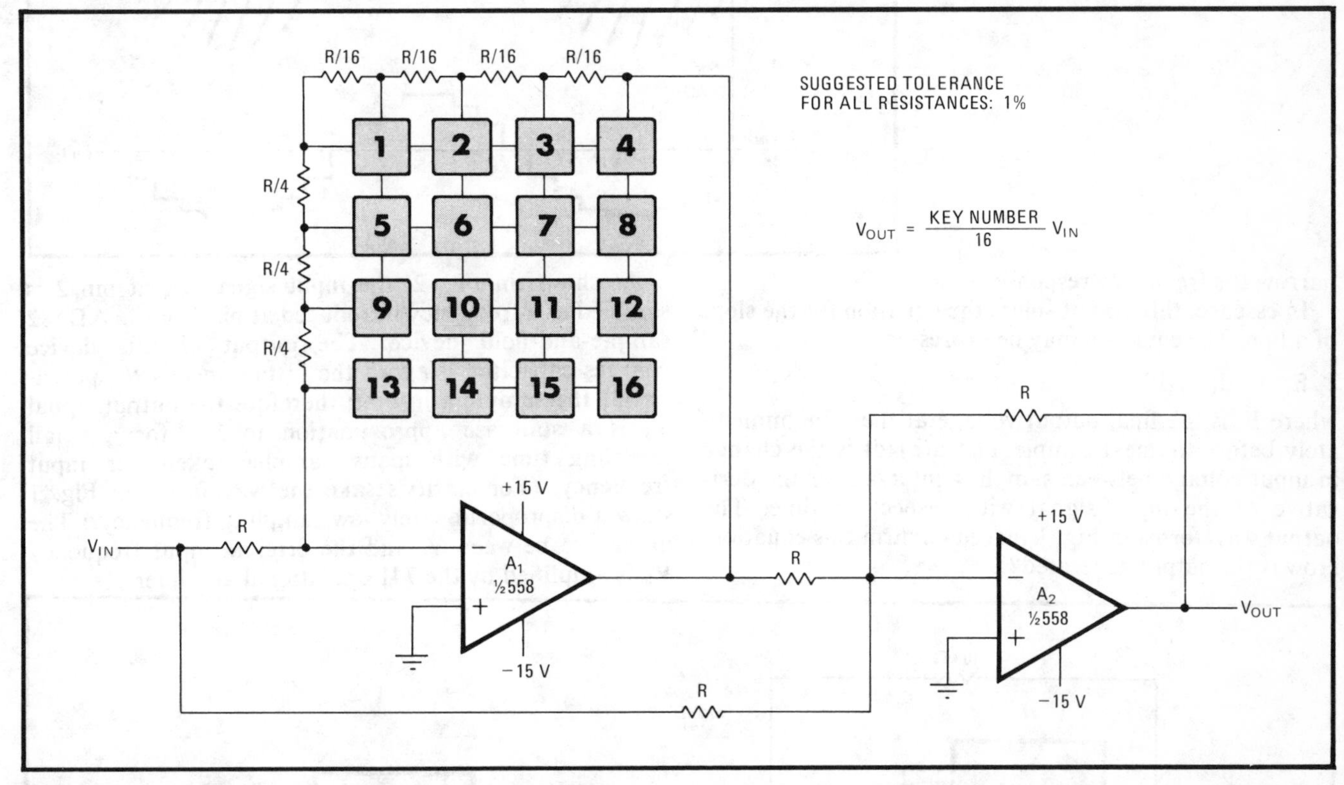

Digital control A standard keyboard and a few precision resistors in op amp's feedback circuit generate an output voltage proportional to the number on the key depressed. Circuit applications vary from component testing to analog-voltage control of computer systems.

Sample-and-hold and op amp form special differentiator

by John Nolte
University of Colorado Medical Center, Denver, Colo.

A sample-and-hold module and an operational amplifier can form a differentiator circuit that is especially useful at very low frequencies. In both accuracy and noise immunity it leaves conventional differentiators far behind, even when they are built with highest-quality components, i.e. low-drift, high-input-impedance amplifiers and high-value, low-leakage capacitors.

High-frequency noise, which can upset the operation of the common RC differentiator, is no problem with this "digital" circuit because its gain is independent of the noise frequency. In addition, its frequency of operation is set simply by adjustment of the sampling frequency. The circuit is therefore capable of a wide frequency response, quite unlike its analog counterpart, in which the components selected to reduce high-frequency noise also

1. Sample waveforms. Input signal V_{in} (shown here as cos *t*) is sampled at rate V_{T1}, and sample-and-hold output V_S is a sawtooth with an envelope that is the derivative of V_{in}, or $-\sin t$. Final sample-and-hold produces staircase of derivative, V_{out}.

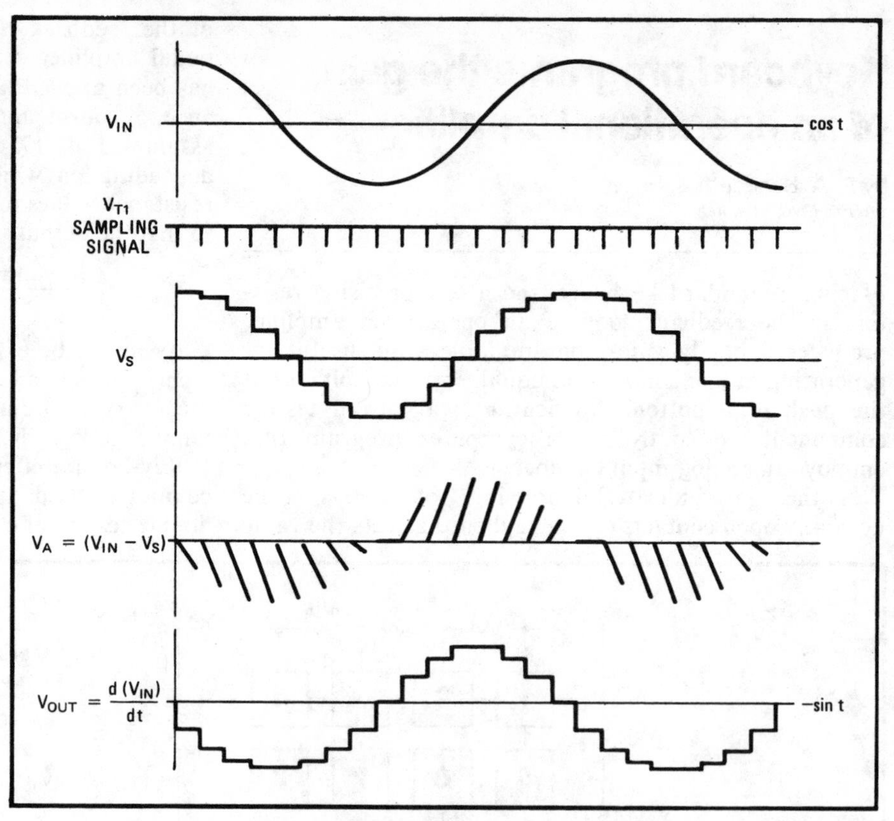

narrow the frequency response.

In essence, this circuit solves the equation for the slope of a line. The equation may be expressed as:

$$E_f = (d(e)/dt)dt$$

where E_f is the final output voltage at the time immediately before the next sample, and d(e)/dt is the change in input voltage between sampling intervals, or the derivative of the input signal with respect to time. The output waveforms in Fig. 1 indeed confirm this equation. How is the output generated?

As shown in Fig. 2, the input signal V_{in} at pin 2 is sampled at a frequency introduced at pin 1 of the AD582 sample-and-hold device. The output of this device charges capacitor C_H with the instantaneous V_{in} potential at the sampling instant; therefore the output signal V_s is a staircase approximation to V_{in}, for a small sampling time with many samples taken per input frequency. (For clarity's sake the waveforms in Fig. 1 show a disproportionately low sampling frequency.) The difference between V_s and the original input frequency V_{in} is amplified by the 741 operational amplifier.

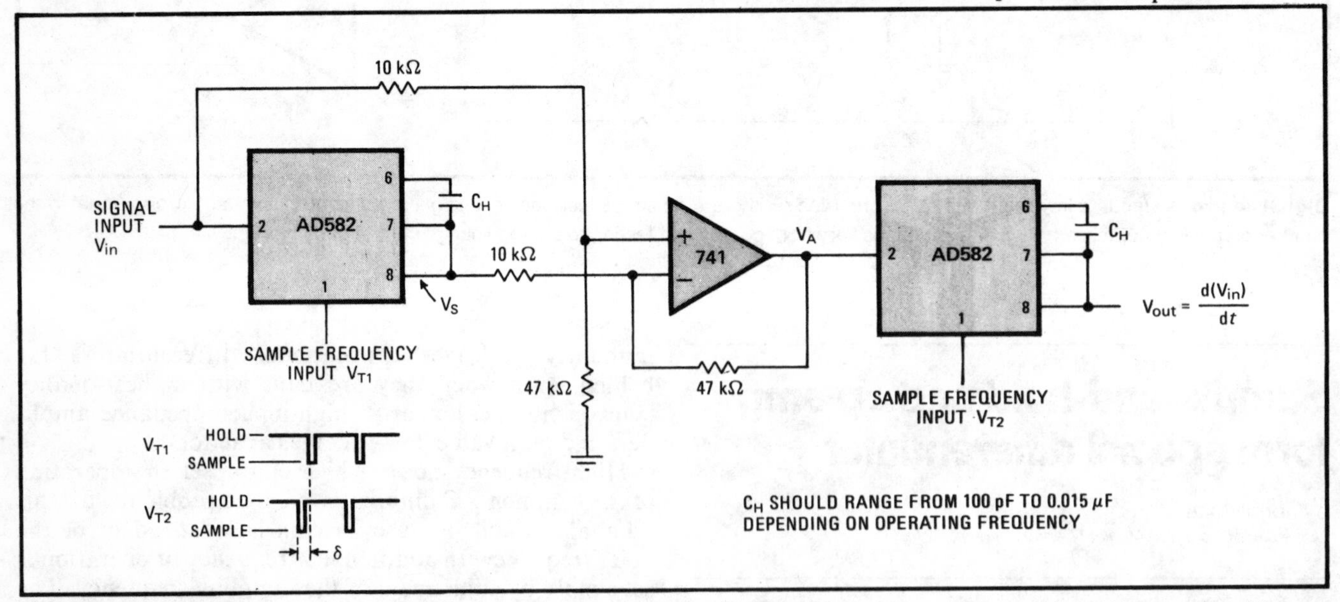

2. Different differentiator. Particularly useful at very-low-frequency signals, this differentiator uses a sample-and-hold and a difference amplifier to obtain the slope of the waveform. Second sample-and-hold recovers peaks of differentiated waveform. Negative pulse train (duty cycle approximately 1%) for sampling signals can be generated by a 555 timer circuit.

During each sampling period, the difference between these two signals is zero, making the output of the 741 (V_A) also zero. However, during each hold period, the output of the op amp rises to a value proportional to the change in input voltage with time. This proportionality is more or less linear, because of the small change in V_{in} versus V_s during a single sample period.

The output of the 741 is therefore a sawtooth waveform with an envelope (formed by the sawtooth peaks) approximating the derivative of V_{in}. To yield a true staircase approximation of the derivative, a second sample-and-hold device can be used to sample the peaks of the sawtooth, thus removing the unwanted components generated by the first sampling module.

For reasonable accuracy in differentiating a signal, the sampling frequency should be at least a few times, but preferably on the order of 100 times, greater than the input frequency. Decreasing the sampling frequency will lower the input frequency that produces a full-scale or maximized-value output. □

FET pair and op amp linearize voltage-controlled resistor

by Thomas L. Clarke
Atlantic Oceanographic and Meteorological Laboratory, Miami, Fla.

A matched field-effect transistor pair can be combined with an operational amplifier and a few resistors to form a circuit in which one FET's drain-to-source resistance (R_{ds}) bears a precisely linear relationship to a control voltage (V_c). Though a single FET can serve as a voltage-controlled resistor, the relationship of R_{ds} to the gate-to-source voltage (V_{gs}) is nonlinear.

The basic idea in this circuit is to control V_{gs} through a feedback loop that senses if the amount of current flowing through the FET, and hence its R_{ds}, is of the proper value. As shown in Fig. 1, this is accomplished by deriving a signal from half of the FET and applying it to a "summing" node at the inverting port of an op amp.

The output of the op amp is connected to the gate of the FET, thus forming a closed loop. The resulting change in V_c causes a proportional change in R_{ds} because the op amp is a linear device, and because input voltages are compared to a fixed voltage (V_{ref}) at the noninverting terminal. Depending on the configuration, R_{ds} can be made proportional to V_c or its reciprocal.

In the circuit to be seen at the left of Fig. 1, R_{ds} varies in proportion to the reciprocal of the control voltage, as is indicated by the following equation:

$$R_{ds} = |V_{ref}|R_c/(|V_c|-|V_{ref}|)$$

where V_{ref} is assumed to be between 0 and V_c. A voltage divider may be used to derive V_{ref}. Moving V_{ref} to the drain of the FET, as in the circuit to the right, the following equation holds:

$$R_{ds} = |V_{ref}|R_c/|V_c|$$

where V_{ref} should be a well-regulated source, since it may have to supply considerable current. These equations are based on the facts that V_c draws current from the negative input of the op amp through R_c and that this voltage drop results in current flow into the terminal by the FET. The application of Kirchhoff's law then

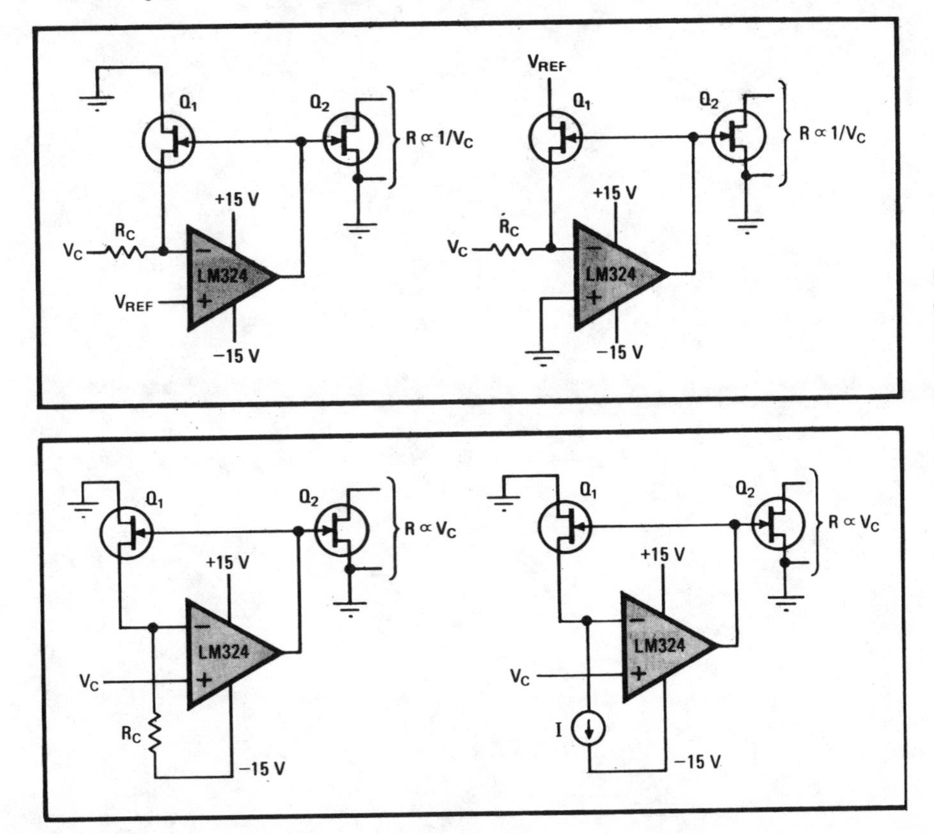

1. Voltage-controlled resistance. Unused half of FET Q_2 can function as voltage-controlled resistor in external circuits. R_{ds} is inversely proportional to control voltage in both circuits. V_c values are negative for n-channel FETs, positive for p-channel FETs.

2. Direct proportional control. R_{ds} varies linearly with V_c in both circuits. If V_c exceeds V_{ref} or breakdown voltages of FET in either figure, a resistor should be inserted between output of operational amplifier and gate of FET to prevent burnout.

yields the above relationships.

As shown in the circuit at the left of Fig. 2, R_{ds} may also vary in direct proportion to the control voltage. Therefore, the relationship becomes:

$$R_{ds} = R_c|V_c|/(V - |V_c|)$$

where V_c is greater than $-V_p$ but less than 0 v. This circuit, while not as linear as those of Fig. 1, can be improved significantly by replacing R_c with a current source, I, as shown at the right of Fig. 2. The relationship then simply becomes:

$$R_{ds} = |V_c|/I$$

The circuits are built with Siliconix 285 dual FET chips and LM324 op amps. Use of high-speed op amps such as the LM318 would permit more rapid variations of resistance. No stability problems are encountered because the FET introduces negligible phase shift in bandpass frequencies of the op amp. Optimal results are obtained, of course, with FETs formed on a common substrate, and if desired, p-channel devices may be used for positive control voltages. □

Chapter 29
OPTOELECTRONIC CIRCUITS

Epitaxial phototransistor with feedback has fast response

by Vernon P. O'Neil
Motorola Inc., Discrete Semiconductor Division, Phoenix, Ariz.

A high-gain negative-feedback loop will reduce the response time of an epitaxial phototransistor to 100 nanoseconds—a significant improvement over several schemes previously suggested.[1,2,3] Because of its construction, the epitaxial device all but eliminates the diffusion of carriers into its depletion region from the bulk collector region, which slows a conventional non-epitaxial phototransistor's operating speed. And added feedback reduces the input-signal swing across the collector-base junction to 1% of what it is normally, further reducing the input-capacitance charge and discharge times.

The MRD 300 phototransistor shown in the circuit has a typical rise time of 2.5 microseconds and a fall time of 4 μs if operated in the conventional emitter-follower configuration. In this modified circuit Q_2 serves as the feedback amplifier that keeps the base of the phototransistor at an almost constant voltage for changes in input-signal level. Thus the effective input capacitance that must be charged and discharged is reduced. Q_3 serves as a buffer. Note that using feedback that is negative enables the switching times to be maximally reduced without fear of creating instability (that is, oscillations can be generated with circuits using positive feedback).

With this circuit, both the rise time and the fall time of the phototransistor are reduced to 100 ns. The output voltage is equal to the product of feedback resistance (10 kilohms) and the collector-base photocurrent. The photograph shows a typical output waveform.

As for the phototransistor itself, it can be hard to determine from data sheets if one is epitaxial or not. The best way to find out is to consult the manufacturer. □

References
1. "Why not a cascode optocoupler?", Electronics, March 2, 1978, p. 132.
2. "Why not a cascode optocoupler? Here's why not", Electronics, April 27, 1978, p. 154.
3. "Bootstrapping a phototransistor improves its pulse response", Electronics, Aug. 17, 1978, p. 105.

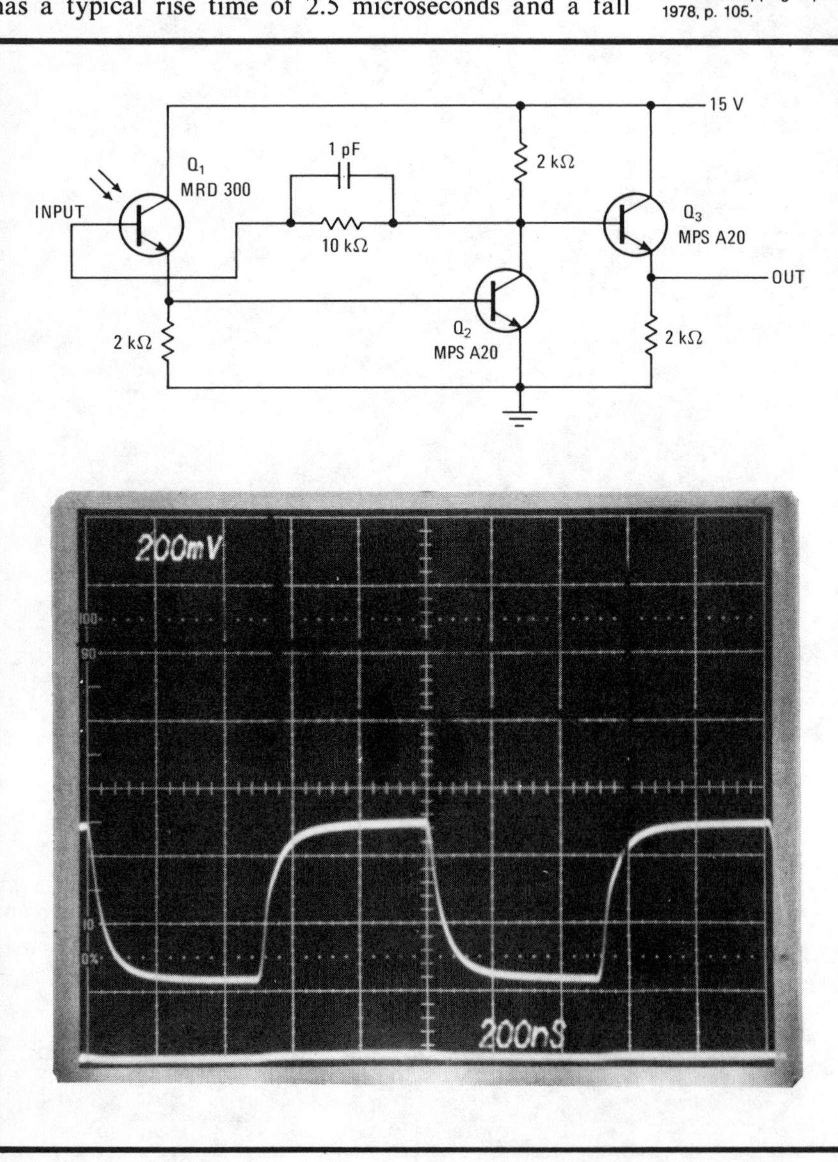

Speedy. Collector-to-base capacitance of phototransistor Q_1 is reduced by employing epitaxial device (MRD 300) and high-gain negative feedback (Q_2), so that operating speed can be increased. Emitter-follower Q_3 provides low-impedance output. Photograph shows typical output response.

Optoisolator initializes signal-averaging circuit

by J. Ross Macdonald, *Department of Physics and Astronomy, University of North Carolina, Chapel Hill*

Long-term averaging circuits require an initializing voltage on their capacitive storage element in order to become almost immediately operational on power up. Here, an optoisolator is used to quickly charge the capacitor with a voltage derived either from the input signal itself or from any dc voltage, the two sources most widely used. The optoisolator circuit is superior to an initializer that uses a relay, which, besides having the disadvantage of being electromechanical, also draws power continuously.

In a circuit that averages a signal over a long period (see figure), the resistor-capacitor (RC) time constant may be on the order of a minute or more. Thus, the output of the averager (V_o) during the time $t = 0-1$ minute is considered to be the circuit's transient response to the input signal, where t is measured from the time that power is applied to the circuit. In most cases, especially when the circuit is part of a more complex system, it is not feasible to wait that long before the RC network starts generating a true average value.

The difficulty may be circumvented by using an optoisolator and a switch, S_1, to charge C on power up. Assume it is desired to charge C from a dc voltage, V_s. When power is applied, C_a, which may be 25 microfarads or more, is charged through R_a. Consequently, as current flows through the photodiode, the value of the photoresistance element in the LM 6000 optoisolator is reduced from more than 10^9 ohms to about 1 kilohm. Thus, in a few tens of milliseconds, C charges to V_s through the element, if S_1 is placed in the V_s position. As C_a becomes fully charged, the resistance of the element quickly increases to at least 10^9 ohms, and the circuit is ready to operate in its intended averaging mode.

When power is removed, C_a discharges through D_1, so that the on-off power cycle can be repeated fairly rapidly. C also discharges slowly through R. This action is of little consequence in circuit operation on a subsequent power up. Note that S_2, a momentary-contact switch, allows the resetting process to be repeated at any time, even while the circuit is active.

To initialize C from the input signal, it is only necessary to connect S_1 to V_{in} prior to power up (or at any time if S_2 is utilized). Otherwise the initializing operation is the same as before. □

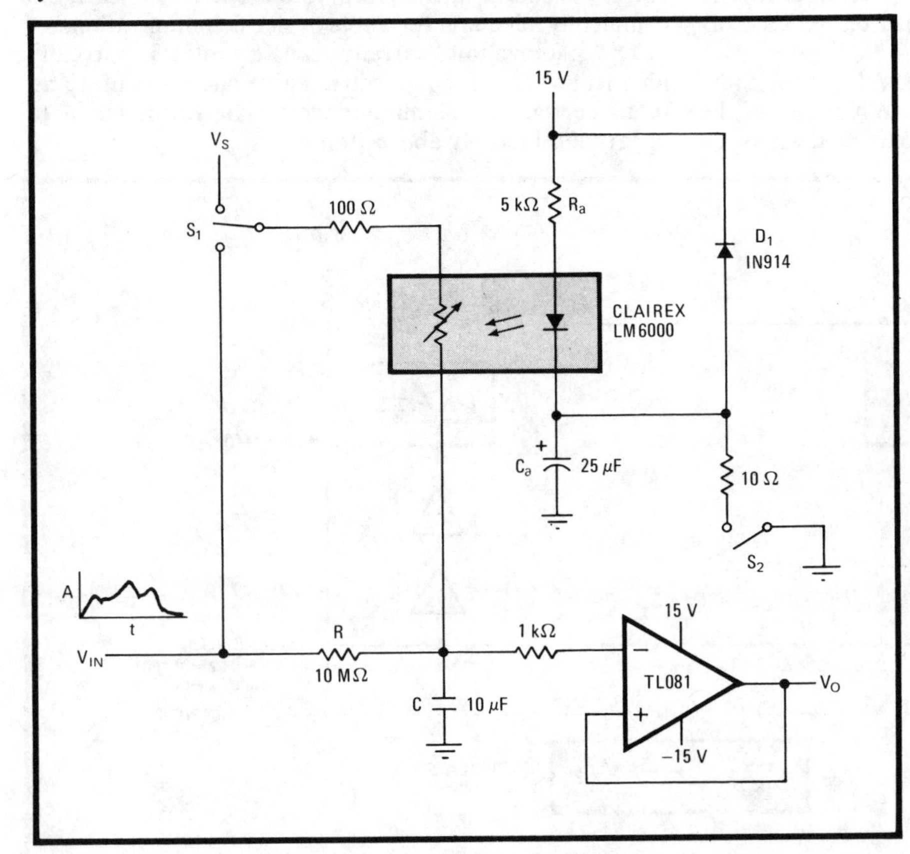

Speedy average. Optoisolator enables long-term averager to operate almost immediately after power up by presenting an initializing voltage to circuit's sampling capacitor, C. Charge is introduced through isolator's low-resistance photoelement. Either a dc voltage or the input signal can be used as the initializing source.

Bidirectional optoisolator puts two LEDs nose to nose

by Forrest M. Mims III
San Marcos, Texas

As conventional optoisolators employ a separate source and sensor, they can transfer current in only one direction. A few photodetectors and electroluminescent diodes can double as both a source and sensor, however, and when they are suitably connected they offer users a convenient way to build a low-cost bidirectional optoisolator, as shown here.

Two OP-195 LEDs, which have gallium-arsenide-silicon infrared emitters, can be made to transfer signals in either direction if they are placed nose to nose in a short length of heat-shrinkable tubing and secured in place by heating the tubing. Alternatively, the LEDs may be quite far apart if they are coupled by a plastic or glass-fiber waveguide.

In either case, the current transfer ratio (I_o/I_{in}) for the pair, with proper biasing, will be 0.06% for an input current of 20 milliamperes. This ratio is far too low for many applications but is good enough for some specialized roles where a bidirectional path is required. In any case, the output signal can be amplified or buffered, as necessary.

A logic-control voltage and two H11A1 optoisolators serve as the input/output port selector. Whichever of the OP-195 devices is designated the output diode may be connected in the reverse-biased photo-conductive mode or the unbiased photovoltaic mode. In the latter case, the output device is not biased. The response of the optocoupler operating in this mode for a given signal-input current is shown in the plot. Note the device linearity is completely adequate for duplex voice communication.

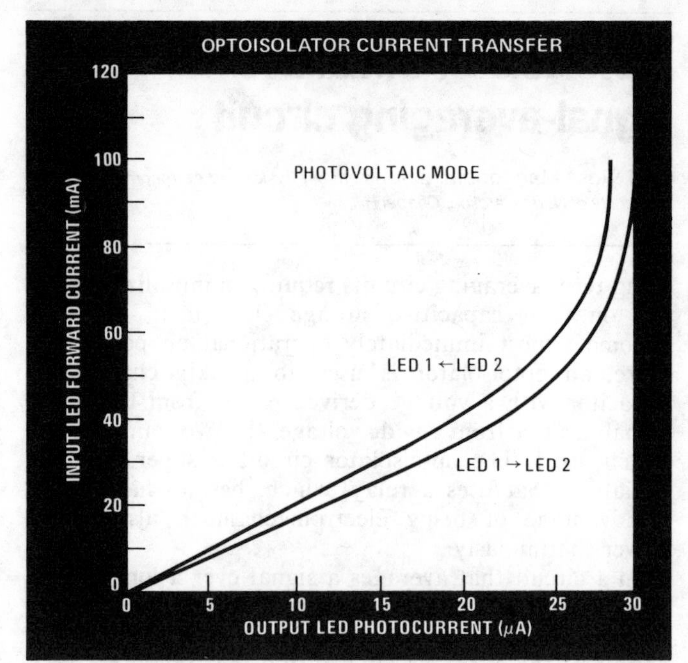

The photovoltaic current transfer ratio is virtually identical to that for photoconductive operation up to an input current of 20 milliamperes. The ratios begin to depart considerably above 40 mA. □

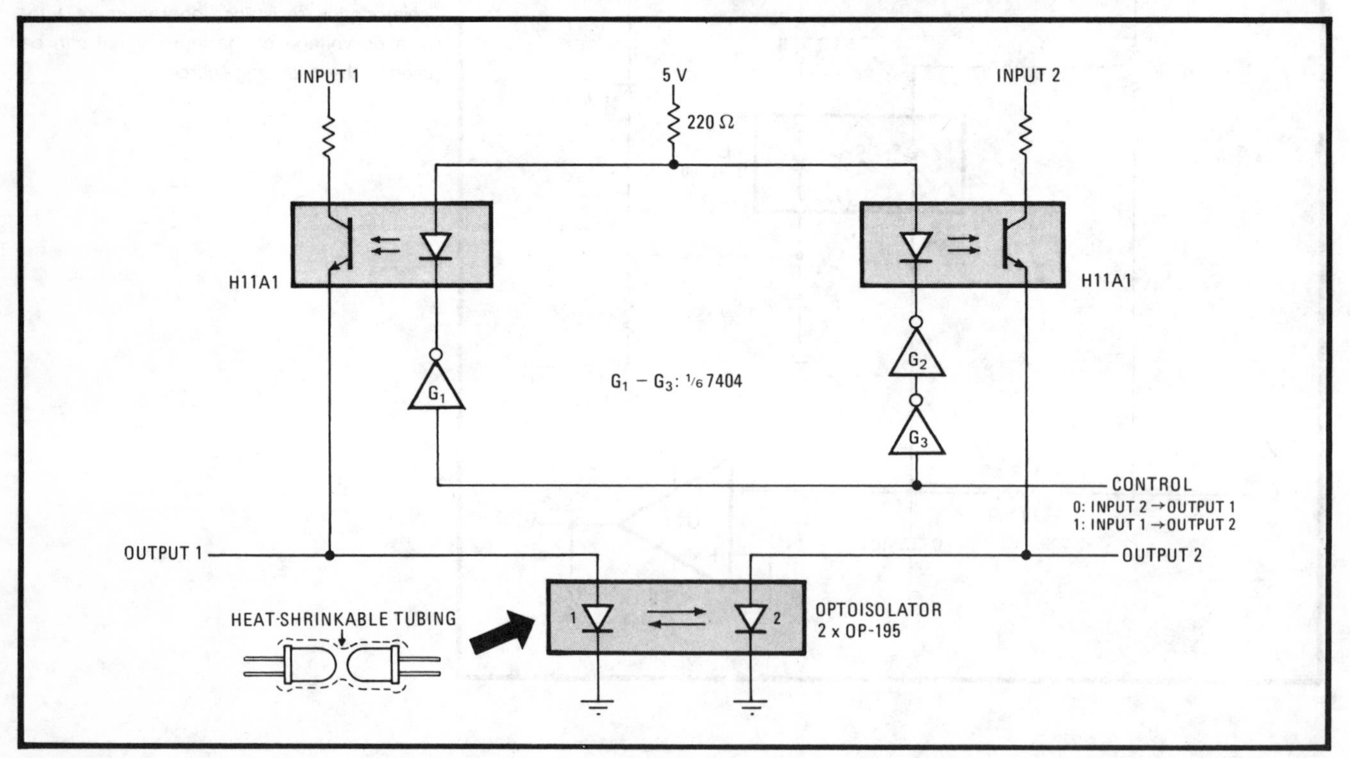

Either way. Standard light-emitting diodes encased in heat-shrink tubing can be made to function as a bidirectional transmission link. Alternatively, LEDs may be coupled through optical fibers. Control circuit for selecting input/output port arrangement is simple, using two optoisolators and three inverters. Circuit's current-transfer ratio suffices for many small-signal applications.

Optoelectronic alarm circuit is time-sensitive

by Forrest M. Mims III
San Marcos, Texas

Using an optoelectronic slot switch and a 556 dual timer operating as both a pulse generator and missing-pulse detector, this circuit generates an alarm when an opaque object blocks the light input for longer than a preset time interval. It has many applications and is especially useful when united with a slotted disk to monitor motor speed stroboscopically, indicating when the steady-state rotation rate is too high or low. It can also be used on the production line for checking the width of materials.

Generally, the output of the pulser periodically activates the light-emitting diode of the H13B1 switch. Other sensors may be used; Darlington photosensing transistors, though, are the most sensitive. In this case, the pulser's operating frequency is set at 1.42 kilohertz, but it may be suitably selected by replacing the 100-kilohm resistor at pin 1 with a potentiometer.

As shown, the H13B1 is built with a slot of several millimeters separating its LED from the output phototransistor so that objects can be placed in the air gap between them. When the slot is not blocked, the phototransistor continuously resets the missing-pulse detector. Should the light path be blocked, pin 8 will remain high and the threshold voltage at pin 12 will fall at a rate determined by the adjustable R_1C_1 time constant.

Depending on the value of this constant, which can be selected for delays from microseconds to seconds, the detector will generate a step voltage if it is not reset within that period. The signal is then inverted by Q_1, which in turn fires the silicon controlled rectifier to drive the load, R_L. □

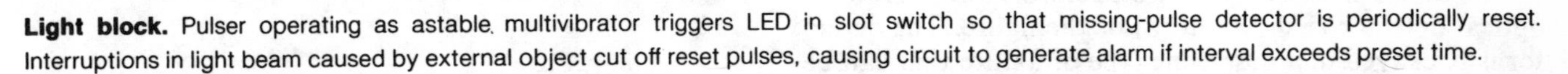

Light block. Pulser operating as astable multivibrator triggers LED in slot switch so that missing-pulse detector is periodically reset. Interruptions in light beam caused by external object cut off reset pulses, causing circuit to generate alarm if interval exceeds preset time.

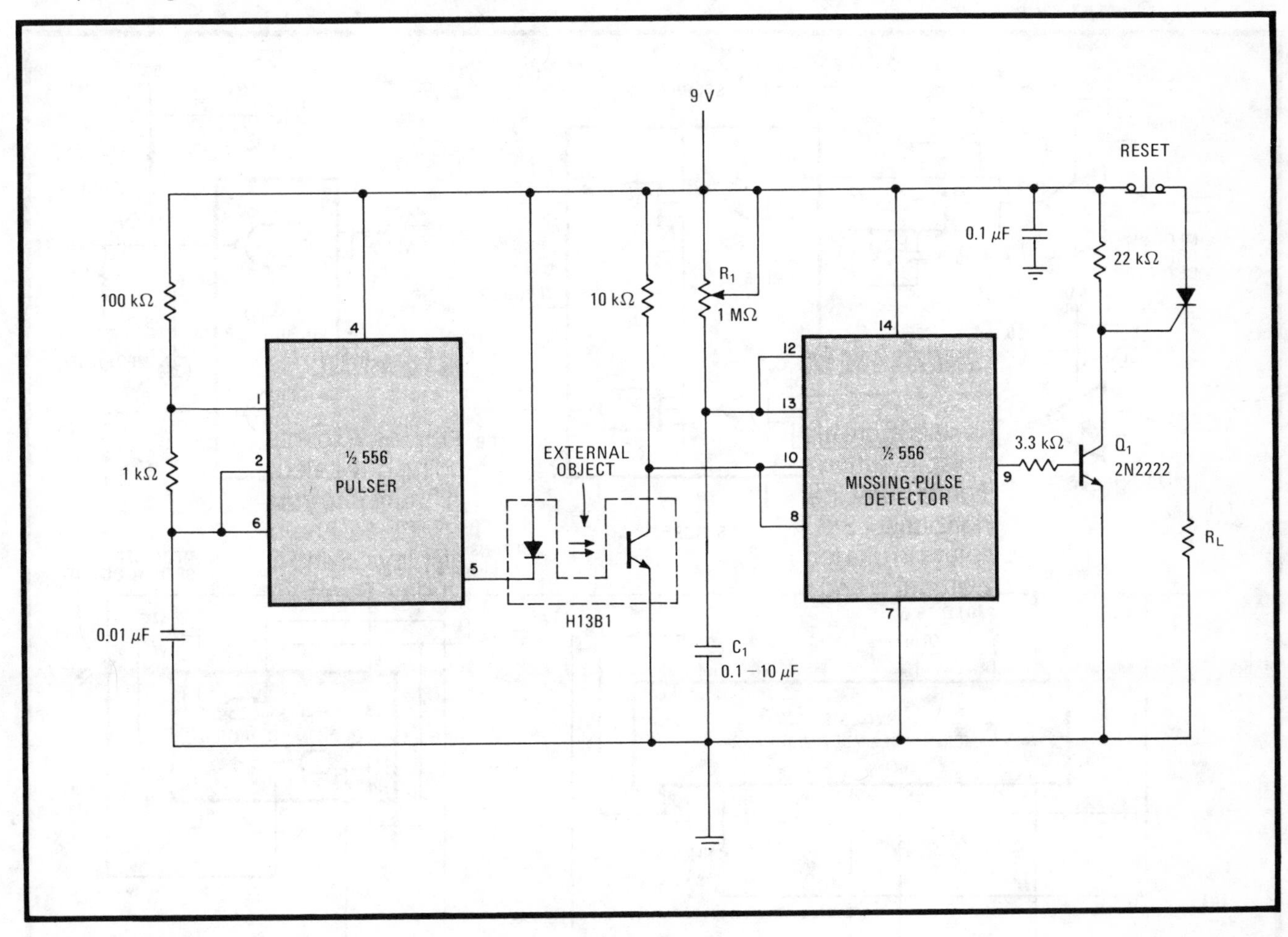

TTL line drivers link fiber optics

by Vernon P. O'Neil and Imre Gorgenyi
Motorola Inc., Discrete Semiconductor Division, Phoenix, Ariz.

Designers who need to convert an existing twisted-pair communications interface into an optical-fiber link can minimize their efforts by simply combining their fiber-optic detector-preamplifiers with TTL line receivers like the industry-standard MC75107 devices. Such an arrangement has other advantages besides simplicity, namely, providing the builder with access to two receivers, complete with strobing inputs, in a single 75107. And although no similar optical line receivers are yet offered as a standard product, this interface will yield good performance at low cost. Building fiber-optic transmitters using TTL line drivers is equally simple.

The union of the 75107 with Motorola's MFOD404F optical detector is shown in (a) of the figure. The detector is packaged in a nose-cone type of fixture that can be directly mounted in standard AMP-connector bushings, making the connection to the optical-fiber cables extremely simple.

The resulting optical receiver will handle data rates of up to 10 megabits per second at a sensitivity as low as 1 microwatt. For even greater data rates, an MFOD405F can be used to extend the data-rate capacity to 50 Mb/s at a sensitivity of 6 μW.

Because the receiver is ac-coupled to the detector, it is necessary to restrict the duty cycle of incoming signals to the range of 40% to 60%. Coupling components between the detector and the 75107 are selected to ensure that the reference level developed at the input of the receiver tracks the average voltage of the input data stream. In this way the circuit self-adjusts to a wide range of input optical power levels.

At the other end of the system (b), a compatible ac-coupled optical transmitter can be constructed from an ordinary 75452 line driver. A 0-to-2-Mb/s fiber-optic transmitter suitable for handling bipolar-pulse (dc-coupled) encoded data (c) is almost as simple. □

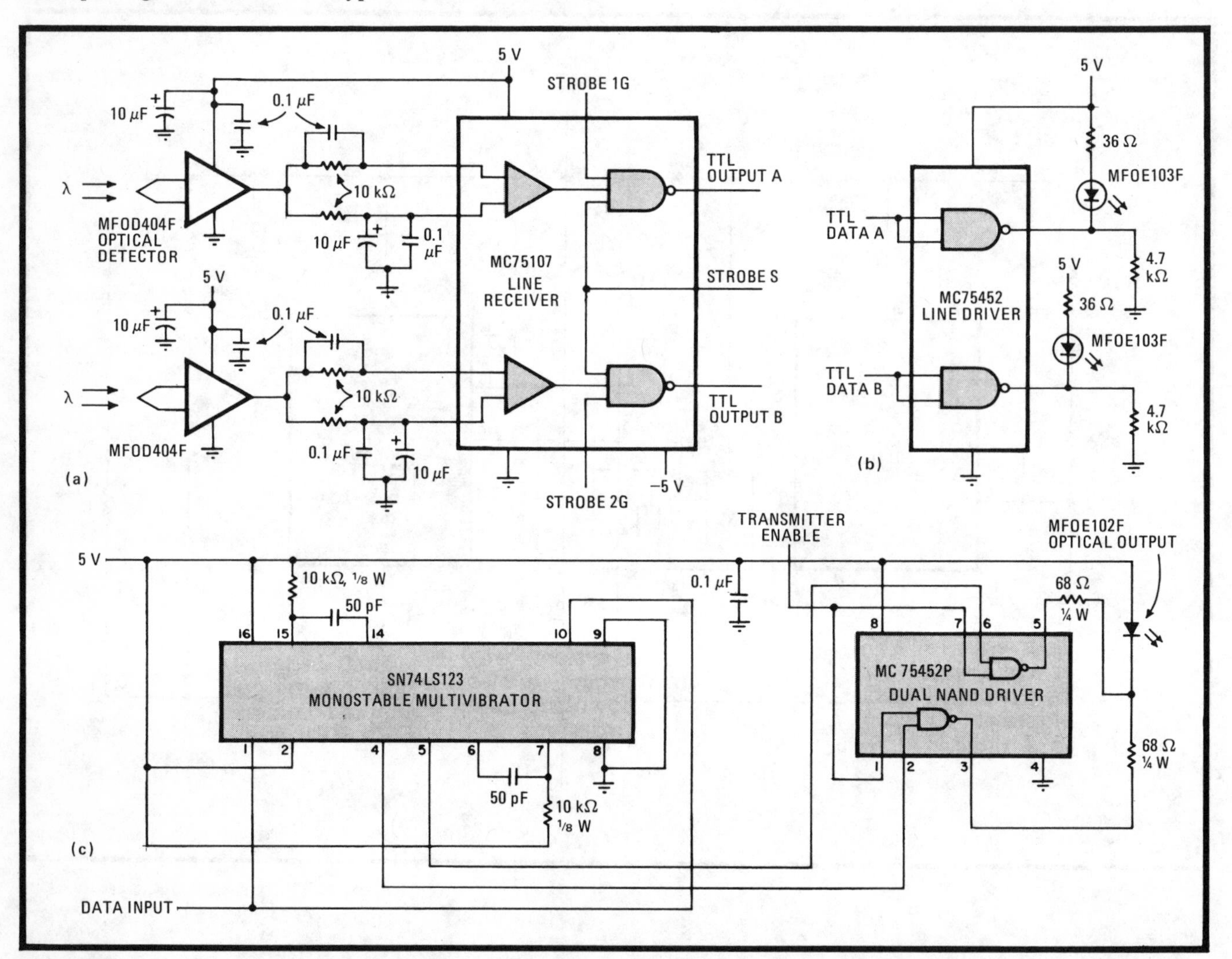

Light line. Standard line receivers such as the 75107 serve well in interface for fiber-optic systems (a). Off-the-shelf TTL line drivers at the transmitting end (b) makes possible low-cost systems. A 0-to-2-Mb/s bipolar-pulse–encoded transmitter (c) is almost as simple to build.

Bootstrapping a phototransistor improves its pulse response

by Peter J. Kindlmann
Engineering and Applied Science Dept. Yale University, New Haven, Conn.

Although the operating speed of a phototransistor cannot be improved simply by connecting a second one in the cascode configuration [*Electronics*, March 2, p. 132, and April 27, p. 154], its response may be improved by employing a standard transistor in a bootstrap circuit in order to reduce the effective value of the phototransistor's junction capacitance. By introducing bootstrap feedback to the base of the input optodevice, the switching speed of a cascode-connected phototransistor can be increased by as much as 10 times over that of a uncompensated one.

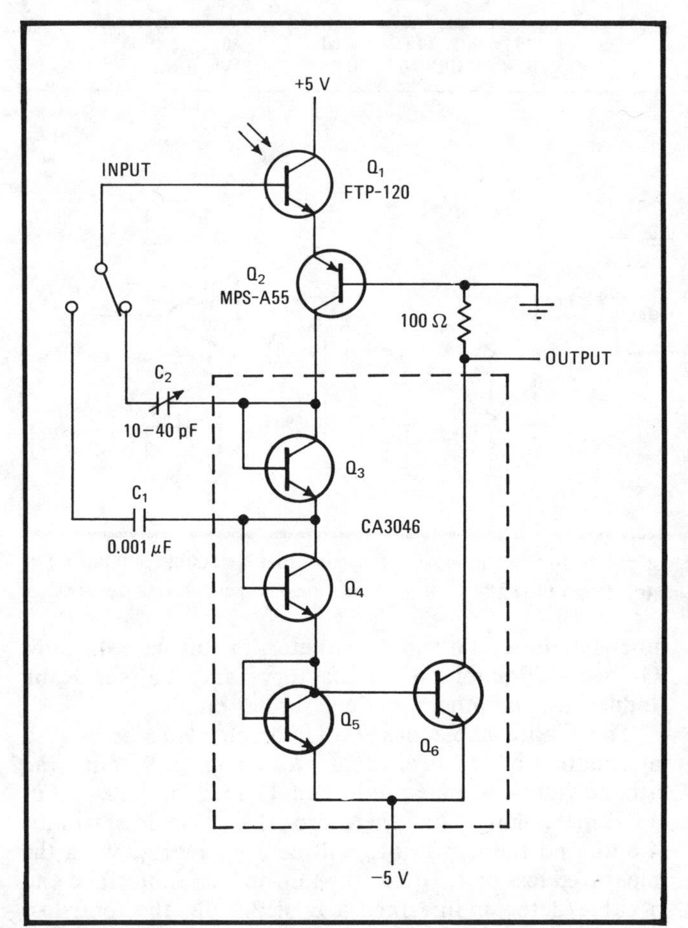

Compensation. Junction capacitance of Q_1, which is not sufficiently reduced despite cascode connection (Q_1, Q_2), is greatly lowered by applying feedback to base. This allows a rapid discharge of Q_1's base-to-emitter capacitance during signal conditions, which acts to increase the phototransistor's high-frequency response.

Phototransistor Q_1 and Q_2, a pnp transistor, form the conventional cascode arrangement, as shown in the figure. Generally, when an input signal is detected, the photocurrent step produced begins to charge the capacitance associated with Q_2's base-emitter and base-collector junctions. The voltage across the base-emitter junction has a magnitude comparable to that across Q_2's base-emitter junction, and therefore a way must be found to compensate for the two V_{be} drops produced, in order to ultimately reduce the effective junction capacitance of the phototransistor.

In theory, the V_{be} drops may be cancelled by making use of the pn drops across two forward-biased diodes of comparable transconductance. Here, diode-connected transistors Q_3–Q_5, which are part of the CA3046 transistor array, are available for use. Using the CA3046 ensures that these transistors will be closely matched.

Feedback from Q_4's collector to Q_1's base through C_1 constitutes the normal bootstrap path, supplying an in-phase current to Q_1's base. This causes a rapid charge of the junction capacitance, and therefore the input photocurrent sees a lower value of capacitance than actually exists. Because Q_1 has a β of several hundred, its base-emitter transconductance is less than that of the lower-β devices, Q_4 and Q_5, used in the feedback path. As a result, the amount of feedback is well below unity loop gain (undercompensated condition).

By using Q_3, however, with feedback applied through C_2, an additional pn drop is gained and compensation becomes almost perfect. For a given quiescent photocurrent, C_2 should be adjusted to a value just above that which will cause oscillation in the circuit.

Fairchild's FTP-120 (Q_1) has a typical rise time and fall time of 18 microseconds when used in the typical emitter-follower configuration specified for a 100-ohm load. With C_1-path compensation, the switching time is about 5 μs. With C_2-path compensation, the switching time is about 2 to 3 μs. □

Low-cost fiber-optic link handles 20-megabit/s data rates

by A. Podell and J. Sanfilippo
Loral Electronic Systems, Yonkers, N. Y.

Providing an inexpensive link for the transmission and detection of digital signals over short distances, this fiber-optic system handles data rates in excess of 20 megabits per second. The system, which can be built for about $90, including cable, processes all types of data—a continuous-wave clock waveform, a burst of N clock cycles, handshaking signals, or a non-return-to-zero (NRZ) stream.

A TTL driver and a light-emitting diode serve well as the transmitter, shown in (a). The 5438 TTL driver is a two-input, open-collector NAND gate selected for its low power dissipation and 48-milliampere current-sinking capability. The LED is a gallium-arsenide device operat-

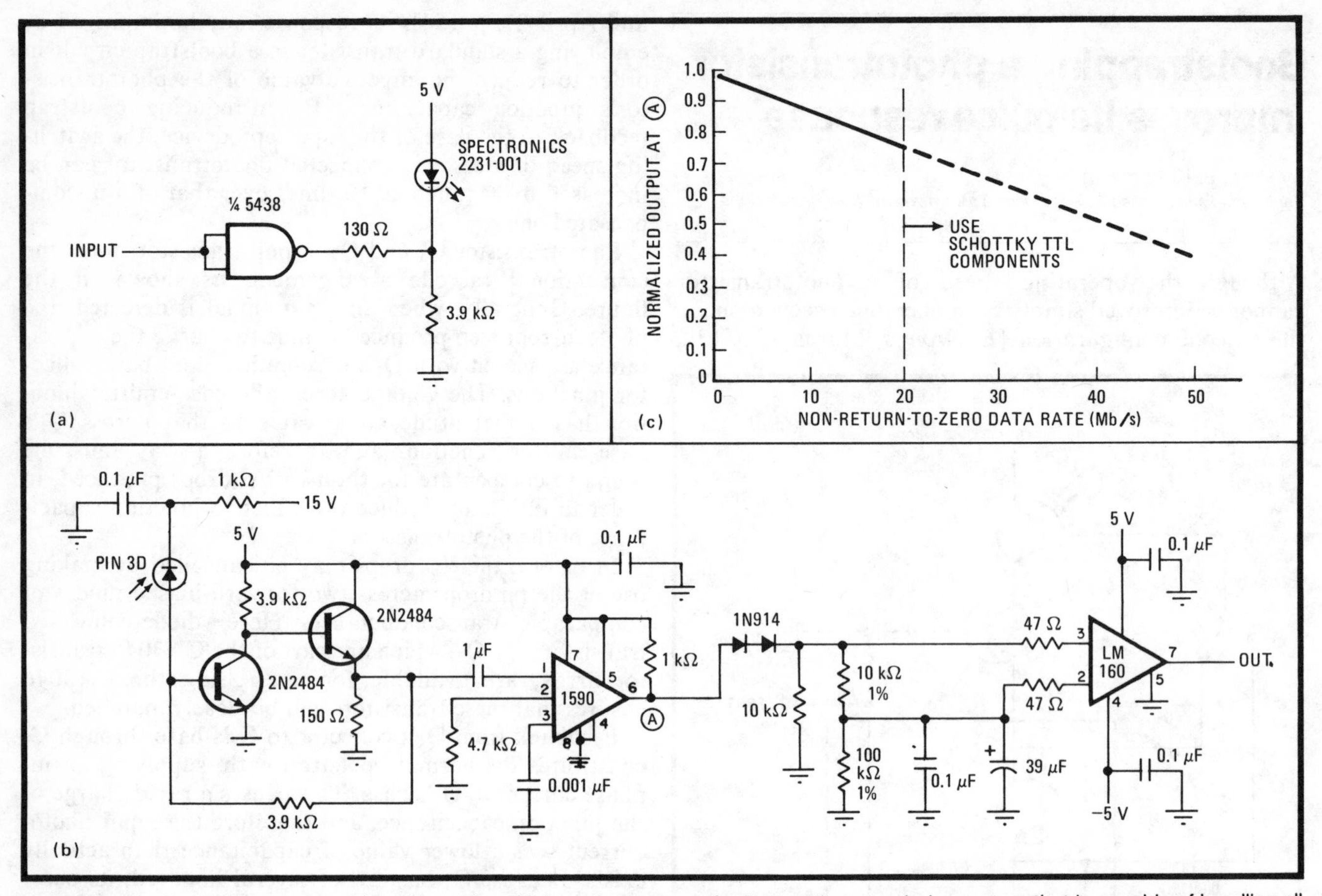

Light bits. Simple data transmitter (a) and a receiver (b) form the nucleus of a fiber-optic transmission system that is capable of handling all types of digital waveforms. Link operates over a wide band of frequencies (c). Cost of the 10-meter-long unit, including cable, is under $90.

ing at 910 nanometers and provides 2 milliwatts of optical power at a forward current of 100 milliamperes. The 130-ohm resistor sets the current through the LED at about 30 mA, and so the output power is about 0.6 mW in this circuit.

The receiver (b) is also simple and sensitive. The output from the p-i-n photodiode (labeled the PIN 3D device) is several microamperes. This current is converted into a voltage by a two-transistor transimpedance amplifier. The 2N2484 transistors selected give low input capacitance, an adequate gain-bandwidth product, and the ability to detect small currents. Amplifier output is about 25 millivolts.

The MC1590 video amplifier that follows greatly boosts signal levels over a wide band (c). Two 1N914 diodes drop the output offset voltage of the single-ended amplifier, nominally at 4 volts, to within the input range of the LM160 comparater. The comparator's threshold is set by a simple voltage divider. The capacitors, across pin 2 and ground, combined with the 100-kilohm resistor, form a low-pass filter providing a threshold that varies with the comparator's supply voltage.

As for the electro-optical interface, the LED, which is contained in a TO-46 package, is easily mounted in an inexpensive window bushing made by AMP, model 530563-1. The PIN 3D photodiode can be mounted in the same type of connector if desired. The need for delicate mounting adjustments is avoided here by using a fiber bundle of sufficient diameter, in this case 45 mils. Galite 2000 cable is satisfactory, and Valtec, Rank Industries, and others produce similar bundles.

The Galite cable has 210 fiber elements having an attenuation of 450 decibels per kilometer at 910 nm and a bandwidth-distance product of 15 megahertz/km. For a 10-meter-long link, therefore, the cable loss will be 4.5 dB and the bandwidth will be 1 gigahertz. With the measured loss of 1.5 dB in the LED-to-cable interface and a cable/detector interface loss of 3.9 dB, the total loss amounts to 10 dB. Thus, the 0.6-mW output of the LED is reduced to 0.06 mW at the receiver.

Transmitter layout is not critical in a one-way link. Duplex operation will require electrical isolation between transmitter and receiver components. There are several precautions to take in constructing the receiver. Notably, the lead from the anode of the detector diode to the transimpedance amp must be kept as short as possible. The output of the receiver should be isolated from all previous stages to prevent unwanted pickup. A ground plane is not a necessity, but is recommended for processing data rates greater than 10 megabits/s.

The link's signal-to-noise ratio is slightly less than 40 dB, implying a bit-error rate above 10^{-8}. The system is operational over a temperature range of −40°C to 100°C, and a supply variation of 4.5 V to 5.5 V. □

Optocoupler in feedback loop aids charger/regulator

by Leonard A. Cherkason
Mt. ISA Mines Ltd., Queensland, Australia

A single, inexpensive circuit that serves either as a voltage regulator or as a charger for low-capacity batteries can be built easily without resorting to usually complex voltage-sensing networks. This circuit uses the diode (emitter) of an optocoupler in a simple feedback loop, permitting easy detection of output voltage changes. It produces a 12.7-volt regulated output at 50 milliamperes or can charge a battery at the same voltage and current limits, and the limits can easily be changed.

The optocoupler is an ideal choice for a detector. The diode senses the output voltage without loading or otherwise disturbing normal operation, and the voltage drop developed across its terminals is constant and relatively small for any charge or load current.

As shown in the figure, the diode bridge and capacitor C_1 rectify and filter an ac input voltage. Assume that the circuit is in the charge mode. If the battery is not fully charged, its voltage will be below 12.7 v ($V_Z + V_D$). This voltage is selected by the choice of an appropriate zener diode, which is placed in series with the optocoupler diode. The 1N2270 pass transistor will therefore be biased on, allowing current to be supplied to the battery. I_A will be limited chiefly by the 220-ohm resistor.

When the battery voltage just exceeds ($V_Z + V_D$), the zener will fire and current I_Z will flow through the optocoupler's diode, switching on its phototransistor and turning off Q. If the battery is removed, thus placing the circuit in the regulator mode, current will be delivered to the load at 12.7 v. Now, of course, the current drain will be determined largely by the load resistance.

The ripple voltage is 25 millivolts in the regulator mode and 1 mv in the charge mode. Line regulation is 30 mv/v, and load regulation is 8 mv/mA over a 5-to-30-mA range. Both may be improved if Q is replaced with a Darlington amplifier.

The output voltage and current may be selected by proper choice of R_1, R_2, and the zener diode. In turn, the resistors may be determined if the battery capacity (C), measured in milliampere-hours, and the input voltage (across C_1) are known.

It has been determined that for best performance, I_A should equal 0.25C. This will occur if the battery is relatively drained. Assuming the input voltage, V_{in}, is much greater than the base-emitter drop across Q, then:

$$R_2 \geq V_{in}/0.25C - R_1/h_{fe}$$

where h_{fe} is the dc gain of Q. To find the minimum value of R_2, assume that I_O is known, from which it can be determined that:

$$R_2 \leq (V_{in} - V_O)/I_O$$

This equation assumes that I_O is greater than I_Z.

The value of R_1 is determined by the minimum value of I_O, which occurs when Q is nearly off. It can be shown that:

$$R_1 \geq (V_{in} - V_O)/0.02C$$

This equation presumes that the minimum value I_O equals about 0.02C, a condition that again has been determined by experiment, rather than by theoretical considerations. □

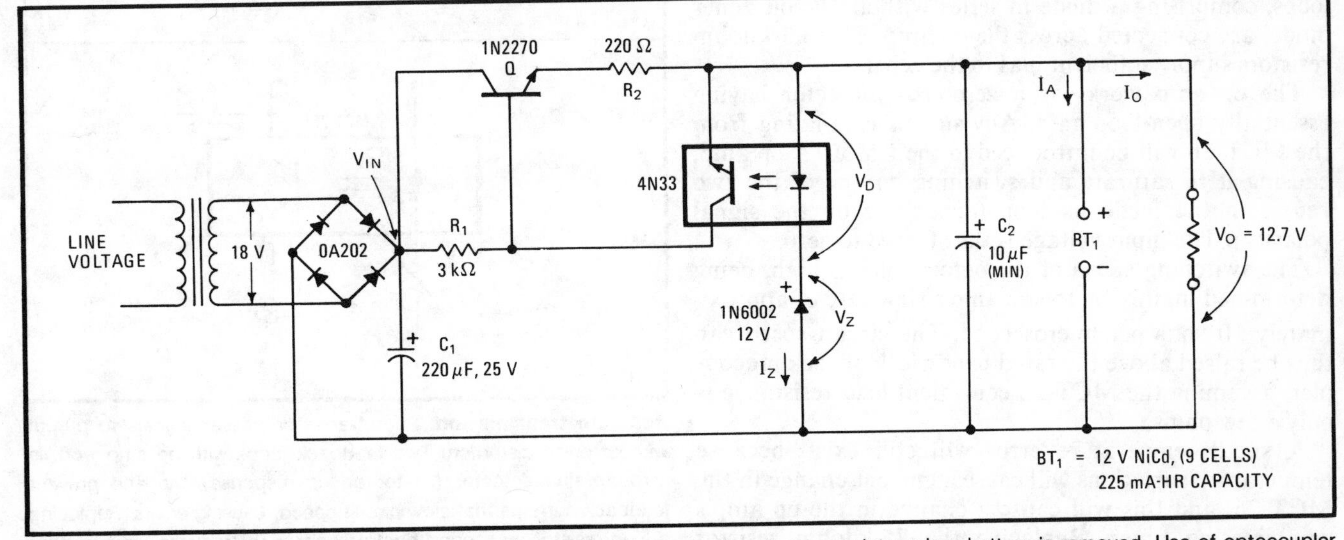

Regulator or charger? Circuit works as low-capacity battery charger or as voltage regulator when battery is removed. Use of optocoupler eliminates complex voltage-sensing networks. Output voltage and load current can be selected by suitable choice of R_1, R_2, and zener diode.

Optocoupler transmits pulse width accurately

by Tadeuz Goszczyński
Industrial Institute of Automation and Measurements, Warsaw, Poland

Though optocouplers work fine in most pulse applications, shortcomings in their switching and temperature characteristics make them poor at such tasks as transmitting pulse-width modulated signals accurately. Adding an operational amplifier to the optocoupler circuit will improve its response time and reduce the effects of temperature on output voltage, enabling it to transmit a pulse width as small as 2 microseconds with an error of only 200 nanoseconds. If a second optocoupler is added to the circuit, temperature problems will be virtually eliminated.

An optocoupler is limited in its ability to transmit pulse width accurately because of two major factors: the response speed of the device is reduced by feedback currents that flow from the output port of the phototransistor to its base, and the current-transfer ratio is highly dependent on temperature. In either the emitter-follower or common-emitter configuration, an output voltage change produces the feedback current and an equal change across the collector-to-base capacitance. A certain time is required for the capacitor to charge to the voltage; this limits the response time and can cause errors in pulse-width transmission.

In addition, the switching times as well as the amplitude of the output pulse generated by this current source vary with temperature. All errors may be greatly reduced if the output voltage of the phototransistor is clamped to a near-zero level for any level of output current, in effect making its load resistance zero so that no feedback current is generated.

Tying a current-limiting resistor and an op amp to the output port of an MCT-26 optocoupler does the job, as shown in Fig. 1a. Two separate bidirectional feedback loops, comprising a diode in series with a 5.6-volt zener diode, are connected across the op amp. The 3.3-kilohm resistors supply sufficient bias to the zeners.

The op amp works as a zero-cross detector having essentially open-loop gain. Any signals emanating from the MCT-26 will be introduced to the LM301A op amp, causing it to saturate and switching on one of the two zeners in the feedback loop (depending on the signal polarity). The input voltage is thus forced to zero.

The switching speed of the optocoupler is high, being determined mainly by the op amp's slew rate of approximately 10 volts per microsecond. The circuit speed can thus be raised above the rated bandwidth of the optocoupler, assuming the MCT-26 equivalent load resistance is only a few ohms.

A small temperature error will still exist, because temperature variations will cause a current change in the MCT-26, and this will cause a change in the op amp's zero-crossing times. Replacing the 75-kilohm resistor with the phototransistor of another optocoupler, as shown in (b), will reduce the temperature error below 3 ns per °C. This ensures equal temperature-dependent voltage drops across both optocouplers, and with them connected as shown, the temperature-generated voltages will cancel. The op amp's temperature coefficient is negligible in comparison and need not be considered. □

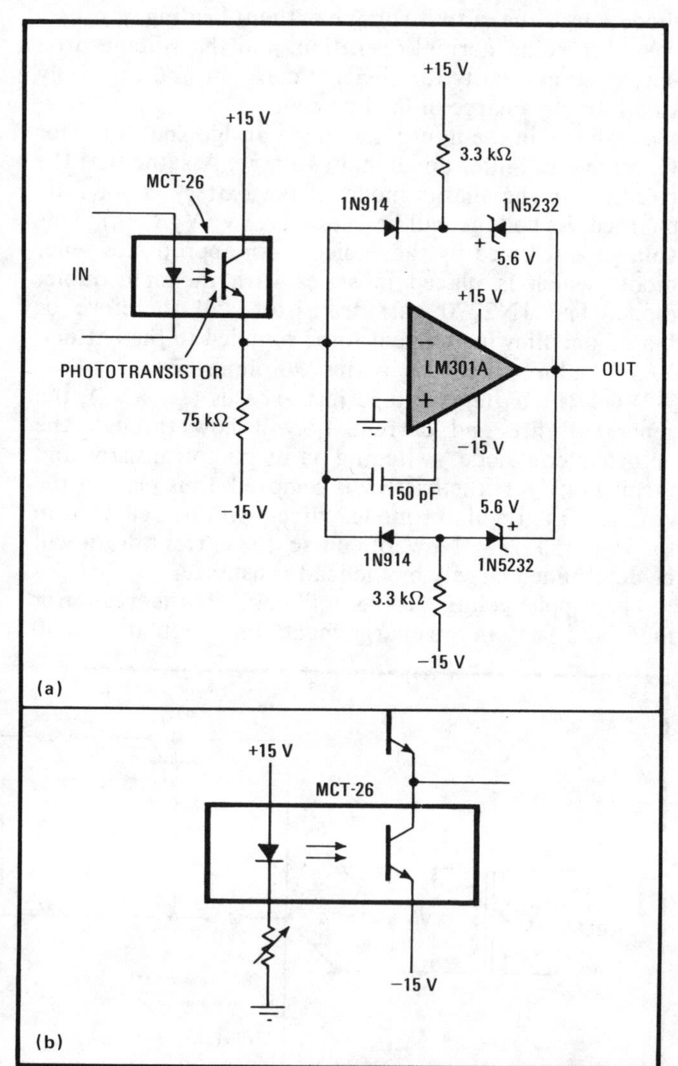

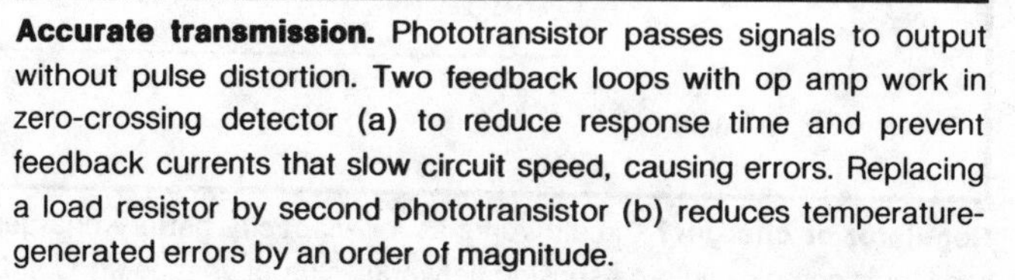

Accurate transmission. Phototransistor passes signals to output without pulse distortion. Two feedback loops with op amp work in zero-crossing detector (a) to reduce response time and prevent feedback currents that slow circuit speed, causing errors. Replacing a load resistor by second phototransistor (b) reduces temperature-generated errors by an order of magnitude.

Chapter 30
OSCILLATORS

Self-gating sample-and-hold controls oscillator frequency

by Peter Reintjes
Research and Design Ltd., Morehead City, N.C.

A phase-locked loop in conjunction with a transmission gate serving as a sample-and-hold device can remember the frequency of a short-duration signal by providing a constant feedback voltage to a voltage-controlled oscillator. The circuit is useful for electronic-music synthesis and tuning radio oscillators by remote push button, as well as other applications.

The circuit can sample and hold frequency bursts without the error introduced when the direct conversion of an input frequency to a voltage or digital quantity is attempted. As shown in the figure, a relatively high-level, short-duration alternating-current input signal is amplified, buffered and compared to a user-determined reference by the three operational amplifiers in the LM3900 device. The output of the last stage drives the 555 timer, which serves as a monostable multivibrator (one-shot). The input signal also drives the CD4046 phase-locked loop.

As the one-shot periodically fires in response to the input signal, the transmission gate conducts, completing the feedback loop in the PLL. This permits the output of a selected comparator in the PLL to feed the voltage-controlled oscillator at pin 9. The output of the comparator is a function of the difference between input and VCO output frequencies; the resulting current charges the RC network in the feedback loop consisting of the 10-kilohm resistor and the 0.1-microfarad capacitor. The final dc voltage, across pin 9, controls the VCO frequency and becomes constant when the input and VCO frequency match.

When the amplitude of the input signal fades, thus signalling the end of the burst, the feedback loop is opened because the 555 no longer fires. The voltage across pin 9 remains, however, because the high-input impedance of the VCO terminal prevents negligible leakage by the RC network. Therefore the VCO continues to oscillate at the same frequency indefinitely. The high input impedance provides a good "hold" characteristic.

Comparator 1 in the PLL, an exclusive-OR gate, should be used for high-noise input sources for optimum performance. Comparator 2 is an edge-detecting device and may be used for most other conditions. The LM555 sampling period is adjustable from 10 milliseconds to 1 second. The supply voltage on all devices may range from 5 to 15 volts. □

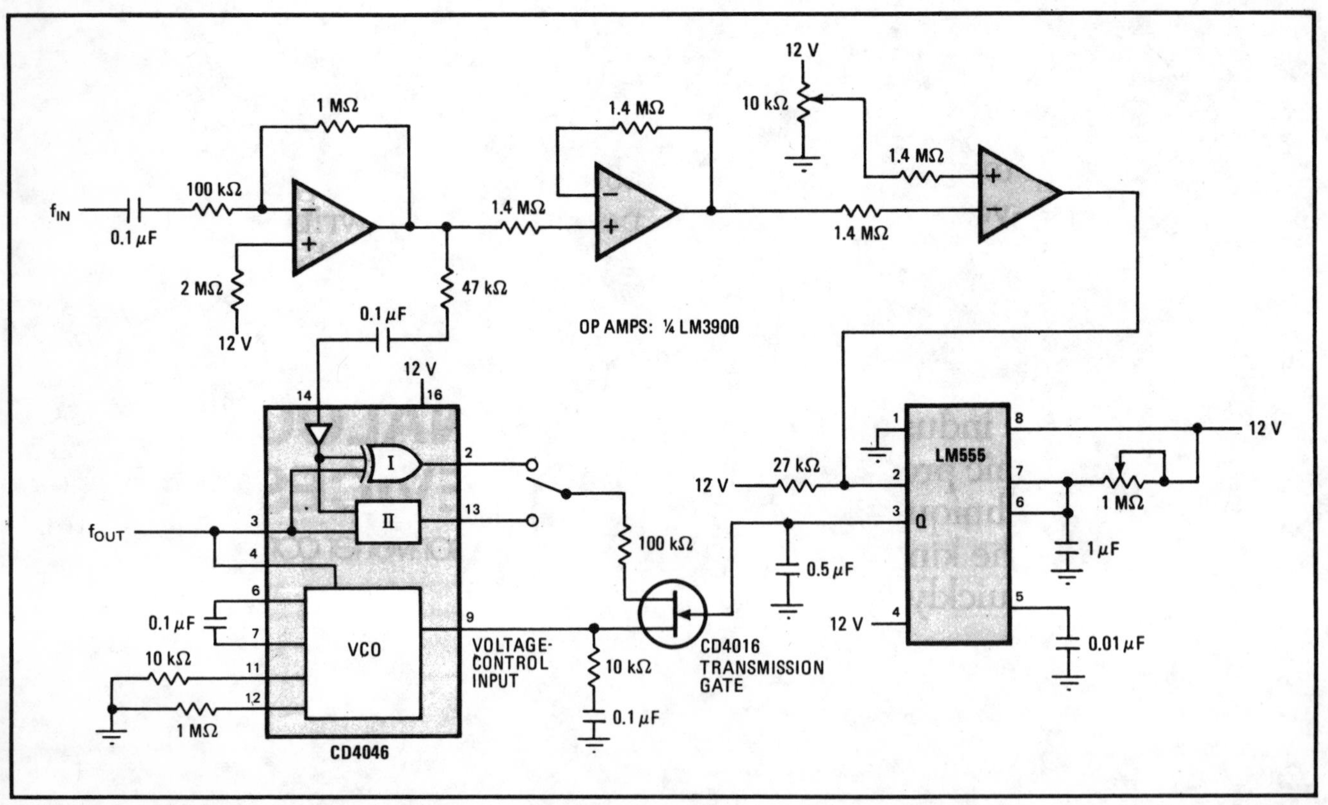

Self-gated sample-and-hold. Input-signal burst closes feedback loop in PLL, locks it to incoming frequency. Removal of input opens loop but causes no change in VCO frequency because of RC network, which stores unchanging voltage to drive oscillator.

C-MOS oscillator has 50% duty cycle

by Bill Olschewski
Burr-Brown Research Corp. Tucson, Ariz.

Astable multivibrators built with complementary-metal-oxide-semiconductor gates suffer one major drawback—their duty cycle may vary from 25% to 75% because of the variations of each gate's switching-threshold voltage (V_{TH}). Variations in the V_{TH} can be canceled and the desired square-wave output therefore attained by adding a C-MOS inverter and three resistors to the basic circuit. The gate-resistor combination uses negative feedback to perform the compensation.

The standard astable multivibrator is shown in (a) of the figure. Running at a frequency ($f = \frac{1}{2}R_TC_T$) that is almost independent of the individual gate used, the circuit nevertheless has an unpredictable duty cycle because of a V_{TH} that can vary by up to 40% on either side of $V_{DD}/2$, where V_{DD} is the supply voltage. If a 50% duty cycle is required, either this circuit must be followed by an edge-triggered flip-flop, or each circuit must be individually adjusted using two trimpots and a diode (see RCA application note ICAN-6267).

The circuit in (b) eliminates these drawbacks. Inverter A_3 creates a second negative-feedback path around A_1 (the signal flow through R_T constitutes the prime path). A_3 is operated at a low closed-loop gain, much like an operational amplifier working in the linear portion of its characteristic. As a result, A_3's inverted threshold voltage can be combined with the negative feedback voltage and injected into A_1. If the ratio R_F/R_I equals the ratio R_C/R_T, complete cancellation of threshold errors between A_1 and A_3 can be obtained. It is assumed that A_1 and A_3 are contained in the same package along with A_2 and that their V_{TH}s are essentially equal.

Since A_3's gain must be set so that its output will not saturate with a ±40% variation of V_{TH}, resistor values must be selected so that $R_I/R_F = 2.33$. At the same time, the correct gain for A_3 is set when $R_CR_I = R_FR_T$. In these circumstances, and ignoring stray and input capacitances, the multivibrator's operating frequency will be $f = 1/R_TC_T$ and the duty cycle will be 50%.

Note that the operating frequency, in this case 20 kHz, is twice that of the standard astable circuit using the same values of C_T, R_T, and R_S because of the second feedback path. □

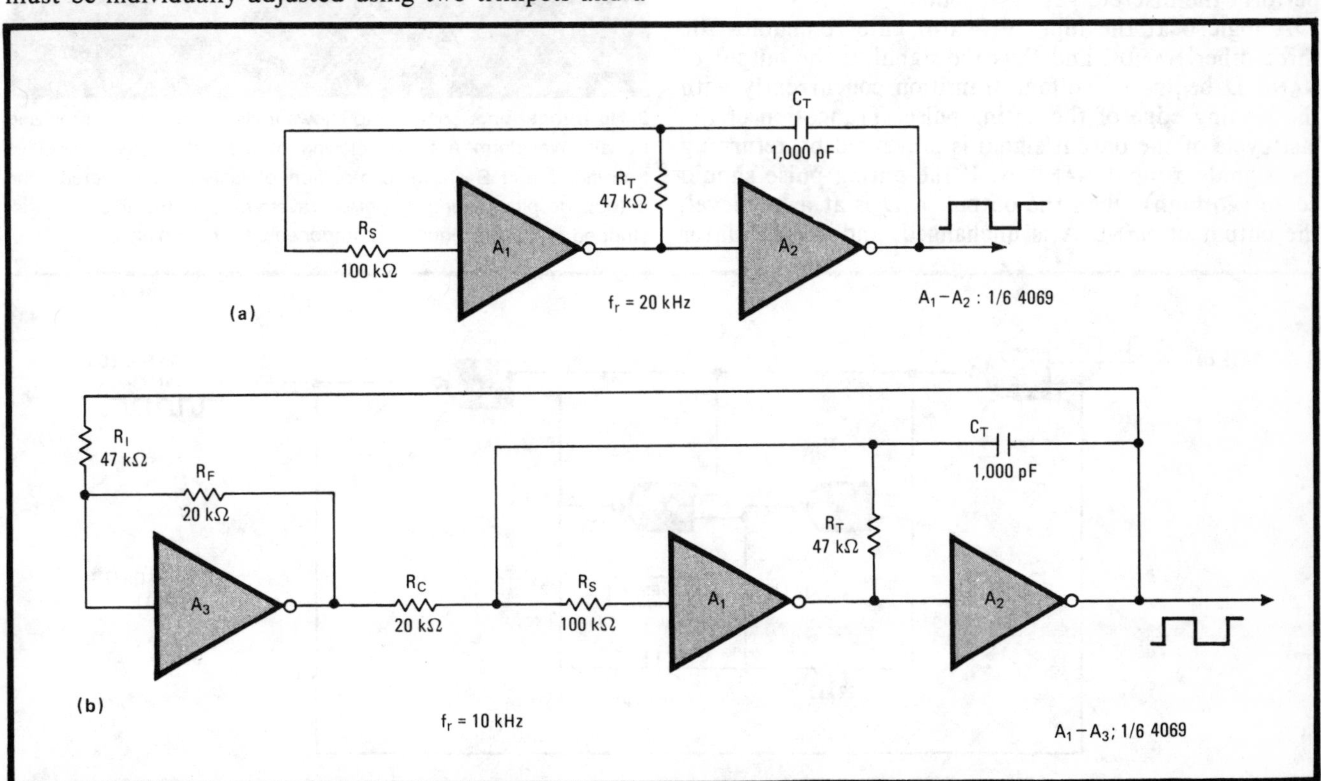

Right on. Standard astable multivibrator using C-MOS gates (a) has unpredictable duty cycle because of variable switching-threshold voltages. Adding inverter and three resistors (b) creates second negative feedback path around A_1, forcing A_1 and A_2's switching point to half the supply voltage, so that 50% duty cycle is attained and square waves are produced.

Fast-starting gated oscillator yields clean tone burst

by Walter C. Marshall
National Oceanic and Atmospheric Administration, Seattle, Wash.

Synchronization pulses required for analyzing recorded data are usually frequency-multiplexed on one channel of a laboratory-type tape recorder. To ensure a high degree of accuracy in the data analysis, the gated tones must exhibit precise start-up and high monotonicity. This gated oscillator, which turns on with the leading edge of the gate pulse, always passes an integral number of cycles. And because the last cycle is never truncated, the oscillator produces no higher-frequency harmonics that could interfere with other decoders.

The complete gated oscillator, shown in Fig. 1, is built with only one complementary-metal-oxide-semiconductor integrated circuit, a 4011 quad two-input NAND gate. Two of the gates, labeled B and C, form the RC square-wave oscillator, which has a frequency that can be adjusted from about 4 to 25 kilohertz by varying the 10-kilohm potentiometer. The remaining two NANDs perform the discrete gating function.

A logic 0 at the input to NAND gate A enables the three other NANDs, and thus the signal at the output of NAND D begins its voltage transition concurrently with the leading edge of the gating pulse. Truncation of the last cycle of the output signal is prevented by returning the signal to input NAND A. If the gating pulse should cease (go high) when the output of D is at a low level, the output of NAND A is unchanged, and the oscillator continues until a cycle is completed. Once the output of NAND D returns high at the end of a cycle, gate A turns off and the tone burst is cleanly terminated.

An oscilloscope photograph of the signals is shown in Fig. 2. Waveform A is the gating pulse, and B is the gated output. At the cessation of the gating pulse (positive transition), the oscillation of the output signal continues until the cycle is completed.

Waveform C is the gated output after it has been passed through a two-pole active bandpass filter to remove higher-frequency components. This tone is then mixed with other synchronization tones and applied to the tape-recorder input. □

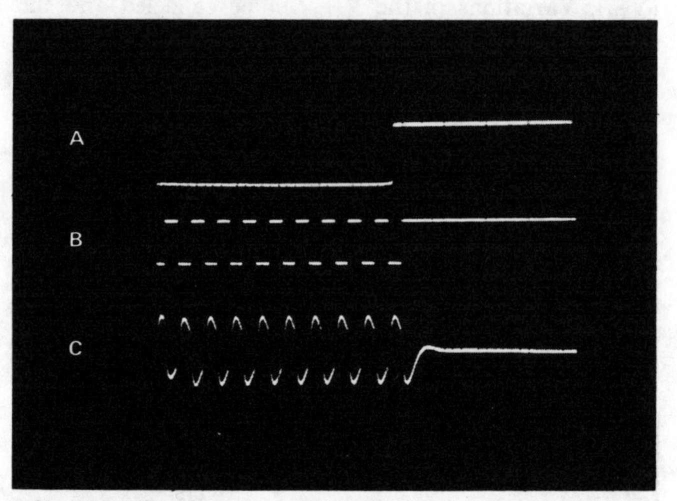

2. No truncation. Scope photo shows trailing edge of synchronizing signals. Waveform A is the closing of the gating pulse (positive transition), and B shows completion of last cycle in gated tone output, despite closing of gate. Waveform C is the filtered burst, stripped of higher-frequency components for recording.

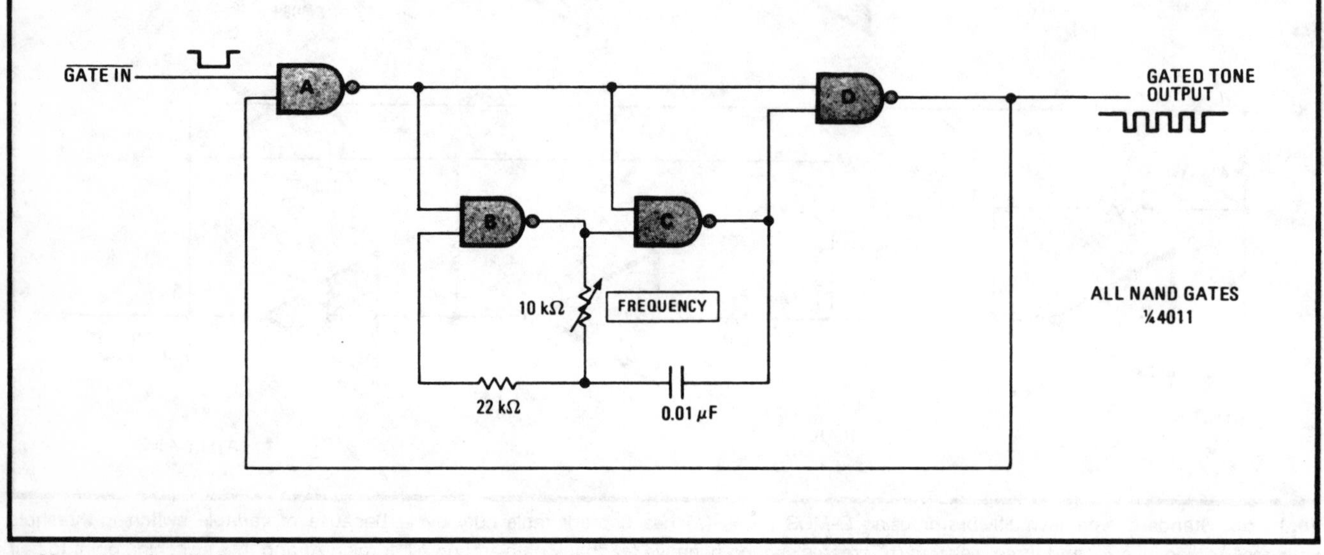

1. Clean tone burst. One-chip circuit generates tone burst of an integral number of cycles when gated by a negative-going input signal. Basic RC square-wave oscillator built around NAND gates B and C is adapted from RCA applications note ICAN-6267.

Divider sets tuning limits of C-MOS oscillator

by Henno Normet
Diversified Electronics Inc., Leesburg, Fla.

Useful as it is, the square-wave RC oscillator implemented in complementary-MOS has one shortcoming—setting its maximum and minimum frequencies of oscillation independently while also maintaining accuracy is extremely difficult. By placing a voltage divider in the feedback loop of the conventional three-gate circuit, however, a one-time trimming adjustment can accurately set the maximum and minimum frequency excursion and will force the ratio of the upper to the lower limit of oscillation to approach a value virtually determined by the resistors used in the same divider.

The standard RC oscillator generates a frequency of $f \approx 0.482/R_1C$, where $R_1 = R_2$, as shown in (a). Generally, it is not practical or economical to use a variable capacitor for C. A potentiometer could be substituted for R_1 to tune the frequency, but slight differences in integrated-circuit parameters will preclude predicting the maximum and minimum frequencies of oscillation with any degree of accuracy for a particular chip. The only other method for setting the upper and lower frequency limits is to parallel several capacitors across C, a tedious procedure at best.

Alternatively, R_1 can be a potentiometer that is placed virtually in parallel with voltage divider R_4–R_5 through C (b). In this way, capacitor C is no longer charged from the fixed-voltage output of the middle gate in (a), but from the voltage divider across the output. R_1 is thus used to change the circuit's time constant without affecting the potential that is applied to C.

The upper and lower limits of oscillation are determined by the position of R_4's wiper arm and by the values of R_4 and R_5. With the tap at point A, the circuit will oscillate at a frequency given by $f = 1/2.2R_1C$. With the wiper at point B, the frequency will be $f = 1/1.39R_1C$. The frequency ratio to be expected is thus $2.2/1.39 = 1.6$. The actual frequency change measured with the particular chip used for breadboarding was 56%, which is thus very close to the intended value. The ratio will increase as R_4 is made larger with respect to R_5.

The circuit has only one small disadvantage—the load presented by R_4 and R_5 does increase the power-supply drain by approximately 0.5 milliampere. □

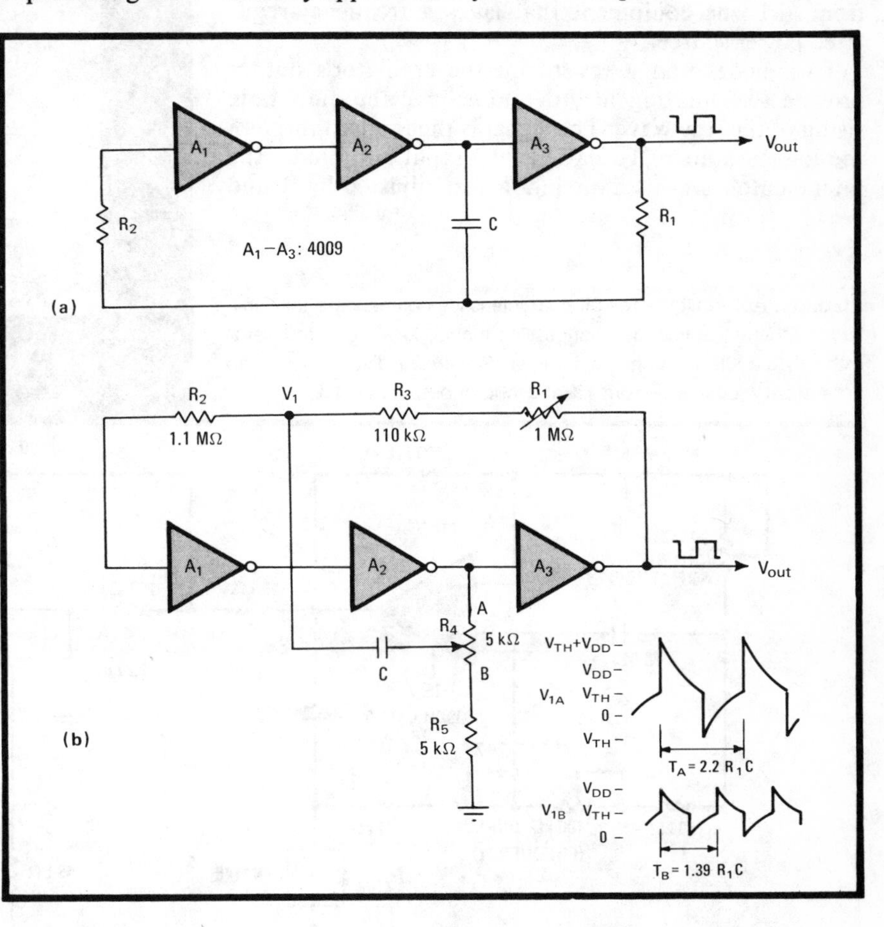

Calibrate. IC anomalies, inherent circuit imbalance, and the expense of making C variable preclude setting upper and lower oscillation limits of typical RC oscillator (a) with any accuracy. Placing R_1 virtually in parallel with voltage divider (b) through C gives circuit one-knob frequency control, with upper-to-lower oscillation ratio in effect determined by R_4 and R_5.

Filter levels output swing of Wien-bridge oscillators

by Maxwell G. Strange
Goddard Space Flight Center, Greenbelt, Md.

Although the output of a tunable Wien-bridge oscillator normally exhibits a large change in amplitude as a function of frequency, a standard active filter will hold it to within ±0.2 decibel over a ±20% frequency range. In this application, the filter's response is set to compensate for the amplitude variation of the oscillator. Most alternative amplitude-stabilization circuits tend to draw high power, create appreciable sine-wave distortion, or stabilize slowly.

The technique can be easily implemented at any frequency over the operating range of the oscillator, since the filter's component values are easy to calculate, being inversely proportional to frequency. The circuit shown was designed to control the speed of a 60-hertz synchronous motor over a range of 48 to 74 Hz. It is used to adjust the tape speed of a recorder in the lab to that of an airplane's recorder so that the data can be recovered from airborne equipment that lacks a frequency-regulated power source.

Two diodes and a resistor at the oscillator's output provide soft limiting in order to confine the amplitude swing of the sine wave. The signal is then passed through the low-pass filter. To flatten the output amplitude, the filter's cutoff and its damping factor, adjusted by R and R′, respectively, are set to compensate for the oscillator's amplitude variations. In general, the slope of the filter's amplitude response is made equal in magnitude but opposite in sign to that of the oscillator's response.

The graph shows the overall output to be expected compared with the individual oscillator and filter responses. In addition to amplitude compensation, the filter provides good rejection of harmonics. Third-harmonic distortion is an order of magnitude below that achieved by the oscillator alone. □

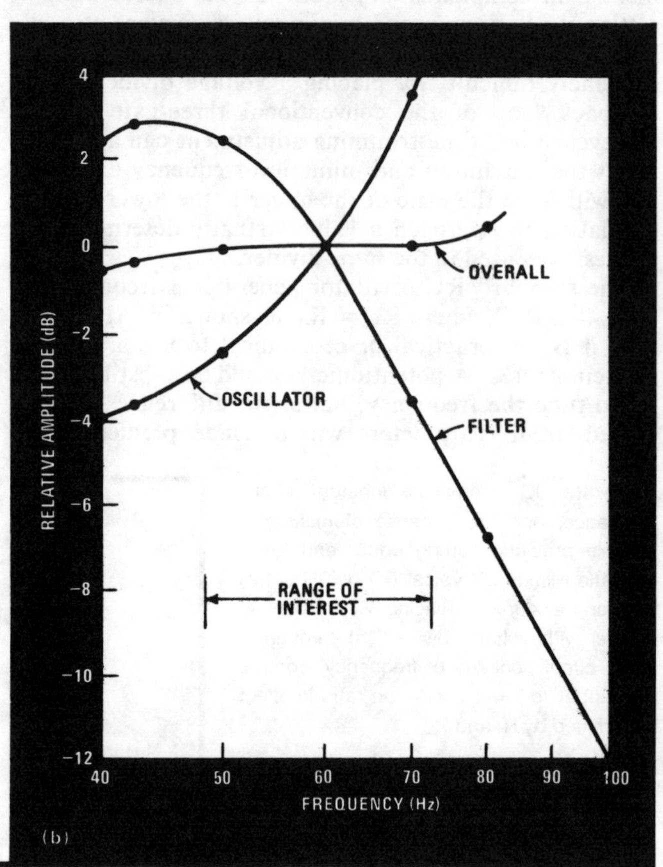

Stability. Active filter's roll-off characteristics compensate for oscillator's inherent amplitude change with tuning, keeping output level within ±0.2 dB in range of interest. Sine-wave distortion is also dramatically reduced—from 1% at oscillator output to 0.1%.

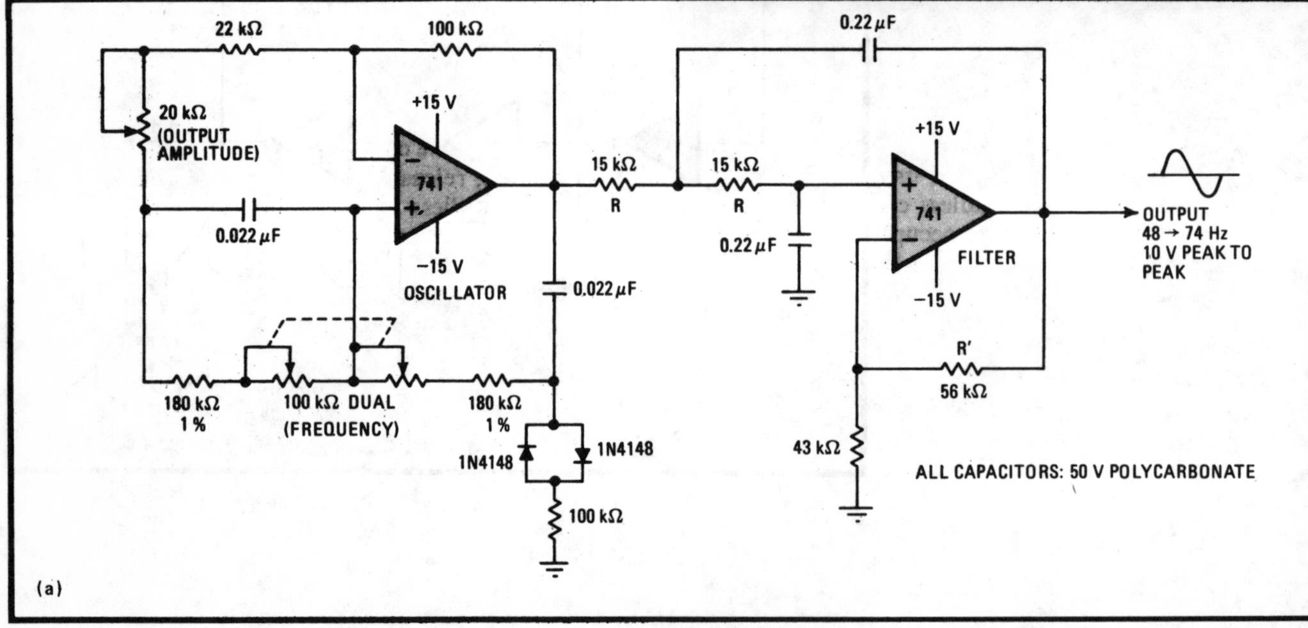

SCRs make serviceable relaxation oscillators

by Larry P. Kahhan
Princo Instruments Inc., Southampton, Pa.

Few designers realize that a silicon controlled rectifier can be used as a relaxation oscillator, just like its sister thyristor, the programmable unijunction transistor. The circuit arrangement is much akin to the PUT oscillator, as shown in the figure.

When power is applied, capacitor C_T charges, and the voltage across resistor R_T decreases exponentially. When the voltage across R_T, which is the cathode voltage of the SCR, drops to 0.6 volt less than the gate voltage, the SCR turns on. This turn-on produces current flow through C_T and a voltage spike across R_3.

Since the large value of R_T prohibits there being sufficient current to maintain conduction, the SCR immediately turns off, and C_T begins its charge cycle again. The period of oscillation is approximately given by $t = -R_T C_T (\ln \eta)$, where η is the fraction of the supply voltage that is applied to the gate, or $R_2/(R_1 + R_2)$. The high-impedance (500-kilohm) sawtooth-waveform output is available at the SCR's cathode, and a pulse waveform appears at its anode. The output impedance at this point is less than 20 ohms. If a high-current pulse is desired, a pulse transformer might replace R_3.

Design equations in the figure outline the criteria for oscillation. Values for SCR on-voltage, holding and forward blocking currents may be obtained from the data sheets. □

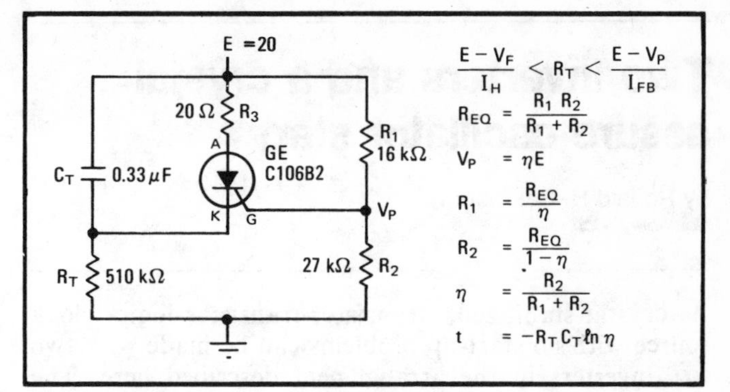

Programmability. An SCR oscillator retains the features of a unijunction-transistor circuit, including programmability of the firing point. Values shown yield an oscillation frequency of about 15 Hz. Gate-to-cathode voltage drop is neglected in the equations.

Two inverters and a crystal assure oscillator start

by Ronald H. Beerbaum
Southbury, Conn.

A crystal-stabilized, transistor-transistor-logic clock source with no start-up problems can be made with two TTL inverters in the arrangement described here. The circuit has two other advantages: it will operate with almost any crystal from any manufacturer, and its stability is dependent on the characteristics of only two components.

TTL circuitry has inherently low impedance at all nodes, necessitating a series-mode oscillator configuration to provide a reasonably high Q. This mode requires a low-impedance path through 360° of phase shift with sufficient gain to overcome the losses in the loop.

In the circuit below, I_1 and I_2 are TTL inverters (or NAND gates with the inputs tied together) that provide the necessary gain and phase inversion. RFC_1 and RFC_2 provide dc feedback at the inverters that forces them into a linear mode. These chokes are chosen to provide enough loop gain to drive a weak crystal, and should have a dc resistance on the order of 100 ohms or less. R_1 and R_2 are Q-swamping resistors inserted to eliminate oscillation of the circuit at the self-resonant frequency of the chokes. If low-Q chokes are used, these resistors may be eliminated, but it is wise to allow for them in the circuit-board layout to permit choke substitution.

Usually serial-mode crystals have three stable modes of oscillation—the one stamped on the can, one above that frequency, and one below that frequency. The higher- and lower-frequency modes are generally determined by both the crystal and distributed circuit parameters. Capacitors C_1 and C_2 are included in the circuit to eliminate these modes; by lowering the impedance of the loop they ensure that the specified frequency occurs. They have very little effect on the operating frequency, so high-quality capacitors are not required.

C_3 can be a trimmer or a fixed capacitor, depending on the exactness of frequency required. A tuning range of about 100 parts in 10^6 can be obtained by varying C_3, but temperature stability and long-term stability are dependent on its characteristics.

This circuit with the values shown will operate in the frequency range of 2 to 5 megahertz. Below this range, the values of RFC_1, RFC_2, C_1, and C_2 should be increased. Above this range C_1 and C_2 should be reduced to 22 picofarads or less. □

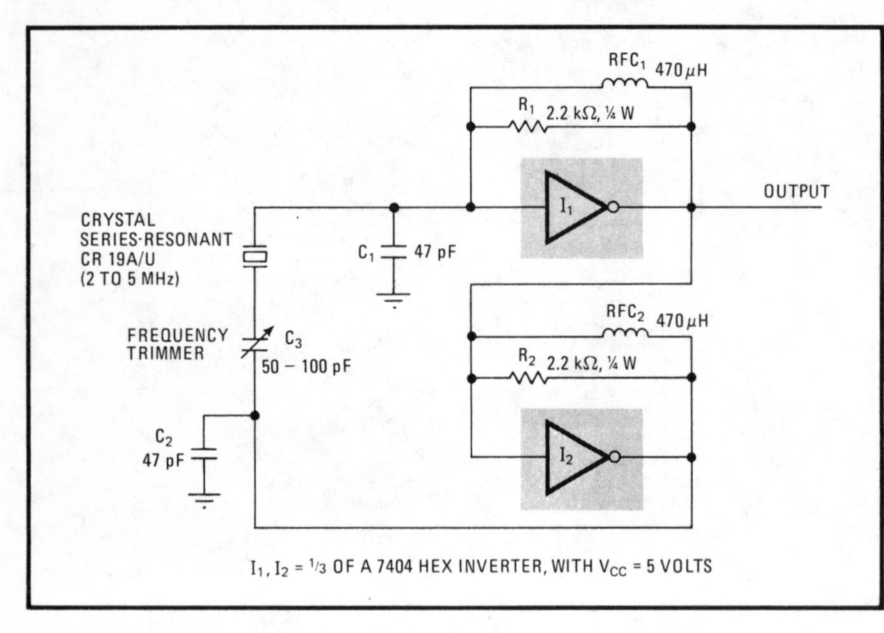

Reliable clock source. When power is applied to this series-mode TTL crystal oscillator, it always starts up at the right frequency. It will operate with most crystals; the components shown here are for crystals in the 2-to-5-MHz range.

One-chip oscillator generates in-quadrature waveforms

by Juan R. Pimentel
Department of Electrical Engineering, University of Virginia, Charlottesville

A quad operational amplifier, working as an oscillator, can generate in-quadrature signals for triangular or square waves, thus eliminating the need for phase-locked loops or other synchronous circuits in applications requiring either wave to be shifted by 90°. The low-cost circuit works over a wide range, thanks to a technique borrowed from Graeme[1] that uses op amps in an oscillator to generate in-quadrature sine waves.

Op amps A_1 and A_3 function as comparators, and A_2 and A_4 as integrators, as shown in the figure. Resistors R_2, R_3, R_6, and R_7 are selected so that 1 volt appears across R_3 and R_7 when A_1 and A_3 are saturated (any other voltage could be chosen).

Circuit operation is simple. If A_3 is saturated at its negative power-supply value, the voltage across R_7 will be -1 v. The voltage at the output of integrator A_4 will then have a positive slope and increase linearly with time, because the input signal is introduced to A_4's inverting port.

When A_4's output passes through zero, the voltage across R_3 will reach -1 v, causing A_2's output to increase linearly (positive slope). Similarly, when the output of A_2 passes through zero, the voltage across R_7 will reach $+1$ v, causing A_4's output to decrease linearly in the negative direction. This operation will repeat, yielding the waveforms shown below the circuit diagram.

The time required to complete one cycle is $T = 4RCV$, and so the frequency of oscillation is $f = 1/4RCV$, where $R = R_4 = R_8$ and V is the power supply voltage. These equations assume a peak output of ± 1 v at points 3 and 4. □

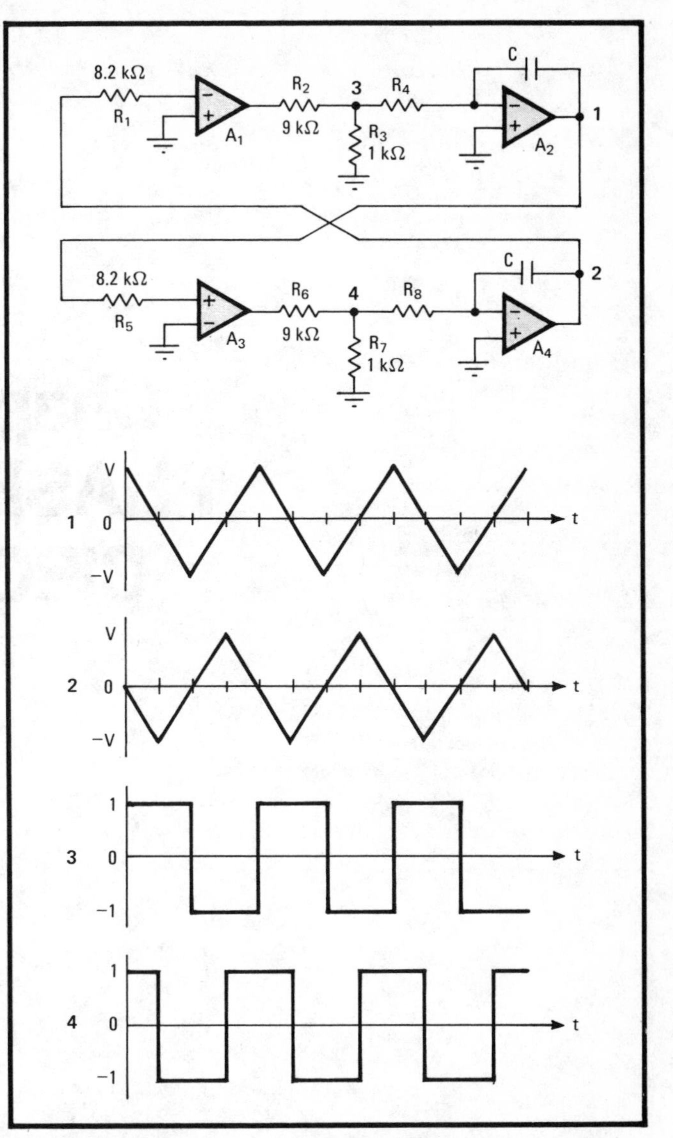

Phasing. Quad op amp, working as oscillator, can generate two triangular- or square-wave outputs that are separated by 90°, thereby eliminating the need for phase-locked loops or other synchronous circuits. Frequency can be controlled by appropriate selection of resistors and capacitors associated with integrators A_2 and A_4.

References
1. J. G. Graeme, G. E. Tobey, and L. P. Huelsman, "Operational Amplifiers—Design and Applications," McGraw-Hill, 1971.

Chapter 31
PHASE-LOCK CIRCUITS

Digital phase shifter covers 0° to 360° range

by J. W. V. Storey
Dept. of Physics, University of California, Berkeley

Offsetting the phase of a signal by digital means over the range of 0° to 89° in any quadrant, this low-power circuit is particularly useful in data-recovery systems that employ synchronous detectors. Unlike most RC phase shifters, the value set is independent of the input frequency.

The input reference signal is first introduced to the 14046 phase-locked loop. Its output, which is set to generate a frequency 360 times that of f_{in}, is then applied to the 14518 binary-coded decimal counter, where it is divided by 10. A second, cascaded counter, A_2, divides A_{2a}'s output by 9, the 89th count of a 90-step cycle being detected by A_{5b}. A_2 is then reset to zero on the 90th count by A_{6a} and A_{5c}.

The signal at the output of A_{6a} is thus at a frequency equal to $4f_{in}$. Flip-flop A_7 performs a divide-by-four operation on this signal, at the same time generating four quadrature outputs. Meanwhile, A_{5a}, which generates one pulse per cycle of f_{in}, locks the PLL in phase with the zero count of A_2. Note that the operation of this divider chain is unaffected by the setting of the digital input lines.

The desired phase is selected by applying the appropriate digital signals D_0–D_7 in binary-coded decimal form. Thus, the output from A_3 and A_4 moves high when A_2 counts to that number and clocks flip-flop A_{6b}. This action occurs four times per each cycle of f_{in}. The appropriate quadrant, available at A_7, is selected by digital inputs D_8–D_{11}, the active quadrant corresponding to which one of the lines is high. Flip-flop A_{6b} thus produces a symmetrical square wave at f_{in} having a phase shift equal to the number of degrees specified plus 0°, 90°, 180°, or 270°. There is an additional phase shift of 0.5° at all settings because of the way the PLL is operated to achieve lock. The error can be eliminated by adding an inverter between the output of A_{5a} and the B input of A_1.

With the component values shown and a 15-volt supply, the circuit will operate over the range of 0.2 hertz to 2 kilohertz. Thumbwheel switches with 1-megohm pull-down resistors are used to set the phase-angle input lines. A four-position switch can be used to select the quadrant. With slight modification, the circuit will find application as a digitally controlled ignition timing system for internal-combustion engines. □

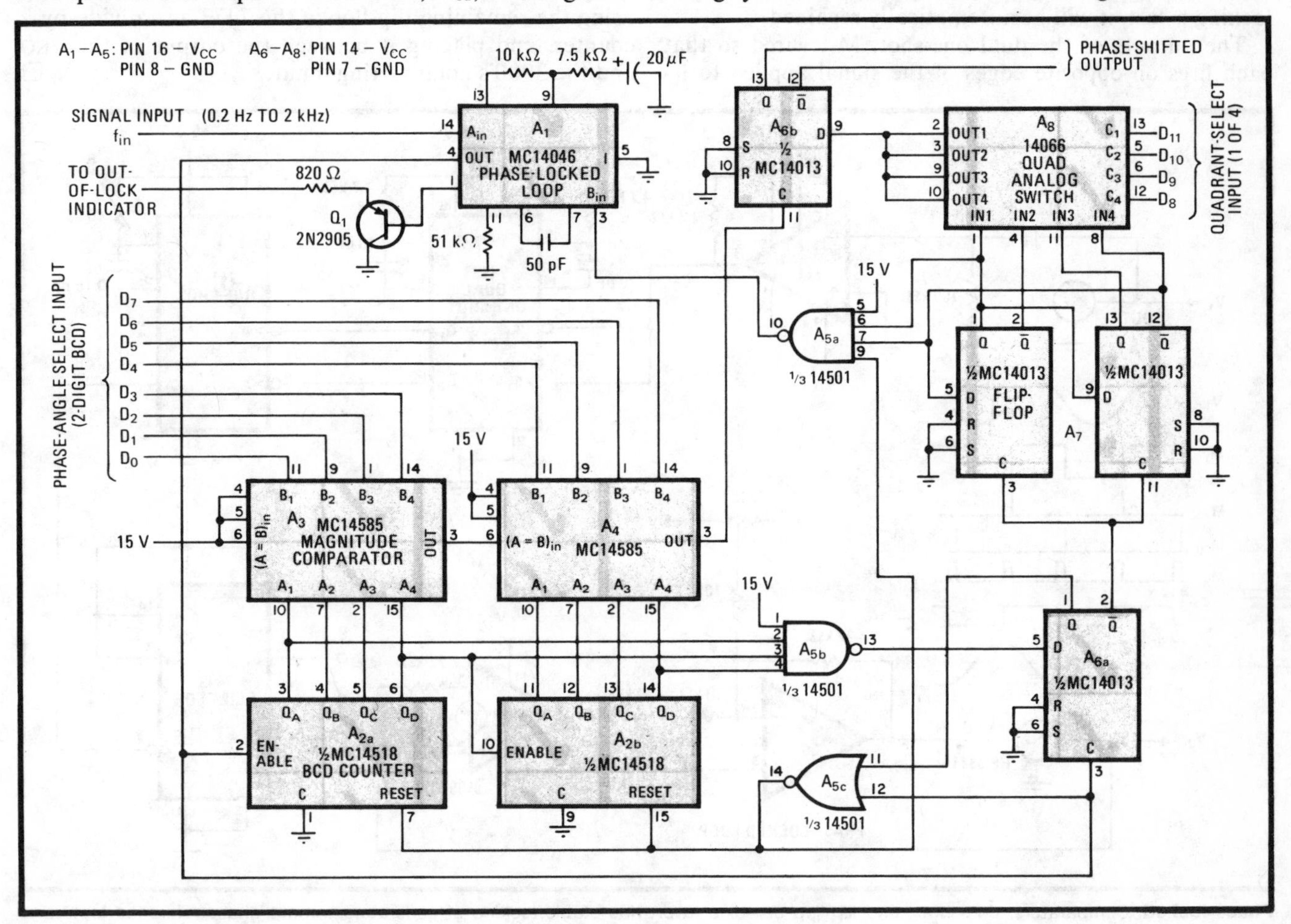

Discrete degrees. Circuit sets 0° to 360° phase shift of reference signal by digital means. Digital inputs D_0–D_7 determine displacement over 0° to 89° range, D_8–D_{11} set quadrant. Output is thus 0° to 89° signal shifted by an additional 0°, 90°, 180°, or 270°.

Circuit phase-locks function generators over 360°

by Lawrence W. Shacklette
Seton Hall University, South Orange, N. J.

This circuit locks a low-cost function generator without a voltage-controlled-oscillator (VCO) input to a second inexpensive generator having such a phase-reference feature. A J-K flip-flop, two one-shots, two comparators, and an operational amplifier are linked in a feedback arrangement that includes the programmable generator as the VCO in a phase-locked loop (PLL). The resultant circuit provides a selectable phase shift between generator outputs over the range of 0° to 360°.

In operation, the output of the low-cost generator, v_1, passes through a high-pass filter to a zero-crossing detector that employs a 311 comparator (M_1), as shown. A dual monostable multivibrator and J-K flip-flop follow. Although these two devices can be eliminated, they enable the phase-locked loop to be operated at the center of its locking range for any phase shift. This arrangement ensures that two often desired phase angles, 0° and 180°, fall within the capture range of the PLL, so that if locking is lost, it will be automatically regained.

The outputs of the dual one-shot, M_2, wired so that each fires on opposite edges of the signal applied to its inputs (see timing diagram), are fed to the OR input of M_3, whose on time is selected by potentiometer R_1. Thus R_1 controls the amount of phase shift.

M_3 produces two pulses for each cycle of v_1 and triggers the J-K flip-flop, M_4, on each negative edge. The flip-flop thus produces a square wave with a frequency equal to v_1, but shifted in phase by up to 180°. An additional shift of 180° can be obtained by using switch S to connect the Q output of M_4 to the inverting input of the 3900 Norton amplifier.

The adjustable-phase square wave serves as a reference signal for the phase-locked loop, which is composed of the 3900, a low-pass filter, a buffer (307), and a VCO (the second function generator, a Hewlett-Packard 3311A). The input signal to the VCO is a negative dc voltage that is the inverted sum of the filtered output of the 3900 and the voltage selected by the offset control, R_2. By turning the generator's front-panel control or R_2, the free-running frequency of the loop can be adjusted.

Because the reference signal is a square wave, the PLL will lock onto either the fundamental of v_1 or its odd harmonics. Selection of a particular harmonic is made by adjusting the free-running frequency to the approximate value of the harmonic desired. Half-multiple harmonics ($\frac{1}{2}f_1$, $\frac{3}{2}f_1$, $\frac{5}{2}f_1$, etc.) can be produced at v_2 by breaking the $\overline{Q}_1$–A_1 connection between M_2 and M_3 and tying A_1 to +5 volts. Even harmonics can be obtained by using the remaining flip-flop in the 7473 as a divide-by-2 counter, and placing it between the output of the VCO and the 3900's noninverting input. □

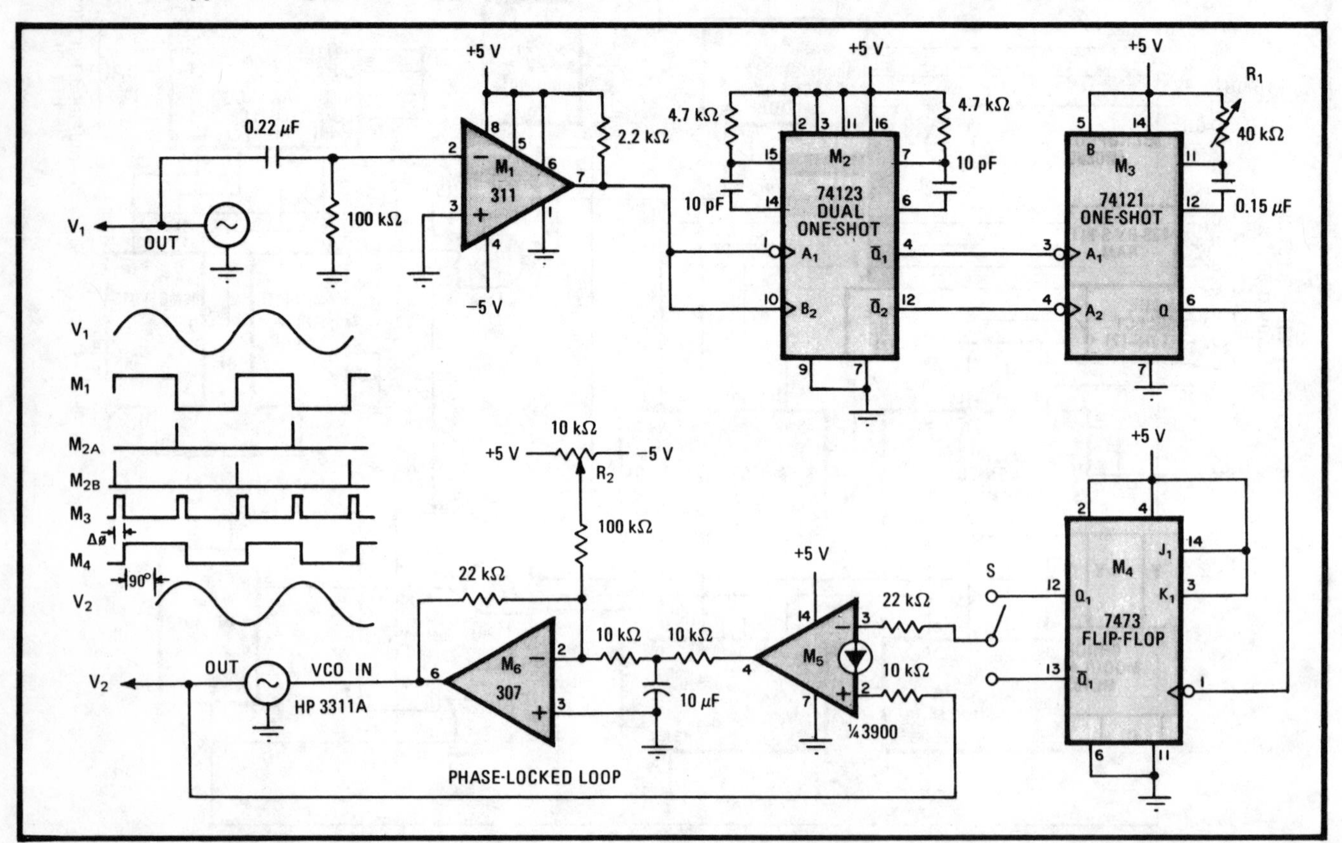

Phase-locked. Comparators, one-shots, and flip-flop combine to provide stable locking of generators without phase-reference feature to those having a VCO input. R_1 and S are used to select phase of v_2 with respect to v_1; phase can be adjusted from 0° to 360°. M_2 and M_4 ensure that locking is regained if it is lost, and R_2 controls lock frequency, which may be set to integer or half-integer harmonics of v_1.

Phase-locked generator converts, filters most inputs

by Peter Reintjes
Research and Design Ltd., Morehead City, N. C.

Replacing the voltage-controlled oscillator in the RCA 4046 phase-locked loop with the popular Intersil 8038 waveform generator forms a circuit that produces sine, square, and triangular wave voltages capable of tracking almost any input signal. Besides performing its prime function of waveform conversion, the circuit serves as a high-Q filter. With a harmonic distortion of only 0.5%, it finds the fundamental frequency of any signal.

The performance of this circuit far exceeds that of a conventional filter, which always adds a phase shift to the incoming signal. Also, traditional filtering methods are often of little use when the fundamental frequency must be recovered from an unpredictable input signal.

Connecting the 8038, which is itself a voltage-controlled oscillator, to the 4046 as shown in the figure does not affect the normal operation of the phase-locked loop. The only difference in the basic PLL circuit is that the 8038 generates sine, triangular, and square waves and drives the 4046 in place of the loop's internal VCO. The output waveshapes are unaffected by the harmonic distortion present on the input signal. Capacitor C sets the center frequency of the 8038 (a value of 0.047 microfarad corresponds to a frequency in the audio range). The frequency-capture range of the circuit, which is determined by the 4046, remains 1,000 to 1. The generator's maximum operating frequency is about 700 kilohertz.

To secure precise locking, the comparators in the 4046 should be driven by the square-wave output of the 8038. If the input waveform is a pulse, phase comparator I should be used. For unpredictable or high-noise signals, phase comparator II is more suitable.

Any phase difference between the square-wave output of the 8038 and the input signal is amplified by two 741 operational amplifiers and then fed back to the VCO to increase or decrease its frequency, as the case may be. Although the internal VCO of the 4046 is not used, it must be enabled by grounding pin 5 of the device so that its voltage-follower will be active. If matched resistors are used at pin 4 and 5 of the 8038, the sine-wave output distortion can be reduced to 0.5%. Potentiometers R_1 and R_2 aid in minimizing the distortion. □

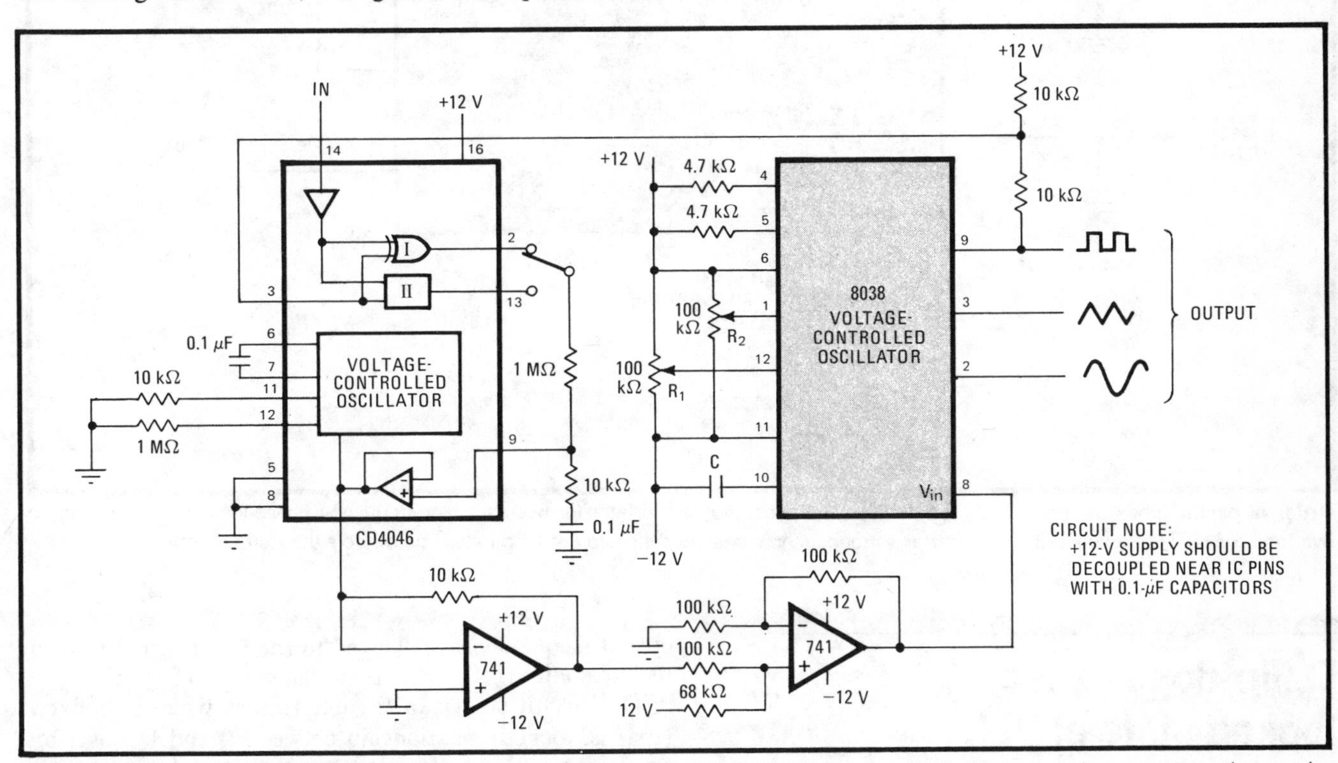

Tracking waveforms. ICL 8038 and two op amps replace internal VCO in 4046 to form phase-locked-loop waveform generator that can be used to recover fundamental frequency of any input signal or to convert the control signal to sine, square, and triangular waves. The generator's maximum output frequency is about 700 kHz. R_1 and R_2 are adjusted for minimum output distortion.

LC network adapts PLL for crystal-overtone operation

by R. J. Athey
National Research Council, Ottawa, Canada

Although Texas Instruments' popular 74S124 oscillator serves reliably in most instances as a crystal-controlled phase-locked loop, problems arise when overtone crystals are utilized for high-frequency (20 megahertz and above) operation. The difficulties may be overcome by adding an LC network in order to retain adequate system gain for oscillation at the required overtone and dampen oscillations at the fundamental frequency. This technique thereby forces the loop to lock onto the crystal's third-order output.

Although the range over which the PLL responds will be limited to about 1 kilohertz for a 0-to-5-V input signal, the method affords repeatable results and will enable use of the 74S124 beyond the normal limits imposed by fundamental-mode crystals. As shown in the figure, L_1 and C_1 are selected to be series-resonant at the desired overtone frequency. Assuming L_1 is 1.5 microhenries, a C_1 value of 2 to 20 picofarads will be adequate for tuning over the range of 38 to 45 MHz required in this particular application. The Q of both L_1 and C_1 should be reasonably high.

C_1, along with R_1, serves as a gross frequency control. Unfortunately, the setting of R_1C_1 will be rather critical. Although the range over which R_1 is effective as a tuning element for a given C_1 is narrow, a miniature carbon potentiometer will have sufficient resolution for making adjustments. C_2 provides an offset for the desired third-overtone frequency. In general, the circuit will work satisfactorily for crystals working to 60 MHz. □

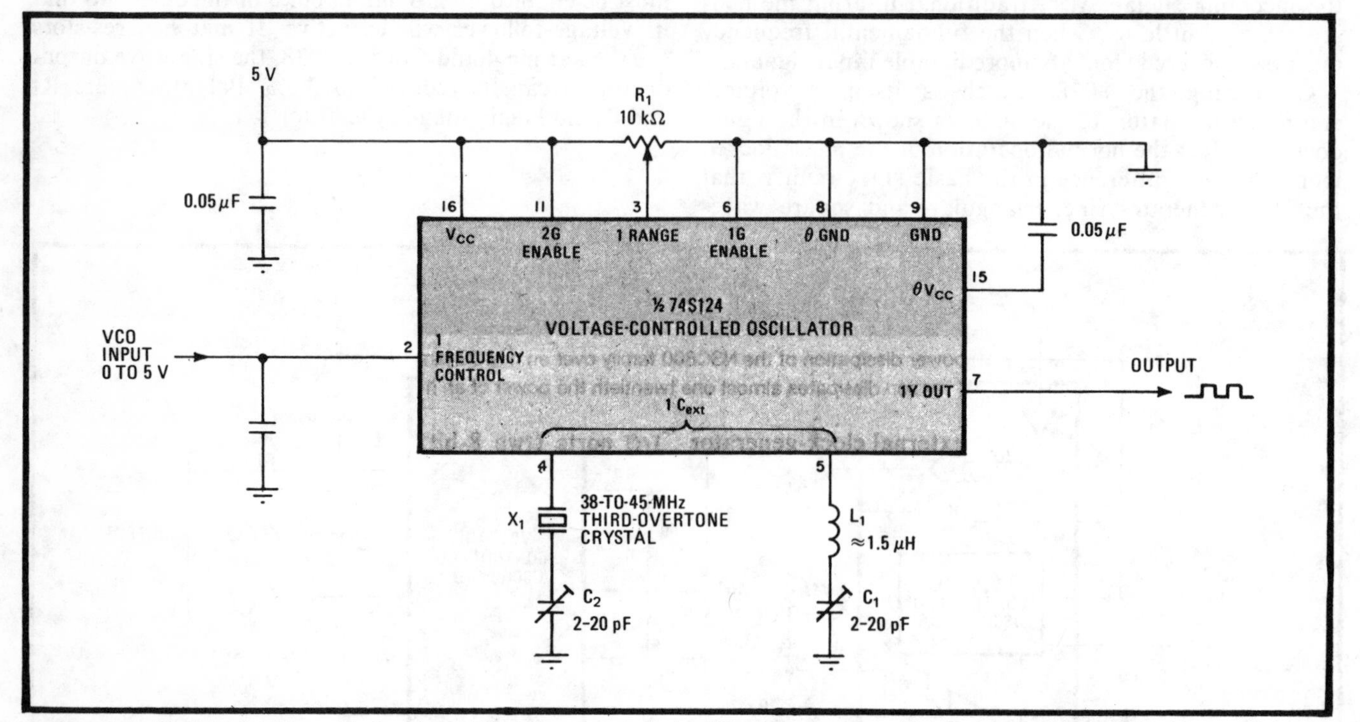

Order of oscillation. Addition of L_1C_1 adapts TI's voltage-controlled oscillator for use as a high-frequency phase-locked loop utilizing an overtone crystal. LC network eliminates crystal's strong fundamental response, forces loop to lock onto slab's third-order output.

D flip-flops sense locked state of PLL

by L. W. Shacklette and H. A. Ashworth
Seton Hall University, Department of Physics, South Orange, N. J.

This circuit uses a dual D flip-flop to sense the locked state of many popular phase-locked loops, such as the Signetics 562 and 565. By adding a dual one-shot–light-emitting-diode combination to the flip-flops, the circuit visually indicates locking for the conditions where the output frequency, f_o, is locked to the input signal (f_s), to its harmonics (Nf_s) or to its subharmonics (f_s/N).

The circuit shown in (a) determines whether a fixed (that is, locked) relationship between f_s and f_o exists by employing both flip-flops in a simple phase detector. The f_s signal drives the D input of flip-flop A_1 and the C input of A_2, and the f_o signal emanating from the voltage-controlled-oscillator output of the PLL drives C_1 and D_2. The design of the phase detector accommodates a PLL having a phase comparator that can generate an upper and lower f_s-to-f_o phase displacement of 180° and 0°, respectively, for the locked condition. The comparator does this by deriving an f_o that is displaced 90° with respect to f_s when the loop is in the center of its range.

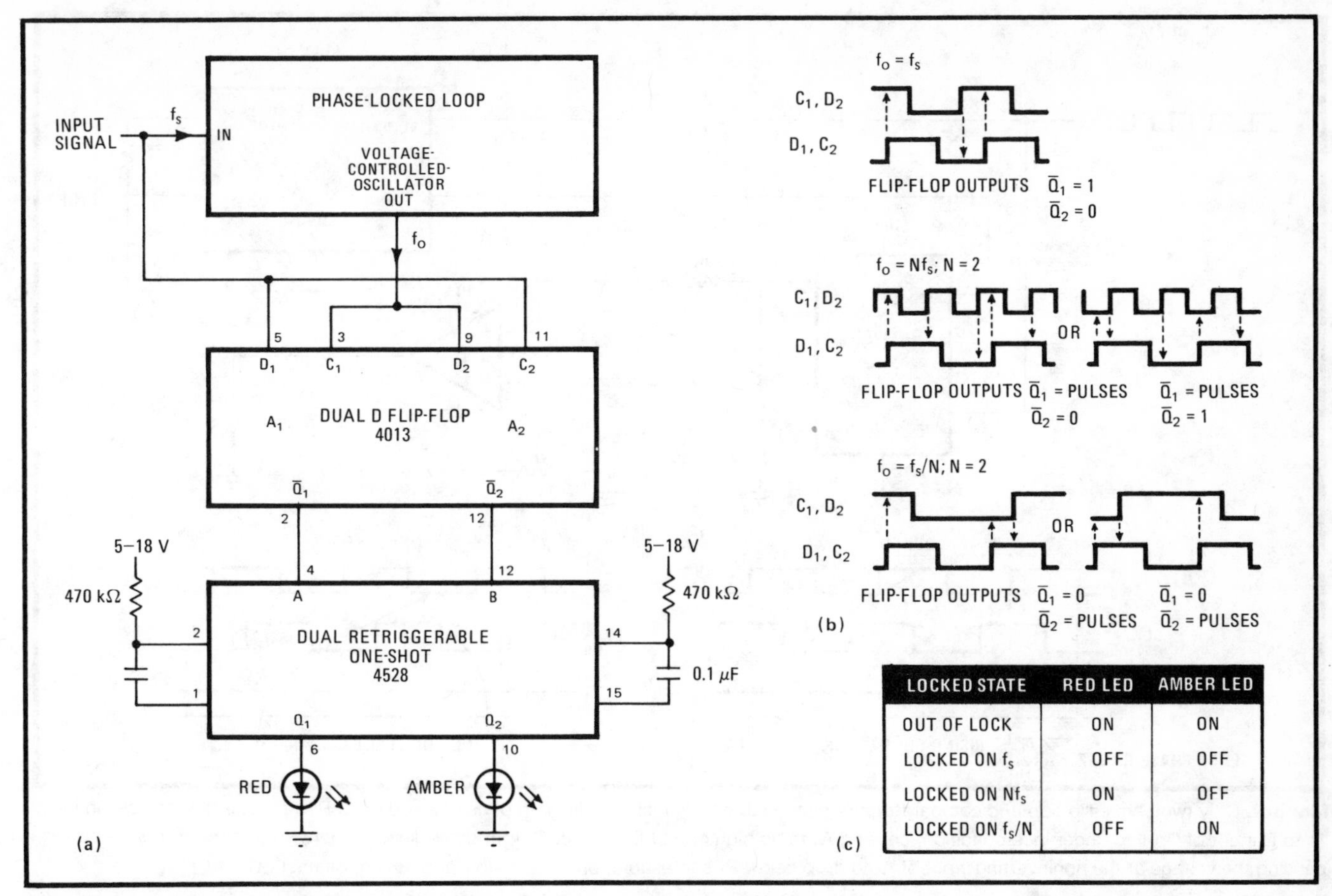

LOCKED STATE	RED LED	AMBER LED
OUT OF LOCK	ON	ON
LOCKED ON f_s	OFF	OFF
LOCKED ON Nf_s	ON	OFF
LOCKED ON f_s/N	OFF	ON

Lock detector. Monitor (a) detects the existence of a phase difference between f_s and f_o and can thus differentiate between three locked conditions, because circuit is also sensitive to ratios f_s/f_o, $f_s/2f_o$, and $2f_s/f_o$ (b). Table (c) summarizes circuit response.

The circuit response for a constant f_o and f_s may be understood with the aid of (b). Because the D flip-flops read the data signals (D_i) on the positive edge of each clock (C_i), whenever the data frequency f_{di} equals the clock frequency, f_{ci}, $\overline{Q}_1$ and $\overline{Q}_2$ of the 4013 remain fixed at either logic 1 or logic 0, depending upon whether the signals at C_i and D_i are in phase or out of phase. In either case, the output from the corresponding edge-triggered one-shot in the 4528 will be zero.

When f_c is an integer multiple of f_d, or f_d is an integer multiple of f_c, there will be a pulsed output signal from one of the output ports of the 4013 and a corresponding signal at the 4528 to light the LED. Note that because the one-shot is retriggerable, its output will be constantly at logic 1 for a pulsed input signal. The output (logic 1 or logic 0) from the other port of the flip-flop will be constant. When f_o and f_s are out of lock, each flip-flop reads random 1s and 0s, causing pulsed output signals to appear at both ports of the 4013. The table (c) summarizes circuit operation.

In cases where it is necessary to detect only the condition $f_o = f_s$, a simpler monitor can be constructed using only a single D flip-flop and one LED that is connected to its $\overline{Q}$ output. The LED will light whenever $f_o \neq f_s$. □

PLL's lock indicator detects latching simply

by Steve Kirby

Department of Electronics, University of York, England

Much less complex than some of the previously described lock indicators for phase-locked loops,[1] with no need to derive and utilize a multiple of the input frequency[2] for phase-comparison purposes, this circuit is easier to set up and use. It sacrifices nothing in the way of accuracy and offers other advantages, such as the ability to lock onto harmonics of the input signal.

The locking technique is illustrated for the C-MOS CD4046 PLL, whose output leads the input by 90° when the lock state is achieved. The loop's capture ratio is such that lock can be maintained for a square-wave input signal no greater than +90° and no less than −90° out of phase with respect to f_{out}. The 4013 D flip-flop detects phase differences by clocking the state of f_{in} at f_{out}'s rising edge. Assuming the PLL and its associated loop filter are working properly, a steady Q = 1 at the output of the flip-flop indicates the PLL is in or will shortly be in the lock state. The noninverting input of the 741 comparator will then rise to 10 volts through integrator

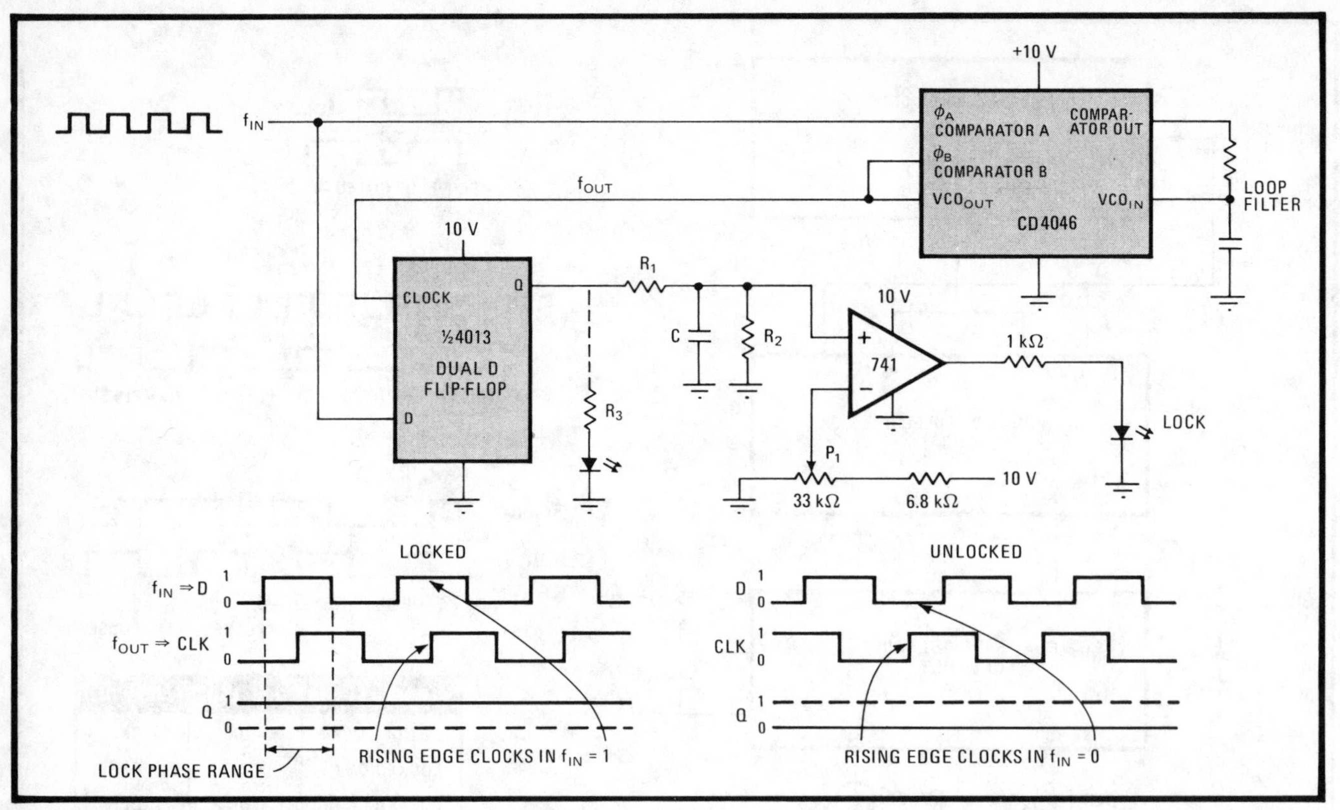

Monitor. Only two chips, flip-flop and comparator, are needed to detect lock condition in phase-locked loop. Rising edge of f_{out} clocks in logic 1s to D input of flip-flop under lock condition, causing A_1 to go high and LED to light. Output of flip-flop is otherwise a random train of pulses, causing the voltage at the noninverting input of A_1 to drop below P_1's threshold, bringing A_1 low and turning off the LED.

R_1R_2C, and its resulting high-going output will light the light-emitting diode.

If the PLL no longer locks on frequency, the phase of f_{in} with respect to f_{out} will be random. The output of the flip-flop will thus be a train of variable-width pulses. The comparator input thus drops to approximately 5 V, and because potentiometer P_1 sets the inverting input at approximately 7 V, A_1 moves low, extinguishing the LED.

The lock detector will lock onto higher harmonics of f_{in}. With a 50/50 mark-to-space square-wave signal, locking has been observed to the fifth harmonic.

If a less precise indication is tolerable, lock detection can be achieved with even fewer parts by placing an LED at the output of the flip-flop and eliminating the comparator circuitry. Resistor R_3 should be selected to hold the LED dim for the out-of-lock condition. □

References

1. J. A. Connelly and G. E. Prescott, "Phase-locked loop includes lock indicator," *Electronics*, Sept. 5, 1974, p. 112.
2. R. P. Leck, "Logic gates and LED indicate phase lock," *Electronics*, May 29, 1975, p. 106.

Chapter 32
POWER SUPPLIES

Modulating the flyback inverter reduces supply's bulk

by Vladimir Brunstein
Nova Electric Manufacturing Co., Nutley, N. J.

The cost, size, and weight of an inverter will be reduced if a flyback transformer, modulated by a high-frequency carrier, is configured in the inverter's output stage. Modulating the flyback transformer is more efficient than attempting to modulate push-pull or bridge circuits, as is sometimes done.

The two subassemblies that are most difficult to miniaturize in the standard inverter are the output filter and transformer. But while the output filter can easily be simplified in a number of ways, the dimensions of the power transformer are determined mainly by the operating frequency.

The obvious solution to reducing the volume and weight of the transformer and the number of filter components, then, is to convert the dc input to a high frequency, such as 20 kilohertz. Here the harmonics of the output signal will be significantly higher than the demodulated frequency, and so a simple low-pass filter can be used to recover the desired 60-Hz waveform.

Using the flyback scheme brings an additional simplification, compared to a push-pull or bridge configuration. The recovered voltage on the secondary of the transformer will be amplitude-modulated, and only a capacitor will be required to obtain the low-frequency component.

As shown, a 60-Hz sinewave voltage is required to drive the flyback transformer via an error amplifier and pulse modulator and also through a power-driver connected to Q_2. Details of these blocks vary with individual requirements and so are not shown in detail here. Picking the right ferrite core is a subject in itself and has been discussed in various papers.

In general, coils T_{2a} and T_{2b} charge during the time interval t_1, and discharge through the load resistance R_L during t_2. Note that an inexpensive optical coupler may be used to replace T_{2b}. The voltage on the transformer secondary will be:

$$V_L = \eta \frac{n_S}{n_p} \frac{\tau}{1-\tau} E \qquad (1)$$

where:

η = efficiency of the power stage
$\tau = t_1/T$
t_1 = turn-on time for transistor Q_1
T = commutation period $t_1 + t_2$
E = dc input voltage

It is seen that the equation lends itself to achieving the end solution, for if the duty cycle is varied by the sine wave while keeping the turn-off time of Q_1 constant, then the peak value of the secondary will follow the sinewave reference as shown in the timing diagram.

This action results if $t_1 = k_1 \sin \omega t$, for in that case Eq. 1 becomes:

$$V_L = K_2 \sin \omega t \qquad (2)$$

where K_1 and K_2 are constants.

In reality, one half of the sine wave would normally be inverted, as shown for V_L. Q_2 and Q_3 act as 60-Hz synchronous commutators serving to restore the original wave shape.

Note that only a relatively small output capacitor, C, is required for filtering—no choke is necessary. Total harmonic distortion at the output is less than 5%. □

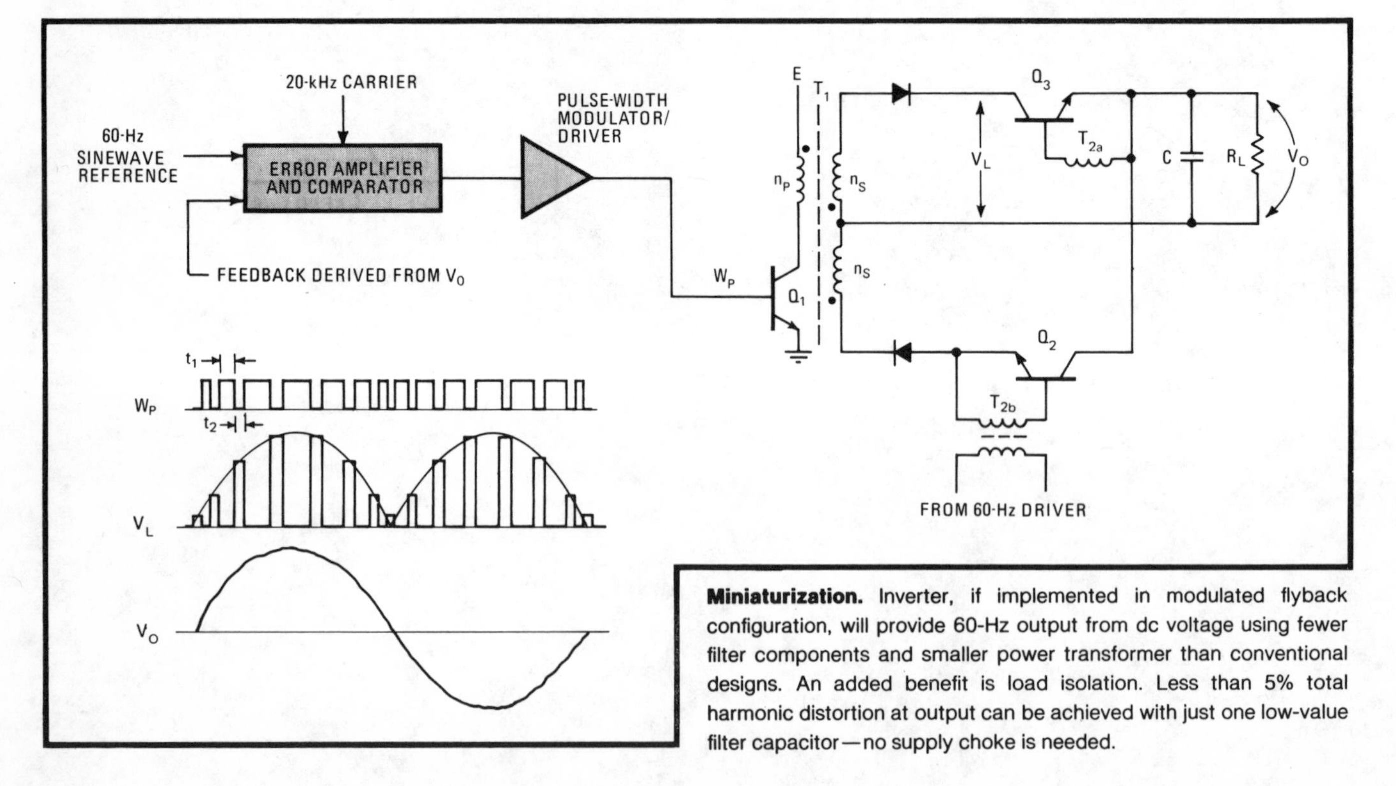

Miniaturization. Inverter, if implemented in modulated flyback configuration, will provide 60-Hz output from dc voltage using fewer filter components and smaller power transformer than conventional designs. An added benefit is load isolation. Less than 5% total harmonic distortion at output can be achieved with just one low-value filter capacitor—no supply choke is needed.

One-chip voltage splitter conserves battery power

by David Bingham
Intersil Inc., Cupertino, Calif.

Positive and negative supply voltages of equal magnitude are usually secured in a battery or other low-power floating-source arrangements by establishing system ground at half the source potential by means of a simple voltage divider. But this scheme more often than not consumes excessive power. Modifying Intersil's ICL7660 positive-to-negative voltage converter to work as a voltage divider, however, will increase the power-conversion efficiency to as high as 98% at an output current of 10 milliamperes.

The conventional voltage divider circuit shown in (a) uses two resistors and a unity-gain operational-amplifier buffer. While this circuit can function at relatively low power, depending upon the op amp used and the value of R, it will suffer from an inherently low power efficiency if the load should be connected between system ground and either V^+ or V^-. In such cases, the current flowing through the load will always be the same as that drawn from the battery, and thus the maximum efficiency can never be greater than 50%.

The ICL7660 can be made to simulate a divide-by-2 voltage converter by simply grounding pin 5 (normally the V^- lead) and using the normally grounded lead at pin 3 as the output, as shown in (b). The voltage distribution on the chip will be unchanged, since pin 3 is midway between V^+ and V^-, as before. And no power is lost in heating up any resistance, as the ICL7660 operates in the switched-capacitor (charge-transfer) mode to derive the output voltage.

With this configuration, an open-circuit output voltage equal to V/2 ±0.1% is achieved. The output impedance is 13 ohms for a supply voltage of 9 V and 0.1 mA $< I_{out} <$ 80 mA, or 17 ohms for V = 6 V and an output current in the same general range.

Because the ICL7660 can source only an output current reliably, difficulty may be encountered if the load is connected between pins 3 and 8 of the device. To ensure startup for current-sinking applications, a 1-MΩ resistor is placed between pin 6 and ground. This step guarantees that there will always be some voltage across the on-chip oscillator and the control circuitry.

As for circuit performance (c), conversion efficiency will be no lower than 80% for V = 6 V and 0.5 mA $< I_{out} <$ 80 mA. In equation form:

$$\eta = (V_{out} I_{out} / V^+ I^+)100$$

where I_{out} is the magnitude of the output current, regardless of sign. □

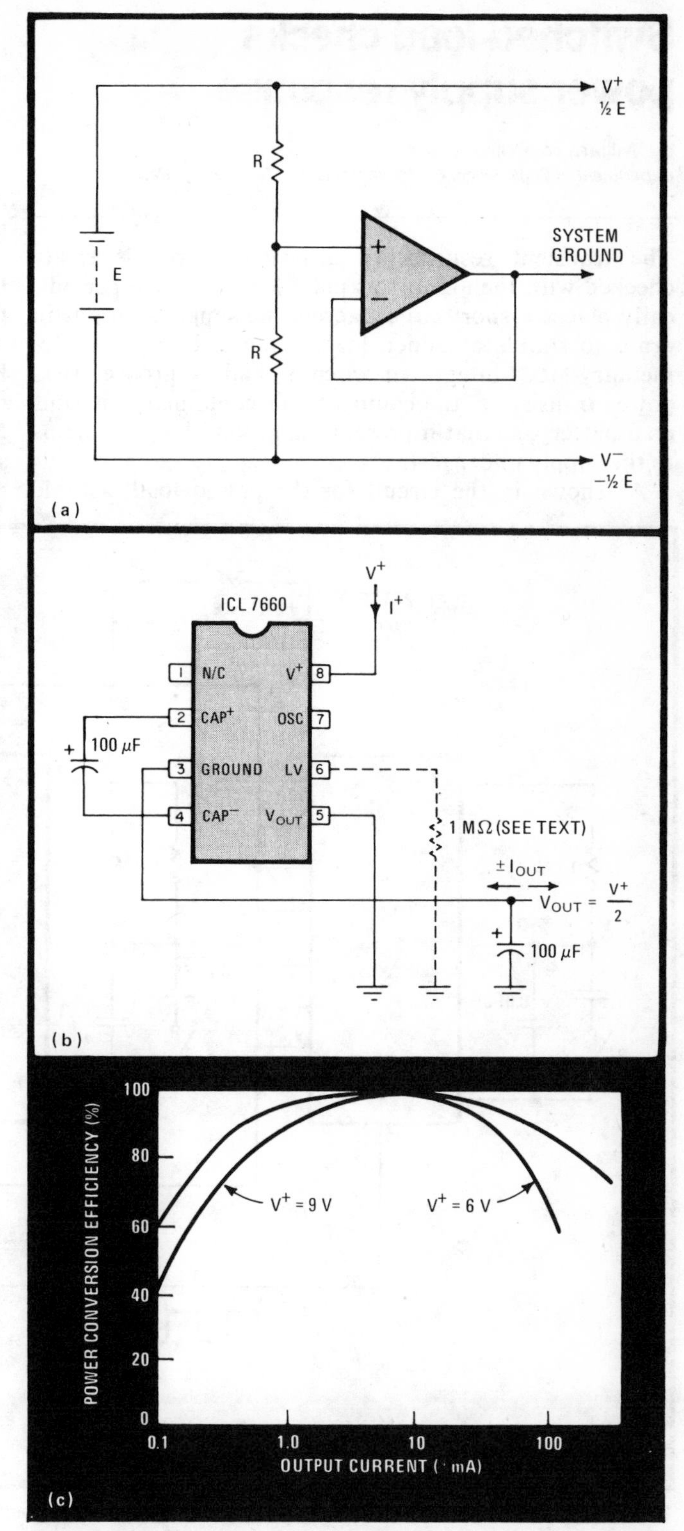

Efficient. Simple resistive voltage divider (a) dissipates excessive power and thus is ordinarily not suited to providing positive and negative supply voltages from floating source. Suitably wired ICL7660 converter (b) provides the function without power loss, yielding conversion efficiencies (c) approaching 98%.

Switched load checks power supply response

by William M. Polivka
Department of Engineering, California Institute of Technology, Pasadena

The transient response of a power supply is easily checked with the aid of this pulse loader, which periodically places a short circuit across the supply's output in order to simulate sudden load changes. Using complementary-MOS integrated circuits and V-groove MOS power transistors, the compact, self-contained unit runs on a battery, so that it presents no ground-loop problems to the supply under test.

As shown in the circuit for the pulsed load, astable multivibrator A_1 and one-shot A_2 set respectively the frequency and the width of the pulses that switch on load transistors Q_1 and Q_2. S_1 selects either of two combinations of frequency and width—in this case, 2 hertz at 25 milliseconds or 20 Hz at 2.5 ms. Note that a low duty cycle is required to reduce heat dissipation in Q_1 and Q_2. A trigger signal for driving an oscilloscope or other instrument to observe the supply's response appears at pin 11 of A_2.

Potentiometer P_1 sets the point at which Q_1 and Q_2 fire, so that the magnitude of the pulsed supply current passing through the load transistors can be selected from zero to the maximum capability of the V-MOS devices. Each field-effect transistor handles 2 amperes at 12 volts, values that derate to 400 milliamperes at 60 V. Moreover, an increase in supply loading may be attained simply by adding transistors in shunt at the output of the unit, as required. □

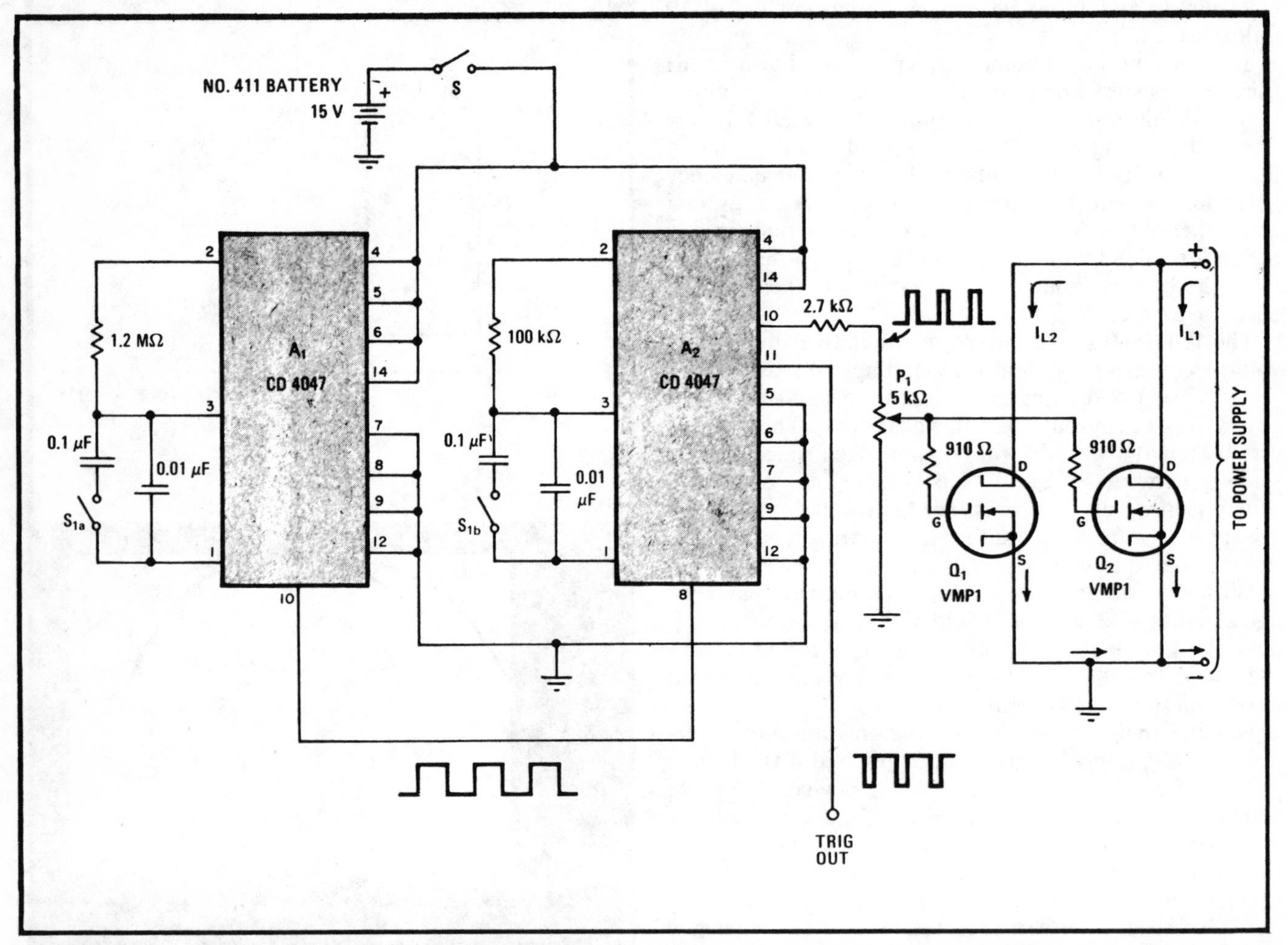

Load dynamics. Low-cost tester, with aid of scope, finds transient response of power supply by ordering periodic increase in supply current to simulate load changes. A_1 and A_2 set frequency and width of switching waveform. $Q_1 - Q_2$ sink current proportional to setting of P_1.

Dc-dc power supply regulates down to 0 volt

by P.R.K. Chetty and A. Barnaba
ISSP, Bangalore, India, and CNES, Toulouse, France

In most dc-input, regulated power supplies, regulation is poor when the desired output voltage is less than the source's internal reference voltage. In addition, circuit considerations usually limit the minimum reference voltage attainable and consequently the minimum regulated output voltage possible. This circuit, however, with a configuration that can bring the reference voltage to virtually zero, overcomes both problems.

The LM723 voltage regulator shown, which provides 12 volts at 1 ampere, must be biased with a negative supply voltage at its $-V_{in}$ port (pin 5) for proper operation. This voltage is provided by the switching inverter shown within the dotted lines.

The LM111 voltage comparator is configured as an astable multivibrator that oscillates at a frequency of about 10 kilohertz. With the aid of the 1-millihenry inductor, which generates the counterelectromotive force required to produce a negative potential from a switched-source voltage, the inverter delivers a well-regulated −7.5 v to the $-V_{in}$ port of the 723.

The magnitude of this voltage is essentially equal to that of the regulator's internal reference voltage, V_{REF}, appearing at pin 4, and properly biases its voltage-reference amplifier. This condition in turn precipitates a condition in the amplifier whereby V_{REF} clamps to ground potential. Thus the output voltage may be adjusted throughout its maximum possible range by potentiometers R_1 and R_2. Although the potential of V_{REF} as measured with respect to ground has been changed, the circuit will retain the regulating properties of the 723. Both the line and the load regulation of the supply are 0.4%. □

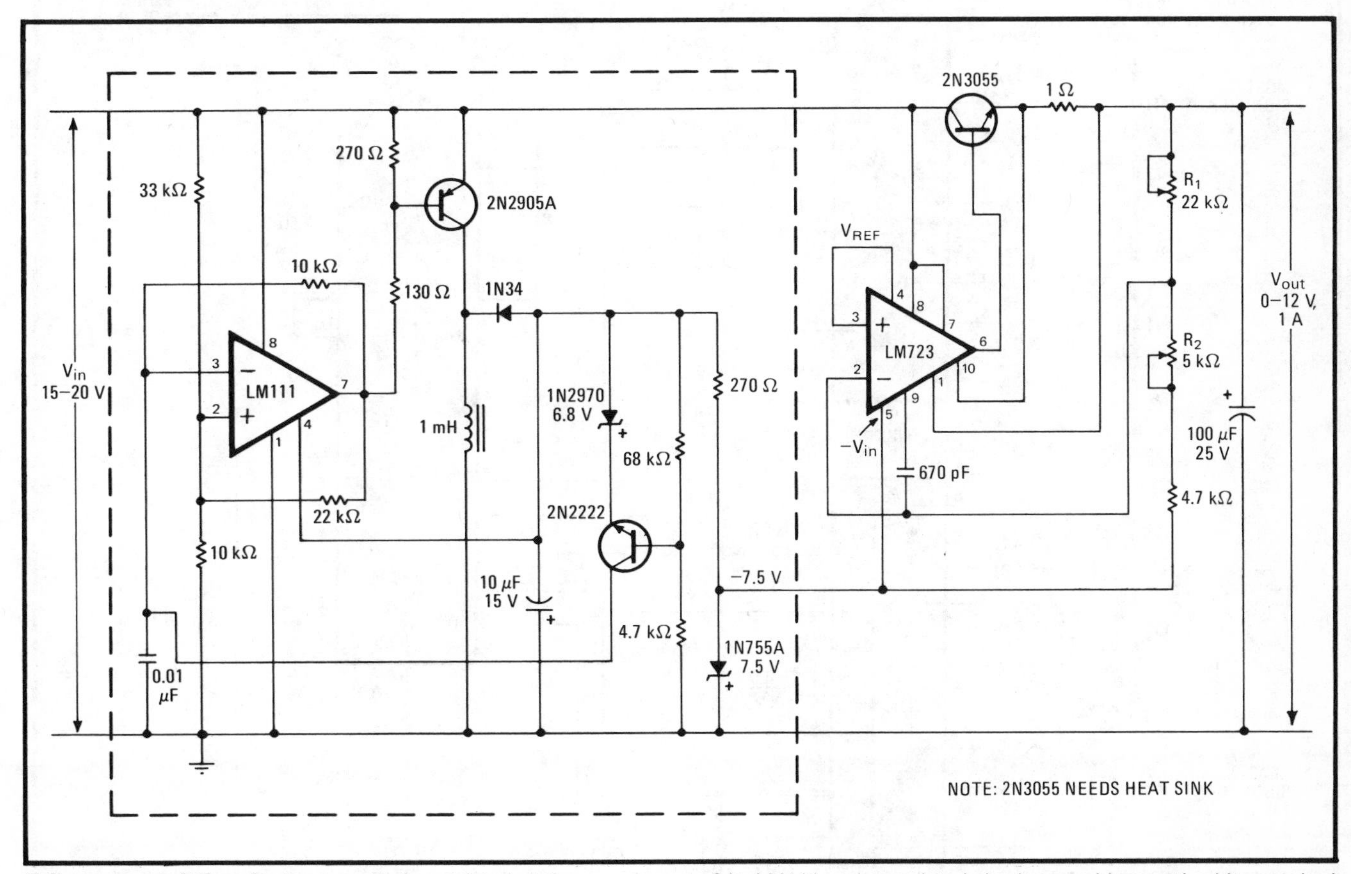

Full-range regulation. Dc-input supply is regulated all the way down to 0 V. LM111 and associated circuitry provide negative bias required for LM723 regulator. Regulator's internal-reference voltage, V_{REF}, is clamped to ground; output voltage is thus adjustable from 0 to 12 V.

Modifying a power supply to add programmability

by Eric Kushnick
Bose Corp., Framington, Mass.

Modifying the commercial bench-type supply is a simple but effective and low-cost solution to the problem of obtaining a programmable power source if the supply's response time is not a primary consideration. Here, members of the popular Hewlett-Packard series (6212A–6218A) are converted by adding opto-isolators and a digital-to-analog converter to their basic circuits so that 8-bit programming capability is achieved. The modification may be completed for less than $50.

The initial changes required are shown in (a), and they are fairly straightforward. The 6.2-volt reference from the supply is disconnected for one end of R_{12} and reconnected to the analog input of A_1, the Precision Monolithics DAC-03-CDX2 d-a converter. C_4 is then mounted directly across R_{12}. The analog output of the converter is reconnected in place of the 6.2-V reference and the supply's front-panel potentiometer, R_{10}, previously used for coarse-voltage adjustment, is replaced by the R_1-R_2-R_4 combination. The digital programming inputs to the converter are isolated from the power supply by A_2 and A_3, the Litronix ILQ-74 optocouplers.

Additions to the section in the reference regulator (not shown) that derives the supply's internal voltages must be made next, as shown in (b), so that the converter can be powered. Note that the components within the dotted line show the standard configuration of the HP supply (this part of the circuit is not modified).

Here, the converter is powered through the circuitry

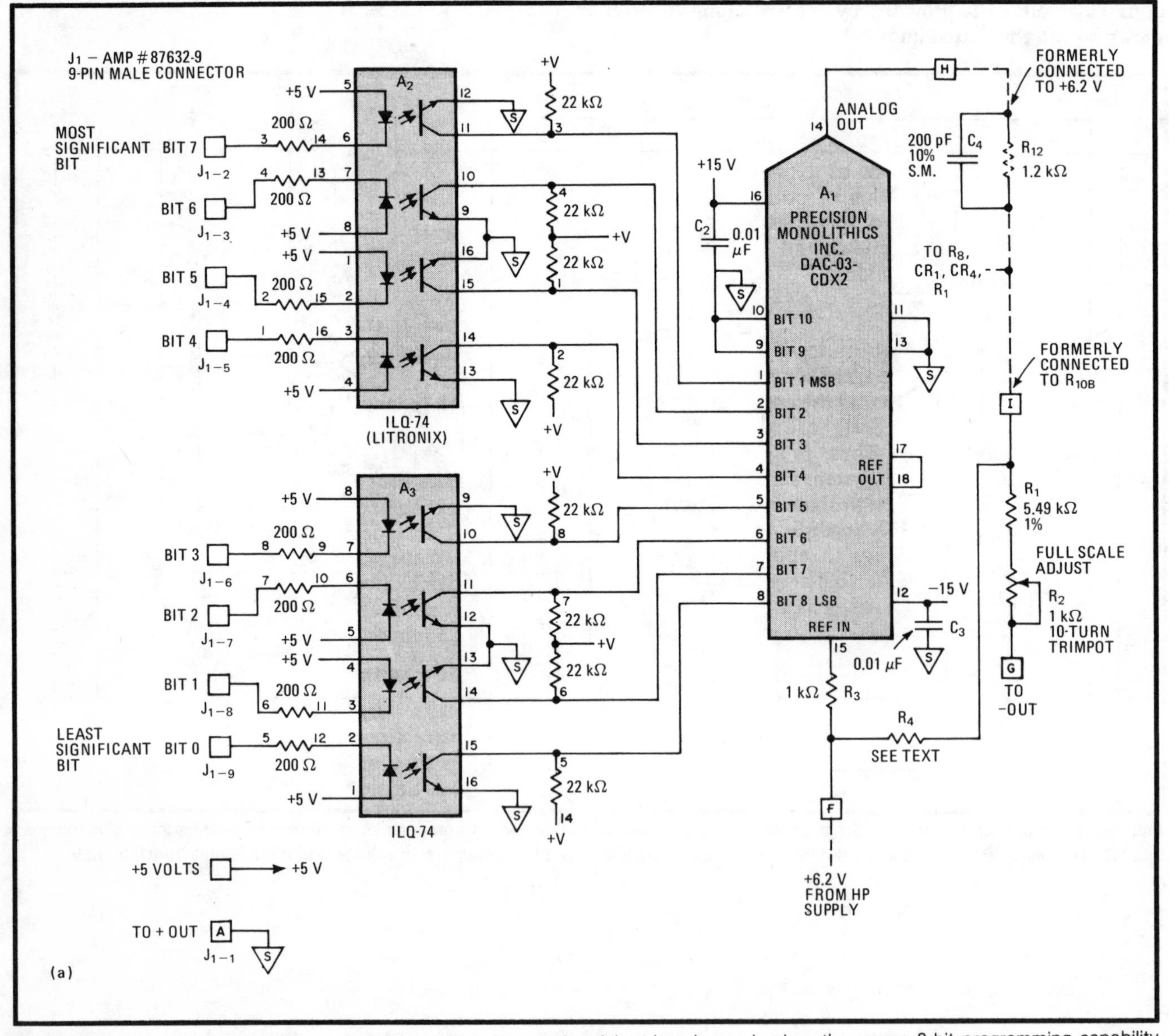

Bits of voltage. Adding digital-to-analog converter and optocouplers (a) to bench supply gives the source 8-bit programming capability. Separate power tap (b) for energizing converter is required to eliminate interaction with supply's regulator (not shown).

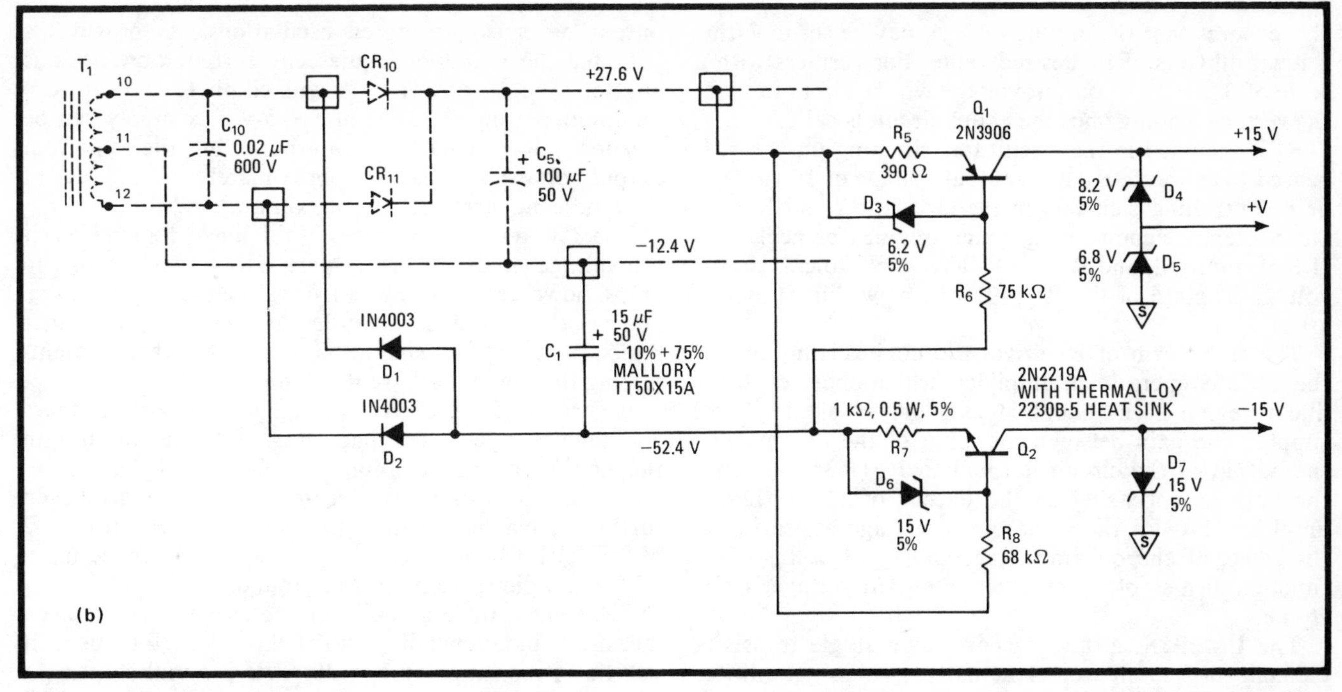

surrounding Q_1 and Q_2. Note that none of the regulator's current can flow into the s terminals, so that regulator operation, which depends on a complex balance of currents to maintain a given output voltage, is not disturbed. C_1 must be in the range of 10 to 30 microfarads to prevent turn-on and turn-off transients from reaching the output of the supply.

The added circuitry can be placed on a printed-circuit board and mounted inside the supply with a small metal bracket bolted to the transformer's mounting screws. As for circuit details, all resistors are ¼-watt carbon-film devices and all capacitors are ceramic-disk types, rated at 25 V, unless otherwise specified. All 200-ohm resistors are contained within a resistor network (Sprague 916C201X5SR). Similarly, the 22-kilohm resistors are contained within the Sprague 914C223X5PE network.

The calibration of the supply is simple. First R_4 is adjusted for a minimum offset voltage at 0 V, and then R_2 is set for a full-scale output voltage of 29.88 V. □

Dc-dc power supply has reference-unit stability

by J. Brian Dance
North Worcestershire College, Worcs., England

The stability of the voltages generated by adjustable dc-dc power supplies is usually no greater than ±100 millivolts, even when the voltages are derived from fixed-voltage regulators. In cases where extreme stability is sought, it is best to design a circuit that utilizes a voltage-reference source instead. Such a circuit, shown in the figure, can provide a 0-to-20-volt output that is within ±5 millivolts of the set value and virtually independent of the current drawn by the load. The supply delivers a maximum of about 1.5 amperes, has a built-in thermal shutdown safeguard, and is protected against short-circuit conditions.

The Precision Monolithics REF-01 voltage reference unit in the circuit provides an extremely stable 10 V across a 10-turn helical potentiometer, R_1, as shown in the figure. This pot, which has a calibrated vernier, sets the output voltage, which will always be equal to twice the value of any voltage derived from the reference source. The linearity of the potentiometer is 0.1%, and this ensures that the output voltage may be set to within a few millivolts of its desired value. For verniers with a scale of 0 to 10, the output voltage will be equal to twice the vernier reading once the entire circuit is calibrated.

R_2 is included in the circuit for trimming purposes. It is used to calibrate R_1 at an output voltage of 10 v. If the gain-controlling elements in the circuit, R_3 and R_4, are close-tolerance components, trimming may be neglected. If R_2 is omitted, and R_3 and R_4 have a 5% tolerance, the voltage at pin 6 of the REF-01 will be within 50 mv of 10 v.

The slider arm of R_1 drives the noninverting port of the LM358 operational amplifier, which operates in the linear region even though it uses only a single-source supply. The gain between the input of the op amp and the output of the circuit is equal to $1+(R3/R4)$. Thus the voltage appearing at the output of the LM295K amplifier is twice the value of the voltage appearing at the input of the op amp; since the input voltage is a function of a stable reference, the output voltage is also stable.

The LM295K, although shown as a single transistor element in the figure, is actually a high-gain linear power amplifier. Its open-loop gain is 1 million, and it is capable of delivering a maximum of 1.5 A to the load.

C_1 and C_2 are added to the circuit to ensure that power-line input-voltage glitches do not appear at the output or cause unwanted oscillations. R_5 provides a path for the amplifier's quiescent current flow; if it is omitted, the output voltage will climb to 9 v at R_1's minimum setting. The use of a −5-v bias supply can be avoided if R_5 is connected to ground, but the minimum output voltage will rise to approximately the product of the quiescent current (5 mA maximum) and R_5.

The 358 amplifier operates in the linear region even if the voltage at the inverting port falls to zero. Most op amps, however, require bias from a dual supply (positive and negative voltages) in order to operate in the linear region, and this fact should be considered when contemplating the use of a different op amp.

A current of 1 A taken from the output port produces a voltage change of less than 1 mv. A change of 10 v on the power-line input voltage results in only a 10-mv change at the output, and this figure can be reduced even further by placing a suitable resistor in series with pin 2 of the REF-01 and the supply voltage and connecting a 15-v zener diode from pin 2 to ground.

The temperature stability of the circuit has not been measured, but, depending on the class of REF-01 used, it will lie in the range of 3 to 20 parts per million per °C. Output noise at very low frequencies is extremely small. □

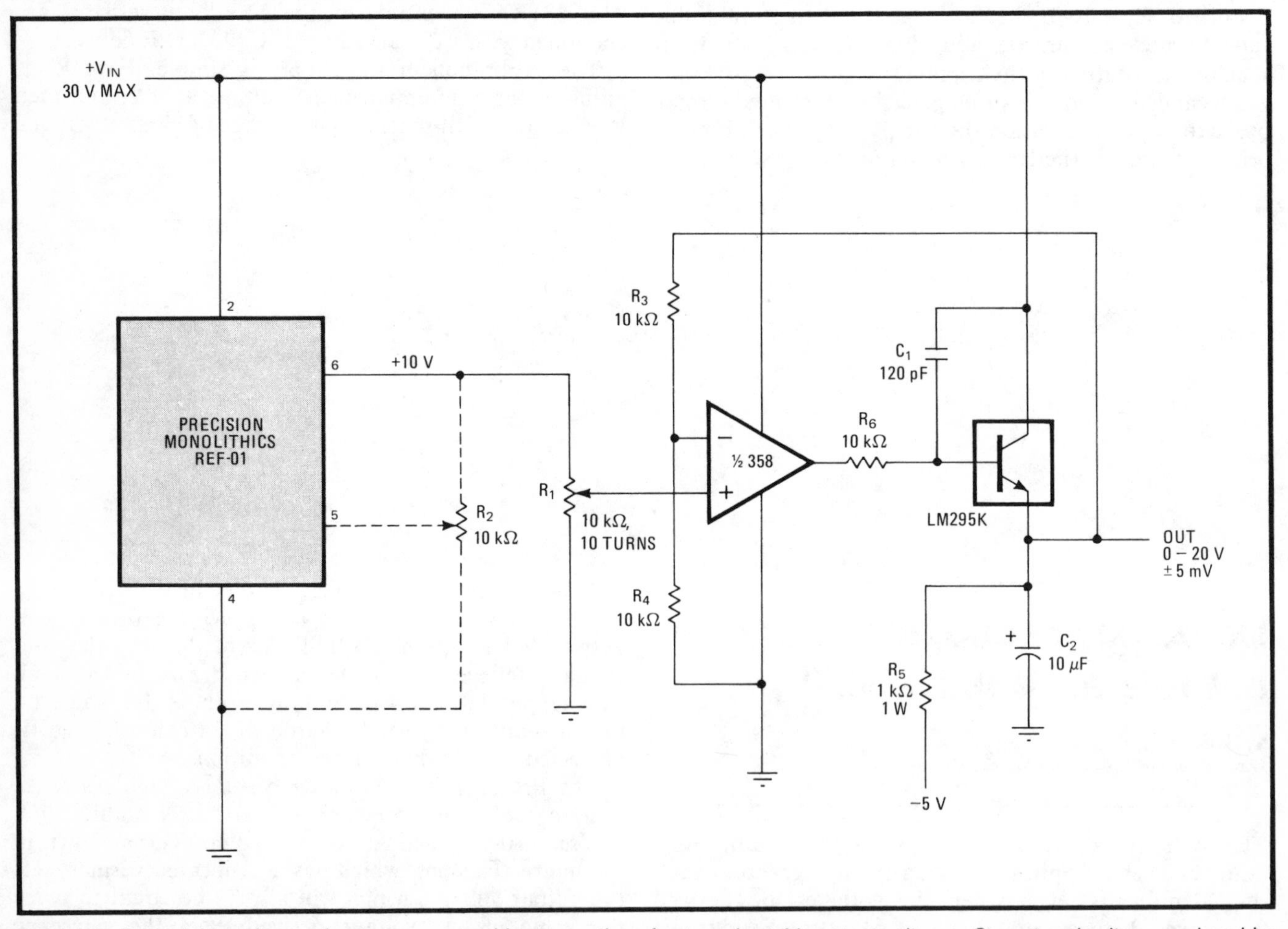

Precision. Rock-stable voltage reference source enables generation of extremely stable output voltages. Current and voltage produced by REF-01 source is independent of load current demands; thus output voltage, once set, will not vary more than ±5 millivolts.

Chapter 33
PROTECTION CIRCUITS

555 timer isolates equipment from excessive line voltage

by R. J. Patel
Tata Institute of Fundamental Research, Bombay, India

Instruments and appliances can be easily damaged when the line voltages that power them become excessively high or low, but a voltage-sensing circuit using the 555 timer will disconnect the equipment from the power lines if the set limits are exceeded. This circuit offers a better alternative for protecting instruments than a voltage stabilizer circuit, which is usually effective for detection and compensation of short-term variations only.

As shown in the figure, the line voltage is converted to approximately 15 volts by the step-down transformer, whose turns ratio is determined by the magnitude of the incoming voltage at the primary winding. This voltage is rectified, then filtered by capacitor C and applied to a 12-v regulator in order to bias the timer and the 2N2222 sense transistor. The magnitude of the unregulated voltage varies proportionally with the line voltage, as is to be expected, and this voltage is continually sampled by potentiometers P_1 and P_2, the upper and lower threshold controls.

The 555 timer is used in the bistable mode, and its state is a direct function of the voltages on its set and reset ports, pins 2 and 4, respectively.

Under normal conditions—that is, when the supply line voltage is within the set limits—the unregulated dc voltage at point A is sufficient to fire zener Z, saturating the transistor. Pin 4 of the timer rises rapidly to 12 v; when this voltage exceeds two thirds of the 12-v bias voltage on the timer, or 8 v, pin 3 moves high and the relay is energized.

If the ac line voltage is below the low set value, the voltage at A is below the value needed to fire the zener, and the relay is de-energized. When the line voltage shoots above the set upper limit and the dc voltage at pin 2 exceeds one third of the 12-v supply voltage, the relay is de-energized as pin 3 moves low.

The upper and lower set limits can be set with an accuracy of ±5 v of the true ac line voltage if precision potentiometers are used. There is no set-point hysteresis, because of the avalanche breakdown characteristics of the zener Z. Any transients generated by the power line are rendered harmless by the large filter capacitor C. □

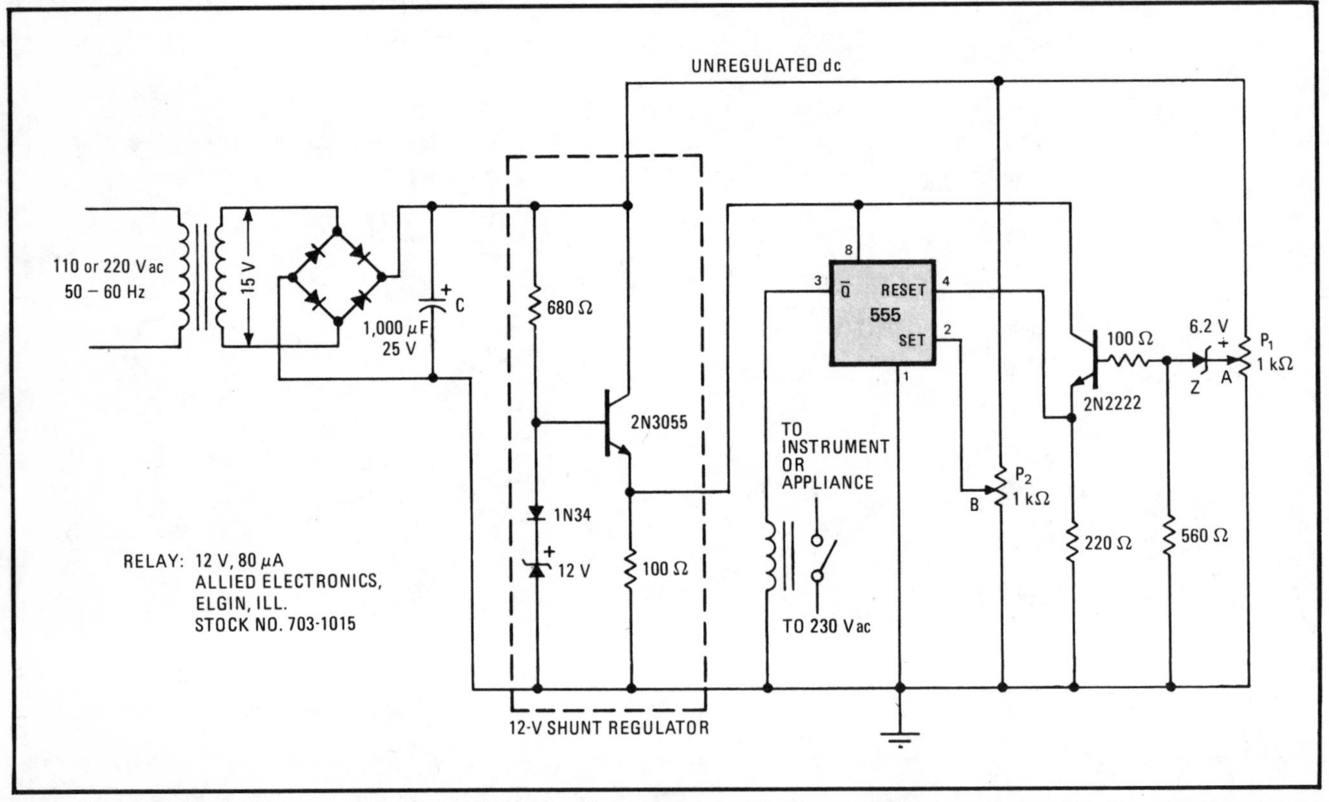

Line-voltage monitor. 555 timer circuit senses if ac line voltage is above or below set limits, then de-energizes line relay if necessary, removing power from equipment. Simple circuit uses set-control potentiometers P_1 and P_2 to monitor unregulated dc voltage, whose value is directly proportional to the ac line voltage. Timer and 2N2222 transistor require the regulated power source.

Acoustic protector damps telephone-line transients

by Gil Marosi
Intech Function Modules Inc., Santa Clara, Calif.

By limiting the transients on telephone lines, this acoustic shock protector prevents those sudden high sound levels that can damage the ear badly enough to cause loss of hearing. It holds the maximum peak-to-peak voltage at the receiver of a telephone headset to 50 millivolts.

A four-terminal device, the shock protector is inserted between the receiver side of the telephone hybrid and the receiver proper. A block diagram of the circuit, which operates from a single 5-volt supply, is shown in (a). Input signals are amplified by a factor of 5 and applied to a voltage-controlled, variable-gain stage. Because this stage also attenuates the signal to the degree indicated by the actual level of a feedback signal, further amplification may be needed, and is available, to retain the loop's gain margin. A voltage doubler then converts the amplifier signal to a dc voltage. This voltage is compared with a preset reference at the inputs of an averaging amplifier.

The output of the averager, which is essentially an integrating network, is connected to the variable-gain stage. As the input voltage from the hybrid becomes greater, so does the feedback voltage, and thus still more attenuation is provided for the variable-gain stage.

As for the actual circuit (b), transformers T_1 and T_2 isolate the protector from the floating telephone line, so that the circuit operates from a 5-V supply referred to ground. A_{1a}, one half of an LM358 operational amplifier, provides the required amplification of the input signal. D_1 and D_2 clamp A_{1a}'s output to 0.7 V and introduce the signal to buffer Q_1. This transistor, along with R_4–R_7, R_{14}, D_4, A_{1b}, and C_3, make up the variable-gain stage.

Zeners Z_1 and Z_2 and op amp A_{2a} bias A_{1a} and A_{1b} so that input signals to those stages swing about a quiescent point of 2 V. The circuit thus provides maximum dynamic range. Note that most op amps require a 12-V supply to achieve a comparable range.

Q_1's output is converted to a current with the aid of R_{14}. Current flows through D_4, which operates as a current-controlled variable resistor. D_4 is biased through

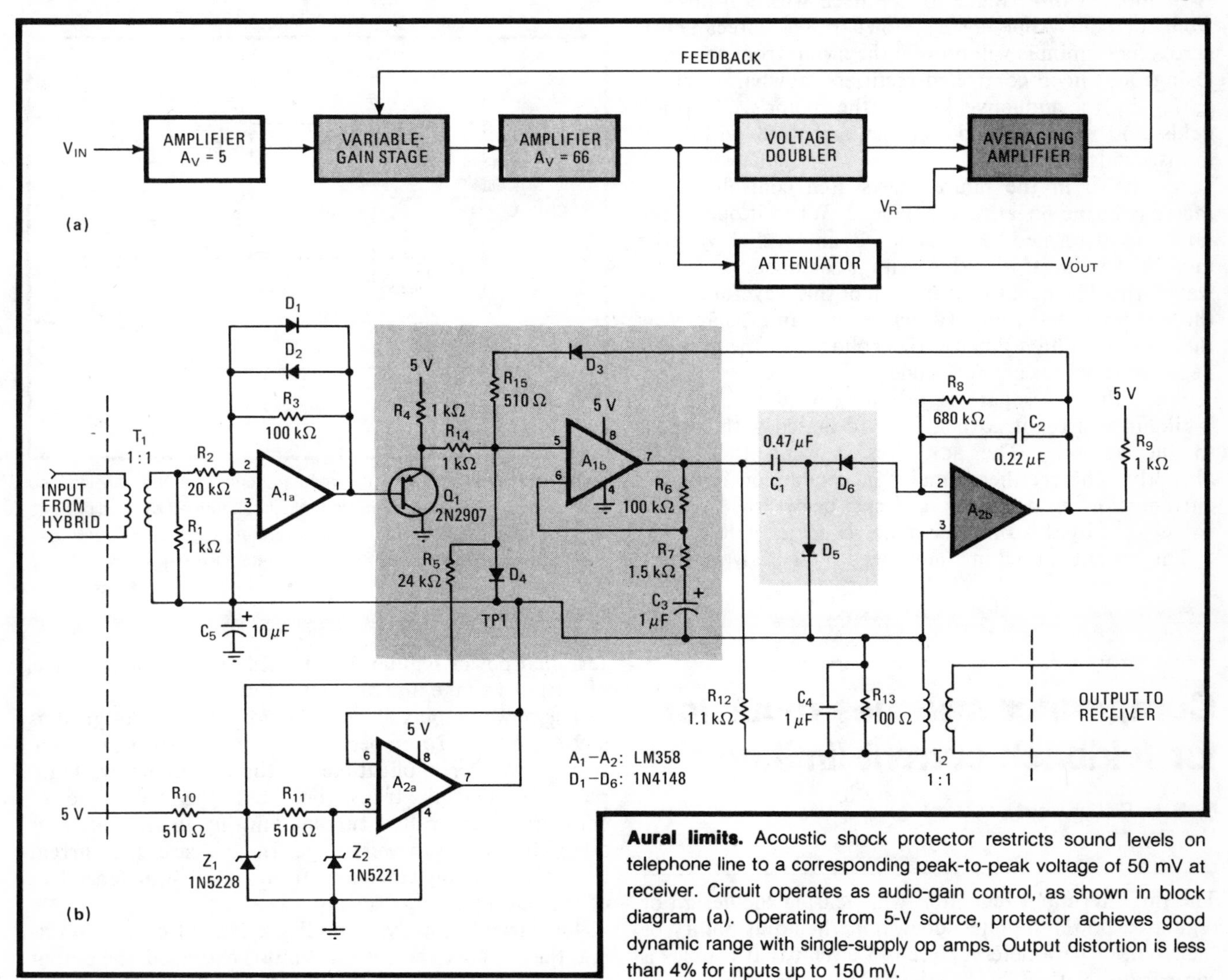

Aural limits. Acoustic shock protector restricts sound levels on telephone line to a corresponding peak-to-peak voltage of 50 mV at receiver. Circuit operates as audio-gain control, as shown in block diagram (a). Operating from 5-V source, protector achieves good dynamic range with single-supply op amps. Output distortion is less than 4% for inputs up to 150 mV.

R_5 such that its nominal resistance is 500 ohms.

The voltage at the noninverting input of A_{1b} is amplified and applied to the voltage doubler (C_1, D_5, and D_6). A_{2b} and its associated circuit perform the averaging function that provides a feedback current to the variable-gain stage. The gain from input to output is unity until the amplifier's input threshold—set at 50 mV—is exceeded. The acoustic shock protector then operates as an automatic gain control for inputs up to 150 mV. The output distortion up to that point does not exceed 4%. Beyond 150 mV, however, the protector simply clamps the output to 50 mV p-p without regard to distortion.

Because the phone receiver is an inductive device, its impedance increases with frequency. C_4 is placed across R_{14} to compensate for this rise in impedance. The overall gain of the acoustic shock protector is thus held flat to within 1 decibel from 300 hertz to 3 kilohertz so long as the output of A_{1b} is below 600 mV or so. □

Inductive kick gates SCR motor crowbar

by Buck Postlewait,
Instrumentation Specialties Co., Lincoln, Neb.

A dc motor can be made to stop itself with a little help from an SCR. Applying a "crowbar," or direct short, across its terminals will prevent the motor from coasting. Using a silicon-controlled-rectifier crowbar permits gating by the inductive kick of the motor's collapsing fields and stops rotation of the armature within 50 milliseconds.

As shown in the figure, the switch controlling the motor is in the negative supply lead. When it opens, the tank circuit formed by capacitor C_1 in parallel with the motor's windings will ring with a damped sinusoidal waveform. The negative transition of this waveform, plus the windings' back emf (which is the same polarity as the supply voltage) drive the voltage at the motor's negative terminal below ground.

Since the SCR's gate is tied to ground through the 1-kilohm resistor it is then forward-biased with respect to the cathode. The SCR fires, slowing the motor abruptly. The rectifier remains in conduction until the current flowing through it drops below the holding value—but by this time the motor is nearly halted.

This circuit is used in a motorized rotary valve system, where the desired valve opening is obtained by running the motor for a specific amount of time; thus, quick stopping is a must. However, the crowbar can brake any small dc motor. The only design considerations would be the SCR current rating, which must be greater than the stalled current drawn by the motor, and the value of C_1, which may have to be adjusted for optimum response. □

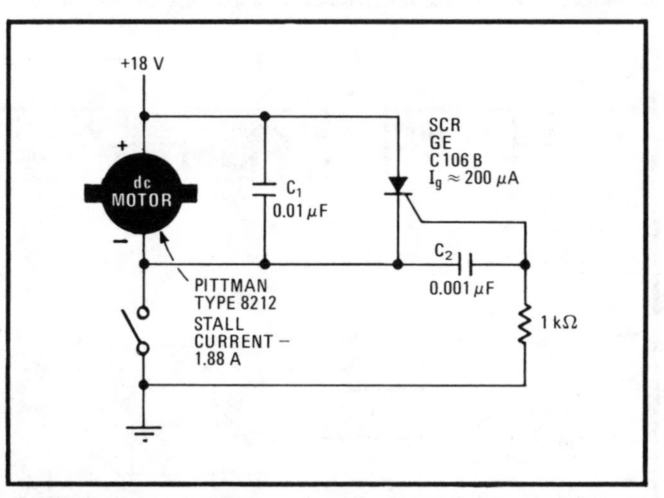

Quick crowbar. Opening the motor circuit generates an induced voltage which, added to signal in tank circuit formed by windings and C_1, gates the SCR. The rotation of the armature is thus crowbarred to a halt within 50 ms. Capacitor C_2 prevents false triggering.

Comparator switches regulator for foldback current limiting

by R. H. Richardson
Melbourne, Fla.

The three-terminal regulator, while easing the design of regulated power supplies, does limit the adjustability of their output parameters. Here is a circuit that couples a comparator with a National LM317 adjustable, three-terminal power regulator to provide independent control of output voltage and current limiting.

The power supply shown in the figure was designed as a 9-to-16-volt, 20-ampere bench regulator for automotive transceiver applications. It therefore employs high-power regulating transistors, but the technique of comparator-controlled current limiting in the output is adaptable to any power range. In this case, the current limit of the supply can be set from 1 to 20 A, independent of the output-voltage setting.

The circuit employs a foldback type of current limiting; that is, once the current limit is exceeded, the output voltage drops nearly to zero. The LM311 comparator, monitoring the current flow, switches the supply into the limiting mode.

The output of the comparator goes high whenever the voltage drop across the current-sense resistor, R_6, exceeds the bias level set by the limit control, R_4. A high level at the output of the comparator turns on transistor Q_1, which in effect shorts out the voltage-adjusting resistors R_1 and R_2, thus dropping down the output voltage.

The design has an inherent hysteresis: in the limiting state, just enough current flows through the sensing resistor R_6 to keep the comparator turned on. Therefore the circuit does not oscillate as many other limiting regulators do—the output voltage remains near zero until the overload is removed. The action of the comparator also provides automatic recovery from the foldback point, since the circuit cannot latch in the current-limiting mode.

Transistor Q_2 turns on an incandescent lamp to indicate the limiting state. Since the value of the current-sensing resistor R_6 is not critical, the device may be replaced with the internal resistance of a 0- to 20-A meter, if desired.

The dial of the current-limiting control can be calibrated with known output loads at a set output voltage. Control of the limiting function is primarily dictated by the biasing resistors surrounding the comparator, $R_3 - R_8$. Designs for output voltages and currents in other ranges may employ the same basic configuration of regulator and comparator, with only the biasing components needing to be changed.

The LM317 adjustable regulator, available in either TO-3 or TO-220 packages, is capable of driving loads up to 1½ amperes without the aid of power-transistor followers. In the supply shown, five power transistors—four 2N5885s and a 2N6053—are required to handle the 20-A loads. All power transistors and diodes, as well as the regulator, must have adequate heat sinking. □

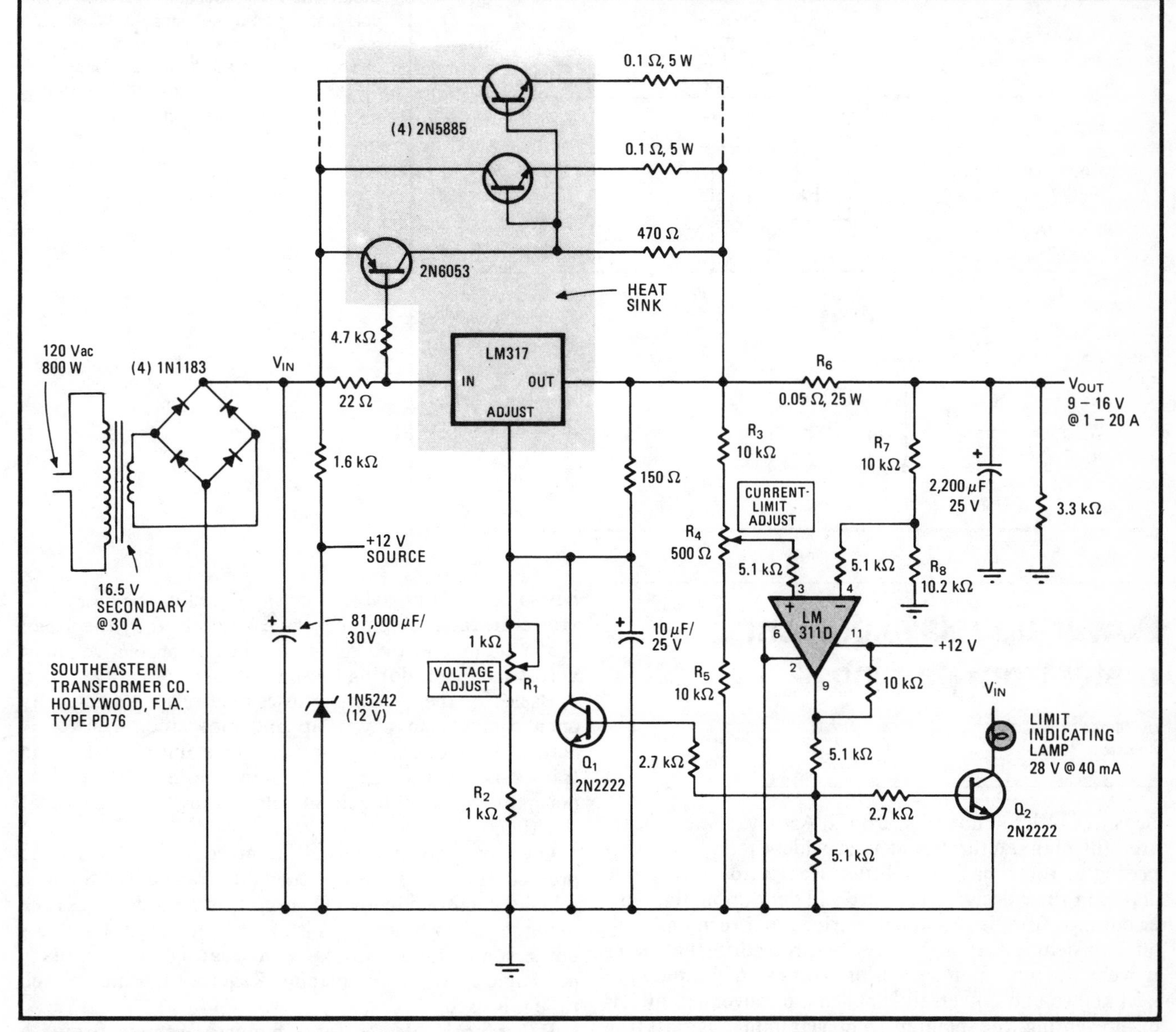

Foldback current-limiting supply. This supply, which has an output voltage adjustable from 9 to 16 V and provides independent current limiting from 1 to 20 A, uses a comparator to drop output voltage when the current limit set by potentiometer R_4 is exceeded.

Resettable electronic fuse consists of SCR and relay

by Russell Quong
Palos Verdes, Calif.

Most direct-current power supplies rely on a circuit breaker, current-sensing circuit, or fuse for current-overload protection, but this simple resettable-fuse circuit has advantages over all three. Built around a silicon controlled rectifier and a line relay, it is faster than a circuit breaker, less complex than most current-sensing circuits, and never in need of replacement.

How the circuit operates is evident from (a). Momentarily depressing S_1 closes the relay so that current flows from the supply to the load. In normal operation, the voltage across points PQ will be equal to the nominal supply voltage, and the normal operating voltage will appear across the relay winding. The relay and resistor R_2 are selected according to the dc supply voltage used and the relay's rated coil voltage, respectively.

Excessive current to the load causes a voltage drop across R_1 greater than 0.65 volt and switches on the SCR. The anode-to-cathode voltage of the SCR in the conducting region is approximately 2 v. This voltage, also across the relay coil, is far below the relay's holding voltage. Consequently, the relay opens, disconnecting the load from the supply. The relay may be reset by depressing S_1 again.

If a variable threshold point for SCR switching is desired, the SCR's gate can be connected to R_1 through potentiometer R_3. Resistor R_1 is calculated as before. □

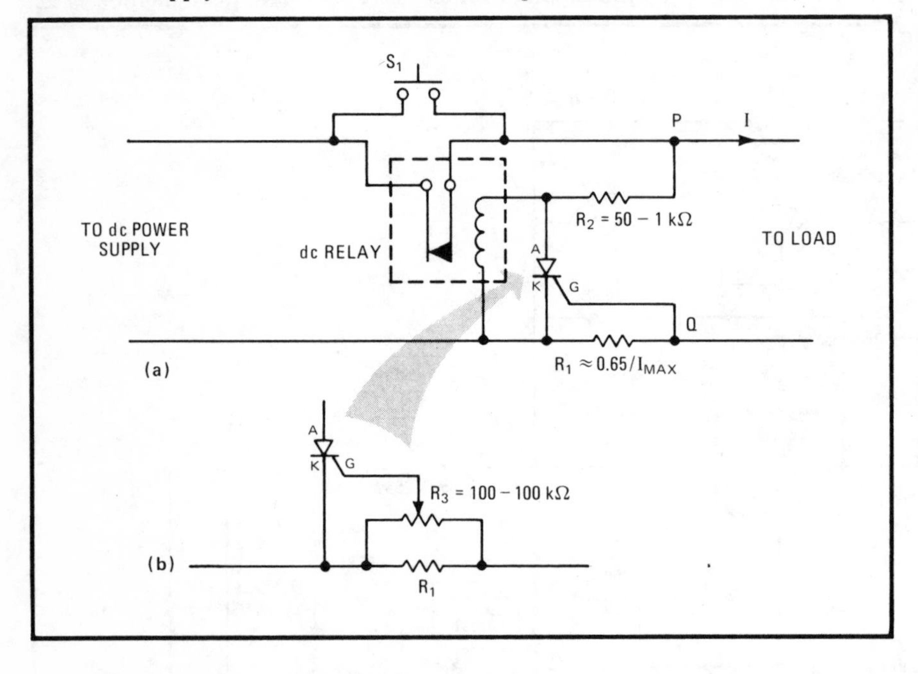

Electronic fuse. SCR and relay form resettable fuse for dc power supplies. When I_{max} is reached, SCR turns on, opening relay and disconnecting power from load. Depressing S_1 reinitializes circuit (a). SCR switching point may be adjusted with R_3 (b).

Power-up relays prevent meter from pinning

by Michael Bozoian
Ann Arbor, Mich.

Sensitive microammeters with d'Arsonval movements are still manufactured and used widely today, but surprisingly, there has been little attempt to correct one defect in their design—they are still very prone to pointer damage from input-signal overload and turn-on/turn-off transients. Although ways of protecting the meter movement from input signals of excessive magnitude are well known and universally applied, no convenient means of preventing the pointer from slamming against the full-scale stop during power-up or power-off conditions has so far been introduced or suggested in the literature. However, the problem may be easily solved by the use of a 555 timer and two relays to place a protective shunt across the meter during these periods.

Basically, the 555 timer closes reed relay A's normally open contacts on power up and puts shunt resistor R across the meter for 5 or 6 seconds until the turn-on transients have subsided, as shown in the figure. The normally closed contacts of relay B are also opened at this time.

On power-down, relay B reintroduces the shunt to protect the meter from turn-off transients. Such a scheme is more effective than placing a diode across the meter, as is often done and is much more elegant and less bothersome than manually activating an auxiliary mechanical switch for placing R across the meter each time it is used.

R has been selected for a meter movement having a full-scale output of 200 microamperes and an internal

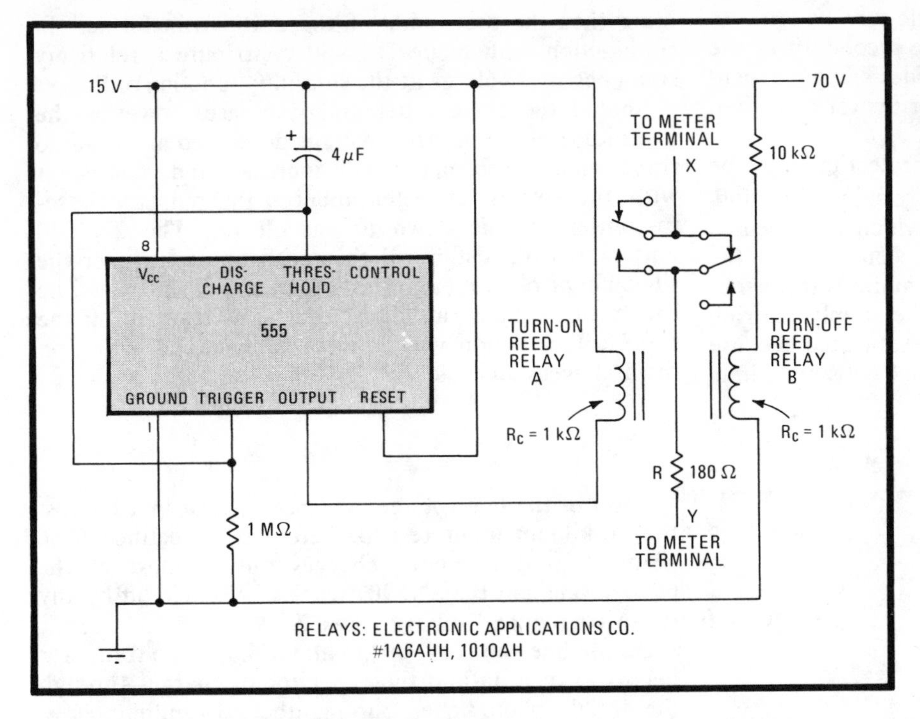

Shunted. Reed relays and 555 timer prevent d'Arsonval movement from slamming against microammeter's full-scale stop during power-up and power-down conditions by introducing shunt resistor across meter terminals until transients die out. Method does not degrade meter's accuracy or its transient response to input signals.

resistance of 1,400 ohms. The complete circuit may be mounted on a 2-by-2¼-inch printed-circuit board. The only design precaution is to ensure that relay **B** is energized from a source that has a fast decay time during power-off conditions. Here, the voltage has been tapped from the meter's power-supply rectifier. □

Low-loss shunt protects high-current supplies

by Roy Hartkopf and Ron Kilgour
Alphington, Victoria, Australia

The usual method for providing short-circuit protection in low-voltage, high-current power supplies is to employ a current-sensing resistor in series with the load. Unfortunately, this scheme develops an appreciable voltage drop across the resistor when large currents flow and may consequently reduce the available output voltage to a great degree. The voltage drop can be virtually eliminated with an alternative method, shown here, which

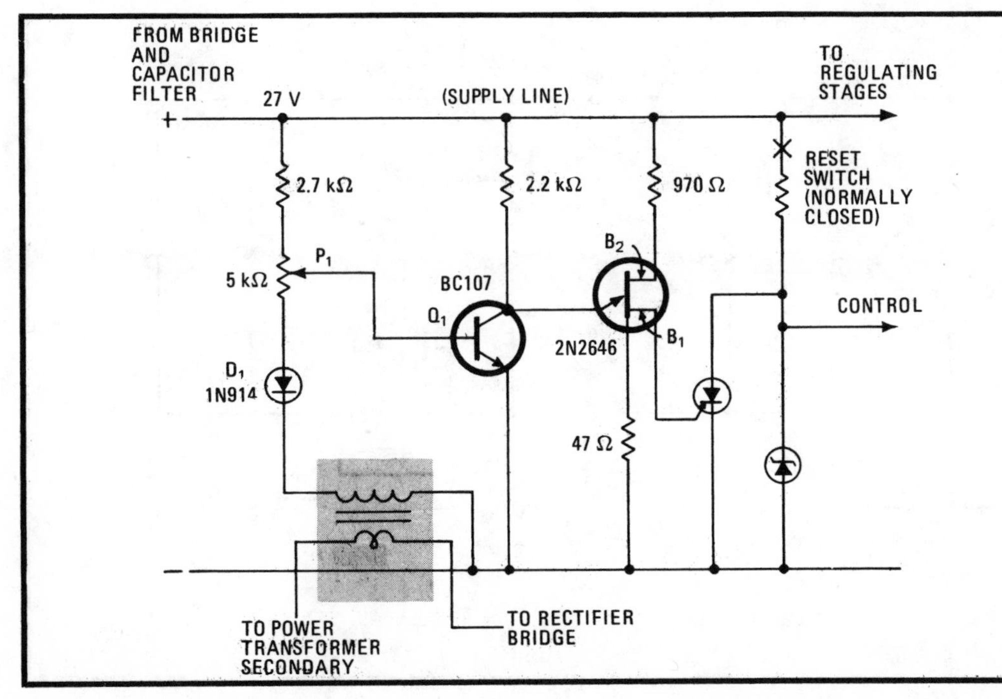

Current gauge. An audio transformer and a single turn of heavy-gauge wire, placed between input transformer's secondary and rectifier, give high-current supplies overload protection without introducing input-to-output voltage drop that occurs with units employing current-sensing resistors. Potentiometer P_1 sets the overload point. Overload detection is instantaneous, occurring on the first positive cycle of input voltage.

uses an audio transformer and a single-turn winding to sense the overcurrent condition at the secondary of the supply's power input transformer. Besides being inexpensive, the current sensor will react faster to overloads than some of the more conventional circuits.

As shown in the figure, current protection may be secured for a typical 27-volt, 20-ampere supply by winding a single turn of 10-gauge wire, which is placed in series with the power transformer's secondary and the supply's rectifier bridge, onto a small audio transformer connected in the control section of the supply. During normal operation, transistor Q_1 will be saturated because current is delivered to its base from the 27-V supply line. Note that the secondary of the audio transformer, in conjunction with diode D_1, will contribute a relatively small negative voltage at the summing junction of P_1.

Should the current demands increase, however, the magnitude of the negative voltage developed at the audio transformer's secondary will increase and, consistent with the setting of potentiometer P_1, pull the base-to-emitter voltage down to cut off Q_1. The 2N2646 unijunction transistor will then turn on and trigger the silicon controlled rectifier, and the control signal will be brought low. Thus this signal can be used to cut off the supply. This action will be instantaneous, occurring on the first overload cycle. □

Opto-isolated detector protects thyristors

by Charles Roudeski
Ohio University, Athens, Ohio

Although gating a thyristor with short pulses greatly reduces the gate and driver dissipation, failure of the driving logic can turn on the thyristor full time, possibly destroying it, the driver, and their supply. Described here is an opto-isolated zero-crossing detector that generates a 100-microsecond pulse each time its 60-hertz power-line input traverses through zero. Besides isolating for the logic element, the circuit terminates the generation of pulses if almost any detector component fails.

Most of the line voltage (see figure) is dropped across the 10-kilohm input resistor before it is rectified. The 25-microfarad capacitor charges during most of the 60-Hz cycle, but the 2N2907 transistor is held off by any full-wave rectified voltage above 2.3 V.

As the line voltage drops to about 4.5 V, the transistor begins to turn on and the capacitor discharges through the 4N26's photodiode, sourcing about 14 milliamperes. This produces a pulse centered about the zero crossing. Wider pulse widths are obtained by reducing the value of the 15-kΩ resistor. If a longer rise time is tolerable, the 33-kΩ resistor in the base lead of the optocoupler's phototransistor can be eliminated.

The 3-V zener establishes the reference voltage for the circuit. □

Protection. Zero-crossing detector uses optocoupler for gating of thyristors by power line. Output pulses, produced 120 times per second as input voltage traverses through zero, last 100 μs. Output of 4N26's phototransistor will be zero if most any element in detector fails, thereby protecting thyristor, driver, and supply from damage that would be caused by activating the thyristor continuously.

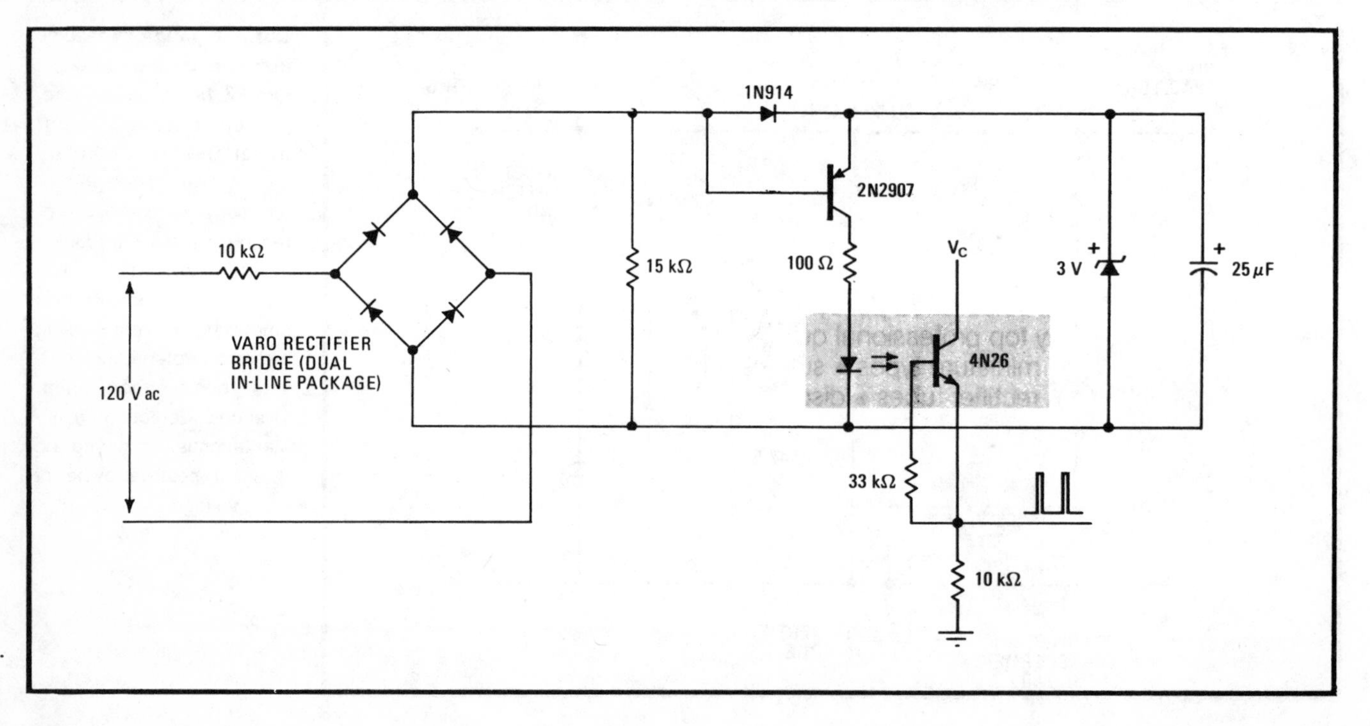

Fast-acting voltage detector protects high-current supplies

by Jorges S. Lucas
Engeletro, Belo Horizonte, Brazil

Protecting a regulated, nonswitching power supply against both short circuits and overvoltages can be difficult, especially if the supply is to deliver high currents. Should either condition occur, this circuit will act quickly to protect the supply, and its load as well, by deactivating the series or shunt pass element in the regulator and thus forcing the output current and voltage to zero.

A typical high-current power supply (5 volts at 5 amperes), which is modified slightly to accommodate the protection circuitry (dotted lines), is shown in the figure. When a short circuit occurs at the output, Q_2 turns on, which in turn disables Q_1. The voltage at the gate of the silicon controlled rectifier then rises at a rate determined by the time constant of elements R_1, D_1, D_2, and C_1.

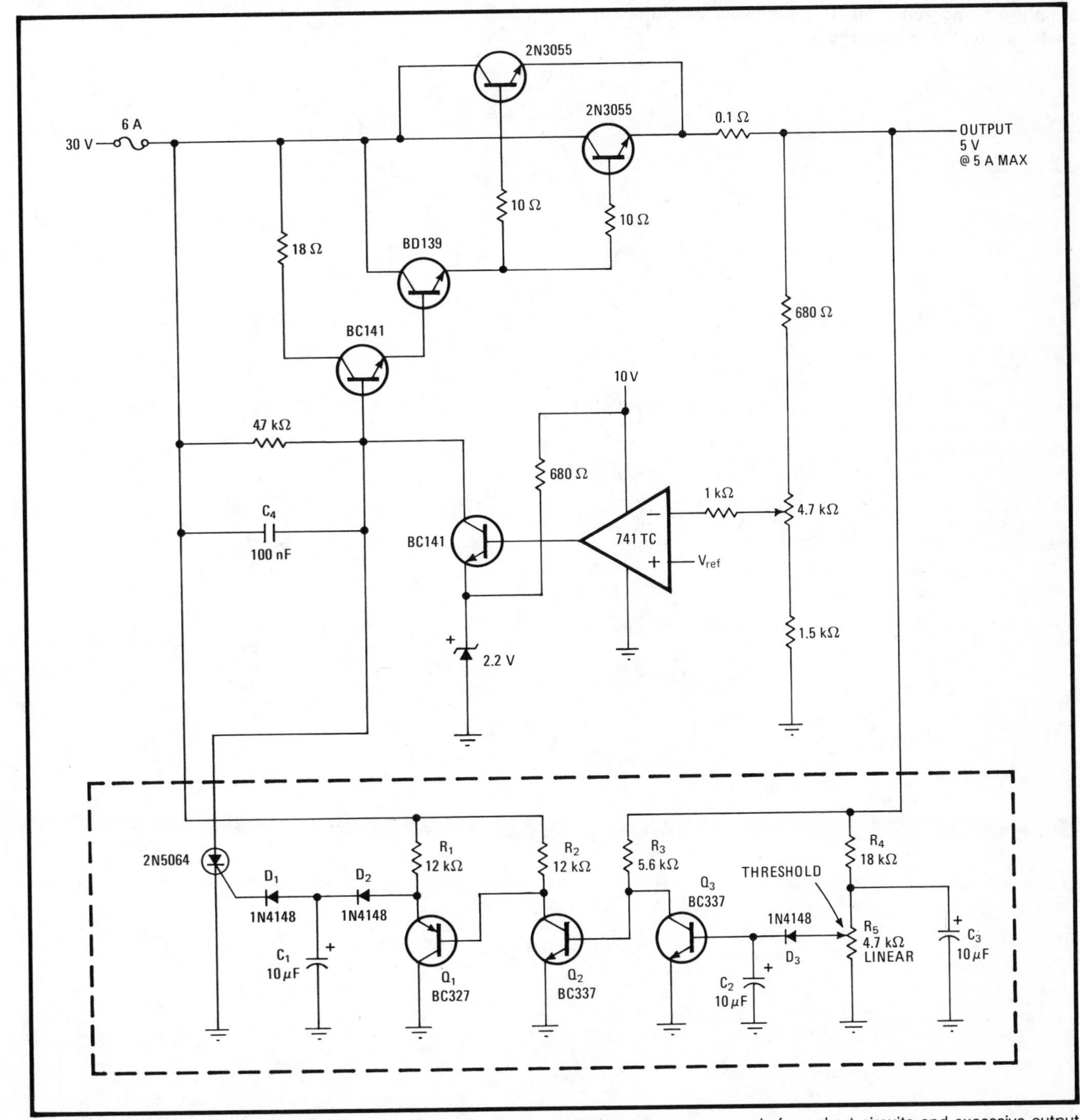

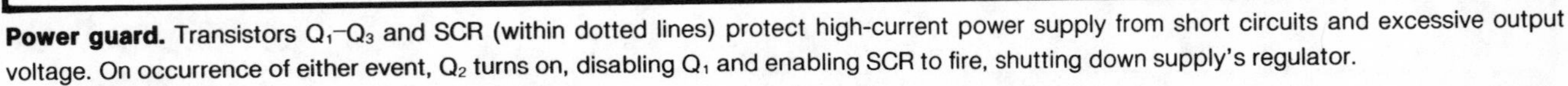

Power guard. Transistors Q_1–Q_3 and SCR (within dotted lines) protect high-current power supply from short circuits and excessive output voltage. On occurrence of either event, Q_2 turns on, disabling Q_1 and enabling SCR to fire, shutting down supply's regulator.

This delay prevents the SCR from triggering when power is first applied to the circuit. The SCR then fires, disabling the BC141/BC139 transistors in the power supply and shutting down the regulator.

Q_3, on the other hand, detects when the output voltage climbs above a user-set threshold. Once the threshold is exceeded, Q_3's base voltage rises at a rate determined by the time constant of elements C_2, R_4, D_3, and threshold potentiometer R_5 (delay must be provided for the reason discussed previously).

Q_3 then turns on. Q_2 and Q_1 react accordingly, and the SCR fires, as it did for the short circuit. Normal circuit operation may be restored simply by turning the power supply off and removing the abnormal condition, then switching on the supply again. □

Chapter 34
PULSE GENERATORS

Delay circuit replicates pulses of variable width

by John H. Davis
Warm Springs, Ga.

Unfortunately, the simple and well-known circuit used to provide true pulse delay—whereupon the first of two one-shots connected in series sets the delay time desired and the second is set to generate a pulse having the same width—cannot be used if the input pulse width is variable. Fortunately, however, pulses of variable width can easily be handled by adding only a quad NOR gate and a few RC differentiators to a modified circuit, as shown here.

Differentiator R_1C_1 provides a positive-going spike from the rising edge of the input pulse to be delayed, in

for the desired delay interval selected by potentiometer R_A. When the Q output returns to its high state, the RS flip-flop at the output, formed by two NOR gates, is set and the I port moves high.

One-shot A_2 is triggered by the falling edge of the input pulse through differentiator R_2C_2, and thus its $\overline{Q}$ output goes low for a time (set by R_B) equal to A_1's delay interval. When A_2's $\overline{Q}$ output returns high, the NOR latch returns to its low state. As long as $\tau_{A1} = \tau_{A2}$, the time during which I is high will always be equal to the width of the input pulse, assuming the delays are equal to or exceed the input pulse width. Pulses that are very much shorter than the set delay time will be reproduced less accurately.

This circuit provides delays over the range of 1 through 20 microseconds, but it is a simple matter to change timing components to achieve times into the millisecond region. Note that the maximum delay that may be set will be limited to the shortest repetition period in the pulse train and in practice should be set to a value less than this to allow for the one-shots to recover.

The circuit is equally suitable for implementation with positive or negative logic. Adjustment is simple. With positive logic, R_A should be set for the desired delay. A train of pulses of nominal width is then introduced at the input, and the I port is monitored with a scope while R_B is adjusted so that the pulse width at the output is equal to that at the input. The circuit will then automatically be calibrated for input pulses having any width. The calibration procedure is similar with negative logic, except that then it is easier to adjust R_B first.

Alternatively, both one-shots may be set for equal delay, but in practice this procedure will cause inaccuracies for very narrow input pulses. In any case, it will be advantageous if circuitry can be configured to program R_A and R_B simultaneously, so that the circuit has only one control. □

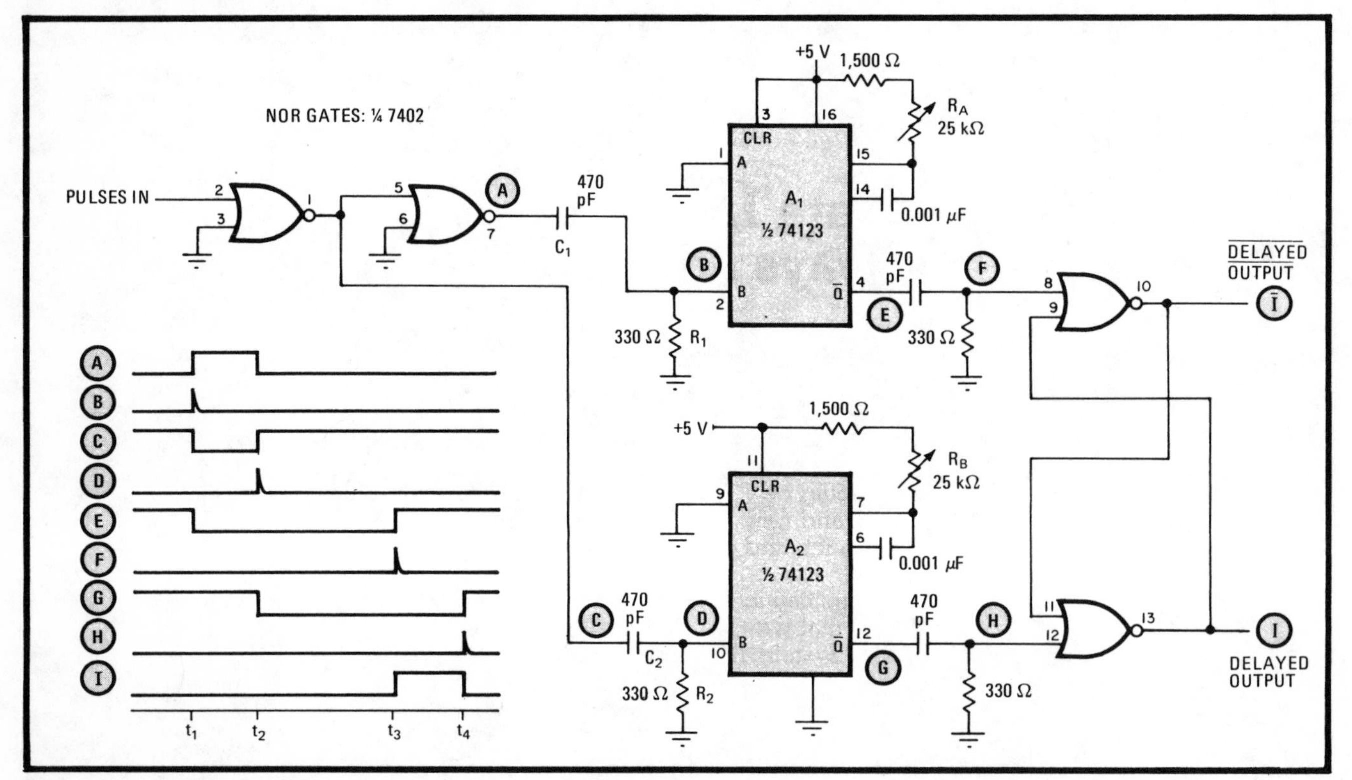

Delayed duplication. Parallel-connected one-shots and NOR gates provide set delay and maintain width of pulse, independent of its value. Low-cost unit will thus be useful for automatically synchronizing blanking pulses in TV systems and for similar applications.

High-speed generator pulses ECL loads

by Andrew M. Hudor Jr.
Department of Physics, University of Arizona, Tucson

Serving the needs of designers and technicians who work with high-speed digital circuits, this inexpensive two-chip pulse generator is invaluable in trouble-shooting emitter-coupled logic. Besides generating signals having frequency and duty cycles that are adjustable to its complementary outputs, the versatile generator also provides two ports at which pulses with a constant 50% duty cycle are available.

One section of a 10116 ECL line receiver, A_1, is used as an RC oscillator[1] whose period is determined by the potentiometer P_1 and the capacitor selected by the frequency-range switch. With the values shown, the oscillator frequency can be varied from a few hundred hertz to more than 50 megahertz. The output of the oscillator is then buffered and squared up by a Schmitt trigger, A_2, which is the second section of the line receiver.

A_2's output toggles both sections of a 10131 dual-D flip-flop, A_3 and A_4. A_3 provides for a 50% duty cycle output, while A_4 and the remaining section of the receiver, A_5, form an adjustable one-shot multivibrator. Here, A_5 serves as a second Schmitt trigger.

The Q output of A_4 is fed to the input of the Schmitt trigger through an RC integrator formed by P_2 and the capacitor selected by the width-range switch. When Q toggles low, the input to the A_5 trigger slowly rises as the capacitor charges. When the trigger level is reached, the Schmitt trigger's output goes high, resetting A_3 and A_4, and the process is repeated.

The time it takes for the integrator to reach the trigger level defines the pulse width. With the values shown, widths from 15 nanoseconds to 10 milliseconds can be selected. Upon resetting of the flip flops, Schottky diode D_1 allows the capacitor to discharge rapidly. The width control allows adjustment of the duty cycle from nearly zero to 50%. For applications where a duty cycle greater than 50% is required, the complementary output should be used.

If desired, a buffer can be easily added at A_3 and/or A_4 in order to drive 50-ohm lines directly. □

References
1. William A. Palm, "ECL IC oscillates from 10 to 50 MHz," *Electronics*, Circuit Designer's Casebook 14D, p. 109.

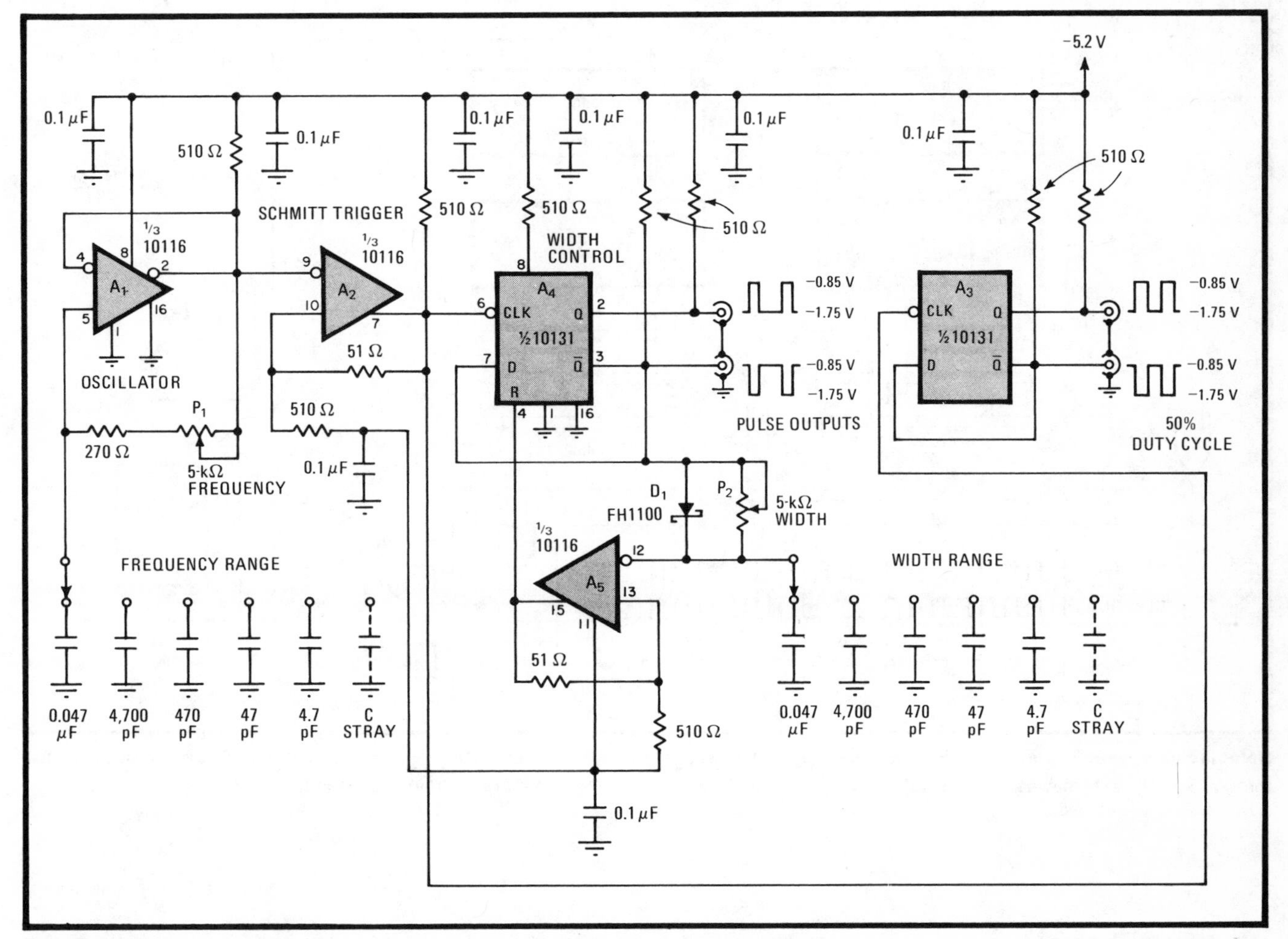

Speedy. Line receiver and dual flip-flop generate high-frequency pulse trains for emitter-coupled logic. Signals to 50 MHz having widths that are adjustable from 15 ns to 10 ms appear at generator's complementary outputs. Circuit also provides output at duty cycle of 50%.

Counter and switches select pulse-train length and dead time

by Héctor Gellón and Enrique Marcoleta
San Luis, Argentina

Only the more expensive pulse generators can repeatedly generate a pulse train of selectable length followed by an off time also of selectable length. But this common requirement is easily met by one cascaded counter, two switches, and some logic. The number of pulses in the train is selectable from 1 to 99, and irrespective of the number of pulses delivered to the output, the off, or dead, time can also be varied between 1 and 99 clock periods. The lengths of both the pulse train and the off time may be extended by adding to the number of stages in the counter.

As shown in the figure, a system clock drives two cascaded 7490 binary-coded-decimal decade counters, and their outputs are converted to a decimal equivalent by the 7442 BCD-to-decimal decoders. The decoded outputs are active low, and when the count reaches the pulse-train length desired (preset in this case to 8 by switches S_{1a} and S_{1b}), gate G_1 moves high.

If the $\overline{Q}$ output of the 7472 flip-flop is high, as it will be once the circuit settles after initialization, G_4 moves low. This causes G_5 to assume a high state, resetting the counter and toggling the flip-flop, which in turn disables G_4 and output gate G_7.

There is no output until the counter reaches the number set by switches S_{2a} and S_{2b}. G_2 moves high, causing G_3 to move low and G_5 to go high. The counter is reset, and the flip-flop is toggled, thus once more enabling G_4 and G_7. The pulse train now appears at the output until the settings of switches S_{1a} and S_{1b} are reached, and the cycle repeats.

The $\overline{Q}$ output of the flip-flop is a signal having a duty cycle that may be anything from 1 to 99 times the period of the system clock. It is essentially an ungated version of the signal at the pulse-output port.

Cascading additional BCD counters and decoders to the circuit will extend its pulse-counting and dead-time limits. Of course, more switches and gates must also be added, to accommodate a greater number of inputs. □

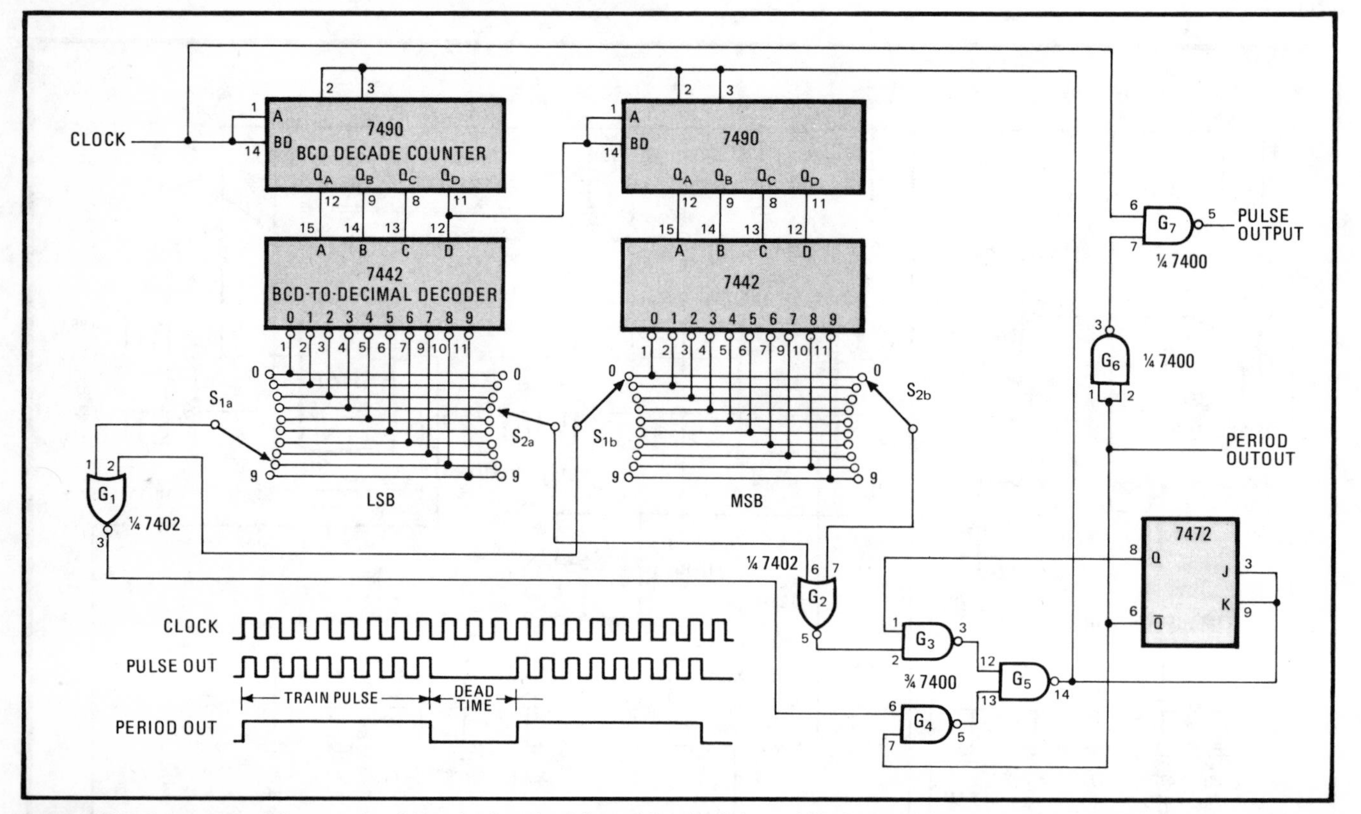

Selectable. Generator produces train of N pulse lengths followed by dead time of M clock pulses, set by 2-pole, 10-position switches S_{1a} through S_{2b}. N and M may assume any value of from 1 to 99. Switches are shown set for pulse train of 8, dead time of 3.

Scanned keyboard activates eight-tone generator

by Albert Helfrick
Aircraft Radio and Control Division of Cessna Aircraft Co., Boonton, N. J.

This keyboard-activated eight-tone generator owes its simplicity to a single oscillator, which makes possible the scanning of the keyboard and simultaneously functions as the tone generator. As a result, its device count is low and its cost is minimal.

Circuit operation is easily understood. The CA3130 operational amplifier, A_1, is configured as a relaxation oscillator, its frequency controlled by R_iC. R_i lies in the 100-to-500-kilohm range, and C is 0.01 microfarad or so for frequencies in the 1-to-10-kilohertz range. The oscillator has excellent frequency stability as a result of the operational amplifier's extremely high input impedance and the complementary-metal-oxide-semiconductor output circuit.

A_1 drives the 4516 4-bit counter, A_2. As the counter increments, it scans each input port of two analog multiplexers, A_3 and A_4. A_3 sequentially places all resistors, R_1 through R_8, in the oscillator circuit, enabling A_1 to generate exactly one cycle of each frequency determined by each R_iC combination. At no time is there any output from G_i, however.

Meanwhile, multiplexer A_4 is scanned to determine whether any keyboard switches are closed. If any switch should be depressed, a logic 1 will emanate from pin 3 of A_4, freezing the counter and enabling G_1. A_1 will then oscillate at the frequency determined by the particular value of R that is in the oscillator circuit when the counter halts. Since the counter cannot advance while the key switch is closed, and simultaneously closing any other key will have no effect on the output frequency, the circuit has in effect a built-in lock-out feature.

The time required for the system to latch to any particular frequency is a function of both the number of frequencies that can be selected and the actual frequencies of oscillation. The maximum acquisition time works out to approximately:

$$t = \frac{1}{f_1} + \frac{1}{f_2} + \ . \ . \ . \ \frac{1}{f_i}$$

where each f_i is equal to $1/0.69\ R_iC$. For eight frequencies in the kilohertz range, t equals about 8 milliseconds, which is an acceptable period of time for manual keystroke applications. □

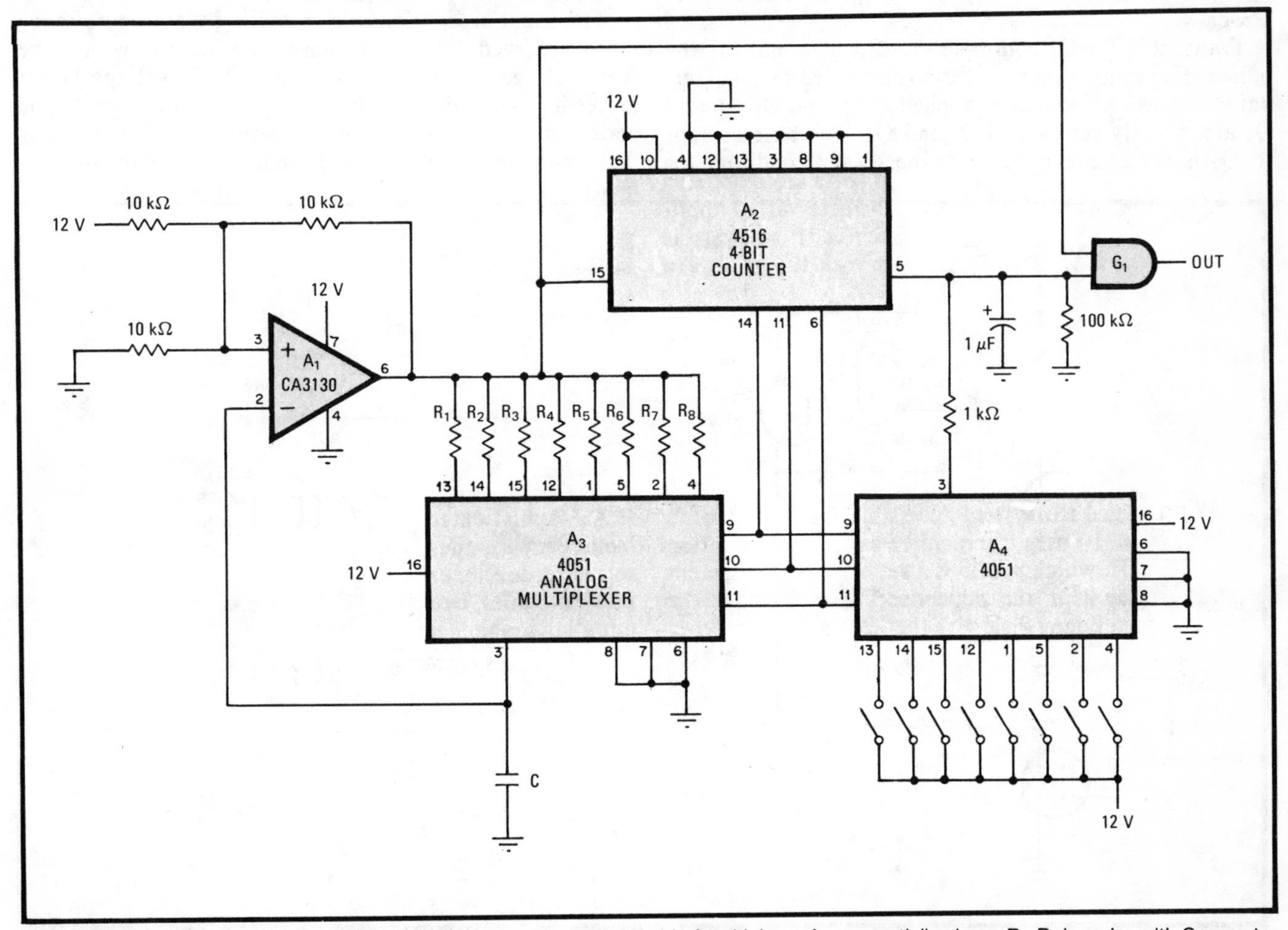

Scanned tones. Self-gating oscillator, A_1, advances counter and with aid of multiplexer A_3 sequentially places R_1–R_8 in series with C so as to control frequency. Op amp's high-input impedance and C-MOS output ensures high oscillator stability. No signal appears at output until a keyboard switch is closed, when A_4 freezes counter and activates G_1, enabling generation of the single desired frequency.

Bipolar and V-MOS hybrid delivers fast power pulses

by Robert H. Hamstra Jr.
Searle Ultrasound, Santa Clara, Calif.

The major advantages afforded by the up-and-coming V-groove MOS transistor—low cost, moderate power capability, high-frequency operation, and immunity to mismatch—are utilized in this relatively inexpensive, compact pulser that, with the aid of two bipolar power transistors, will deliver a peak power of 5 kilowatts at widths as narrow as 20 nanoseconds. In terms of voltage and current, the unit can supply an output pulse of as much as 250 volts or 30 amperes.

This pulser overcomes the size, weight, and standby power limitations of vacuum tubes, the bulkiness and limited frequency response of charged delay lines, and the exponential fall rate and lack of constant output impedance that occur with silicon controlled rectifiers. Bipolar transistors that are fast enough cannot handle the current or voltage. The drawbacks are overcome by combining the bipolar transistors and V-MOS devices, which are fast but cannot deliver great amounts of power.

Transistors Q_1–Q_4 comprise the driver circuit for the pulser. The energizing waveform is provided by an external low-power pulse source applied at Q_1 and Q_2. Q_1 and Q_2 are directly coupled to Q_3 and Q_4. The base current to Q_3 is set at about 1.5 A, a necessary condition for achieving rise times of 10 ns across the V-MOS input capacitance of 500 picofarads.

D_1–D_3 serve as clamping diodes to prevent Q_3 from saturating on the rising edge of the pulse, and the 5.6-ohm resistor helps turn off Q_3 quickly. The amplitude of the output pulse is determined by the control voltage, which is applied to Q_3's emitter. The circuit could be further simplified if a p-channel V-MOS device were to replace Q_3, but unfortunately the p-channel units are not yet readily available.

In general, the basic output circuit is in a cascode arrangement using 10 V-MOS transistors Q_i and two bipolar transistors Q_j, with the gates of Q_i driven by the pulse. The base of Q_j is at ac ground. In this way, the current delivered to the load Z_L is approximately proportional to the voltage applied to Q_i.

The cascode circuit helps minimize the effect of the gate-to-source and gate-to-drain (input) capacitances of the V-MOS transistors, thereby enhancing the response time or speed of the circuit. Note that the output voltage of Q_i overcomes the drop across Q_j's (internal) emitter inductance (10 to 20 nH) when the stage is made active, and that Q_i is selected to withstand the inductive kick (10 to 20 V) that occurs when the stage is switched at slew rates as fast as 2 A/ns. In addition, Q_j is never driven into saturation, and as a consequence, fast response times are maintained.

The usual precautions for protecting V-MOS gates have been observed. The input-pulse amplitude is within the gate-voltage ratings. The source-to-drain voltage is not exceeded when the device is off. No attempt to bias the gate near its turn-off point to improve device linearity has been made, either, as a small misadjustment may

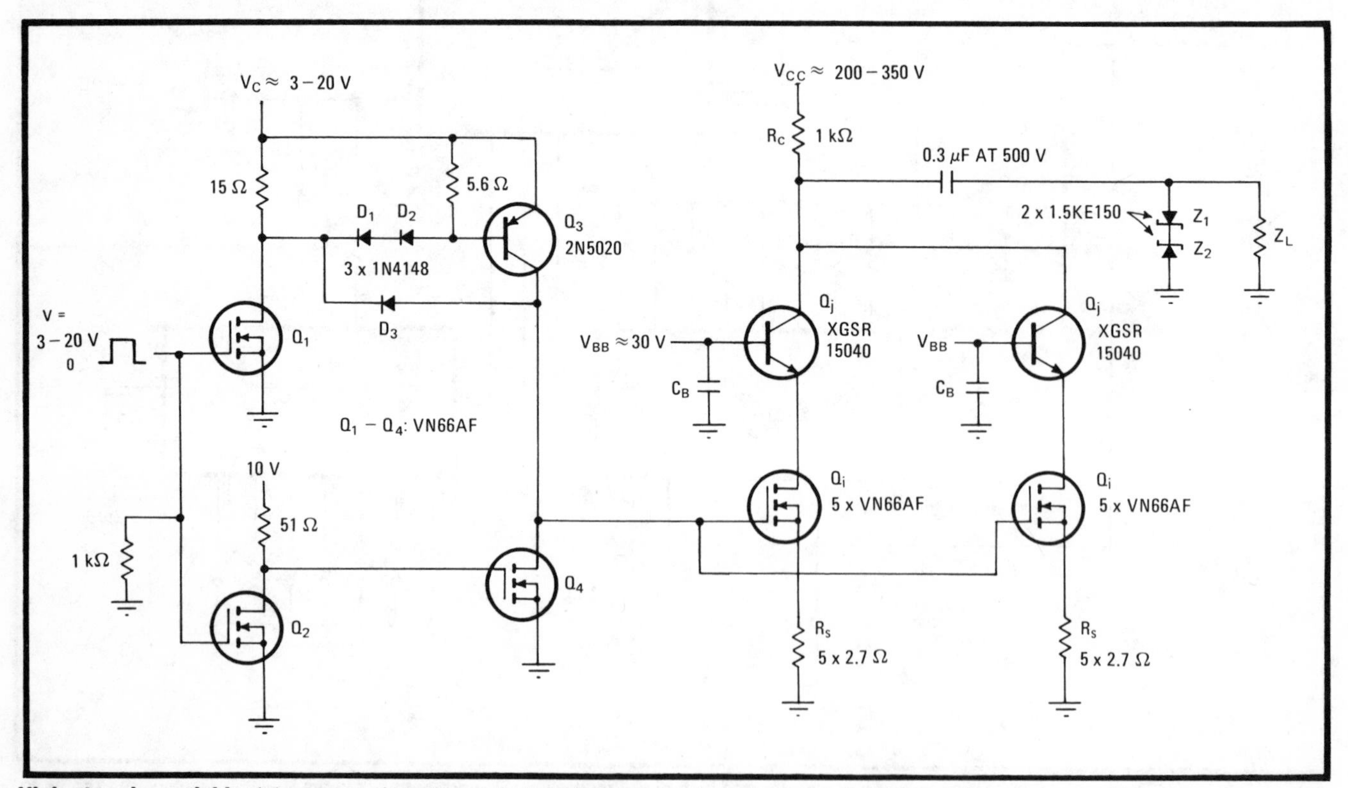

High-stepping quickly. Advantages of bipolar and V-MOS transistors are combined in circuit that delivers pulses to 5 kilowatts at a nominal width of 20 ns, if load is resistive. Output pulse width is limited by the 0.3-μF coupling capacitor.

cause excessive dissipation and device failures.

The bipolar transistor is a current source and so short circuits at the output of the pulser cause no trouble. Protecting the output transistors from voltage spikes is important, however, and fast transient suppressors Z_1 and Z_2 have been added to avoid transistor damage. Z_1 and Z_2 work well for low-impedance loads (10 Ω or less) but their high capacitance may otherwise serve to slow circuit speed.

The circuit has been built on a brass plate, to avoid parasitic oscillation problems in V-MOS that would be manifest if the circuit were constructed on a printed-circuit board. Very low-impedance bypass capacitors (monolithic ceramic) are connected between the base of the bipolar transistors and ground. Note that the gates of each V-MOS device are connected directly in parallel without a resistor or ferrite bead. This arrangement is equally effective in improving the suppression of parasitic oscillations. □

Dual charge-flow paths extend pulse repetition rate

by J. Klimek
Pretoria, South Africa

Although the basic, one-gate pulse generator shown in part (a) of the figure cannot be beaten for convenience in general test applications, it has a relatively narrow repetition-rate range, typically only a few tens of kilohertz. But with a few modifications (b), the repetition rate for a narrow-width pulse train can be extended from dc to 1 megahertz or so.

The range of the pulse generator is increased because the timing capacitor is charged and discharged through separate paths. This operation decreases circuit-switching times and enables the circuit to oscillate over a wide band of frequencies. Gate T_1 is one sixth of the 4007 chip, which contains three n-channel and three p-channel enhancement-mode transistors; it charges capacitor C for as long as pin 1 is high. For the circuit configuration shown, the charging period is a fraction of a microsecond.

When T_1 is switched low by pin 4 of the 4093, C discharges through T_2, the current source-sink whose value is controlled by R. Once C is discharged, T_1 switches high again and the process repeats.

In this instance, when 0.7 volt $\leq V_{in} \leq$ 3.4 volts, the corresponding repetition rate varies from dc to 1 MHz. The pulse width, which is about 0.5 microsecond, may vary by as much as a factor of 2, depending on the particular 4007 used. But whatever the value, it will be constant throughout the 0-to-1-MHz range.

To minimize the phase jitter that may occur at low frequencies because of the small charge current involved, the circuit should be placed inside a metal enclosure. In line with the low-frequency consideration, a low-leakage capacitor is also recommended. □

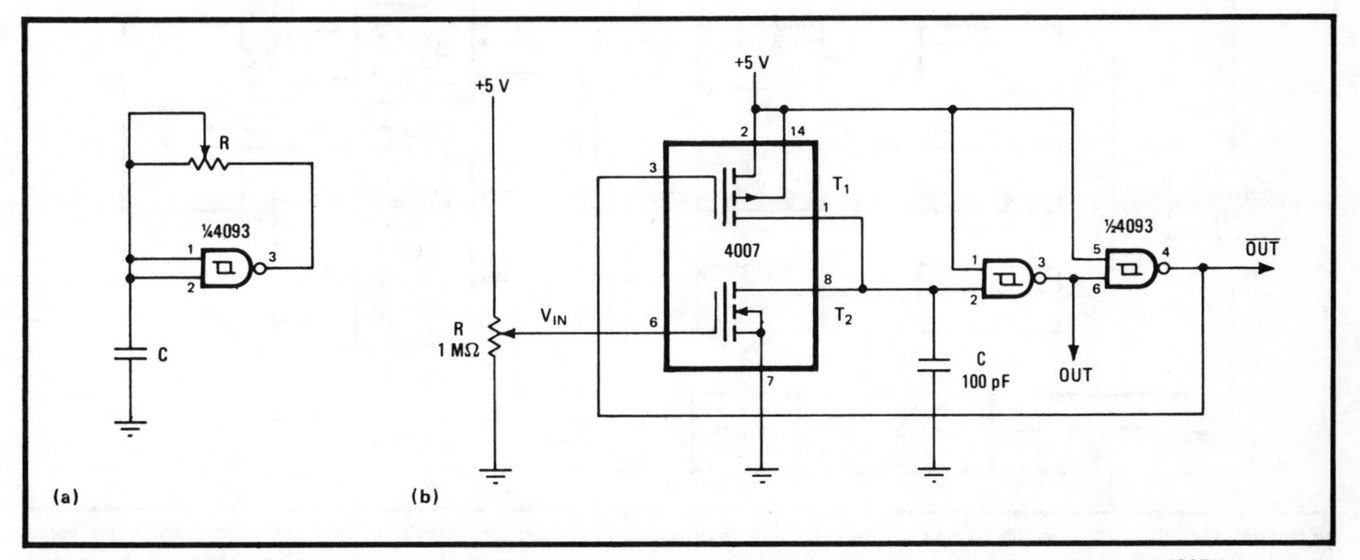

Simple improvement. The range of a standard Schmitt-trigger pulse generator (a) can be easily extended by adding one 4007 gate array to circuit (b). New circuit charges and discharges timing capacitor C through separate paths, enabling the circuit to generate a narrow-width pulse train (0.5 microsecond) over 0 to 1 MHz for an input voltage ranging from 0.7 to 3.4 V.

VCOs generate selectable pseudo-random noise

by James D. Long
Aerojet ElectroSystems Co., Azusa, Calif.

In this circuit, several voltage-controlled oscillators, whose outputs are summed in order to generate a suitable feedback voltage to their inputs, are used to generate pseudo-random noise over band limits that can be selected by the user. Using VCOs makes possible a transfer function that is closer to the ideal and ensures that the circuit has better amplitude-versus-temperature stability than conventional generators, which rely on special and often expensive diodes to produce noise over a wide band. This generator will be suitable for many applications, producing random noise over a bandwidth of three octaves, with a crest factor (ratio of peak to true root-mean-square voltage) of three.

The key to circuit operation is to generate a random feedback voltage to the bank of VCOs so that their output frequencies vary randomly, thus in effect producing noise. This task can be accomplished with the circuit shown in the block diagram. Note that the frequency of the feedback (modulating) voltage is unimportant so long as the maximum modulating frequency is less than any oscillator's output frequency.

A single VCO usually retains its linear input-voltage-to-output frequency characteristic over less than one octave. The three-octave bandwidth can be realized by operating several VCOs over adjacent frequency bands (staggered tuning). Low-pass filters remove the higher-order harmonics contained in the square-wave outputs of the oscillators and the signal appearing at the output of the summing amplifier is similar to random noise.

The feedback signal required to generate the random noise is derived by first combining the output of all VCOs in a resistive summing network. The output from the summer is a signal that contains zero crossings occurring at random intervals.

A divide-by-64 circuit detects threshold crossings and brings the summer signal down to a frequency range consistent with the requirement that the highest modulation frequency be much less than the total output frequency of the VCOs. A narrowband ($Q=5$) LC filter then smooths these random amplitudes into a continuous signal for the VCOs' inputs. With proper scaling factors provided by the amplifier, the desired range of random frequencies can be generated.

Each VCO is biased to its appropriate geometric-mean frequency and operates there if the modulation amplitude is zero. Attenuators a_1 through a_n allow independent control of the deviation range of each VCO, and the adjustable limiter permits control of the peak deviation frequency. The function of the adjustable limiter is to ensure that the oscillators do not deviate beyond their prescribed frequency bands. If the VCOs' band edges are aligned so that no overlaps nor gaps occur between adjacent bands, the distribution of frequencies at the output of the summing amplifier will be uniform. Alignment of the band edges is not difficult.

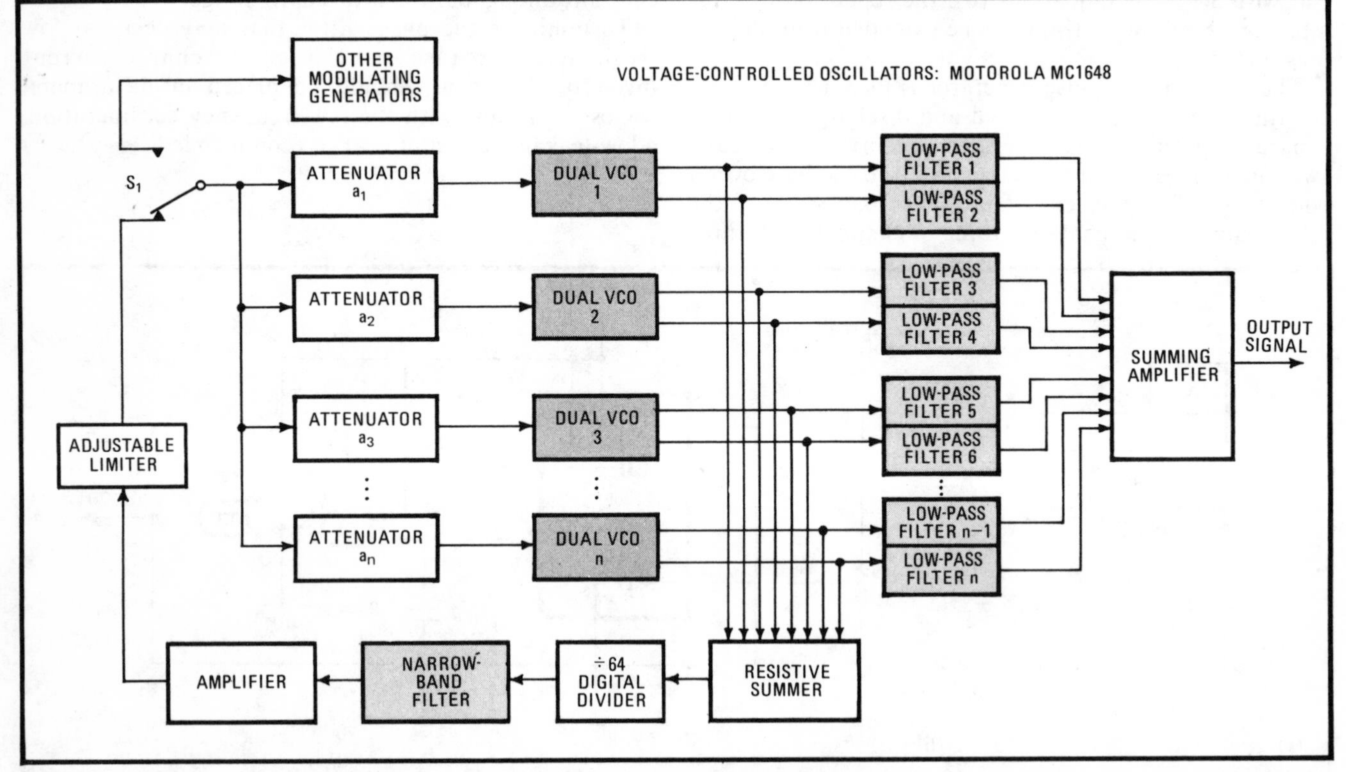

In the noise. Method for generating noise uses summed outputs of voltage-controlled oscillators to generate feedback voltage that varies randomly with time, thereby causing frequencies of oscillators to vary in the same manner. System provides uniform noise output over three octaves, with a noise-voltage crest factor of three. Generator can produce other waveforms if appropriate modulating signals are applied.

This generator can produce signals other than noise. Switch S_1 may be used to apply any modulating signal to the VCOs' inputs. For example, a sawtooth input signal will generate a swept-tone output. If the VCO has an inhibit input, it can be selectively activated so that pulsed tones can be generated. □

Ultrasonic pulser needs no step-up transformer

by Paul M. Gammell
Jet Propulsion Laboratory, Pasadena, Calif.

When a transducer is used in a pulse-echo ultrasonic system of the kind that is excited by high-voltage impulses at a low duty cycle, the device often requires the services of a bulky, step-up line transformer and switching circuits that must withstand the full supply voltage. The ultrasonic pulser shown in the figure, however, generates a 300-volt pulse train at a low-duty cycle (about 1/20,000 of a cycle rise time at 2 kilohertz) without the need for a transformer and without placing an excessive voltage on the switching devices.

Used in place of the transformer supply is a voltage tripler circuit that arranges for each of three capacitors, in effect arranged in parallel, to be charged by a 100-v dc input voltage. The circuit then places the capacitors in series during discharge so that the output voltage is three times the input voltage. Moderately priced ($6) silicon controlled rectifiers with a fast, 10-nanosecond rise time serve as the switching elements and need withstand only 100 v each. Furthermore, since the power drain is modest, merely a few tens of milliamperes, a small power supply of the kind intended for glow-discharge displays is suitable.

As for the operation of the circuit, at times when no signal is applied to the gate of S_1, as during a power-up, S_1–S_3 are off and C_1–C_3 charge up through their respective 100-kilohm resistors and diodes, D_1–D_3.

A positive-going signal applied to the gate of S_1 by the 555 timer, which operates as an astable multivibrator at 2 kHz, turns S_1 on, pulling point A from +100 v to ground. The 555, with the aid of the Q_4–Q_5 line driver/buffer, generates a waveform at the sync output that precedes the pulse to S_1 by 2 microseconds or so and thus is suitable for triggering a scope.

Because the voltage across C_1 cannot change the instant S_1 switches, point B also changes by 100 v, moving from ground to −100 v. Point B was clamped close to ground potential by D_1 during the charge period but is free to make negative excursions after the period ends. S_2 then turns on. The RC network connected to the gate of S_2 limits gate current to a safe value and allows the SCR to return to the off condition when required. With S_2 on, points B and C assume the same potential of −100 v. Because C_2 is charged to 100 v, point D is pulled to −200 v.

The cathode of S_3, which is connected to point D, also assumes a value of −200 v, while the gate is at −100 v for a brief instant. S_3 thus turns on. In a like manner described above, point F is pulled to −300 v. The fall in voltage from 0 to −300 v happens in 20 to 30 μs, then decays back to zero at a rate determined by the output load resistance and capacitance. D_4 has been incorporated to isolate the particular receiver used in the system from any noise generated in the pulser's high-voltage

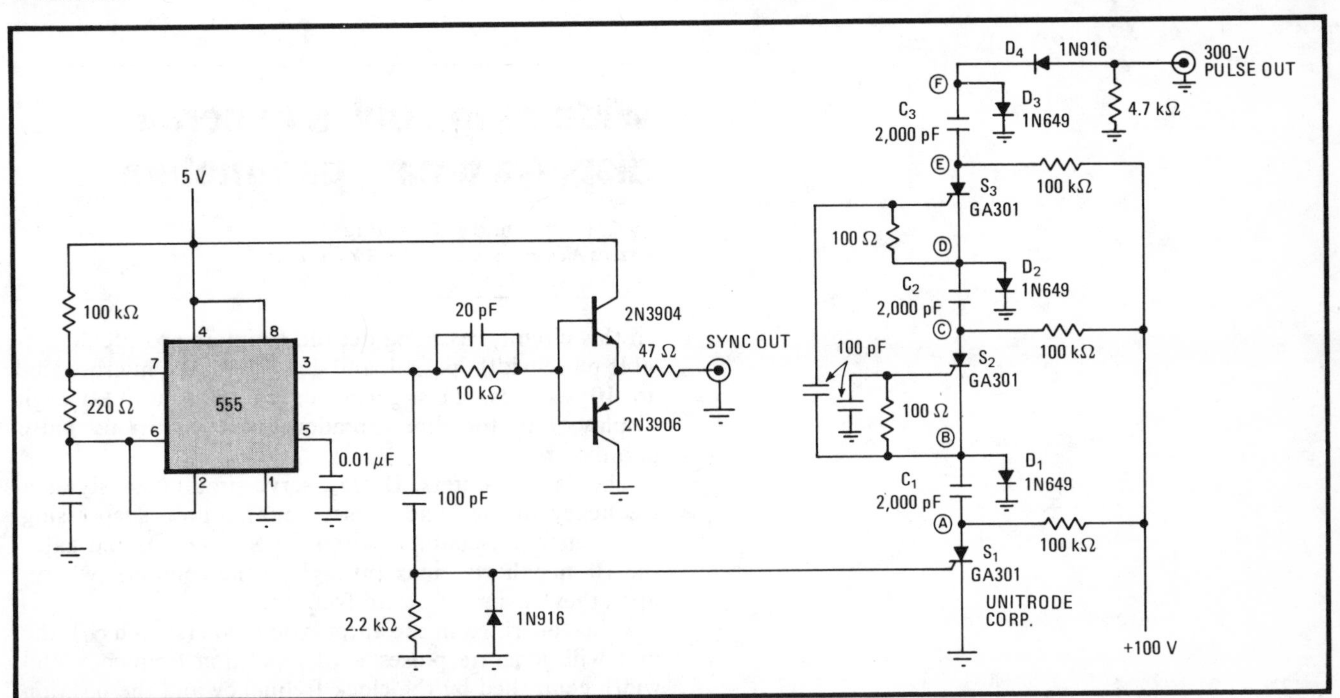

Pulsed tripler. Circuit generates 300-V pulse without the benefit of step-up transformer by charging capacitors C_1–C_3 in parallel from 100-V source and then discharging them in series. Output has repetition rate of about 2 kHz and a rise time of only 20 to 30 ns.

supply and also to minimize the loading of the receiver by the pulser during receive (echo) time slots. The high-speed, low-voltage diode used for D_4 is adequate in most cases, since it never needs to stand off more than a few volts unless a highly reactive echo signal is reflected back into the pulser. The resistor at the output of the circuit provides a dc return path for the output pulse.

The use of slow diodes for D_1–D_3 does not preclude obtaining pulses of fast rise times at the output. Because C_1–C_3 are fully charged when a pulse is commanded, there is practically no current flowing through D_1–D_3; hence, there are no stored carriers.

Additional voltage-multiplying stages may be added as required, using the technique described here. The rise time at the output will increase slightly for each new stage added. □

Wide-range pulse generator displays timing parameters

by C. L. Bhat and R. C. Yadav
Bhabha Atomic Research Center, Srinigar, India

In this circuit, cascaded decade counters provide adjustable pulse width, period, and delay from 0.1 microsecond to 10 seconds. This generator contains an LED digit display, too, for direct readout of the various pulse parameters.

The 7490 counters, D_1–D_8, serve simultaneously as a frequency divider and preset counter unit. Depressing the momentary-contact switch S resets D_1–D_8 and gates the 10-megahertz clock through to the counters, whereupon they advance upward from zero.

When operated in the automatic mode (switch S_1), the unit will generate pulses with repetition frequency and width controlled by the clock frequency and the position of two sets of taps at the output of the 7442 4-to-10-line decoders. Here taps a′–h′ bring output flip-flop F_1 low

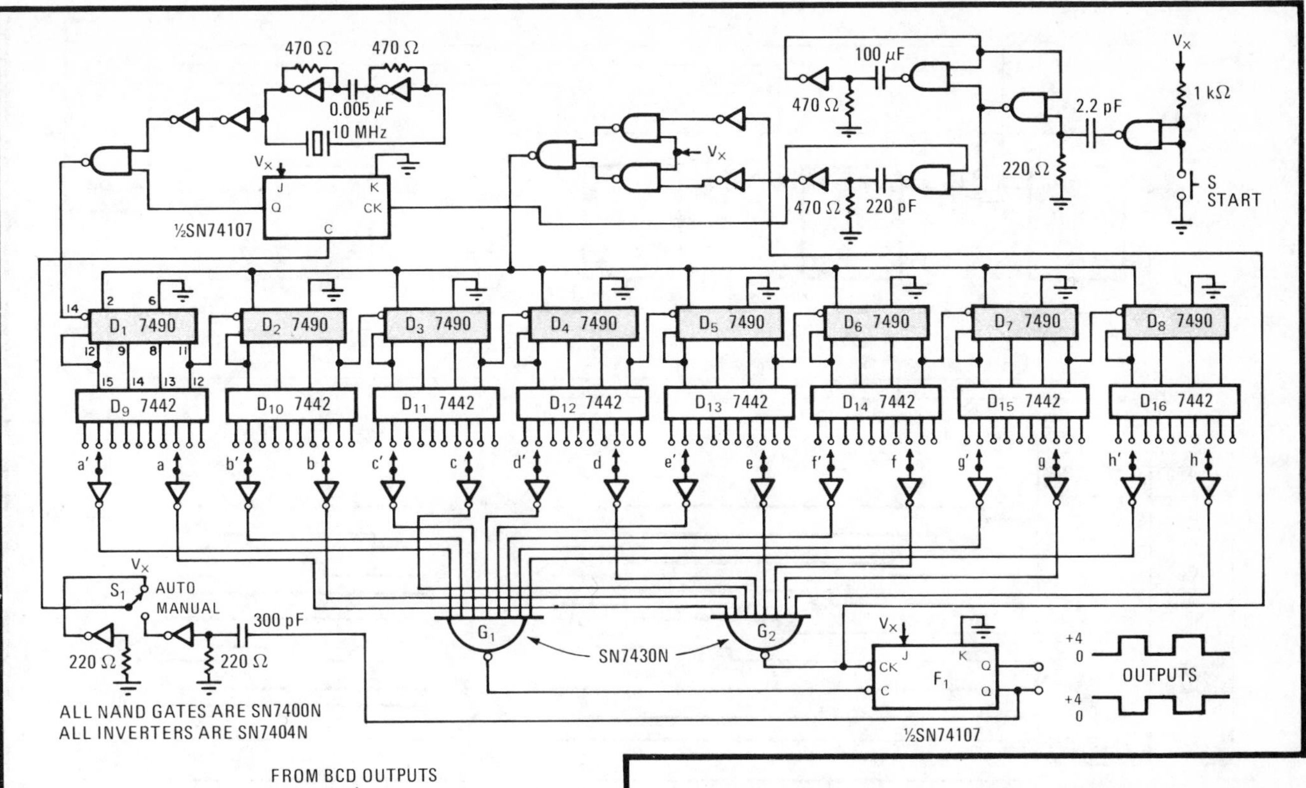

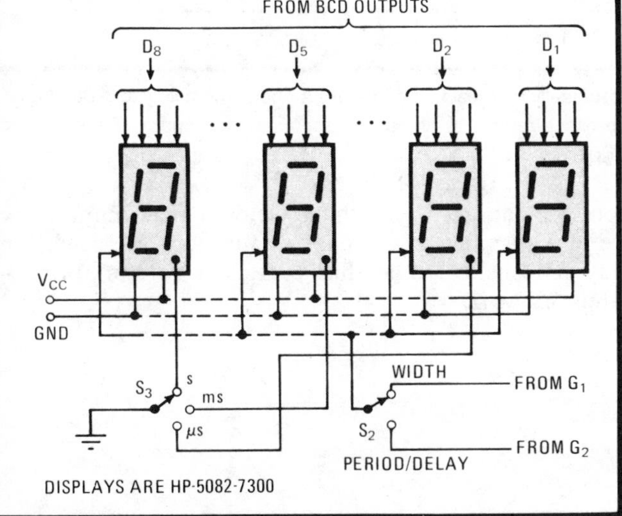

Watching width. Pulse generator provides adjustable width and period over a 0.1-microsecond-to-10-second range. A single pulse of specified delay may also be generated. LED-digit display gives direct readout. Wiring of display-switching circuitry (inset) is simple.

through gate G_1 when D_1–D_8 reaches some preset number. The counter continues to advance, reaching a number determined by taps a–h, which are set to activate G_2 and clock F_1 high. The counters are then reset, and the process repeats.

In the manual mode, a single pulse having a specified delay is generated, with the pulse width again selected by both sets of taps. The basic difference in manual operation is that the resetting of F_1 results in the resetting of F_2 and the disabling of the clock signal to D_1–D_8. Note that taps a–h are set below taps a′–h′.

As for the display circuitry (see inset), the outputs of all counters drive their respective light-emitting diodes. The decimal points of displays two, five, and eight are wired to time-base switch S_3 as shown. The pulse's width and period/delay will be displayed by appropriately setting switch S_2. S_3 orders up the readout time in seconds, milliseconds, or microseconds. □

Cascaded flip-flops set periodic-sequence generator

by Carlos Correia and Cidálio Cruz
University of Coimbra, Portugal

This circuit, which generates a periodic sequence of nonconsecutive binary numbers, will serve well as an address generator in multiplexed data-communications systems. Using several J-K flip-flops whose outputs drive a priority encoder, the circuit produces a selectable, monotonically increasing output code having zero dead time (no lag) between numbers. Implementing this circuit is far simpler than modifying a standard binary-

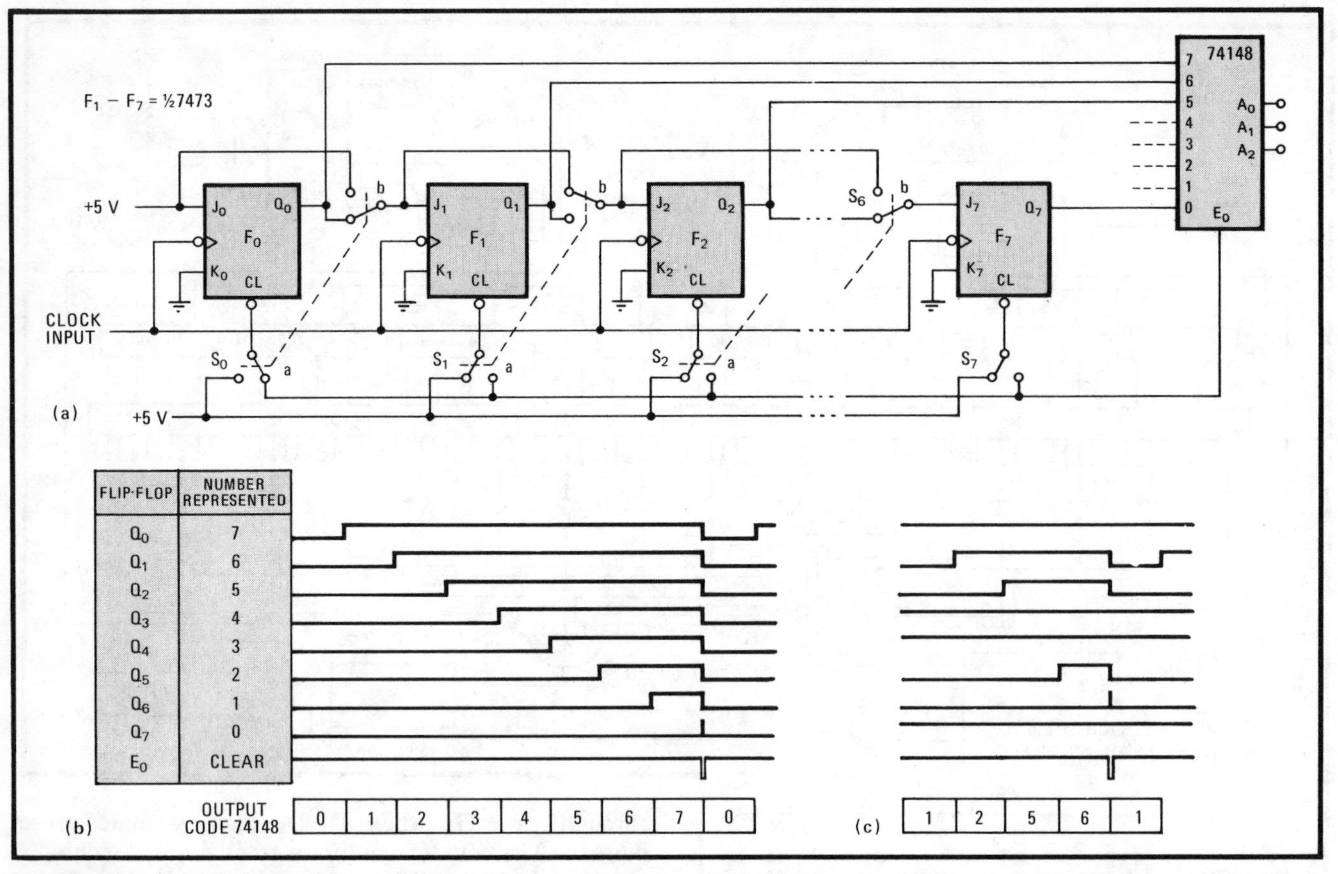

Chosen order. Periodic-sequence generator (a) produces selectable, monotonically increasing binary output code (b) having zero dead time. Any number can be omitted from the sequence (c) by using double-pole, double-throw switches to disconnect Q output of the flip-flop corresponding to that number from succeeding flip-flop and bringing its clear port to 5 V.

counter circuit, which is more useful in applications where the numbers to be generated are consecutive.

As shown in (a), a double-pole, double-throw switch is required for all but the last flip-flop desired, which requires a single-pole, double-throw switch. In this case, seven flip-flops are used—thus the numbers 0 through 7 can be generated.

When each flip-flop is active (Q disconnected from the J port of its succeeding 7473 flip-flop and its clear port connected to E_0 of the 74148), the sequence generator will advance in order from 0 through 7, as shown in (b). At the end of the sequence, when all inputs to the 74148 are high, E_0 moves high, clearing all flip-flops and initializing the sequence at its first number.

Note that any number can be omitted from the sequence by connecting the Q output of the flip-flop that corresponds to that number to the J port of the next flip-flop and then connecting its clear port to 5 volts. For instance, the sequence 1, 2, 5, 6, seen in (c), is generated by disabling flip-flops 3, 4, and 7.

A sequence having a maximum word length of 10 can be generated if a 74147 is used in place of the 74148. However, the E_0 signal is not available in the former device, and therefore to derive that signal each input of a 10-lead NAND gate must be connected to all flip-flops, with its output connected to the switches.

The maximum clock rate is determined by the propagation delay times encountered by the signal that clears the flip-flops. For the circuit shown, the delay time is about 95 nanoseconds, which yields a maximum clock rate of 1 megahertz. If Schottky flip-flops are used, the delay is 44 ns, corresponding to a clock rate of approximately 2.2 MHz. □

Chapter 35
SENSORS

Programmed module automates transducer's linearization

by C. Viswanath
Indian Institute of Science, Bangalore, India

As a consequence of its unusual transfer function, the Analog Devices 433 J/B programmable multifunction module finds wide use in performing vector operations, generating trigonometric functions, raising a number to an arbitrary power, and linearizing the response of transducers used in medical and industrial electronics.

The module's transfer function is:

$$e_o = \frac{10}{9} V_y \left(\frac{V_z}{V_x}\right)^m = P \qquad 0.2 \leq m \leq 5.0$$

Programming of the exponent, m, contained in the transfer function, which is necessary to generate the required operations, is done more quickly and accurately with a digital-to-analog converter and two field-effect transistors than with a potentiometer, the component most often used. Digital selection of the exponent is particularly useful where an automatic test sequence must be generated from a microprocessor to multiplex several transducers, each requiring a different m. With this circuit, the value of m may be adjusted throughout the entire specified range, in increments of 0.1.

A circuit used for transducer linearization is shown in the figure. The technique used for linearizing a transducer's transfer function (Q) is to control m so that it varies inversely with the known exponent (n) contained within the transducer's characteristic equation. Thus, when the output voltage from the transducer $(V_z)^n$ has been processed by the 433 J/B, the effects of m and n on the output voltage will cancel each other (as a result of multiplying P and Q to obtain e_o), and the entire transfer function is then simply expressed by:

$$e_o = K V_z$$

The value of m is programmed by controlling the resistance between pins 1, 7, and 8 of the 433 J/B by means of the AD3954 dual-FET stage and the 10-bit AD7520 d-a converter. The combination of the converter and the dual FETs is thus intended to serve as a digitally controlled potentiometer.

Two binary-weighted current sources, the magnitudes of which are dependent on the 10-bit input and the sum of which is constant (equal to $I_{out\,1}$ plus $I_{out\,2}$), drive operational amplifiers A_1 and A_2. The magnitude of $I_{out\,1}$ and $I_{out\,2}$ are determined by the reference voltage at pin 15 of the converter.

A_1 and A_2 convert the currents to voltages $V_{out\,1}$ and $V_{out\,2}$, respectively, and drive the gates of the dual FET. The FETs operate as voltage-controlled resistors and are selected to provide good tracking throughout the 0-to-−3.5-volt input-voltage range.

If all 10 bits of the AD7520 are set to logic 1, corresponding to an m value of 5.0, $I_{out\,1}$ will equal 1 milliampere, and $I_{out\,2}$ will eqûal 0. Thus $V_{out\,1}$ should be

PROGRAMMING OF EXPONENT M

M	DIGITAL INPUT MSB LSB
0.2	0000000000
0.3	0000010100
0.4	0000101010
0.5	0001000000
0.6	0001010100
0.7	0001101010
0.8	0010000000
0.9	0010010100
1.0	0010101010
1.1	0011000000
1.2	0011010100
1.3	0011101010
1.4	0100000000
1.5	0100010100
1.6	0100101010
1.7	0101000000
1.8	0101010100
1.9	0101101010
2.0	0110000000
2.1	0110010100
2.2	0110101000
2.3	0111000000
2.4	0111010100
2.5	0111101010
2.6	1000000000
2.7	1000010100
2.8	1000101010
2.9	1001000000
3.0	1001010100
3.1	1001101010
3.2	1010000000
3.3	1010010100
3.4	1010101010
3.5	1011000000
3.6	1011010100
3.7	1011101010
3.8	1100000000
3.9	1100010100
4.0	1100101010
4.1	1101000000
4.2	1101010100
4.3	1101101010
4.4	1110000000
4.5	1110010100
4.6	1110101010
4.7	1111000000
4.8	1111010100
4.9	1111101010
5.0	1111111111

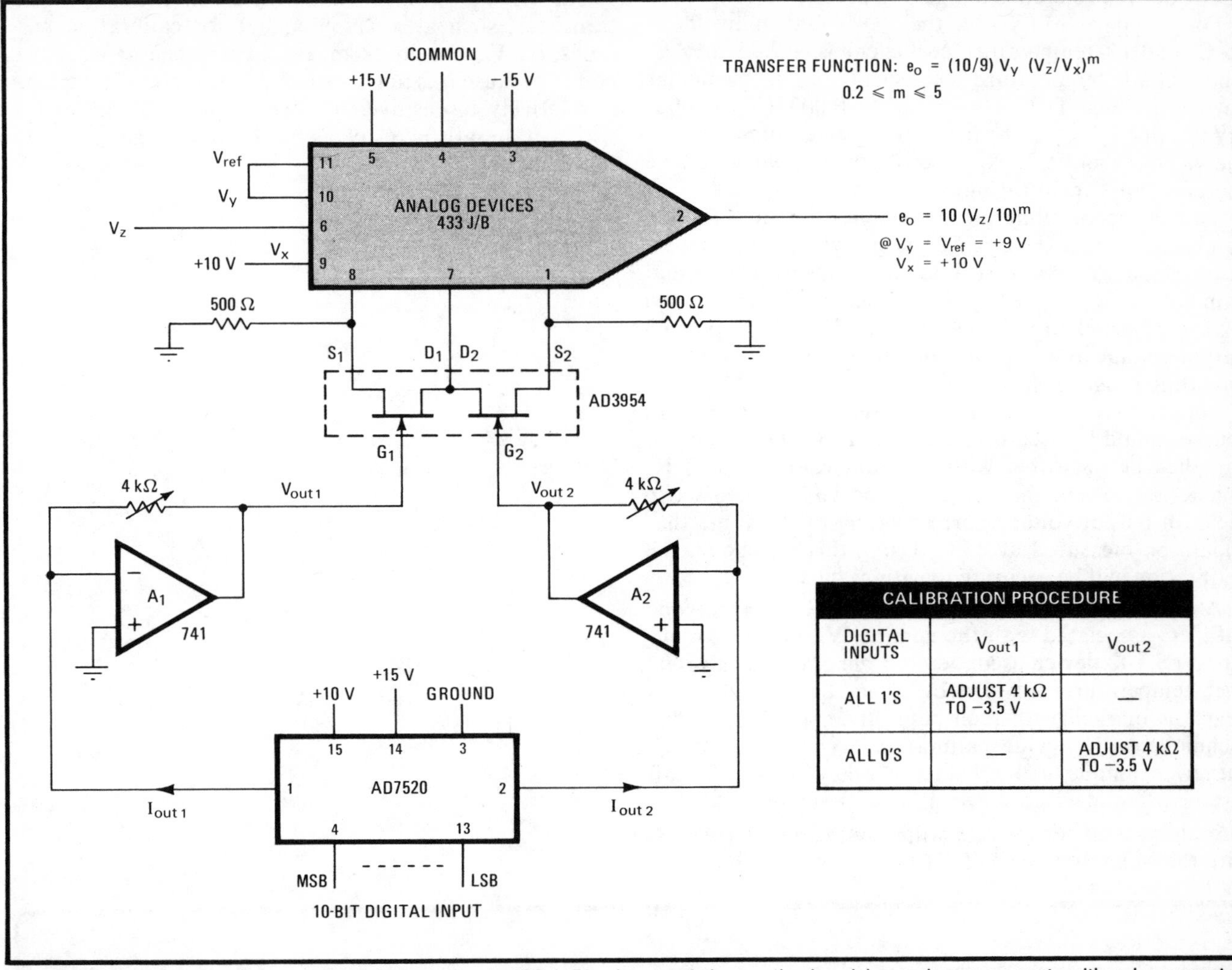

DIGITAL INPUTS	$V_{out\,1}$	$V_{out\,2}$
ALL 1'S	ADJUST 4 kΩ TO −3.5 V	—
ALL 0'S	—	ADJUST 4 kΩ TO −3.5 V

Exponent programming. Programming of constant m in 433 J/B's characteristic equation is quicker and more accurate with a d-a converter and FETs that operate as voltage-controlled resistors than with a potentiometer. This circuit linearizes the response of transducer voltage V_z if the 433 J/B is programmed so that m is the inverse of exponent value n contained in the transducer's transfer function.

set to a full-scale output (−3.5 v), and $V_{out\,2}$ should be set to 0. Similarly, $V_{out\,1}$ must be set to 0, and $V_{out\,2}$ to −3.5 v, when all inputs are set to logic 0, corresponding to an m of 0.2. The 4-kilohm potentiometers are provided for calibration purposes.

The table outlines the inputs to the d-a converter required for any value of m from 0.2 to 5. □

Semiconductor thermometer is accurate over wide range

by Larry G. Smeins
Hewlett-Packard Co., Loveland Instrument Division, Loveland, Colo.

Reducing the circuit complexity and simplifying the calibration of thermometers that use the base-emitter drop of a transistor to detect temperature changes, this circuit provides readings accurate to within ±2°C over the range of −40°C to +150°C. Using the low-offset, low-drift characteristics of National's micropower LM-10 operational amplifier and Motorola's MTS-102 temperature sensor, the thermometer will compete with the more expensive platinum-resistance units.

The LM-10 is used to bias, scale, and zero the MTS-102 sensing transistor, thereby removing the need for a separate constant-current bias source for the sensor and an additional scaling and nulling circuit. Further, the well-defined characteristics of the MTS-102 permit single-point calibration of the thermometer.

The MTS-102, Q_1, is placed in the feedback loop of the LM-10 reference amplifier. A constant-current bias is created by V_{ref} across resistor R_1 giving $I_{bias} = 0.2$ v/2kΩ = 100 μA. The voltage at the output of the reference amplifier will thus be $V_{ref} + V_{be}$. The second stage of the LM-10 is configured to provide a gain that is constant with respect to the temperature coefficient of the V_{be} drop and to subtract V_{ref} and V_{be} from the output at any desired reference temperature.

For a typical MTS-102, the V_{be} is 600 millivolts at 25°C, and its temperature coefficient is −2.25 mV/°C. The actual temperature coefficient for a particular device is thus TC = $-2.25 + 0.0033(V_{be}' - 600)$ mV/C, where V_{be}' is the measured base-emitter drop for the given sensor at 25°C. A corresponding offset voltage therefore appears at the output of A_1.

The gain-controlling elements of A_2, resistors R_2–R_4, can be set so that the circuit's output-voltage-to-temperature slope will be correct for any sensor. The actual gain will be $R_4/(R_2+R_3)$. Once the gain is set, R_5 and R_6 are adjusted to null the offset and yield the desired output voltage at any calibration temperature within the operating range of the circuit.

For very accurate calibration, a reference temperature source should be used to keep Q_1 at 25°C. The V_{be} of Q_1 can then be measured with a digital voltmeter and R_2 and R_3 set to null the offset. R_5 and R_6 are then set to yield an output voltage corresponding to the value that should be measured at 25°C. Using this technique, the calibration will be accurate to within ±1°C.

A simpler alternative is to set R_2 and R_3 to correspond with TC associated with the nominal V_{be} of the particular MTS-102 device used, zeroing the circuit at a reference temperature provided by an ice bath. Each MTS device is marked with their respective $V_{be} \pm 2$ mV. This technique will provide calibration accuracy to ±2°C. Either technique will provide accuracy to 4°C for all interchanged devices marked with the same V_{be}. But interchangeability accuracy will vary to a greater degree with the MTS-103 and MTS-105 devices.

The circuit will also work with a conventional silicon transistor, such as a 2N3904, but for calibration purposes, its V_{be} should be measured between at least two points because it is not a specified parameter. The circuit is relatively insensitive to power supply voltage and it will operate satisfactorily over 2-to-40-V range. □

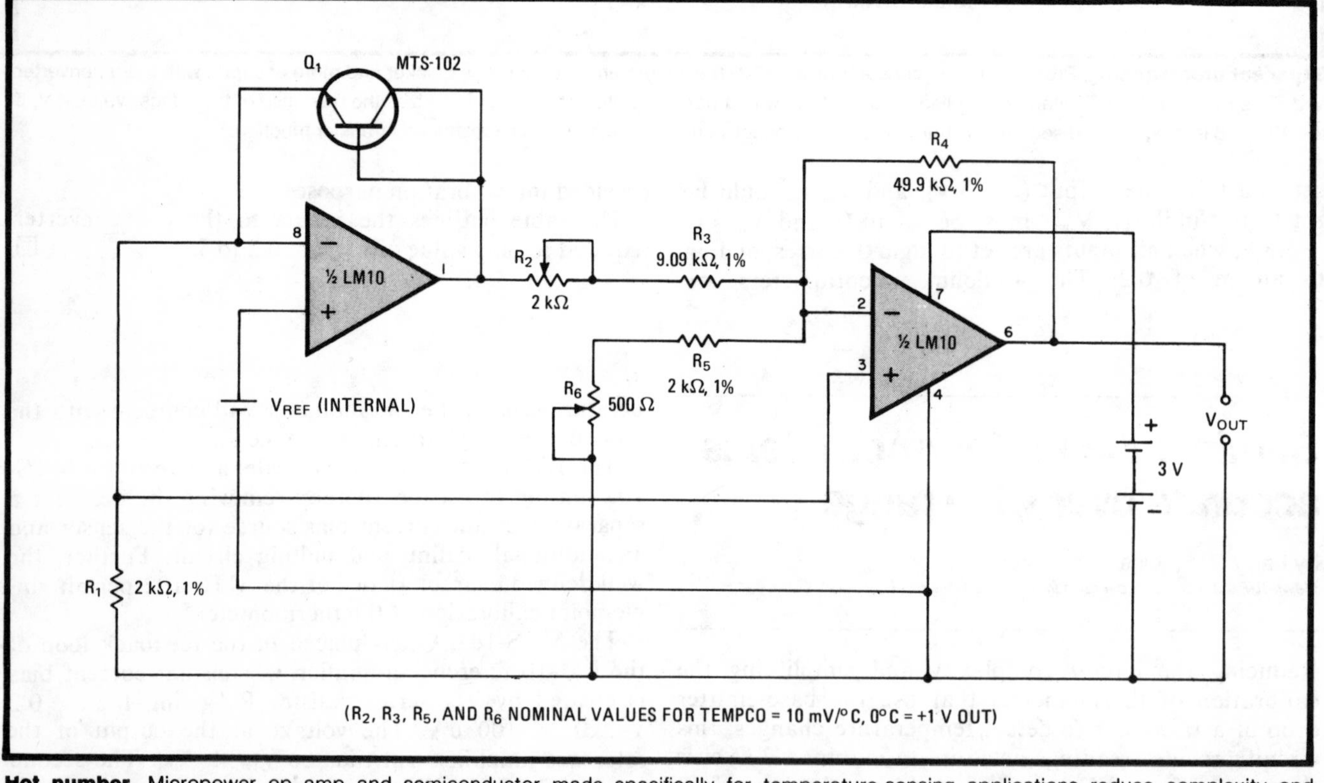

Hot number. Micropower op amp and semiconductor made specifically for temperature-sensing applications reduce complexity and calibration procedure of thermometers that use V_{be} of transistors to detect temperature changes. Accuracy of device, no worse than ±2°C over the range of −40°C to +150°C, and its simplicity enable it to compete with much more expensive units.

Diode sensor and Norton amp control liquid-nitrogen level

by V. J. H. Chiu
National Research Council, Ottawa, Canada

In parametric amplifier and other cryogenic applications, it would be handy to have an inexpensive sensor and controller of the level of liquid nitrogen. One can be built around a standard silicon diode and a Norton (current) amplifier.

The circuit's operation is based on the principle that the diode's junction voltage increases from 0.7 volt at room temperature to 1.05 v in liquid nitrogen (liquefaction temperature: −196°C). This voltage change is used to activate the amplifier, which controls a solenoid valve. The valve regulates a nitrogen-gas supply, which pumps liquid nitrogen from a reservoir to the desired container that houses the sensor.

The controller is shown in the figure. The sensor is placed in the container at any desired level. When liquid nitrogen rises to this level, voltage V_S reaches the preset voltage V_R almost instantly, and the output of the 3900 Norton amp becomes zero, closing the valve. When the liquid nitrogen falls below the desired level, V_S drops below V_R, and the valve opens.

Circuit sensitivity is adjusted by R. The diode need not be completely immersed in the liquid nitrogen, for its range is such that liquid as much as 2 inches below it will start the refilling of the container. Frequent cycling is thus avoided. The state of the solenoid valve may be determined by observing the light-emitting diode. □

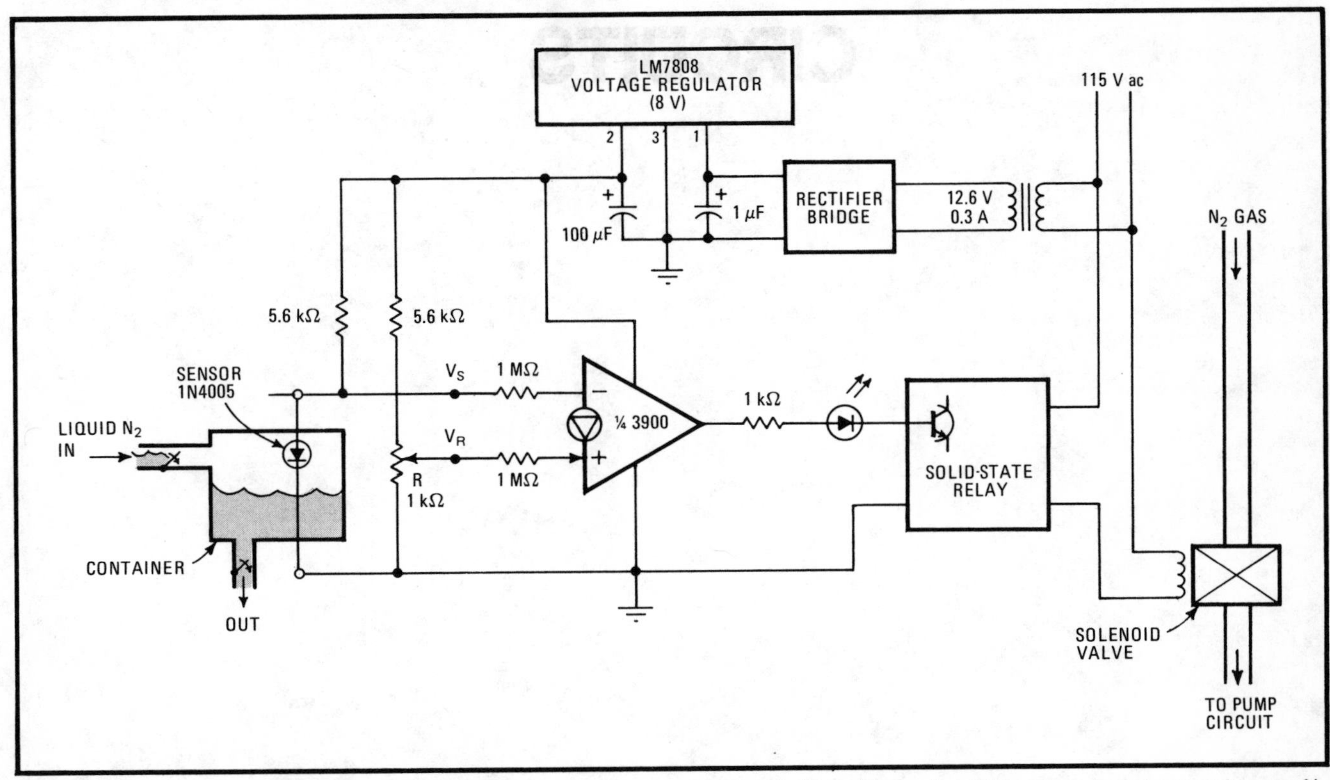

Fixing the nitrogen level. When liquid-nitrogen level in container is below the diode position, solenoid is turned on and pumps in more N_2. When level reaches that of diode, its junction voltage jumps from 0.7 to 1.05 V, turning off solenoid and stopping N_2 inflow.

Chapter 36
SWITCHING CIRCUITS

Removing the constraints of C-MOS bilateral switches

by W. Chomik and A. J. Cousin
Department of Electrical Engineering, University of Toronto, Canada

Two major limitations imposed on the popular complementary-metal-oxide-semiconductor 4016 switch may be overcome with this circuit. As well as allowing the signal magnitude to exceed the power-supply voltage, it enables unipolar control signals to switch bipolar input signals. Only a second switch and an inverter need be added to a standard circuit to remove these operating constraints on the signal-handling gate.

Usually, the signal voltage to be passed through a single switch must be limited to between $V_{DD}+0.7$ volt and $V_{SS}-0.7$ v, where V_{DD} is the positive supply (drain) voltage and V_{SS} is the minus supply (source) voltage. Otherwise, the signal voltage will cause the forward biasing of the diode between the substrate and channel of one MOS field-effect transistor, and the gate may be destroyed.

This problem might arise if the power-supply value applied to some active element in a circuit happened to lie outside the voltage range that could be applied to the switch ($V_{DD}-V_{SS}$), dictating that the gate must be protected from input and control signals that saturate to the supply level.

Furthermore, many circuits, especially those containing operational amplifiers, use bipolar supplies. The resulting signals to be processed are likely to be bipolar as well. Yet the channel-voltage constraints inherent in the design of the 4016 (that is, the fact that the logic 0 control voltage, V_{SS}, must be at or below the most negative signal voltage, and the logic 1 control voltage, V_{DD}, must be at or above the most positive signal voltage) means that bipolar supplies and control signals must also be applied to the switch if these bipolar signals are to be passed. Unfortunately, too, many systems use digital control signals that are unipolar, and so logic-level shifters are needed also, to make this signal symmetrical with respect to ground.

With the addition of a second bilateral switch and an inverter to a standard op-amp circuit, as shown, the signal-handling switch can operate from a single power supply and be driven by unipolar logic at the control input in order to pass bipolar signals. Moreover, the signal can lie outside the $V_{DD}-V_{SS}$ limit of the switches.

The channel voltage of both switches is set by fixing their drain potentials at the virtual ground of the op amp or to circuit ground, depending on which switch is on. Because the virtual ground never strays from true ground by more than a few millivolts, the switches will be protected from burn out, as their channel-voltage limit will never be exceeded.

When switch A is on and switch B is off, node n will be essentially at ground potential. When B is on and A is off, the signal is removed from the op amp's input, but node n will still be at ground (through B), and the same channel-voltage conditions will prevail.

Note that the actual input voltage to the gate at node n will never drop more than a few millivolts below the minimum control voltage, even if the input signal is negative. Thus, the gate's channel-voltage constraint is always met. □

No limitations. Inverter and gate B enable switching of bipolar input signals by unipolar control signals at gate A and also allow magnitude of input to exceed gate's supply voltage. Node n is held near to ground at all times, so that channel-voltage limit of gate is never exceeded. Magnitude of signal at node n never exceeds control-signal potential, enabling gate to switch properly.

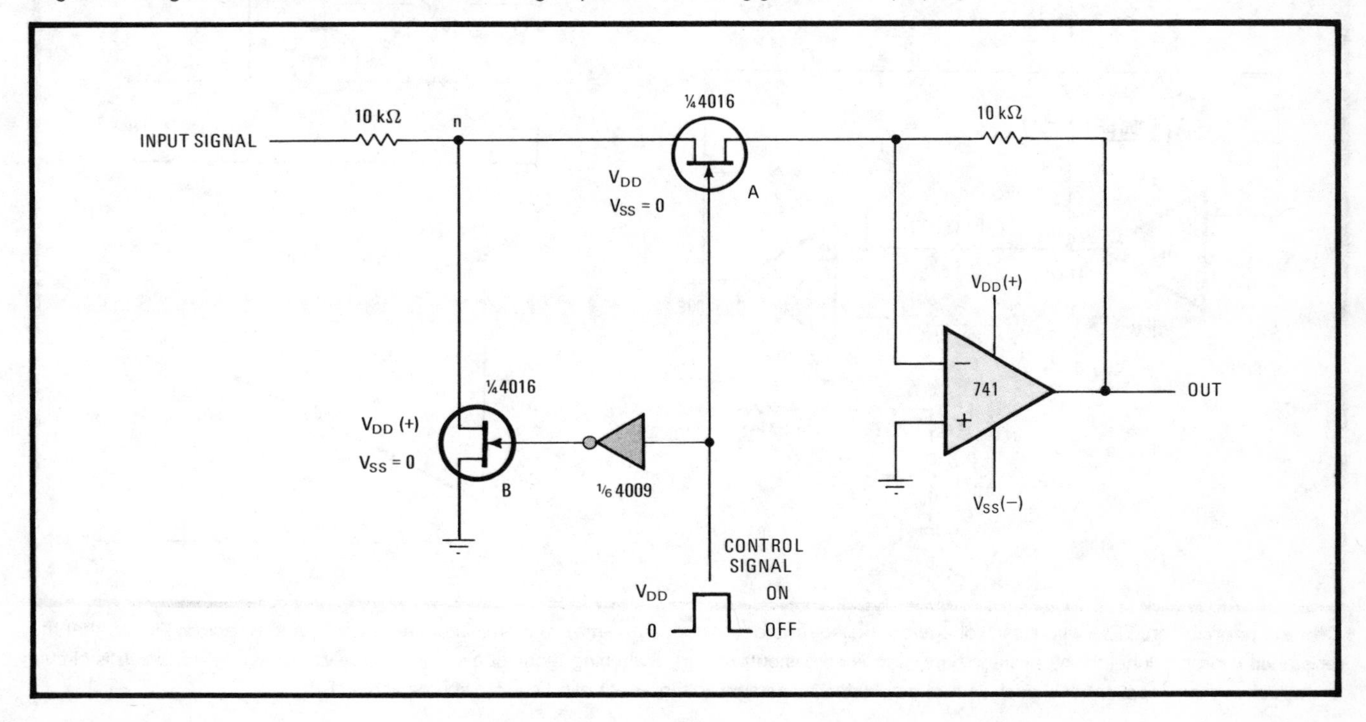

Voltage-controlled resistance switches over preset limits

by Chris Tocci
Halifax, Mass.

Using two field-effect transistors as switches, this voltage-controlled resistor network can order up any value of resistance between two preselected limits. It is unlike other circuits in that it does not employ the drain-to-source resistance of matched FETs, whose R_{ds} characteristics are usually proportional to a control voltage. As for circuit linearity, it will far exceed that of conventional networks using a single FET in various feedback configurations.[1]

In operation, oscillator A_1–A_2 generates a 0-to-10-volt triangle wave at 100 kilohertz, which is then compared with the control signal, V_c, at A_3. During the time that the control exceeds switching voltage V_T, FET Q_1 is turned on, and resistor R_1 is placed across resistance R_{out} (disregarding the R_{ds} of Q_1). At all other times, FET Q_2 is on and resistor R_2 is placed across R_{out}. Thus R_{out} is equal to an average value proportional to the time each resistor is placed across the output terminals, with the actual resistance given by $R_{out} = (R_1 - R_2)V_c/10 + R_2$, for $R_1 > R_2$. This relationship will hold provided any potential applied to the R_{out} port from an external device is less in magnitude than the supply voltages; that any signal processing at R_{out} be done at a frequency at least one decade below the 100-kHz switching frequency; and that the upper and lower resistance limits, R_1 and R_2, are much greater than the on-resistance of Q_1 and Q_2, respectively.

Potentiometer R_{11} adjusts the baseline of V_T to zero so that with $V_c = 0$, $R_{out}n = R_2$, where n is a constant. Further calibration can be carried out by trimming R_1 and R_2 to precise values.

This circuit is readily adapted to many applications, such as a one-quadrant multiplier. This is achieved by connecting a voltage-controlled current source into the R_{out} port to build a dc-shift amplitude modulator whose carrier frequency is the switching frequency. The audio information or data is taken from V_c, but with the signal offset by 5 volts. Thus the dynamic range of the circuit will be 10 V. □

References
1. Thomas L. Clarke, "FET pair and op amp linearize voltage-controlled resistor," p. 251

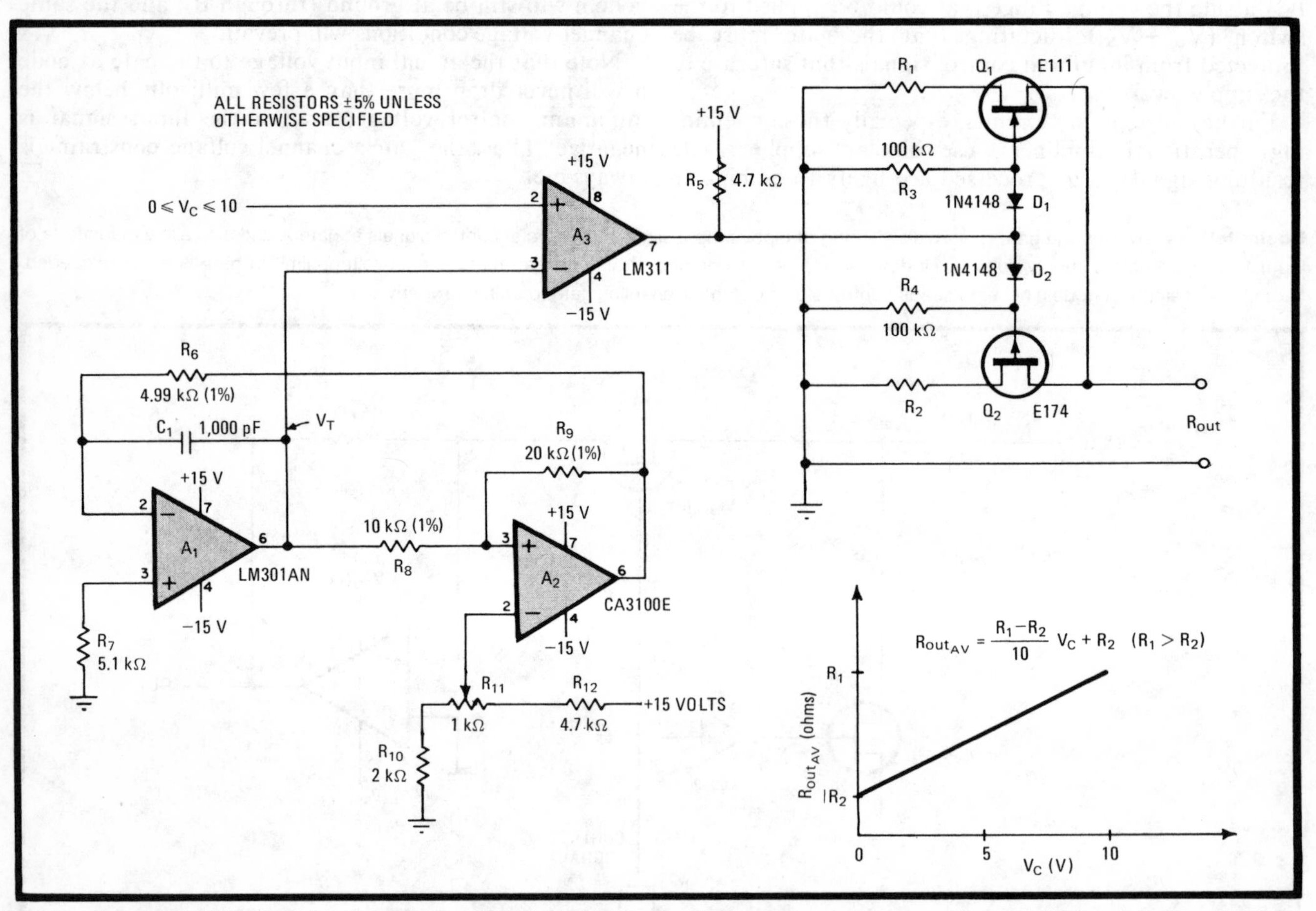

Ohmic linearization. FET switches in voltage-controlled resistor network place maximum-minimum resistors R_1 and R_2 across R_{out} so that the resistance is proportional to the average time each is across output port. Switching technique ensures piecewise-linear operation. This circuit lends itself to many applications, such as a-m modulator, by placing voltage-controlled current source across R_{out}.

Open-collector logic switches rf signals

by W. B. Warren
TRW Subsea Petroleum Systems Inc., Houston, Texas

The open-collector outputs of transistor-transistor logic can provide a simple way of switching low-level radio-frequency signals and thus digitally selecting signal sources in test equipment or filters in a communications receiver. With a 50-ohm source and load, the rf attenuation through the switch at 10 megahertz is only 1.3 decibels when the switch is active and greater than 40 dB when it is open.

Shown in (a) is the basic rf switching element. When the logic control signal is low, the open collector output of the NAND gate is high. Thus diode D_1 is back-biased, D_2 conducts, and the switch is turned on. The capacitance of the open-collector output of the logic element will not affect circuit operation, since the reverse-biased D_1 prevents the element from shunting the rf path.

When the control signal is high, the open-collector output is low and D_1 conducts, forcing the dc voltage at the junction of the diodes to a low value. The voltage across D_1 is therefore not sufficient to keep it conducting (that is, it is reverse-biased), and the rf signal cannot appear at the output.

The application of the basic element to a single-pole double-throw rf switch is shown in (b). Logic control for the SPDT switch is provided by two 7401 open-collector NAND gates. □

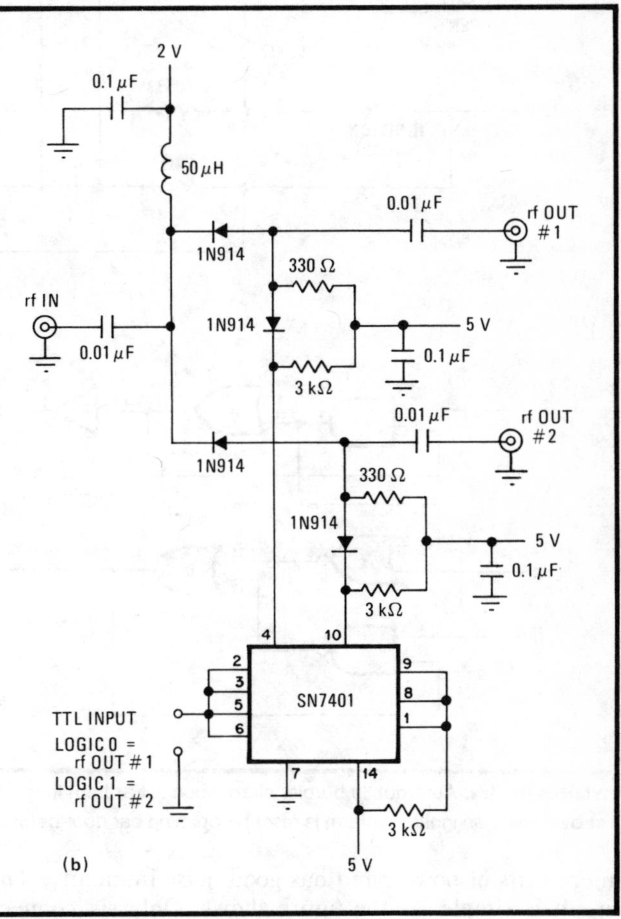

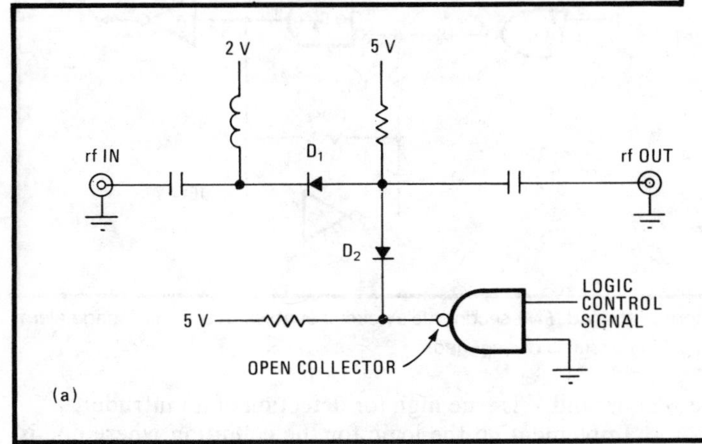

Digital rf switch. Logic signal is capable of switching low-level radio-frequency signals without significant attenuation, if open-collector transitor-transistor-logic element is used (a). Basic idea can easily be extended to single-pole, double-throw rf switch (b).

Vehicle-intruder alarm has automatic set/reset switching

by M. B. Horan
Derby, England

Providing fully automatic operation at negligible standby current, this low-cost burglar alarm for automobiles will sound the horn if any door is opened. No alarm-set switch is required on the body of the automobile: there is a built-in time delay between the opening of the door and the sounding of the alarm, which gives the driver time to activate the ignition circuit and so disengage the alarm. Resetting the circuit requires only that the driver open a car door before the ignition key is removed, thereby engaging the alarm circuit.

The alarm has been designed using complementary-metal-oxide-semiconductor logic, because it is inexpensive, rugged, reliable, and available, requires only a few

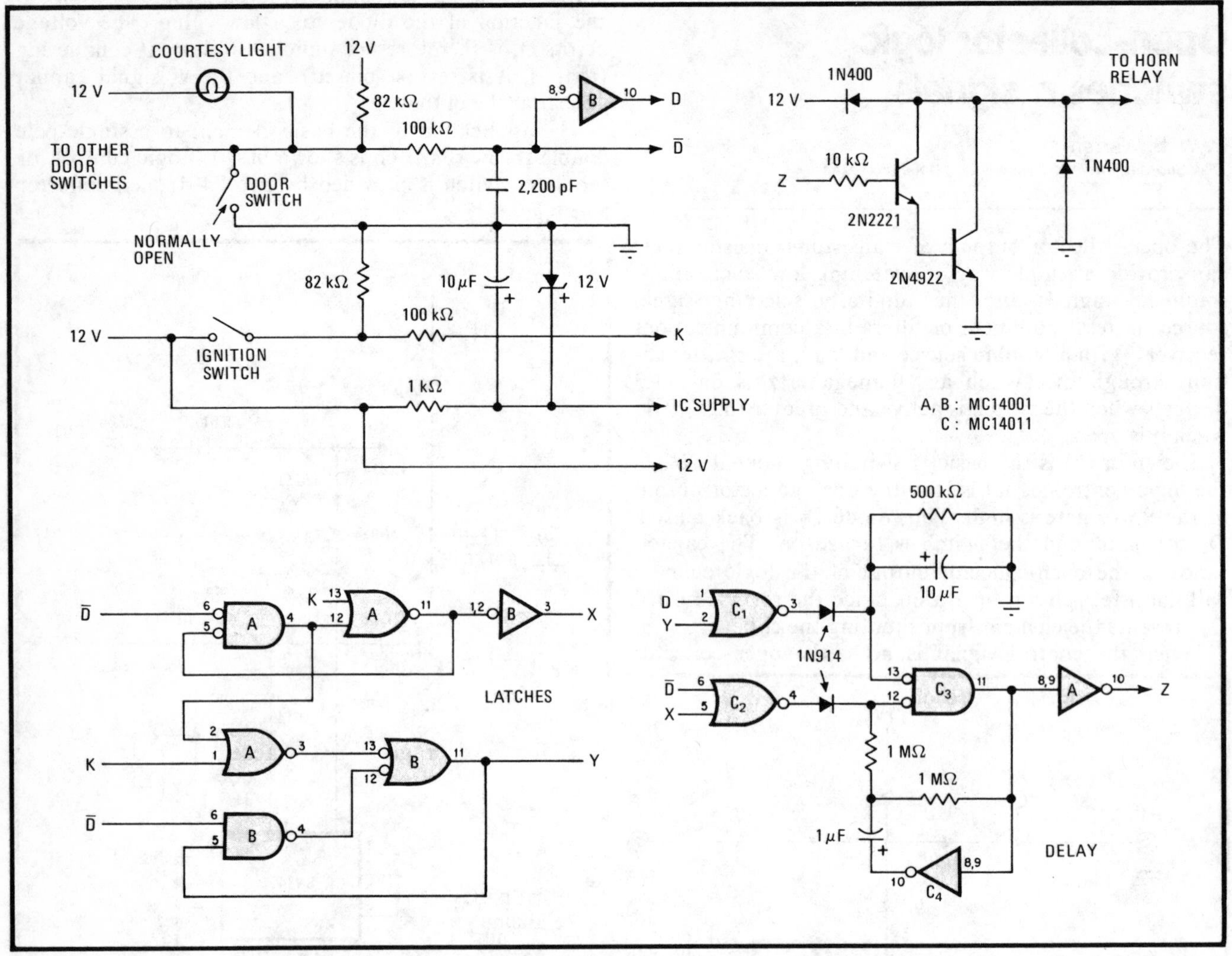

Invisible sentry. Automobile burglar alarm sounds car horn if any door is opened. Five-second-delay circuit enables user to disengage alarm just by turning on ignition. Alarm is reset by opening car door before ignition circuit is disengaged.

microwatts of power, and has good noise immunity. The circuit is simple, as the figure shows. Only six connections are required to interface it with the auto—two for the ignition, two for the door switches, and two for the horn relay.

The circuit must distinguish between several asynchronous events encountered in normal operation and store their present states so that:

- The horn will sound approximately 5 seconds after the time any door is opened, provided the ignition switch is not engaged.
- Once the horn sounds, it will continue to do so, independent of the position of any door.
- The alarm can always be reset by engaging the ignition switch.
- If the door is opened with the ignition on, and then the ignition is turned off and the door subsequently closed within 5 seconds, the horn will not sound.

In order to perform these tasks, the circuit implements the function:

$$Z = (D+Y)(\overline{D}+X)$$

where $X = K+DX$, $Y = K+DX+\overline{D}Y$, and K is true high for ignition switch-on, D is true high for an open door, and Z is true high for detection of an intruder.

Implementing the logic for the condition where one is entering the vehicle is simple. The equation given becomes more involved, however, because the circuit must allow the operator to leave the auto while setting the alarm without triggering the horn. The logic to implement this fourth condition is controlled by two latches, one of which generates the secondary variable X and the other generating Y. X is set high, also allowing Y to be set high when the condition occurs. Latch Y remains high, ready to reset on the opening of any door. Latches X and Y and the door signal are then gated to preset the alarm signal.

The alarm-gating signal (Z) is actually generated once the 1-microfarad capacitor in the delay oscillator discharges below gate C_3's logic-1 point, about 5 seconds after the X signal arrives at C_2. The inverting-gate astable multivibrator, C_4, modulates the horn at 1 hertz to enhance effectiveness as an alarm signal. Other components are included in the circuit for protection against switching transients. □

Chapter 37
TIMING CIRCUITS

Sample-and-hold modules shrink delay-line cost

by T. G. Barnett
London Hospital Medical College, Department of Physiology, England

Providing a selectable time lag while faithfully reconstituting an analog signal, this circuit, which uses a 555 timer and three one-shots to clock two sample-and-hold modules, serves as a low-cost substitute for the conventional analog delay line in some applications. Although its spectral response is not nearly so great as that found in commercial charge-coupled-device–type delay lines, the circuit is suitable for use in the analysis of low-frequency physiological data.

Components R_a, R_b, and C_t of the 555 (wired in the astable mode) set the desired sampling frequency $f = 1.44/(R_a + 2R_b)C_t$. In line with the sampling theorem, this frequency should be equal to or greater than twice the highest analog frequency to be processed.

The output of the 555 drives one-shot A_1, which provides a sampling pulse for the first LF398 sample-and-hold module. Similarly, the 555 drives A_2, which (through A_3) drives the second sample-and-hold device.

Timing components R_{t1} and C_{t1} should be set to provide an output pulse of a minimum of 10 microseconds from A_1. Components R_{t2} and C_{t2} surrounding one-shot A_2 should be set so that its output pulse width is less than $t = t_i - t_3$, where t_i is the sample-interval time and t_3 is the width of the pulses emanating from one-shot A_3, which is set to give a minimum width of 10 μs.

Thus, assuming A_1 and A_3 are set for the minimum width possible, analog signals appearing at the input will be sampled and transferred to the output of the second sample-and-hold device after a delay time approximately equal to t, the time set by A_2. □

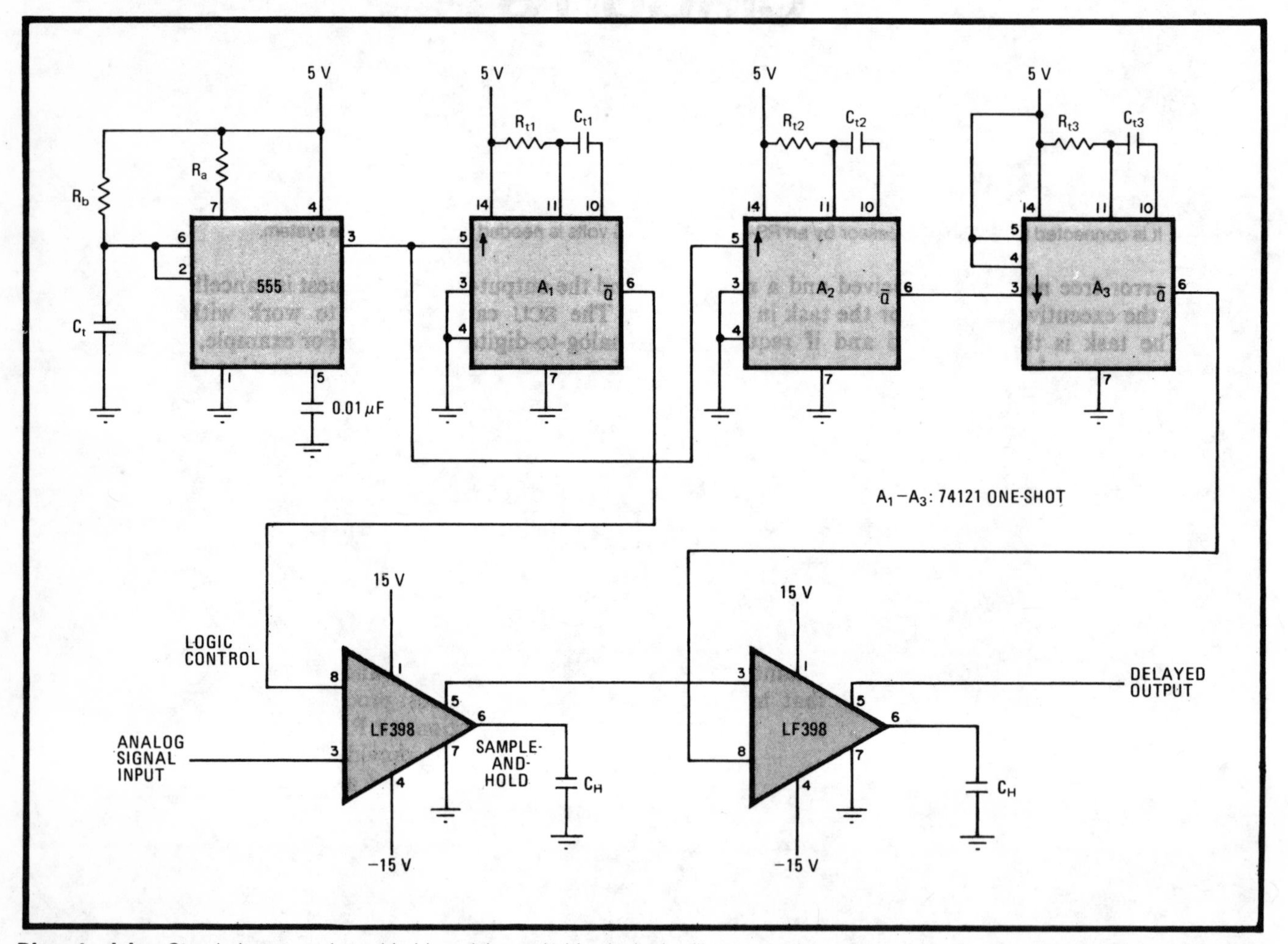

Discrete delay. Stand-alone sample-and-hold modules, suitably clocked, will serve to delay analog signals. Sampling frequency set by 555 should be at least twice the highest analog frequency to be processed. One-shot multivibrator A_2 sets delay time.

Clock module supplies chart-recorder time markers

by G. J. Millard
Volcanological Observatory, Rabaul, Papua, New Guinea

Utilizing the display segments of a standard digital clock module—or more precisely, the signals that drive them—this circuit enables a chart recorder to mark intervals of 1 minute or 1 hour or both. Use of an already built electronic clock, such as National Semiconductor's popular MA1012C, guarantees a chronometer that is accurate, simple to construct, and low in cost ($25).

The 1-minute markers are developed from the signals that drive segments d and e of the minutes-unit display in the MA1012C. These signals, which are readily accessible, drive two frequency doublers, A_1–A_4, that in turn produce a pulse at point C every minute. The pulse then triggers a one-shot (A_5, A_6, R_1, C_1), which switches the relay on for 2 seconds.

If desired, markers can be generated at 1-hour intervals by connecting segments d and e of the hours-unit display to a similar circuit, the output of which is connected to point D. To differentiate between the minute and hour markers, the relay on-time should be set at 4 seconds by making R_1 in the corresponding circuit 680 kilohms. □

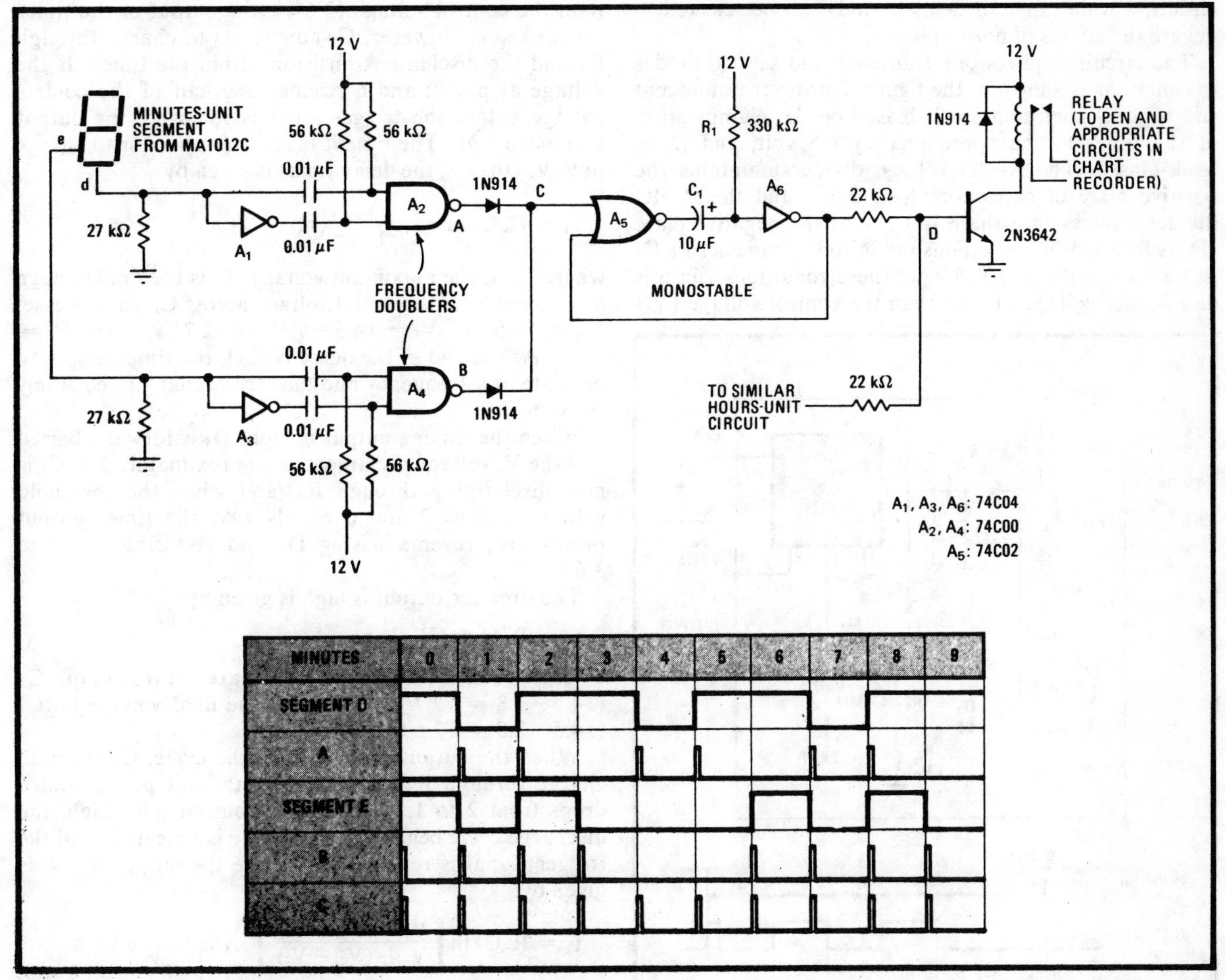

On time. Markers at 1 minute and 1 hour are derived from signals that drive segments d and e of minutes-unit and hours-unit display, respectively, in MA1012C clock module. Relay's on-time is controlled by R_1C_1. Timing diagram details operation for minute markers.

Timer's built-in delay circumvents false alarms

by Kamalakar D. Dighe
American Instrument Co., Silver Spring, Md.

A single 555 timer operating as an astable multivibrator can be used to set its own delay time before turning on, thereby circumventing the most annoying problem encountered with threshold-detector and industrial-alarm circuits—how to distinguish between an actual trigger signal and one that is generated by noise. With this circuit, a wide range of delay periods can be selected to negate the effects of noise spikes.

The circuit requires one transistor and several diodes to function, as shown in the figure. During the quiescent (no-trigger) condition, Q_1 is biased on, D_1 clamps pin 5 of the timer to the approximately 0.8 volt, and D_2 is back-biased. The R_1–R_2 voltage divider maintains the positive plate of capacitor C_1 at 2.5 v and the R_3–R_4 divider ensures that there is 1.5 v at the negative plate (D_3 is forward-biased). Thus the initial voltage across C_1 is 2.5 − 1.5 volts, or 1 v. Under these conditions, pin 6 is at a higher voltage (1.5 v) than the control voltage (V_c) at pin 5, and so the output of the timer at pin 3 is driven low, back-biasing D_4.

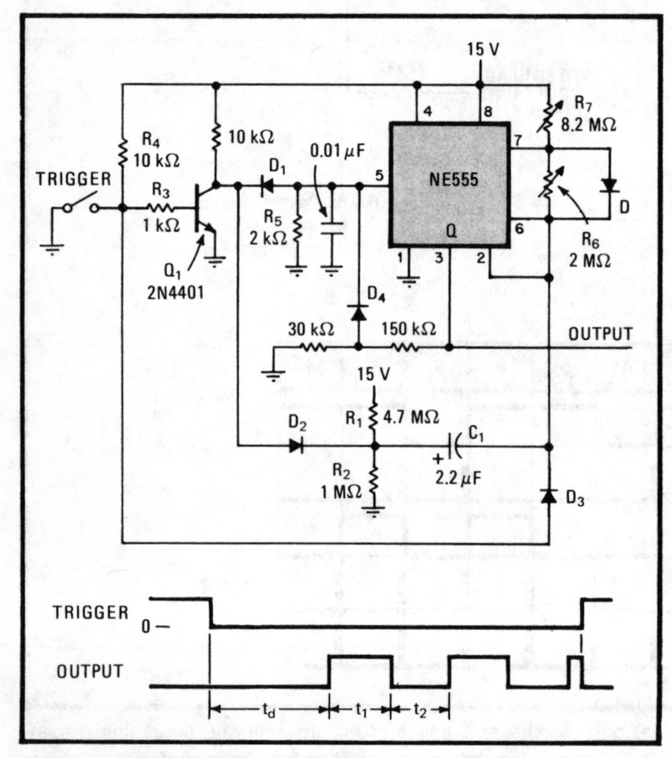

Sure-fire. Circuit can differentiate between real and noise-generated alarm signal by waiting a user-specified time before turning on. Frequency and duty cycle of astable multivibrator can be tailored for optimum response from loudspeaker or other audio-alarm monitor.

The delay period begins when the trigger or alarm input, represented by a switch, is grounded. Q_1 turns off, and the collector voltage climbs to 15 v, back-biasing D_1 and bringing pin 5 to 3.8 v (the voltage depends on R_5, which is in parallel with two 5-kilohm resistors internal to the 555).

Meanwhile, D_2 becomes forward-biased, with the result that approximately 14.5 v is applied at the positive plate of C_1, a step rise of approximately 12 v above the previous voltage. D_3 is now back-biased; the step voltage is therefore transferred to the negative plate of C_1, and pins 2 and 6 of the timer jump to 13.5 v.

Because the threshold voltage (13.5 v) is still larger than the control voltage (3.6 v), the output of the timer remains low. However, C_1 now starts to charge through R_6 and the discharge transistor within the timer. If the voltage at pins 2 and 6 reaches one half of the control voltage before the trigger input is removed, the output will move high. The time it takes the capacitor to charge to ½ V_c (that is, the delay time) is given by:

$$t_d = R_6C_1 \ln \frac{V_m}{V_m - (V_f - V_i)}$$

where V_m is the maximum voltage, V_f is the final voltage on C_1, and V_i is the initial voltage across C_1. In this case, $V_m = 14.5$ v, $V_f = 14.5 - \frac{1}{2}V_c = 12.7$ v, and $V_i = 1.0$ v. At the end of the delay period, the timer begins to oscillate at a frequency and duty cycle that can be set by the user.

When the timer's output is high, D_4 is forward-biased and the V_c voltage is clamped to approximately 2 v. C_1 is now discharging through R_7, and when the threshold voltage at pins 2 and 6 equals 2 v, the timer output moves low, reverse-biasing D_4 and restoring pin 5 to 3.6 v.

The time the output is high is given by:

$$t_1 = R_7C_1 \ln(V_o/V_f)$$

where V_o is the initial on-state voltage of C_1 (14.5 − 1.8 = 12.7 v), and V_f is the final voltage on C_1 (14.5 − 2.0 = 12.5 v).

When the output moves low in the cycle, C_1 starts to charge through R_6. When the voltage at pins 2 and 6 drops from 2 to 1.8 v, the timer output goes high, the discharge cycle begins, and the cycle is repeated until the trigger signal is removed. The time the output is low is given by:

$$t_2 = R_6C_1 \ln \left[\frac{V_m - (V_{f2} - V_i)}{V_m - (V_{f1} - V_i)}\right]$$

where $V_{f1} = 14.5 - 1.8 = 12.7$ v and $V_{f2} = 14.5 - 2.0 = 12.5$ v. □

Four-function calculator times long intervals accurately

by Steve Newman
Hospital of the Good Samaritan, Los Angeles, Calif.

Time intervals extending up to several years can be set with the aid of an inexpensive four-function calculator and an integrated-circuit divider. This timer is accurate to within a few seconds per year. The only requirements for the calculator are that it have at least an eight-digit display, a minus-sign indicator at the left-most position of the display, and a continued-operation mode, which allows subtraction of a given number repetitively by depressing a single key (usually the equal-sign button).

The time interval is set by keying in the desired value, expressed in seconds. For instance, if the timer is set to go off in 3 days, 10 hours, 54 minutes, and 33 seconds, the number 298,473 is entered. (With an eight-digit display, an interval of 10^8 seconds, or 38 months, can be set.) Next, the -1 command is keyed in. The calculator is now programmed to subtract 1 from the displayed value each time the equal-sign key is activated and then to display the new value.

To decrement the count at proper intervals, a precise time base is required, and the calculator's keyboard pins must be made accessible. In this case, a divide-by-60 counter, the HEP C4055P, is used to derive a 1-hertz signal from the 60-Hz power line, as shown in the figure. This clock drives the transistor switch connected across the pins corresponding to the equal-sign key, which are closed when that key is depressed in normal operation.

The timing interval begins when the time base is activated. The displayed number is decremented for each pulse emanating from the C4055P, so that after N seconds the calculator will display zero, and after $N+1$ seconds it will display -1.

The minus-sign output, which is part of the display circuitry, provides a convenient way to detect the end of the set timing interval. Here, the g-segment pin corresponding to the minus-sign key is brought to one input of a NAND gate, G_1, as is the lead corresponding to the left-most digit of the display. An identical gate, G_2, follows, so that the entire logic function is that of an AND gate. This function is suitable for driving positive-logic circuits or devices.

Pulses emanate simultaneously from both the minus-sign pin and the last-digit driver port when the calculator first displays a minus value, and for every result thereafter. The first pulse may thus be used to turn on an alarm. Because many calculators scan the display at a frequency of a few hundred hertz, an audible output can be obtained if a loudspeaker is connected to G_2. □

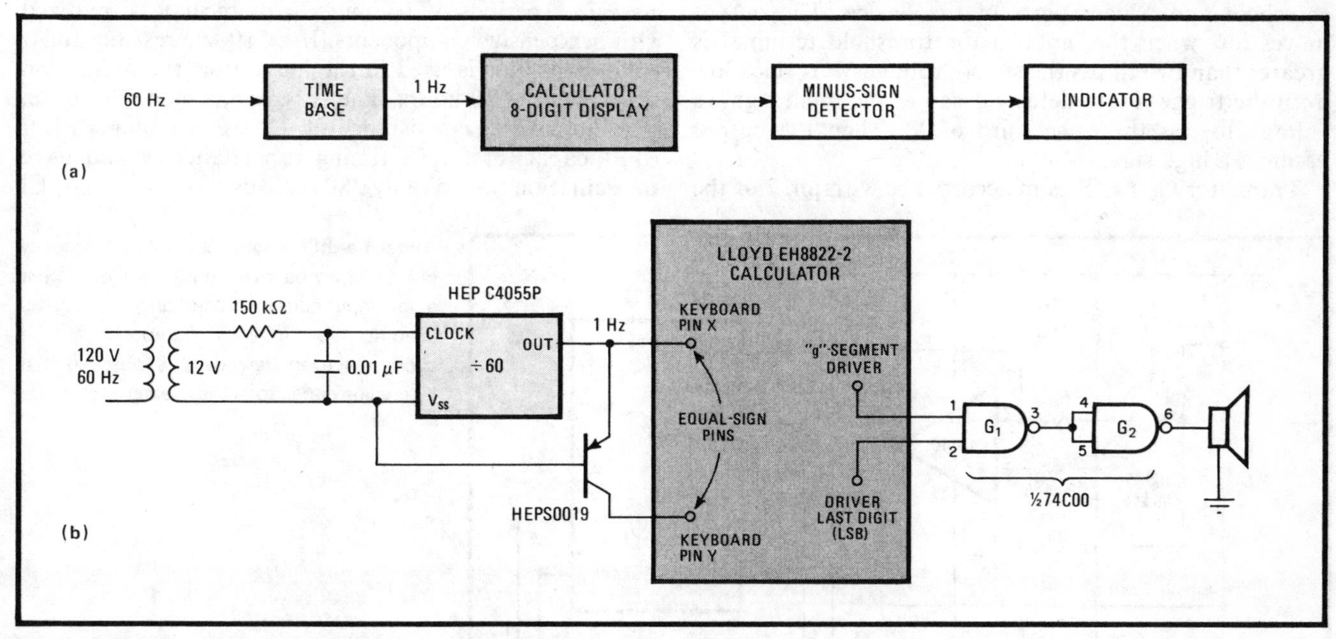

On the button. Precision, low-cost timer (a) can be built with most any four-function calculator and integrated-circuit divider (b). Keyboard pins corresponding to equal-sign button must be accessible. Circuit is accurate to within a few seconds per year.

High-impedance op amp extends 555 timer's range

by Ronald Zane
University of California, Los Angeles, Calif.

The period of oscillation of the 555 timer can be increased 20 times or more if the timing components are replaced by a feedback loop containing a transistor and a very-high-impedance input operational amplifier configured as an integrator. The circuit is an inexpensive way of generating timing periods of hours or days to control industrial processes, to turn on lights in the home for burglar protection, and for like applications.

As shown in the figure, resistor R_5 and capacitor C combine with the CA3140T op amp to make the integrator that controls the period of oscillation in the 555. The very low offset current of this op amp (typically 3 picoamperes but no greater than 30 pA) permits accurate integration of very low input currents (100 pA). Thus it ensures excellent control over the actual oscillation times.

The timer, operating in its astable-multivibrator mode, produces a change of state at pins 3 and 7 each time the input signal requirements are met at the threshold and trigger ports of the device. The output moves low when the input at the threshold terminal is greater than two thirds the supply voltage V_s. It stays low until the trigger input detects decay of the input signal's voltage to less than one third of V_s. Then the output assumes a high state.

Transistor Q_1 switches in accordance with pin 7 of the 555; point E_1 will be at ground when Q_1 is on, and at $V_s/2$ when Q_1 is off, because of the voltage divider made up of R_1 and R_2. The voltage at E_2, $V_s/4$, is determined by the divider made up of R_6 and R_7 (R_1, R_3, and R_4 in series hardly affect the calculation).

When Q_1 is on, the current through R_5 at the inverting input of the op amp is thus:

$$I = \frac{(V_s/2 - V_s/4)\,R_4}{R_5(R_3+R_4)} = \frac{V_s R_4}{4R_5\,(R_3+R_4)}$$

and when it is off, the current is:

$$I = \frac{(0 - V_s/4)\,R_4}{R_5\,(R_3+R_4)} = \frac{-V_s R_4}{4\,R_5\,(R_3+R_4)}$$

The magnitudes of these currents are the same. The integrator then operates on this input current. Because its output is $E_3 = dV_s/dt = I/C$ by definition, and because E_3 switches between one third and two thirds of the supply voltage each cycle, the oscillator's period is:

$$t = \int_{1/3}^{2/3} \frac{C}{I}\,dV_s \frac{8(R_3+R_4)\,R_5C}{3R_4} = 50\,R_5C$$

The low offset current of the op amp allows R_5 to assume values in the hundreds of megohms. Even when R_5 equals 110 MΩ, the current through it is 1.8 nanoamperes, far exceeding the op amp's offset current.

If R_5 is 110 MΩ and C is 1 microfarad, the 555's period is 5,500 seconds. Since R_5 and C can be increased, periods of 10 hours or more may be realized with inexpensive components. If a 510-MΩ resistor and a 10-μF capacitor is used in the integrator, the oscillation period will be 70 hours. The bare 555, which may use a maximum timing resistance of 20 MΩ, coupled with a 10-μF capacitor for its timing capacitance, would have an oscillation time of only 280 seconds. □

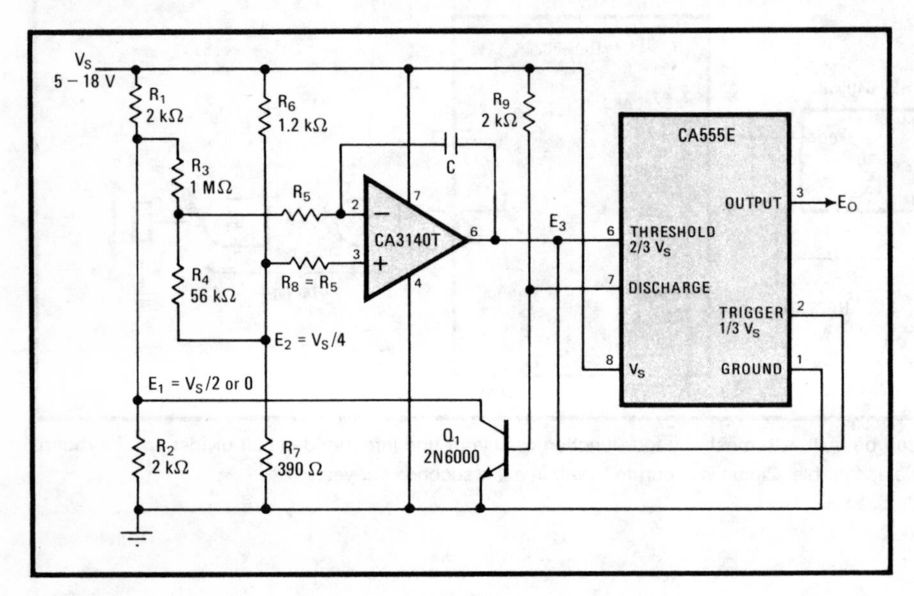

Time-magnification. Oscillation frequency of a 555 may be lowered 20 times or more if a low-input-current, integrating op amp is used to replace timing components. One oscillation every few days is possible if a high-value capacitor, C, is used in integrator.

Timer IC circuit separates rep rate and duty cycle control

by Arturo Sancholuz
Laboratorio Nacional de Hidraulica, Caracas, Venezuela

Combining both halves of a 556 dual timer with an operational amplifier in this simple circuit enables independent control of the output frequency and the duty cycle. The frequency is adjustable throughout the normal 10-hertz-to-10-kilohertz range of the 556, and the duty cycle is selectable from 1% to 99% of the total waveform period.

As shown in the figure, one half of the 556 (A_1) is connected as an astable multivibrator, oscillating at a frequency given by $f = 1.4/(R_1 + R_2)C$. This oscillator is the frequency-governing element in the circuit.

The negative-going edge of signal v_1 periodically triggers timer A_2, which operates as a monostable multivibrator. An exponential ramp emanating from the threshold port of A_1 drives A_2 through the 531 op amp.

The duty cycle in this timer is determined not by external resistance-capacitance elements, but by the voltage on the threshold port. The output of A_2 will remain high if the threshold voltage stays below two thirds of the supply voltage, V_{cc}. This circuit can generate a dc offset voltage at the port to modify the threshold-switching time.

The voltage at the threshold port is determined by the two input voltages, v_2 and v_3, at the summing junction. Thus:

$$v_4 = v_2\left(\frac{R_5}{R_5+R_6}\right) + v_3\left(\frac{R_6}{R_5+R_6}\right)$$

Voltage v_2 is an exponential ramp resulting from charging C through resistances R_1 and R_2. The boundaries of the signal, determined by the internal comparators of A_1, lie between $\frac{1}{3}V_{cc}$ and $\frac{2}{3}V_{cc}$.

The 531 op amp is a buffer for the high-impedance A_2 signal and prevents current from flowing into the timing port, which could charge C from V_{cc} through R_5 and R_6. Dc voltage v_3 can be varied from $\frac{2}{3}V_{cc}$ to V_{cc}. Thus it can be seen that R_3 will determine how large a dc voltage is superimposed on v_2, thereby controlling the duty cycle. Since there are no feedback loops linking A_1 and A_2, it is clear that frequency and duty cycle adjustments are independent. □

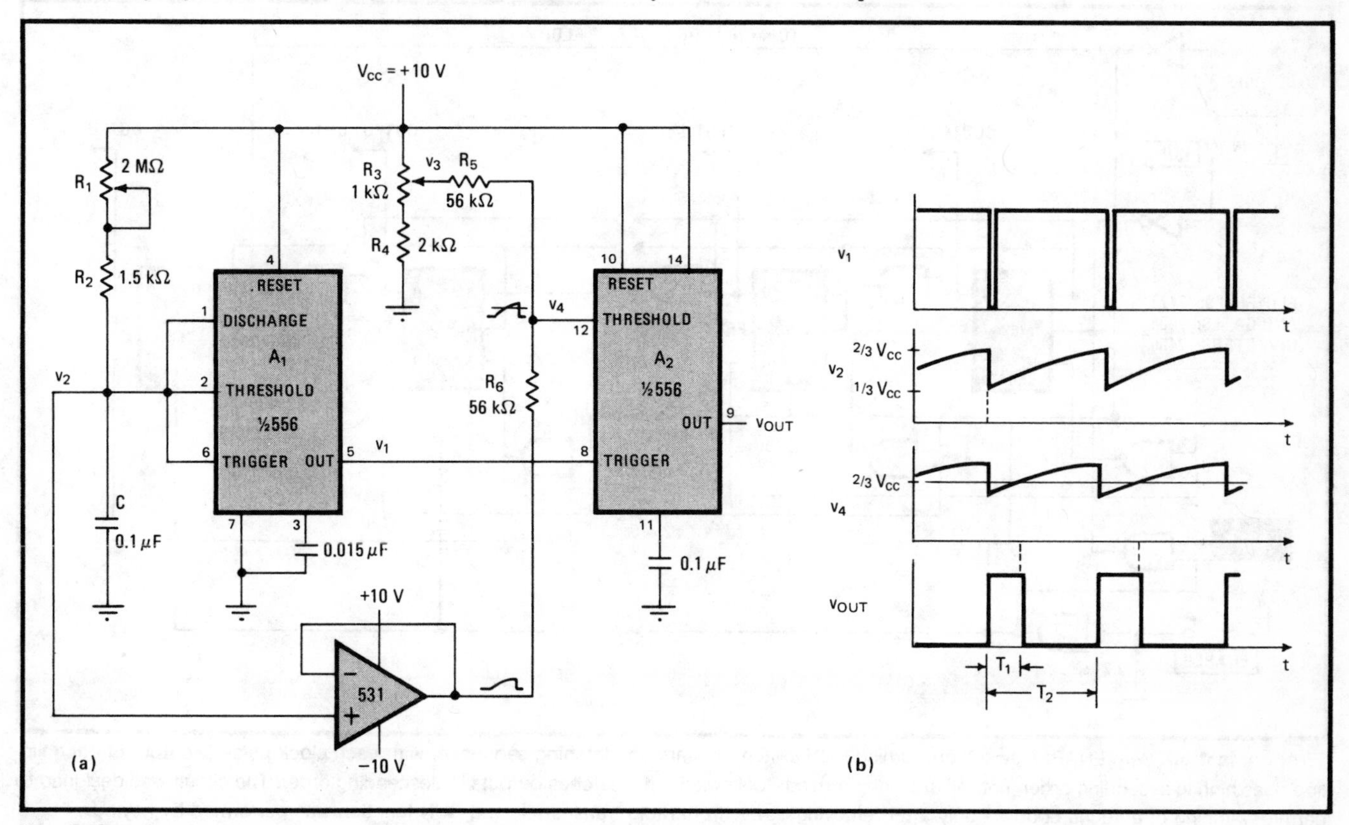

No relation. A_1 runs at frequency set by R_1. But duty cycle is selected by R_3, which controls signal offset at threshold port of A_2. No feedback loops link A_1 and A_2, thereby ensuring independent adjustment of rep rate and duty cycle. Timing diagram details operation.

Up/down latching sequencer keeps order

by Marc D. Williams
Standard Telephone Co., Cornelia, Ga.

Circuits sometimes need energizing in a prescribed order and de-energizing in the reverse order, on a first-on, last-off basis. At a high-power radio station, for instance, the antennas are switched for transmitting and receiving, and the linear amplifier must be turned off before the switching and on afterwards if its output section is not to be destroyed. A sequencer that performs this task can be built with transistor-transistor-logic circuits, having a sequence rate controlled by an external TTL clock and a completely expandable number of outputs.

The circuit uses AND gating and inhibiting inputs on the control flip-flops to execute the proper sequence. As shown in the schematic, closing the ENABLE switch allows clock pulses to pass to the flip-flops, which are inhibited by the low level at their J inputs. Turning on the RUN switch allows the first flip-flop, FF_1, to toggle upon receipt of the next clock pulse, and the Q output of FF_1 enables FF_2 to toggle with the following clock pulse. The $\overline{Q}$ output of FF_2 inhibits FF_1, while the Q output of FF_2 allows FF_3 to toggle, and so on down the line to the last flip-flop, FF_n, as long as the RUN switch remains on. The circuit now is stable with all the Q outputs high.

Turning the RUN switch off starts the unlatch sequence by enabling FF_n to return to its off state due to the high K and low J input levels. The $\overline{Q}$ output of FF_n enables the preceding flip flop FF_{n-1} to return to its low state, and so on back to FF_1. Opening the ENABLE switch simultaneously resets all the flip-flops and prevents further toggling.

A three-stage sequencer will require five TTL packages, excluding the clock circuit. Driving displays or relays requires buffering of the Q outputs on the flip-flops.

Although the sequencing may be extended to many stages, the inverters may have to be buffered to drive the AND gates for many stages. □

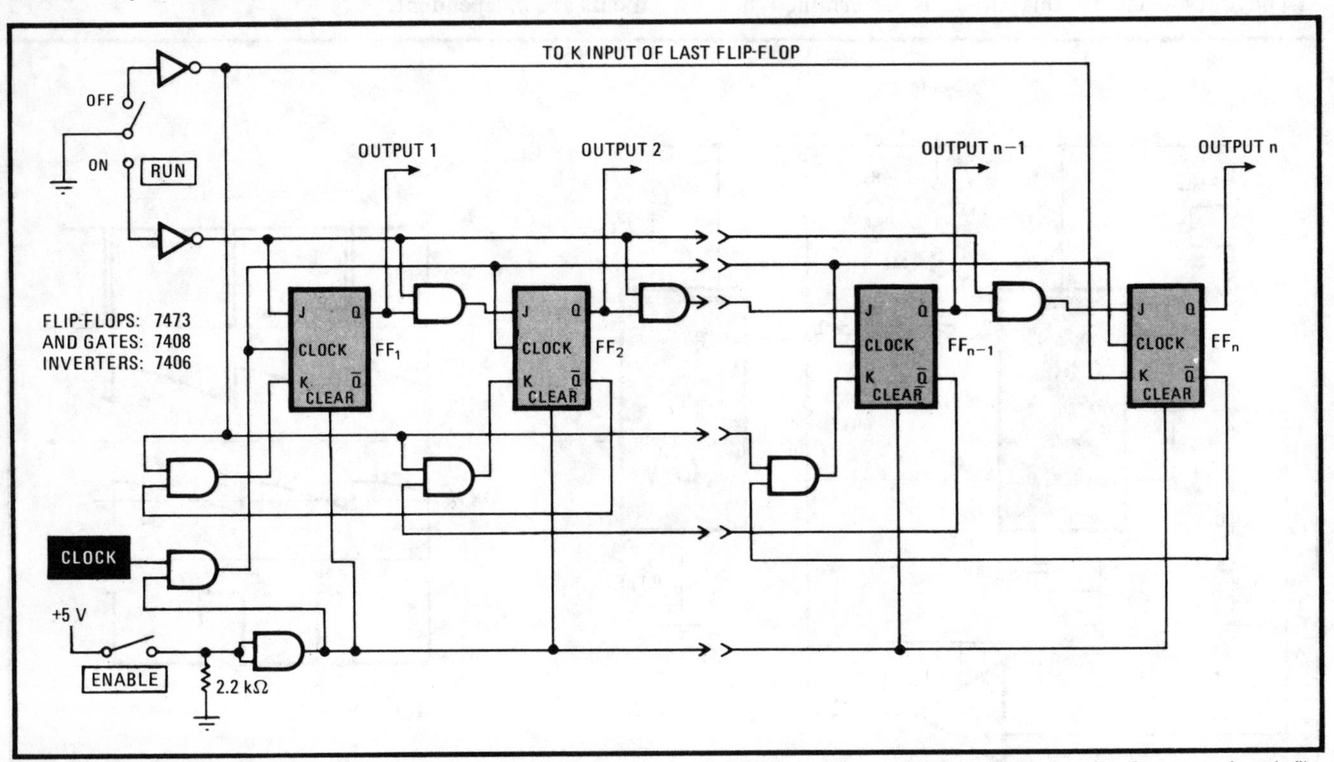

First on, last off. With ENABLE switch on, turning RUN switch on starts the latching sequence: with each clock pulse Q output of each flip-flop goes high in ascending order, until all are high. Turning RUN switch off unlatches outputs in descending order. The circuit was designed to permit switching of antenna couplers only after removing power from radio transmitter, and RUN function was performed by keying.

Timer extends life of teletypewriter

by Bob Piankian
Massachusetts General Hospital, Boston, Mass.

The mean time between failures of a teletypewriter terminal can be extended by a circuit that turns on the motor only when information is being received or transmitted. In that way, the mechanical wear and destructive temperatures are minimized in a motor that is continuously powered-up but seldom active. Moreover, the cost of power to the terminal is significantly reduced. A teletypewriter motor consumes about $50 worth of electricity a year. If it is only active 10% of the time, this circuit could save $45 each year for each machine.

A logic circuit for this application is shown in Fig. 1. When no characters are being transmitted over the current loop, the MCT2 opto-isolator keeps the trigger input of the 74121 monostable multivibrator at logic 0. If characters are sent to the teletypewriter, or if the teletypewriter BREAK key is depressed, the current loop is broken, and the 74121 is triggered. This triggering fires the 555 timer, activating relay K through Darlington pair Q_1 and Q_2, thereby turning on the motor of the teletypewriter.

The time the motor remains energized, T, is given in seconds by $1.1(R_5 + R_6)C_4$, where C_4 is expressed in microfarads and R_5 and R_6 in megohms. Resistor R_6 adjusts the time period to between 2.5 seconds and 20 minutes.

If another character is received or the keyboard is used during the time span that the motor is energized, the monostable resets the timer and then retriggers it, keeping the teletypewriter on for another time period, T. To discharge C_4 completely during the reset process, R_4 C_3 should be greater than 2.4 C_4, where R_4 is in kilohms and C_3 and C_4 in microfarads. D_3, R_7, and C_6 prevent the relay from turning off during the 0.3-second reset operation.

Standard current loops of 20 or 60 milliamperes can be used. Diode D_2 protects against accidental reversal of line polarity, and C_2 provides immunity from noise on the line. Defeat switch S permits normal use of the teletypewriter terminal.

Figure 2 shows the power supply for the circuit, which uses a standard transformer already in a KSR-33 Teletype. The entire circuit was built on a 3-by-3½-inch printed-circuit board and installed in many terminals. Total parts cost is about $15 for each unit—and they pay for themselves in just a few months.

With this device installed, the power switch is left in the ON position. The system software should be changed to send a nonprinting character to the teletypewriter 1 second before actual information output so that the motor can come up to speed. In a half-duplex system, hitting the BREAK key starts up the teletypewriter locally. For this feature to work in a full-duplex system, the software must echo the break to the machine. Turn-off time-delay T can be changed as desired to avoid needless turn-on/turn-off cycles. □

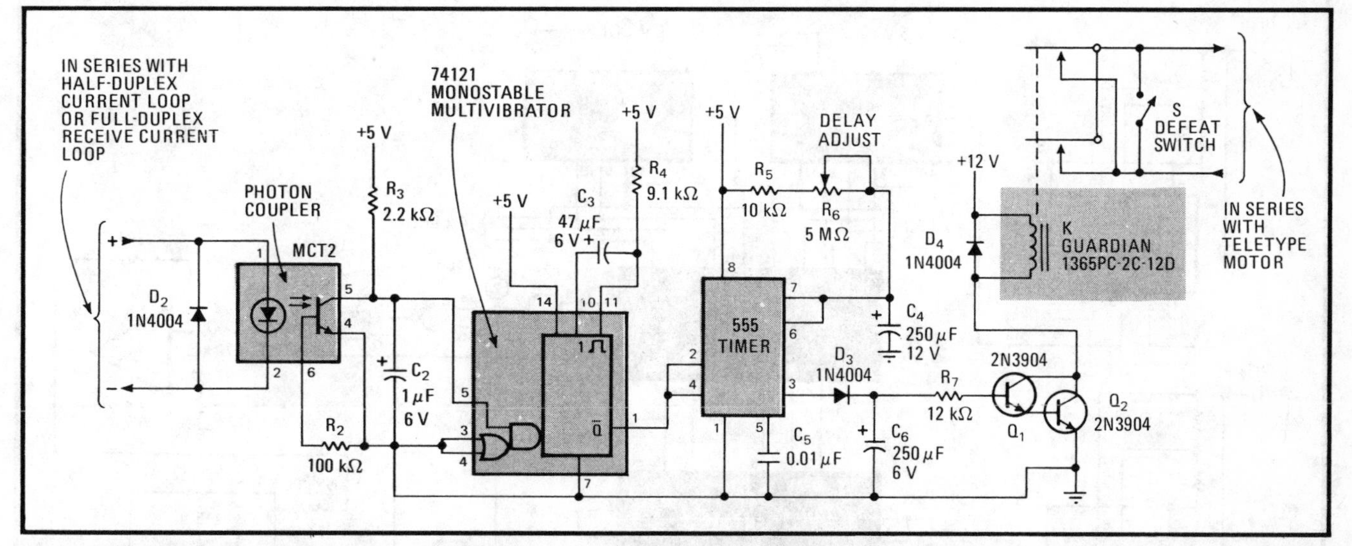

1. Turn-off circuit. Teletypewriter motor remains off unless information is being received or transmitted by terminal. When current loop is broken by transmission or reception of a character, the monostable fires and triggers the timer, which turns on the motor and keeps it on for a drop-out time of up to 20 minutes. Use of this $15 circuit saves power and prolongs teletypewriter life.

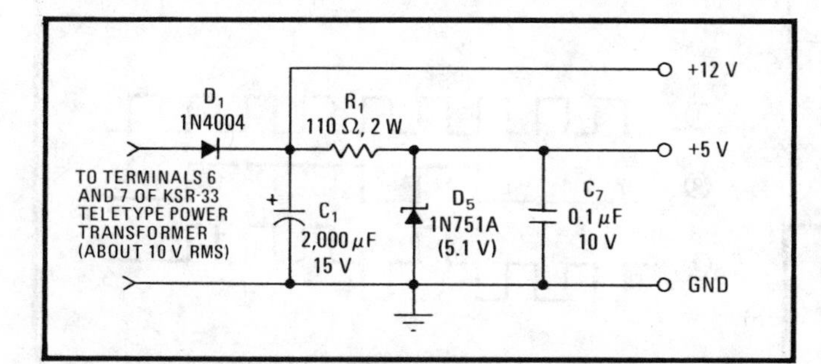

2. Power supply. By using a few additional parts, the 5-volt and 12-volt supplies required for the circuit of Fig. 1 can be obtained from a power transformer already present in the KSR-33 Teletype terminal.

Time-encoding system formats cassette tapes

by Fathi Saleh and Benham Bavarian
Abadan Institute of Technology, Abadan, Iran

Using operational amplifiers and standard digital elements to place 1-minute markers on a stereo cassette tape, this formatter is extremely useful when implemented in simple data-recording and control systems. Also shown here is a rudimentary decoder circuit, and the information required to utilize the markers for controlling any cassette's stop, play, fast forward or rewind functions for more advanced systems.

Figure 1a gives the block diagram of the tape formatter, which was used with the JVC MC-1820R recorder. The timing signals are recorded at the expense of one track of the tape, leaving the other track for monaural sound or data recording. It might have been possible to add a thin head to the recorder in order to record marker signals and at the same time make possible two-channel data recording, but this was not tried.

As shown in Fig. 1b, the recording system uses a 555 timer to generate a pulse every 60 seconds, two 7490 decade counters that are stepped by the timer to yield required 8-bit patterns, eight light-emitting diodes to monitor the number generated, a parallel-to-serial converter (74151) that is clocked at 5 kilohertz by a second timer, a 4-bit counter (7493), and shaping circuitry.

Pulses emanating from the 1-minute timer advance the 7490s. The output of these devices is introduced to the 74151, which is being clocked at a 5-kHz rate

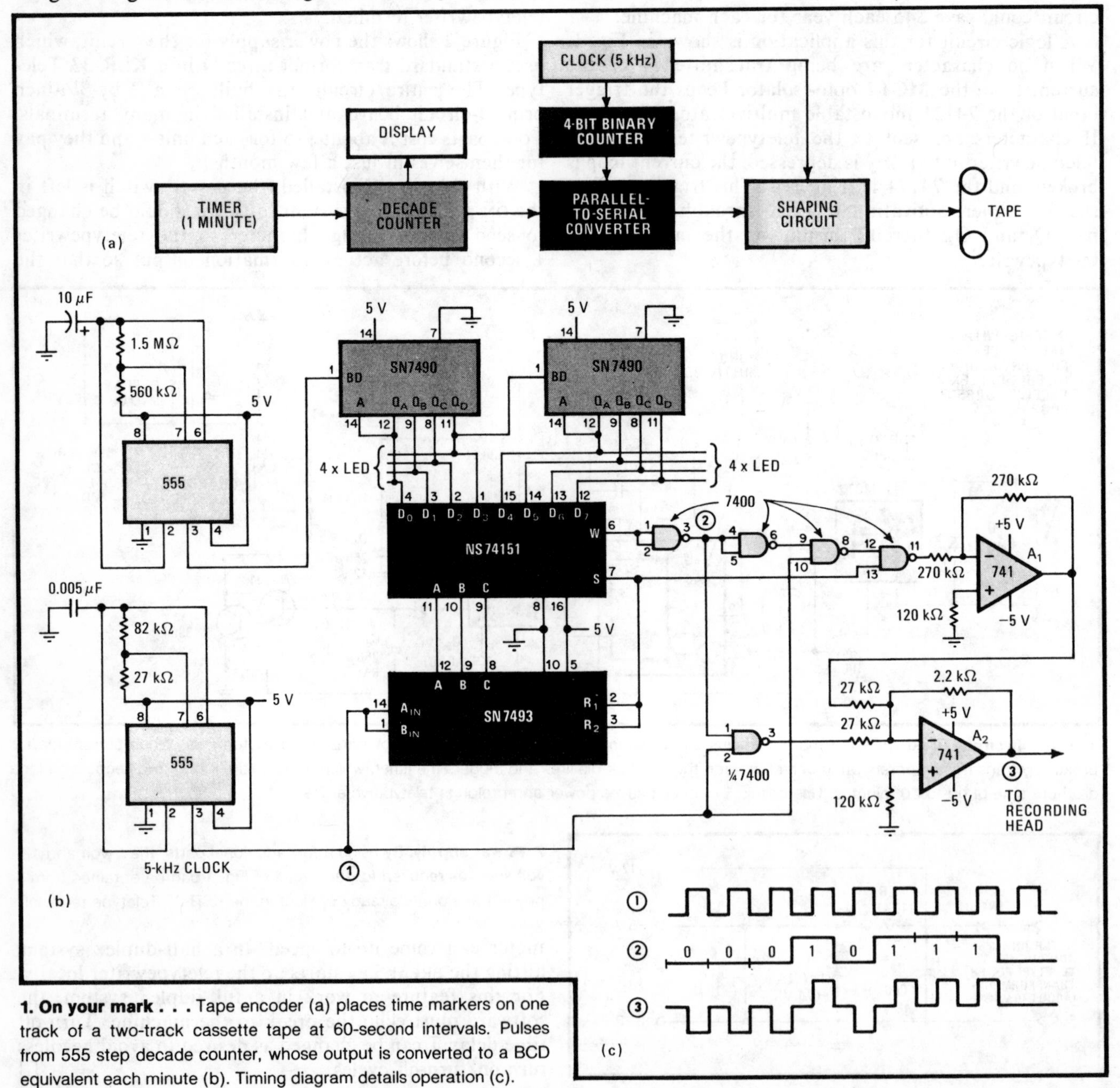

1. On your mark . . . Tape encoder (a) places time markers on one track of a two-track cassette tape at 60-second intervals. Pulses from 555 step decade counter, whose output is converted to a BCD equivalent each minute (b). Timing diagram details operation (c).

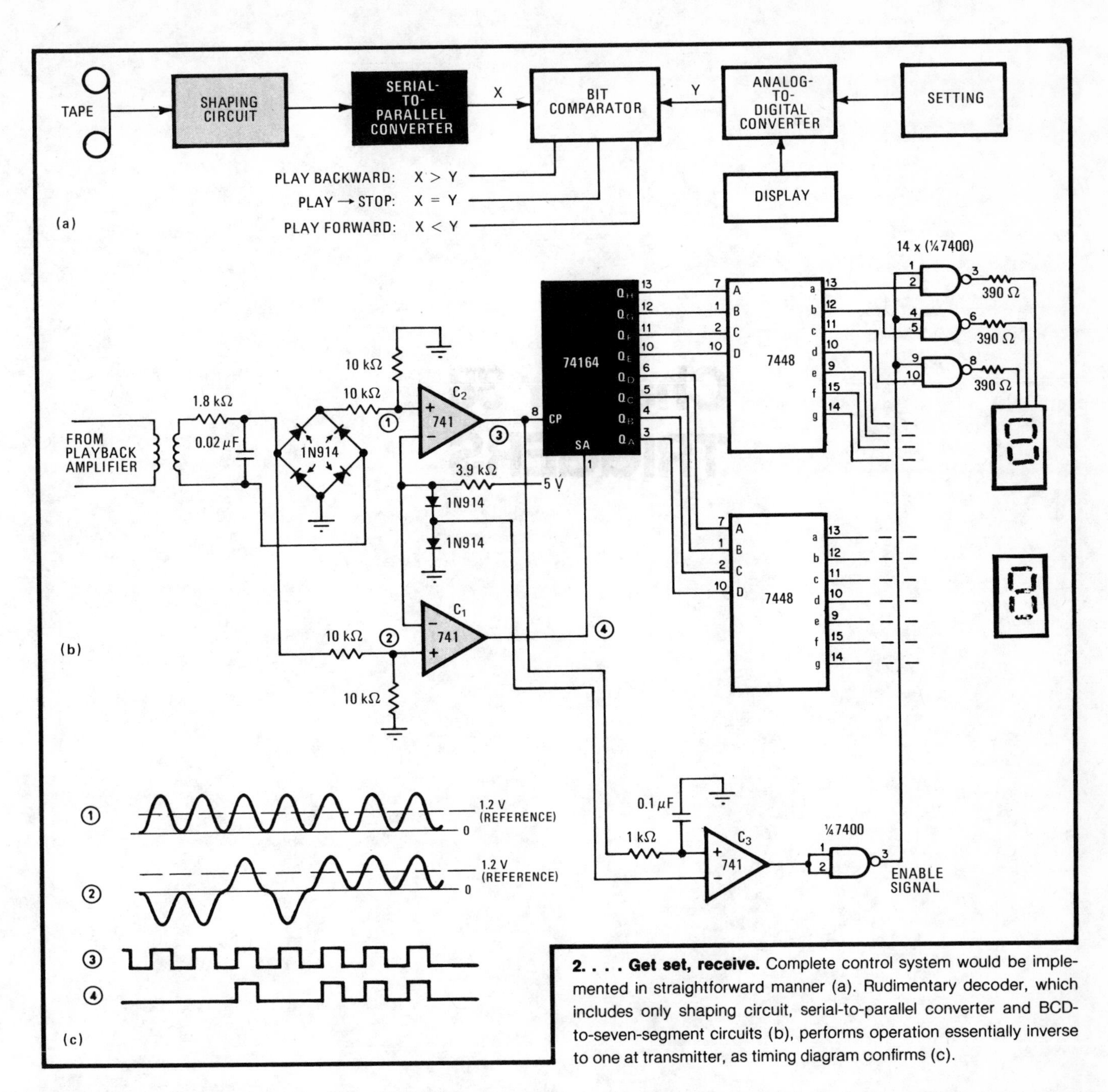

2. . . . Get set, receive. Complete control system would be implemented in straightforward manner (a). Rudimentary decoder, which includes only shaping circuit, serial-to-parallel converter and BCD-to-seven-segment circuits (b), performs operation essentially inverse to one at transmitter, as timing diagram confirms (c).

through the 7493. The output of the 74151, after passing through a NAND gate, appears as shown in the timing diagram (Fig. 1c).

Recombining the 5-kHz clock with other summing-circuit (A_1) signals yields an output at A_2 having a 50% duty cycle. This output returns to a logic 0 when the output from the 74151 is a logic 1 and moves high for a logic 0 output from the 74151. The signal is a two-digit binary-coded-decimal character which advances from 0 to 60 minutes. The return-to-zero format is used here so that one may differentiate between the zero-state and the no-signal levels. This format also simplifies the design of the decoding circuits.

The playback (decoder) circuit is shown in Fig. 2. The block diagram (Fig. 2a) shows a complete control system, but the system, although straightforward, encompass more circuitry than can be shown here.

The basic decoder is shown in Fig. 2b. Passing through the shaping circuit is the output from the tape recorder. This input signal is compared with a 1.2-volt reference at C_1, enabling the original bit pattern recorded to be recovered at its output. Meanwhile, the original clock pulses are regenerated at C_2 by comparing the rectified 1s and 0s of the bipolar return-to-zero input signal with the 1.2-v reference. The bit pattern output from C_1 is connected to the 74164 serial-to-parallel converter, which is stepped by the output of C_2.

Thus the output of the 74164 is a binary-coded-decimal equivalent of the marker-input's value. The seven-segment displays that follow indicate the time, in minutes, corresponding to the last marker detected. □

Chapter 38
TRIGGERS

Dc-coupled trigger updates quickly

by Andrzej M. Cisek
Electronics for Medicine, Honeywell Inc., Pleasantville, N. Y.

Two operational amplifiers and a few components build this broadband trigger, which is absolutely immune to the problems created by dc offset and base-line wandering. A self-adapting stage for maintaining a constant level of hysteresis, as well as dc coupling at the input, permits a low-cost yet superior performance over a wide range of input signals that makes the circuit very attractive for biomedical applications.

Input signals are applied to the LM307 amplifier, A_1, through charge-storage network $D_1D_2R_1R_2C$ (a). Diodes D_1 and D_2 quickly charge capacitor C to the peak of the input voltage (less one diode drop) and minimize C's discharge until the signal falls more than two diode drops below the peak voltage.

The output of A_1 is then compared with the original signal at A_2. The amount of hysteresis is set by constant-current source Q_1 and Q_2, two p-channel field-effect transistors wired as diodes in A_2's positive feedback loop, and R_3. Q_1 and Q_2 ensure that the level of hysteresis is maintained virtually constant for any dc offset at the input. Resistor R_1 provides a zero level for input signals whose amplitude is smaller than one diode drop, and R_2 protects the input-signal source.

Such an arrangement has a reliable triggering level and responds fast, its speed being limited only by time constant $(R_{D1,2}+R_2)C$. This time constant can be easily adjusted to meet the requirements of practically any application.

As for the circuit's use in medical electronics, consider the two cases illustrated in (b) and (c), where a cardiac signal representing heart rate is superimposed on the respiratory (breath) signal. Depending on the trigger level, either heart rate (b) or respiration rate (c) may be counted. □

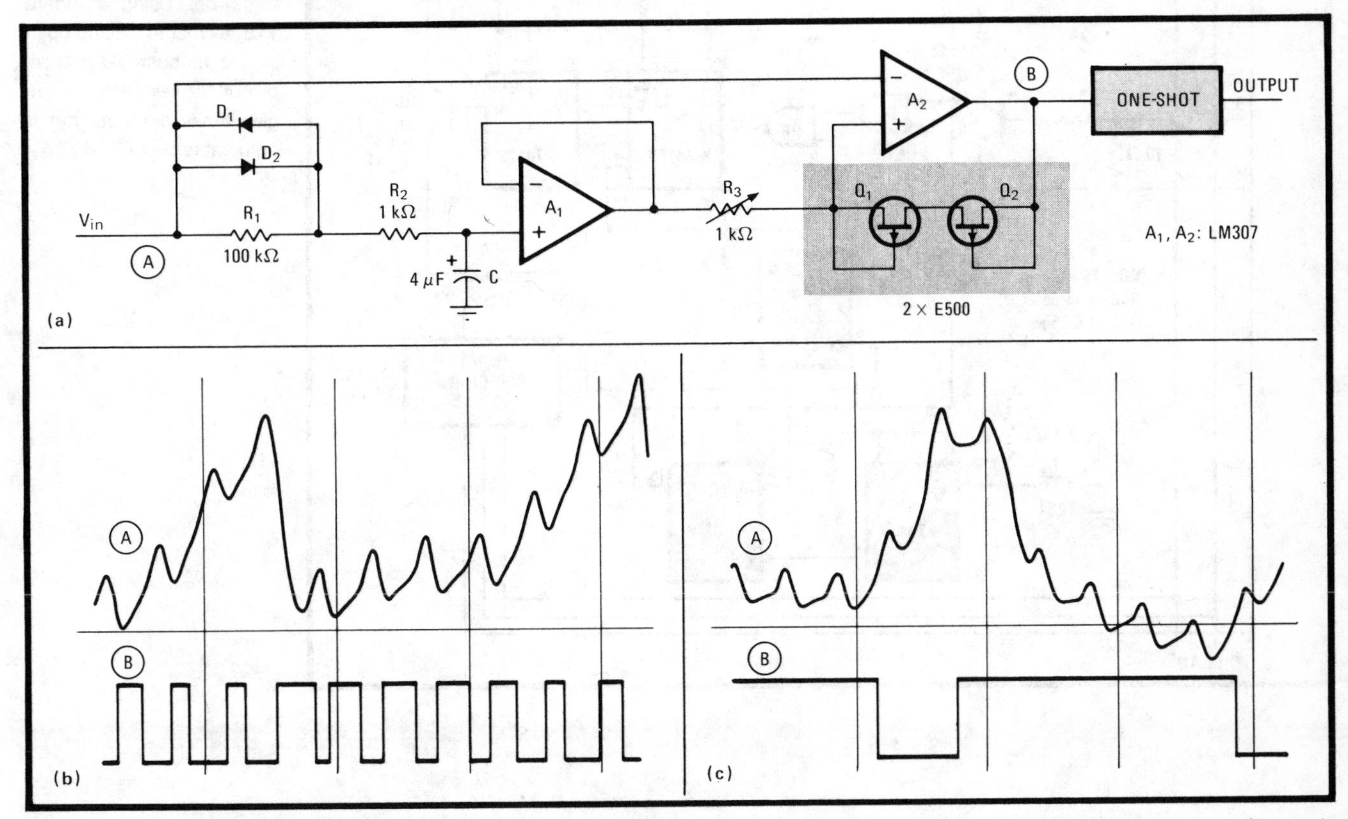

Flat. Trigger (a) maintains switching level over a wide band of frequencies and input-signal amplitudes. Operation is independent of input dc offset and problems created by baseline shift. Directly coupled arrangement at input contributes to circuit's high-speed response. In biomedical application, trigger switches on superimposed signals produced by heart rate (b) and respiration rate (c).

Triac trigger sets firing point over 0° to 360°

by Tagore J. John
Meerut, Uttar Pradesh, India

Because it selects the point on the cycle at which a 60-hertz line input will fire a triac, this circuit is very useful for checking the transient response of overload protection systems and in many other power-control applications. Standard TTL elements are used here to switch the triac on when commanded, over any point in the range of 0° through 360°.

Generally, the 60-Hz voltage on the secondary winding (T_1) of the current transformer turns transistor Q_1 on and off at every positive and negative transition of the power-line input, respectively, as shown in the figure. During these times, the output of gate G_2 is always high.

When the test button is depressed, one-shot A_1 fires to set flip-flop A_2. The resulting low output from G_3 that occurs when Q_1 is off then sets A_3, so that Q_1's output passes through to G_2 and one-shot A_4.

A_5 turns on when the Q output of A_4 moves low. The delay between the positive-going transition of the input wave and the firing of the fault-switching device may be selected by adjusting R. Note that because the time constant of A_4 is slightly greater than the period of the input waveform, the triac will not be pulsed until after the passage of one cycle.

A calibrated potentiometer for R is suggested. The trigger-point markings will be accurate to within a few degrees throughout the range if an oscilloscope is used to calibrate the circuit. □

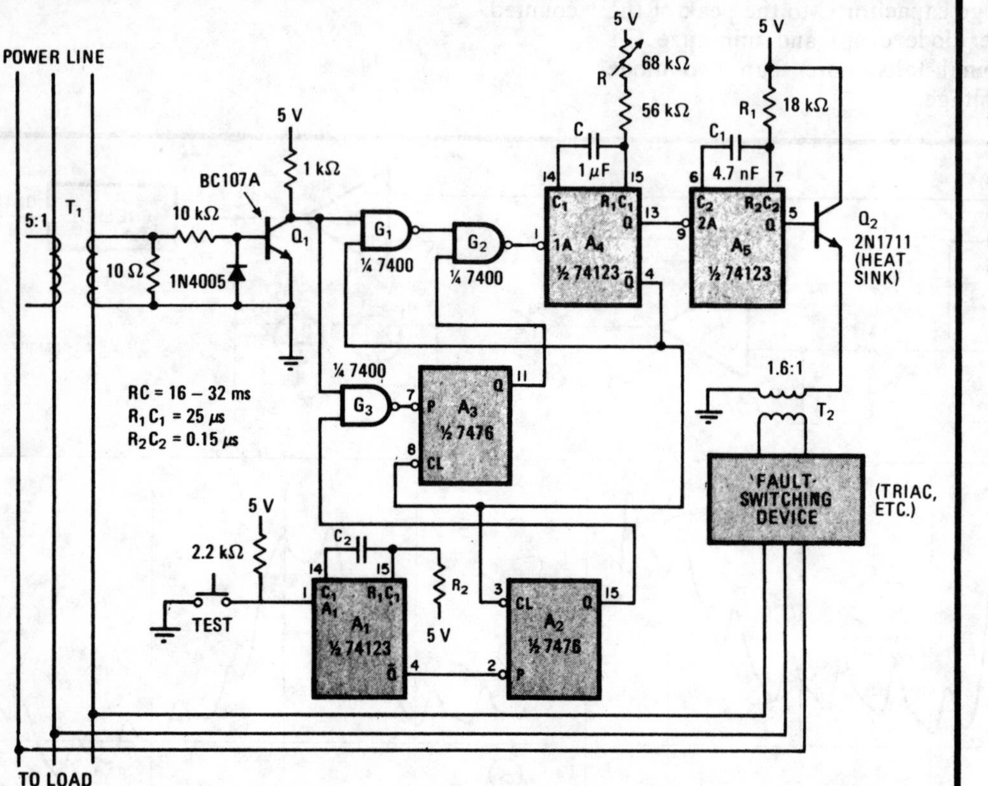

Power angle. Trigger circuit sets point on ac cycle at which triac or other power-controlling device turns on. Using standard TTL elements, including a simple monostable multivibrator for selecting fire angle, it has a range that is adjustable over 0°-360°.

C-MOS triac trigger cuts parts count

by Hul Tytus
Tytus & Co., Cincinnati, Ohio

Integrated circuits made to trigger triacs often consume large amounts of power and require biasing schemes substantially increasing the number of components in the circuit facing the ac load. But by using a C-MOS operational amplifier, the circuit shown can deliver a peak current of 100 milliamperes to the triac, draws an average current of only a few milliamperes and uses a minimum of parts.

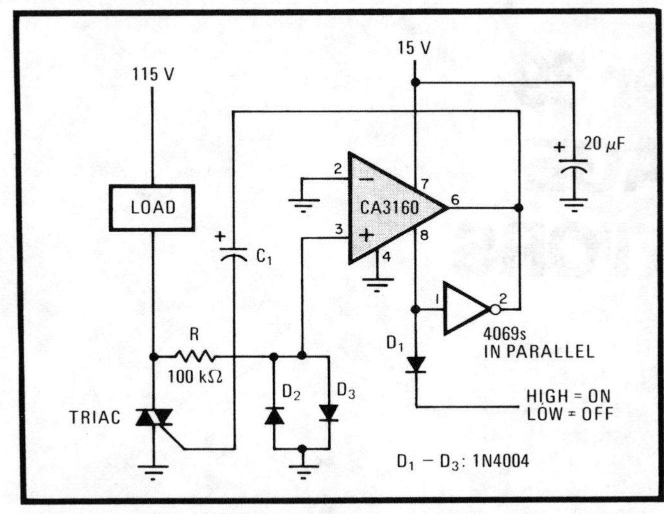

Diminutive driver. C-MOS operational amplifier, operated from a single supply, minimizes power drain and parts count in triac trigger that can source 100 milliamperes. C-MOS inverters may be placed in parallel with op amp for circuit to drive heavier loads.

Circuit operation is based on the fact that the output of a C-MOS device acts as a current source or sink. As the ac-input voltage rises up through zero, the CA3160 generates a positive-going current pulse of a preset magnitude and charges C_1, which determines the triac's pulsing time. As the ac line voltage completes a half cycle and drops through zero, C_1 discharges through the op amp, which now serves as a current sink, so that the triac is again fired.

Note that several 4069 inverters can be placed in parallel with the op amp's strobe (pin 8) and normal outputs to drive heavier loads. Pin 8 and D_1 may also serve as a control terminal, enabling operation with 4049s permanently wired into the circuit.

The maximum rated trigger current of the triac should equal the minimum current the 3160 and 4069s are capable of sourcing. The pulse time for the triac—typically measured in microseconds—should equal the time necessary for the ac supply to generate the necessary conduction voltage across the triac, plus t_T max, the maximum trigger time for the triac. The value of capacitance required for a given trigger current, I, triac pulse time, t, and maximum trigger voltage, V_T, is given by:

$$C_1 = I\,t/(0.75\,V_{dd} - V_T)$$

where V_{dd} is the op amp's supply voltage. □

Chapter 39
VOLTAGE REGULATORS

Low-drain regulator extends battery life

by T. C. Penn
Texas Instruments Inc., Dallas, Texas

Stable regulation of battery power supplies once required dumping current into a zener diode or sacrificing several volts and milliamperes to a three-terminal regulator. Powering up voltage-controlled oscillators and other supply-sensitive circuits with batteries becomes practical when the circuit described here regulates the supply voltage. Providing ±1-millivolt regulation with loads ranging from 5 to 55 milliamperes, the circuit maintains accuracy even when the battery voltage is only 50 mV higher than the regulated output, and it draws rather less than 1 mA of quiescent current.

As evident in the schematic, the simplicity of the circuit belies its performance. The regulating operational amplifier A gets its reference voltage from the bridge made up of R_1 to R_3 and a light-emitting diode. The drop across the LED, which varies from device to device, is about 1.4 V and it remains relatively stable over a wide temperature range. Changing the temperature from 20°C to 40°C varies the regulated-output voltage only about 1 mV. If greater stability is required, a low-voltage reference diode or even a two-terminal regulator may be substituted for the LED in the reference leg of the bridge.

The op amp controls Q_2 and Q_1 to maintain a zero offset between its inverting and noninverting inputs, thus establishing that the voltage drop across R_2 is equal to the drop across the LED. The regulated-output voltage is the sum of the drops across R_1 and R_2; this can be shown to be $V_{REG} = V_{LED}(R_1/R_2 + 1)$. With the values that appear in the schematic and if $V_{LED} = 1.4$ V, the regulated-output voltage V_{REG} by substitution becomes 1.4(22 kilohms/10 kilohms + 1) = 4.48 V. If the voltage must be trimmed to a particular value, a 5-kilohm potentiometer can be inserted between R_1 and R_2, with its wiper connected to the op amp.

Although the op amp is powered by the bus it regulates, the circuit exhibits no start-up problems. As soon as power is applied, Q_2, biased on by the current through R_6, turns on Q_1, supplying power to the op amp and reference bridge. Q_1 and Q_2, in conjunction with R_4 and R_5, form a feedback pair with a voltage gain of about 3. For that reason, a high-gain op amp may be used with no danger of oscillation.

The op amp is half of a dual package, although it could equally well be one fourth of a quad package. In either case, the regulated bus can power all the op amps in the package, so long as the total current required does not exceed 55 mA. The current drawn by the TLO 22 dual op amp alone is about 800 microamperes. Also pictured in the schematic are optional filtering and decoupling components, C_1, C_2, and R_7. □

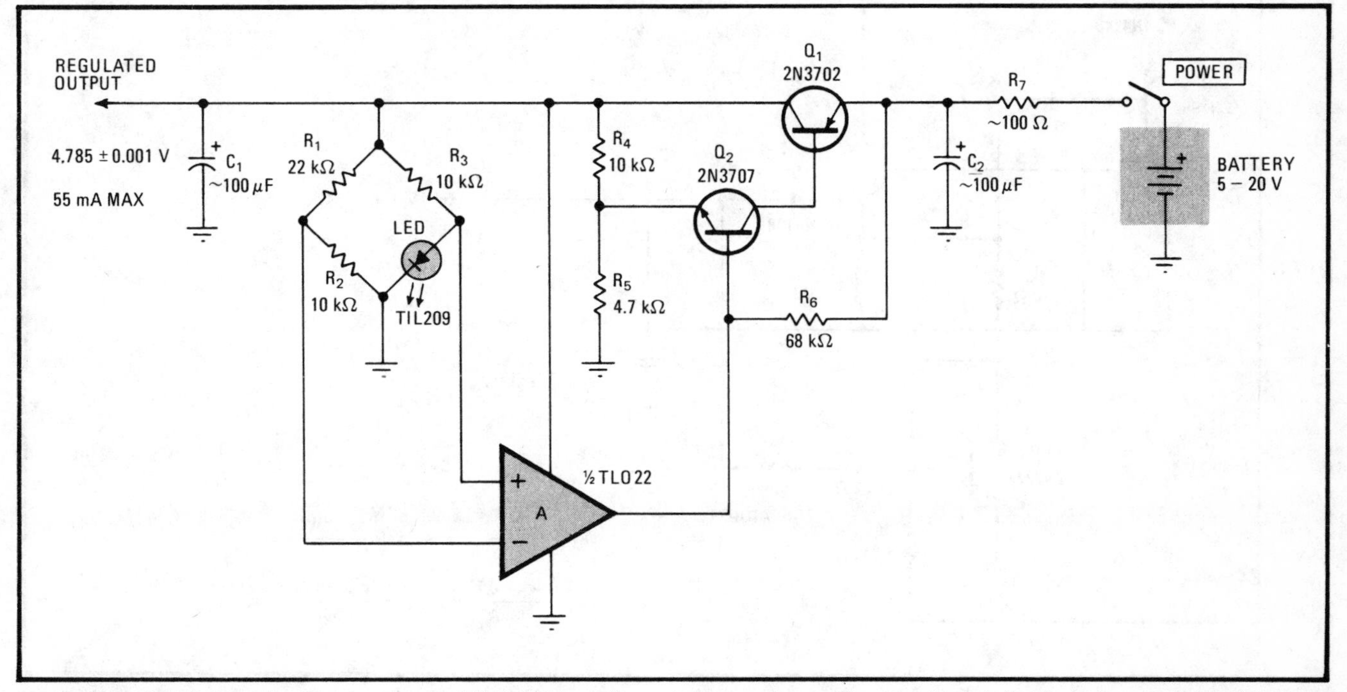

Battery supply regulator. Voltage drop across LED in bridge references op amp, controlling transistors in feedback loop. Under 5-milliampere load current, regulated output of 4.785 ± .0005 V is achieved with supply voltages ranging from 4.835 to 20.00 V. With a 9.000-V supply, the regulated output is 4.785 ± .001 V with loads ranging from 5 to 55 mA. Current through LED is barely enough to make it glow.

Negative-output regulator tracks input voltage

by Gil Marosi
Intech Function Modules Inc., Santa Clara, Calif.

By using an astable multivibrator in a flyback arrangement to develop negative voltages from positive ones, this regulator ensures that its output tracks the input, such that $V_O = -V_{in}$. The voltage-controlled circuit requires only three active devices, all of them transistors.

Q_1 and Q_2 form the free-running oscillator, as shown in the figure. With Q_2 on, the V_{in} voltage is impressed across resistor L, causing the current through L to increase linearly. The peak value of current reached before Q_2 turns off will be directly proportional to the magnitude of the output voltage developed across capacitor C_4.

During the time the current through the inductor increases, no voltage can be developed across C_4 because diode D is back-biased. When Q_2 switches off, however, the collector voltage drops from V_{in}, and the capacitor charges to a negative voltage. This occurs because the charging current through the coil makes D turn on, thereby causing a negative voltage at the output.

The field across L then begins to collapse and D is biased on, placing the output voltage ($V_o + V_d$) across L, where V_d is the diode drop. The current through L must then fall linearly to zero. This completes one cycle of the flyback operation.

The input voltage is next compared with the output voltage at the summing node, at the base of Q_3. This transistor amplifies the voltage difference and transforms it into a current that is used to control Q_2's turn-on time. Thus if V_{out} should fall, the control current will act to increase the on time of Q_2, thereby increasing the peak current through L and so raising the output voltage. This analysis assumes that V_{in} emanates from a stiff source—that is, an increased current demand will not cause a drop in V_{in} because of an increased voltage drop across the source's internal impedance.

Without Q_3, the load regulation would be directly proportional to a change in load current (I_L) and so a 10% change in I_L would cause a 10% change in load voltage V_L. Q_3 ensures that such a change in I_L causes only a 0.2% change.

Component values are given for a circuit that operates with an I_L = 20 milliamperes, a V_o = −5 volts, and an astable multivibrator operating at 50 kilohertz (τ = 20 microseconds). Equations are given in order to facilitate the design of regulators for specific parameters. □

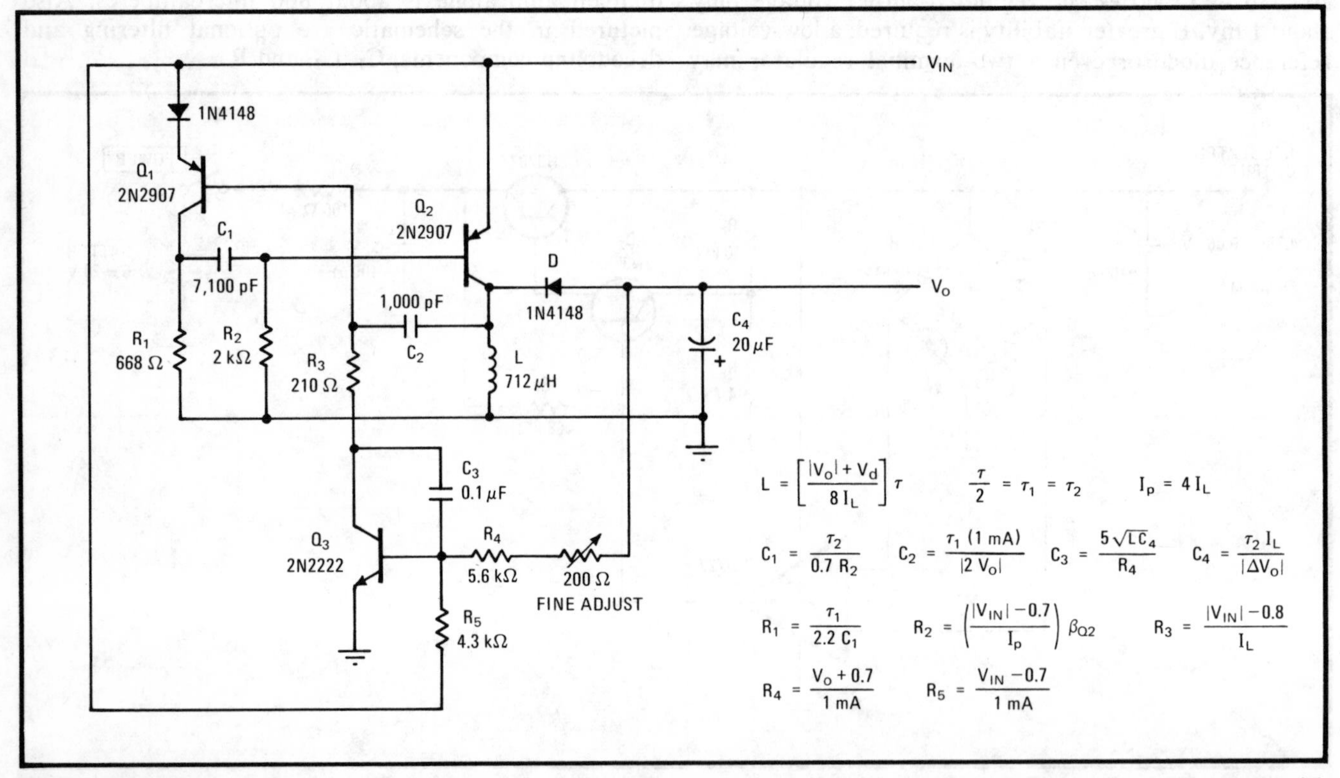

$$L = \left[\frac{|V_o| + V_d}{8 I_L}\right]\tau \qquad \frac{\tau}{2} = \tau_1 = \tau_2 \qquad I_p = 4 I_L$$

$$C_1 = \frac{\tau_2}{0.7 R_2} \qquad C_2 = \frac{\tau_1 (1\ \text{mA})}{|2 V_o|} \qquad C_3 = \frac{5\sqrt{LC_4}}{R_4} \qquad C_4 = \frac{\tau_2 I_L}{|\Delta V_o|}$$

$$R_1 = \frac{\tau_1}{2.2 C_1} \qquad R_2 = \left(\frac{|V_{IN}| - 0.7}{I_p}\right)\beta_{Q2} \qquad R_3 = \frac{|V_{IN}| - 0.8}{I_L}$$

$$R_4 = \frac{V_O + 0.7}{1\ \text{mA}} \qquad R_5 = \frac{V_{IN} - 0.7}{1\ \text{mA}}$$

Flyback follower. Regulator uses astable multivibrator Q_1–Q_2 and inductor to generate negative output voltages from positive inputs while also ensuring that $V_{out} = -V_{in}$. Differential amplifier Q_3 serves to develop feedback control voltage to readjust on time of Q_2 and thus voltage developed across L and C_4 when $V_{out} \neq -V_{in}$. Component values are given for I_L = 20 mA, V_o = −5 V, and f = 50 kHz.

Solar-powered regulator charges batteries efficiently

by G. J. Millard
Volcanological Observatory, Rabaul, Papua New Guinea

For use with solar panels, this simple and efficient regulator circuit provides an energy-saving solution to charging batteries of the lead-acid type commonly found in automobiles. Not considering the cost of the solar cells, assumed to be at hand for use in other projects, the regulator alone is under $10.

Unlike many other shunt regulators that divert current into a resistor when the battery is fully charged, this circuit opens the charging path so that the resistors can be eliminated. This method is extremely advantageous when solar panels are used, for large resistors would otherwise be required to dissipate the high power levels typically encountered.

When the battery voltage, e_o, is below 13.5 volts (normally the open-circuit potential of a 12-V battery), transistors Q_1, Q_2, and Q_3 turn on and charging current flows from the solar panels as required. The active green light-emitting diode indicates the battery is taking charge.

As e_o approaches the open-circuit voltage, op amp A_{1a} switches Q_1–Q_3 off. This condition is maintained until such time as the battery voltage drops to 13.2 V, whereupon the charge cycle repeats.

If the battery voltage should continue to fall from 13.2 to approximately 11.4 V, indicating a flat battery, A_{1b} switches low, causing a red LED to flash at a rate determined by the astable multivibrator A_{1c}, in this case oscillating at a frequency of 2 hertz. A_{1d} provides a reference of 6 V to maintain the switching points at the 11.4- and the 13.2-V levels.

The circuit will handle currents to 3 amperes. To draw larger currents, it is necessary to increase the base currents of Q_2 and Q_3 so that these transistors will remain in saturation during the charging periods. □

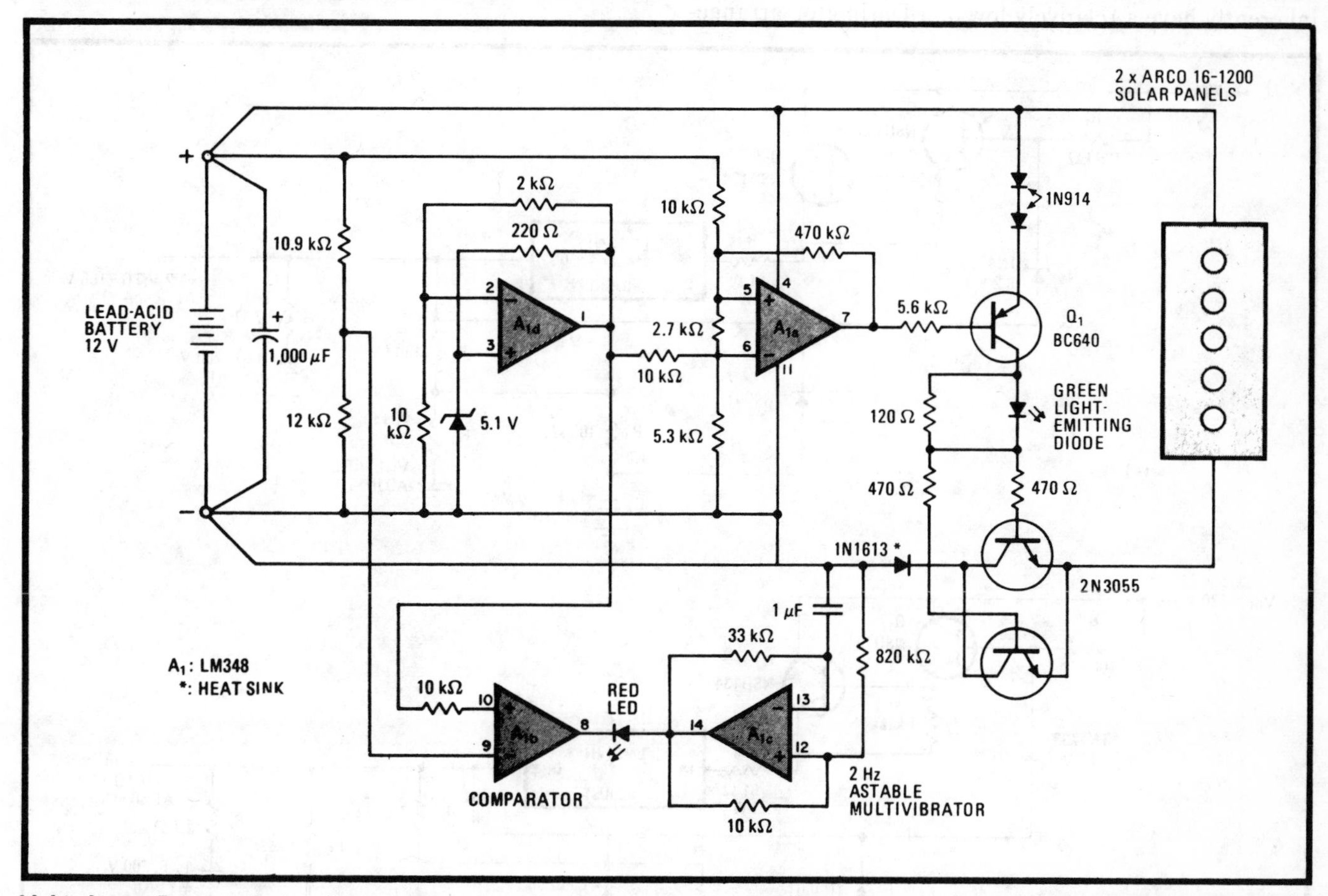

Light charge. Regulator for handling currents produced by solar panels charges lead-acid batteries without wasting excessive power. Circuit cuts off current to battery when its open-circuit voltage is greater than 13.5 V, eliminating need for dissipating power in resistors. Green LED indicates battery is charging. Flashing red LED indicates battery is flat (battery voltage below 11.4 V) and refuses to take charge.

High-voltage regulator is immune to burnout

by Michael Maida
National Semiconductor Corp., Santa Clara, Calif.

The floating-mode operation of adjustable three-terminal regulators in the LM117 family make them ideal for high-voltage service. Because the regulator sees only the input-output differential—40 volts for the LM117—its voltage rating will not be exceeded for outputs in the hundreds of volts. But the device may break down if the output is shorted unless a circuit can be developed for withstanding the high voltage typically encountered and the output current is limited to a safe value in the event of a dead short.

The circuit surrounding the regulator will serve to solve the problem. Zener diode D_1 maintains a 5-V input output differential over the entire range of output voltages from 1.2 to 160 V. Because high-voltage transistors inherently have a relatively low β, a Darlington arrangement is used to stand off the high input potentials.

The zener diode's impedance will be low, so that no bypass capacitor is required directly at the regulator's input. In fact, no capacitor should be used if the circuit is to survive a short at the output. Resistor R_3 limits the short-circuit current to 100 milliamperes. The RC network at the output improves the circuit's transient response, as does bypassing the adjustment pin. R_4 and D_2 protect the adjustment input from breakdown, if there should be a short circuit at the output.

The approach shown in (b) will serve well in precision regulator applications. Here a LM329B 6.9-V zener reference has been stacked in series with the LM317's internal reference to improve temperature stability and regulation.

These techniques can be employed for higher output voltages and/or currents by either using better high-voltage transistors or cascoded or paralleled transistors. In any event, the output short-circuit current determined by R_2 must be within Q_2's safe area of operations so that secondary breakdown cannot occur. □

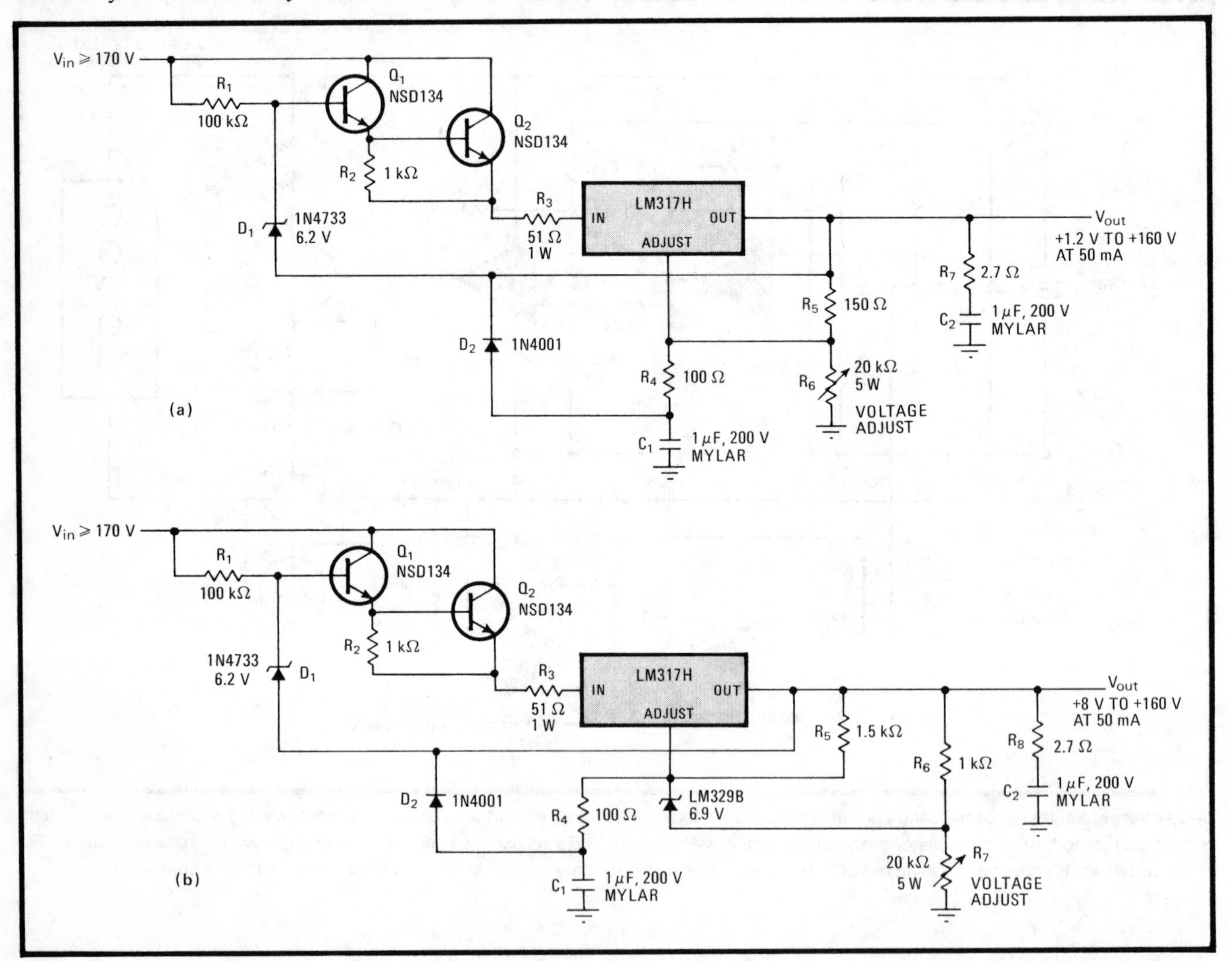

Skirting shorts. Three-terminal regulator (a), configured for high-voltage duties as a consequence of operating in the floating mode, is protected by appropriate circuitry against burnout due to shorts. LM329B zener (b) and minor changes improve stability and regulation.

Twin regulators deliver constant voltage and current

by Ladislav Grýgera and Milena Králová
Tesla-Popov Research Institute, Prague, Czechoslovakia

Cascading two μA723 precision voltage regulators in such a way as to enable them to monitor both output voltage and load current yields a circuit that can generate a constant-voltage, constant-current output. Output voltage can be adjusted over a range of 0 to 15 volts at a load current that is selectable from 0 to 3 amperes with the configuration shown.

In this circuit, output voltage is controlled by A_1, which monitors load voltage with the aid of the associated network that is connected to the ports of its error-voltage amplifier (pins 4, 5, and 6). Any change in load current is detected by A_2, which acts to generate a signal at its V_{out} port.

The current-limit input of A_1 is then activated, and thus the current-limiting transistor internal to A_1, which acts as a shunt across the error amplifier, can control the amount of driving current supplied to pass transistor Q_1 at the set output voltage.

The output voltage is adjusted by R_1. The value of R_1 required is approximately 1 kilohm for each volt appearing at the load R_L. Because $R_2 = R_3$,

$$V_{out} = K_1 R_1 V_{ref1}$$

where $K_1 = 1/2R_4$.

Output current is adjusted by R_5, the value of which will increase 1 kilohm for every 100-milliampere increase in load current. Assuming $R_6 = R_7$, then it can be shown that:

$$I_{out} = K_2 R_5 V_{ref2}$$

where $K_2 = 1/2R_8R_{sc}$.

Zener diode D_1 allows selection of the output current down to zero. Diodes D_2 and D_3 provide well-defined switching thresholds for A_1 and Q_1 to enhance the circuit's response time to a changing output current and/or voltage. □

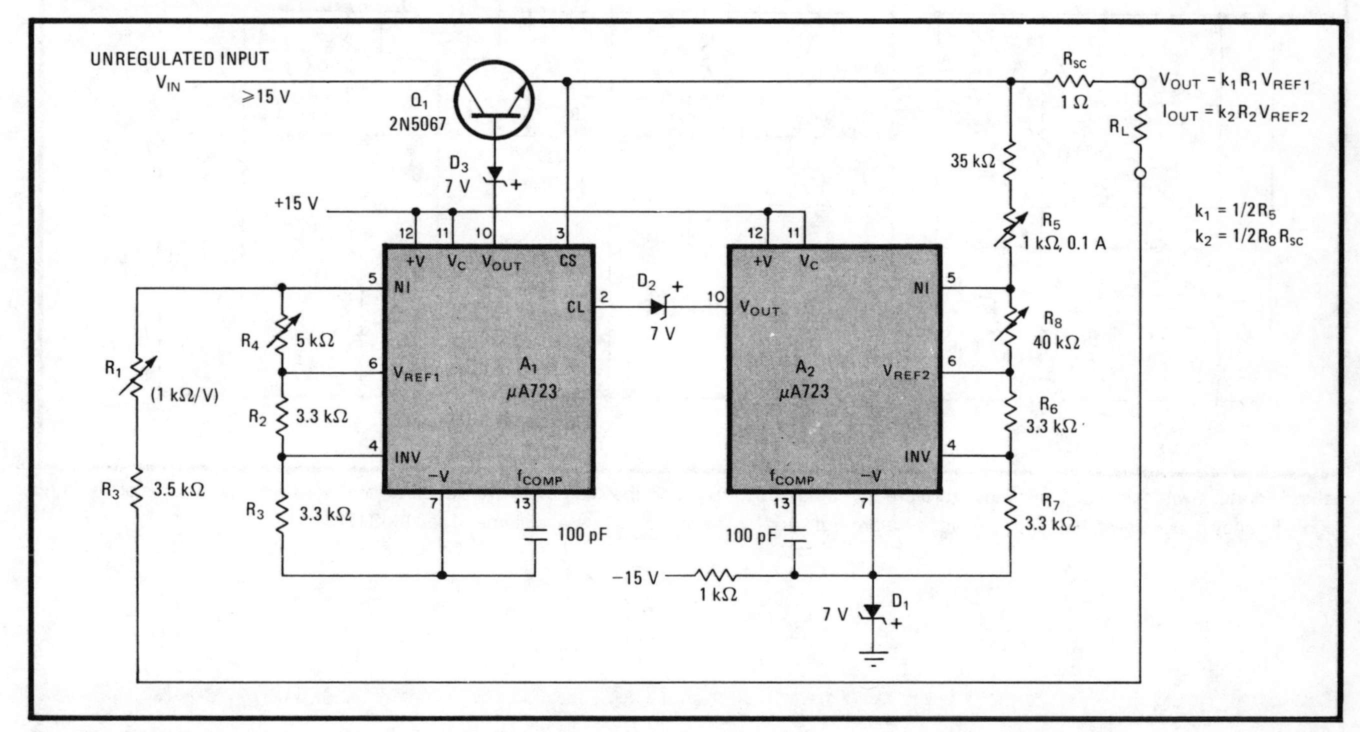

Steady. Cascaded regulators, one for monitoring load voltage, the other for current, form circuit that generates constant-voltage, constant-current output. Current and voltage are adjustable from zero. Maximum output voltage is 15 V; current limit depends on Q_1's rating.

Switching converter raises linear regulator's efficiency

by Sadeddin Kulturel
Istanbul, Turkey

The low ripple and fast recovery of a series-pass voltage regulator can be attained at the high efficiency of a switching regulator if both are combined. In this circuit, the performance is achieved by using the switching circuit as a preregulator for the linear element.

As shown in the illustration of the general circuit, which is designed to transform the 35-volt raw input into a well-regulated output, heat dissipation across the LM317K series element can be reduced if it is made to handle a switched, rather than a continuous, input. Here, the switching regulator is formed by transistors Q_1–Q_4, D_1, and L_1. During power up, Q_1, driven through R_1–R_3, is brought into saturation. Q_2 remains off and Q_3 is turned on.

Switching occurs when V_d equals 3.6 volts, which is D_2's zener voltage. Q_4 then turns on, as does Q_2, and Q_3 is turned off.

As Q_2 turns on, Q_1 switches off, and because of the positive voltage spike created by L_1, load current is momentarily forced through D_1 as V_d decreases. When V_d reaches the lower hysteresis threshold of Q_3 as established by R_5 and R_6, Q_2 and Q_4 turn off, and Q_1 turns on, completing the switching cycle. With the supply's negative path restored, V_d rises until it reaches V_2, and the process is repeated.

The linear regulator can be of any type, including a three-terminal, nonadjustable device. Note that a switching current regulator can be formed if the regulator is replaced by a resistor. In that case, the switching current will be $I_S = V_Z/R$. □

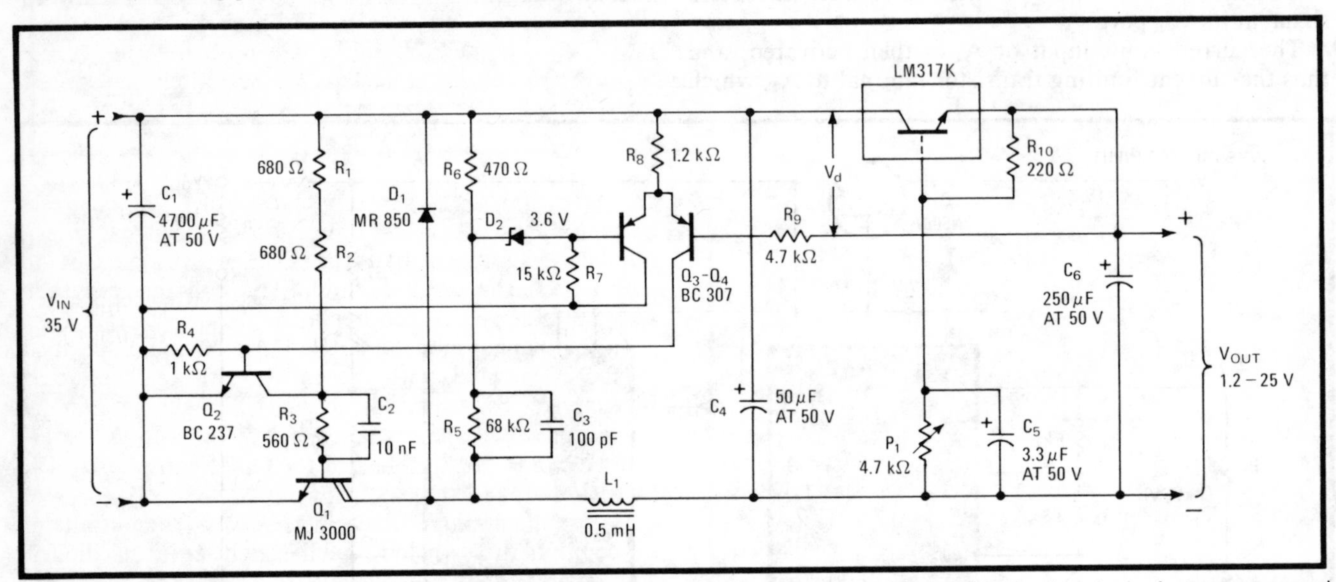

Mixed mode. Switched and linear regulators are combined to form a unit that has the advantages of both—low ripple, fast response, and high efficiency. Here a switched circuit serves as a preregulator for the linear series-pass element, the LM317K.

Foldback limiter protects high-current regulators

by A. D. V. N. Kularatna
Ratmalana, Sri Lanka

This circuit provides foldback protection for a series-regulated source that has to deliver high current. Because it requires no current-monitoring resistor, the circuit achieves wide dynamic response at good efficiency. It draws only 2% of maximum load current and its cost is reasonable.

Here, a low-current shunt-regulated module (a) provides the overload protection. This module is configured into the conventional regulator system to work as a switch, in which role it quickly turns off a series-pass transistor when the load current exceeds some predetermined value.

The circuit details are explained with the aid of the diagram (b) for a representative regulator designed to deliver 12 volts at 4 amperes. Transistors Q_1 and Q_2 form a differential amplifier, which compares a 6.2-V reference to a potential derived from the 12-V output through potentiometer R_V. Shunt elements Q_5–Q_6 act to maintain the potential at the base of Q constant for any load condition by taking up the difference between the set and the actual base drive.

It is necessary that the current source Q_3-Q_4 be set to I_L/h_{fe} for proper tracking, where I_L is the maximum load current and h_{fe} is the current gain of Q. The value of the constant current, I, is $h_{fe\,Q4}\,(V_{Z5} - V_{be\,Q3})/R_7$, so that the current is most easily set by adjusting resistor R_7.

The module requires a current of 70 to 80 milliamperes under maximum load conditions. The short-circuit output current is less than 200 mA, because the drop in output voltage switches transistor Q_2 off. The voltage across zener diode Z_5 is then reduced to a very low value, and this action in turn lowers the voltage at Q_6 and cuts down the base drive to Q.

Zener diodes Z_1–Z_3 were added to improve the ripple characteristics of the supply. As configured, the source has an output ripple of 6 mV peak to peak.

The shunt regulator module can be easily configured for any output voltage mainly by selecting the appropriate zener-diode values. □

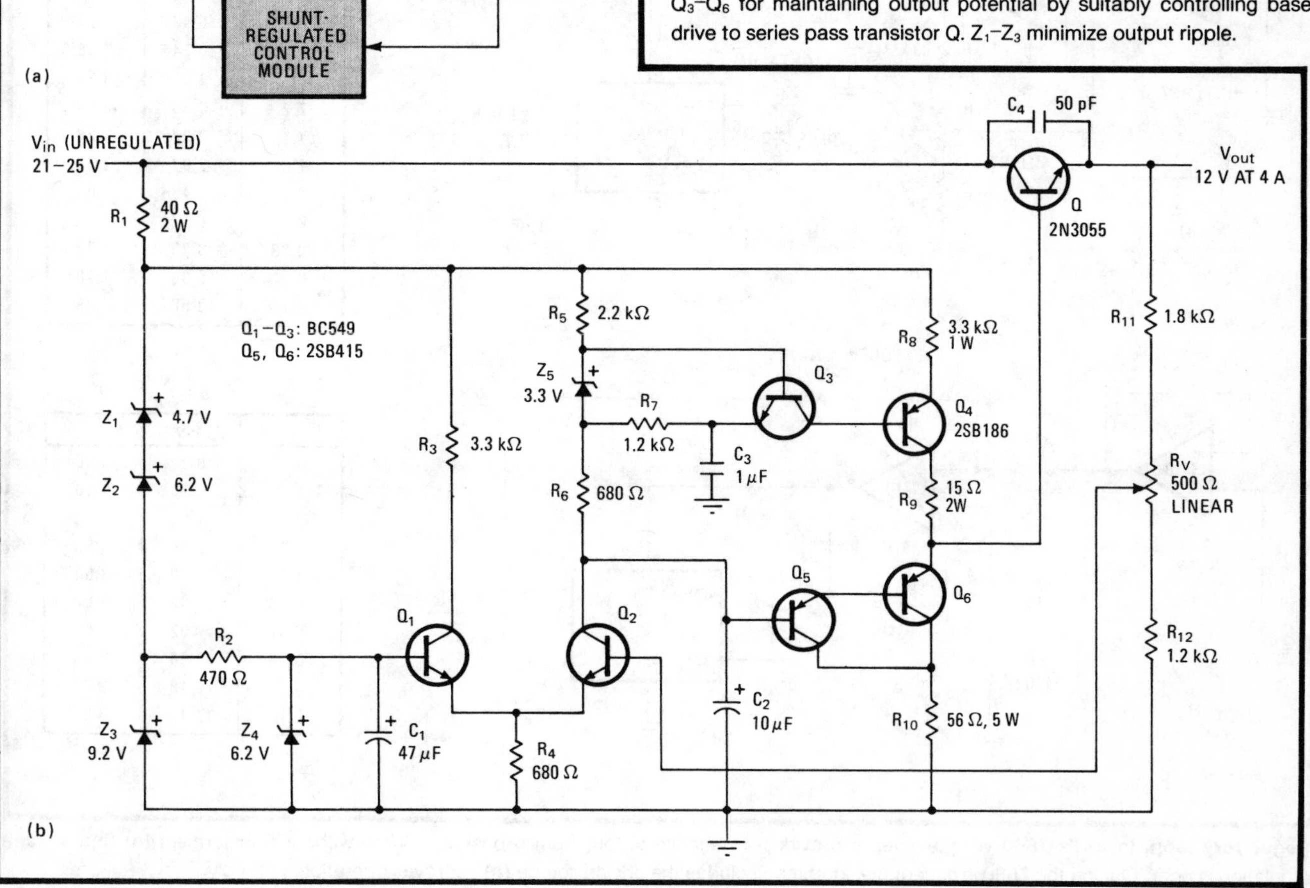

High handling. Low-current shunt regulator (a) provides foldback limiting for high-current power sources at good efficiency and reasonable cost. Circuit (b) for 12-V regulator uses differential pair Q_1–Q_2 for detecting differences in reference and output voltage, Q_3–Q_6 for maintaining output potential by suitably controlling base drive to series pass transistor Q. Z_1–Z_3 minimize output ripple.

Converter in feedback loop improves voltage regulation

by David Abrams
Winchester, Mass.

One of the most frustrating experiences a designer faces is to discover that his TTL or complementary-MOS circuit, which he intended for single-supply operation, actually requires a minus potential at some miniscule current for one or two of its integrated circuits. A new chip, Intersil's 7660 voltage converter, now enables the designer to obtain the minus voltage at low currents from a positive supply without the need for a transformer or other complicated inverter circuitry, and at low cost. In addition, placing the converter in a feedback loop that includes the chip's power—or driving—source permits a degree of voltage regulation that is not possible with the conventional stand-alone driver configuration.

As shown in (a), the 7660 can supply −3.5 volts to a single chip in a C-MOS or TTL system. The chip requires +3.5 V, which is generated by the LM10 operational amplifier from the +5-V supply. Although some other low-voltage op amp and an external reference could be substituted, the LM10 will run off a single supply, has its own reference, and has an output stage that can swing within ½ V of the supply while delivering −20 milliamperes to the 7660.

Though this circuit performs well at very low load currents, its output voltage drops rapidly as load currents increase (see table) because its output impedance is fairly high. At a no-load output voltage of −3.5 V the converter exhibits an output resistance of about 100 ohms, but it will increase 50% for V_{out} = 2 V. This value will render the 7660 useless in systems where more then a few milliamperes are required.

By adding a single resistor and configuring the circuit to the topology in (b), however, the converter can be made to perform much as an ideal voltage source for loads of 1 kilohm or greater. The regulation for loads less than 1 kΩ will be much superior to that in (a), as seen in the table.

Here the circuit works as an inverting amplifier with a gain of −17.5, which is set by $(R_1+R_2)/R_3$. The converter provides a gain of −1, requiring that the noninverting input of the op amp be used as the summing junction. Thus the circuit can still be run from a single supply because the LM10's input common-mode range includes the negative supply (ground, in this case).

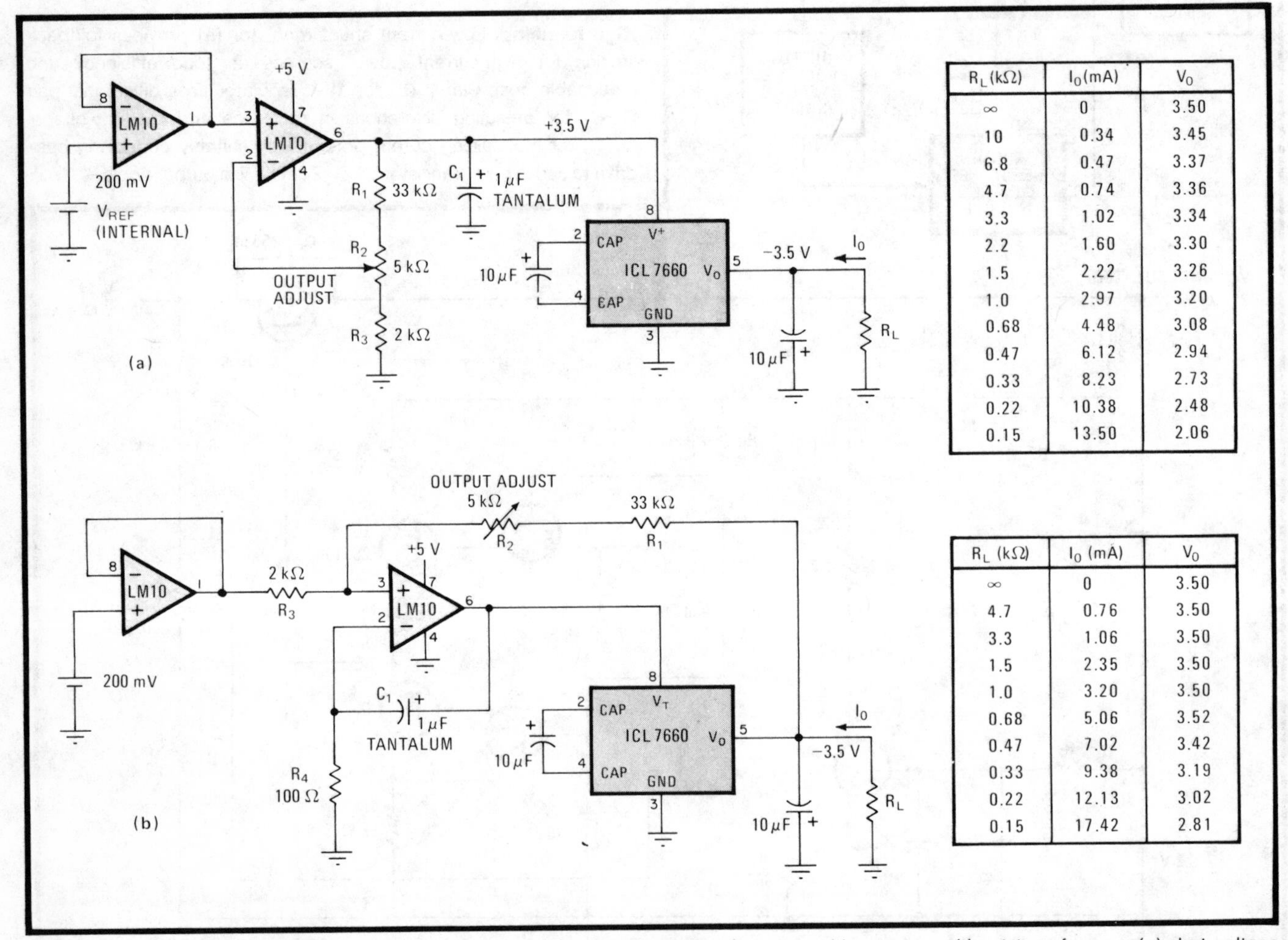

R_L (kΩ)	I_O (mA)	V_O
∞	0	3.50
10	0.34	3.45
6.8	0.47	3.37
4.7	0.74	3.36
3.3	1.02	3.34
2.2	1.60	3.30
1.5	2.22	3.26
1.0	2.97	3.20
0.68	4.48	3.08
0.47	6.12	2.94
0.33	8.23	2.73
0.22	10.38	2.48
0.15	13.50	2.06

R_L (kΩ)	I_O (mA)	V_O
∞	0	3.50
4.7	0.76	3.50
3.3	1.06	3.50
1.5	2.35	3.50
1.0	3.20	3.50
0.68	5.06	3.52
0.47	7.02	3.42
0.33	9.38	3.19
0.22	12.13	3.02
0.15	17.42	2.81

Regulatory loop. Intersil's 7660 voltage inverter provides a negative output from a positive source without transformers (a), but voltage regulation is poor. Placing the 7660 in a feedback loop that includes the driving source (b) improves operation markedly.

R_4 and C_1 provide local feedback around the op amp to stabilize the loop. Without these components, the delay between input and output voltage changes of the 7660 would cause the output of the LM10 to oscillate between ground and +5 V.

In operation, the feedback loop will force the op amp to try to hold the negative output voltage constant. Even at the higher currents, the output resistance is half of what it is in (a).

The circuit may also be used to supply negative voltages other than −3.5 V. If higher voltages are desired, it is necessary to choose a supply voltage for the LM10 that will provide sufficient output from the op amp under the expected load conditions. In this case, the effective voltage gain of the 7660 drops from −0.99 V to zero, and so the output voltage of the op amp must rise as the load current increases in order to compensate for the loss of gain. □

Micropower regulator has low dropout voltage

by Kelvin Shih
General Motors Proving Ground, Milford, Mich.

Designed specifically to regulate the output of lithium batteries, which have a low terminal voltage at low temperatures, this circuit provides a stable 5.0 volts at 10 milliamperes for an input voltage as low as 5.2 V.

The low dropout voltage of the regulator (5.2−5.0 = 0.2 V) is attained in part by operating the circuit's output transistor in the common-emitter mode. As a result, its collector-to-emitter voltage drop is much lower than the base-to-emitter drop of transistors operated as emitter followers in standard regulators. And, because it uses a low-power operational amplifier operating from a single supply, and a low-current, low-voltage zener diode for the voltage reference, the regulator's idle current is only 250 microamperes.

Three lithium batteries drive the regulator shown in the figure. Their terminal voltage is usually 3 V per cell at room temperature, but it will drop to 2 V at −40°C.

Z_1 provides a low-voltage reference (1.22 V) to the noninverting input of the LM224 op amp, A_1. The Intersil ICL 8069CMQ zener has been selected because it requires only 50 μA of bias current and has a temperature coefficient of better than 50 parts per million/°C.

The 1.22-V reference is compared to the output voltage from a divider network (R_1,R_2, P_1), which is used to trim the output voltage to the desired value. Any voltage difference appearing at the output of A_1 drives transistor Q_1, and thus determines the drive current to Q_2. As a result, Q_2 conducts more heavily if the output voltage is low, or limits the application of battery voltage to the load if the output voltage is high.

There will be no observable change of output voltage for an input voltage variation between 5.2 and 10 V, over the temperature range of −40°C to +70°C. □

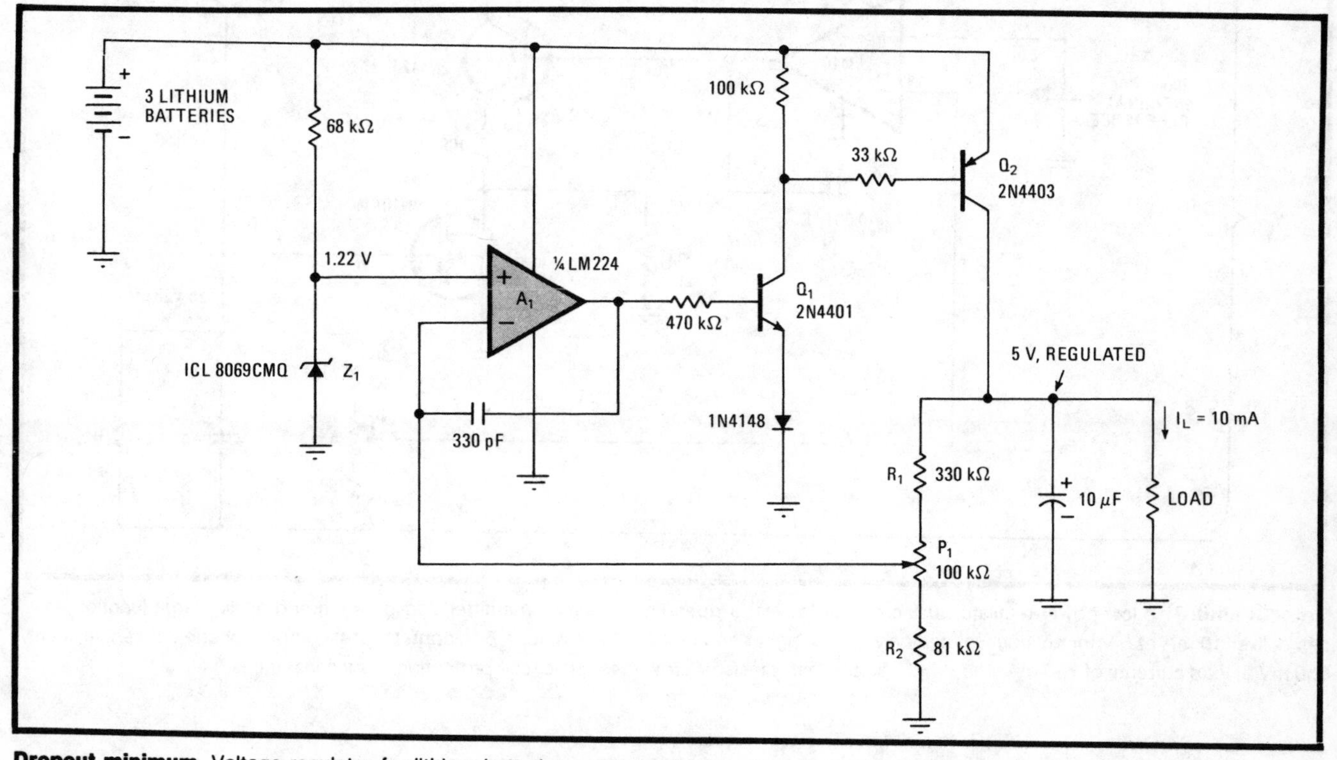

Dropout minimum. Voltage regulator for lithium batteries maintains 5-volt output for a minimum input voltage of 5.2 V. Output voltage is constant over the temperature range −40°C to +70°C. Using a low-power op amp operating from a single-ended supply, and a low-current zener, the circuit holds the idle current to 250 μA, well below the 2 to 10 mA required by standard regulators.

Protected regulator has lowest dropout voltage

by Thomas Valone, *A-T-O Inc., Scott Aviation Division, Lancaster, N. Y.*
and Kelvin Shih, *General Motors Proving Ground, Milford, Mich.*

Providing an output of 5 volts at 10 milliamperes for an input of only 5.012 V, this regulator is ideal for use in many micropower applications, such as regulating the output of lithium batteries that drive low-power detection and recording instruments in the field. The circuit is useful in high-current situations also, as it can deliver up to 1 ampere at 5 V for an input of only 6.0 V. Short-circuit protection in this instance is provided by a single V-groove MOS field-effect transistor.

Contributing to the low-dropout characteristic of the circuit is the 2N6726 output transistor, which has a large junction area that allows a lower emitter-to-collector drop than most other devices, including Darlington arrangements. Thus the input-to-output voltage differential, 12 millivolts, is 6% that of one of the best low-dropout regulators reported to date.[1]

The input-to-output differential is only 350 mV at a load current of 500 mA. The 2N6726 is physically a small transistor but can dissipate 1 watt safely without a heat sink.

Short-circuit protection is provided by a Siliconix VN10KM, which presents a resistance of less than 10 ohms to the emitter circuit of the 2N4424 drive transistor under normal conditions. However, when the output is shorted to ground or excessive current is demanded, the drain-to-source resistance of the FET rises, safely shutting down the pass transistor. This characteristic can be used to advantage in adjustable current limiters, where the trip point is set by the input voltage. This method, incidentally, is more effective than any transistor foldback technique.

In operation, the LM10CH reference amplifier compares the voltage set by potentiometer P_1 to its internal 200-mV reference and through Q_1 acts to minimize voltage differences at the amplifier's input. With suitable selection of the component values in divider network R_1–R_2, the circuit will regulate over any voltage from 1 to 40 V. The operational-amplifier half of the LM10CH is available for other uses.

The load regulation is to within 0.3% for the range 0 to 100 mA and to within 1% for the range 100 mA to 1 A. The regulator's idle current is 320 μA. □

References
1. Kelvin Shih, "Micropower regulator has low dropout voltage," p. 343

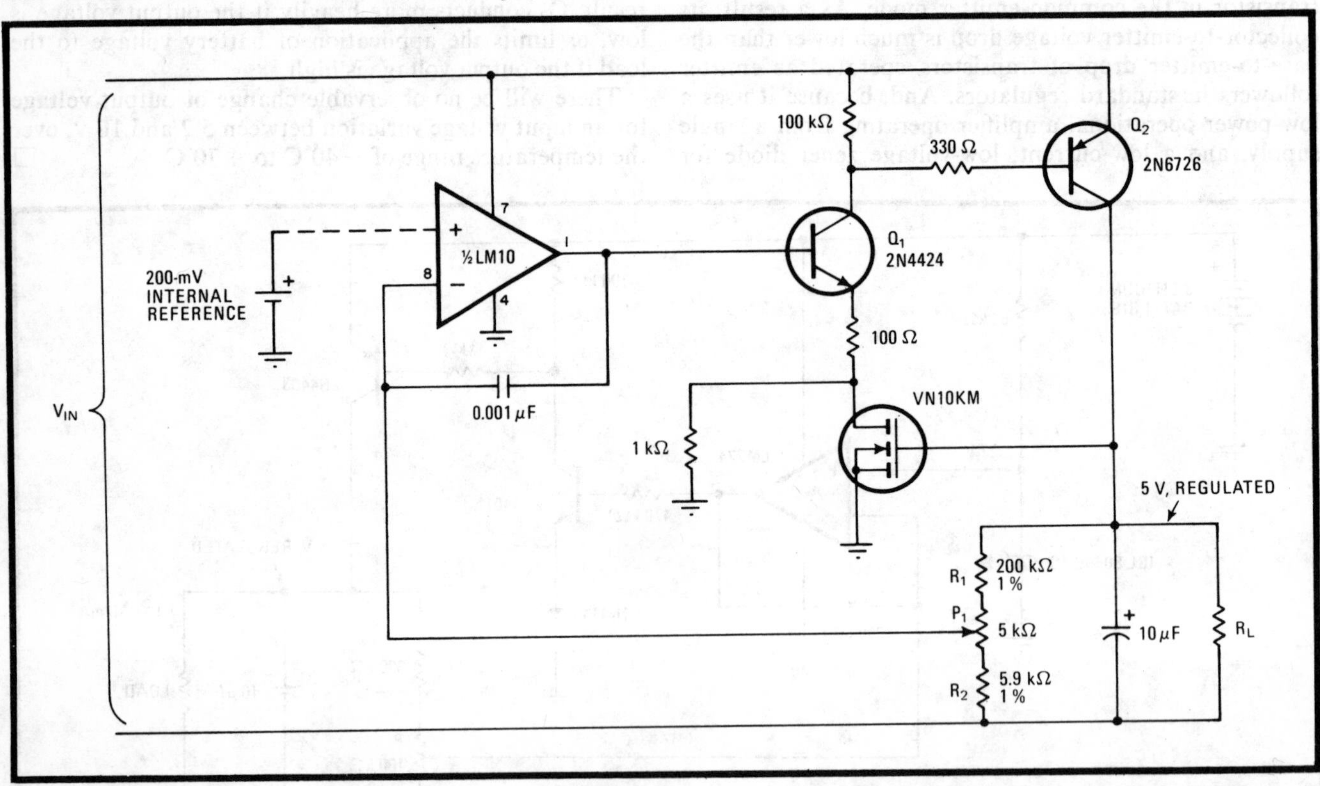

Dropout limit. This low-power regulator, using output transistor operating in common-emitter configuration and having large junction area, can deliver 10 mA at 5 V for an input voltage only 12 mV higher and up to 1 A at 5 V for a 6-V input. Input-to-output voltage differential is only 650 mV at load currents of 750 mA. The V-MOS field-effect transistor provides short-circuit protection in such instances.